Andreas Bikfalvi

Encyclopedic Reference of
Vascular Biology & Pathology

Springer-Verlag Berlin Heidelberg GmbH

Andreas Bikfalvi

Encyclopedic Reference of
Vascular Biology
& Pathology

Springer

Professor Andreas Bikfalvi
Laboratoire des Facteurs de Croissance et
de la Différenciation cellulaire (LFCDC)
Université Bordeaux I
Avenue des Facultés
33405 Talence
France

Additional material to this book can be downloaded from http://extras.springer.com

Die Deutsche Bibliothek – CIP-Einheitsaufnahme
Comprehensive vascular biology and pathology : an encyclopedic reference / Andreas Bikfalvi (ed.). –
Berlin ; Heidelberg ; New York ; Barcelona ; Hong Kong ; London ; Milan ; Paris ; Singapore ; Tokyo :
Springer, 2000

ISBN 978-3-642-62983-9   ISBN 978-3-642-57063-6 (eBook)
DOI 10.1007/978-3-642-57063-6

Typesetting: MEDIO GmbH, Berlin
Cover-Design: design & production GmbH, Heidelberg

Printed on acid-free paper   SPIN: 10551176   27/3136KM - 5 4 3 2 1 0

Preface

Vascular biology has become one of the most exciting fields in the biomedical sciences. The development of molecular biology and of genetic approaches in the mouse embryo has largely contributed to our current understanding of the biology of the vascular cell. Major advances have been achieved in the understanding of vascular development and in the role of the vasculature in various physiological or pathological processes.

The aim of the present book is to provide the reader with a reference in which information can be looked-up quickly or to spark interest in a topic for later research. It should be valuable not only for scientists working actively in vascular biology or in related fields but also to clinicians because it will provide both with the necessary information about the physiopathological mechanisms encountered in their daily work. In addition, the book should also be of great help to teachers and to students in the life sciences.

We did not want to organize this book in a textbook fashion. Instead, we chose to organize the book alphabetically, thus providing the reader with rapid access to information. However, we also wanted the various topics dealt with in enough depth for it not to be so condensed and short as in a lexicon. Thus, the book lies somewhere between the two.

Major areas in the vascular biology field are covered and the reader is also provided with specialized information on various topics. The book will surely be a success and find its place on the shelf of almost everyone interested in vascular biology.

I would like to thank all the contributing authors for their effort in writing the various book chapters. They are truly outstanding. I would also like to thank Dr. Rolf Lange and Ms. Hiltrud Wilbertz from Springer-Verlag for their effort and support.

Finally, I would like to dedicate this book to the memory of Professor Werner Risau, who died recently. Professor Risau was head of the Max Planck Institute for Physiological and Clinical Research – W.G. Kerckhoff-Institute in Bad Nauheim (Germany) and was one of the leading european scientists in the vascular biology field. As a highly respected and esteemed scientist, Professor Risau made important contributions in the area of vascular development. His death not only represents a great loss to his family but also to the entire vascular biology community. Although Werner is no longer with us, his legacy will certainly guide us towards the elucidation of the mechanisms of vascular development and to the establishment of angiogenesis therapy.

Bordeaux, August 1999 *Andreas Bikfalvi*

List of Contributors

Austin, Sandra C.
Center for Experimental Therapeutics, University of Pennsylvania
Stellar Chance Laboratories Room 905, 422 Curie Blvd., Philadelphia, PA 19104-6100, USA

Bachelot, Christilla
U428 INSERM, "Risque Thrombotique et Mécanismes Cellulaires et Moléculaires de l'Hémostase", Faculté de Pharmacie, 4 avenue de l'Observatoire, 75006 Paris, France

Badet, Josette
INSERM U427, Développement Humain: Croissance et Différenciation, Université René Descartes, 4, avenue de l'Observatoire, 75270 Paris, France

Barritault, Denis
Laboratoire de Recherche sur la Croissance Cellulaire, la Réparation et la Régénération Tissulaires (CRRET), UPRESA 7053 CNRS, Université Paris-XII Val de Marne, Avenue du Général de Gaulle, 94010 Créteil Cedex, France

Baruch, Dominique
INSERM U143, Hôpital de Bicêtre, 94276 Bicêtre Cedex, France

Bicknell, Roy
Molecular Angiogenesis Laboratory, Imperial Cancer Research Fund, Institute of Molecular Medicine, University of Oxford, John Radcliffe Hospital, Oxford, OX3 9DU, UK

Bikfalvi, Andréas
Laboratoire des Facteurs de Croissance et de la Differenciation cellulaire (LFCDC) Université Bordeaux I, Avenue des Facultés, 33405 Talence, France

Bini, Alessandra
Laboratory of Blood Coagulation Biochemistry, Lindsley F. Kimball Research Institute, New York Blood Center, 310 East 67th Street, New York, NY 10021, USA

Birnbaum, Daniel
Laboratoire d'Oncologie Moléculaire, U.119 INSERM, 27 Bd. Leï Roure, 13009 Marseille, France

Borg, Jean-Paul
Laboratoire d'Oncologie Moléculaire, U.119 INSERM, 27 Bd. Leï Roure, 13009 Marseille, France

Bouton, Marie-Christine
Laboratoire de Recherche sur l'Hémostase et la Thrombose, Faculté Xavier Bichat, Université Paris 7, 16, rue Henri Huchard, BP 416, 75870 Paris cedex 18, France

Bussolino, Federico
Institute for Cancer Research and Treatment, Division of Molecular Angiogenesis, School of Medicine, University of Torino, Strada Provinciale 142, Km 3.95, 10060 Candiolo, Italy

Caen, Jacques
Institut des Vaisseaux et du Sang, Hôpital Lariboisière, 8, rue Guy Patin, 75 010 Paris, France

Carmeliet, Peter
Center of Transgene Technology and Gene Therapy, Campus Gasthuisberg, University of Leuven, Herestraat 49, 3000 Leuven, Belgium

Caruelle, Danièle
Laboratoire de Recherche sur la Croissance Cellulaire, la Réparation et la Régénération Tissulaires (CRRET), UPRESA 7053 CNRS, Université Paris-XII Val de Marne, Avenue du Général de Gaulle, 94010 Créteil Cedex, France

Chassagne, Catherine
INSERM U127, IFR "Circulation", Hôpital Lariboisière, 41, Boulevard de la Chapelle, 75475 Paris cedex 10, France

Collen, Désiré
Center of Transgene Technology and Gene Therapy, Campus Gasthuisberg, University of Leuven, Herestraat 49, 3000 Leuven, Belgium

Comoglio, Paolo
Institute for Cancer Research and Treatment, Division of Molecular Oncology, School of Medicine, University of Torino, Strada Provinciale 142, 10060 Candiolo, Italy

Courty, José
Laboratoire de Recherche sur la Croissance Cellulaire, la Réparation et la Régénération Tissulaires (CRRET), UPRESA 7053 CNRS, Université Paris-XII Val de Marne, Avenue du Général de Gaulle, 94010 Créteil Cedex, France

Cuevas, Pedro
Servicio de Histología, Hopital Ramon Y Cajal, Carretera de Colmenar Km. 9,100, 28034 Madrid, Spain

Delbé, Jean
Laboratoire de Recherche sur la Croissance Cellulaire, la Réparation et la Régénération Tissulaires (CRRET), UPRESA 7053 CNRS, Université Paris-XII Val de Marne, Avenue du Général de Gaulle, 94010 Créteil Cedex, France

Dieterlen-Lievre, Françoise
Institut d'Embryologie Cellulaire et Moleculaire, Collège de France, 49 bis, avenue de la Belle Gabrielle, 94736 Nogent sur Marne Cedex, France

Drouet, Ludovic
Angio-hematologie Biologique et Moléculaire, Hôpital Lariboisiere, 8, rue Guy Patin, 75475 Paris Cedex 10, France

Dupuy, Evelyne
Service d'Angio-Hèmatologie, Hôpital Lariboisière, 2, rue Ambroise Parè, 75475 Paris Cedex 10, France

Duriez, Patrick
INSERM U 325, Institut Pasteur de Lille, 1, rue du Professeur Calmette, 59019 Lille cedex, France

Eichmann, Anne
Institut d'Embryologie Cellulaire et Moleculaire, Collège de France, 49 bis, avenue de la Belle Gabrielle, 94736 Nogent sur Marne Cedex, France

Emonard, Hervé
Laboratoire de Biochimie, "Signalisation Cellulaire et Matrice Extracellulaire" (SIME), CNRS UPRESA 6021, IFR 53 "Biomolécules", Faculté de Médecine, 51 rue Cognacq Jay, 51095 Reims cedex, France

Feige, Jean-Jacques
INSERM U244 - DBMS/BRCE, CEA Grenoble, 17 rue des Martyrs, 38054 Grenoble cedex 9, France

Fischer, Elizabeth
INSERM U 430, Immunopathologie Humaine, Hôpital Broussais, 96, rue Didot, 75074 Paris cedex 14, France

FitzGerald, Garret A.
Department of Pharmacology, University of Pennsylvania, Medical Center, 153 Johnson Pavillion, 3620 Hamilton Walk, Philadelphia, PA 19104-6084, USA

Fons, Pierre
Laboratoire de Biologie Moléculaire Eucaryote, CNRS UPR 9006, 118 Route de Narbonne, 31062 Toulouse, France

Frémeaux-Bacchi, Véronique
INSERM U 430, Immunopathologie Humaine, Hôpital Broussais, 96, rue Didot, 75074 Paris cedex 14, France

Fruchart, Jean-Charles
INSERM U 325, Institut Pasteur de Lille, 1, rue du Professeur Calmette, 59019 Lille cedex, France

Giansily, Muriel
Laboratoire Central d'Hématologie, CHU Montpellier, 2, avenue Bertin Sans, 34295 Montpellier Cedex 5, France

Fournier, Emmanuel
Laboratoire d'Oncologie Moléculaire, U.119 INSERM, 27 Bd. Leï Roure, 13009 Marseille, France

Guillin, Marie-Claude
Laboratoire de Recherche sur l'Hémostase et la Thrombose, Faculté Xavier Bichat, Université Paris 7, 16, rue Henri Huchard, BP 416, 75870 Paris cedex 18, France

Haidaris, Patricia J.
Department of Medicine/Hematology Unit Box 610, University of Rochester, 601 Elmwood Avenue, Rochester, NY 14642, USA

Han, Zhong Chao
Institute of Hematology and National Laboratory of Experimental Hematology, Chinese Academy of Medical Sciences and Pekin Union Medical College, 300020 Tianjin, China

Hansson, Göran K.
Cardiovascular Research Unit, Department of Medicine, Center for Molecular Medicine, Karolinska Hospital, 171 76 Stockholm, Sweden

Heldin, Carl Henrik
Ludwig Institute of Cancer Research, Biomedical Center, Box 595, 75124 Uppsala, Sweden

Hornebeck, William
Laboratoire de Biochimie, "Signalisation Cellulaire et Matrice Extracellulaire" (SIME), CNRS UPRESA 6021, IFR 53 "Biomolécules", Faculté de Médecine, 51 rue Cognacq Jay, 51095 Reims cedex, France

Hornych, Antonin
9, avenue Charles Bras, 77184 Emerainville, France

Huminiecki, Lucas
Molecular Angiogenesis Laboratory, Imperial Cancer Research Fund, Institute of Molecular Medicine, University of Oxford, John Radcliffe Hospital, Oxford, OX3 9DU, UK

Hutchings, Helen
Laboratoire de Biologie Moléculaire Eucaryote, CNRS UPR 9006, 118, route de Narbonne, 31062 Toulouse cedex, France

Jandrot, Martine -Perrus
Laboratoire de Recherche sur l'Hémostase et la Thrombose, Faculté Xavier Bichat, Université Paris 7, 16, rue Henri Huchard, BP 416, 75870 Paris cedex 18, France

Kalafatis, Michael
Department of Chemistry, Cleveland State University, Euclid Avenue at East 24th Street, Cleveland, OH 44115, USA and Department of Molecular Cardiology, Lerner Research Institute, Cleveland Clinic Foundation, Cleveland OH 44195, USA

Kazatchkine, Michel D.
INSERM U 430, Immunopathologie Humaine, Hôpital Broussais, 96, rue Didot, 75074 Paris cedex 14, France

Kerbiriou-Nabias, Danièle
INSERM U143, Hôpital de Bicêtre, 94276 Bicêtre Cedex, France

Kieffer, Nelly
Laboratoire Franco-Luxembourgeois de Recherche Biomédicale, CNRS, France - CRP-Santé, Centre Universitaire, 162A, avenue de la Faäencerie, 1511 Luxembourg, Luxembourg

Kleinman, Hynda K.
Cell Biology Section, National Institute of Dental Research, NIH, 30 Convent Drive, MSC 43 70, Bethesda, MD 20892, USA

Kudryk, Bohdan J.
Laboratory of Blood Coagulation Biochemistry, Lindsley F. Kimball Research Institute, New York Blood Center, 310 East 67th Street, New York, NY 10021, USA

Lawrence, David A.
UPR 9079 du CNRS, Groupe des Laboratoires de Villejuif, 7, rue Guy Moquet, 94801 Villejuif, France

Le Douarin, Nicole
Institut d'Embryologie Cellulaire et Moleculaire, Collège de France, 49 bis, avenue de la Belle Gabrielle, 94736 Nogent sur Marne Cedex, France

Malavaud, Bernard
Laboratoire de Biologie Moléculaire Eucaryote, CNRS UPR 9006, 118 Route de Narbonne, 31062 Toulouse, France

Malinda, Katherine M.
Cell Biology Section, National Institute of Dental Research, NIH, 30 Convent Drive, MSC 43 70, Bethesda, MD 20892, USA

Mann, Kenneth G.
Department of Biochemistry, College of Medicine, University of Vermont, Given Building, Burlington VT 05405, USA

Mendelsohn, Laurane G.
Lilly Research Laboratories, Lilly Corporate Center, Indianapolis, Indiana 46285, USA

Meyer, Dominique
INSERM U143, Hôpital de Bicêtre, 94276 Bicêtre Cedex, France

Michel, Jean-Baptiste
INSERM U460, Faculté de Médecine Xavier Bichat, 16, rue Henri Huchard, 75018 Paris, France

Milhiet, Pierre Emmanuel
Laboratoire de Recherche sur la Croissance Cellulaire, la Réparation et la Régénération
Tissulaires (CRRET), UPRESA 7053 CNRS, Université Paris-XII Val de Marne, Avenue du
Général de Gaulle, 94010 Créteil Cedex, France

Ortéga, Nathalie
Laboratoire de Biologie Moléculaire Eucaryote, CNRS UPR 9006, 118, route de Narbonne,
31062 Toulouse cedex, France

Moscatelli, David
Department of Cell Biology, New York University Medical Center, 550 First Avenue, New
York, NY 10016, USA

Perrot-Applanat, Martine
INSERM U460, CHU Xavier Bichat, 16, rue Henri Huchart, 75870 Paris cedex 18, France

Plouët, Jean
Laboratoire de Biologie Moléculaire Eucaryote, CNRS UPR 9006, 118, route de Narbonne,
31062 Toulouse cedex, France

Ponce, M. Lourdes
Cell Biology Section, National Institute of Dental Research, NIH, 30 Convent Drive, MSC 43
70, Bethesda, MD 20892, USA

Pouysségur, Jacques
Centre de Biochimie and Centre Antoine Lacassagne, CNRS-UMR 6543, Université de Nice,
Faculté des Sciences, Parc Valrose, 06108 Nice cedex 2, France

Preissner, Klaus T.
Institut für Biochemie, Fachbereich Humanmedizin, Justus-Liebig-Universität,
Friedrichstrasse 24, 35392 Giessen, Germany

Rappaport, Lydie
INSERM U127, IFR "Circulation", Hôpital Lariboisière, 41, Boulevard de la Chapelle, 75475
Paris cedex 10, France

Rendu, Francine
INSERM U428, "Risque Thrombotique et Mécanismes Cellulaires et Moléculaires de
l'Hémostase", Faculté de Pharmacie, 4 avenue de l'Observatoire, 75006 Paris, France

Samuel, Jane-Lyse
INSERM U127, IFR "Circulation", Hôpital Lariboisière, 41, Boulevard de la Chapelle, 75475
Paris cedex 10, France

Schved, Jean-Francois
Laboratoire Central d'Hématologie, CHU Montpellier, 2, avenue Bertin Sans, 34295
Montpellier Cedex 5, France

Seiffert, Dietmar
DuPont Merck Research Laboratories, Experimental Station, Wilmington, DE 198800-0400,
USA

Singh, Jai Pal
Lilly Research Laboratories, Lilly Corporate Center, Indianapolis, Indiana 46285, USA

Sordello, Sylvie
Laboratoire de Biologie Moléculaire Eucaryote, CNRS UPR 9006, 118 Route de Narbonne,
31062 Toulouse, France

Tournier, Jean-François
Laboratoire de Biologie Moléculaire Eucaryote, CNRS UPR 9006, 118, route de Narbonne, 31062 Toulouse cedex, France

van't Veer, Cornelis
University of Maastricht, General Surgery, P.O. Box 616, 6200 MD Maastricht, The Netherlands

Viñals, Francesco
Centre de Biochimie and Centre Antoine Lacassagne, CNRS-UMR 6543, Université de Nice, Faculté des Sciences, Parc Valrose, 06108 Nice cedex 2, France

Wautier, Jean-Luc
Institut National de la Transfusion Sanguine, 6, rue Alexandre-Cabanel, 75739 Paris Cedex 15, France

Yan, Zhong-Qun
Cardiovascular Research Unit, Department of Medicine, Center for Molecular Medicine, Karolinska Hospital, 171 76 Stockholm, Sweden

Ziche, Marina
Institute of Pharmacology, University of Siena, Medical School, Vie E.S. Piccolomini 170, 53100 Siena, Italy

Zini, Jean-Marc
Service d'Angio-Hèmatologie, Hôpital Lariboisiëre, 2, rue Ambroise Parè, 75475 Paris Cedex 10, France

AA

Definition   *Arachidonic acid*

See: →Bleeding disorders; →Complement system (interaction of vascular cells with); →Vasomotor tone regulation, molecular mechanisms of; →Platelet stimulus-response coupling; →Prostacyclin; →Prostaglandins; →Thromboxanes

AC

Definition   *Adenylate cyclase*

See: →Hormonal regulation of vascular cell function in angiogenesis; →Vasomotor tone regulation, molecular mechanisms of

ACE

Definition   *Adrenal gland capillary endothelial cell*

See: →Endothelial cells

AchE

Definition   *Acetylcholinesterase*

See: →Megakaryocytes

Actin/Myosin

Definition   *Cytoskeleton proteins involved in contractility, motility and cell division. Two myosin-heavy chains are smooth muscle specific (MHC SM-1 and SM-2) and have contributed to the understanding of development and disease.*

See: →Smooth muscle cells

Adaptor Molecules

Definition   *Molecules of the signal transduction cascade with no intrinsic kinase activity that bind phosphate containing tyrosine kinases domains via their SH-2 domain.*

See: →Signal transduction mechanisms in vascular biology; →FGF-1 and -2; →Tyrosine Kinase Receptors for Factors of the VEGF family; →Platelet stimulus-response coupling

Adhesion

Definition   *Fundamental process in biology. It implies adhesion of cells to the extracellular matrix or to other cells (intercellular adhesion). Adhesion molecules involved in vascular biology include intercellular adhesion molecule (ICAM-1), vascular cell adhesion molecule (VCAM-1), selectins, VE-cadherins and integrins.*

See: →Extracellular matrix; →Signal transduction mechanisms in vascular biology; →Thrombosis; →Vascular integrins; →Cytokines in vascular biology and disease; →Fibrin/fibrinogen; →Blood Cells, interaction with vascular cells

ADP

Definition   *Adenosine diphosphate*

See: →Bleeding disorders

aFGF

Definition   *Acidic fibroblast growth factor*

See: →Fibroblast growth factors; →FGF receptors

AFP

Definition   *Alpha-fetoprotein*

See: →Angiogenesis inhibitors; →Transforming growth factor β

AGE

Definition   *Advanced glycosylation end product*

See: →Blood cells, interaction with vascular cells

Age-Related Macular Dystrophy (AMD)

Definition   *Ocular disease in the elderly characterized by abnormal choroid blood vessels, hemorrhage and retinal lesion*

See: →Endothelial cells

AMD

Definition   *Age-related macular dystrophy*

See: →Age-related macular dystrophy; →Endothelial cells

Ang-1

Definition   *Angiopoietin-1*

See: →Angiopoietins

Ang-2

Definition   *Angiopoietin-2*

See: →Angiopoietins

Angioblasts/Hemangioblasts

Definition *Putative precursor cells for cells of the endothelial and hematopoietic lineage*

See: →Ontogeny of the vascular system; →Endothelial cells

Angiogenesis

Synonym: Neovascularisation

Definition *Angiogenesis describes the formation of new blood vessels from preexisting vessels. Although angiogenesis focuses mainly on the formation of capillaries, it includes the formation of small and large blood vessels. In contrast, vasculogenesis (the other recognised mechanism of vessel development) involves the differentiation of new vessels from embryonic structures known as blood islands.*

See also: →Ontogeny of the vascular system; →Hormonal regulation of vascular cell function; →Vascular endothelial growth factor; FGF-1 and -2; →FGF receptors; →Angiogenesis inhibitors; →Vascular integrins; →Fibrinolytic, hemostatic and matrix metalloproteinases, role of

Introduction   In adults all new vessels are formed by angiogenesis, in contrast during embryonic development some organs are vascularised by either vasculogenesis, angiogenesis, or both (for example in the brain, larger vessels initially form by vasculogenesis but the interior is subsequently vascularised by three distinct and successive waves of angiogenesis [1]).

Extensive angiogenesis occurs during embryonic development, in contrast, active angiogenesis is an exception in healthy adults. Physiological processes of which angiogenesis is a component include wound healing and the female reproductive cycle.

It follows that since angiogenesis is the only mechanism in adults by which new blood vessels are formed, diseases in which new vessel formation occurs must be angiogenesis-dependent. Among such disorders are the two major killers of western populations namely atherosclerosis and cancer. In contrast, some pathologies are characterised by insufficient angiogenesis (e.g. impaired wound healing, some female reproductive failures) and enhancing new vessel formation could potentially be a successful approach to treatment.

Angiogenesis is currently a focus of major scientific interest. In particular, extensive effort has been directed towards elucidation of its physiological control. A driving force being the wealth of potential clinical applications of both proangiogenic and antiangiogenic therapies.

Characteristics   Angiogenesis is a multistep process subject to complex regulation. Numerous proangiogenic and antiangiogenic factors have been isolated. In many cases, the activity of factors depends on the local concentration and/or the microenvironment.

Although, endothelial cells are the primary focus of angiogenic research, several other cell types are also involved. Chief among these are vascular smooth muscle cells, pericytes and fibroblasts. Moreover, non-cellular structures, such as the basal lamina and the extracellular matrix also play an important part in angiogenesis.

Endothelial cells are usually quiescent and it has been shown that an endothelial cell will undergo only a few divisions in an individual's adult life. In contrast, angiogenic endothelium undergoes rapid proliferation. Such proliferation is accompanied by endothelial cell migration and both processes contribute to new vessel formation.

As was mentioned above, angiogenesis comprises several stages, starting with the release of proteases, followed by endothelial cell proliferation, invasion and tube formation. A possible sequence of events follows:

a. retraction of pericytes from the ablumenal surface of the capillary;
b. release of proteases by endothelial cells;
c. degradation of basal lamina and the extracellular matrix surrounding the vessel;
d. endothelial cell migration and proliferation;
e. formation of tube-like structures;
f. anastomoses (fusion) of newly formed tubes;
g. initiation of blood flow.

Migration usually involves invasion, since endothelial cells must penetrate the tissue to be vascularised. Thus, activated endothelial cells locally degrade the underlying basal lamina and invade the surrounding stroma. This is achieved by the release of proteases (e.g. serine proteases such as urokinase [2] (uPA) or tissue plasminogen activator [3] (tPA) and members of a large family of matrix metalloproteinases (MMPs)), and down-regulation of the expression of protease inhibitory factors (e.g. tissue inhibitors of MMP – TIMPs [4]). The net result is partial extracellular matrix (ECM) degradation (for a review of the role ECM in angiogenesis see [5]). This, in turn, enables cellular movement that involves repeated adhesion/degradation cycles, and pseudopodia-based crawling. There exist numerous heparin-binding ECM-stored endothelial cell growth factors and chemoatractants that are released as a consequence of ECM degradation. Since proteolytic activity is often highly localised (spatial control by means of soluble inhibitors, cell surface anchoring receptors, etc; for a comprehensive review see [6]), subtle chemical gradients are formed that directly contribute to endothelial cell alignment and tube formation.

The angiogenic factor known as endothelial-cell-stimulating angiogenesis factor (ESAF) has been shown to directly activate proteinase activity by dissociation of collagenase and gelatinase A complexes from tissue inhibitor of metalloprotease type 1 (TIMP-1) [7].The molecular identity of ESAF is as yet to be elucidated.

Regulation   Angiogenesis is a process that is regulated by a combination of paracrine, autocrine, and localised environmental stimuli. This regulation involves the endothelial and other cellular components of a vessel, the extracellular matrix and cells of the target tissue.

Undoubtedly, the two environmental factors of primary significance in the regulation of angiogenesis are the local concentrations of glucose and of oxygen. This points to the central physiological role of angiogenesis, which is to ensure that the blood flow through a given tissue is adequate to supply nutrition and to enable cells to be rid of metabolic wastes.

Traditionally, it was a shift in metabolic pathways that was viewed as the primary cellular response to glucose or oxygen depletion. Enhanced glycolysis via allosteric regulation of various metabolic enzymes is a classic example. Such activity clearly has relevance in physiology (particularly in the case of muscle fibers) but the effect is short term when compared to the angiogenic response. In development, such hypoxia- and hypoglycemia-driven pro-angiogenic stimulation could be sufficient for angiogenesis-mediated vascularisation of developing organs, particularly where no vasculogenesis is involved (e.g.: brain [8] and kidney in an avian chimeric embryo model [9]).

A major effort is directed towards understanding exactly how conditions of stress such as hypoxia and hypoglycemia act at the molecular level to initiate angiogenesis, for example, what changes in gene expression patterns they cause and which specific transcription factors are involved.

The best-known hypoxic response pathway is that of hypoxia-inducible factor 1 (HIF1-alpha). HIF1 (for a review see [10]) specifically recognises an 8 bp motif (the hypoxia-response element – HRE), first identified in the erythropoietin gene promoter. Subsequently, HREs have been found in the promoters of several other hypoxia-stimulated genes including various glycolytic enzymes and of interest to angiogenesis research, both in vascular endothelial growth factor (VEGF) [11] and one of its receptors, flt-1 [13].

Hypoxia-inducible factor 1 (HIF1) is a heterodimeric complex composed of two basic-helix-loop-helix Per-AHR-ARNT-Sim (PAS) proteins (HIF-1-alpha and -1-beta). HIF1-beta had been previously described as the aryl hydrocarbon receptor nuclear translocator, hence the acronym: AHR-ARNT. The pathway has been shown to be activated not only by hypoxia but also in response to Co^{2+} ions and to iron chelation.

Recently the cDNA for another putative hypoxia-response factor was identified with a (predicted) amino acid sequence similarity to HIF1-alpha [12]. The factor, termed HIF1-alpha-like factor (HLF), is similar to HIF1-alpha in that it binds hypoxic-response elements (HREs); however, its tissue expression was markedly different to that of HIF1-alpha.

Not only VEGF but also its receptors flt-1 [13] and KDR are induced under hypoxia. However, the expression of KDR has also been reported to be down-regulated by hypoxia. Studies in some models have also shown a more complicated pattern, with an initial decrease followed by a subsequent increase in expression [14].

Clearly, gene expression in response to hypoxia is complex, differs between models and may be significantly different in *in vitro* cell culture compared to the *in vivo* environment [15]. The angiogenic enzyme platelet-derived endothelial cell growth factor/thymidine phosphorylase was recently shown to be up-regulated by hypoxia [16] but by an as yet unidentified mechanism. A major effort is now being directed towards the identification of other genes up-regulated by hypoxia.

The abundance of both pro- and antiangiogenic factors and the fact that none appear to be the uniquely controlling factor shows that the control of angiogenesis is complex. Indeed, the concept of an angiogenic switch (for a review see [17]) assumes that the final outcome of pro- and antiangiogenic action depends on a net balance between them. One cannot predict the response of endothelial cells *in vivo* to a particular stimulus without an appreciation of the microenvironment in which the cell is situated.

Genes Vascular endothelial growth factor/vascular permeability factor (VEGF/VPF) increases vessel permeability and is a specific growth factor for endothelial cells. VEGF and its receptors are considered to be one of the most crucial regulatory pathways in angiogenesis and as a result have been intensely studied.

VEGF is induced in a wide range of normal and cancer cells under conditions of hypoxia [18] and hypoglycemia [19], and in the presence of some growth factors [20], oncogenes [21] and tumour suppressor gene mutations [22].

Four spliced-isoforms of VEGF, termed VEGF121, VEGF165, VEGF189, and VEGF206 (named after the number of amino acids in the mature protein) have been described. VEGF121 and VEGF165 are the most abundant isoforms in most tissues. VEGF189 is also fairly abundant and its transcript can be detected in most tissues in which VEGF is expressed. In contrast, VEGF206 is rare and has so far been detected only as a message in a human foetal liver cDNA library.

As mentioned above there are currently two recognised VEGF receptors, both tyrosine kinases: Flt-1 (fms-like-tyrosine kinase) and KDR (kinase insert domain region). Flk-1 (fetal liver kinase-1) is the murine homologue (85% sequence identity) of human KDR. Although similar (seven immunoglobulin-like domains, a transmembrane region, and a tyrosine kinase motif), Flt-4 has not been shown to be a VEGF receptor. In contrast to Flt-1, the KDR promoter does not contain an 8 bp HRE (hypoxia-response element). Thus, Flt-1 expression is upregulated in a HIF-1 driven hypoxia-response pattern, whereas KDR is not. However, it cannot be excluded that another as yet unidentified pathway, regulates KDR expression in hypoxia [23] possibly at the post-transcriptional as opposed to the transcriptional level.

Recently two new genes with homology to VEGF have been identified (termed VEGF-B [24], and VEGF-C [25] – localising to chromosome 11q13 and 4q34 respectively) and the existence of a VEGF protein family (of at least 3-members) postulated. VEGF-C is a ligand for Flt-4 (or VEGFR-3), and judging from its expression in tissues [26], and transgenic studies, it plays a crucial role in the development of the lymphatics. For VEGF-B, two isoforms with completely different C-terminal domains have been described and different functional properties postulated. Both VEGF-B and VEGF-C have been found to be intensely expressed in the human placenta. Despite having sequence similarity, the members of the VEGF family show differential regulation of expression and different patterns of expression in tissues [27].

The fibroblast growth factor protein family now comprises eleven distinct but related growth factors. They are known to possess diverse biological activities and to be involved in numerous physiological and pathological processes. Not only are there numerous members of the family, but also assorted sets of receptors that operate via different signal transduction pathways (for a recent review see [28]).

Two of the eleven are considered to be key angiogenic factors, namely FGF-1 (or acidic FGF) and FGF-2 (or basic FGF), but others may also be involved. Both FGF-1 and FGF-2 are growth and chemotactic factors for endothelial cells. They also induce release of proteinases from the endothelial cell which are capable of degrading basal lamina and the extracellular matrix. FGFs bind strongly to heparin, and associate with the basement membrane and extracellular matrix. Endothelial cells produce large quantities of FGF-2 in culture suggesting a possible autocrine mode of action.

Both FGF-1 and FGF-2 lack classic secretion signals and the mechanism of their transport outside a cell remains unclear. It has been proposed that these factors associate with other protein components and are released as complexes with them under conditions of stress (for example hypoxia).

Tumour necrosis factor alpha has multiple effects on endothelial cells. It can induce both apoptosis and inflammation. Apoptosis prevails if the TNF-alpha concentration is high, inflammation if it is comparatively low [29]. There are two receptors for TNF-alpha: termed p55 and p75. In general, p75 is associated with the induction of apoptosis, whereas p55 is mainly active in the context of a proliferative/proinflammatory response.

There is substantial evidence that TNF-alpha *in vitro* causes apoptosis of proliferating endothelium [30, 31]. This points to a potential use of TNF-alpha in vascular-targeted anticancer experimental therapies[32]. It should be noted that production of TNF-alpha protects against the effects of extracellular TNF-alpha, possibly by the induction of manganous superoxide dismutase that neutralises the free oxygen radicals which mediate exogenous TNF-alpha toxicity. This activity can also protect against certain chemotherapeutic agents: e.g. doxorubicin [33] and adriamycin [34].Thus, TNF-alpha exerts its cytotoxic effect only in a paracrine fashion. However, in gene therapy applications this need not pose a problem and may even be advantageous. Since

the transduction efficiency of most delivery systems is low, a few endothelial cells producing large amounts of TNF-alpha could induce apoptosis in a large area of surrounding tumour endothelium [32].

Low doses of TNF-alpha induce several adhesion molecules on endothelial cells, including E-selectin, ICAM-1 (intercellular adhesion molecule-1) and VCAM-1 (vascular (1)cell adhesion molecule-1). These promote leukocyte adhesion to the endothelium. Possibly due to the enhanced expression of such adhesion molecules, administration of low doses of TNF-alpha enhances tumour xenografting efficiency [35].TNF-alpha also exerts a procoagulant effect [36 – 38]. This, at least in part, accounts for haemorrhagic necrosis that occurs in tumours treated with high doses of TNF-alpha and melphalan in an isolated limb perfusion model [39].

Thymidine phosphorylase (or platelet-derived endothelial cell growth factor – TP) is an angiogenic enzyme overexpressed in many tumour types (for a review see [40]). The mechanism by which TP stimulates angiogenesis is rather unusual. TP is not a classic growth factor in that it has no receptor and it is not directly mitogenic on endothelial cells. Rather, the product of TP's enzymatic action on thymidine, 2-deoxy-D-ribose, is chemotactic to endothelial cells. The mechanism of this activity is unknown. 2-deoxy-D-ribose is secreted from cells in which it is released from thymidine by TP [41]. Necrosing tumour cells release DNA that is hydrolysed to thymidine that is then in turn converted by TP to 2-deoxy-D-ribose.

Angiopoietin-1 [42] is a ligand for the tyrosine kinase receptor Tie-2. Tie-2 was described prior to identification of its ligands, the latter of which were expected to play a key role in angiogenesis in view of the complete specificity of Tie-2 expression in the endothelium [43]. It was a significant advance when the angiopoietin-1 cDNA was isolated [44], and found to stimulate Tie-2 phosphorylation although unexpectedly without direct proliferative or chemotactic action on the endothelial cell.

A critical role for angiopoietin-1 in physiological angiogenesis has come from studies of transgenic mice. Thus, it was demonstrated that both Tie-2 and angiopoietin-1 are indispensable for embryonic vascular development. Both Tie-2 and angiopoietin-1 mouse gene knock-outs exhibit a similar vascular phenotype with poorly differentiated pericytes and vascular smooth muscles. It is now thought that rather than exerting a direct proliferative effect on endothelial cells, angiopoietin-1 plays a role in the recruitment of perivascular cells (e.g. pericytes, smooth muscle cells, andmyocardiocytes) and elaboration of the newly formed vascular tree. Interestingly, an activating mutation in Tie-2 is associated with inherited venous malformations [45].

Both Tie-2 and angiopoietin-1 have close homologues termed Tie-1 and angiopoietin-2 respectively. Angiopoietin-2 acts as a competitive inhibitor of angiopoietin-1 binding to Tie-2 but does not induce its phosphorylation. No ligands for Tie-1 have as yet been identified.

It has been demonstrated that a xenografted primary Lewis lung carcinoma produces a circulating inhibitor of angiogenesis termed: angiostatin [46].Removal of the primary tumour results in rapid vascularisation and growth of metastases. Angiostatin is a 38 kDa internal proteolytic fragment of plasminogen. More recently another tumour-derived angiogenesis inhibitor called endostatin, secreted by a haemangioendothelioma has been identified [47]. Endostatin is, a 20 kDa C-terminal fragment of collagen XVIII, and like angiostatin, has the capacity to selectively inhibit tumour angiogenesis and growth in animal models. Thrombospondin-1 (and its fragments) are yet another tumour derived angiogenesis inhibitor [48]. The p53 tumour suppressor gene induces expression of thrombospondin-1 in some cell types (e.g. Li-Fraumeni fibroblasts) [49].

The characterisation of endogenous circulating inhibitors of angiogenesis provide not only new therapeutic opportunities but also identify a novel physiological angiogenesis control mechanism. Thus, both angiostatin and endostatin are fragments of larger proteins devoid of anti-angiogenic activity. An N-terminal fragment of prolactin, an internal fragment of platelet factor 4 and laminin fragments have also been shown to exhibit antiangiogenic properties. It could be that proteases released by activated endothelial cells are also engaged in the proteolytic production of angiogenesis inhibitors, constituting a feedback control mechanism.

Molecular Interactions   Angiogenesis is so complex a process that is clearly impossible to describe in this article all known angiogenic molecules and the molecular interactions involved. However, an important area of molecular interactions in angiogenesis is cell adhesion. avb3 and avb5 integrins have been shown to play a critical role in angiogenesis. avb3 is involved in VEGF induced neovascularisation, while avb5 is involved in bFGF mediated angiogenesis. Furthermore, specific antibodies to avb3 [50, 51] and avb5 [52] integrins block respectively VEGF and bFGF mediated angiogenesis. The integrins are most likely to have a role in endothelial cell attachment to the extracellular matrix and in cell migration. NO is also involved in VEGF but not bFGF induced angiogenesis as administration to rabbits of the NO synthase inhibitor, L-NAME, completely blocks VEGF but not bFGF induced corneal angiogenesis [53]. Curiously, both soluble E-selectin and soluble VCAM have been shown to be angiogenic [54].

Cells and Cellular Interactions   Traditionally endothelial cells have been a major focus of angiogenesis research. However, other cells such as pericytes, fibroblasts and macrophages are clearly involved. Pericytes are thought to act mainly as angiogenesis inhibitors and their retraction is one of the first steps in angiogenesis. It is possible that minor paracrine angiogenic stimuli prevail in the body and pericytes act as a key factor restricting spontaneous endothelial cell proliferation [55]. However, the topic of pericyte sensing of angiogenic stimuli has been

as yet rather neglected and little data is available. The exact origin and fate of pericytes is uncertain, however, some evidence suggests that they are derived from fibroblasts and also that pericytes themselves could give rise to other mesenchymal-type cells (for a review see [56]). Aminopeptidase A (specifically the subtype recognised by the monoclonal antibody RC38) has been reported to be upregulated in activated pericytes [57]. Platelet-derived growth factor-beta and VEGF receptors are present in pericytes [58, 59]. Thus, it is possible that VEGF may directly activate pericytes and induce their retraction. Finally, the recently described angiopoeitin is a strong candidate to be a key regulator of pericyte/endothelial cell interactions during vessel formation.

Tumor-associated macrophages (TAMs) are also thought to play a role in tumour angiogenesis. TAMs are abundant in many tumours, for example, in some breast carcinomas they may contribute up to 50 % of the overall tumour mass. Tumour cells can produce a range of monotactic factors: MCP-1, MCP-2, MCP-3, GM-CSF, G-CSF and M-CSF that attract macrophages into the tumour. Interestingly VEGF has also been shown to act as a monocyte chemoattractant. Previously TAMs were considered to be a part of the host immune response to tumours. However, recently TAMs have also been shown to be pro-angiogenic and may therefore enhance rather than inhibit tumour growth. It appears that TAMs produce several angiogenic factors (including VEGF, bFGF, EGF, TNF-alpha, TP, HGF/SF, IGF-1, IL-8). Studies have indicated that the degree of macrophage infiltration in a series of invasive breast carcinomas correlates with high vascular grade, reduced relapse-free survival and reduced overall survival [60].

Additional Features   Mathematical modelling of angiogenesis has been attempted, particularly with regard to wound healing [61] and tumourigenesis [59]. Mathematicians have attempted to analyse space and time interactions between the different cells and molecules involved in angiogenesis. Parameters examined include pre-existing vessels, capillary sprouts, pericytes, fibroblasts, oxygen tension, angiogenic factors and extracellular matrix. Future computer modelling could give insight into the dynamic balance of factors controlling the angiogenic switch.

A successful attempt to model tumour angiogenesis [62] has involved reaction-diffusion theory (Turing models). This demonstrated that small vascularised tumours secreting both angiogenesis stimulating and inhibiting factors should, in agreement with experimental data, invade their surroundings as columns of cells spreading from the central tumour mass.

There is striking correlation between activated endothelial cell behaviour and cancer invasion (for a review see [63]). The invasive phenotype, which in the case of endothelial cells is an essential component of their physiological activity, in cancer is purely a pathological consequence of cell deregulation and dedifferentiation. It is notable that the same set of proteolytic enzymes is involved in pathological invasion by carcinoma cells as in physiological invasion by endothelial cells (e.g.: metalloproteinases, urokinase and tissue-specific plasminogen activator). Both these protease-dependent processes (i.e. angiogenesis, and cancer-cell invasion) are necessary for metastatic spread and this adds appeal to protease inhibition based anti-metastatic therapies. Indeed some protease inhibitors (e.g. peptide derived 'Marimastat') have proved quite successful in pre-clinical and early clinical trials. For a review on MMP inhibitors in clinical trials see [64].

A recent report has questioned whether new vessels arise in the adult solely by angiogenesis. Thus, researches claim [65] to have isolated endothelial cell progenitors from peripheral blood by means of magnetic bead mediated selection for expression of the surface antigens CD34 and Flk-1. These putative progenitors have been reported to integrate into the existing vasculature on re-injection preferentially at sites of active angiogenesis in animal models, allegedly taking part in angiogenesis itself. If confirmed, these observations would strongly suggest that a mechanism similar to embryonic vasculogenesis might exist in adults. In such a case, peripheral blood endothelial cell progenitors would clearly be the adult equivalent of the cells that comprise the embryonic blood islands.

Clinical Significance   Angiogenesis constitutes the only mechanism by which new vessels develop in adults. Delayed angiogenesis severely impairs wound healing. Insufficient, angiogenesis may be involved in some female reproductive failures since the menstrual cycle involves periods of intensive angiogensis [66]. Angiogenesis is indispensable for successful skin grafting. On the other hand, there are a host of conditions where it is excessive angiogenesis that contributes to the pathology. The following are some examples

a. cancer;
b. atherosclerosis;
c. arthritis;
d. psoriasis;
e. diabetic retinopathy;
f. endometriosis;
g. menorrhagia;
h. haemangiomas and venous malformations.

Much attention has focussed on anti-angiogenesis as a means of antimetastasic therapy. The concept was formulated as early as 1971 when Judah Folkman proposed [67] that angiogenesis is indispensable for the growth of solid tumours beyond a diameter of a few millimetres. In 1990 in the *Journal of the National Cancer Institute* [68] Folkman collected early evidence in the form of 14 points which will be quoted herein verbatim from this important paper:

A. The growth rate of tumors implanted in subcutaneous transplant chambers in mice is slow and linear before vascularization and rapid and nearly exponential after vascularization.

B. Tumors grown in isolated perfused organs where blood vessels do not proliferate are limited to 1-2mm^3 but expand rapidly to 1-2cm^3 after vascularization on transplantation to mice.

C. Tumor growth in the avascular cornea proceeds slowly and at a linear rate but switches to exponential growth after vascularization.

D. Tumors suspended in the aqueous fluid of the anterior chamber of the eye remain viable, avascular, and limited in size (<1 mm^3). Once they are implanted on the iris vessels, however, they induce neovascularization and grow rapidly, reaching 16,000 times their original volume within 2 weeks.

E. Human retinoblastomas metastic to the vitreous or the anterior chamber are similarly avascular, viable and growth restricted.

F. Within a solid tumor, the [3H] thymidine labeling index of tumor cells decreases with increasing distance from the nearest open capillary. The mean labeling index for a given tumor is a function of the labeling index of the vascular endothelial cells in that tumor.

G. Tumors implanted on the choriollantoic membrane of the chick embryo are often restricted in growth during the avascular phase (=72 hr.), but rapid growth begins within 24 hours after vascularization. In one study, tumors did not exceed a mean diameter of 0.93 ± 0.29 (SD) mm during the avascular phase, but after vascularization, tumors reached a mean diameter of 8.0 ± 205 mm by day 7.

H. The chorioallantoic membrane appears on day 5 in the chick embryo, and the [3H] thymidine labeling index of its vascular endothelial cells decreases with age, with an abrupt reduction on day 11. Tumors implanted on the chorioallantoic membrane in successively older embryos grow at slower rates in parallel with the reduced rates of endothelial cell growth.

I. Vascular casts of metastases to the rabbit liver show that these tumors are avascular up to 1 mm in diameter. Beyond that size, the tumors are vascularized.

J. Carcinoma of the ovary metastasizes to the peritoneal membrane as tiny avascular seeds, which rarely grow beyond a limited size until after vascularization.

K. Angiogenesis inhibitors that are not cytostatic to tumor cells *in vitro* inhibit tumor growth *in vivo*.

L. The appearance of neovascularization at the base of a human melanoma is associated with increased growth and metastasis. Metastasis is rare prior to neovascularization.

M. In one study of transgenic mice that develop carcinomas of the pancreatic islet (beta cells), large tumours arose from a subset of preneoplastic hyperplastic islets that had become vascularized.

N. After subcutaneous injection of tumour cells into mice, tumours have become vascularized at about 0.4mm^3. With increasing tumour size, the blood vessels occupied approximately 1.5% of the tumour volume, a 400% increase over normal subcutaneous tissue. The tumour infiltrated surrounding connective tissue and expanded into the newly formed vessels in that tissue.

Antiangiogenesis should be viewed as a field rapidly moving from preclinical to clinical trials. There are a number of small molecular-weight compounds already in phase I or II clinical trials, including: thalidomide, AGM-1470, carboxyamido-triazole (CAI), interleukin-12 (IL-12), and a group B *Streptococcus* toxin (for detailed reviews see [69, 70]). In general, the preliminary results suggest that disease stabilisation may be achievable with only moderate side-effects.

Thalidomide is a well-known drug with a bad press as the causative agent of a number of birth defects in the late 50s. The teratogenic effects of the drug, including impaired limb development, are now thought to be results of its antiangiogenic activity. Angiogenesis appears to be essential for normal prenatal development, particularly in the case of limbs where long bones must form. There are ongoing phase II clinical trials involving thalidomide for patients with breast cancer, prostate cancer, Kaposi's sarcoma and glioblastoma [71, 72].

AGM-1470, probably the best known antiangiogenic drug in clinical trials, is a fumagillin derivative which shows less toxicity and more potent antiangiogenic properties than the maternal compound. AGM-1470 toxicity observed in phase I and II clinical trials included neurotoxicity with anxiety, asthenia, dysphoria, agitation and gait disturbance. Prolonged periods of disease stabilisation in some patients were observed.

Carboxyamido-triazole (CAI) is a calcium-mediated signal transduction pathway inhibitor. CAI exhibits not only antiangiogenic, but also inhibits endothelial and cancer-cell migration and invasion. The mechanism of pleiotropic activity of this kind is probably the universal block of calcium-mediated signal transduction. Such a broad effect could be the cause of major toxicity of the compound. Fortunately, clinical trials in patients with melanoma, colorectal, ovarian, and lung cancer demonstrated that the drug is quite well tolerated, with nausea, vomiting and fatigue, or cerebellar ataxia and confusion (depending on the route of administration) being the dose-limiting toxicities [73].

Animal experiments with a panel of immunodeficient mouse strains revealed that the presence of T cells is necessary for the action of IL-12. Furthermore, administration of anti-interferon-gamma antibodies inhibited the antiangiogenic response, and interferon-gamma itself was able to abrogate proangiogenic stimuli in a similar fashion to IL-12. Thus, IL-12 is thought to act indirectly through the upregulation of interferon-gama and consequently interferon-inducible protein 10 which, in turn, exerts direct antiangiogenic action. Since the pharmacokinetic characteristics of IL-12 are clinically preferable to those of interferon-gamma, it is reasonable to use IL-12 as a physiological interferon-gamma inducer in antiangiogenic treatment.

Group B *Streptococcus* toxin is the agent mediating pulmonary vessel toxicity in a respiratory distress syndrome in human noenates infected with group B β-hemolytic *Streptococcus*. It was suggested that abnormal tumour vasculature might be subject to the toxic effects in a similar way to the immature pulmonary endothelium. In phase I clinical trials dose limiting toxicities turned out to be dysponea and cardiac arrhythmia. Tumour pain, bronchial spasm and hypotension were also observed. There have, however, been some very encouraging observations of tumour regression in these trials [74, 75].

It should be noted that if antiangiogenic compounds with sufficiently low toxicity are identified, they would be of clinical use not only in the treatment of cancer but also in the many other pathologies of which angiogenesis is a component.

There are biological therapies involving the vasculature that are alternatives to antiangiogenesis, for example, vascular targeted immunotoxins or gene therapy. The potential of such therapies has been demonstrated in models developed by Burrows and Thorpe [76]. In their study, antibodies recognising mouse major histocompatibility complex (MHC) class II (which is only expressed in mouse endothelium when it is inflamed) were employed. Following xenografting of an interferon-gamma expressing neuroblastoma cell line only endothelium in the tumour expresses MHC class II, and the antibody coupled to ricin-toxin is administered into the tail vein. A single administration of the immunotoxin brought about complete eradication of large solid tumours. It was shown that endothelial destruction induced extensive thrombosis followed by hypoxic tumour necrosis.

The key to success in vascular targeting is specific killing of tumour, i.e. proliferating, migrating, and frequently also inflamed endothelium. Systemic therapy constitutes the ultimate goal. Selective killing could be ensured by either selective delivery of a toxic particle (e.g. immunotoxins targeted to tumour endothelial specific antigens), selective transcription of a toxin encoding a prodrug activating gene, or by means of a combined approach (e.g. adenovirus coupled to an antibody delivering a gene under the control of a specific promoter). The use of retroviruses in gene therapy strategies offers an additional level of targeting (all retroviruses with the exception of lentiviruses will not incorporate into non-dividing cells); however, the transduction efficiency in retroviral delivery systems is usually very low.

Lucas Huminiecki and Roy Bicknell

References

1. Breier G, Risau W, Trends Exp Clin Med (in press)
2. Iwasaka C (1996) J Cell Physiol 169:522-531
3. Tyagi SC et al (1996) Canadian J Physiol Pharmacol 74: 983-995
4. Arbiser JL et al (1997) PNAS 94:861-866
5. Vlodavsky et al (1997) In: Bicknell R, Lewis CE, Ferrara N (eds) Tumour Angiogenesis. Oxford, Oxford University Press, pp 125-140

6. Werb Z (1997) Cell 91:439-442
7. McLaughlin B, Weiss JB (1996) Biochem J 317:739-745
8. Stewart PA et al (1981) Dev Biol 84:183-192
9. Pardanaud L et al (1989) Development 105:473-485
10. Guillemin K, Krasnov MA (1997) Cell 89:9-12
11. Liu Y et al (1995) Circulation Res 77:638-643
12. Ema M et al (1997) PNAS 94:4273-4278
13. Detmar M et al (1997) J Invest Dermatol 108:263-268
14. Takagi H et al (1996) Invest Ophthalm Visual Sci 37:1311-321
15. Sandner P et al (1997) Kidney International 51:448-453
16. Griffiths L et al (1997) Cancer Res 57:570-572
17. Hanahan D, Folkman J (1996) Cell 86:352-364
18. Shweiki D et al (1992) Nature (London) 359:843-845
19. Shweiki D et al (1995) PNAS 92:768-772
20. Horiuchi T et al (1997) Am Journal Resp Cell Mol Biol 17:70-77
21. Mukhopadhyay D et al (1995) Cancer Research 55:6161-6165
22. Fontanini G et al (1997) British J Cancer 75:1295-1301
23. Waltenberger J et al (1996) Circulation 94:1647-1654
24. Olofsson B et al (1996) PNAS 93:2576-2581
25. Paavonen K et al (1996) Circulation 93:1079-1082
26. Kukk E et al (1996) Development (Cambridge) 122:3829-3837
27. Enholm B et al (1997) Oncogene 14:2475-2483
28. Christofori G (1997) In: Bicknell R, Lewis CE, Ferrara N (eds) Tumour Angiogenesis. Oxford, Oxford University Press, pp 201-237
29. Fajardo LF et al (1992) Am J Pathol 140:539-544
30. Meyrick B et al (1991) Am J Pathol 138:92-101
31. Sato N et al (1987) J Natl Cancer Inst 79:1383-1391
32. Jaggar RT et al (1997) Hum Gene Ther 8:2239-2247
33. Kobayashi D et al (1997) Blood 89:2472-2479
34. Maeda M et al (1994) Int J Cancer 58:376-379
35. Batten P et al (1996) Immunology 87:127-133
36. Poll T (1990) N Engl J Med 322:1622-1629
37. Ten Cate JW et al (1997) Thrombosis and Haemostasis 78:415-419
38. De Benedetti F et al (1997) British J Rheumatology 36:581-588
39. Nooijen PTGA et al (1996) British J Cancer 74:1908-1915
40. Moghaddam A et al (1997) In: Bicknell R, Lewis CE, Ferrara N (eds) Tumour Angiogenesis. Oxford, Oxford University Press, pp 251-260
41. Shaw T et al (1988) Mutation Research 200:99-116
42. Suri C (1996) Cell 87:1171-1180
43. Schlaeger TM et al (1995) Development 121:1089-1098
44. Davis S (1996) Cell 87:1161-1169
45. Vikkula M (1996) Cell 87:1181-1190
46. O'Reilly MS et al (1994) Cell 79:315-328
47. O'Reilly MS et al (1997) Cell 88:277-285
48. Good DJ et al (1990) PNAS 87:6624-6628
49. Dameron KM et al (1994) Science 265:1582-1584
50. Brooks P et al (1994) Cell 79:1157-1164
51. Brooks P et al (1995) J Clin Invest 1815-1822
52. Friedlander M et al (1995) Science 270:1500-1502
53. Ziche M et al (1997) J Clin Invest 99:2625-2634
54. Koch AE et al (1995) Nature 376:517-519
55. Egginton S et al (1996) Microvascular Research 51:213-228
56. Hirschi KK et al (1996) Cardiovascular Res 32:687-698
57. Schlingemann RO et al (1996) J Pathol 179:436-442
58. Sundberg C et al (1993) American J Pathol 143:1377-1388
59. Takagi H et al (1996) Diabetes 45:1016-1023
60. Leek-R-D et al (1996) Cancer Research 56:4625-4629
61. Pettet G et al (1996) Proceedings Royal Soc London Series B Biol Sci 263:1487-1493
62. Chaplain MAJ (1995) J Biol Systems 3:929-936
63. Kohn EC, Liotta LA (1995) Cancer Research 55:1856-1862
64. Wojtowicz-Praga SM et al (1997) Investigational New Drugs 15:61-75
65. Asahara T et al (1997) Science 275:964-967
66. Reest MCP, Bicknell R (1998) Angiogenesis (in press)
67. Folkman J (1971) New Engl J Med 285:1182-1186
68. Folkman J (1990) J Natl Cancer Inst 82:4-6
69. Masiero L et al (1997) Angiogenesis 1:23-35
70. Harris AL (1997) Angiogenesis 1:36-37
71. Fine HA et al (1997) Proc Am Soc Clin Onc 16:385a
72. Figg WD et al (1997) Proc Am Soc Clin Onc 16:333a
73. Kohn EC et al (1996) Cancer Res 56:569-573
74. DeVore RF et al (1997) Clin Cancer Res 3:365-372
75. Wamil BD et al (1997) J Cancer Res Clin Oncol 123:173-179
76. Burrows FJ et al (1993) PNAS 90:8996-9000

Angiogenesis Inhibitors

Definition *Factors that inhibit the formation of blood vessels in vitro and/or in vivo*

See also: →Angiogenesis; →Signal transduction mechanisms in vascular cells

Introduction Angiogenesis, the formation of new blood vessels, is a multi-step process involving endothelial cell activation, migration, proliferation, and tube formation and maturation. In adult animals, physiological angiogenesis is largely limited to reproductive system (ovary, uterus, placenta) and the tissues undergoing wound healing. A high degree of control of angiogenic processes is maintained by multiple endogenous positive and negative regulators. Several cytokines and growth factors have been identified that initiate and promote angiogenic responses in endothelial and supporting cells in a paracrine and autocrine manner [1-2]. Among the endogenous stimulators described to date, the basic fibroblast growth factor (bFGF) and vascular endothelial growth factor (VEGF) family of growth factors have emerged as the primary and direct stimulators of angiogenesis. The expression of bFGF and VEGF correlates with active angiogenesis observed under physiological and pathological situations. VEGF or bFGF stimulate endothelial cell proliferation and migration *in vitro* and promote angiogenesis *in vivo*. Endothelial cells express specific receptors for VEGF and bFGF which are coupled to the receptor tyrosine kinase signal transduction system. Animals deficient in VEGF receptors exhibit severe defects in blood vessel formation (see reviews [1-4]). In addition, the recent heightened interest in angiogenesis research has led to the identification of other novel endogenous mediators of the angiogenesis pathway.

The discovery of Tie 2 tyrosine kinase receptor ligands angiopoietin 1 and 2, and the elucidation of their role in angiogenesis underscore the importance of endogenous regulators of angiogenesis still to be identified. Perturbation of angiogenic regulators and the associated cellular responses could lead to pathologic angiogenesis as observed in several disease states including, diabetic retinopathy, arthritis, psoriasis, and cancer. On the other hand, deficient angiogenesis may contribute to insufficient circulation following ischemia, and poor wound healing as it occurs in decubital and venous stasis ulcers. Several reviews have been published on the current understanding of the cellular and molecular mechanisms of angiogenesis. The nature of the direct and indirect stimulators of angiogenesis, and their proposed roles in disease states have also been reviewed [1-4]. In this article, we will focus on the endogenous negative regulators of angiogenesis and the pharmacological inhibitors that have been targeted for the prevention of angiogenesis in diseases such as retinopathy and solid tumors. The challenges inherent in the clinical development of the antiangiogenic therapy for cancer are also discussed.

Role in Vascular Biology

Physiological Function In this section the physiological and pharmacological inhibitors of angiogenesis will be discussed. This section also includes the roles of these inhibitors in pathology.

Endogenous Inhibitors of Angiogenesis

In recent years several endogenous polypeptide inhibitors affecting endothelial cell activities *in vitro* and angiogenesis *in vivo* have been identified. Characterization of the endogenous inhibitors have revealed that a local generation of the polypeptide inhibitors from precursor proteins by proteolytic modification may be an important mechanism controlling angiogenesis. Listed in Table 1 are some of the known endogenous antiangiogenic peptides and their precursor proteins. These proteins are known to serve other functions in the regulation of vascular homeostasis or thrombosis. The proteolytic cleavage of the propeptide may serve as

an activation mechanism of antiangiogenic response. Proteolytic enzymes have been shown to facilitate angiogenesis by degrading extracellular matrix, promoting endothelial cell migration, tube formation, and vessel remodelling [5]. The observations summarized in Table 1 suggest that proteolytic enzymes produced at the site of injury, thrombosis or inflammation may also play an important role in the generation of mediators that exert negative control over angiogenesis. Like the endogenous stimulators of angiogenesis, the ability to localize and concentrate at the site of angiogenesis is also an important characteristic of the endogenous inhibitors. Thus, a site specific activation and localization of angiogenesis inhibitors may be fundamental biological mechanisms contributing to a local control of angiogenesis after tissue injury or pathogenic stimulation.

Thrombospondin-1 (TSP-1) TSP-1 is a large glycoprotein (MW=420,000) initially isolated form platelets. TSP-1 is released from alpha granules during platelet aggregation and vascular injury *in vivo*. Subsequent studies have shown that TSP-1 is also produced by other cells, and becomes incorporated into the extracellular matrix [6]. The trimeric protein consists of several functional domains including an N-terminal heparin binding domain, collagen type V binding region, properdin-like and EGF-like repeats, and a calcium-binding domain [6]. The antiangiogenic activity of TSP-1 resides in a 70 kDa fragment containing the collagen homology region and the properdin-like type-1 repeats [7]. The cellular expression of TSP-1 inversely correlates with the angiogenic activity of various tissues. For example, the expression of TSP-1 is greatly increased during the vascular regression phase in endometrium and mammary gland [8]. To the contrary, reduced expression of TSP-1 is found in cells derived from growing tumor or endothelial cells from hemangioma [9]. High level expression of TSP-1 in transformed endothelial cells reverts their phenotype to normal and reduces their ability to form hemangiomas *in vivo* [10]. TSP-1 and the 70 kDa peptide inhibit endothelial cell proliferation *in*

Table 1. Proteolytic Processing and Extracellular Matrix Binding of Endogenous Angiogenesis Inhibitors

Pro-peptide	Modification	Active-peptide	Heparin/ ECM binding
Fibronectin	Proteolytic	N&C-truncated, 29 kDa	yes
Thrombospondin	Proteolytic	N&C-truncated, 70 kDa	yes
Prolactin	Proteolytic	N-truncated, 16 kDa	na
Plasmin	Proteolytic (Angiostatin)	N-truncated, 43 kDa	yes
PF-4	Proteolytic ?	N-truncated, 6kDa (MPF-4)	yes
Collagen XVIII	Proteolytic	20 kDa fragment (Endostatin)	yes

na: data not available
N or C-truncated: N-terminal or C-terminal truncated

vitro and bFGF induced angiogenesis *in vivo* [7]. The *in vitro* activities of TSP-1 are not only limited to endothelial cells. TSP-1 also promotes adhesion and proliferation of smooth muscle cells and fibroblasts. Some of the actions of TSP-1 may be secondary to the induction of growth factors by TSP-1 in the target cells [11]. Despite the demonstration of the *in vitro* and *in vivo* antiangiogenic activities, clinical utility of TSP-1 may be limited because of its multiple activities on normal cells, the difficulties associated with developing an optimum therapeutic candidate from a complex protein, and the difficulties of delivery of such a protein *in vivo*.

Platelet Factor-4 (PF-4)   PF-4 is a 7.8 kDa member of CXC chemokine supergene family. PF-4 is released from alpha granules during platelet aggregation. PF-4 binds heparin and cell surface heparin sulfate. The inhibition of heparin-antithrombin III-thrombin complex formation and thus promotion of coagulation was identified as the primary physiological function of PF-4 [12]. More recently, PF-4 was shown to inhibit endothelial cell proliferation and migration [13]. Recombinant PF-4 inhibits angiogenesis and tumor growth in experimental animals when injected intralesionally [14]. Viral vector mediated gene transfer of a secreted form of PF-4 also produced a marked reduction in vascularization and tumor growth [15]. Compared to other chemokines which are known to exert their cellular responses at low concentration (nM range), endothelial cell growth inhibition by PF-4 requires micromolar concentrations. An N-terminal truncated form of PF-4 has been isolated from mixed lymphocyte/platelet culture supernatant that was 30-100 fold more potent as an inhibitor of endothelial cell proliferation than PF-4 [16]. PF-4 deposited at the site of vascular injury may initially promote the desired coagulation pathway by preventing the association of antithrombin III with heparin sulfate. At a later stage, cleavage of PF-4 by proteases and the dissociation of the N-terminal peptide by still unknown reductive mechanisms may produce a potent antiangiogenic activity. PF-4 binds to heparin sulfate with high affinity, however, the role of the cell surface heparin binding or the existence of other signal transducing receptors for PF-4 is not known. Cell cycle studies have demonstrated that PF-4 blocks endothelial cells in the S phase [17].

Clinical studies with recombinant PF-4 in patients with Kaposi's sarcoma, melanoma, renal cell carcinoma, colon and prostate cancer have been initiated. In phase 1 trials, PF-4 was well tolerated. Efficacy in Kaposi's sarcoma patients receiving intralesional injections has been reported [1]. However, early data from the clinical trials against colon carcinoma using intravenous administration have not shown antitumor activity [18]. The lack of efficacy in these trials could have been due to an inadequate dose. Further studies with higher doses or a more potent form of PF-4 are required to determine the therapeutic potential of PF-4 for the treatment of angiogenic disease.

Prolactin (PRL)   PRL has been shown to promote breast carcinoma in rodents. In hypophysectomized rats, reduced circulating PRL levels were associated with a significant reduction in tumor progression. Prolactin naturally exists as multiple forms generated by proteolytic modifications of the 23 kDa PRL. A 16 kDa N-terminal fragment of prolactin acts as a potent inhibitor of endothelial cell proliferation *in vitro* and of angiogenesis in animal models [19]. The intact 23 kDa prolactin was ineffective in blocking angiogenesis. Although the16 kDa PRL is distinct from other endogenous antiangiogenic molecules, the local generation of active antiangiogenic PRL from a precursor protein further lends support for site specific control of angiogenesis. The effects of 16 kDa PRL are mediated by binding to a specific high affinity receptor on endothelial cells [20]. The intact PRL did not exhibit binding to the 16 kDa PRL receptor. The 16 kDa PRL also inhibited VEGF and bFGF induced MAP kinase activity in endothelial cells. These antiangiogenic activities of 16 kDa PRL are intriguing. Further validation of antiangiogenic activity of 16 kDa PRL and its potential use in the pathogenesis of angiogenesis are still to be determined.

Somatostatin   Somatostatin is an endogenous cyclic tetradecapeptide originally known as somatomedin release-inhibiting factor. Somatostatin inhibits release of peptides such as gastrin, insulin, motilin, cholecystokinin, and substance P [21]. At least five receptor subtypes, belonging to the family of seven transmembrane domain receptors, have been identified. Receptor subtype 2 (SST-2) is believed to control secretory processes in human tumors. Nuclear scanning studies using the SST-2 selective analog, octreotide acetate (SMS 201995), have shown the presence of somatostatin receptors in astrocytomas, meningiomas, melanomas, breast cancer, renal cell carcinomas and small cell carcinomas of the lung. Antiproliferative activity of octreotide acetate on cultured tumor cells has been reported [21]. Somatostatin also inhibits endothelial cell proliferation *in vitro* and angiogenesis *in vivo* as determined using the chick chorioallantoic membrane (CAM) assay and rat corneal neovascularization assay. In phase I studies, SMS -2010995 was found safe up to 8 mg/day [22]. Disease stabilization was observed in patients with diabetic retinopathy [23].

Angiostatin   Angiostatin is a 38 kDa fragment of plasminogen initially isolated from serum and urine of tumor bearing mice [24]. Subsequent studies have demonstrated that angiostatin can be generated by proteolytic cleavage of human plaminogen. The active peptide is apparently generated from plasmin by a unique mechanism whereby the disulfide bonds in plasmin are first reduced by extracellular reductases. The reduced peptide then undergoes autoproteolysis to generate active angiostatin [25]. The murine tumor derived protein or the peptide purified from the proteolytic digest of human plasminogen inhibited endothelial cell proliferation *in vitro* and angiogenesis *in vivo* [24]. The antiproliferative action of angiostatin was selective for endothelial cells [24]. Daily intraperitoneal injection of angiostatin inhibited growth and metastasis of primary murine and human tumors in mice [26]. Angiostatin treatment was accompanied by an increased tumor cell apoptosis *in vivo* [26]. The *in vitro* endothelial selectivity, a marked reduction of human breast, colon, and prostate tumors in mice, and a favorable toxicity profile suggest that angiostatin may be an attractive candidate for antiangiogenic therapy. However, because of its short half-life and subsequent requirement of daily injections of high doses to achieve efficacy, a derivative with improved pharmacokinetics may be a more suitable candidate for clinical development.

Endostatin   Endostatin is a 20 kDa C-terminal fragment of collagen XVIII isolated from conditioned medium of murine hemangioendothelioma cell line EOMA [27]. Endostatin derived from hemangioendothelioma cells or produced by gene cloning in *E. coli* inhibited endothelial cell proliferation *in vitro* and angiogenesis in the CAM assay [27]. Systemic administration of recombinant endostatin suppressed tumor growth and regressed primary tumors [26]. Further studies to determine the therapeutic potential of the recombinant angiostatin and endostatin for the treatment of human cancers are expected.

Interferons   Initially characterized for their antiviral activity, interferons have been shown to exhibit multiple cellular activities, including the inhibition of cell proliferation. Interferon-α is the first endogenous mediator that has demonstrated therapeutic efficacy for the treatment of angioproliferative disease such as malignant melanoma and Kaposi's sarcoma [28]. Furthermore, the angiomatous condition in children, hemangioma, is reduced by interferon-α. The mechanism of interferon's effect on angiogenesis appears to be multifaceted. Interferons are know to inhibit growth factor signal transduction, bFGF production in fibroblasts and human tumor cell lines, and endothelial cell proliferation and migration *in vitro*. Recent studies have shown that expression of interferon inducible protein (IP-10), an antiangiogenic chemokine may be one of the mechanisms contributing to the antitumor activities of interferons [29]. Despite the observed efficacy in melanoma and Kaposi's sarcoma, interferons have not yet proven a viable therapy for the treatment of other angiogenic diseases or solid tumors. The multiple activities of interferons, their indirect antiangiogenic action which may vary in different tumor types, the ability to deliver interferons into the tumor, and the undesired effects are some of the limitations for a wider use of interferons as antiangiogenesis therapy.

Tissue Inhibitors of Metalloproteinases (TIMPs) Degradation of extracellular matrix allowing cell migration, proliferation, and vessel remodelling is considered crucial for angiogenesis and metastasis. A family of tissue proteases known as matrix metalloproteinases (MMPs) catalyze the degradation of matrix proteins [30]. Overexpression and activation of MMPs is commonly observed in tumors. Inhibition of MMPs by TIMPS is one of the major endogenous mechanisms for regulating MMP activity which may also contribute to the local control of angiogenesis. Based on the structure and activities of MMPs and TIMPS, peptide mimetic small molecules have been synthesized that act as potent inhibitors of MMPs. The preclinical and clinical studies on the synthetic MMP inhibitors are discussed in the section on pharmacological inhibitors of angiogenesis.

Angiopoietin-2   Angiopoietin-1 and 2 are the two recently identified ligands for the receptor tyrosine kinase (RTK) Tie-2. Angiopoietin-1 (Ang-1) induces Tie-2 RTK activation in endothelial cells whereas Angiopoietin-2 (Ang-2) acts as an antagonist of Tie-2 activation. Ang-1 is a 75 kDa protein with structural homology to fibrinogen. The functional outcome of Tie-2 activation by Ang-1 is different from VEGF or bFGF receptor tyrosine kinase activation. The former does not mediate endothelial cell proliferation or migration [31]. The action of Tie-2 or Ang-1 resides at latter stages of angiogenesis pathway involving tube maturation. These conclusions are derived from transgenic mice overproducing Ang-2. The Ang-2 transgenic exhibited phenotype similar to that of Ang-1 knockout mice wherein the vascular defects reside in the recruitment of perivascular cells, maturation of tubule structures and their association with the tissue matrices [31]. The antiangiogenic mechanism of angiopoietin-2 may serve as a template for the development of a new class of antiangiogenic agents.

Pharmacological Inhibitors of Angiogenesis
The identification of endogenous inhibitors and a greater understanding of the cellular and molecular mechanisms of angiogenesis have led to the identification of several pharmacological agents that inhibit endothelial cell functions *in vitro* and angiogenesis *in vivo*. Some of the compounds have shown promising activity in models of angiogenic diseases, including cancer, diabetic retinopathy and rheumatoid arthritis [2, 32-35]. The nature of the pharmacological inhibitors, their putative mechanisms of action, and the key preclinical and clinical findings are discussed in the following section.

Inhibitors of Heparin Binding Growth Factors:
Pentosan polysulfate
Pentosan, a sulfated polysaccharide with anticoagulant activity inhibits endothelial cell proliferation and migration *in vitro*. Inhibition of prostate tumor associated angiogenesis has been demonstrated *in vivo*. In addition, large vascular tumors developing from implanted adrenal tumors engineered to secrete FGF were inhibited by pentosan polyphosphate [33]. Stabilization of tumor growth was observed in Phase I trials in AIDS-related Kaposiís sarcoma [36]. Disease stabilization and one response were noted in a phase II trial.

Tecogalan sodium

Tecogalan, a sulfated polysaccharide isolated from the cell walls of Arthrobacter sp AT-25, inhibits endothelial cell growth and chemotaxis by blocking binding of growth factors eg bFGF to their receptors. It inhibits bFGF-induced angiogenesis in the CAM assay. Two Phase I studies have recently reported a minor response and one sustained clinical benefit response in cancer patients [33, 37].

Suramin

Suramin is a polysulfonated napthylurea with multiple cellular activities. Originally utilized as an antiparasitic agent, suramin has been shown to inhibit binding of several growth factors to their respective receptors (bFGF, TGFa and b, IGF, EGF, PDGF and VEGF). Suramin inhibits angiogenesis in the CAM assay and rat corneal assay. It has also shown growth inhibitory activity against tumor cell lines including breast, prostate, sarcoma and colorectal carcinoma. Suramin has demonstrated activity against hormone refractory prostate cancer, Kaposiís sarcoma, renal carcinoma, adrenal carcinoma and non-Hodgkin's lymphoma [33].

Inhibitors of Signal Transduction

Genistein

Genistein is a naturally occurring phytoestrogen found in soybeans. Genistein inhibits protein tyrosine kinases such as the EGF receptor tyrosine kinase, and endothelial cell proliferation and angiogenesis. Genistein also inhibits ATP-induced calcium influx which may account for its antiangiogenic activity. Other activities of genistein including the inhibition of topoisomerase II and tumor promoter-induced H_2O_2 production may also contribute to its antitumor activity [38].

LY333531

LY333531, a macrocyclic bis(indolyl)maleimide, is a selective inhibitor of protein kinase C-β2 (PKC-β2) [39]. LY333531 inhibits the VEGF activated signal transduction cascade and endothelial cell proliferation [40]. The potency of LY333531 to inhibit PKC-β correlates well with *in vivo* plasma concentrations that reduce the glomerular filtration rate, albumin excretion rate and retinal circulation in a model of diabetic retinopathy. LY333531 is currently in Phase I clinical trials for the treatment of diabetic complications including retinopathy.

Bryostatin-1

Bryostatin is a naturally occurring macrocyclic lactone that causes transient activation of PKC followed by its down-regulation. *In vitro*, bryostatin inhibits cell growth and induces differentiation of tumor cell lines. Antitumor activity in a number of models including melanoma have led to clinical trials. Recent studies on the structure activity-relationships of 26-epi-bryostatin suggest that the antitumor activity of the bryostatin may be dissociated from its PKC-mediated effects [41]. However, other studies suggest that PKC plays an important role in regulation of MMP production and that the modulation of PKC by bryostatin-1 is one of the likely mechanisms of the antiangiogenic activity of bryostatin [42].

Carboxyamidotriazol (CAI)

CAI inhibits basal or stimulated calcium uptake and consequently influences Ca^{++}-dependent signal transduction including: release of second messengers, protein phosphorylation and gene transcription. Treatment with CAI inhibits endothelial cell adhesion, spreading, migration, expression of proteolytic enzymes, *in vitro* and *in vivo* tube formation. In phase I studies, CAI was cytostatic, achieving disease stabilization in nearly half of the treated patients [43]. A micronized formulation has been developed for Phase II administration [44].

Nitric oxide synthase (NOS) inhibitors

NO is a free radical gas generated from L-arginine by the action of oxidoreductase enzymes NO synthases (NOS). NO plays roles in the regulation of vascular tone, platelet aggregation and inflammation. High concentrations of NO and its byproducts such as proxynitrite and OH are toxic to cells. Increased NOS expression and NO production have been reported in a variety of human tumors, including breast, uterine, ovarian, melanoma, and brain tumors. NOS activity in tumors correlates with tumor grade [45]. In a recent study with 22 patients with primary breast tumors, a strong correlation was found between NOS expression and metastatic potential [46]. The expression of inducible NOS (iNOS) in inflammatory cells as well as tumor cells may contribute to a high concentration of NO in tumors.

High concentration of NO may have two major consequences in the pathogenesis of cancer. First, NO may produce direct effects on the growth and survival of tumor cells through its effects on protooncogene Ras or suppresser gene p53. NO mediated S-nitrosylation of protooncogene Ras results in the accumulation of activated Ras-GTP and thus favors increased cell growth signals and decreased apoptosis signals [47]. NO mediated S-nitrosylation of the tumor suppression protein P53 reduces the ability of p53 to bind DNA [48]. The modification of p53 by NO may lead to reduced tumor cell apoptosis. Thus, the post translational modification of Ras and p53 by NO produces functional alterations similar to those produced by genetic mutations that favor tumor growth.

The second major consequence of excessive NO production is the stimulation of angiogenesis. Vasodilation of preexisting microvessels is one of the early events in the initiation of angiogenesis. The angiogenic factors, bFGF and VEGF, induce vasodilation by activation of NOS as well as induction of NOS gene expression. The rate of tumor growth and tumor vascularization was markedly increased by transfection of iNOS in colon adenocarcinoma cell line [49]. Evidence for the stimulation of angiogenesis and accelerated wound healing by NO has also been obtained [53]. L-arginine treatment increases the gastric blood flow, angiogenesis and accelerated healing of acute gastric lesions [50]. These effects of L-arginine are blocked by inhibitors of NOS [50]. Treatment of endothelial cells with NO-donating compounds increases their proliferation and migration [51]. NO donors have been shown to induce VEGF production by tumor cells derived from glioblastoma and hepatocarcinoma [52]. Thus, in addition to its direct effects on signal tranduction pathways regulating cell growth and apoptosis, the induction of VEGF and bFGF expression by NO may promote angiogenesis. The role of NO in angiogenesis is further substantiated by pharmacological inhibition of NOS. NOS inhibitor L-nitro-arginine methyl ester (L-NAME) or L-monomethyl-Nitro- arginine inhibit angiogenesis in rabbit cornea model and reduce growth of xenografted tumors [53]. The compounds tested to date have been non-selective inhibitors of the three isoforms of NOS. Inhibition of endothelial NOS by non-selective agents is associated with undesired hemodynamic effects. In this regard, the recently synthesized iNOS selective inhibitor, 1400W may allow the evaluation of the therapeutic potential of NOS inhibitors without the undesired hemodynamic side effects [54].

Inhibitors of Endothelial Cell-Matrix Interactions

Batimastat (BB-94) and Marimastat (BB-2516)

Matrix metalloproteinases (MMPs) are Zn^{++}-dependent proteases that degrade basement membrane proteins including collagen, laminin, geletin and fibronectin. MMPs are regulated at transcriptional and post translational levels. Tissues secrete MMPs in a latent form and also secrete peptide inhibitors known as tissue inhibitors of metalloproteinases (TIMPs). Batimastat (BB-94), a low-molecular weight synthetic hydroxamate peptide mimetic, binds to most MMPs at the active site Zn^{++} atom resulting in potent but reversible inhibition. Antitumor and anti-metastatic activity of batimastat have been reported in a number of animal models [55]. Phase I/II clinical trials have demonstrated a delay in ascites accumulation in patients with malignant effusions, however, a lack of oral bioavailability has hampered further development of BB-94. A second generation analog, marimastat (BB-2516) with greater oral bioavailability has been evaluated in clinical settings. Patients were dosed twice daily with 25 or 50 mg of BB-2516. Dose-limiting studies have shown toxicities related to musculoskeletal symptoms with pain and tenderness in joints, muscles, and tendons of the hands and shoulders appearing in most patients within 3–4 months. Pharmacokinetics of marimastat in cancer patients was different from that in normal volunteers; doses of 10 mg twice daily resulted in trough plasma concentrations 2–3 times the levels obtained with Phase I volunteers. Other MMP inhibitors currently in development include, AM6001, AA3340, CGS27023A, and 12-9566.

Integrin avb3 antagonists

Expression of endothelial cell integrin avb3 is induced by angiogenic stimuli such as bFGF. Integrin $\alpha_v\beta_3$ recognizes several extracellular matrix proteins including fibronectin, vitronectin, osteopontin, von Willebrand factor, fibrinogen, proteolyzed collagen and matrix metalloprotease II. Ligation of $\alpha_v\beta_3$ reduces p53 activity and p21^{WAF} increases the cellular bcl-2/bax ratio, and stimulates expression of an adhesion-dependent cell survival. Treatment with the $\alpha_v\beta_3$-selective monoclonal antibody, LM609, inhibits endothelial cell migration and angiogenesis, and induces unscheduled programmed cell death. Clinical trials with a humanized form of LM609 are planned [56]. Other peptides and small molecule mimetics that antagonize $\alpha_v\beta_3$ are being considered for development [4].

Inhibitors with Multiple Activities:

Linomide

Linomide is a quinolone-3-carboxamide which has demonstrated immunomodulatory activity *in vivo*. Linomide treatment of rats bearing prostatic tumors have shown a 37 % reduction in tumor blood vessels and a reduction in lung tumor metastases. Antitumor activity was not observed in phase I/II clinical trials in patients with renal cell carcinoma, melanoma and colon cancer [33].

Thalidomide

Thalidomide was developed in the late 1950's as a sedative but was removed from the market because of the severe deformities induced in developing human limbs. Investigations into its teratogenic mechanism have suggested that thalidomide may inhibit angiogenesis [57]. *In vivo*, thalidomide is activated by conversion in liver to an epoxide [34] and as many as twelve other metabolites. Thalidomide inhibits bFGF and VEGF-induced neovascularization in the mouse corneal assay. A recent phase I study demonstrated activity in AIDS-related Kaposiís sarcoma [58]. Phase II studies are in progress in breast and prostate cancer, glioblastoma multiforme, Kaposiís sarcoma, macular degeneration, and diabetic retinopathy [43].

TNP-470 /AGM-1470

TNP-470 is a semisynthetic analog of fumagillin with antiproliferative and antimigratory activities on endothelial cells. TNP-470 arrests cells in the late G1 phase of the cell cycle, inhibits cyclin dependent kinase activation (cdk2/4) and cyclin E expression. Recently it was shown that fumagillin and TNP-470 covalently bind and inactivate methionine aminopeptidase-2, an enzyme thought to be important for protein myristoylation [59]. TNP-470 also exhibits cytotoxic activity against breast and prostate tumors lines (see [33] for review). *In vivo*, antitumor activity against several xenograft tumors including ovarian cancer, endometrial tumors, choriocarcinoma, gastric cancers and human neuroblastoma has been demonstrated. Phase I trials are ongoing in AIDS related Kaposiís sarcoma, prostate cancer and squamous cell carcinoma of the cervix [43]. Phase II trials have been initiated in glioblastoma multiforme. A Phase III trial is in progress in locally advanced non-metastatic, nonresectable pancreatic cancer [60].

Inhibitors Selectively Targeted to Tumor Vasculature

Vascular targeting which exploits the differences between mature endothelial cells and tumor neovasculature in order to selectively induce necrosis is a unique therapeutic approach for the treatment of cancer [61]. The group B streptococcus polysaccharide toxin, CM101, has been shown to selectively target tumor vasculature. It preferentially binds to capillary endothelium in tumors and induces inflammatory responses including necrosis, hemorrhage, thrombosis and release of cytokines. Treatment with CM101 was associated with reduction in tumor size and increased survival time in tumor bearing mice [62]. In phase I studies, all patients experienced time- and dose-dependent elevations of cytokines; three of the 15 patients experienced tumor shrinkage [63]. Other studies suggest that vascular targeting is more effective in the treatment of melanoma limb metastases when combined with regional arterial perfusion with tumor necrosis factor-a [64].

Pathology   see above

Clinical Relevance and Therapeutic Implications

The lack of clinical success of several agents initially demonstrating antiangiogenic activity in preclinical studies (Table 2) may have been due to their non-selective mechanism of action, poor bioavailability, and a lack of an appropriate strategy for clinical testing. With advances in the understanding of the molecular mechanisms of angiogenesis, more selective antiangiogenic agents are expected to become available. Along with the development of selective agents, the challenges inherent in their clinical testing also need to be addressed. Antiangiogenic agents may be useful in treating cancer patients in several clinical situations: 1) the adjuvant setting; 2) as maintenance therapy for patients needing chronic treatment because of high risk of relapse; 3) maintenance therapy for patients with advanced metastatic inoperable disease, where disease stabilization is desirable; 4) in combination with chemotherapy or radiation therapy. For each of these clinical situations, establishing appropriate therapeutic doses and treatment schedules are essential. In cancer clinical trials, therapeutic doses are determined by escalating doses in Phase I studies to establish the maximum tolerated dose. Therapeutic doses are then chosen slightly below the toxic levels. This strategy is inappropriate for developing drugs to be used in an adjuvant setting or on a chronic basis. Instead, therapeutic doses will need to be determined based on novel pharmacokinetic measurement such as plasma concentrations and half-life that are consistent with efficacy based on *in vivo* preclinical models, and surrogate indicators of pharmacodynamic activity. The definition of efficacy or clinical response expected in antiangiogenic trials is another important consideration. Although tumor shrinkage with antiangiogenic agents in preclinical models has been observed, it is not clear whether reduction in tumor mass will be a measurable clinical outcome in cancer patients. Thus, classical observations of objective responses will not be suit-

Table 2. Antiangiogenic Agents in Clinical Development

Compound	Cellular or Molecular Action*	References
Inhibitors of heparin binding growth factors		
Pentosan Sulfate	EC migration and proliferation	[33, 36]
Tecogalan sodium	EC migration and proliferation	[33, 37]
Suramin	EC proliferation	[33]
Inhibitors of signal transduction		
LY333531	PKCβ2; VEGF signal transduction in EC	[39, 40]
Bryostatin-1	stimulates PKC, followed by down regulation MMP production	[41, 42]
Carboxyamidotriazol	Ca^{++} channels, protein phosphorylation	
and gene transcription	[43, 44]	
L-NAME, 1400W	Nitric oxide synthase	[53, 54]
Inhibitors of matrix-endothelial cell interactions		
Batimastat (BB-94)	MMP by binding to Zn^{++} site	[55]
Marimastat (BB-2516)	MMP by binding to Zn^{++} site	[55]
LM609	$\alpha_v\beta_3$ integrin receptor antibody, EC migration and induces apoptosis	[4, 56]
Inhibitors with multiple activities		
Linomide	FGF induced angiogenesis	[33]
Thalidomide	early stage neovascularization	[57, 58]
TNP-470 (AMG-1470)	methionine aminopeptidase-2, EC proliferation, cyclin kinase, cyclin E expression	[43, 60]
Vascular targeted agents		
CM101	induces hemorrhage, thrombosis and cytokine release	[63, 64]

*Compounds inhibit the listed activities except in case of Bryostatin and CM101 as indicated.
EC: endothelial cells

able endpoints. Chronic therapy with an antiangiogenic agent may instead result in disease stabilization or tumor dormancy [65], a desirable outcome after bulk tumor reduction following surgery, or in patients with metastatic inoperable cancers. Assessing stable disease will require evaluation of novel clinical indicators such as: blood flow, measured using non-invasive techniques such as color doppler; tumor metabolism, measured using positron emission tomography; angiogenesis, assessed using magnetic resonance imaging and tumor proliferative index (Ki67 antigen) assessed on fine needle tumor biopsies. Tumor vascularity has been reported to be a useful prognostic indicator of survival and for predicting patient response to chemotherapy in breast cancer patients. Assessment of tumor vascularity prior to patient enrollment will be important to ensure appropriate interpretation of response data. Stable disease for a minimum of three months or more has been suggested as one criteria. For assessment of disease stabilization, tumors with good serum markers e.g., αFP for hepatoma, PSA for prostate cancer, CEA and erbB2 for breast cancer, may be most suitable [66]. Preclinical studies have demonstrated synergistic effects resulting form the combination of chemotherapy and antiangiogenic therapy [67]. Randomized phase II/III trials of standard chemotherapies with or without the angiogenic agent may be the most expedient way to capture measures of antiangiogenic efficacy.

Finally, adjuvant trials with antiangiogenic agents are likely to take longer, requiring several years to demonstrate stable diseases, decrease in metastatic events, or enhanced survival. Surrogate markers such as PCR to amplify PSA in order to detect circulating micrometastases in prostate cancer are being explored. Early diagnostics for relapse and progression will allow more rapid decision making for efficacy of antiangiogenic therapies. Table 2 shows a list of molecules that are in clinical development as antiangiogenesis agents.

Jai Pal Singh and Laurane G. Mendelsohn

References

1. Bussolino F (1996) Eur J Can 32A:2401-2412
2. Gastl G et al (1997) Oncology 54:177-184
3. Iruela-Arispe ML, Dvorak HF (1997) Throm Hemo 78:672-677
4. Hanahan D, Folkman J (1996) Cell 66:353-364
5. Vassalli JD, Pepper MS (1994) Nature 370:14-15
6. Bornstein P (1992) FASEB J 6:3290-3299
7. Tolsma SS et al (1993) J Cell Biol 122:497-511
8. Iruela-Arispe ML et al (1996) J Clin Invt 97:403-412
9. Zabrenetzsky et al (1994) Int J Cancer 59:191-195
10. Sheibani N, Frazier (1995) Proc Natl Acad Sci (USA) 92:6788-6792
11. Schultz-Cherry S et al (1994) J Biol Chem 269:26783-26788
12. Lane DA et al (1984) Biochem J 218:725-732
13. Maione TE et al (1990) Science 247:77-79
14. Maione TE et al (1991) Can Res 51:2077-2083
15. Tanaka T et al (1997) Nature Med 3:437-442
16. Gupta SK et al (1995) Proc Natl Acad Sci (USA) 92:7799-7803
17. Gupta SK, Singh JP (1994) J Cell Biol 127:1121-1127
18. Belman N et al (1996) Invest New Drug 14:387-389
19. Clapp C et al (1993) Endocrinology 133:1292-1299
20. Clapp C, Weiner RI (1992) Endocrinology 130:1380-1386
21. Woltering EA et al (1997) Invest New Drugs 15:77-86
22. Antony L et al (1993) Acta Oncol 32:217-233
23. Mallett B et al (1992) Diabetic Metabol 18:438-444
24. OíReilly MS et al (1994) Cell 79:315-328
25. Stathakis P et al (1997) J Biol Chem 272:20641-20645
26. OíReilly MS et al (1996) Nature Med 2:689-692
27. OíReilly MS et al (1997) Cell 88:277-285
28. Gutterman JU (1994) Proc Natl Acad Sci (USA) 91:1198-1205
29. Arenberg DA (1996) J Exp Med 184:981-992
30. Matrisian LM (1992) BioEassay 14:455-463
31. Maisonpierre PC et al (1997) Science 277:55-60
32. Gasparini G, Harris AL (1995) J Clin Oncol 13 :765-782
33. Gradishar WJ (1997) Lancet 349 (suppl 2):13-15
34. Marshall JL et al (1995) Breast Can Res and Treatment 36:253-261
35. Pluda JM (1997) Seminar in Oncology 24:203-218
36. Pluda JM et al (1993) J Natíl Cancer Inst 85:1585-1592
37. Eckhardt SG et al (1996) Annals of Oncology 7:491-496
38. Teicher BA (1995) Critical Reviews in Oncology/ Hematology 20:9-39
39. Ishii H et al (1996) Sci 272:728-731
40. Xia P et al (1996) J Clin Invest 98:2018-2026
41. Szallasi Z et al (1996) Can Res 56:2105-2111
42. Wojtowicz-Praga SM et al (1997) Invest New Drugs 15:61-75
43. Price JT et al (1997) Crit Rev Biochem Mol Biol 32:175-253
44. Kohn EC et al (1997) J Clin Oncology 15:1985-1993
45. Thomsen LL et al (1995) Br J Can 27:41-44
46. Duenas-Gonzalez A et al (1997) Mod Pathol 10:645-649
47. Lander HM et al (1997) J Biol Chem 272:4323-4326
48. Calmels S et al (1997) Can Res 57:3365-3369
49. Jenkins DC et al (1995) Proc Natl Acad Sci 92:4392-4396
50. Brzozowski T et al (1997) J Gastroentrol 32:442-452
51. Ziche M et al (1994) J Clin Invt 94:2036-2044
52. Chin K et al (1997) Oncogene 15:437-442
53. Tozer GM et al (1997) Can Res 57:948-955
54. Thomsen LL et al (1997) Can Res 57:3300-3304
55. Talbot DC, Brown PD (1996) Eur J Cancer 32A:2528-2533
56. Stromblad S, Cheresh DA (1996) Trends in Cell Bio 6:462-468
57. DíAmato RJ et al (1994) Proc Natíl Acad Sci USA 91:4082-4085
58. Soler R, Howard R (1996) Clin Infect Dis 23:501-503
59. Sin N et al (1997) Proc Natl Acad Sci (USA) 94:6099-6103
60. Castronovo V, Belotti D (1996) Eur J Cancer 32A:2520-2627
61. Bicknell R (1994) Ann Oncol 5:45-50
62. Thurman GB et al (1994) J Cancer Res Clin Oncol 120:470-484
63. DeVore RF et al (1997) Clinic Can Res 3:365-372
64. Harris AL (1997) The Lancet 349:13-15
65. Folkman J (1996) Eur J Can 32A:2534-2539
66. Fox SB, Harris AL (1997) Investigational New Drugs 15:15-28
67. Kakeji Y, Teicher BA (1997) Investigational New Drugs 15:39-48

Angiogenin

Synonym: DIP, degranulation inhibiting protein; RNase 5, pancreatic-type ribonuclease 5.

Definition *Ribonuclease with angiogenic activity*

Introduction Angiogenin is a plasma protein [1] with angiogenic [2] and ribonucleolytic [3] activities. It was originally purified from 2000 litres of serum-free medium conditioned by HT-29 human adenocarcinoma cells (yield: 500 ng/L), on the basis of its ability to induce neovascularization [2]. The strategy of purification used by Professor Vallee and his co-workers was based on the observation by Professor Folkman that tumour growth depends on neovascularization, and their project was inspired by the hypothesis that anti-angiogenesis might be used as a therapeutic approach [4].

Angiogenin elicits new blood vessel formation in the chick chorioallantoic membrane, where only femtomole amounts were needed to induce angiogenesis [2], and in the rabbit cornea [2] and meniscus [5].

Angiogenin is also present in normal human tissues and fluids such as plasma [1, 6] and amniotic fluid [7]. Angiogenin is secreted in culture by vascular endothelial cells, aortic smooth muscle cells (SMC), fibroblasts and tumour cells [2, 8].

Angiogenin is a ribonuclease-related molecule. This cationic single-chain protein has 35% amino acid sequence identity with human pancreatic ribonuclease (RNase), many of the remaining residues being conservatively replaced [9]. Angiogenin also displays ribonuclease activity, albeit markedly different in both its specificity and magnitude [3]. Angiogenin has ribonucleolytic specificity for ribosomal and transfer RNA. The limited cleavage of 18S rRNA seems to be responsible for the inhibition of cell-free protein synthesis by specific inactivation of the 40S ribosomal subunit [10]. Angiogenin blocks protein synthesis when injected into *Xenopus* oocytes [11]. Its physiological substrate remains to be identified. In addition, angiogenin binds to a protein RNase inhibitor (RI) first isolated from placenta, which abolishes both its ribonucleolytic and angiogenic activities ([12, 13], for reviews).

The integrity of the catalytic site and a cell-binding domain are required for its capacity to induce neovascularization [14, 15]. Indeed, angiogenin binds to high-affinity receptors on subconfluent endothelial cells [16-18], activates cell-associated proteases [19, 20] and triggers several intracellular events [21, 22]. It has been reported to stimulate proliferation of endothelial cells [17,18]. In aortic smooth muscle cells, angiogenin activates phospholipase C and induces cholesterol esterification [23].

Angiogenin binds copper [24]. The presence of copper, a modulator of angiogenesis *in vivo*, enhances angiogenin binding to endothelial cells *in vitro* [16, 24].

Angiogenin suppresses, *in vitro*, the proliferation of stimulated human lymphocytes [25] and, under the name DIP, at concentrations in the nanomolar range, inhibits degranulation of polymorphonuclear leukocytes (PMNL) [26].

Angiogenin supports the adhesion of endothelial cells, fibroblasts and tumour cells [27, 28].

Angiogenin is a heparin-binding protein [29].

Besides its angiogenic potency, the possible involvement of angiogenin in the development of tumours is suggested by the demonstration that angiogenin antagonists prevent the growth of human tumour xenografts in athymic mice [30]. In addition, angiogenin expression increases in pancreatic cancer [31] and serum angiogenin concentrations increase in cancer patients [31-35]. As angiogenin is not a tumour-specific product, these data point to potential modulatory mechanisms of angiogenin functions.

Characteristics

Molecular Weight Calculated relative molecular mass are 14 124, 14 595, 14 059 and 14 362 for the human [1], bovine [36], pig and rabbit [37] angiogenins, respectively.

Domains The similarity of angiogenin with ribonuclease has been used to define structure/function relationships through existing information on RNase. However, only angiogenin is able to induce angiogenesis, which suggests that its biological activities result from structural characteristics. A tremendous effort has been made, mainly by Professor Vallee's group, to identify the regions of the molecule that are critical for its activities. Domains or residues corresponding to those known to be important for the enzymatic activity of RNase have been modified chemically or by mutagenesis. Only part of this work is reported here. The reader is also referred to recent reviews [38, 39].

Receptor binding domains: The putative receptor binding domain (Figures 1, 2, 4; see conformation §§) includes two segments on adjacent loops (segment 58-70: loops 4 and 5 and β strands B_2 and B_3; segment 107-110: loop 9) containing residues 60-68 [15] and Asn-109 [40] respectively. This location was deduced from studies on proteolysis [15], deamidation [40] and mutagenesis [41]. The segment 58-70 in angiogenin contains two fewer residues than RNase, and angiogenin lacks the two cysteines (replaced by Pro-64 and Leu-69 in human angiogenin) that form the disulphide loop in RNase A (positions 65 and 72) involved in purine binding. Replacement of the segment 58-70 by the corresponding sequence of RNase A by means of regional mutagenesis causes a reduction in angiogenic potency [41]. Substituting the surface loop of RNase A (residues 59-73) with residues 58-70 of angiogenin endows ribonuclease with angiogenic activity [42]. Furthermore, peptide ANG(58-70) inhibits endogenous angiogenesis in mice [42]. The segment of bovine angiogenin includes an Arg-Gly-Asp sequence (67-69) which is replaced by Arg-Glu-Asn in the human molecule. Arg-66 has been identified as an essential component of this site, as its mutation reduces the angiogenic potency of angiogenin [43].

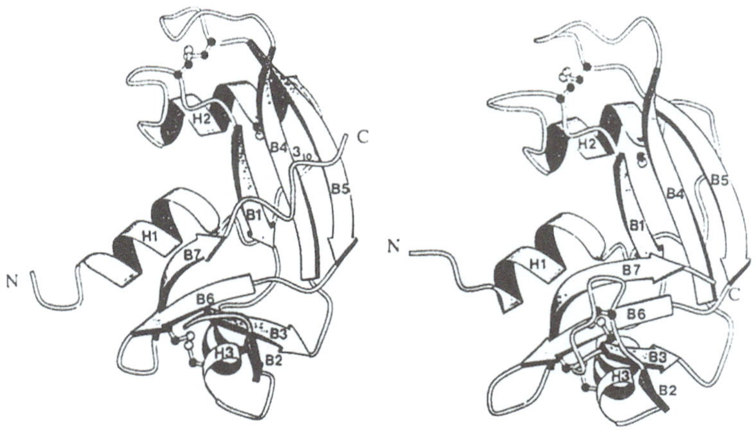

Figure 1. Three-dimensional structure of human angiogenin (Left) and RNase A (Right). MOLSRIPT diagramm reproduced with permission from [44] (Copyright 1994, Proc. Natl. Acad. Sci. USA).

Angiogenin inhibits the degranulation of PMNL [26]. A similar but weaker inhibitory effect has been obtained with the synthetic peptide Leu[83]-His-Gly-Gly-Ser-Pro-Trp-Phe-Phe-Cys-Gln-Tyr-Arg[95], which might be the domain of angiogenin involved in this biological activity. Indeed, the segment 83-95 includes loop 7 (connecting β sheets B4 and B5), a region that diverges from that of RNase [44]. Moreover, this domain is exposed to solvent, and nuclear magnetic resonance (NMR) studies of bovine angiogenin have shown increased disorder in this domain, which is likely to reflect greater flexibility [45]. Epitope: Loop 7 (residues 85-92) forms part of the epitope for a monoclonal antibody that has been shown to prevent the establishment of human tumours in athymic mice [30].

Cell adhesion domains: The endothelial cell adhesion domain might include the Arg[67]-Gly-Asp[69] segment of bovine angiogenin, which is replaced by Arg[66]-Glu-Asn[68] in human angiogenin [27]. Although these segments do not have a conformation typical of an integrin recognition site when analysed by X-ray crystallography [46], in solution the Arg-Gly-Asp sequence of bovine angiogenin forms a short loop at the apex of the B_2-B_3 sheet exposed to solvent, with signs of flexibility in proton NMR. These characteristics are shared with many integrin binding proteins [45]. Human angiogenin also supports human adenocarcinoma HT-29 cell adhesion. The domains involved might include Arg[31]-Arg-Arg[33], Arg-[66] and Arg[70], as mutation of these basic amino acids decreases the capacity of angiogenin to mediate cell adhesion [28].
The heparin-binding domain includes the segment Arg[31]-Arg-Arg[33] and Arg[70] [29].
Nuclear localisation signal: Segment Arg[31]-Arg-Arg-Gly-Leu[35] of human angiogenin has been identified as the nuclear localisation signal responsible for nucleolar targeting of angiogenin in calf pulmonary and human umbilical artery endothelial cells. Arg[-33] is the essential amino acid [47].

Actin-binding domain: The actin-binding domain of angiogenin might involve the segment 60-68, as the two derivatives, cleaved at residues 60-61 and 67-68 respectively, fail to bind to angiogenin-binding protein [48], a dissociable cell-surface component of endothelial cells and a member of the actin family [49].

Enzymatic active site: As in RNase A, the ribonucleolytic active site of angiogenin consists of several subsites [44]. The catalytic centre (P_1), at which phosphodiester bond cleavage occurs, involves the three catalytic residues: His-13, Lys-40 and His 114. The B_1 site, for binding the pyrimidine, whose ribose donates its 3' oxygen to the scissile bond, corresponds to Gln-12, Thr-44, Ser-118; and the B_2 site, that preferentially binds a purine, corresponds to Glu-108. The side-chain of Gln-117 forms two hydrogen bonds with Thr-44 and obstructs the pyrimidine binding site B_1. Ile-129 and Phe-120, in the middle of the C-terminal helix, make intramolecular hydrophobic

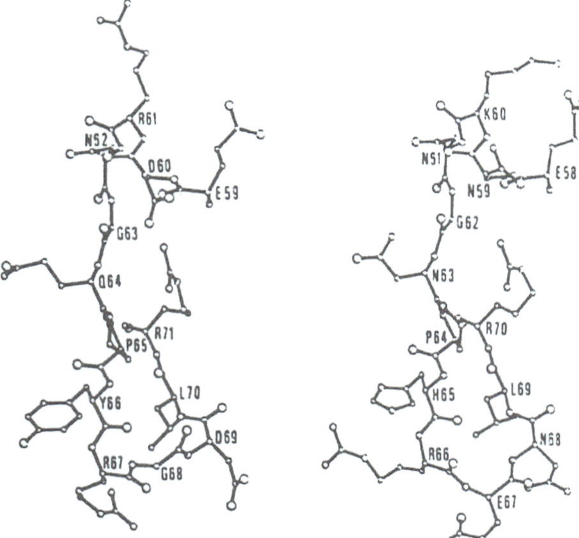

Figure 2. View of the putative receptor binding site of bovine angiogenin (Left) and human angiogenin (Right). Reproduced with permission from [46] (Copyright 1995, Proc. Natl. Acad. Sci. USA).

interactions that stabilise the conformation. The terminal segment of angiogenin, Arg[121]-Arg-Pro[123], does not seem to form contacts with the rest of the molecule and might be a subsite for binding polymeric substrates [50]. Thus, the C-terminal region might play a key role in angiogenin activity. The C-terminal peptide ANG(108-121) inhibits both the enzymatic and angiogenic activities of angiogenin. It is devoid of activity by itself.
The N-terminal region of angiogenin ANG(1-21) is highly conserved in angiogenins from five mammalian species and differs from the corresponding region of RNase A. It might be involved in the biological activities of angiogenin (Figure 4).

RI binding site: Human angiogenin binds human placental RNase inhibitor with the extremely low Ki value of 0.7 fM ([12, 13], for reviews). The tight interaction results from a large contact surface involving 24 residues distributed over seven segments of angiogenin [51]. The domains of the molecule involved in this complex include the catalytic site (mostly residue Lys-40) and the surface loop 84-89 [12, 13, 51].

Binding Sites and Affinity

Cell-surface receptors: Specific binding sites of angiogenin are present on endothelial cells from calf pulmonary artery, and bovine aorta, cornea and adrenal cortex capillary [16], bovine brain capillary [17], human umbilical artery and vein, and human foreskin dermal tissue [18]. They were not detected on Chinese hamster lung fibroblasts [16]. Cell binding of iodinated angiogenin at 4 °C is time- and concentration-dependent, reversible, saturable and specific, whereas iodinated RNase does not bind to endothelial cells [16, 17]. Scatchard analyses of binding data reveal two apparent types of specific interaction with CPAE cells. High-affinity binding sites with an apparent dissociation constant (Kd_1) of 1-5 10^{-9} M bind about 2×10^5 molecules per cell [16, 24]. Low-affinity/high-capacity binding sites with a Kd_2 of 0.2×10^{-6} M are associated with pericellular components and bind several million molecules [16, 24]. Angiogenin that is bound with low affinity to extracellular matrix and cells is released by a wash at 0.6 M NaCl. On bovine brain capillary endothelial cells, high-affinity binding sites with an apparent dissociation constant of 0.5×10^{-9} M bind about 11×10^3 molecules of bovine angiogenin per cell [17]. The high-affinity binding sites are likely cell-surface receptors, as angiogenin triggers a series of intracellular events when added at 1 ng/ml to cultured CPAE cells [21], and a concentration of 100-1000 ng/ml enhances [3H]thymidine incorporation and proliferation in human umbilical venous and microvascular endothelial cells by up to 50 % [18]. Furthermore, bovine angiogenin at 100 ng/ml induces [3H]thymidine incorporation and proliferation in bovine brain capillary endothelial cells [17].
Density-dependent regulation of angiogenin receptors has been observed on endothelial cells. High-affinity binding sites decrease as cell density increases and are not detected at confluence [16, 18].

Characterisation of angiogenin binding sites on bovine brain capillary endothelial cells by ligand-blotting of iodinated-angiogenin to solubilized membrane proteins separated by sodium dodecyl sulphate-polyacrylamide gel electrophoresis (SDS-PAGE) and transferred to nitrocellulose identifies a labelled component with an apparent molecular mass of 49 kDa [17]. In contrast, affinity labelling of human umbilical venous, artery and microvascular endothelial cells with iodinated angiogenin using the homobifunctional cross-linker bi(sulfosuccinimidyl)suberate yields a cross-linked band of 180 kDa on SDS-PAGE autoradiography [18]. When biotinylated surface molecules are purified by affinity on an angiogenin-Sepharose column, and the final product is analysed by SDS-PAGE and western-blotting with alkaline phosphatase-labelled streptavidine, a band is detected at 170 kDa [18].
Among the factors that modulate angiogenesis, protamine, an inhibitor of angiogenesis, competes for angiogenin binding to CPAE cell receptors, whereas heparin interacts to a lesser extent [16]. Placental RI, a tight-binding inhibitor of both the ribonucleolytic and angiogenic activities of angiogenin ([12, 13], for reviews), abolishes its binding to CPAE cells at a molar ANG/placental RI ratio of 1 [16]. Cu^{2+}, a modulator of angiogenesis *in vivo*, increases by 4-fold the number of angiogenin molecules bound to high-affinity receptors on CPAE cells at saturation, at a concentration of 0.1 mM [16, 24]. Specific binding is also increased in the presence of Zn^{2+}, whereas Ni^{2+}, Co^{2+} and Li^+ have no effect. Specific angiogenin binding to the low-affinity/high-capacity sites is increased about 2-fold in the presence of Cu^{2+} and Zn^{2+} respectively [24]. Furthermore, in the presence of Cu^{2+} or Zn^{2+}, no inhibitory effect of RI on cell binding was observed. Metal ions do not irreversibly inactivate placental RI, as their action can be reversed by adding diethyldithiocarbamate, a metal chelator [16, 52].

Copper binding: Metal ion affinity chromatography and atomic absorption spectrometry have been used to show the direct interaction of angiogenin with copper and zinc ions. Angiogenin binds 2.4 mol of copper per mol protein, *in vitro* [24]. Cu^{2+} inhibits angiogenin-catalysed cleavage of tRNA, with an IC50 of 0.03 mM [53].

Interaction with PMNL: Angiogenin inhibits granule discharge from PMNL at concentrations between 7 and 70 nM [26].

Cell adhesion: Angiogenin supports the adhesion of endothelial cells, fibroblasts and tumour cells when coated at concentrations > 100 ng/cm². It has no effect on cell adhesion when in solution. Both human and bovine angiogenins promote adhesion of CPAE cells and Chinese hamster lung fibroblasts, whereas RNase A does not. Endothelial cell adhesion on bovine and human angiogenin is inhibited by the peptide Arg-Gly-Asp-Ser. Adhesion of endothelial cells is Ca^{2+} and Mg^{2+}-dependent but the addition of collagens I and IV, fibrinogen, fibronectin or vitronectin in solution at concentrations up to 0.01 mg/ml has no effect [27]. In contrast, adhesion of HT-29 human adenocarcinoma cells on human angiogenin is not inhibited by the Arg-Gly-Asp-Ser peptide, does not require Ca^{2+} or Mg^{2+} but is affected by heparin, which has no effect on endothelial cell adhesion at concentrations up to 0.01 mg/ml [28]. Platelet factor-4 also reduces tumour cell adhesion by 60% at 0.01 mg/ml, but protamine has no effect at concentrations up to 0.5 mg/ml. Finally, placental RI prevents tumour cell adhesion. Adhesion of tumour cells to angiogenin is sensitive to heparinase and heparitinase treatment, and a cell-surface proteoglycan of apparent molecular mass > 200 kDa has been isolated by angiogenin-affinity chromatography [28].

Heparin: Angiogenin binds to heparin-Sepharose and is eluted by 0.78 M NaCl [28]. The stoichiometry of the angiogenin-heparin interaction, estimated by light-scattering measurements, is 1 heparin chain (molecular mass 16.5 kDa): ≈ 9 angiogenin molecules [29]. Heparin partially protects angiogenin from tryptic cleavage at Arg-31, Lys-60 and Arg-101 [29]. It inhibits angiogenin-catalysed cleavage of tRNA at pH 5.5 with an IC_{50} value of 0.7 mg/ml [53]. Adhesion of HT-29 human adenocarcinoma cells to human angiogenin is reduced by 60% in the presence of 0.05 mg/ml heparin [28]. The minimum heparin fragment required for inhibition is the dodecamer [29].

Actin: An angiogenin-binding protein of 42 kDa is released from CPAE and GM7373 fetal bovine aortic endothelial cells by incubating starved subconfluent monolayers with 1 mg/ml heparan sulphate. It has been cross-linked to ^{125}I-angiogenin in a crude cell lysate and in heparan sulphate-released material by using 10 mM EDC (1-ethyl-3 (3-dimethylaminopropyl) carbodiimide). The formation of a 58-kDa complex is inhibited by a 50-fold molar excess of either bovine or human unlabelled angiogenin, and is reduced by a factor of 3.4 in the presence of RNase A. Pre-incubation of ^{125}I-angiogenin with placental RI prevents the formation of the complex, whereas co-incubation does not [48]. Among the angiogenin-binding molecules obtained by a purification procedure comprising angiogenin-affinity chromatography, a 42-kDa protein has been purified and further identified as a member of the actin family [49]. A 42-kDa band has also been revealed by immunoblotting with a monoclonal antibody to smooth-muscle alpha-actin of CPAE cell-surface proteins selected by biotinylation of the cell surface and further isolation by avidin affinity chromatography. In addition, immunoreactivity has been detected at the surface of CPAE cells by immunofluorescence with a monoclonal antibody specific for smooth muscle alpha-actin. Angiogenin binds to bovine muscle actin and induces its polymerisation *in vitro*. The cross-linking of ^{125}I-angiogenin to actin is inhibited by platelet factor-4 (400 ng/ml) and protamine (0.01 mg/ml) [49]. Finally, both actin at a 100-fold molar excess and anti-actin antibody at a 10-fold molar excess inhibit the angiogenesis induced by 10 ng of bovine angiogenin in the chick chorioallantoic membrane assay [49].

Ribonuclease inhibitor: Placental RI, a tight-binding inhibitor of both the ribonucleolytic and angiogenic activities of angiogenin ([12, 13], for reviews), abolishes angiogenin binding to CPAE cells [16] and angiogenin-induced increase of diacylglycerol in cultured CPAE cells [21]. It also prevents tumour cell adhesion to angiogenin [28]. The stoichiometry of binding between angiogenin and placental RI is 1:1. The apparent association rate constant of 1.8 x 10^8 $M^{-1}s^{-1}$ and the dissociation rate constant of the complex of 1.3 10^{-7} s^{-1} result in an extremely low calculated Ki of 7.1 x 10^{-16} M. The half-life of the complex is about 60 days. Inhibition is competitive and reversible, and 1 mM p-(hydroxymercuri) benzoate dissociates the complex to yield active angiogenin. The human RI gene (RNH) is located on the terminal part of the short arm of chromosome 11, subband 11p15.5, within 90 kb of the Harvey-ras protooncogene (HRAS). Human placental RI is an acidic protein (pI 4.7) composed of 460 amino acids with a calculated molecular mass of 49 847 Da; it is detected in western blot analysis as a single polypeptide chain of 51 kDa ([12, 13], for reviews). Human RI has a high leucine (92 residues, 20%) and cysteine (32 residues, 7%) content organised in fifteen alternate homologous leucine-rich repeats. Each repeat corresponds to a single right-handed β-α structural unit [54]. The amino acid sequences of RIs are highly conserved, as the human, pig and rat species are 75-77% identical, with no insertions or deletions, except for a short insertion at the N-terminus of human RI. Crystal structure studies of porcine RI at a resolution of 2.5 Å reveals a horseshoe structure with overall dimensions 70 Å x 62 Å x 32 Å (Figure 3, [54]). The 16 helices align on the outer circumference and 17 β-strands form a curved parallel β-sheet on its inner circumference exposed to the solvent. Extensive mutagenesis studies of the human RI-angiogenin complex [12] and X-ray crystallographic analysis at 2.0 Å reveal that the tight interaction results from a large binding interface involving 26 human RI residues from 13 of the 15 repeat units of RI and a total of 124 contacts [51]. The interaction is predominantly electrostatic with a high chemical complementarity, mainly in the C-terminal segment 434-460 of human RI. Angiogenin binds to RI with its catalytic site covering the C-terminal part of the inhibitor; the ε-amino group of Lys-40 forming two salt bridges with the carboxylate of Asp-435. About one-third of angiogenin is located inside the central cavity of RI and the other part of the molecule lies over it. Finally, the complex crystallises as a dimer [51]. RI mRNA is ubiquitously expressed at ≈ 2 kb in the 16 normal human tissues so far tested [55]. Immunoreactivity, using a monoclonal antibody against placental RI, has been detected in normal human serum and would correspond to 2-3 mg/L, but there is no evidence that an active form is present. The absence of disulphide bonds in RI is consistent with its cytoplasmic location; it is irreversibly inactivated by sulfhydryl reagents.

RNA: Angiogenin has the same general catalytic properties as RNase A. It cleaves RNA preferentially on the 3'

(a)

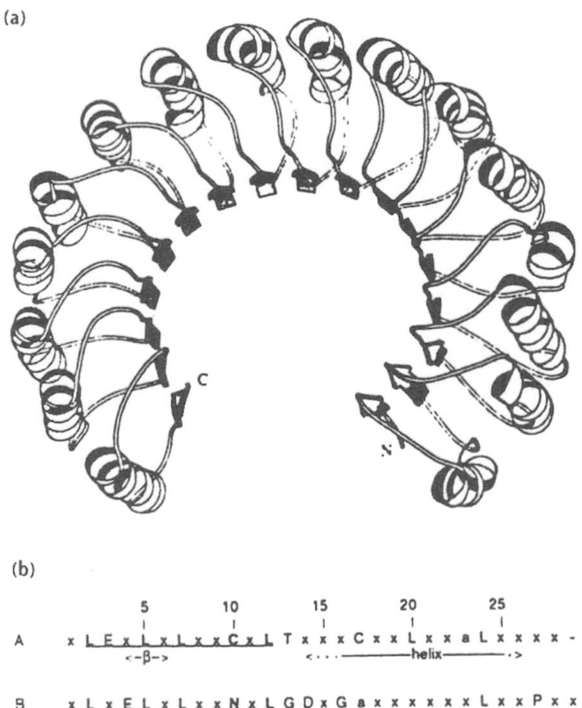

(b)

Figure 3. Structure of ribonuclease inhibitor (RI). (a) Ribbon diagram of the structure of porcine RI generated using the program MOLSCRIPT. (b) Consensus sequences and secondary structure of leucine-rich repeats (LRRs) of porcine RI. The sequence of RI was aligned so that two types of repeats (A and B) alternate in the sequence. One-letter amino acid code is used. 'x' indicates any amino acid and 'a' denotes an aliphatic amino acid. The part of the repeat that is strongly conserved in all LLR proteins is underlined, and the conserved residues are shown in bold. Below the sequence, solid lines mark the core region of b-sheet and helix; dots denote extensions of helix in different repeats. Reproduced with permission from [54].

side of pyrimidine by a transphosphorylation/hydrolysis mechanism. However, its activity differs markedly in both its magnitude and specificity [3, 41]. Angiogenin is 10^5 to 10^6-fold less active than RNase A on mono-, di- and polynucleotide substrates [41, 53]. It is inactive on various dinucleotides and polynucleotides used as substrates for RNase [3]. The order of reactivity for angiogenin is CpA>CpG>UpA>UpG [41]. The base-cleavage specificity towards RNA has been determined with *Saccharomyces cerevisiae* and *Escherichia coli* 5S RNAs [56]. Like RNase A, phosphodiester bond cleavage occurs at the 3' side of cytidylic or uridylic acid residues when the pyrimidine is followed by adenine, but not at all the potential cleavage sites [56]. Angiogenin cleaves 28S and 18S rRNAs to give a mixture of products 100 to 500 nucleotides long. The overall pattern is different from that of RNase, and it requires 10^4-10^5 as much angiogenin to degrade the same amount of rRNA as with RNase A [3, 14]. Like RNase A, angiogenin prefers

single-stranded RNA as substrate. It does not cleave single-stranded DNA. However, angiogenin has been reported to bind to single-stranded DNA, double-stranded plasmid DNA and high-molecular-weight DNA [56]. The optimal parameters for yeast tRNA cleavage by angiogenin are pH ≈ 6.8, 15-30 mM NaCl, and ≈ 55°C [53].

Additional Features   Both the ribonucleolytic and angiogenic activities of angiogenin are abolished by reagents that modify histidine, lysine, or arginine residues, but not those that modify tyrosine, aspartate, glutamate or methionine residues. The activities are also abolished by reduction of the three disulphide bonds, and are restored by reoxidation [14].

Fluorescence changes on binding of angiogenin to placental RI: Placental RI has six tryptophans and a fluorescence spectrum maximum at 338 nm. Human and rabbit angiogenins with one tryptophan (Trp-89) exhibit a weaker fluorescence spectrum with a maximum at 343 nm when excited at 285 nm. An enhancement of fluorescence is observed upon binding of angiogenin to placental RI. This property has been used as a probe to study placental RI binding to angiogenin ([12, 13], for reviews).

Adsorption of angiogenin to surfaces: Loss of angiogenin due to adsorption to container surfaces is significant. Conditions used to minimise this interaction are as follows: polypropylene containers and an angiogenin concentration above 300 nM (≈ 4 mg/L). Lysozyme or serum albumin should be added as carriers at 0.1 g/L to more dilute angiogenin solutions.

The sequence data are available in the Swiss-Prot database under the accession number placental RI, P13489. The nucleotide sequence accession number in the EMBL/Genbank Data Library is M22414.

Structure

Sequence and Size   Human angiogenin is a single-chain protein of 123 amino acids and has a calculated relative molecular mass of 14 124. The amino-acid composition of the mature protein is 15 Asp/Asn, 7 Thr, 9 Ser, 10 Glu/Gln, 8 Pro, 8 Gly, 5 Ala, 6 Cys, 5 Val, 1 Met, 7 Ile, 6 Leu, 4 Tyr, 5 Phe, 7 Lys, 6 His, 13 Arg, 1 Trp. It has the following sequence:

Glu[1]-Asp-Asn-Ser-Arg-Tyr-Thr-His-Phe-Leu[10]-Thr-Gln-His-Tyr-Asp-Ala-Lys-Pro-Gln-Gly[20]-Arg-Asp-Asp-Arg-Tyr-Cys-Glu-Ser-Ile-Met[30]-Arg-Arg-Arg-Gly-Leu-Thr-Ser-Pro-Cys-Lys[40]-Asp-Ile-Asn-Thr-Phe-Ile-His-Gly-Asn-Lys[50]-Arg-Ser-Ile-Ile-Lys-Ala-Ile-Cys-Glu-Asn-Lys[60]-Asn-Gly-Asn-Pro-His-Arg-Glu-Asn-Leu-Arg[70]-Ile-Ser-Lys-Ser-Ser-Phe-Gln-Val-Thr-Thr[80]-Cys-Lys-Leu-His-Gly-Gly-Ser-Pro-Trp-Pro[90]-Pro-Cys-Gln-Tyr-Arg-Ala-Thr-Ala-Gly-Phe[100]-Arg-Asn-Val-Val-Val-Ala-Cys-Glu-Asn-Gly[110]-Leu-Pro-Val-His-Leu-Asp-Gln-Ser-Ile-Phe[120]-Arg-Arg-Pro[123]-OH. Three disulphide bonds link Cys[26]-Cys[81], Cys[39]-Cys[92] and Cys[57]-Cys[107]. The protein is free of the glycosylation signal sequence Asn-X-Ser/Thr [9].

Angiogenin isolated from normal human plasma and from medium conditioned by HT-29 human adenocarcinoma cells has identical chromatographic behaviour, molecular weight, amino-acid composition, ribonucleolytic and angiogenic activities, and immunoreactivity [1].

Homologies

Angiogenins: Angiogenin, that can be referred to as angiogenin-1, has been isolated from HT-29 human adenocarcinoma cells [2] and from human plasma [1]. It has also been purified from mouse, rabbit and pig sera [37] and from bovine serum and milk [57, 58]. The mouse protein sequence has been deduced from the genomic DNA sequence. Mouse, rabbit, pig and bovine angiogenins have 75, 73, 66 and 64% sequence identity to human angiogenin, respectively (Figure 4). Most of the differences are the result of conservative substitutions. Apart from cyclization of the N-terminal glutamyl residue in the human, mouse and rabbit proteins, there is no evidence of post-translational modification. There is no Asn-X-Thr/Ser site for potential N-linked glycosylation and no evidence of O-linked glycosylation. All five angiogenins contain the essential catalytic residues His-13, Lys-40 and His-114 (human angiogenin-1 numbering), and three disulphide bonds. They induce neovascularization *in vivo* and display very low ribonucleolytic activities [1, 37, 59]. Human, rabbit, pig and bovine angiogenins all bind human placental RI with 1:1 stoichiometry [1, 12, 37, 59].

Angiogenin-2 isolated from bovine serum and milk is 57% identical to bovine angiogenin-1, with an overall similarity of 71%. It is glycosylated at Asn-33 and contains 2-3 glucosamine, 5-6 mannose, 1-2 galactosamine and 0-1 xylose. Its apparent molecular mass is 20 kDa [60]. Its ribonucleolytic activity is lower than that of bovine angiogenin-1. Angiogenin-2 is a less potent inducer of angiogenesis than angiogenin-1. The conserved Asn-109 of angiogenin-1 is replaced by Asp-108 in bovine angiogenin-2. The same replacement by site-directed mutagenesis in human angiogenin-1 (replacement of Asn-109 by Asp-109) abolishes angiogenic activity, and the aspartyl derivative inhibits angiogenin-induced angiogenesis [40].

Angiogenin-related family: By screening a mouse genomic library with an angiogenin-1 gene probe, an « angiogenin-related protein » gene with 88% nucleotide sequence identity to the BALB/c mouse angiogenin gene has been identified [59]. The mouse angiogenin-related protein (Angrp), produced in *E. coli*, is 78% identical to mouse angiogenin-1. It is free of consensus sequences for N- and O- linked carbohydrate chains. It has higher ribonucleolytic activity than angiogenin-1 and is inhibited by human placental RI. Angiogenin-related protein lacks angiogenic activity in the chick embryo chorioallantoic membrane assay, which could result from poor conservation of the receptor-binding domain. Angiogenin-related protein does not inhibit angiogenin-1-induced angiogenesis. It has not been detected in mouse serum [59].

EF-5 induced by E2a-Pbx1 in mouse NIH 3T3 fibroblasts encodes another member of the angiogenin gene family (EF5/angiogenin-3) that is 74% identical to mouse angiogenin-1 and 81% identical to mouse angiogenin-related protein [61]. Its transcript is expressed in adult liver and on day 7 of development. It is not expressed in

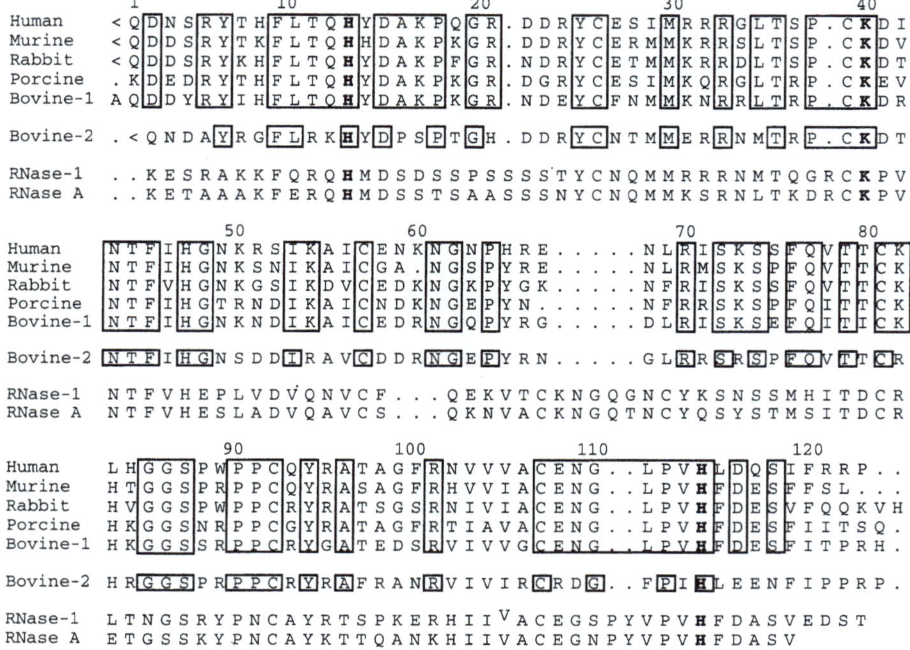

```
            1               10                  20                  30                  40
Human     < Q D N S R Y T H F L T Q H Y D A K P Q G R . D D R Y C E S I M R R R G L T S P . C K D I
Murine    < Q D D S R Y T K F L T Q H H D A K P K G R . D D R Y C E R M M K R R S L T S P . C K D V
Rabbit    < Q D D S R Y K H F L T Q H Y D A K P F G R . N D R Y C E T M M K R R D L T S P . C K D T
Porcine   . K D E D R Y T H F L T Q H Y D A K P K G R . D G R Y C E S I M K Q R G L T R P . C K E V
Bovine-1  A Q D D Y R Y I H F L T Q H Y D A K P K G R . N D E Y C F N M M K N R R L T R P . C K D R

Bovine-2  . < Q N D A Y R G F L R K H Y D P S P T G H . D D R Y C N T M M E R R N M T R P . C K D T

RNase-1   . . K E S R A K K F Q R Q H M D S D S S P S S S S T Y C N Q M M R R R N M T Q G R C K P V
RNase A   . . K E T A A A K F E R Q H M D S S T S A A S S S N Y C N Q M M K S R N L T K D R C K P V

            50                  60                  70                  80
Human     N T F I H G N K R S I K A I C E N K N G N P H R E . . . . . N L R I S K S S F Q V T T C K
Murine    N T F I H G N K S N I K A I C G A . N G S P Y R E . . . . . N L R M S K S P F Q V T T C K
Rabbit    N T F V H G N K G S I K D V C E D K N G K P Y G K . . . . . N F R I S K S S F Q V T T C K
Porcine   N T F I H G T R N D I K A I C N D K N G E P Y N . . . . . . N F R R S K S P F Q I T T C K
Bovine-1  N T F I H G N K N D I K A I C E D R N G Q P Y R G . . . . . D L R I S K S E F Q I T I C K

Bovine-2  N T F I H G N S D D I R A V C D D R N G E P Y R N . . . . . G L R R S R S P F Q V T T C R

RNase-1   N T F V H E P L V D V Q N V C F . . . Q E K V T C K N G Q G N C Y K S N S S M H I T D C R
RNase A   N T F V H E S L A D V Q A V C S . . . Q K N V A C K N G Q T N C Y Q S Y S T M S I T D C R

            90                  100                 110                 120
Human     L H G G S P W P P C Q Y R A T A G F R N V V V A C E N G . . L P V H L D Q S I F R R P . .
Murine    H T G G S P R P P C Q Y R A S A G F R H V V I A C E N G . . L P V H F D E S F F S L . . .
Rabbit    H V G G S P W P P C R Y R A T S G S R N I V I A C E N G . . L P V H F D E S V F Q Q K V H
Porcine   H K G G S N R P P C G Y R A T A G F R T I A V A C E N G . . L P V H F D E S F I I T S Q .
Bovine-1  H K G G S S R P P C R Y G A T E D S R V I V V G C E N G . . L P V H F D E S F I T P R H .

Bovine-2  H R G G S P R P P C R Y R A F R A N R V I V I R C R D G . . F P I H L E E N F I P P R P .

RNase-1   L T N G S R Y P N C A Y R T S P K E R H I I V A C E G S P Y V P V H F D A S V E D S T
RNase A   E T G S S K Y P N C A Y K T T Q A N K H I I V A C E G N P Y V P V H F D A S V
```

Figure 4. Protein sequence alignment of human, murine, rabbit, porcine and bovine (Bovine-1) angiogenins, bovine angiogenin-2 (Bovine-2), human pancreatic ribonuclease (RNase-1), and bovine pancreatic ribonuclease (RNase A). The alignment is numbered according to human angiogenin. Amino acids conserved in all angiogenins are boxed. The active site residues His-13, Lys-40, and His-114 are shown in bold.

NIH 3T3 fibroblasts, exhibits dose-dependent expression in response to E2a-Pbx1, and slight upregulation in Abl-transformed cells. It is not induced by oncogenic Lck, Ras, Neu, Src or Sis, or in myeloblasts immortalised by E2a-Pbx1 [61]. It has the residues required for ribonucleolytic activity but diverges at residues in the receptor-binding domain.

RNase family: The human RNase family is composed of related structurally proteins that cleave ribonucleic acids on the 3' side of pyrimidine and have a variety of distinct biological activities ([62, 63], for reviews). In human angiogenin (RNase-5), 43 of the 123 amino acids are identical to those of human pancreatic RNase (RNase-1) at the corresponding positions, and many of the remaining residues are conservatively replaced [9]. Thus, angiogenin has 35% sequence identity and a similarity of 68% with human pancreatic RNase-1. RNase-1 has also been purified from urine, seminal plasma, brain and kidney ([62, 63], for reviews). Angiogenin has 28% sequence identity with RNase-2 (also named EDN, for eosinophil-derived neurotoxin). RNase-2 occurs predominantly in eosinophils, liver, spleen and placenta. It is known for its neurotoxicity, due to its ribonuclease activity. RNase-2 is 35% identical to RNase-1. Angiogenin is 27% identical to RNase-3 (also named eosinophil cationic protein or ECP). RNase-3, isolated from granulocytes, is highly similar to RNase-2 (70% identity) but less neurotoxic than RNase-2. It is a helminthotoxin with antibacterial activity as well as cytotoxic activity for mammalian cells *in vitro*. Angiogenin is 40% identical to RNase-4. First isolated from tumour-cell-conditioned medium, RNase-4 has been purified from plasma and is highly similar to bovine and porcine liver RNase (9 ≈ 90% identity). RNase k6 has been amplified from human genomic DNA and its mRNA transcript has been detected in all the human tissues so far tested (predominantly in lung). The deduced amino acid sequence of RNase k6 is 30% identical to that of angiogenin ([62, 63], for reviews). None of the ribonucleases tested (human RNase-1, -2 and -4 and bovine RNase A) are angiogenic, emphasising the peculiarity of angiogenin.

"Fibroblast growth factor receptor ligand 2" cloned *Xenopus* gene (FRL2) encodes a protein that is 21% identical to angiogenin/RNase, with no homology with typical FGF family members. FRL2 increases protein-tyrosine phosphorylation in yeast cells expressing FGF receptors and binds to the extracellular domain of the FGF receptor [64].

A gene differentially expressed in two types of v-myb-transformed avian myelomonocytic cells (clone 462) encodes an RNase/angiogenin-related protein that shows 34% sequence identity with angiogenin [65].

The RNase-superfamily-related gene (RSFR), highly expressed in chicken bone marrow cells, codes for a protein that shows 32% identity with angiogenin and differs fully in the angiogenin receptor-binding domain [66].

Several RNases were first isolated on the basis of their biological properties and subsequently identified as RNases [67]. Angiogenin, EDN (RNase-2) and ECP (RNase-3) belong to this family, together with bovine seminal ribonuclease (BS-RNase), frog onconase and bullfrog lectin. The three last are 32%, 27% and 30% identical to human angiogenin, respectively. BS-RNase exhibits antitumour, immunosuppressive and antispermatogenic actions. Onconase from *Rana pipiens* and the sialic-acid-binding lectin purified from *Rana catesbeiana* oocytes possess antitumour activity [67].

Other homologies: Bovine prochymosin has 6 residues identical to residues 103-108 of human angiogenin and 9 of the 14 in positions 103 to 116 are also identical. Residues 1136-1149 of DNA-dependent RNA polymerase align with 7 of the 14 C-terminal residues of human angiogenin [9].

Conformation Human Met-(-1) angiogenin crystallises in the orthorhombic system, space group C222$_1$, with unit cell dimensions a=83.4 Å, b=120.6 Å and c=37.7 Å (one molecule per asymmetric unit, specific volume 3.3 Å^3/Da, 63% solvent content). When determined at 2.4-Å resolution, the three-dimensional structure of human angiogenin is kidney-shaped with dimensions 38 Å x 43 Å x 34 Å, and consists of three helical regions and seven β-strands (Figure 1, [44]). The secondary-structure elements in human angiogenin are helix 1 (H1, residues 3-14), helix 2 (H2, residues 21-33), β-sheet 1 (B1, residues 41-47), helix 3 (H3, residues 49-58), β-sheet 2 (B2, residues 62-65), β-sheet 3 (B3, residues 69-73), β-sheet 4 (B4, residues 76-84), β-sheet 5 (B5, residues 93-101), β-sheet 6 (B6, residues 103-108), β-sheet 7 (B7, residues 111-116) and 310-helix (residues 117-121). The central core of angiogenin is organised around the two antiparallel twisted β-strands B3-B4 and B5-B6. It is completed by two additional strands on either side (B1 and B7) and the short strand B2. The N-terminal helix H1 is close to the short C-terminal 3$_{10}$-helix. Helix 2 and H3 are oriented at ≈ 70° relative to the plane of the β-sheet. Three disulphide bridges are present in angiogenin, linking H2 to B4 (Cys^{26}-Cys^{81}), B1 to B5 (Cys^{39}-Cys^{82}) and H3 to B6 (Cys^{57}-Cys^{107}) [44]. The overall structure of human angiogenin is similar to that of RNase A but differs in the putative receptor binding site and the ribonucleolytic active centre. The 1.5-Å-resolution crystal structure of bovine angiogenin confirms that the site spatially analogous to that for pyrimidine binding in RNase A is obstructed by Gln-117 and Glu-118 in human and bovine angiogenin, respectively [44, 46]. The properties of crystalline angiogenin are conserved in solution, as shown by ^{1}H NMR spectroscopy of bovine angiogenin [45]. Five loops and one helix show a larger dispersion, *viz.* the N- and C- terminal segments, the Arg^{67}-Gly-Asp^{69} loop covering part of the putative receptor binding domain and the endothelial cell adhesion domain [27], and the 35-42 and 86-94 segments covering part of the domains involved in the inhibition of PMNL degranulation by angiogenin [26].

Additional Features

Isoelectric point: Angiogenin elutes after lysozyme (pI 10.5) in cation-exchange high-performance liquid chromatography [2] suggesting a pI > 10.5.

Ultraviolet absorption: Angiogenin has a molar absorptivity of 12,500 M^{-1} cm^{-1} at 280 nm in 20 mM Tris, pH 7.5 [68].

Metal content: Atomic absorption spectrometry of plasma-derived angiogenin at a concentration of 0.07 mM showed that less than 0.01 mol/mol of copper, iron, manganese, and zinc is present [1].

Deamidation of asparagine residues is a non enzymatic post-translational protein modification that occurs in angiogenin, BS-RNase, RNase A, human RNase-4 and other proteins such as human growth hormone, mouse epidermal growth factor, interleukin-1β, calmodulin, adrenocorticotropin and lysozyme. In human angiogenin, treatment at alkaline pH or long-term storage results in deamidation of Asn-61 and Asn-109 [40]. The desamido derivatives mainly contain isoaspartic acid, exhibit nearly full enzymatic activity, have low angiogenic activity on the chick embryo chorioallantoic membrane, and do not inhibit angiogenin-induced neovascularization. The aspartic acid derivatives, obtained by site-directed mutagenesis, differ from the isoaspartic derivatives by their inhibition of angiogenin-induced angiogenesis. This underlines the importance of Asn-61 and Asn-109 for the angiogenic activity of human angiogenin [40].

The sequence data are available in the Swiss-Prot database under the following accession numbers: human RNase-1, P07998; bovine RNase-1, RNase A, P00656; human RNase-2, EDN, P10153; human RNase-3, ECP, P12724; RNase-4, P34096; RNase-5, human angiogenin, P03950; rabbit angiogenin, P31347; pig angiogenin, P31346; bovine angiogenin-1, P10152; bovine angiogenin-2, P80929; mouse angiogenin-1, P21570; mouse angiogenin-related protein, Q64438; EF5/mouse angiogenin-3, P97802; chicken clone 462 RNase/angiogenin, P27043; chicken G. gallus RSFR, P30374; RNase k6, Q93091; BS-RNase, P00669; onconase, p30 (*Rana pipiens*), P22069; sialic acid-binding lectin, SBL-C (*Rana catesbeiana*), P14626.

The atomic coordinates are available in the Protein Data bank, Chemistry Department, Brookhaven National Laboratory, Upton, NY 11973, under the following entry codes: human angiogenin, 1ANG; bovine angiogenin, 1AGI; RNase A, 7RSA.

The proton NMR coordinates of bovine angiogenin are available from the Brookhaven Protein Data bank under the file name 1GIO.

Gene

Gene Structure

The gene for angiogenin is devoid of introns in the coding and 3'-noncoding regions of the gene ([69], Figure 5). A consensus sequence for a 3' boundary suggests the presence of an intron that exceeds 1700 base pairs in the 5' untranslated region [69, 70]. The presence of an intron in the 5' untranslated region and the absence of introns in the coding sequence are common features in many members of the RNase A family. The angiogenin gene contains a leader sequence coding for a signal peptide of 24 (or 22) amino acids, indicating that angiogenin is a secreted protein, 369 nucleotides coding for the mature protein of 123 amino acids, a stop codon, a 3'-noncoding sequence of 175 nucleotides and a poly(A) tail of 36 nucleotides. A potential TATA box is present. Two Alu sequences flank the gene, 400 base pairs upstream and 300 base pairs downstream of the coding region, respectively. These two Alu repeats are in the same inverted orientation. At 1100 base pairs downstream of the gene, a third Alu sequence is present in the typical orientation. The coding region of the mouse angiogenin gene is 82% identical to the human region.

Southern blotting of human leukocyte DNA has suggested that only one copy of the angiogenin gene is present in human DNA [69]. However, 2-3 copies of the gene have been detected in various experimental conditions [70].

Chromosomal Location

The human angiogenin gene is located on chromosome band 14q11 [71]. It resides proximal to a translocation breakpoint within the T cell receptor α/δ locus, upstream of this locus. Chromosome translocations in peripheral T lymphocytes are frequently observed in patients with the autosomal recessive disease ataxia telangiectasia. The excessive vascularization observed in this disease points to the involvement of an angiogenic factor [71].

In mice, the angiogenin gene is assigned to chromosome 14.

Gene Expression

Angiogenin mRNA is widely distributed, in both tissues and cultured cells. Angiogenin mRNA is expressed predominantly in human and rat liver [55, 72]. The ≈ 1-kb angiogenin messenger has also been detected by northern blotting in human pancreas, lung, prostate, testis, ovary, small intestine, colon, heart, kidney, skeletal muscle and peripheral blood leukocytes [55]. In the rat, it is also present in small intestine, colon, heart, kidney, adrenal, spleen, ovary, brain, lung and skeletal muscle [72]. Larger transcripts have been detected in human liver [55] and in HT-29 human colon adenocarcinoma cells, at 1.6, 2.7, 3.5, 5.2 and 6 kb [73].

A 0.8- to 1.1-kb angiogenin mRNA is present in normal and tumour cells [8, 72, 73]. Angiogenin transcripts have been detected in human tumour cell lines such as lung carcinoma A549 cells [8, 73], the colon lines HT-29, SW620, SW480 and WiDr, the medulloblastoma line TE-671, the fibrosarcoma line HT-1080, SK-HEP hepatoma cells, embryonic tumour cells from rhabdomyosarcoma, and bladder carcinoma HT-1376 cells [72, 73], as well as in T-cell leukaemia CEM cells [73] and MT4 cells [8]. They are also present in normal cells such as epithelial cells from colon [72, 73] and liver [72, 73], mesothelial cells [73] and fibroblasts from embryonic lung, foreskin [73], adult saphenous vein [8], human umbilical endothelial cells and endothelium-derived EA.hy926

```
-1000 AACAGAGAGGTCCCCAAATCCCGGTCTGTGGCCTGTCCGCCTAAGCTCTGCCTCCTGCCAGATCAGCAGGCAGCATTAGATTCTCATAGGAGCTGGACGC

 -900 CTATTGTGAACTGCGCATGTGCGGGATCCAGATTGTGCACTCTTTATGAGAATCTAACTAATGCTTGATGATCTATCTGAACCAGAACAATTTCATCCTG

 -800 AAACCATCCCCCACCAATCCATAGAAATACTGTCTTCCACAAAAATGATCCCTGGTGCCAAAAATGTTAGAGACCACTCCCCTAAAACTCTCTTCTTAGC

 -700 TCTCACCTCCTGTATTACTATCTCATCTCAGTACATTGAAGCCCCCATCTTTTCCCCATGGATGCCTCATTTCCTATTAGGGAGGCATTTTTTTTATTTTT

 -600 TGTTTTTATTTTTTTTCCGAGACGGAGTCTCGCTCTGTCGCCAAGGCTGGAGTGCAGTGGCGCGATCTCGGCTCACTGCAAGCTCCGCCTCCCGGGTTCAC

 -500 GCCATTCTCCTGCCTCAGCCTCCCAAGTAGCTGGGACTACAGGCGCCCGCACTACGCCCGGCTAATTTTTTGTATTTTTAGTAGAGACGGGGTTTCACCG

 -400 TGGTAGCCAGGATGGTCTCGATCTCCTGACCTCGTGATCCGCCCGCCTTGGCCTCCCAAAGTGCTGGGATTACAGGCGTGAGACCGCGCCCGGCCGTCAT

 -300 TTGGTATGTCTTAATGTGCCTCAGGACCTAGCACAGTCCCTGGTACCCAGTAGAGACCTATGTAATGTTCGTTATTCAATAATAAATACATGAATTAAAG

 -200 AGTGAGAGTGGATTTTGTAATGTTACGACTGATAGAGAAATACTCAGTGATTCTAAGGGATGGGGAAGAACGGTTGGAGCTAGAGGTTGTGCTCAGGAAA

 -100 CTATTAAATAGACGTTCCGCAGGAAGGGATTGACGAAGTGTGAGGTTAATGAGGAAGGGAAAATAGAA[TATAAA]ATTTGGTGGTGGAAAAGATCTGATTC

    1 ATGATGCCGTGTCAGAGAGCAAAGCTCCTGTCCTTTTGGCCTAATTTGGTGATGCTGTTCTTGGGTCTACCACACCTCCTTTTGCCCTCCGCAGGAGCCT

      ......↓.......-24.........-20.......................-10.....................+1.........
      ........M..V..M..G..L..G..V..L..L..L..V..F..V..L..G..L..G..L..T..P..P..T..L..A..Q..D..N..S..R..
  101 GTGTTGGAAGAGATGGTGATGGGCCTGGGCGTTTTGTTGTTGGTCTTCGTGCTGGGTCTGGGTCTGACCCCACCGACCCTGGCTCAGGATAACTCCAGGT

      ..........10.....................20.......................30...........
      Y..T..H..F..L..T..Q..H..Y..D..A..K..P..Q..G..R..D..D..R..Y..C..E..S..I..M..R..R..R..G..L..T..S..P..C
  201 ACACACACTTCCTGACCCAGCACTATGATGCCAAACCACAGGGCCGGGATGACAGATACTGTGAAAGCATCATGAGGAGACGGGGCCTGACCTCACCCTG

      ..40.................50.....................60.......................70.....
      ..K..D..I..N..T..F..I..H..G..N..K..R..S..I..K..A..I..C..E..N..K..N..G..N..P..H..R..E..N..L..R..I..S.
  301 CAAAGACATCAACACATTTATTCATGGCAACAAGCGCAGCATCAAGGCCATCTGTGAAAACAAGAATGGAAACCCTCACAGAGAAAACCTAAGAATAAGC

      ..........80...................90.....................100..............
      .K..S..S..F..Q..V..T..T..C..K..L..H..G..G..S..P..W..P..P..C..Q..Y..R..A..T..A..G..F..R..N..V..V..V..
  401 AAGTCTTCTTTCCAGGTCACCACTTGCAAGCTACATGGAGGTTCCCCCTGGCCTCCATGCCAGTACCGAGCCACAGCGGGGTTCAGAAACGTTGTTGTTG

      ...........110.....................120......123
      A..C..E..N..G..L..P..V..H..L..D..Q..S..I..F..R..R..P
  501 CTTGTGAAAATGGCTTACCTGTCCACTTGGATCAGTCAATTTTCCGTCGTCCGTAACCAGCGGGCCCCTGGTCAAGTGCTGGCTCTGCTGTCCTTGCCT

  600 TCCATTTCCCCTCTGCACCCAGAACAGTGGTGGCAACATTCATTGCCAAGGGCCCAAAGAAAGAGCTACCTGGACCTTTTGTTTTCTGTTTGACAACATG

  700 TTTAATAAATAAAAATGTCTTGATATCAGTAAGAATCAGAGTCTTCTCACTGATTCTGGGCATATTGATCTTTCCCCCATTTTCTCTACTTGGCTGCTCC

  800 CTGAGAGGACTGCATAGGATAGAAATGCCTTTTTCTTTTCTTTTCGTTTTTTTTTTTTTTTTTTTTTTTGAGATGGAGTCTCACTCTGTCGCCCAGGCTTAA

  900 GTGCAATGGCACAATCTCGGCTCACTGCAACCTCTCTCTCCTGGGTTCAAGTGATTCTCCTGCCTCAGCCTCCCAAATAGCTGAGATTACAGGCATGCAC

 1000 CACCACACCTGGCTAATTTTTGTGTTTTTAGTAGAGACAGGGTTTCACCGTTTTGGCCAGGTTGGTCTTGAACTCCTGACCTCGGGAGATCCGCCCACCT

 1100 TGGCCTCTCTTTGTGCTGGGATTACAGGCATGAGCCACTGAGCCGGGCCACTTTTTCCTTATCAGTCAGTTTTTACAAGTCATTAGGGAGGTAGACTTTA

 1200 CCTCTCTGTGAAGGAAAGTATGGTATGTTGATCTACAGAGAGAGATGGAAAAATTCCAGGGCTCGTAGCTACTAAGCAGAATTTCCAAGATAGGCAAATT

 1300 GTTTTTTCTGTCAAATAATAAGCTAATATTACTTCTACAAATATGAGACCTTGGAGAGAAGTTTCCAAGGACCAAGTACCAACATACCAACAGATTATTA

 1400 TAGTTTCTCTCACTCTTACACACACACACACACATATACACATATGTAATCCAGCATGAATACCAAAATTCATTCAGGGTAGCCACCTTTTGTCTTAATC

 1500 GAGAGATAATTTTGATGTTTGAATGGAATGCTCCCAGGATATTCTCTTGTCATGGTTATTTTATATAAAATTCAAAAACCAATTACATTATTTCCTCTGT

 1600 AATCTTTTACTTTATCAACTAATGTCTGGCAAGTGTGATGTTTTGGGGAAGTTATAGAAGATTCCGGCCAGGCGCTTATCTCACGCTTGTAATCCAGCAC

 1700 TTTGGGAAGCTGAGGCGGACAGATCACGAGGTCAAGAGATCAAGACCATCCTGGACAACATGGTGAAACCTTGTCTCTACTAAAAATGTGAAAATTAGCT

 1800 GGGCGTGGTGGCACACACCTATAGTCCCAGCTACTCGGGAGGCTGAGGCAGGAGAATCGCTTGAACCTAGGAGGCGGAGGTTGCACTGAGCCGAGATCAC

 1900 GCCACTGCACTCCAGCCTGGGCGACAGAGCGAGACTCCATCTCAAAAAAAAAAAAAAAAGAAAGATCCCAGTTTATCCCAGTTTATCCCTTATTCTTCCT
```

Figure 5. Nucleotide sequence of the human angiogenin gene and inferred amino acid sequence. Nucleotides are numbered beginning with +1 at the proposed transcription initiation site; TATA box is boxed; AATAAA and CACTG sequences are underlined with solid bars. The two arrows limit the cDNA sequence. Underlining shows the three Alu repeat sequences in the 5'- and 3'- flanking regions of the gene; dashed underlining indicates the direct repeat sequences flanking each Alu repeat (modified from [69]).

cells [8]. The abundance of angiogenin RNA transcripts is increased in stimulated peripheral blood lymphocytes [73]. Angiogenin transcripts have not been detected in promyelomonocytic HL-60 and U-937 cells, which is consistent with the absence of immunoreactivity in cell-conditioned media [8].

Angiogenin immunoreactivity is widely and differentially found in anchorage-dependent growing cells such as vascular endothelial cells from saphenous and umbilical veins, SMC, fibroblasts from embryos (WS1 and WI-38 cells), new-borns (AG1523) and adults (cells from saphenous vein), and tumour cells (A-431, A549, HT-29 and HeLa). The secretion of angiogenin can reach ng/10^6 cells/day. As compared to anchored cells, normal peripheral blood cells and tumour cells such as myelomonocytes (HL-60, U-937) and megakaryocytes (Dami) do not secrete angiogenin or secrete low levels ($<$30 pg/10^6 cells/day), but myeloma RPMI 8226 cells produce as much angiogenin cross-reactivity as anchored cells. Among the tumour T-cells tested, Jurkat and MT4 cells express cross-reactivity, while H9 and HuT 78 cells do not [8].

In vivo, angiogenin is present in normal human plasma at 110-380 ng/ml (n=65) [6] and in human amniotic fluid [7]. The bovine angiogenin concentration has been estimated at 100-180 ng/ml of serum and 4-8 mg/L of milk [74]. Immunohistochemical studies have identified angiogenin in epithelial and secretory cells in sections of cow mammary gland, visceral peritoneum and gall-bladder [74].

Gene Regulation   The angiogenin concentration in human serum increases in the perinatal period. A statistically significant increase in angiogenin concentrations has been reported in day 4 neonatal serum relative to umbilical cord serum, with values reaching maternal levels [75]. Similarly, expression of the angiogenin gene in rat liver is developmentally regulated: the mRNA level is low in the foetus, increases in the neonate and is predominantly expressed in the adult [72].

The distribution of 5-methyldeoxycytidine in 5'-CG-3'-rich sequences can be related to the degree of transcriptional activity, or to replication or recombination in the genome. The human angiogenin gene has three Alu sequences (see Gene Structure §). The Alu element located upstream of the transcription start site in the angiogenin gene is unmethylated in the DNA of haploid spermatozoa, while it is highly methylated in the DNA of differentiated diploid human cells. This is surprising, because DNA from sperm is usually highly methylated. Cell-free transcription experiments suggest that 5'-CG-3' methylation can lead to transcriptional inactivation [76].

Additional Features   Pseudogenes: Screening of a 129-strain mouse genomic DNA library with a BALB/c angiogenin gene probe yielded angiogenin pseudogene 1 (Ang-ps1) [77]. Ang-ps1 contains a single insertion resulting in a frameshift mutation in the early part of the coding region. Polymerase chain reaction cloning

has also yielded a second angiogenin-like pseudogene, Ang-ps2 [77].

Nucleotide sequence accession numbers (EMBL/Genbank Data Library): human RNase-1, X62946; human RNase-2, EDN, M24157; human RNase-3, ECP, X15161; human RNase-4, U36775; RNase-5: human angiogenin, M11567; mouse angiogenin-1, U22516; mouse angiogenin-related protein, U22519; EF5/mouse angiogenin-3, U72672; chicken clone 462 RNase/angiogenin, X 61192; chicken G gallus RSFR, X64743; mouse Ang-ps1, U22517; mouse Ang-ps2, U22518; human RNase k6, U64998.

National Biomedical Research Foundation accession numbers of sequences: angiogenin, NRHUAG; RNase-2, EDN, A35328; RNase-3, ECP, B35328; RNase A, A32471; onconase, p30 (*Rana pipiens*), A39035; sialic acid-binding lectin, SBL-C (*Rana catesbeiana*), A27121.

Processing and Fate   Nuclear translocation of angiogenin has been studied by means of indirect immunofluorescence microscopy in CPAE cells grown for one day in minimal essential medium containing 20% heat-inactivated fetal bovine serum [78]. Bright staining was observed in the nucleoli of cells exposed to 100 or 1000 ng/ml ($\approx$7 or 70 nM) of human Met-(-1)angiogenin and native $<$Glu-angiogenin for 30 min at 37 °C in serum-free medium. Immunostaining was not observed in the presence of a 10-fold excess of bovine angiogenin with mAb 26-2F, a monoclonal antibody against human angiogenin which does not react with the bovine protein. Neither lysozyme nor RNase A affects nuclear translocation of human angiogenin in endothelial cells. Exogenous smooth muscle alpha-actin (0.001 or 0.01 mg/ml), a monoclonal antibody to smooth muscle cell alpha-actin (0.1 mg/ml), heparin (0.0001, 0.001 or 0.01 mg/ml), and an 8-fold molar excess of FGF-2 all inhibit this process, but chondroitin sulphate does not. CPAE cells do not internalise enzymatically active angiogenin derivatives with an altered cell-binding site, such as angiogenin K (Met-(-1)angiogenin cleaved at residues 60-61), ARH-1 (Met-(-1)angiogenin/RNase hybrid protein in which the angiogenin segment 58-73 is replaced by residues 59-73 of RNase A), R66A and N109D. In contrast, two enzymatically inactive mutants, K40Q and H13A, whose cell-binding site is intact, are translocated to the nucleus. Like the angiogenin derivatives angiogenin K, ARH-1, R66A and N109D, the R33A mutant is not translocated to the nucleus and lacks angiogenic activity [78]. In digitonin-permeabilised endothelial cells, angiogenin mutants with an altered cell-binding domain (angiogenin K, ARH-1 and R66A) are imported into the nucleus and targeted to the nucleolus, whereas the R33A derivative is not. Segment Arg[31]-Arg-Arg-Gly-Leu[35] of human angiogenin has been identified as the nuclear localisation signal responsible for nucleolar targeting of angiogenin. The corresponding peptide Arg-Arg-Arg-Gly-Leu targets non nuclear proteins and the R33A derivative to the nucleolus of digitonin-permabilised endothelial cells, while the peptide Arg-Arg-Ala-Gly-Leu does not [47].

Nuclear translocation has also been observed in GM7373 and human umbilical artery endothelial cells [47, 78]. No staining was observed in confluent CPAE cells [78].

Biological Activity   Angiogenin is a potent blood vessel-inducing factor: 0.5 ng (35 fmol) induces angiogenesis in the chick embryo chorioallantoic membrane assay [2], 50 ng (3.5 pmol) in the rabbit corneal assay [2], 100 ng (7 pmol) in the rabbit meniscus [5], and 1 nmol in the disc angiogenesis assay in mice [42]. Placental RI blocks angiogenin-induced angiogenesis in the chick embryo chorioallantoic membrane assay [12]. In the disc angiogenesis assay in mice, placental RI also inhibits the angiogenesis induced by FGF-2 and sodium orthovanadate. In the mouse cornea assay, placental RI inhibits FGF-2-induced angiogenesis [79].

During angiogenesis, quiescent endothelial cells lining microvessels are induced to invade the extracellular matrix. In an experimental model, microvascular endothelial cells grown as a monolayer on a three-dimensional matrix can be stimulated to invade the underlying matrix and to form capillary-like tubular structures. Unlike FGF-2 and VEGF, human angiogenin does not induce cloned bovine microvascular endothelial cells from the renal cortex to form capillary-like tubules in a collagen invasion assay (R. Montesano, M. Moenner and J. Badet, unpublished observation). However, bovine angiogenin stimulates the cell-associated proteolytic activity of CPAE cells and GM 7373 cells, a foetal bovine aortic endothelial cell type transformed by benzo[α]pyrene, as determined by using a tissue-type plasminogen activator (t-PA)-specific peptide as substrate. Stimulation is maximal at an angiogenin concentration of 500 ng/ml [19]. Angiogenin induces fibrinolytic activity in GM 7373 cells grown on fibrin gels. Maximal stimulation was observed in the presence of angiogenin at ≈ 1000 ng/ml and resulted in small focal defects of the fibrin gel [19]. Angiogenin also induces endothelial cells to invade Matrigel. Invasion is inhibited by a rabbit polyclonal anti-angiogenin antibody and α_2-antiplasmin [19]. Although actin has been shown to inhibit angiogenin-induced angiogenesis *in vivo* [49], it does not inhibit angiogenin-induced invasion of endothelial cells into Matrigel [19]. On the contrary, the presence of both angiogenin and actin induces more invasion than does angiogenin alone [19]. In fact, angiogenin interacts with actin. It enhances actin acceleration of plasmin generation from plasminogen by t-PA and abolishes actin inhibition of plasmin [48, 49].

When confluent bovine aortic endothelial cells cultured on collagen-coated dishes are wounded, bovine angiogenin at 10 or 100 ng/ml induces cell migration [20]. Angiogenin upregulates the mRNA of both urokinase-type plasminogen activator (u-PA) and plasminogen activator inhibitor 1 (PAI-1). Both cell lysates and conditioned media express u-PA activity. It has been reported that angiogenin at 10 and 100 ng/ml induces the formation of tube-like structures by endothelial cells on type-I collagen gel, an effect blocked by aprotinin, an inhibitor of serine proteases [20]. Finally, bovine angiogenin at 100 ng/ml induces an increase in c-fos and c-jun mRNA levels 30 min after stimulation [20].

Angiogenin has been reported to stimulate the proliferation of endothelial cells [17, 18]. Human angiogenin at concentrations ranging from 100 to 1000 ng/ml enhances [³H]thymidine incorporation and cell proliferation in human umbilical venous and microvascular endothelial cells by up to 50% [17, 18]. Bovine angiogenin at 100 ng/ml induces [³H]thymidine incorporation and cell proliferation in bovine brain capillary endothelial cells [17].

Angiogenin interacts with cultured endothelial cells to activate second messengers. Angiogenin at concentrations in the ng/ml range, is a weak inducer of diacylglycerol formation in CPAE and BACE cells [21]. Higher concentrations of angiogenin (> 100 ng/ml) are required to induce diacylglycerol in HUVE cells [21]. Angiogenin seems to activate phospholipase C; however, the concomitant 20% increase in inositol triphosphate is small compared to the 500% increase induced by the endothelial cell agonist bradykinin [21]. Chemical modification of one or both His residues in the catalytic site (His-13 and His-114) abolishes the ability of angiogenin to increase CPAE cellular diacylglycerol. Placental RI also completely abolishes the response induced by angiogenin. Angiogenin at 1 ng/ml activates BACE and HUVE cells but not CPAE cells to secrete prostacyclin, a potent vasodilator and inhibitor of platelet aggregation [22]. Angiogenin does not stimulate the secretion of the angiogenic prostaglandins PGE_1 and PGE_2. The secretion of prostacyclin by BACE cells is blocked by pre-treating the cells with indomethacin and tranylcypromine (inhibitors of prostacyclin synthesis), quinacrine (a specific inhibitor of phospholipase A_2), RHC 80267 (a diglyceryl and monoglyceryl lipase inhibitor), H7 (a protein kinase inhibitor) and also phorbol 12-myristate 13-acetate (that down-regulates protein kinase C). Angiogenin-induced prostacyclin secretion by BACE cells is also blocked by pre-treatment with pertussis toxin, which has no effect on the induction of diacylglycerol by angiogenin. This suggests that a putative angiogenin receptor is coupled by a pertussis-sensitive G protein to phospholipase A_2. No calcium mobilisation has been detected in BACE cells after exposure to angiogenin, either by fura-2 labelling and fluorescent measurements or by determining $^{45}Ca^{2+}$ efflux [22].

In cultured rat SMC, angiogenin activates phospholipase C [23]. Angiogenin at 1 ng/ml induces a transient increase in inositol triphosphate to a maximum of 400% of control, and also generates a transient increase in 1,2-diacylglycerol. However, the authors observed no detectable internal Ca^{2+} release by fura-2-labelled cells [23]. Angiogenin transiently depressed rat aortic smooth muscle cell cAMP levels by a pertussis toxin-sensitive mechanism. It has no effect on cellular cGMP. In addition, angiogenin stimulates rapid incorporation of fatty acids into cholesterol esters, which might have important physiological implications [23].

Angiogenin supports cell adhesion (see "Binding sites and affinity") [27, 28].

A degranulation inhibiting protein (DIP) purified from ultrafiltrates of plasma from patients with uraemia, has been shown to be identical to angiogenin by amino acid sequence determination, immunoblotting and assays of inhibitory effects on leukocyte degranulation [26]. DIP at concentrations in the nanomolar range inhibits spontaneous degranulation by 40 % in PMNL, and by 70 % in cells previously stimulated with the chemotactic peptide formyl-norLeu-Leu-Phe-norLeu-Tyr-Leu [26].

Human angiogenin suppresses [3H]thymidine incorporation in human lymphocytes stimulated by phytohemagglutinin or concanavalin A, and in allogenic human lymphocytes in mixed lymphocyte culture. This immunosuppressive activity has been observed with an angiogenin concentration of 0.02 mg/ml [25].

Angiogenin has ribonucleolytic activity on both 18S and 28S ribosomal RNAs, as well as transfer RNA [3]. At a concentration of 40-60 nM, angiogenin abolishes cell-free protein synthesis by ribonucleolytic inactivation of the 40S ribosomal subunit. It generates limited cleavage products from reticulocyte RNAs. Kinetic studies of protein synthesis indicate that angiogenin inhibits either the chain elongation or termination step of protein synthesis, not its initiation [10]. Angiogenin blocks protein synthesis when injected into *Xenopus* oocytes, by acting as a tRNA-specific ribonuclease [11]. Protein synthesis, *in vitro* and *in vivo*, is restored in the presence of RI [10,11].

Role in Vascular Biology

Physiological Function   Angiogenin is a potent inducer of blood vessel formation in experimental models, *in vivo* [2, 5, 42]. However, the widespread expression of angiogenin suggests a physiological function not restricted to the neovascularization process [8, 72]. The presence of angiogenin in normal plasma [1] suggests that it may be involved in vascular homeostasis.

The angiogenin concentration in plasma is in the range 8-30 nM [6, 8]. At such concentrations, angiogenin inhibits the degranulation of PMNL *in vitro*, as described by Tschesche and colleagues, who suggested that angiogenin might participate in an endogenous inhibitory mechanism to counterbalance plasma-derived molecules released during inflammatory responses [26].

Angiogenin expression is developmentally regulated in rat liver [72]. In human serum, the angiogenin concentration increases in the perinatal period [75]. These two observations might provide clues to the physiological function(s) of angiogenin.

Pathology   By *in situ* hybridisation, increased angiogenin mRNA expression has been detected in human colonic adenocarcinoma, gastric adenocarcinoma and pancreatic cancer tissues compared to corresponding normal tissues [31, 80]. Over-expression of angiogenin

in pancreatic cancer tissue is associated with cancer aggressiveness [31]. There is a significant increase in serum levels of angiogenin in patients with pancreatic [31], breast [35], ovarian [32], endometrial [33] and cervical [34] cancer. The possible involvement of angiogenin in tumour development is suggested by the demonstration that angiogenin antagonists prevent the growth of human tumour xenografts in athymic mice [30].

An increase in serum levels of angiogenin has also been reported among children and adolescent with insulin-dependent diabetes mellitus [81] and women with severe ovarian hyperstimulation syndrome [82].

Angiogenin is regulated as an acute-phase protein in experimental models *in vivo* [83].

Clinical Relevance and Therapeutic Implications   Correlation between angiogenin concentrations in biological fluids and disease processes might provide prognostic information. For example, increased angiogenin concentrations in midtrimester amniotic fluid has been shown to be a marker of preterm delivery [7]. Similarly, progression of endometrial and pancreatic cancer is associated with increased serum levels of angiogenin [31, 33].

Inhibition of angiogenesis might be a therapeutic target in cancer. Angiogenin-induced angiogenesis is inhibited by angiogenin derivatives such as angiogenin H13A and H114A, N61D and N109D, and the C-terminal peptide ANG(108-121). The peptide ANG(58-70), corresponding to the putative receptor binding domain, and two peptides complementary to the receptor-binding domain of angiogenin are also inhibitors [42, 84]. The amino acid sequences of these latter two peptides have been deduced from the antisense RNA sequence corresponding to the putative receptor binding site of angiogenin (residues 58-70) in either the 5'→3' (chANG) or the 3'→5' (chGNA) direction [84]. They have the following sequences: Val-Phe-Ser-Val-Arg-Val-Ser-Ile-Leu-Val-Phe (chANG) and Leu-Leu-Phe-Leu-Pro-Leu-Gly-Val-Ser-Leu-Leu-Asp-Ser (chGNA).

The monoclonal antibody mAb 26-2F, that recognises both Trp-89 and the 38-41 segment of human angiogenin, neutralises the angiogenic activity of angiogenin. It has been shown to prevent or delay the appearance of HT-29 tumours in athymic mice. Similarly, mAb 36u, that interacts with segment 58-73, prevents the appearance of tumours [30].

The activity of angiogenin (10 ng) in the chick embryo chorioallantoic membrane assay is also inhibited by actin (3000 ng) and anti-actin antibody (1000 ng) [48, 49]. On the basis of the angiogenin-actin interaction, angiogenin antagonists have been screened from a phage-displayed peptide library. ANI-E peptide (sequence Ala-Gln-Leu-Ala-Gly-Glu-Cys-Arg-Glu-Asn-Val-Cys-Met-Gly-Ile-Glu-Gly-Arg), that contains a disulphide bond, blocks the neovascularisation induced by angiogenin in the chick chorioallantoic membrane assay [85]. The disulphide bond and the glutamic acid inside the disulphide ring are both required for the activity of the peptide. The peptide alone does not induce or inhibit embryonic angiogenesis. ANI-E peptide does not inhibit the adhesion of angiogenin-secreting PC3 human prostate adenocarcinoma cells to angiogenin, but does block the angiogenesis induced by these cells [85].

Because the integrity of the catalytic site is required for angiogenin to express its angiogenic properties, the homology of angiogenin with RNases generates novel approaches to antitumoral therapy. Homology modelling can serve as a guide for the design of tight-binding inhibitors of angiogenin. For example, kinetic and modelling studies of the catalytic centre subsites of angiogenin suggest that 5'-diphosphoadenosine 2'-phosphate might be a potent inhibitor [86].

Specific ligands for angiogenin have been selected by using the procedure of systematic evolution of ligands by exponential enrichment. An oligodeoxynucleotide aptamer inhibits both the enzymatic and biological activities of angiogenin [87].

The cytotoxic potential of RNases has been explored with a view to designing targeted drugs, named immunotoxins, by coupling non cytotoxic RNases to cell-binding ligands. The RNase chimeras display receptor-mediated cytotoxicity. On the basis of the ability of angiogenin to block protein synthesis, angiogenin fused to a single-chain antibody against the human transferrin receptor has been expressed as a targeted immunofusion protein. Angiogenin single-chain immunofusion proteins that retain both enzymatic and biological activity might form a new class of therapeutic agents [88].

Abbreviations: ANG, angiogenin; BACE cells, bovine adrenal capillary endothelial cells; BS-RNase, bovine seminal ribonuclease; CPAE cells, calf pulmonary artery endothelial cells; DIP, degranulation inhibiting protein; ECP, eosinophil cationic protein; EDN, eosinophil-derived neurotoxin; FGF, fibroblast growth factor; FRL2, FGF receptor ligand 2; HUVE cells, human umbilical vein endothelial cells; NMR, nuclear magnetic resonance; PA, plasminogen activator; t-PA, tissue-type PA; u-PA, urokinase-type PA; PMNL, polymorphonuclear leukocytes; PG, prostaglandin; RI, ribonuclease inhibitor; RNase, ribonuclease; RNase A, bovine RNase-1; SDS-PAGE, sodium dodecyl sulphate – polyacrylamide gel electrophoresis; RSFR, RNase super-family-related gene; SMC, aortic smooth muscle cells; VEGF, vascular endothelial growth factor.

Mutant proteins are designated by the single-letter code for the original amino acid followed by its position in the sequence and the single letter code for the new amino acid. "Protein(n-n')" refers to a peptide whose N- and C- terminal residues, denoted by n and n' respectively, are the positions in the primary structure of the protein.

Acknowledgements: I apologise to colleagues whose work has not been cited or been cited indirectly through other articles, due to the space limitation. I would also like to thank David young for editorial assistance and Marina Yassenko for translating papers published in Russian.

Josette Badet

References

1. Shapiro R et al (1987) Biochemistry 26: 5141-5146
2. Fett JW et al (1985) Biochemistry 24: 5480-5486
3. Shapiro R et al (1986) Biochemistry 25: 3527-3532
4. Folkman J et al (1971) J Exp Med 133: 275-288
5. King TV et al (1991) J Bone Joint Surg (Br) 73-B: 587-590
6. Bläser J et al (1993) Eur J Clin Chem Clin Biochem 31: 513-516
7. Spong CY et al (1997) Am J Obstet Gynecol 176: 415-418
8. Moenner M et al (1994) Eur J Biochem 226: 483-490
9. Strydom DJ et al (1985) Biochemistry 24: 5486-5494
10. St. Clair DK et al (1988) Biochemistry 27: 7263-7268
11. Saxena SK et al (1992) J Biol Chem 267:21982-21986
12. Lee FS et al (1993) Prog Nucl Ac Res Mol Biol 44: 1-30
13. Hofsteenge J (1997) Ribonuclease inhibitor. Academic Press, New York, pp 1-621-658
14. Shapiro R et al (1987) Proc Natl Acad Sci USA 84: 8783-8787
15. Hallahan TW et al (1991) Proc Natl Acad Sci USA 88: 2222-2226
16. Badet J et al (1989) Proc Natl Acad Sci USA 86: 8427-8431
17. Chamoux M et al (1991) Biochem Biophys Res Commun 176: 833-839
18. Hu G-F et al (1997) Proc Natl Acad Sci USA 94: 2204-2209
19. Hu G-F et al (1994) Proc Natl Acad Sci USA 91: 12096-12100
20. Jimi S-I et al (1995) Biochem Biophys Res Commun 211: 476-483
21. Bicknell R et al (1988) Proc Natl Acad Sci USA 85: 5961-5965
22. Bicknell R et al (1989) Proc Natl Acad Sci USA 86: 1573-1577
23. Moore F et al (1990) Biochemistry 29: 228-233
24. Soncin F et al (1997) Biochem Biophys Res Commun 236: 604-610
25. Matousek J et al (1995) Comp Biochem Physiol 112B: 235-241
26. Tschesche H et al (1994) J Biol Chem 269:30274-30280
27. Soncin F (1992) Proc Natl Acad Sci USA 89: 2232-2236
28. Soncin F et al (1994) J Biol Chem 269: 8999-9005
29. Soncin F et al (1997) J Biol Chem 272: 9818-9824
30. Olson KA et al (1995) Proc Natl Acad Sci USA 92: 442-446
31. Shimoyama S et al (1995) Cancer Res 56: 2703-2706
32. Chopra V et al (1996) Cancer J Sci Am 2: 279-285
33. Chopra V et al (1997) J Cancer Res Clin Oncol 123: 167-172
34. Chopra V et al (1998) Cancer Invest 16: 152-159
35. Montero S et al (1998) Clin Cancer Res 4: 2161-2168
36. Bond MD et al (1989) Biochemistry 28: 6110-6113
37. Bond MD et al (1993) Biochim Biophys Acta 1162: 177-186
38. Strydom DJ (1998) Cell Mol Life Sci 54: 811-824
39. Riordan JF (1997) In: D'Alessio G, Riordan JF (eds) Ribonucleases: Structures and functions. Academic Press, New York, pp 445-489
40. Hallahan TW et al (1992) Biochemistry 31: 8022-8029
41. Harper JW et al (1989) Biochemistry 28: 1875-1884
42. Raines RT et al (1995) J Biol Chem 270: 17180-17184
43. Shapiro RT et al (1992) Biochemistry 31: 12477-12485
44. Acharya KR et al (1994) Proc Natl Acad Sci USA 91: 2915-2919
45. Lequin O et al (1996) Biochemistry 35: 8870-8880
46. Acharya KR et al (1995) Proc Natl Acad Sci USA 92: 2949-2953

47. Moroianu J et al (1994) Biochem Biophys Res Commun 203: 1765-1772
48. Hu G-F et al (1991) Proc Natl Acad Sci USA 88: 2227-2231
49. Hu G-F et al (1993) Proc Natl Acad Sci USA 90: 1217-1221
50. Russo N et al (1996) Proc Natl Acad Sci USA 93: 3243-3247
51. Papageorgiou AC et al (1997) EMBO J 16: 5162-5177
52. Badet J et al (1990) Blood Coag Fibrin 1: 721-724
53. Lee FS et al (1989) Biochem Biophys Res Commun 161: 121-126
54. Kobe B et al (1995) Curr Opin Struct Biol 5: 409-416
55. Futami J et al (1997) DNA Cell Biol 16: 413-419
56. Rybak SM et al (1988) Biochemistry 27: 2288-2294
57. Bond MD et al (1988) Biochemistry 27: 6282-6287
58. Maes P et al (1988) FEBS Lett 241: 41-45
59. Nobile V et al (1996) Proc Natl Acad Sci USA 93: 4331-4335
60. Strydom DJ et al (1997) Eur J Biochem 247: 535-544
61. Fu X et al (1997) Mol Cell Biol 17: 1503-1512
62. Sorrentino S (1998) Cell Mol Life Sci 54: 785-794
63. Beintema JJ et al (1998) Cell Mol Life Sci 54: 825-832
64. Kinoshita N et al (1995) Cell 83: 621-630
65. Nakano T et al (1992) Oncogene 7: 527-534
66. Klenova EM et al (1992) Biochem Biophys Res Commun 185: 231-239
67. D'Alessio G (1993) Trends Cell Biol 3:106-109
68. Shapiro R et al (1991) Biochemistry 30: 2246-2255
69. Kurachi K et al (1985) Biochemistry 24: 5494-5499
70. Neznanov NS et al (1990) Molekulyarnaya Biologiya 24: 709-715
71. Weremowicz S et al (1990) Am J Hum Genet 47: 973-981
72. Weiner HL et al (1987) Science 237: 280-282
73. Rybak SM et al (1987) Biochem Biophys Res Commun 146: 1240-1248
74. Chang S-I et al (1997) Biochem Biophys Res Commun 232: 323-327
75. Malamitsi-Puchner A et al (1997) Pediatr Res 41: 909-911
76. Kochanek S et al (1993) EMBO J 12: 1141-1151
77. Brown WE et al (1995) Genomics 29: 200-206
78. Moroianu J et al (1994) Proc Natl Acad Sci USA 91: 1677-1681
79. Polakowski IJ et al (1993) Am J Pathol 143: 507-517
80. Li D et al (1994) J Pathol 172: 171-175
81. Malamitsi-Puchner A et al (1998) Pediatr Res 43: 798-800
82. Aboulghar MA et al (1998) Human Reproduction 13: 2068-2071
83. Olson KA et al (1998) Biochem Biophys Res Commun 242: 480-483
84. Gho YS et al (1997) J Biol Chem 272: 24294-24299
85. Gho YS et al (1997) Cancer Res 57: 3733-3740
86. Russo N et al (1996) Proc Natl Acad Sci USA 93: 804-808
87. Nobile V et al (1998) Biochemistry 37: 6857-6863
88. Newton DL et al (1996) Biochemistry 35: 545-553

Angiopoietins

Definition *Angiopoietins are ligands for the Tie-2 tyrosine kinase containing receptor expressed in angiogenic endothelial cells. Angiopoietin-1 (Ang-1) is essential in embryonic vascular development and promotes vessel maturity. It is also important in repair-associated angiogenesis. Angiopoietin-2 antagonizes the effect of Ang-1.*

See: →Angiogenesis; →Angiogenesis inhibitors; →Ontogeny of the vascular system

Angiostatin

Definition *Fragment derived from plasminogen that inhibits angiogenesis in vitro and in vivo*

See: →Angiogenesis; →Angiogenesis inhibitors

Anticoagulation Factors

Same as inhibitors of coagulation. These include Antithrombin III, Heparin-cofactor-II, Thrombomodulin, Protein C, Protein S, proteases nexins or heparin.

See: →Coagulation factors; →Fibrinolytic, hemostatic and matrix metalloproteinases, role of; →Thrombin

α2-Antiplasmin

Definition *Protease inhibitor that inhibits plasmin activity*

See: →Fibrinolytic, hemostatic and matrix metalloproteinases, role of

AP-1, -2 etc

Definition *Activating protein 1, 2 etc.*

See: →Coagulation factors; →Tissue inhibitors of metalloproteinases; →Matrix Metalloproteinases; →Transforming growth factor β

aPTT

Definition *Activated partial thromboplastin time*

See: →Bleeding disorders; →Coagulation factors

ARNT

Definition *Aryl hydrocarbon-receptor nuclear translocator*

See: →Angiogenesis; →Vascular endothelial growth factor family

Arteriosclerosis

See: →Atherosclerosis

AT

Definition *Angiotensin*

See: →Renin-angiotensin system; →Prostaglandins; →Vasomotor tone regulation, molecular mechanisms of

ATF-2

Definition *Activating transcription factor-2*

See: →Signal transduction mechanisms in vascular biology

Atherosclerosis

Synonym: Arteriosclerosis

Definition *Lesion of large arteries that includes thickening and remodeling of the vessel wall leading progressively to blood vessel occlusion through plaque formation and thrombosis.*

See also: →Thrombosis; →Cytokines in vascular biology and disease; →FGF-1 and -2; →Fibrin/fibrinogen; →Fibrinolytic, hemostatic and matrix metalloproteinases, role of

Introduction Atherosclerosis is characterized by focal, slow and progressive accumulation of cells, extracellular matrix and lipids in the intima of medium-sized and large arteries. Basically, the hallmark of atherosclerosis disease are intimal plaques called *atheromas* that protrude into the arterial lumen. The resulting vessel narrowing leads to intermittent impairment of blood flow supply to the end-organ causing tissue ischemia and end-organ damage. Although atherosclerosis was considered a degenerative disease that was an inevitable consequence of aging, a growing body of experimental and clinical data has shown that atherosclerosis is a chronic and focal inflammatory arterial disease that is converted to an acute clinical event by the induction of plaque rupture which undergoes a series of complications that predispose to overlying thrombosis [1].

Characteristics Hemodynamic variables outside of the normal range, raised serum lipoproteins (LDL), oxidized lipoproteins (ox-LDL), oxygen free radicals, infectious agents, homocysteine and cytokines lead to endothelial cell dysfunction, which then continues through the growth of atherosclerotic lesion. Atherosclerotic lesions evolve over time. The earliest recognizable lesion of atherosclerosis is the so-called fatty streak, an accumulation of lipid-rich macrophages or foam cells and T lymphocytes within the intima [2]. These subendothelial lesions beginning as 1-mm yellow protrusion into the arterial lumen are found in the coronary arteries of half of the necropsy specimens from children aged 10-14 [3]. Although fatty streaks are not clinically significant, these lesions are responsible for the latter steps that lead to a clinical event. Fatty streaks develop into the occlusive lesions termed fibrous plaque, which increases in size and protrudes into the arterial lumen affecting blood flow by vessel narrowing and by endothelial dysfunction that produces abnormal vasoconstriction.

Histologically the composition of fibrous plaque has essentially three components: (a) cells, including blood-derived monocytes/macrophages, smooth muscle cells, activated T lymphocytes and mast cells; (b) connective tissue fibers and matrix; and (c) lipids [4]. Calcification of fibrous plaque leads to complex plaque in which process a highly acidic glycoprotein (osteopontin) has been implicated [5]. These plaques are prone to rupture leading to occlusive platelet thrombi. In the presence of plaque and arterial stenosis, constriction and thrombosis lead to impairment of blood supply that can cause

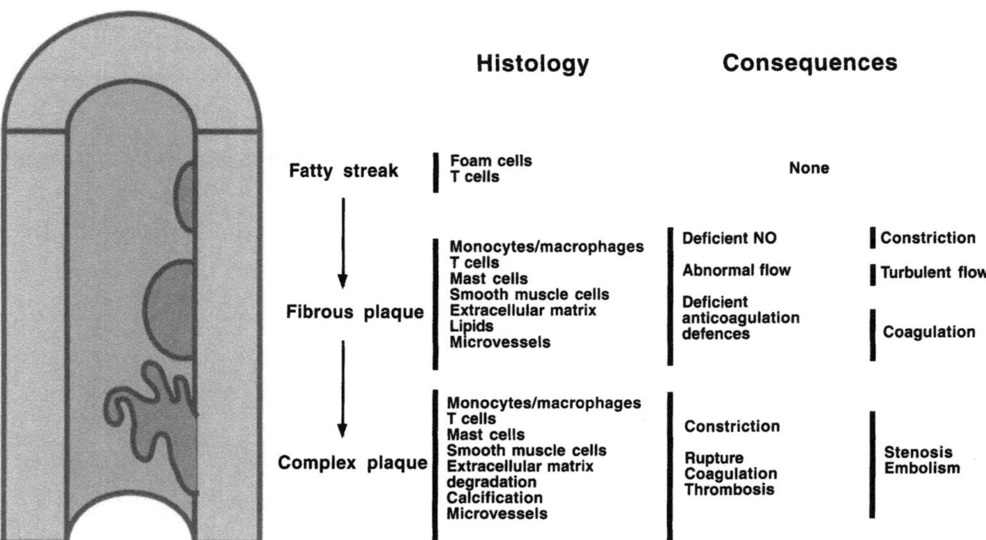

Figure 1. Plaque progression. The fatty streak contains only foam cells and T cells. Extracellular lipids, smooth muscle cells, and collagen appear successively and lead to a complex plaque on which the thrombotic complication can arise.

stroke, acute myocardial infarction, and sudden death. A summary of histological findings of plaque growth and its functional disturbance are shown in Figure 1.

Regulation Atherogenesis starts with the loss of a wide range of healthy functions by the endothelium. Healthy endothelial cells that are oriented in the direction of laminar flow, exhibit slow turnover, specific lipoprotein permeability and several mechanisms to maintain an anticoagulant surface [6]. Endothelial cells are continuously subjected to hemodynamic forces, particularly fluid shear stress. Although atheromas can be localized through the body, usually areas subjected to increased hemodynamic stress (bifurcations) are particularly prone to development of atherosclerotic lesions. Experimental evidence shows that low shear and turbulent flow induce, in endothelial as well as in smooth muscle cells and in macrophages, selective upregulation of gene expression of adhesion molecules, coagulation factors, fibrinolytic factors, growth factors, and cytokines [7,8]. This selective regulation of these products suggests that hemodynamic forces may represent pathophysiologically regulatory stimuli in inflammation and atherosclerosis [9], and could explain the preferential localization of atheromas at anatomic sites of the arterial tree subjected to hemodynamic stress.

Genes The ability of the endothelium to change from one phenotype to an entirely opposite phenotype is a key event in atherogenesis. Anticoagulant changes to procoagulant (by expression of tissue factor (TF) and down regulation of thrombomodulin), fibrinolytic to anti-fibrinolytic (by changing the relative proportion of expressed tissue-type plasminogen activator (t-PA) to its plasminogen activator inhibitor-1 (PAI-1)) and anti-inflammatory to pro-inflammatory (by the expression of adhesion molecules and the synthesis of chemokines). All of these changes require alteration in levels of gene transcription. It is currently accepted that atherosclerotic lesions contain an immune-mediated inflammatory reaction [10]. Active expression of cell adhesion molecules genes that have a role in inflammatory cell recruitment which occurs during early atherosclerotic lesions has been reported [11]. Among these, the expression of the cell adhesion molecules (CAMs) may have an important role in the selective recruitment of inflammatory cells from the circulation into the lesion [12]. Experimental evidence suggests that ox-LDL is a potent inducer of inflammatory molecules [13]. Ox-LDL induces these inflammatory molecules both by inducing increased rates of gene transcription and by stabilizing the mRNA for these genes [14]. In addition, upregulation of genes that code for the production of inducible nitric oxide synthase, matrix metalloproteinases, cytokines and growth factors has been suggested to have an important role in atherogenesis [15]. The endothelium plays a critical role in the regulation of coagulation through the constitutive expression and release of anticoagulants and the inducible expression of procoagulant substances [16]. Atherogenesis dysregulates this process by endothelial cell activation in

response to circulating inflammatory mediators. This activation leads to gene expression of TF into atherosclerotic plaques [17]. Within the plaque, TF is particularly abundant in the relatively acellular lipid-rich core and is also located in macrophage foam cells, monocytes adjacent to the cholesterol clefts and smooth muscle cells [18]. TF expression in atherosclerotic plaques is involved in the initiation of thrombus formation upon spontaneous or induced plaque rupture.

Molecular Interactions Abnormalities of endothelial cell function are likely to be major determinants of atherosclerotic disease. The endothelium performs several important physiological functions. It forms a non-thrombogenic surface, serves as a barrier between extravascular and intravascular compartments, and produces several vasoactive mediators. Dysfunctional endothelial cells, as it occurs in atherogenesis, display properties opposite to those normally expressed in the normal quiescent state. In atherosclerotic disease, endothelial cells tend to balance opposing actions: anti-versus procoagulant, vasoconstrictor versus vasodilator, and growth promoting versus inhibiting. Endothelial cells are important components of the immune system in that they regulate leukocyte adhesion and diapedesis into vascular wall. The accumulation of leukocytes in the intima and transformation into lipid-engorged foam cells result in the formation of atherosclerotic plaques.

It has been established that elevated levels of plasma LDL cholesterol is a risk factor for atherosclerosis. LDL transports cholesterol from the liver to peripheral cells via interaction with LDL receptors [19]. Initiation of atherosclerotic lesion formation involves alteration of the functional integrity of the endothelium, allowing an influx of LDL from the circulating plasma into the subendothelium. LDL can be oxidized by endothelial and smooth muscle cells to form ox-LDL [20]. ox-LDL activates endothelial cells to produce chemoattractants and adhesion molecules that attract circulating monocytes to the arterial wall and induce their migration through endothelial cell junctions [21]. Thereafter, the monocytes differentiate into macrophages that secrete chemoattractants for monocytes [22]. Furthermore, the macrophages accumulate ox-LDL via scavenger receptors, resulting in foam cell generation [23]. ox-LDL stimulates the migration of smooth muscle cells from the media into the intima via induction of PDGF in macrophages and smooth muscle cells [15] and aFGF and bFGF in endothelial cells, macrophages and smooth muscle cells [15, 24]. Additionally, ox-LDL induces the expression of scavenger receptors in smooth muscle cells, resulting in uptake of ox-LDL, and subsequent smooth muscle foam cell generation. The effects of ox-LDL on endothelial cells, monocytes/macrophages and smooth muscle cells are shown in Table 1.

Endothelial cells synthetize nitric oxide (NO) from the terminal guanidino nitrogen atom of the amino acid L-arginine by the NO synthase enzyme [25]. Endothelium-derived NO exerts vasodilatory effects on underlying vas-

Table 1. Effects of ox-LDL

	Direct effects	Indirect effects
Endothelial cells	Synthesis of adhesion molecules (ICAM-1, ELAM-1, VCAM-1)	Adhesion of monocytes
	Synthesis of chemoattractants (MCP-1)	Migration of monocytes
	Synthesis of colony stimulating factors (M-CSF, GM-CSF)	Proliferation of monocytes
	Synthesis of growth factors (aFGF, bFGF, PDGF, VEGF, TGF-β)	Proliferation of smooth muscle and endothelial cells
	Impaired NO secretion	Vasoconstriction and platelet adhesion
	Secretion of prostaglandins	Platelet aggregation
	Synthesis of TF	Coagulation
	Inhibition of synthesis of t-PA and stimulation of synthesis of PAI-1	Defective fibrinolysis
Monocytes/macrophages	Synthesis of MCP-1	Chemoattraction of monocytes
	Direct uptake	Foam cell generation
	Synthesis of growth factors (aFGF, bFGF, PDGF, TGF-β)	Proliferation of smooth muscle and endothelial cells
	Synthesis of TF	Coagulation
Smooth muscle cells	Chemoattraction	Migration of smooth muscle cells
	Direct uptake	Foam cell generation
	Synthesis of growth factors (aFGF, bFGF, PDGF, TGF-β)	Proliferation of smooth muscle and endothelial cells
	Synthesis of TF	Coagulation

ELAM-1, endothelium leukocyte adhesion molecule-1; aFGF, acidic fibroblast growth factor; bFGF, basic fibroblast growth factor; GM-CSF, granulocyte-macrophage colony stimulating factor; ICAM-1, intercellular adhesion molecule-1; M-CSF, macrophage colony stimulating factor; MCP-1, monocyte chemoattractant protein-1; NO, nitric oxide; PAI-1, plasminogen activator inhibitor 1; PDGF, platelet-derived growth factor; TF, tissue factor; TGF-β, transforming growth factor β; t-PA, tissue-type plasminogen activator; VCAM.1, vascular adhesion molecule 1; VEGF, vascular endothelial growth factor

cular smooth muscle cells by stimulating soluble guanylate cyclase and increasing intracellular cyclic GMP (cGMP) [26]. In addition, endothelium-derived NO inhibits the aggregation of platelets [27] and attenuates neutrophil adhesion to endothelium [28]. In atherosclerotic vessels, reduced endothelium-derived NO has been reported [29] and may promote vasoconstriction and facilitate platelet aggregation and release of platelet-activating mediators (e.g. thromboxane A2, serotonin, ADP, platelet-activating factor (PAF) and PDGF), which may contribute to the progression of the disease. Furthermore, it has been reported that decreased NO release in hypercholesterolemia is associated with enhanced neutrophil adherence to the endothelium [30]. These findings have pathophysiological significance, since activated neutrophils may influence platelet aggregation and may produce superoxide radicals, arachidonic acid metabolites (leukotrienes A4 and B4) and proteases [31], all of which may amplify endothelial dysfunction and promote arterial damage. In addition to the loss of the protective actions of NO, endothelial dysfunction may result in a number of abnormalities which may enhance progression of atherosclerosis including increased release of vasoconstrictor substances, expression of surface adhesion molecules which facilitate monocyte recruitment and egress into the intima, production of growth factors which promote vascular smooth muscle cell proliferation and migration, and enhanced thrombogenicity mediated by increased platelet activation, PAI-1, and expression of TF [23].

Cells and Cellular Interactions Recent data have provided evidence that the first steps of atherogenesis are inflammatory in nature. The discovery of activated T lymphocytes, macrophages and mast cells in atherosclerotic lesions, detection of HLA class II antigen expression and cytokines have been implicated in the immune and inflammatory mechanisms of the pathogenesis of atherosclerosis [10]. These cells are capable of adhering to the injured endothelium and migrating into subendothelial space and releasing substances that ultimately affect arterial lumen. Although the inflammatory cells interact in specific patterns to form the atherosclerotic plaque, the special features of these cells and their interrelationships are described separately below and summarized in Figure 2.

Activated T lymphocytes T lymphocytes are present from a very early stage of atherosclerosis and are detectable in the fatty streak [32]. During atherogenesis activated lymphocytes release into the intima cytokines that result in macrophage, B-cell, and cytotoxic T cells activation, and expression of neutrophil adhesion molecules [10]. Interferon-γ (IF-γ), their principal cytokine secreted by activated T cells, may play a role in lesion progression or regression through its effect on macrophage (inhibits their transformation into a foam cell) or smooth muscle cells (inhibits their proliferation) [33].

Monocytes and Macrophages Monocytes and macrophages participate in all stages of the formation of the atherosclerotic plaque. Ox-LDL activates endothelial cells to produce chemoattractants and adhesion molecules that attract circulating monocytes to arterial wall. Activated endothelium produces monocyte chemoattractant protein-1 (MCP-1), monocyte colony-stimulating factor (M-CSF) and granulocyte-macrophage-colony stimulating factor (GM-CSF) [34]. These adhesion molecules serve to recruit more mononuclear cells which, in turn, activate adjacent endothelial cells through the secretion of growth factors and cytokines in a paracrine fashion. Once they adhere, monocytes migrate into the intima and transform into macrophage and scavenge ox-LDL [35]. Once engorged with lipids, the macrophages are referred to as foam cells, the characteristic infiltrating cells in the early atherosclerotic lesion, which play an important role in the attraction of smooth muscle cells to the intima. Macrophages release cytokines and growth factors, including transforming growth factor α and β (TGF-α and β), both forms of platelet-derived growth factors (PDGF) and acidic and basic fibroblast growth factors (aFGF and bFGF) [23]. These growth factors induce smooth muscle cell migration and proliferation into the subendothelial space. Smooth muscle cells in the intima change the contractile phenotype to synthetic phenotype and secrete extracellular matrix. In late lesions, macrophages release proteases that contribute to the rupture of complicated plaques leading to acute thrombosis [36]. Degradation of extracellular matrix by proteases is associated with a decrease in the number of smooth muscle cells. The decrease in the smooth muscle cell population in theses regions might result from growth inhibition due to IF-γ secreted by T cells. In addition, necrosis and apoptosis contribute to decrease the number of smooth muscle cells in plaques showing matrix degradation [37]. Although the precise signals that regulate apoptosis in atheroma are unknown, reduced oxygen species induced by ox-LDL might play an important role [38].

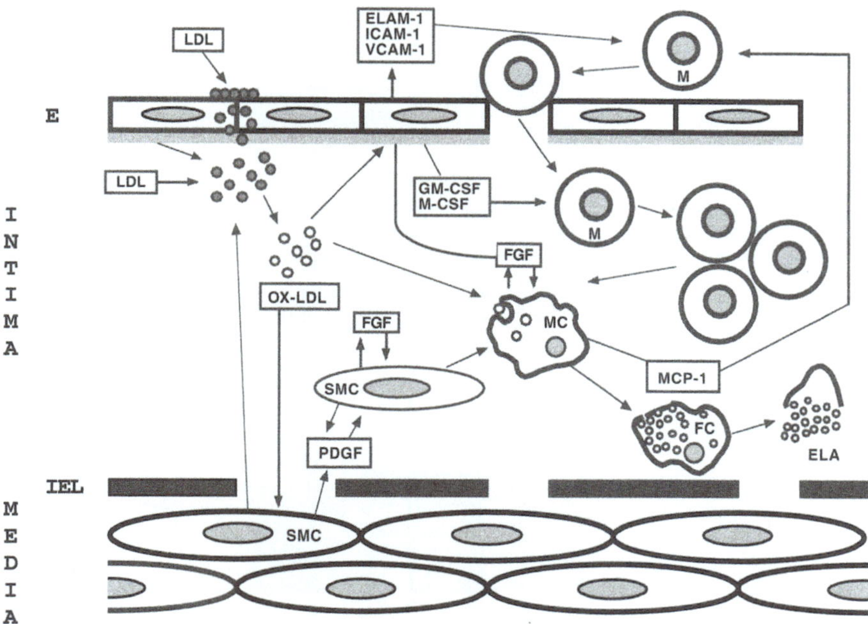

Figure 2. This figure shows a network between pro-inflammatory, chemoattractant and mitogen molecules and cells in the atheroma. LDL is oxidized by SMC from the media and by endothelial cells. Ox-LDL stimulates the secretion of adhesion molecules (ELAM-1, ICAM-1 and VCAM-1) by endothelial cells, resulting in increased monocyte adhesion and their transendothelial migration. The proliferation of intimal monocytes is induced by specific colony stimulating factors (GM-CSF and M-CSF). Monocytes differentiate into macrophages that secrete specific chemoattractants (MCP-1) for circulating monocytes. Simultaneously, ox-LDL stimulates the migration of SMC from the media into the intima. Activated SMC segregates mitogens (PDGF and FGF), by an autocrine and paracrine fashion, which induces the migration and proliferation of SMC. Several of these cells differentiate into macrophages by FGF-induced expression of scavenger receptors. Via these receptors, SMC accumulates ox-LDL, resulting in foam cell generation. Finally, the foam cell death results in extracellular lipid accumulation into the intima. ELAM-1, endothelium leukocyte adhesion molecule-1; E, endothelium; ELA, extracellular lipid accumulation; FC, foam cell; FGF, fibroblast growth factor; GM-CSF, granulocyte-macrophage colony stimulating factor; ICAM-1 intercellular adhesion molecule-1; IEL, internal elastic lamina; M, monocyte; MC, macrophage; MCP-1, monocyte chemoattractant protein-1; M-CSF, monocyte colony stimulating factor; PDGF, platelet-derived growth factor; SMC, smooth muscle cell; VCAM-1, vascular adhesion molecule-1.

Mast Cells   Activated mast cells in the arterial intima release cytokines (IL-5, TNF-α), growth factors (FGF, PAF, TGF-β) and other mediators. They could serve as chemotactic molecules for mononuclear cells and mitogens for smooth muscle cells [39]. Following mast cell activation, the two neutral proteases of mast cells, tryptase and chymase are released [40] and are able to activate matrix metalloproteinases secreted by macrophages and smooth muscle cells [41]. This can lead to collagen degradation, which may weaken the fibrous cap with subsequent plaque rupture. Tryptase, histamine and platelet-activating factor (PAF) released by mast cells contribute to smooth muscle cell proliferation and to the formation of microthrombi in complex plaque. Finally, mast cell degranulation products promote procollagenase production by other cells, resulting in collagen degradation and subsequent atheroma rupture. Other non-inflammatory cell types that participate in the atherogenesis steps are platelets and vascular smooth muscle cells.

Platelets   Platelets play a central role in atherogenesis. Endothelial activation or denudation promote platelets adhesion to the vessel wall. In these conditions activated platelets release molecules that affect coagulation, arterial tone, and migration and proliferation of medial smooth muscle cells in arterial intima [42]. Activated platelets release a potent mitogen (PDGF) for smooth muscle cells. In addition, platelets release thromboxane A2, serotonin, ADP and histamine. These products promote platelet aggregation, thrombosis and vasoconstriction. Platelets play a critical role in the development of the acute obstructive thrombosis that occurs after plaque rupture. Ox-LDL may change the nonthrombogenic vascular phenotype to a prothrombotic phenotype that could induce thrombosis by activation of platelets and by stimulating coagulation and inhibiting the fibrinolytic pathway [43,44] (Figure 3).

Smooth Muscle Cells   Smooth muscle cells accumulation in the arterial intima is a key event in the growth of atherosclerotic lesions, and results from a combination of proliferation and migration of smooth muscle cells from the media into the intima. Both of these activities are induced by cytokines and growth factors. Growth factors, which can originate in endothelial cells, monocytes, macrophages, mast cells, lymphocytes, platelets or smooth muscle cells themselves, change the contractile phenotype of smooth muscle cells to promote cellular division and secretion of large amounts of connective tissue matrix [2]. The combination of extracellular matrix accumulation, smooth muscle cell

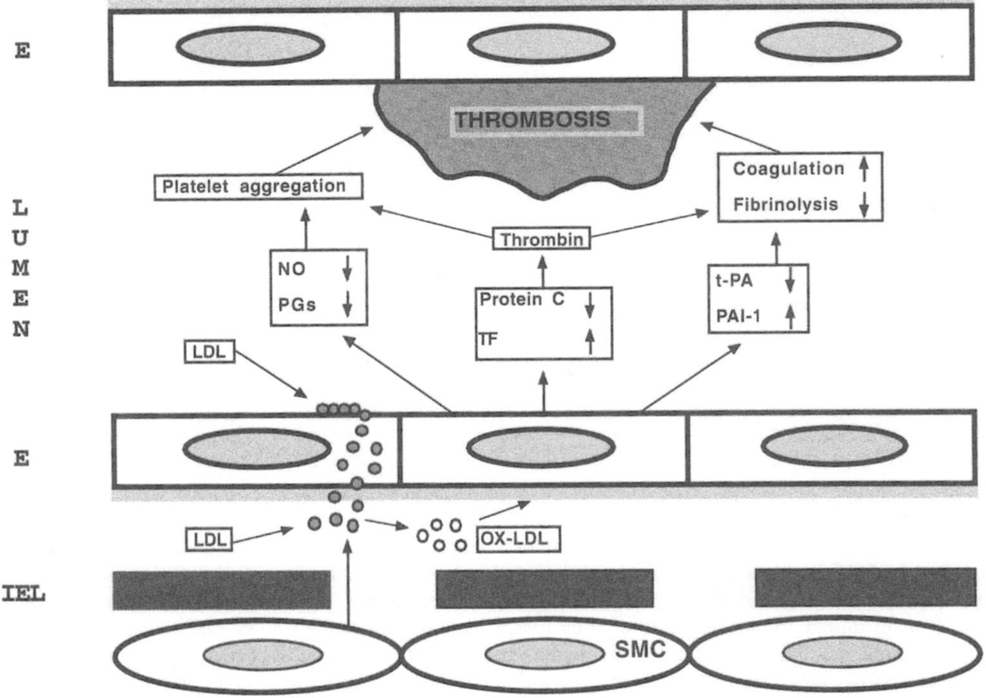

Figure 3. Ox-LDL induces platelets activation by inhibiting the synthesis of nitric oxide (NO) and stimulating the synthesis of prostaglandins (PGs). Furthermore, ox-LDL inhibits the synthesis of protein C and induces synthesis of tissue factor (TF), resulting in thrombin generation that induces platelet aggregation and triggers coagulation. Ox-LDL impairs the fibrinolytic capacity of the endothelial cells, resulting in defective clot lysis. E, endothelium, IEL, internal elastic lamina; SMC, smooth muscle cells.

Table 2. Growth promoters of vascular smooth muscle cells

Substance	Source
Epidermal growth factor (EGF)α	P
Acidic fibroblast growth factor (aFGF)	EC,M,MC,SMC
Basic fibroblast growth factor (bFGF)	EC,M,MC,SMC
Insulin growth factor-1 (IGF-1)	EC,M,P,SMC
Interleukin-1 (IL-1)	EC,M,P,SMC,T
Interleukin-6 (IL-6)	EC,M,P,SMC,T
Platelet derived growth factor (PDGF)	EC,M,P,SMC
Transforming growth factor-α (TGF-α)	M
Transforming growth factor-β (TGF-β)	EC,M,MC,P,SMC,T
Tumor necrosis factor-α (TGF-α)	EC,M,MC,,SMC,T
Platelet-activating factor (PAF)	MC
Interferon-γ (IFN-γ)	M,T
Thrombin	PI

EC, endothelial cell; M, monocyte/macrophage; MC, mast cell; P, platelet; PI, plasma; SMC, smooth muscle cell; T, lymphocyte T

replication, and increased vasoconstriction contributes to the focal arterial stenosis.

The most important growth promoters of vascular smooth muscle cells are listed in Table 2.

Additional Features Atherosclerotic lesions arise as a consequence of the inflammatory response to ox-LDL incorporated by macrophages and deposited in the intima. This is associated with the accumulation of medial smooth muscle cells in the neointima. In the stable plaque, a fibrous cap excludes material contained within the lesion from the circulation. Breakdown of this cap, such as plaque fissure or rupture, causes distal embolization or vessel occlusion. Plaque neovascularization is a candidate process for intraplaque haemorrhages and supraplaque thrombosis at the site of plaque rupture [45]. Rupture of plaque microvessels may be produced by wall stress and pulsatil flow as the plaque enlarges.

In normal arteries, the media is dependent on diffusion of oxygen from the lumen and from vessels of the adventitial vasa vasorum. Wall thickening associated with atherosclerosis impairs oxygen delivery from the lumen, requiring a compensatory increase from the vasa vasorum [45]. Intraplaque hypoxia causes an increased expression of angiogenic peptides such as bFGF and VEGF in infiltrating cells [46] that promote the neovascularization of the plaque from vessels arising from vasa vasorum. Intimal neovascularization plays a role in the growth of plaques, because it contributes to the rapid growth and complications of plaque through edema, inflammation, small vessel rupture, haemorrhage, thrombosis and necrosis.

Despite the implication of FGF in atherogenesis as growth promoters for medial smooth muscle and for endothelial cells [47], the non-mitogenic activities of these peptides could play a role in platelet activation, inflammation, procoagulation, or paradoxical constriction and apoptosis that are associated with the atherosclerotic plaque. Indeed, the capacity of FGF to attenuate platelet activation [48], and to suppress transcriptional activation of TF and several inflammatory genes [49],

and its vasodilatory [50] and anti-apoptotic effects [51] suggest potential mechanisms by which these peptides may interfere with atherosclerotic plaque complications.

Clinical Relevance   Atherosclerosis is the leading cause of death in the western world [52] as the main source of myocardial infarction, stroke and peripheral vascular disease [53]. Myocardial infarction often strikes at the height of productive capacity. Stroke causes disability and death. Peripheral arterial disease robs the ability to walk. Taken together, these events are costly in terms of human suffering and economic burden. Atherogenesis starts with the loss of a wide range of defensive and healthy functions by the endothelium. This results in an increase in inflammatory, procoagulant, vasoactive and proliferative response into the arterial wall. These changes promote neointimal formation, vasospasm, and thrombus formation, all of which play an important role in the progression and clinical manifestations of atherosclerosis. Understanding the molecular interactions between injured endothelium, circulating immune cells and vascular smooth cells may lead to therapeutic strategies for the prevention and treatment of atherosclerotic disease.

Pedro Cuevas

References

1. Fuster V et al (1992) N Engl J Med 326:242-250
2. Ross R (1995) Annu Rev Physiol 57:791-804
3. Stary HC (1989) Arteriosclerosis 9:19-32
4. Kovanen PT (1995) Chem Immunol 62:132-170
5. Giachelli CM (1993) J Clin Invest 92:1686-1696
6. Ginbrone MA Jr et al (1997) Ann NY Acad Sci 811:1-10
7. Davis PF, Tripathi SC (1993) Circ Res 72:234-245
8. Endres M et al (1997) Stroke 28:77-82
9. Nagel T et al (1994) J Clin Invest 94:885-891
10. Hansson GK (1993) Br Heart J 69(Suppl):S38-S41
11. Gimbrone MA Jr (1995) Monogr Pathol 37:1-11
12. Wood KM et al (1993) Histopathology 22:437-444
13. Berliner JA et al (1995) Circulation 91:2488-2496
14. Bork RW et al (1992) Arterioscler Thromb 12:800-806
15. Clinton SK, Libby P (1992) Arch Pathol Lab Med 116:1292-1300
16. Rubanyi GM (1993) J Cardiovasc Pharmacol 22 (suppl4) S1-S14
17. Wilcox JM et al (1989) Proc Natl Acad Sci USA 86:2839-2843
18. Moreno PR et al (1996) Circulation 94:3090-3097
19. Goldstein JL et al (1979) Proc Natl Acad Sci USA 76:333-337
20. Steinberg DS et al (1989) N Engl J Med 320:915-924
21. Leake DS (1991) Current Opinion Lipidology 2:301-305
22. Osterud B (1997) Thromb Res 85:1-22
23. Boyle EM et al (1997) Ann Thorac Surg 64:S47-S56
24. Ananyava NM et al (1997) Arterioscler Thromb Vasc Biol 17: 445-453
25. Palmer RMJ et al (1988) Nature 333:664-666
26. Waldman SA, Murad F (1987) Pharmacol Rev 39:163-196
27. Radomski MW et al (1987) Br J Pharmacol 92:181-187
28. Kubes P et al (1991) Proc Natl Acad Sci USA 88:4651-4655
29. Vanhoute PM, Shimokawa H (1989) Circulation 80:1-9
30. Lefer AM, Ma X-I (1993) Arterioscler Thromb 13:771-776
31. Weiss SJ (1989) N Engl J Med 320:265-276

32. Munro JM et al (1987) Hum Pathol 18:375-380
33. Hanson GK et al (1988) Cir Res 63:712-719
34. Rollins BJ et al (1990) Am J Pathol 136:1229-1233
35. Matstumoto A et al (1990) Proc Natl Acad Sci USA 87:9133-9137
36. Galis ZS et al (1994) Proc Natl Acad Sci USA 92:402-406
37. Bennett MR et al (1995) J Clin Invest 95:2266-2274
38. Dimmeler S et al (1997) Circulation 95:1760-1763
39. Metzler B, Xu Q (1997) Int Arch Allergy Immunol 114:10-14
40. Irani AA et al (1986) Proc Natl Acad Sci USA 83: 4464-4468
41. Hibbs MS et al (1987) J Clin Invest 80: 1644-1650
42. Badimon L, Badimon JJ (1996) In: Fuster V et al (eds) Artherosclerosis and coronary disease. Lippincott-Raven, Philadelphia, pp 639-656
43. Weiss JR ett al (1991) FASEB J 5:2459-2465
44. Kugiyama K et al (1993) Circ Res 73:335-343
45. Fryer JA et al (1987) J Vasc Surg 6:341-349
46. Winlaw D (1997) Ann Thorac Surg 64:1204-1211
47. Cuevas P et al (1991) Circ Res 69:360-369
48. Zarge JI et al (1997) Am J Surg 174:188-192
49. Pendurthi UR et al (1997) Arterioscler Thromb Vasc Biol 17:940-946
50. Cuevas P et al (1991) Science 254:1208-1210
51. Cuevas P et al (1997) Eur J Med Res 2: 465-468
52. WHO Tech Rep Serv (1985) 143:1-20
53. Ross R (1993) Am J Pathol 143:987-1002

Atherothrombosis

Definition   *Atherosclerotic lesion of arteries that contain in their lumen thrombi.*

See: →Atherosclerosis; →Thrombosis

ATP

Definition   *Adenosine triphosphate*

See: →FGF receptors

β-TG

Definition   *β-thromboglobulin*

See: →Megakaryocytes

BACE

Definition   *Bovine aortic-cross endothelial cell*

See: →Angiogenin

BAE

Definition   *Bovine aortic endothelial cell*

See: →Angiogenin; →Endothelial cells; →von Willebrand factor

BAG-1

Definition   *Bradyzoite-specifically expressed gene-1*

See: →Hepatocyte growth factor

Basal Lamina

See: →Extracellular matrix (basement membranes)

Basement Membranes

Definition   *Membrane underneath epithelial or endothelial cells composed of extracellular matrix proteins. Endothelial cells are separated from the subendothelium by basement membranes.*

See: →Extracellular matrix

BCE

Definition   *Brain capillary endothelial cells*

See: →Angiogenin; →Endothelial cells; →Heparin-affin regulatory protein; →Signal transduction mechanisms in vascular biology

Bernard-Soulier Syndrom

Definition   *Autosomal recessive platelet disorder characterized by abnormal platelet glycoprotein Ib or IX*

See: →Bleeding disorders; →Vascular integrins

bFGF

Definition   *Basic fibroblast growth factor*

See: →Fibroblast growth factors; →FGF receptors

BFU

Definition   *Burst-forming unit*

See: →Megakaryocytes

BFU-MK

Definition   *Megakaryocytic burst forming unit*

See: →Megakaryocytes

Bleeding Disorders (Congenital)

Synonym: Congenital bleeding disorders, including: $\alpha 2$ anti-plasmin deficiency; Abnormal responses to arachidonic acid: cyclooxygenase deficiency and thromboxane synthetase deficiency; ADP receptor abnormality; Afibrinogenemia/hypofibrinogenemia; Bernard-Soulier syndrome (BBS); Combined factor V-VIII deficiency; Defect in calcium mobilization or utilization; Defect in platelet G-protein; Deficiency of factors II, VII, IX and X or vitamin K-dependent factors deficiency; Dysfibrinogenemia (congenital); Ehlers-Danlos syndrome (EDS); Epstein syndrome or Alport syndrome; Factor V deficiency; Factor V Quebec or Quebec platelet disorder or multimerin deficiency; Factor X deficiency; Factor XIII deficiency; Fanconi syndrome (FA); Glanzmann's thrombasthenia (GT); GPIα deficiency; Gray platelet syndrome or alpha-granule deficiency or alpha-storage pool disease; Hemophilia A; Hemophilia B; Hereditary factor XI deficiency; Hereditary hemorrhagic telengiectasia or Rendu-Osler-Weber syndrome; May-Hegglin anomaly; Montreal platelet syndrome; Platelet-type von Willebrand disease or "pseudo" von Willebrand disease; Prothrombin deficiency or hypoprothrombinemia and dysprothrombinemia; Scott syndrome; Storage pool disease or dense granules deficiency; 'Swiss-cheese' platelet disorder; Thrombocytopenia with absence of radii or TAR syndrome; Thromboxane A_2 receptor deficiency; von Willebrand disease (vWD).

Definition   *Congenital disorders of primary hemostasis, blood coagulation or fibrinolysis that lead to bleeding.*

See also: →Bleeding disorders; →Fibrin/fibrinogen

Introduction

α2 anti-plasmin deficiency: Autosomal recessive bleeding disorder. Homozygous patients have a complete defect in α2 anti-plasmin, heterozygous have α2 anti-plasmin level between 40% and 60% of normal values in both antigenic and functional assays.

Abnormal responses to arachidonic acid: cyclooxygenase deficiency and thromboxane synthetase deficiency: Mild hemorrhagic syndrome with a prolonged bleeding time, normal platelet count and normal platelet dense-granule content.

ADP receptor abnormality: Rare platelet disorder characterized by an abnormal ADP platelet response.

Afibrinogenemia/hypofibrinogenemia: Congenital autosomal recessive bleeding disorder.

Bernard-Soulier syndrome (BBS): Rare autosomal recessive platelet disorder described in 1948 by J. Bernard and J.P. Soulier characterized by a prolonged bleeding time, thrombocytopenia, very large platelets and abnormal prothrombin consumption.

Combined factor V-VIII deficiency: Autosomal recessive disorder due to a single genetic defect.

Defect in calcium mobilization or utilization: Characterized by a mild bleeding disorder.

Deficiency of factors II, VII, IX and X or vitamin K-dependent factors deficiency: Autosomal recessive bleeding disorder concerning all vitamin K-dependent factors (Factor II, VII, IX, X, protein C and S).

Dysfibrinogenemia (congenital): Characterized by the biosynthesis of a structurally abnormal fibrinogen. The functional abnormalities are usually reflected as abnormalities in one or more phases of the fibrinoformation process.

Ehlers-Danlos syndrome (EDS): EDS represents various genetic connective tissue disorders which affect the skin, ligaments, joints, blood vessels and visceral organs. The current classification recognizes nine subtypes. The types VI and VIIc are autosomal recessive, the other types are autosomal dominant.

Epstein syndrome or Alport syndrome: Rare mild autosomal dominant disease with thrombocytopenia, giant platelets, hereditary nephritis and deafness.

Factor V deficiency: Autosomal recessive trait with an incidence approximately of 1/1,000,000.

Factor V Quebec or Quebec platelet disorder or multimerin deficiency: Autosomal-dominant qualitative platelet disorder.

Factor X deficiency: Autosomal recessive bleeding deficiency with an incidence of 1/500,000.

Factor XIII deficiency: Rare autosomal bleeding disorder with an incidence of 1/3,000,000.

Fanconi syndrome (FA): Rare autosomal recessive disease characterized by multiple congenital abnormalities, bone marrow failure, and cancer susceptibility. Patients can be severely affected, with multiple congenital anomalies, or may have mild phenotype with no major malformations.

Glanzmann's thrombasthenia (GT): Rare autosomal recessive platelet disorder characterized by a prolonged bleeding time, normal platelet count and abnormal clot retraction described by Glanzmann in 1918.

GPIα deficiency: Exceptional bleeding disorder characterized by a prolonged bleeding time, normal platelet count and no platelet response to collagen.

Gray platelet syndrome: Rare inherited disorder characterized by thrombocytopenia with large and gray platelet due to the absence of normal alpha granules.

Hemophilia A: X-linked bleeding disorder with an incidence of 1/5000.

Hemophilia B: X-linked bleeding disorder with an incidence of 1/30,000.

Hereditary factor XI deficiency: Rare autosomal recessive bleeding disorder with an incidence less than 1/1,000,000 occurring predominantly in the Ashkenazi Jews.

May-Hegglin anomaly: Autosomal dominant thrombocytopenia with giant platelets and Dôhle bodies in the leukocytes.

Montreal platelet syndrome: Hereditary autosomal dominant bleeding disorder.

Platelet-type von Willebrand disease: Rare platelet disorder with a prolonged bleeding time, thrombocytopenia and low plasmatic level of vWF.

Prothrombin deficiency: Autosomal recessive disorder characterized by a lack of production of prothrombin, or production of a mutant prothrombin or a combination of both.

Rendu-Osler-Weber syndrome: Hereditary hemorrhagic telangiectasia (HHT) is an autosomal heterogeneous dominant bleeding disorder characterized by telangiectasic lesions of the nose, lips and visceral organs.

Scott syndrome: Extremely rare bleeding disorder, only one family is described.

Storage pool disease: Heterogeneous thrombopathy characterized by either a granule dense deficiency or abnormal dense-granule functions.

'Swiss-cheese' platelet disorder: Autosomal dominant mild bleeding disorder with giant vacuolated platelets.

Thrombocytopenia with absence of radii or TAR syndrome: Severe bleeding disorder related to thrombocytopenia at the level of megacaryocyte cells.

Thromboxane A$_2$ receptor deficiency: Rare bleeding disorder with prolonged bleeding time.

von Willebrand disease (vWD): The most common inherited bleeding disorder. vWD is clinically and genetically heterogenous due to qualitative or quantitative abnormalities of von Willebrand factor (vWF). Phenotypic classification of vWD defines quantitative defects that are divided into partial deficiency (1) and severe deficiency (3). Qualitative defects (2) are divided into 4 sub-categories.

Characteristics *α2 anti-plasmin deficiency:* α2 anti-plasmin is an inhibitor which reacts very rapidly to form a stochiometric complex with plasmin and inhibits plasmin activity. Bleeding tendency may be due to premature lysis of the hemostatic plugs. The half-life of plasmin molecules generated on the fibrin surface may be considerably prolonged. PT, aPTT and plasminogen and fibrinogen levels are normals. Euglobulin fibrinolytic activity is slightly increased. α2 anti-plasmin level must be determined by antigenic and functional assays.

Abnormal responses to arachidonic acid: cyclooxygenase deficiency and thromboxane synthetase deficiency Two cyclooxygenases (COX) are identified COX-1 and COX-2. Cox-1 is constitutively expressed in cells and COX-2 is inducible. Platelets contain only COX-1 and therefore platelet cyclooxygenase deficiency concerns only COX-1 deficiency. This disorder is characterized by an abnormal platelet response to collagen and arachidonic acid with a decreased thromboxane A2 generation. Thromboxane synthetase deficiency is characterized by an abnormal platelet response to arachidonic acid and a decrease thromboxane A2 generation.

ADP receptor abnormality: an impaired platelet aggregation induced by ADP with normal GPIIb/IIIa complex.

Afibrinogenemia/hypofibrinogenemia: Afibrinogenemia is marked by clotting tests incoagulable and are prolonged in hypofibrinogemia. Hypofibrinogenemia is characterized by a level of fibrinogen below 1 g/l (moderate) or 0.5 g/l (severe). Bleeding time is prolonged and abnormal platelet aggregation is observed in severe hypofibrinogenemia and afibrinogenemia. The defect of platelet aggregation is corrected by the addition of fibrinogen.

Bernard-Soulier syndrome (BBS): Bernard Soulier platelets are unable to bind to von Willebrand factor (vWf) localized in the subendothelial matrix. The biochemical defect is the deficiency of the membrane glycoprotein Ib-IX complex and glycoprotein V.

Combined factor V-VIII deficiency: characterized by a prolonged APTT and PT and reduced level of factor V and VIII of 5%–15% of normal value. Protein C and protein C inhibitor levels are normals.

Defect in calcium mobilization or utilization: The disease is related to a defect of calcium mobilization with impaired platelet response to ionophore A 23187. As a consequence, an abnormal phosphatidyl inositol hydrolysis has been found suggesting an impaired internal mobilization or utilization of calcium. A platelet defect of calcium influx from external medium has also been described.

Defect in platelet G-protein: characterized by a diminished platelet aggregation and secretion in response to multiple agonists despite the presence of normal dense granules stores.

Deficiency of factors II, VII, IX and X or vitamin K-dependent factors deficiency: characterized by a prolonged APTT and PT and combined deficiency of factors II, VII, IX, X, protein C and S activities with normal antigen levels. The defect stems from abnormal carboxylation of the vitamin K-dependent proteins.

Dysfibrinogenemia (congenital): Dysfibrinogenemia is defined by prolonged thrombin and reptilase times. The fibrinogen concentration appears low when measured as clottable protein, but is normal when measured immunologically. Different types of dysfibrinogenemia are described:

- Congenital dysfibrinogenemia with abnormal fibrinopeptides release. More than thirty mutations of the fibrinogen A chain have been described resulting in abnormal fibrinopeptide A release. The most common mutation site in the fibrinogen molecule is Arg[16] of the A α chain.

- Abnormal release of fibrinopeptide B. The mostly involved mutations is Arg 14 with a decrease release of fibrinopeptide B. In fibrinogen New-York, a deletion 9–72 of the B β chain results in a decrease release of fibrinopeptide B and slow release of fibrinopeptide A.

- Congenital dysfibrinogenemia with polymerization defects. There are the most commonly functional abnormalities described in dysfibrinogenemia. More than fifteen types are described and the most common abnormality is found on the Arg[75] of the γ chain.

- Congenital dysfibrinogenemia with defective cross-linking. Several mutants exhibit defective cross-linking in addition of abnormalities in fibrinopeptides release or fibrin polymerization.

- Congenital hypo/dysfibrinogenemia. They include cases in which abnormal fibrinogen are present with total plasma fibrinogen concentration below 1.5 g/l. About eleven cases are described.

Ehlers-Danlos syndrome (EDS): EDS is an abnormal inherited collagen disorder characterized by a prolonged bleeding time. Platelets count and functions are normal. Bleeding tendencies are observed in all EDS types but are major in type IV (vascular sub-type), therefore we will focus on this sub-type.

Epstein syndrome or Alport syndrome: characterized by thrombocytopenia with large platelets. Fechtner syndrome is a variant of Alport syndrome with leukocyte inclusions.

Factor V deficiency: characterized by a prolonged PT and aPTT with normal thrombin time. Bleeding time is prolonged in severe factor V deficiency. Two forms are described, severe (<1%) and moderate (<50%).

Factor V Quebec or Quebec platelet disorder or multimerin deficiency: The moderate to severe bleeding disorder is characterized by:

- mild thrombocytopenia,
- an epinephrine aggregation defect,
- a quantitative deficiency of platelet multimerin,
- a quantitative and qualitative abnormalities of platelet factor V (2%–4% of the normal value) and moderate decreased plasma factor V level (50% of the normal value),
- and abnormal proteolysis of platelet thrombospondin, fibrinogen, von Willebrand factor, osteonectin, fibronectin and P-selectin. By contrast, albumin and IgG are not degraded.

Factor X deficiency: characterized by a prolonged APTT, PT, and Russel's viper venom-based assay. Antigenic and functional specific assay reveals variants of factor X deficiency.

Factor XIII deficiency: coagulation screening tests are normal despite a convincing history of bleeding. Only specific antigenic assay of factor XIII is able to make the diagnosis.

Fanconi syndrome (FA): defined by cellular hypersensitivity to DNA cross-linking agents.

Glanzmann's thrombasthenia (GT): GT is a defect of platelet aggregation. Biochemical defect is a deficiency of the platelet membrane glycoprotein IIb/IIIa complex. There are two types of Glanzmann's thrombasthenia, in type I clot retraction, platelet fibrinogen and GPIIb/IIIa are absent, in type II clot retraction and platelet fibrinogen are decreased and the amount of platelet GPIIb/IIIa is between 20%–30%. Several variants of GT are described with 50%–100% of GPIIb/IIIa and abnormal functions.

GPIα deficiency: characterized by the decrease of the expression of platelet membrane GPIa.

Gray platelet syndrom: absence of α granules detected by electron microscopy. The basic abnormality in this disorder is the inability of megakaryocytes to transfer secretory proteins into α granules.

Hemophilia A: characterized by a factor VIII deficiency and detected by prolonged APTT and decrease of factor VIII level. One third of the hemophilia A defect is due to spontaneous genetic mutations.

Hemophilia B: characterized by factor IX deficiency and detected by a prolonged APTT and decrease of factor IX level.

Hereditary factor XI deficiency: characterized by a prolonged APTT with normal PT with a decrease of factor XI. Homozygous or composed heterozygous have factor XI activity level less than 15% of the normal and heterozygous have factor XI activity level more than 15% of normal.

May-Hegglin anomaly: characterized by giant platelets with a spherical form in relation to unorganized microtubules.

Montreal platelet syndrome: characterized by thrombocytopenia with large platelet with spontaneous platelet aggregation and prolonged bleeding time. Platelet aggregation is normal in response to all inducers except thrombin.

Platelet-type von Willebrand disease: characterized by the absence of high molecular weight von Willebrand multimers due to the increase affinity of vWf for platelet GPIbα.

Prothrombin deficiency: characterized by a prolongation of PT and APTT with decrease prothrombin level. Dysprothrombinemia results of decrease prothrombin activity with a prothrombin antigen level normal or not.

Rendu-Osler-Weber syndrome: Two types of HHT are described involving members of the family of the transforming growth factor-β1 (TGF-β1) receptor complex endoglin (CD105) and activin receptor-like kinase 1 (ALK-1). A reduction in the level of functional endoglin is involved in the generation of HHT type 1 (HHT1), and is associated with arteriovenous malformations.

Scott syndrome: characterized by an impaired binding of coagulation factors Va and VIIIa by activated platelets, reflecting a dimished surface exposure of phosphatidylserine (PS) associated with reduced shedding of micro-particles from the platelet surface.

Storage pool disease: Dense-granule deficiency or δ-storage pool is characterized either by an absence of dense granules which correlates with the depletion of the specific dense-granule components or by the presence of small abnormal granules which an inability to package secretory contents within the dense granules. Bleeding time is usually but not always prolonged with impaired platelet aggregation in response to weak inducers. Uptake and secretion of serotonin are decreased. Electron microscopy or mepacrine test demonstrate the absence of dense granules. Recently, a platelet dense granule protein was described (granulophysin or CD63) that is deficient in Hermansky-Pudlack syndrome.

'Swiss-cheese' platelet disorder: Patients have prolonged bleeding time, absence of secondary wave of platelet aggregation in response to epinephrin, collagen and adenosine diphosphate (ADP), and defective ^{14}C-serotonin release and platelet factor 3 availability. 30%–70% of the platelets in any given sample exhibited a prominent membrane complex and dilated, tortuous surface-connected canalicular system ('Swiss-cheese' platelet).

Thrombocytopenia with absence of radii or TAR syndrome: an autosomal recessive syndrome with amegacaryocytopenia, abnormalities of the skeletal system, microcephaly, micrognathia and abnormalities of the hearth.
Thromboxane A₂ receptor deficiency: characterized by a low responses to collagen, normal response to high doses of thrombin and platelet do not aggregate in the presence of arachidonic acid and thromboxane A_2 (TXA_2).
von Willebrand disease (vWD):

- vWD–type 1, the most common form (70 % of the patients), is autosomal dominant. It is caused by partial quantitative deficiency of vWF and characterized by prolonged APTT, decreased ristocetin cofactor activity (vWF:RCo) and vWF antigen (vWF:Ag).
- vWD type 3 is a recessive disorder with virtually no detectable vW protein. Absence of vWF causes a secondary deficiency of factor VIII and severe combined defect in platelet adhesion and blood clotting. Screening assays show absence of vWF:RCo, vWF:Ag, prolonged aPTT and mark decrease of factor VIII.
- Type 2A vWD is a autosomal dominant qualitative variant with decreased platelet-dependent function that is associated with the absence of high molecular weight vWF multimers. Disproportionately low vWF:RCo relative to vWF:Ag is found.
- Type 2B vWD is characterized by increased affinity of the variant vWF for platelets leading to spontaneous binding of large vWF multimers to platelets in vivo, followed by clearance of both vWF and platelets. Platelet agglutination occurs at low concentrations of ristocetin in patient's platelet-rich plasma. Thrombocytopenia is frequent with decreased vWF level.
- Type 2M vWD refers to qualitative variants with decreased platelet-dependent function that is not caused by the absence of high molecular weight vWF multimers. Laboratory results are similar to those in vWD type 2A, but large multimers are present.
- Type 2N vWD refers to qualitative recessive variants with markedly decreased affinity for factor VIII. Screening tests typically show normal vWF levels and a prolonged aPTT. The factor VIII level was low and vWF multimers were normal in appearance.

Regulation

Genes

Abnormal responses to arachidonic acid: cyclooxygenase deficiency and thromboxane synthetase deficiency: COX-1 gene (22 kB) is located on chromosome 9.
Afibrinogenemia/hypofibrinogenemia: The synthesis of the three chains of fibrinogen is under the coordinated control of three separate genes localized on chromosome 4.
Bernard-Soulier syndrome (BBS): Several point mutations and a deletion mutation in the genes for GPIbα and GPIX have been described in BBS. A variant type of BBS has been related to an abnormality of the GPIbβ gene.

Ehlers-Danlos syndrome: EDS type IV is a deficiency of collagen type III resulting from structural mutations in the COL3AI gene. Range of mutations in the COL3AI gene have been identified. The majority of molecular defects are point mutations which cause substitution of a glycine residue by a bulkier amino-acid.
Epstein syndrome or Alport syndrome: The 3' half of the human type IV collagen αV gene is affected in the Alport's syndrome.
Factor V deficiency: Human factor V is located on chromosome 1 with 25 exons (300 Kbases).
Factor X deficiency: Factor X is located on chromosome XIII (13q34). Molecular defect is charaterized by single point mutation. All mutations involved CG dinucleotides in the DNA sequence.
Factor XIII deficiency: Factor XIII is located on chromosome 6 for the A chain but precise location of B chain is not actually known. In the majority of cases, inherited factor XIII deficiency is due to a mutations in the gene for the factor XIII A subunit. The factor XIII A gene contains 15 exons and extend over greater 160 kb.
Fanconi syndrome (FA): Using cell-cell fusion techniques, it has been established that there are at least five FA complementation groups FA (A) to FA (E). The genes, encoding the abnormal or missing protein in patients with FA (C) has been cloned. This gene encodes a protein of 558 amino acids and has been localized at chromosome 9q22.3. Cells from FA (C) patients have abnormalities of cell cycle regulation directly related to the genetic mutation. FA (A) and FA (D) genes have been recently cloned.
Glanzmann's thrombasthenia (GT): More than 20 abnormalities are described in both GPIIb and GPIIIa genes.
Hereditary factor XI deficiency: The type I mutation interferes with the normal exon splicing. The type II mutation introduces a premature termination codon. The type III mutation results in the substitution of leucine for phenylalanine at position 283. This substitution causes a reduction in the amount of factor XI secreted from the cell. The type II and type III mutations together account for most of the factor XI deficiency in Ashkenazi Jews, but account for only 12 % of the factor XI deficiency in non-Jewish patients. The remainder of cases in Ashkenazis and most of the other cases are caused by a number of different mutation.
Hemophilia A: Factor VIII gene is localized to chromosomes X q28. The molecular defects are heterogeneous. Large deletions are associated with less than 5 % of the severe patients. Non-sense mutations cause severe hemophilia. More than 80 mutations have been identified and are associated with wide range of clinical manifestations. Recently, a new mechanism was described involving a recombination between sequences in intron 22 that are homologous with sequences up-stream of the factor VIII gene leading to an inversion of all intervening DNA and a disruption of the gene.
Hemophilia B: The gene is localized at Xq26. Approximately 800 variants have been described. Large deletions are associated with severe hemophilia B, but small-

er deletions can also be the source of severe bleeding due to an abnormal protein synthesis. Punctual mutations are the main abnormalities found in hemophilia B (nonsense 16%, anti-sense 70%, or abnormalities of alternative splicing).

Platelet-type von Willebrand disease: Two mutations in the GPIbα gene resulting in amino acid substitution (Gly^{233}→Val; Met^{239}→Val) have been described.

Prothrombin deficiency: The prothrombin gene is located on chromosome 11.

Rendu-Osler-Weber syndrome: Endoglin is mapped to chromosome 9 and ALK-1 to chromosome 12. Several mutations are described for HHT1.

Scott syndrome: This disorder involved a deletion or a mutation of a gene implicated in the transmigration of PS to the cell surface.

Thromboxane A$_2$ receptor deficiency: Recently, a single amino-acid substitution (Arg^{60} to Leu) has been described in the first cytoplasmic loop of the TXA$_2$ receptor in two unrelated families with a defect in platelet response to TXA$_2$. This mutation impairs the coupling of the receptor to the G protein linked to phospholipase C.

von Willebrand disease (vWD): Quantitative deficiencies will correlate with promoter, non-sense and frameshift mutations and with large deletions. No genetic distinctions can be made between type 1 and type 3. Qualitative deficiencies will correlate with mis-sense mutations and small in-frame deletions or insertions.

Molecular Interactions

α2 anti-plasmin deficiency: α2 anti-plasmin binds to lysine-binding sites I of plasmin and inhibits its active site and the binding of plasminogen to fibrin.

ADP receptor abnormality: A defective interaction between ADP and its receptor(s) on platelets characterize this platelet disorder.

Bernard-Soulier syndrome (BBS): GPIb-IX is a 1:1 non covalent complex. Both GPIb and IX are embedded in the platelet membrane. The cytoplasmic portion of GPIb is associated with the platelet cytoskeleton through an interaction with the actin-binding protein. GPIb is composed of two subunits (GPIbα and GPIbβ) linked by a disulfide bridge. The complex GPIb-IX is one of the platelet membrane receptors of exposed sub-endothelium vWf.

Dysfibrinogenemia (congenital): Fibrinogen is converted into fibrin after thrombin cleavage. Congenital dysfibrinogenemia results in abnormal clotting.

Defect in platelet G-protein: characterized by an abnormality in Gα subunit function.

Deficiency of factors II, VII, IX and X or vitamin K-dependent factors deficiency: The abnormal carboxylation results in protein-induced vitamin K absence (PIVKA).

Ehlers-Danlos syndrome (EDS): Type III collagen is secreted by fibroblasts and smooth muscle cells and is common in a variety of tissues (skin, blood vessel wall and visceral organs). It is also involved in platelet adhesion to sub-endothelium.

Factor V Quebec or Quebec platelet disorder or multimerin deficiency: It is related to a pathologic proteolysis of α-granular contents, rather than a defect in targeting proteins to α-granules.

Factor X deficiency: Factor Xa is involved in the tenase complex (factor VIIa, factor IXa, factor Xa and phospholipids).

Factor XIII deficiency: Factor XIII is activated by thrombin. Activated factor XIIIa catalyse e (γ-glutamyl) lysine cross-links between fibrin molecules which increases the mechanical strength of blood clots. Factor XIIIa cross-links fibronectin and thrombospondin to anchor the clot to the site of injury and fibrin to α2-anti-plasmin to increase clot resistance to plasmin. Factor XIIIa also cross-links fibronectin, vitronectin, collagen in the extracellular matrix.

Glanzmann's thrombasthenia (GT): GPIIb/IIIa is an integrin receptor which binds fibrinogen and vWF on activated platelets. The GPIIb/IIIa complex is a calcium dependent heterodimer. Following fibrinogen binding, the cytoplasmic portion of GPIIb/IIIa interacts with actin filaments involving a cytoskeletal protein (talin). This interaction plays a role in clot retraction.

GPIα deficiency: GPIa/IIa complex (α2/β1) forms a heterodimeric complex that appears to be part of the collagen-binding sites on platelets. This disorder is characterized by little or no GPIα and reduced amounts of GPIIα.

Montreal platelet syndrome: The platelet-defective membrane reorganization during shape change is related to a defect in the proteinase calpain.

Rendu-Osler-Weber syndrome: Endoglin and ALK-1 genes encode transmembrane proteins expressed on endothelial cells which bind TGFβ1 and 3 for endoglin and TGFβ1 and activin A for ALK-1.

Scott syndrome: characterized by an isolated defect in the expression of the membrane catalytic tenase activity (complex of coagulation factors VIIIa and IXa) and prothrombinase activity (complex of coagulation factors Va and Xa) in stimulated platelets.

Thromboxane A$_2$ receptor deficiency: TXA$_2$ receptor is a member of the family of the G protein-coupled receptors.

von Willebrand disease (vWD): caused by an abnormal platelet adhesion to sub-endothelial vWF with the exception of type 2N vWD.

Cells and Cellular Interactions

Defect in platelet G-protein: Receptor-mediated release of arachidonic acid from phospholipids and calcium mobilization are impaired upon platelet activation.

Fanconi syndrome (FA): Cells from individuals with FA arrest excessively in the G2/M phase of the cell cycle after exposure to low doses of DNA cross-linking agents. The relationship of this abnormality to the fundamental genetic defect in such cells is unknown, but many investigators have speculated that the various FA genes directly regulate cell cycle checkpoints.

Gray platelet syndrome: A stable bone marrow fibrosis is always observed, due to a leakage of platelet-derived growth factor (PDGF) into the marrow stroma.

Clinical Relevance

α2 anti-plasmin deficiency: Homozygous patients have severe hemorrhagic diathesis with hemarthrosis and hematomas after minor trauma. Heterozygous patients show no or moderate bleeding tendencies.

Abnormal responses to arachidonic acid: cyclooxygenase deficiency and thromboxane synthetase deficiency are an extremely rare mild bleeding disorder with a prolonged bleeding time.

ADP receptor abnormality: Moderate hemorrhagic syndrome are described but most of them occurred after surgery or trauma.

Afibrinogenemia/hypofibrinogenemia: The hemorrhagic syndrome is related to the plasma fibrinogen concentration. Patients with afibrinogenemia exhibit a very high incidence of hemorragic complications (umbilical cord bleeding, gastrointestinal hemorrhage, hemarthrosis, hematoma, intra cranial bleeding). Bleeding is the cause of death in approximately one-third of patients in the neonatal period. Bleeding in hypofibrinogenemia occurs usually in patients with fibrinogen concentrations below 0.5 g/l and the hemorrhagic manifestations are associated with trauma or surgical procedures. Thrombotic episodes might be observed and induced by fibrinogen infusion.

Bernard-Soulier syndrome (BBS): a severe bleeding disorder characterized by ecchymosis, epistaxis, gingival bleeding, menorrhagia, gastrointestinal hemorrhage. Hemarthrosis and hematomas are unusual. In one report of 59 cases of BBS, there were 10 deaths from hemorrhage.

Combined factor V-VIII deficiency: This bleeding disorder is mild, mostly mucosal bleeding and post surgery.

Defect in calcium mobilization or utilization: This bleeding disorder is moderate (mostly cutaneous).

Defect in platelet G-protein: a disorder associated mild mucocutaneous bleeding diathesis associated with prolonged bleeding time and normal platelet counts.

Deficiency of factors II, VII, IX and X or vitamin K-dependent factors deficiency: This severe bleeding diathesis occurs in infancy with spontaneous intracerebral hemorrhage, multiple bruises and hematomas but not hemarthrosis.

Dysfibrinogenemia (congenital): Some mutants are completely asymptomatic, others have thrombotic disorders, and others have bleeding disorders. Parma fibrinogen is probably the dysfibrinogenemia associated most frequently with bleeding complications.

Ehlers-Danlos syndrome (EDS): EDS type IV is characterized by excessive bruising and bleeding, delayed wound healing, spontaneous arterial rupture and thin, transparent but not hyperelastic skin.

Epstein syndrome or Alport syndrome: The bleeding disorder is function of the severity of the thrombocytopenia.

Factor V deficiency: Bleeding disorders are characterized by umbilical cord bleeding, menorraghia, epistaxis, hemarthrosis but bleeding is not as frequent as in hemophilia A and most of the time diagnosis is made in adulthood (half of the patients). Variants of factor V have been described with a decrease in V_a function and normal antigenic level.

Factor V Quebec or Quebec platelet disorder or multimerin deficiency: The bleeding is moderate to severe and irresponsive to platelet transfusions.

Factor X deficiency: The bleeding disorder is characterized by hemarthrosis, menorrhagia and hematuria. Soft tissue hemorrhages are observed in patients with severe factor X deficiency. Patients with more than 15% of factor X level have very few hemorrhagic episodes except after surgery or trauma.

Factor XIII deficiency: The hallmark of severe factor XIII deficiency is umbilical stump bleeding. Intra-cranial hemorrhage is relatively frequent. Recurrence of tissue hemorrhage after trauma, recurrent abortions and poor wound healing are frequent. Typically, a clot appears to form normally in the wound site, which then breaks down approximately 24 hours later, and bleeding resumes. Heterozygous patients show no or middle bleeding tendencies but they might have abnormalities of wound healing and recurrent spontaneous abortion.

Fanconi syndrome (FA): The bleeding disorder is correlated to the severity of the thrombocytopenia.

Glanzmann's thrombasthenia (GT): GT typically presents with muco-cutaneous bleeding beginning from the neonatal period.

GPIα deficiency: The few patients described exhibited a severe muco-cutaneous bleeding disorder.

Gray platelet syndrome: The patients exhibit a mild bleeding diathesis.

Hemophilia A: It is a clinically heterogenous bleeding disorder as would be expected from the large number of different defects in the factor VIII gene. Bleeding disorder is characterized by spontaneous hematoma and hemarthrosis. It is important to distinguish severe hemophilia (factor VIII <1%), from moderate (1%–4%) and mild hemophilia (5%–25%). Mild hemophiliacs bleed only after surgery or trauma. Exceptional hemophilia has been described in female patients.

The incidence of the development of inhibitors in patients with hemophilia is about 20%. The risk in patients with mild or moderate hemophilia is low.

The risk to develop inhibitors is strongly associated with non-sense mutations and large deletions in the factor VIII gene.

Hemophilia B: It is a clinically heterogeneous bleeding disorder as would be expected from the large number of different defects in the factor IX gene. The bleeding disorder is characterized by spontaneous hematoma and hemarthrosis. It is important to distinguish severe hemophilia (factor IX <1%), from moderate (1%–4%) and mild hemophilia (5%–25%). Mild hemophilia B sufferers bleed only after surgery or trauma. Exceptional hemophilia B has been described in female patients. The factor IX level in female carriers ranges between 15% and 50% of normal level. The incidence of the development of inhibitors in patients with hemophilia is about 20%. The risk to develop inhibitors is strongly associated with non-sense mutations and large deletions in the factor IX gene. Hemophilia B-Leyden is characterized by the gradual amelioration of bleeding after the onset of puberty.

Hereditary factor XI deficiency: Spontaneous bleeding is rare. Clinical bleeding after trauma or surgery is extremely variable and unpredictable in patients with factor XI deficiency and across the entire group of patients (homozygous or heterozygous). Excessive bleeding has only a rough correlation with the factor XI level. The risk of bleeding in homozygous and composed heterozygous patients has clearly been shown to be greater than in heterozygous patients.

May-Hegglin anomaly: a rare minor bleeding disorder.

Platelet-type von Willebrand disease: characterized by hereditary muco-cutaneous bleeding.

Prothrombin deficiency: This bleeding disorder is characterized by epistaxis, post surgery bleeding but hemarthrosis are uncommon.

Rendu-Osler-Weber syndrome: A bleeding disorder characterized by severe recurrent epistaxis (>90%), mucosal (50%–80%), gastrointestinal hemorrhages (10%–40%), cerebral hemorrhages (venous malformation) and rare pulmonary hemorrhages.

Scott syndrome: a moderately severe bleeding disorder.

Storage pool disease: Different forms are described: Hermansky-Pudlack syndrome, an autosomal recessive disease, associate oculo cutaneous albinism, dense-granule deficiency and the presence of ceroid-like inclusions in cells of the reticulo-endothelial system; Chediak-Higashi syndrome, an autosomal recessive disease, consists of partial oculo-cutaneous albinism, frequent pyogenic infections and giant lysosomal granules in many cell types. The bleeding disorder is generally moderate.

'Swiss-cheese' platelet disorder: This disorder is characterized by a mild bleeding diathesis.

Thrombocytopenia with absence of radii or TAR syndrome: a mild to life-threatening bleeding. The prognosis is related to the severity and duration of thrombocytopenia.

Thromboxane A$_2$ receptor deficiency: a moderate bleeding disorder.

von Willebrand disease (vWD): vWD is a mucocutaneous bleeding disorder. The mean vWF:Ag level in people with blood type O is approximately 25% lower than that found in people with other types. Prolonged bleeding time is not specific for vWD and does not predict bleeding at surgery. In type 3 vWD the bleeding disorder is very severe with spontaneous muco-cutaneous bleeding, hematoma and hemarthrosis.

Evelyne Dupuy and Jean-Marc Zini

References

1. al-Mondhiry H et al (1994) Am J Hematol 46:343-347
2. Anwar R et al (1998) Blood 91:149-153
3. Brenner B et al (1990) Br J Haematol 75:537-542
4. Canfield WM et al (1982) J Clin Invest 70:1260-1272
5. De Paepe A (1996) Thromb Haemost 75:379-386
6. Epstein CJ et al (1972) Am J Med:52:299-310
7. Everse SJ et al (1998) Thromb Haemost 80:1-9
8. Gabbeta J et al (1997) Proc Natl Acad Sci 94:8750-8755
9. George GN et al (1984) N Eng J Med 311:1084-1098
10. Gerard JM et al (1991) Blood 77:107-112
11. Giannelli F et al (1993) Nucleic Acids Res 21:3075-3083
12. Green D et al (1981) Br J Haematol 48:595-600
13. Guillin MC (1986) Ann NY Acd Sci 485:56-65
14. Hayward CPM et al (1997) Blood 89:1243-1253
15. Heinrich MC et al (1998) Blood 91:275-287
16. Hirata T et al (1994) J Clin Invest 94:1662-1667
17. Homans AC et al (1988) Br J Haematol 70:205-210
18. Hoyer LW (1994) Hemophilia A N Eng J Med:330:38-47
19. Kluft C et al (1982) Blood 59 1169-1180
20. Kurachi S et al (1994) Biochemistry 33:1580-1591
21. Lages B et al (1988) Thromb Haemost 59:175-179
22. Levy-Toledano S et al (1981) J Lab Clin Med 98:831-848
23. Lopez JA et al (1998) Blood 91:4397-4418
24. Miller et al (1991) Proc Natl Acad Sci USA 88:4761-4765
25. Miller JL et al (1982) Blood 60:790-794
26. Nieuwenhuis MK (1985) Nature 318:470-472
27. Nurden AT et al (1975) Nature 225:720-722
28. Nurden P et al (1995) In: Sampol J et al (eds) Manuel d'hèmostase. Elsevier, Paris, pp 277-302
29. Nurden P et al (1995) J Clin Invest 95:1612-1622
30. Okita JR et al (1989) Blood 74:715-721
31. Pugh R E et al (1995) Blood 85:1509-1516
32. Rao AK et al (1989) Blood 74:664-672
33. Roberts HR et al (1991) In: Hoffman R et al (eds) Hematology:basic principles and practice. Churchill Livingstone, NewYork, pp 1336-1337
34. Sadler JE et al (1994) Blood 84:676-679
35. Seeler RA (1972) Med Clin North Am 56:119-125
36. Shovlin CL (1997) Thromb Haemost 78:145-150
37. Soria J et al (1995) In: Sampol J et al (eds) Manuel d'hèmostase. Elsevier, Paris, pp 489-506
38. Weiss HG (1994) Semin Hematol 31:312-319
39. Weiss HJ et al (1993) Br J Haematol 83:282-295
40. Wu KK et al (1981) J Clin Invest 67:1801-1804

Blood Cells, Interaction with Vascular Cells

Introduction The endothelium is a large surface of exchange which is in direct contact with the blood. Blood cell-endothelium contacts are driven by mechanical forces but also by specific molecular mechanisms. Blood flow generates velocity profiles which are different according to blood cell microrheology but also to the cell size. Beside the physical forces chemoattraction is one of the factors directing the cell movement. More recently the molecular specificity was found to be a determinant factor modulating blood cell-endothelial cell interactions.

A review of leukocyte and erythrocyte interactions with the endothelium indicate that the consequences for the physiology of the endothelium and the rheology of the blood are of considerable importance in the general physiology of the organism. The perturbation of blood cell-endothelial cell interaction may initiate vascular dysfunction, stimulate an inflammatory response, induce thrombosis or favor the defense mechanism against infection but also be modulated by drugs [1, 2].

Characteristics   Since the observation of Metchnikoff describing leukocyte margination, several techniques have been used [3]. In animals intravital observation of leukocyte behavior was made on the vessel of the rat cremaster muscle or the hamster cheek pouch. In vitro systems have been improved by culturing human endothelial cells on plastic dishes or capillaries and the blood cell adhesion measured on these supports in static or dynamic flow conditions [4]. The adhering blood cells are quantified by direct visual counting, radiometric or enzymatic (peroxidase) techniques [5].

The leukocyte adhesion has been divided in three steps: "rolling" which consists of a slowing down of leukocytes, "firm adhesion" and "transmigration" through the endothelium. The molecular interaction which results in rolling consists of a binding between a leukocyte and an endothelial cell molecule; on the leukocyte side, sialylated proteins, sLex which is analogous to the blood group substance Lewis and L-selectin have been demonstrated to support rolling. On the endothelium P-selectin, L-selectin ligand, E-selectin, CD34 participate in the reaction resulting in rolling. Activation which leads to firm adhesion is mediated by chemokines, PAF, PECAM, E-selectin and chemoattractant receptors. Adhesion to the endothelium occurs when leukocyte integrins (β1, β2, β7) bind to ICAM, VCAM molecules.

In contrast to leukocytes, normal red blood (RBC) does not adhere to endothelial cells when adhesion is measured with endothelial cells in culture.

Regulation

Molecular Interactions

Leukocyte Adhesion   The first information regarding the molecular basis of leukocyte adhesion was obtained from the observation of patients with recurrent infections and leukocyte adhesion deficiency (LAD).

Three molecules were identified to be involved in leukocyte adhesion: LFA1 on lymphocyte, P150-95 on granulocyte and Mac 1 on monocyte [3]. These three molecules were shown to belong to a large family of molecules, the integrins. This is a large family of molecules found on different cell types, commonly possessing a α and β subunit. Five molecules belonging to the integrins are involved in the leukocyte-endothelial cell interactions. LeuCAM (Leukocyte Adhesion Molecules) or β2 integrins are the first three recognized molecules but now named CD11a/CD18 (LFA1), CD11b/CD18 (Mac1), CD11c/CD18 (P150/95). The group includes a β1 integrin VLA4: CD49d/CD29 and a β7 integrin (a1β7) which is a homing receptor for lymphocytes.

The development of endothelial cell culture allows the direct study of leukocyte adhesion to the endothelium and the characterization of the structures present on endothelial cells which mediate the attachment of leukocytes. Three major groups of molecules are expressed by endothelial cells and involved in leukocyte adhesion. They are members of the selectin, integrin and immunoglobulin families.

Intercellular adhesion molecule (ICAM-1) was first isolated and cloned [6]. ICAM-1 (CD54) with ICAM-2 (CD102) vascular cell adhesion molecule VCAM-1 (CD106) and platelet endothelial cell adhesion molecule PECAM-1 (CD31) are members of the immunoglobulin superfamily. ICAM-1 and ICAM-2 are constantly present on endothelial cells but can be overexpressed after cytokine stimulation. VCAM-1 is only detectable after induction by cytokines such as tumor necrosis factor (TNF) or interleukine-1 (IL-1) but can be found on dendritic cells.

PECAM-1 is present on various cell types: endothelial cells, platelets, leukocytes and can bind in homotypic reaction to PECAM-1. It plays a role in endothelial cell cohesion and is important for granulocyte and monocyte diapedesis [6].

Endothelial cell adhesion molecule-1 (ELAM-1) was identified on TNF and IL-1 stimulated endothelial cells and, when cloned, was shown to belong to a family of molecules which includes molecules present on platelet, leukocyte and complement regulatory molecules. This family of molecules is known as selectins. The endothelial cells expressed two selectins, E-selectin (ELAM-1, CD62E, endothelial) and P-selectin (CD62P, platelet). The third selectin L-selectin (CD62L, leukocyte) is present on leukocytes [4].

Antigens of the major histocompatibility complex (MHC) class I are present on endothelial cells. MHC class II is also expressed in endothelial cells after interferon stimulation [7] (Figure 1).

Erythrocyte Adhesion   see clinical relevance

Clinical Relevance

Leukocyte Adhesion Leukocyte adhesion deficiency (LAD) was known in patients with integrin deficiency. A second type of LAD was recently described in patients with sLex deficiency. These patients suffered from recurrent infections and had a Bombay blood group phenotype.

After the adhesion molecules were discovered, their possible alteration in various pathological situations was postulated and several investigations were conducted in inflammatory conditions, sepsis and cardiovascular disease.

The ability to bind via selectins is characteristic of all polymorphonuclear (PMN) neutrophils without stimulation, except in rare individual lacking enzymes required for production of ligands for endothelial selectins [5]. Attention, therefore, focuses on the ability to bind via integrins, which is activation dependent, both because constitutive integrins require conformational activation to bind efficiently and because de novo expression of integrins underlies prolonged attachment [4]. As an alternative to adhesion assays, measurement of the level of surface expression of integrins, CD11b and CD18 can be used to monitor adhesive potential. After activation,

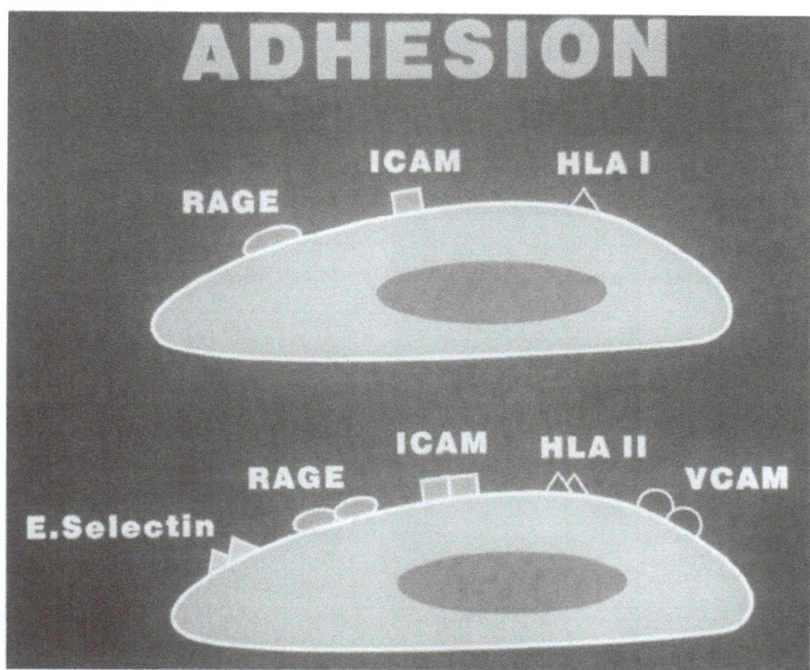

Figure 1. Adhesion molecule on resting (up) and activated (down) endothelial cells. RAGE, receptor for advanced glycated end products; ICAM, intercellular adhesion molecule; HLA, human leukocyte antigen; VCAM, vascular cell adhesion molecule.

CD11b rises more in proportion than CD18 (which exists as a heterodimer with other non-increasing integrin components CD11a and CD11c). In addition, the method can be applied to whole blood for clinical studies [8], where expression may be lower than on isolated cells because of the lack of preparative trauma. It should be noted, however, that expression itself is not equal with adhesion, because of the need for conformational activation and because functional activity is not strictly dependent on newly expressed CD11b/CD18.

Since it is known that L-selectin is shed from PMN upon activation, reduction in L-selectin expression should also influence activation. In fact, plasma contains quite a high level of the presumably-shed molecule, and an alternative to measuring CD11b in neutrophils has been to measure the level of soluble L-selectin. Perhaps unexpectedly, this level has been shown to correlate with a positive prognosis in septic shock. Much attention has been paid to the possibility that the plasma concentration of shed endothelial cell receptors for neutrophils may correlate with adhesive processes and hence reflect disease state or outcome. It is not clear at this stage whether this is a more specific or useful marker than other acute phase proteins such as the von Willebrand factor.

Erythrocyte Adhesion   Increased erythrocyte adhesion to the endothelium was first described in sickle cell anemia and diabetes mellitus [9, 10]. In these diseases, the extent of erythrocyte adhesion was correlated to the severity of the vascular complications [10]. Enhanced erythrocyte adhesion was found in malaria and was considered to be an important factor in the cerebral dissemination of the parasite (Figure 2).

Diabetes Mellitus   In diabetic patients, one of the biological consequences of the high plasma glucose concentration is the non-enzymatic glycosylation (glycation) of a variety of intra- and extracellular proteins. Erythrocyte membrane proteins are glycated and the membrane content in glucose is two-fold higher in diabetics as compared to non- diabetic subjects. Two major erythrocyte membrane proteins have been studied: the glycophorins and membrane protein band 3. Band 3, the major transmembrane protein, is easily accessible to glycation. Another protein, band 4.1, can also be glycated. Band 4 is a membrane skeletal protein which consists of two peptides, 4.1a and 4.1b. During the life span of erythrocytes, the ratio of 4.1a to 4.1b increases, thus constituting a marker of cell senescence. Some studies have suggested that 4.1 is synthesized as 4.1b and later converted to 4.1a. It has been estimated that 20 to 40 % of band 4.1 is glycated, although this modification has no apparent effect on the electrophoretic mobilities of the bands 4.1a and 4.1b. Whether the biological functions of the glycated band 4.1 protein are different from those of non-glycated band 4.1 remains as yet unknown [11].

Diabetic erythrocytes adhere abnormally to cultured human endothelial cells. The increase in adhesion is correlated with the extent of vascular complications as evaluated by the severity of retinopathy, neuropathy, peripheral vascular disease and myocardial ischemia. Fibrinogen and to a lesser extent fibronectin potentiate this adhesion of erythrocytes to human endothelial cells. The abnormal interaction also affects the endothelial cell arachidonic acid pathway, erythrocyte adhesion being correlated with an increase in PGI2 as assessed by the increase in release of 6-keto-PGF1 from the endothe-

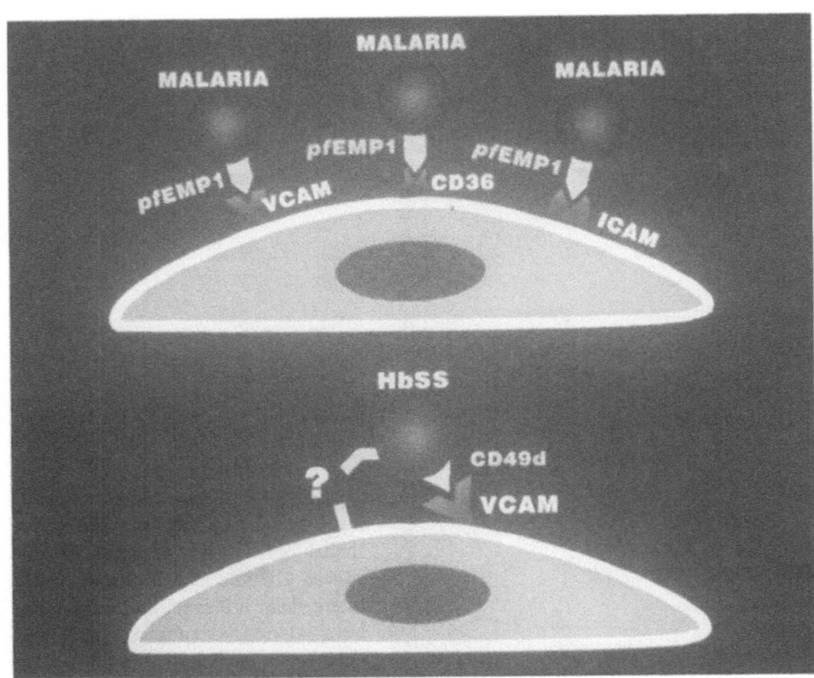

Figure 2. Erythrocyte adhesion to endothelium in Malaria and sickle cell anemia (HbSS). PfEMP, Plasmodium falciparum external membrane protein 1; ICAM, intercellular adhesion molecule; VCAM, vascular cell adhesion molecule.

lium. Recently, Schmidt et al [12] identified two receptors for the advanced glycosylation end product (AGE). These receptors are expressed on the macrophage and endothelial cell surfaces.

Recently, the molecular basis of increased erythrocyte adhesion to endothelium in diabetes mellitus was identified. Special structures present on endothelial cells were found to bind advanced glycated end products (AGE). One of the structures was found to be identical to Lactoferrin. A 35kDa molecule binds AGE protein and diabetic RBC(s). This molecule was purified, characterized and cloned and considered to be a receptor for AGE (RAGE). RAGE belongs to the immunoglobulin superfamily of membrane molecules. Antibodies directed against AGE or RAGE block the interaction between diabetic RBC and endothelium suggesting that RAGE is the receptor for the diabetic RBCs and that AGE present on the cell membranes are the counterpart [13]. A truncated form of the receptor (sRAGE) when preincubated with diabetic RBCs inhibited the adherence to endothelium in culture.

RBCs obtained from diabetic rats when injected in normal syngeneic rats had a reduced life span which was corrected when the animal received anti-RAGE antibodies while antithrombomodulin antibodies which recognize an endothelial cell component were without effect [14].

Diabetic RBCs bearing AGE when bound to endothelial cell RAGE induced an oxidant stress as evidenced by an increase in thiobarbituric reactive substances (TBARS) similarly when injected into liver TBARS. In addition to the TBARS level, AGE-RAGE interaction induces an activation of the nuclear factor Kappa B (NFKB). The endothelium barrier function was altered by AGE-RBC binding to RAGE. The vascular hyperpermeability observed in diabetic rats can be prevented by blocking RAGE with specific anti-RAGE antibodies. The vascular permeability measured by radiolabelled albumin transfer to the extravascular compartment is augmented when diabetic RBCs are infused in normal rats. This effect was blocked when diabetics were incubated with sRAGE before infusion in the animal.

To further evaluate whether oxidant stress was involved in the transmission of the intracellular message responsible for the enhanced permeability, we incubated the endothelial cells with antioxidant (Probucol, Vitamin E) or treated the animal with the two drugs. Both drugs reduced in vitro and in vivo the hyperpermeability produced by RBC AGE-RAGE interaction [15].

Several endothelial cell functions are modified by RAGE occupation. Vascular cell adhesion molecule expression is induced after AGE binding to RAGE, as is the case for tissue factor. AGE-RBC binding to endothelial cell potentiates Interleukin 6 production. All these alterations may play a determinant role in the genesis of vascular dysfunctions observed in diabetic vasculopathy [15].

Sickle cell anemia In sickle cell anemia the erythrocyte population is extremely heterogeneous and characterized by an increase in the proportion of the reticulocytes, a subpopulation of abnormally dense, less deformable cells showing extruding spicules at the cell surface. These spicules are composed of Band 3 proteins. In deoxygenated erythrocytes, Hb S polymers penetrate the membrane near the basal region of the spicules and after erythrocyte reoxygenation in vitro the spicules become round and detach from the cell membrane [16].

Erythrocytes from patients with sickle cell anemia exhibit an abnormal propensity to adhere to vascular endothelium or monocytes. As red cells from HbSS patients are widely heterogeneous, the abnormalities already considered may be partly responsible for their increased adhesion to the endothelium. However, plasmatic and vascular factors are also involved. Erythrocyte adhesion to endothelium may be studied in static systems or under dynamic conditions [17] with defined shear stress parameters. Sugihara et al. have recently proposed several distinct hypotheses to explain the interaction of sickle erythrocytes with the endothelium [17].

Sickle erythrocytes differ in their ability to adhere to large and microvascular vessels. Since HbSS erythrocytes adhere mainly to post capillary venules, obliteration starts with adhesion of young sickle cells to post-capillary vessels and occlusion is achieved by recruitment of irreversibly sickled erythrocytes. Flow modifications favor increased adherence and promote local hypoxia and acidosis. These biochemical and hemodynamic changes induce further sickling, adhesion and trapping of dense sickle erythrocytes, thus creating a self-maintained cycle of sickling. Such a cycle was in fact already proposed but without clear identification of the initiating factors. Erythrocytes from patients with HbSS adhere to endothelial cells of large vessel and microvascular sites by quantitatively and qualitatively different mechanisms and autologous plasma differently promotes adherence to microvascular and venous endothelial cells. High molecular weight von Willebrand factor, which is synthesized by endothelial cells, did not increase HbSS erythrocyte adherence to human microvascular or umbilical vein endothelial cells in culture, but increased their adhesion to bovine arterial endothelial cells. Recently, Swerlick et al. [18] postulated that the interactions between reticulocytes of patients with sickle cell anemia and endothelium may initiate vascular occlusion. Binding is mediated by the 4ß1-integrin complex (VLA4) which is expressed on the membrane of the reticulocytes and by VCAM-1 (vascular cell adhesion molecule-1) on the "activated" endothelial cell surface. This phenomenon could be important in cases of diseases with a high reticulocyte count.

Under static conditions, dense erythrocytes including irreversibly sickled cells are trapped in microvessels. Several factors are involved in the interaction between the endothelium and sickled erythrocytes. Plasma depleted in divalent cations or collagen binding proteins do not enhance the adherence of erythrocytes as well as a normal plasma, while fibrinogen and fibronectin have been identified as adhesion promoting factors and may represent the required collagen binding proteins [19]. Adhesion of sickle cells is enhanced by thrombospondin and abolished using specific antibodies against either thrombospondin or its receptor CD36. Finally, the abnormal interaction with sickle erythrocytes stimulates the production and release of prostacyclin by endothelial cells, as revealed by a significant rise of prostacyclin in plasma [20]. The increase in prostacyclin synthesis is correlated to the extent of sickle erythrocyte adhesion.

Plasmodium falciparum malaria   Clinical complications are favored by mechanical intravascular obstruction with erythrocytes containing asexual stages of the parasite. Post-mortem analyses of patients dying from cerebral malaria have revealed parasitized erythrocytes sequestrated within capillaries and post-capillary venules. The pathophysiological mechanism of parasite dissemination includes the abnormal erythrocyte adhesion and cell deformability.

The adhesion of erythrocytes to the endothelium is associated with the occurrence of knobs on the cell membrane. As observed by electron microscopy, these knobs appear regular and symmetrical and are thought to be the exclusive site at which the parasitized erythrocytes bind to the endothelium. However, parasitized erythrocytes without knobs have been shown to be capable of cytoadhesion *in vitro*. The knobs are composed of modified proteins, in particular of Histidine-Rich Protein which induces the excrescence below the lipid bilayer and binds to the submembranous cytoskeletal proteins spectrin and actin. High concentrations of band 3 protein, glycophorin and spectrin have also been detected in the knob regions. The human band 3 protein contains at least two exofacial regions that can serve as putative adhesins. The adhesive sequences of band 3 involve amino-acids 506–555 and amino-acids 821–834. Synthetic peptides based on these sequences inhibit the adhesion of *Plasmodium falciparum*-infected erythrocytes in vitro and affect sequestration [21]. *Plasmodium falciparum* Erythrocyte Membrane Protein 1 (PfEMP 1), a family of high molecular weight proteins (220–350 kDa) specifically located at knob sites on the outer membrane surface, presents a wide range of phenotypic antigens and is probably responsible in part for the immune response. Sera from immunized patients precipitate these antigens. The diversity in molecular weight is due to the addition of repetitive peptidic sequences involved in cytoadherence. PfEMP 1, considered as a high affinity adhesion molecule, is a candidate for a number of ligands such as CD36, ICAM-1 and thrombospondin. Recently, a recognition protein for CD36 termed sequestrin, was identified as a mediator in the adhesion of infected erythrocytes to the endothelium. This 270 kDa protein is a member of the high molecular weight antigens expressed on the surface of parasitized erythrocytes, and it seems to be involved solely in cytoadhesion and is only detected in wild-type *Plasmodium falciparum*. PfEMP 2 (MESA) appears in the knob regions at later stages of parasite maturation. Binding of this entity to the membrane skeleton involves band 4.1 protein [22], which is also associated with another protein located at knob sites, a phosphoprotein of about 85 kDa. *Plasmodium falciparum* induced alterations such as protein dephosphorylation inhibit new parasite invasion, as this process requires high levels of phosphoproteins.

The ability of parasitized cells to adhere *in vitro* to C32 melanoma cells, human umbilical vein endothelial cells or the macrophage cell line U937 does not appear to correlate with the severity of disease. Nash et al. [23] have found that adhesion, which is initially weak and shear reversible, is subsequently stabilized by cooperative effects of multiple receptor interactions. Three endothelial proteins have been identified *in vitro* as receptors for infected erythrocytes: thrombospondin, CD36 and ICAM-1 [24]. The precise role of these receptors in the pathogenesis of the disease is not yet clear. Thrombospondin is a complex glycoprotein of high molecular weight (420 kDa) composed of three identical chains linked by disulfide bonds. As a major component of platelet α-granules, it is synthesized by many adherent cell types such as endothelial cells and macrophages. It is a multifunctional protein involved in cell-cell interactions, carrying binding sites for fibrinogen and fibronectin. CD36 is a glycoprotein of 85 kDa expressed on the endothelial cell membrane and also on circulating monocytes macrophages and some tumor cell lines (leukemias and melanomas). This molecule is a receptor for thrombospondin and collagen type I. ICAM-1 (CD54) is a 95-112 kDa glycoprotein belonging to the immunoglobulin superfamily, which is widely distributed on cells of the immune system and on the endothelium. Its specific ligands are the integrins LFA-1 (Leukocyte Function-associated Antigen-1, CD11a/CD18) and Mac-1 (CD11b/CD18). However, ICAM-1 is ubiquitous and binds to the human rhinovirus. Using monoclonal antibodies (MAbs) which recognize different epitopic sites of ICAM-1, the molecular domain-1 has been identified as the binding site for infected erythrocytes and been shown to be distinct from the LFA-1 sites [25], while the domain-2 appears to differ from this binding site but to play an essential role in the structural conformation of the first domain. LFA-1 recognizes a molecular region bordering domains 1 and 2 but on the opposite side to the binding site for infected erythrocytes [25]. Ockenhouse et al. [26] have further demonstrated the importance of domains 1 and 2 in the interaction of ICAM-1 with infected erythrocytes by means of COS cells transfected with a cDNA encoding wild type or mutant ICAM-1. These transfected cells have been tested for the adhesion of malaria infected erythrocytes. Soluble ICAM and synthetic peptides derived from the amino-acid sequence of ICAM bind to infected erythrocytes. These results suggest that such molecules could be employed in the treatment of severe and cerebral malaria.

Among other endothelial cell molecules constituting putative receptors for malaria infected erythrocytes, adhesion molecules induced by cytokines may also mediate the cytoadhesion of parasitized cells to the endothelium. Adhesion is inhibited by antibodies against VCAM-1 and E-selectin (ELAM-1). E-selectin and VCAM-1 are present on the endothelium of post-mortem brain tissue from patients dying of cerebral malaria [26], while during clinical episodes of malaria, increases in the plasma levels of soluble ICAM-1 and E-selectin are positively correlated with plasma concentrations of soluble IL-2. The association of acute-phase proteins with inflammation marked by rises in cytokines such as IL-1ß and TNF, may create conditions which favor the adhesion of parasitized erythrocytes to the endothelium. Blood levels of

TNF may be increased in patients with malaria. This cytokine is frequently elevated in cerebral complications [27]. Its systemic effects include effects on the vasculative which may participate in the pathogenesis of the disease. Thus, cytokines activating endothelial cells induce the expression of receptors which can potentiate the adhesion of Plasmodium falciparum infested erythrocytes.

Jean-Luc Wautier

References

1. Dosquet C et al (1995) Journal of Cardiovascular Pharmacology 25:13-19
2. Zoukourian C et al (1996) International Angiology 3:195-200
3. Dosquet C, Wautier JL (1992) Clinical Hemorheology 12:817-829
4. Rainger GE et al (1996) Brit J Haematol 92:192-199
5. Weil D et al (1995) J Lab Clin Med 6:768-774
6. Carlos TM, Harlan JM (1994) Blood 84:2068-2101
7. Pober JS, Gimbrone MA (1982) Proc Natl Acad Sci (USA) 79:6641-6645
8. Setiadi H et al (1987) J Immunol 138:2330-2334
9. Hebbel RP et al (1980) N Engl J Med 302:992-995
10. Wautier JL et al (1981) N Engl J Med 305:237-242
11. Chappey O et al (1994) Nouv Rev Fr Hématol 36:281-288
12. Schmidt AM et al (1994) Arterioscler Thromb 14:1521-1528
13. Wautier JL et al (1994) Proc Natl Acad Sci (USA) 91:7742-7746
14. Wautier JL et al (1996) J Clin Invest 97:238-243
15. Chappey O et al (1997) Europ J Clin Invest 27:97-108
16. Hebbel RP et al (1989) Am J Physiol 256:C579
17. Sugihara K, Hebbel RP (1992) Clin Hemoreol 12:185-189
18. Swerlick KA et al (1993) Blood 82:1891-1899
19. Wautier JL et al (1983) J Lab Clin Med 101:911-918
20. Wautier JL et al (1986) J Lab Clin Med 107:210-215
21. Crandall I et al (1993) Proc Natl Acad Sci 90:4703-4707
22. Coppel RL (1992) Mol Biochem Parasitol 50:335-348
23. Nash GB et al (1992) Blood 79:798-807
24. Berendt AR et al (1989) Nature 341:57-59
25. Berendt AR et al (1990) Cell 68:71-81
26. Ockenhouse CF et al (1992) J Exp Med 176:1183-1189
27. Grau GE et al (1989) N Engl J Med 320:1586-1591

Blood Coagulation

Definition *Process that leads to clot formation. The blood coagulation system is constituted of a series of biochemical reactions that lead to the formation of fibrin.*

See: →Coagulation factors; →Fibrinolytic, hemostatic and matrix metalloproteinases, role of; →Thrombosis

BSS

Definition *Bernard-Soulier syndrom*

See: →Bleeding disorders

Cadherins

Calcium-dependent intercellular adhesion molecules. Three main types have been described, Epithelial

Cadherin (E-cadherin), Neural cadherin (N-cadherin) and vascular endothelial cadherin (VE-cadherin).

See: →Signal transduction mechanisms in vascular biology

CAI

Definition *Carboxyamido-triazol*

See: →Angiogenesis; →Angiogenesis inhibitors

CAM

Definition *Chorioallantoic membrane*

See: →Angiogenesis; →Angiogenesis inhibitors; →Atherosclerosis

cAMP

Definition *Cyclic adenosine monophosphate*

See: →Fibrinolytic, hemostatic and matrix metalloproteinases, role of; →Nitric oxide; →Platelet stimulus-response coupling; →Vasomotor tone regulation, molecular mechanisms of

CD142

See: →Procoagulant activities

Cell Cycle

Definition *In order to divide, cells pass through different phases that include G1, S (DNA synthesis), G2 and M (mitosis) which together form the cell cycle. The major regulators of the cell cycle are cyclins/cyclin-dependent kinases which are connected to the signalling machinery.*

See: →Signal transduction mechanisms in vascular biology

CETP

Definition *Cholesteryl ester transfer protein*

See: →Lipoproteins

CFU

Definition *Colony-forming unit*

See: →Megakaryocytes

CFU-MK

Definition *Megakaryocytic colony forming unit*

See: →Megakaryocytes

cGMP

Definition *Cyclic guanidine monophosphate*

See: →Cyclic nucleotides; →Nitric oxide; →Platelet stimulus-response coupling; →Vasomotor tone regulation, molecular mechanisms of

Chemokines

Definition *Family of regulatory molecules. Different subtypes have been described based on the occurrence of the cysteine motif. Among CXC chemokines, some contain an ELR motif and have stimulatory properties while others are ELR negative and are inhibitory.*

See: →Angiogenesis; →Angiogenesis inhibitors

CHO

Definition *Chinese hamster ovary*

See: →FGF receptors; →FGF-1, FGF-2; →Vascular integrins; →von Willebrand factor

Chronic Inflammatory Disorders

Definition *Family of chronic diseases characterized by a non-resolved process of inflammation*

For mouse models see: →Fibrinolytic, hemostatic and matrix metalloproteinases, role of

CL

Definition *Corpus luteum*

See: →Hormonal regulation of vascular cell function in angiogenesis

cNOS

Definition *Constitutive nitric oxide synthase*

See: →Nitric oxide

Coagulation Factors

Roman numbers identify the majority of coagulation factors However, several coagulation factors have a synonym that in some cases is associated with a disease state or the first individual identified with the deficiency of the corresponding factor. The synonym for each coagulation factor (when applied) will be always given below in parenthesis.

Definition *Serine proteases that constitute the extrinsic and intrinsic blood coagulation pathway.*

See also: →Fibrin/fibrinogen; →Fibrinolytic, hemostatic and matrix metalloproteinases, role of; →Procoagulant activities; →Thrombin; →Thrombosis

Introduction   Following vascular injury, the blood clotting process is initiated in order to stop blood from leaking outside the vasculature. Blood clotting involves a multitude of proteins, which act in concert to produce the procoagulant enzyme α-thrombin, which in turn is responsible for the generation of the fibrin plug. However, while generation of the fibrin plug is required for the arrest of excessive bleeding, unregulated clotting will result in the occlusion of the vessels and thrombosis. Thus, the regulation of the delicate balance between the procoagulant and anticoagulant mechanisms is of extreme importance for survival. The maintenance of hemostasis in human blood is a complex event, requiring the controlled interaction of protease, zymogens, cofactors and inhibitors on surfaces presented to the blood by the platelets, blood cells and vessel wall. Procoagulant events must be regulated and are localized to damaged regions of vessels; otherwise, disseminated coagulation throughout the vasculature will result in the occlusion of blood flow to tissues and organs. Blood clotting occurs when platelets adhere and aggregate at sites of injury, leading to the exposure of procoagulant membrane surface which localizes and concentrates the proteins required for normal hemostasis and formation of polymeric fibrin strands which stabilize the initial platelet mass. These events are dependent upon the conversion of prothrombin to α-thrombin, the central enzyme of blood coagulation. The factors, which regulate thrombin generation, are manifold, and diseases of abnormal coagulation are often associated either with defects in the pathway leading to thrombin or in the processes, which regulate thrombin production. Deficient or unrestricted thrombin generation can lead to bleeding (hemorrhage) or vessel occlusion (thrombosis, infarction, etc.), respectively. Thus, an understanding of the origins of hemorrhagic and thrombotic disease must be rooted in an understanding of the factors regulating the thrombin response.

Theories of blood coagulation have recognized that initiation of thrombin generation and clotting of plasma *in vitro* arises from two distinct sources: the contact (or intrinsic) pathway and the tissue factor (or extrinsic) pathway. First Paul Morawitz in 1905 [1] proposed that tissue thromboplastin is required for the initiation of blood clotting and established the basis for the extrinsic pathway of blood coagulation. Davie and Ratnoff [2] later outlined the classical "cascade" models of the intrinsic pathway of blood coagulation. Both pathways are characterized by a series of reactions initiated by the presentation of a procoagulant surface to the blood flow. No procoagulant complex can by assembled in the absence of an adequate membrane surface. Every reaction in the series involves the activation of an inactive plasma protease precursor (zymogens) via limited proteolysis to an activated plasma protease. Each newly activated protease activates other zymogens later in the sequence, producing a string of activation processes, which culminate in thrombin generation and subsequent clot formation. However, the activity of all of the complexes is also limited by the availability of an adequate membrane surface provided by endothelial cells, platelets, and monocytes. The cell surface provides a site for the recruitment of the appropriate proteins and allows for fast and efficient clot formation. Subsequent investigations detailed the architecture of each active enzyme at the procoagulant surface as an assembly, requiring calcium, a proenzyme activated by proteolysis with a cofactor also activated by proteolysis. Thus, timely exposure of the adequate membrane surface is an important step in the regulation of α-thrombin formation and since each reaction in the series requires this surface-dependent assembly for efficient expression of activity, coagulation can be modulated in vivo by regulation of either the surface presented to the blood, as in platelet activation; the protease component, by inhibitors, or the cofactor component.

Structure and Characteristics

Molecular Weight; Domains

Antithrombin III (AT-III)   AT-III contains an NH_2-terminal heparin binding domain, a carbohydrate rich domain, and a COOH-terminal serine protease-binding domain (Figure 1). AT-III is a member of the serine protease inhibitor (SERPIN) family which inhibit serine proteases. AT-III (M_r 58,000) circulates in plasma at a concentration of 150-400 µg/ml (2.6 µM-3.4 µM).
AT-III contains a COOH-terminal arginine-serine reactive site that interacts with the serine proteases of the blood coagulation process as well as two positively charged clusters of amino acid residues that bind sulfated polysaccharides (Figure 1).

Factor II (Prothrombin)   Prothrombin circulates in plasma at a concentration of 1.4 µM as an inactive zymogen of M_r 72,000 [3]. The NH_2-terminal portion of the molecule contains the γ-carboxyglutamate residues, which are involved in the metal binding properties of the molecule and the two kringle domains (Figure 2). The entire catalytic domain is located at the COOH-terminus of the molecule. Prothrombin is a vitamin-K dependent protein. The light chain includes 46 amino acid residues containing the γ-carboxyglutamate residues, which are required for the metal binding and the proper interaction of the molecule with the membrane surface. Two kringle domains of 79 amino acids each follow this region. The serine protease domain is localized at the COOH-terminal domain, is composed of 259 amino acids and contains the catalytic triad which is responsible for the enzymatic activity of the molecule (Figure 2).

Factor V (Proaccelerin)   Human factor V circulates in plasma at a concentration of 20 nM [4], as a large single chain procofactor with a M_r 330,000 [5]. The cDNA and deduced amino acid sequence shows that the molecule

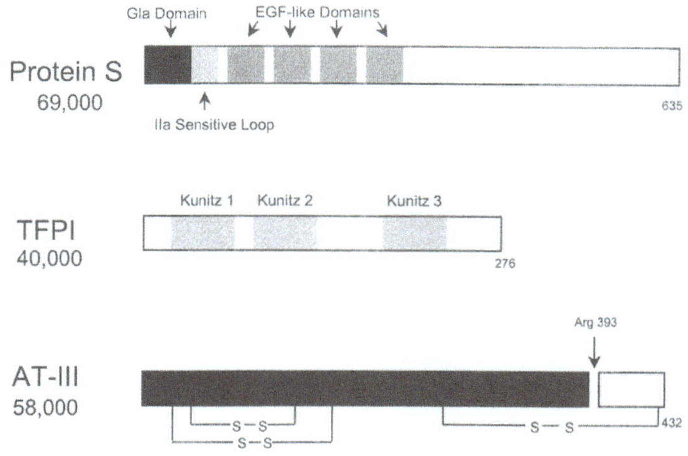

Figure 1. *Schematic representation of the organization of the protein S, TFPI, and ATIII molecules.* The specific domains are indicated. (Reprinted with permission from The Regulation of Clotting Factors by Kalafatis et al. In: Critical Reviews in Eukaryotic Gene Expression, 7(3):241-280 (1987), Ed Begell House Inc).

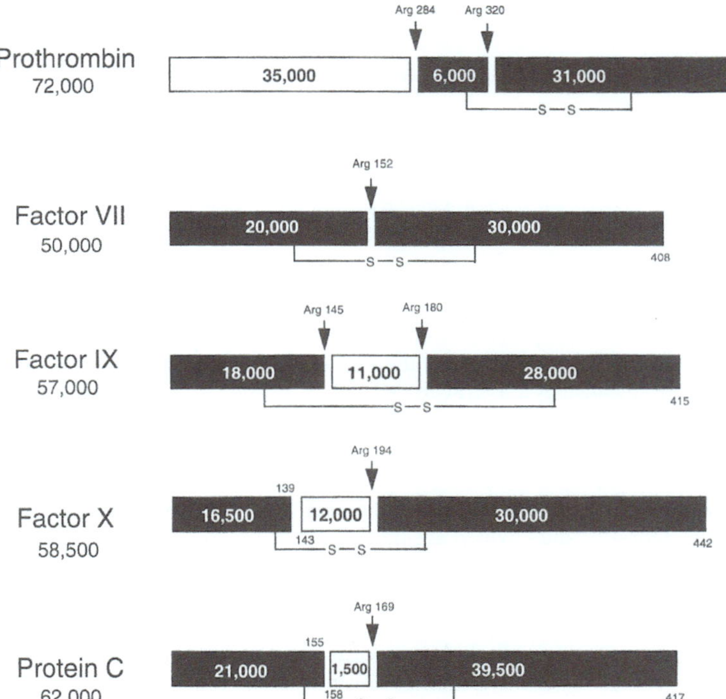

Figure 2. *Schematic representation of the structure of the vitamin K-dependent proteins.* The active molecule is identified in each case by the black shading. The specific activating cleavage sites are also shown. The NH_2-terminal portion of each protein contains the γ-carboxyglutamic acids which are required for the metal binding of the proteins (not shown). (Reprinted with permission from The Regulation of Clotting Factors by Kalafatis et al. In: Critical Reviews in Eukaryotic Gene Expression, 7(3):241-280 (1987), Ed Begell House Inc).

contains triplicate "A" domains which share a high degree of homology with ceruloplasmin, duplicated "C" domains which share homology with discoidin and a connecting B region which functions as an activation peptide (Figure 3).

Factor V is cleaved by α-thrombin to generate the active cofactor, and two heavily glycosylated activation peptides [6] (Figure 3). The factor Va molecule is a heterodimer composed of a heavy chain of M_r 105,000 and a light chain of M_r 74,000. Both subunits are non-covalently associated via a divalent metal ion process (Figure 3). The heavy chain contains the NH_2-terminus of the procofactor (residues 1-709) and is composed of two "A" domains (residues 1-303, and 317-656), connected by an amino acid region containing basic amino acids (residues 304-316). The COOH-terminal portion of the heavy chain (residues 657-709) is rich in acidic amino acids. The light chain of the cofactor contains the COOH-terminus of the factor V molecule (residues 1546-2196) and is composed of one "A" domain (residues 1546-1877) and two "C" domains (residues 1878-2036, and 2037-2196). Factor Va has 18 cysteines residues (Fig 3). Four of these cysteines are present as free -SH while the remaining fourteen are involved in disulfide bridges as follows (Figure 3): three 26 amino acid residue α-loops are present, one in each "A" domain; the "A1" and "A2" domains also contain one

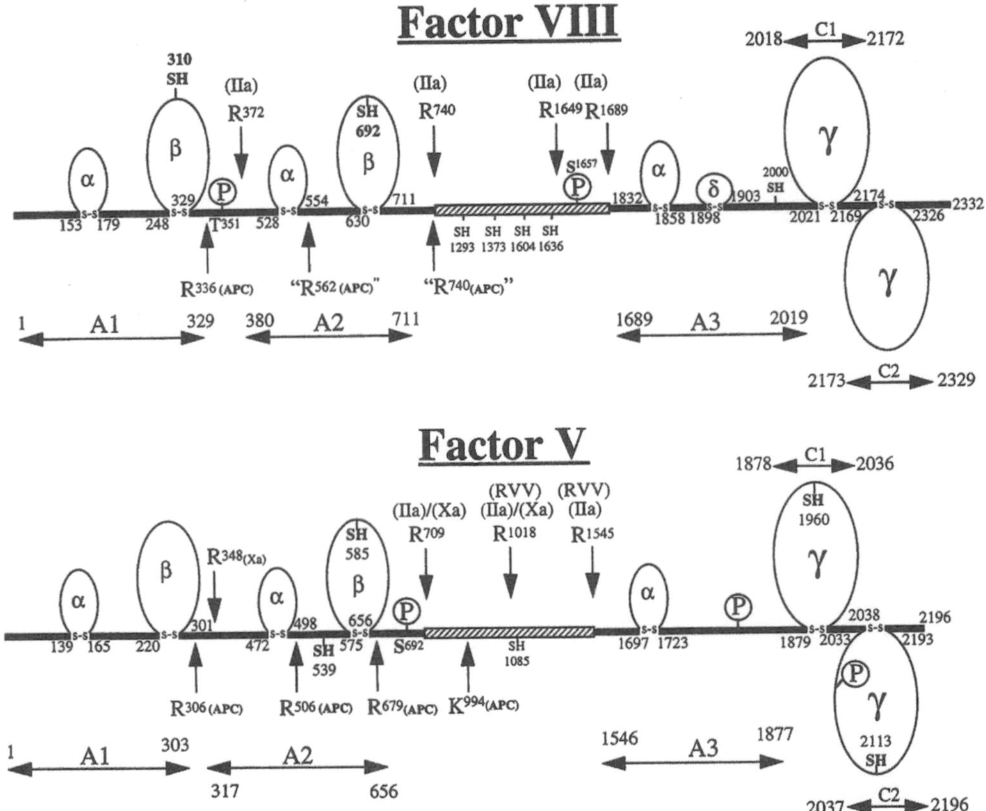

Figure 3. *Structural features of factor VIII and factor V.* The activating cleavage sites of the procofactors are shown on top. The APC-inactivating cleavage sites that are required for factor V/Va inactivation by APC are shown at the bottom. The APC cleavage sites for factor VIII are also shown. The location of the free cysteines (indicated by SH) and of the disulfide bridges in each molecule are also shown. The phosphorylation sites of both procofactors are identified by a "P". (Reprinted with permission from The Regulation of Clotting Factors by Kalafatis et al. In: Critical Reviews in Eukaryotic Gene Expression, 7(3):241-280 (1987), Ed Begell House Inc).

β-loop each of which composed of 82 amino acids; two γ-loops which are composed of 154 and 155 residues represent the "C1" and "C2" domains. The factor V molecule has multiple potential N-linked glycosylation sites in the B region and on the heavy and light chains that are required for secretion. The molecule is also phosphorylated by a membrane associated casein kinase II-like enzyme on the heavy chain (at Ser^{692}) and by a protein kinase C isoform on 2 sites of the light chain (Figure 3). Phosphorylation at Ser^{692} is required for fast and efficient inactivation of the cofactor by APC. Finally, Tyr^{665}, Tyr^{696}, Tyr^{698}, Tyr^{1494}, Tyr^{1510}, Tyr^{1515}, and Tyr^{1565} are believed to be sulfated. Inhibition of sulfation by sodium chlorate results in a cofactor molecule 5-fold less active that the native molecule. It has been thus suggested that sulfation of factor V is important for α-thrombin activation of the procofactor. It has been also suggested that sulfation may by important for the interaction of factor Va with factor Xa.

Factor VII Factor VII circulates in plasma predominantly as a single chain inactive zymogen (M_r 50,000) at a concentration of 10 nM [7]. The NH_2-terminal part of the molecule contains the γ-carboxyglutamate residues, which are involved in the metal binding properties of the protein and a β-hydroxyaspartic acid at position 63 (Figure 2).

Factor VII is a vitamin-K dependent protein. The light chain includes 49 amino acid residues containing the γ-carboxyglutamate residues, which are required for the metal binding and the proper interaction of the molecule with the membrane surface (Figure 2). Two epidermal growth factor (EGF) domains of 32 and 37 amino acids respectively follow this region. The serine protease domain is localized at the COOH-terminal domain, is composed of 254 amino acids and contains the catalytic triad, which is responsible for the enzymatic activity of the molecule.

Factor VIII (Antihemophilic Factor) Human factor VIII (M_r 280,000) circulates in plasma at 0.7 nM [8] in non-covalent association with von Willebrand factor (vWf) (M_r~15x10^6). Deficiency of factor VIII constitutes one of the most common inherited coagulation disorders (hemophilia A). The cDNA and deduced amino acid sequence of factor VIII shows that the molecule is composed of triplicated "A" domains, duplicated "C" domains and a B region (Figure 3).

Human factor VIII (M_r 285,000) circulates in plasma as a dimer composed of a multitude of heavy chain fragments (M_r 90-200,000) and a light chain (M_r 80,000) [9]. α-Thrombin converts factor VIII to an heterotrimer. Factor VIIIa is a composed of fragments of M_r 50,000 (A1 domain) containing a COOH-terminal portion rich in acidic amino acids (residues 337-372), M_r 40,000 (A2 domain), and M_r 74,000 (A3C1C2) containing the carboxyl-terminal part of the factor VIII molecule (Figure 3). The three subunits are non-covalently associated via divalent metal ions. Factor VIIIa contains 19 cysteines residues (Figure 3). Three of these cysteines are present as free -SH while the remaining sixteen are involved in disulfide bridges. Three γ-loops are present, one in each "A" domain. The "A1" and "A2" domains also contain β-loops. Two γ-loops encompass the "C1" and "C2" domains of the light chain of factor VIIIa. A small δ-loop is also present in the A3 domain of the cofactor (Figure 3).

Factor VIII is subject to several post-translational modifications. The molecule contains several potential N-linked glycosylation sites that are necessary for secretion and is phosphorylated by a platelet membrane casein kinase II at Thr^{351} and Ser^{1657}. Tyr^{346}, Tyr^{718}, Tyr^{719}, Tyr^{723}, Tyr^{1664}, and Tyr^{1680} are sulfated and sulfation at residues 346 and 1664 increases the rate of factor VIII activation by α-thrombin, whereas sulfation at residues 718, 719, and 723 is required for optimum interaction of factor VIIIa with the components of the intrinsic tenase. Finally, sulfation of Tyr^{1680} is necessary for the proper interaction of factor VIII with vWf.

Factor IX (Christmas Factor)   Human factor IX circulates in plasma as a single chain zymogen (M_r 55,000) at a concentration of 90 nM [10]. Deficiency of this glyco-protein constitutes an inherited coagulation disorder called hemophilia B. The NH_2-terminal part of the molecule contains the γ-carboxyglutamate residues, which are involved in the metal and membrane binding properties of the protein (Figure 2).

Factor IX is a vitamin-K dependent protein. The light chain includes 50 amino acid residues containing the γ-carboxyglutamate residues which are required for metal ion binding and the proper interaction of the molecule with the membrane surface. This region is followed by two EGF domains of 32 and 37 amino acids respectively. The serine protease domain is localized at the COOH-terminal region, is composed of 235 amino acids and contains the catalytic triad which is responsible for the enzymatic activity of the molecule (Figure 2).

Factor X (Stuart Factor)   Human factor X circulates in plasma as an inactive zymogen (M_r 59,000) [11] composed of a heavy chain (M_r 42,000) and a light chain (M_r 16,500) which are covalently associated through a disulfide bond at a concentration of 170 nM. The light chain contains the γ-carboxyglutamate residues, which are required for the metal binding and the proper interaction of the molecule with the membrane surface (Figure 2).

Factor X is a vitamin-K dependent protein. The light chain includes 49 amino acid residues containing the γ-carboxyglutamate residues which are required for metal ion binding and the proper interaction of the molecule with the membrane surface. This region is followed by two EGF domains of 32 and 36 amino acids respectively. The serine protease domain is localized at the COOH-terminal domain, is composed of 254 amino acids and contains the catalytic triad, which is responsible for the enzymatic activity of the molecule (Figure 2).

Factor XI   Factor XI circulates in plasma at a concentration of 30 nM [12] in a non-covalent complex with high molecular weight kininogen. Factor XI (M_r 160,000) contains two identical polypeptide chains (M_r 80,000) associated through disulfide bridges.

The mature factor XI molecule is a two-chain homodimer that are linked by disulfide bonds. A monomer is composed of four tandem amino acid sequence repeats, which constitute the heavy chain of factor XIa. The catalytic domain, which is located at the carboxyl terminal portion of the molecule, will constitute the light chain of factor XIa. Following activation the proenzyme will give rise to an enzyme with two active sites per molecule.

Factor XII (Hageman factor)   Factor XII has a M_r 85,000 and circulates in plasma at a concentration of 375 nM. Factor XII is single chain protein of 596 amino acids. The protein was found to contain multiple domains, which have considerably homology with the EGF and kringle regions of tissue-type plasminogen activator (tPA) and fibronectin.

Factor XIII (Plasma Transglutaminase)   Factor XIII is a heterodimer of M_r 320,000 composed of two subunits: the α subunit (M_r 82,000) and the β subunit (M_r 76,500). Factor XIII ($α_2β_2$) circulates in plasma at a concentration of 70 nM. Factor XIII is a dimer composed of two non-identical subunits α and β with the overall structure $α_2β_2$. The active site is on the α chain whereas the β chain functions as a carrier protein.

Fibrinogen   Fibrinogen is a M_r=340,000 protein which is present in plasma at a concentration of 7 μM [13]. Fibrinogen is composed of three pairs of polypeptides, which are linked via disulfide bridges. The three polypeptides are: Aα (M_r 66,500), Bβ (M_r 52,000) and γ (M_r 46,500).

Fibrinogen is a disulfide linked dimer composed of 3 pairs of disulfide linked non-identical polypeptide chains (Aα, Bβ, and γ). The two subunits are aligned in an antiparallel manner forming a trinodular arrangement of the six chains. The nodes are formed by disulfide bridges between the three chains. The Aα chain contain an NH_2-terminal fibrinopeptide A [amino acid residues 1-16], the factor XIIIa cross-linking sites and 2 phosphorylation sites. The Bβ chain contains the fibrinopeptide B [amino acid residues 1-14] one N-linked carbohydrate moiety and an NH_2-terminal pyroglutamic acid. The γ chain contains all the other N-linked glycosylation sites and a factor XIIIa cross-linking site.

High molecular weight kininogen (Fitzgerald factor)
Two forms of kininogens are found in plasma. The high molecular weight kininogen circulates in plasma at 670 nM as a single chain protein with a M_r 120,000 whereas, the low molecular weight kininogen (M_r 66,000) circulates in plasma at a concentration of 2.4 μM.

The kininogens are proteins which are composed of multiple domains (D1–D6) and can be divided into three structural portions: a heavy chain that is common to the high and low molecular weight kininogens, the bradykinin portion, and the light chains that are unique to each of the two forms of kininogens. Domains 1–3 constitute the kininogens' heavy chain and domain 4 is the bradykinin moiety. Domain 5 is a unique domain that represent the light chain of low molecular weight kininogen, whereas domains 5 and 6 are unique to the light chain of the high molecular weight kininogen.

Prekallikrein (Fletcher Factor)   Prekallikrein, which is a single chain protein with a M_r 85,000, circulates in plasma at a concentration of approximately 500 nM. Prekallikrein is composed of a 371 amino acid heavy chain and a 248 amino acids light chain. Both chains are held together by a disulfide bond. The amino acid sequence of prekallikrein is 58 % homologous to the amino acid sequence of factor XI. Prekallikrein has four tandem repeats located at the amino terminal portion of the molecule. Because of the intrachain disulfide bond arrangement the molecule has four "apple" domains each composed of 90 or 91 amino acids. These structures have been also observed in factor XI.

Protein C   Protein C which is a vitamin K-dependent protein of M_r 62,000 (Figure 2) [14] circulates in plasma at a concentration of 60 nM and is composed of a heavy chain (M_r 41,000) and a light chain (M_r 21,000) which are associated covalently through a disulfide bond. The light chain contains the γ-carboxyglutamate residues, which are required for the metal binding and the proper interaction of the molecule with the membrane surface. Single chain protein C represents approximately 20 % of the circulating plasma zymogen suggesting that the protein is initially generated as a single chain protein which is converted to a two chain form by an as yet unidentified protease. Protein C is a vitamin K-dependent zymogen with γ-carboxyglutamate residues located on the NH_2-terminal portion of the light chain. These residues are required for the proper interaction of the protein with the membrane surface and its physiological function. After these residues there is a stack of hydrophobic residues, two EGF repeats (40 and 36 amino acids respectively) and a protease domain (250 amino acids). A β-hydroxyaspartic acid residue is produced by a post-translational modification in the EGF domain. This residue is required for Ca^{2+}-dependent alteration in the molecule (Figure 2).

Protein S   Plasma protein S which is another vitamin K-dependent protein (300 nM, M_r 69,000) exists in two states, free-protein S (~150 nM) and in complex with the C4B-binding protein (C4BP) of complement (150 nM). Protein S is also present at low concentrations in platelets (Fig 1).

Protein S is composed of an NH_2-terminal "Gla"-domain, a thrombin sensitive region, four EGF-like structures and a COOH-terminal portion, containing 60 % of the whole molecule (Figure 1). Protein S was initially identified due to the presence of the γ-carboxyglutamic acid residues [15]. Plasma protein S exists in two states, free-protein S and in complex with the C4B binding protein (C4BP) of complement. Protein S is also present at low concentrations in platelets. The anticoagulant properties of protein S are only observed for the free fraction of the protein and are the object of intense investigation.

Thrombomodulin   Human thrombomodulin (M_r~ 100,000) is an endothelial cell surface glycoprotein that exhibits similarity to the low-density lipoprotein receptor [16]. Thrombomodulin is composed of 559 amino acids and contains five domains: an NH_2-terminal lectin-like domain (residues 1–224) containing a potential α-thrombin binding site, six EGF-like domains (residues 225-461), a 34 amino acid segment which contains eight hydroxyl-amino acids, several potential glycosylation sites for O-linked carbohydrate in a region rich in threonine and serine residues (residues 462–496), a hydrophobic membrane spanning domain (residues 497-520), and a cytoplasmic tail (residues 521-559).

Tissue Factor (TF)   TF (has M_r 29,600) is a membrane-bound glycoprotein, which is exposed at the site of vascular injury and, in association with the serine protease factor VIIa, initiates the blood coagulation cascade [17]. Tissue factor is selectively expressed by a variety of vascular and extravascular cells including monocytes, macrophages and endothelial cells. Human tissue factor is composed of 263 amino acids and contains three domains: a putative extracellular domain which contains residues 1–219 (soluble tissue factor), an extremely hydrophobic (region residues 220–242) which is the transmembrane domain, and a cytoplasmic domain (amino acids 243–263) which contains a potential N-linked glycosylation site and one cysteine covalently associated with a palmityl or stearyl thioester.

Tissue Factor Pathway Inhibitor (TFPI)   The normal plasma concentration of TFPI is approximately 2.5 nM [18]. TFPI in human plasma (M_r 34,000) is very heterogeneous. (Figure 1). The full-length TFPI has an acidic amino-terminal region, three tandem repeated serine protease inhibitor domains which are structuraly homologous to the Kunitz domain of the trypsin inhibitor, and a basic COOH-terminal domain.

Gene

Chromosomal Localization; Gene Structure, Expression, and Regulation

Antithrombin III (AT-III)   The 15 kb gene encoding AT-III is located on chromosome 1 and has seven exons. The promoter region does not possess a TATA box at -25.

There is an 8-bp segment at the 5' flanking region that exhibits high homology with the J_κ-C_κ enhancer element of the human Igκ chain. This element appears to be critical for the efficient synthesis of AT-III. Transcription of the AT-III gene is initiated at a site located 72-bp upstream of the ATG initiation codon and results in a mRNA of 1.8 kb.

Factor II (Prothrombin)   Prothrombin is the product of a 21 kb gene, which is located on chromosome 11. Transcription of the prothrombin gene gives rise to a 2 kb mRNA transcript [19]. Prothrombin is primarily synthesized in the liver. The 5' flanking region of the prothrombin gene contains potential binding sites for several transcription factors including EBP20, Sp1, Ap-1, and CTF/NF-1. Low level expression of prothrombin has also been reported in the brain, diaphragm, stomach, kidney, spleen, intestine, uterus, placenta and adrenal tissues.

Factor V (Proaccelerin)   The factor V gene is located on chromosome 1 at q21-25. The factor V gene is 80 kb in length, contains 24 introns and following transcription gives rise to a 6.8 kb mRNA [20,21]. Synthesis of factor V has been demonstrated by bovine aortic endothelial cells, a human hepatocarcinoma cell line, HepG2, and by human and guinea pig megakaryocytes, the precursor cells of circulating platelets. The factor V antigen level in megakaryocytes has been shown to increase following treatment with PMA. This increase in antigen is at the protein level and is not associated with an increase in mRNA levels. Platelet factor V (20% of the circulating factor V) is also localized in the platelet α-granules [4].

Factor VII   The factor VII gene is located on chromosome 13q34 [22]. The 12.8 kb gene for factor VII is located 3 kb upstream of the gene for factor X and encodes a 2.4 kb mRNA transcript. The factor VII gene is expressed only in the liver. The 5' flanking region of the factor VII gene contains binding sites for the liver enriched transcription factor HNF-4. This region also contains 3 potential AP-1 binding sites. The upstream regulatory region also contains a putative C/EBP-β/NF-IL6 binding site. This site is not important for expression of the factor VII gene because IL6, IL1-β, and IL6 plus dexamethasone do not affect mRNA levels.

Factor VIII (Antihemophilic Factor)   The factor VIII gene is present on the X chromosome at q28 [23]. The factor VIII gene is 187 kb in length, contains 25 introns and gives rise to a 9 kb mRNA transcript [17]. While the introns present in the factor VIII gene are larger than the introns in factor V with the exception of intron 6, all of the intron exon boundaries are at approximately the same amino acid residue position. Hepatocytes are the major source of plasma factor VIII. Expression of factor VIII can also be detected in a variety of other cell types. DNase I footprint analysis revealed 19 protein binding sites in the -1175 to -9 region of the factor VIII promoter. This region contains all of the required elements for maximal promoter activity. Cis acting elements in this region include binding sites for HNF-1, NFκB, C/EBPα, C/EBPβ, and HNF-4.

Factor IX (Christmas Factor)   The 34 kb factor IX gene is located on the X chromosome and is transcribed into a mRNA of 2.8 kb [24]. Five cis acting elements have been identified in the proximal promoter region of the factor IX gene. The 5' and 3' cis acting sites are liver specific binding sites which bind the transcription factor C/EBP α. The other cis acting sites in the proximal promoter have been shown to bind the transcription factors NF-1 and HNF-4. The factor IX proximal promoter region also contains a consensus androgen response element sequence.

Factor X (Stuart Factor)   The gene encoding factor X is on chromosome 13 [25]. The gene is 27 kb in length and encodes a 1.5 kb mRNA. The gene for Factor X is located near the gene for factor VII. Factor X is expressed primarily in the liver. The sequences present in the 279 bp upstream from the AUG site are sufficient for maximal promoter activity in Hep G2 cells.

Factor XI   The factor XI gene is 23 kb in length and is located on chromosome 4 at q35. The factor XI gene encodes a 2.1 kb mRNA transcript [26].

Factor XII (Hageman factor)   The gene for factor XII is located on chromosome 5, is 12 kb long and encodes for a 2.4 kb mRNA. Human liver and rat hepatocytes were found to synthesize factor XII. In postmenopausal and pregnant women estrogens enhance the concentration of the protein.

Factor XIII (Plasma Transglutaminase)   The 160 kb gene for the α chain of factor XIII is located on chromosome 6 at p24-25 and encodes a 3.8 kb mRNA transcript [27]. The gene for the β chain is 28 kb in length and is located on chromosome 1 at q31-32 and encodes a mRNA transcript of 2.3 kb.

Fibrinogen   The three genes which code for the three fibrinogen chains are located within a 50 kb length of DNA on chromosome 4 at q23-q32. The gene for the α-chain is 5.5 kb in length and encodes a 2.2 kb mRNA transcript. The 8 kb β-chain gene encodes a 1.9 kb mRNA, and the 8.5 kb γ-gene gives rise to a 1.6 kb mRNA transcript [28].

High molecular weight kininogen (Fitzgerald factor) The two forms of kininogens circulating in plasma are products of a single gene, which is localized to chromosome 3q26-qter producing an mRNA of 27 kb. The gene consisting of 11 exons produces the mRNA for both high and low molecular weight kininogen by alternative splicing. Both molecules share the coding sequence of the first nine exons. Exon 10 contains the coding sequence for the M_r 56,000 light chain of high molecular weight kininogen whereas exon 11 encodes for the M_r 4,000 light chain of low molecular weight kininogen.

Prekallikrein (Fletcher Factor)   The gene for prekallikrein is localized on chromosome 4 and is 22 kb long. The gene is composed of 15 exons and 14 introns. Computer analysis of the 5'-promoter region of the gene revealed that plasma kallikrein and factor XI genes may be derived from a common ancestor.

Protein C   The 11.2 kb protein C gene is located on chromosome 2q13-q14. The protein C gene encodes mRNA transcripts of 1.8 and 1.6 kb [29]. The presence of the two transcripts is believed to be a result of alternative polyadenylation signals present in the transcript. The promoter region contains binding sites for the liver enriched transcription factors HNF-3 and HNF-1.

Protein S   The protein S gene is located near the centromere on chromosome 3 [30]. The protein S gene produces mRNA transcripts of 4, 3.1, and 2.6 kb. Protein S is synthesized in the liver and endothelial cells, testicular lining cells, and megakaryocytes.

Thrombomodulin   The intronless gene for thrombomodulin is on chromosome 20. Thrombomodulin is constitutively expressed on endothelial cells [31]. Expression of thrombomodulin in contrast to TF is downregulated by exposure of endothelial cells to IL-1, TNF-α, endotoxins or hypoxia, thus contributing to the establishment of a procoagulant environment.

Tissue Factor (TF)   The 12 kb TF gene located on chromosome 1 contains 6 exons and is transcribed into a 2.3 kb mRNA [32]. The endothelial cells align the luminal side of the blood vessels forming a barrier between the flowing blood and the underlying cells such as smooth muscle cells and fibroblasts, which constitutively express TF on their membrane. TF is selectively expressed by a variety of vascular and extravascular cell types including monocytes, macrophages and endothelial cells. Induction of TF expression occurs in response to expression of a multitude of transcription factors which are constitutively expressed in differentiated cells or as a result of activation by growth factors, and inflammatory cytokines such as tumor necrosis factor α (TNFα) and interleukin 1β (IL-1β) as well as bacterial lipopolysaccharide (LPS). Vessel wall injury or the induction of TF expression results in the establishment of a procoagulant "milieu". Transcriptional expression of the TF gene is regulated in extravascular and vascular cells by three Sp1 binding sites in the TF promoter (–111 to +14 bp). Transient expression of the TF gene is regulated by a distal enhancer (–227 to –172 bp) which contains two AP-1 binding sites and aκB binding site. This enhancer region mediates LPS and cytokine induced expression of TF in cultured monocytes and endothelial cells. The proximal enhancer (–111 to +14 bp) will induce transcription following exposure of cultured epithelial cells to PMA or serum.

Tissue Factor Pathway Inhibitor (TFPI)   Under normal conditions, TFPI is expressed in megakaryocytes and the endothelium. Adherent monocytes and macrophages will express TFPI under pathologic conditions. TFPI is encoded by a gene located on chromosome 2 and contains nine exons and eight introns resulting in a 4.0 kb mRNA. The gene was mapped to the region 2q31-2q32.1. The 5' DNA does not contain a TATAA box and several studies demonstrated multiple transcription initiation sites. The three Kunitz domains are encoded by separate exons.

Processing and Fate

Antithrombin III   No proteolysis is necessary for activation.

Factor II (Prothrombin)   Factor Xa with its cofactor, factor Va, a phospholipid surface in the presence of Ca^{2+} ions, activates prothrombin after proteolysis at two positions to yield α-thrombin. Cleavage at Arg^{320} generates a two-chain intermediate of M_r 72,000 (meizothrombin) which has esterase activity and is very unstable [33]. Meizothrombin is cleaved at Arg^{284} to generate α-thrombin containing the COOH portion of the prothrombin molecule and fragment 1·2 representing the NH_2-terminal part of prothrombin (Figure 2). Fragment 1·2 is further cleaved by α-thrombin to prothrombin fragment 1 plus prothrombin fragment 2.
α-Thrombin is a two chain enzyme which is composed of an A-chain (M_r 6,000) and a B chain (M_r 31,000) associated covalently through a disulfide bond. Purified α-thrombin undergo autolysis after long term storage. The resulting products are less active than α-thrombin. Similar results were found following limited digestion of α-thrombin by trypsin. β-Thrombin is produced following cleavage of α-thrombin at Arg^{70} or/and Arg^{73} in the B-chain of α-thrombin, whereas β'-thrombin is produced following cleavage of α-thrombin at Arg^{154} in the B chain. Cleavage of α-thrombin at both sites will result in the formation of γ-thrombin.

Factor V (Proaccelerin)   Factor V is cleaved by α-thrombin at Arg^{709}, Arg^{1018}, and Arg^{1545} to generate the active cofactor, and two heavily glycosylated activation peptides [34] (Figure 3). The factor Va molecule is a heterodimer composed of a heavy chain of M_r 105,000 and a light chain of M_r 74,000. The two subunits are non-covalently associated via a divalent metal ion process.
The inactivation of human factor Va is membrane-dependent and occurs because of limited proteolysis by APC of the heavy chain at Arg^{506}, Arg^{306}, and Arg^{679} (Figure 3) [35]. The light chain of the cofactor is not cleaved by APC. When factor Va is inactivated by physiologically relevant concentrations of APC in the presence of a membrane surface, the inactivation proceeds in a sequential manner. The first cleavage at Arg^{506} appears necessary for optimum exposure of the cleavage sites at Arg^{306} and Arg^{679}. The membrane dependent cleavage at Arg^{306} is required for complete inactivation. Phosphorylation of the cofactor by a platelet-membrane associated casein kinase II enzyme at the COOH-terminal portion of the heavy chain (at Ser^{692}) increases the rate of cofactor inactivation by APC because of acceleration of cleavage at Arg^{506}. It has been recently shown that upon cleavage at Arg^{306}/Arg^{506}, the A2 domain of the cofactor is dissociated from the rest of the molecule.

The procofactor factor V is also inactivated by APC because of cleavages at Arg^{306}, Arg^{506}, Arg^{679}, and Lys^{994} [35]. Cleavage at Arg^{306} in factor V is the inactivating cleavage site of the procofactor.

Factor VII   Factor VII is converted to its active form, factor VIIa, by a variety of blood coagulation enzymes: α-thrombin, factor IXa, factor Xa, factor VIIa and factor XIIa through cleavage of a single peptide bond (Arg^{152}-Ile^{153}) (Figure 2). The resulting factor VIIa (M_r 50,000) is composed of a light chain (M_r 20,000) and a heavy chain (M_r 30,000) covalently associated through a disulfide bond.

Factor VIII (Antihemophilic Factor)   α-Thrombin converts factor VIII to a heterotrimer following cleavage at Arg^{372}, Arg^{740}, and Arg^{1648}. Cleavage by α-thrombin at Arg^{1689} releases the cofactor from vWf and generates the active cofactor, factor VIIIa (Figure 3). Factor VIIIa is a composed of fragments of M_r 50,000 (A1 domain) containing a COOH-terminal portion rich in acidic amino acids (residues 337-372), M_r 40,000 (A2 domain), and M_r 74,000 (A3C1C2) containing the carboxyl-terminal part of the factor VIII molecule. The three subunits are non-covalently associated via divalent metal ions.

The majority of factor VIIIa cofactor activity (70-80 %) is lost at physiological pH following dissociation of the A2 domain from the rest of the molecule [36]. The remaining cofactor activity is lost because of cleavage by APC at Arg^{336}. The K_d for the dissociation of the A2 domain of factor VIIIa from the A1/light chain dimer is 270 nM [37]. Thus at physiological protein concentrations (0.7 nM), factor VIIIa inactivation is mostly the result of dissociation of the A2 domain rather than of proteolysis at Arg^{336}. Spontaneous loss in factor VIIIa cofactor activity because of dissociation of the A2 domain explains the difficulty of working with factor VIIIa.

Factor VIII is inactivated following cleavages at Arg^{336} and Arg^{562}. The first cleavage at Arg^{336} in factor VIII appears to be the inactivating cleavage site. Cleavage by APC at Arg^{562} and Arg^{740} in factor VIII have been suggested to be necessary for complete inactivation of the factor VIII latent cofactor activity.

Factor IX (Christmas Factor)   During activation, factor IX is converted to an inactive intermediate, factor IXα, and then to the active enzyme, factor IXa, by two sequential cleavages at Arg^{145} and Arg^{180} (Figure 3) resulting in the release of a M_r 11,000 activation peptide. These cleavages are catalyzed by either the extrinsic tenase complex (TF/factor VIIa) or factor XIa. Factor Xa on an acidic membrane surface will also activate factor IX. Proteolysis of factor IX at Arg^{145} will first convert factor IX to the inactive intermediate, factor IXα. Generation of this obligate intermediate enhances the overall rate of production of fully active enzyme. The resulting factor IXa (M_r 45,000), which has coagulant and esterase activities, is composed of a heavy chain (M_r 28,000) and a light chain (M_r 18,000) covalently associated through a disulfide bond.

Factor X (Stuart Factor)   Factor X is converted to its active form, factor Xa, following cleavage of the heavy chain at Arg^{194}. This cleavage is catalyzed by either the intrinsic tenase (factor IXa/factor VIIIa) or the extrinsic tenase (factor VIIa/TF) and results in the release of a M_r 12,000 peptide. The resulting factor Xa (M_r 48,000) is composed of a heavy chain (M_r 30,000) and a light chain (M_r 18,000) covalently associated through a disulfide bond (Figure 2).

Factor XI   Factor XIIa as well as α-thrombin will cleave factor XI at Arg^{369} to generate an active molecule with two active sites. The resulting active enzyme is a homodimer composed of two subunits, each made of a heavy chain and a light chain.

Factor XII (Hageman factor)   Factor XII binding to a negatively charged surface result in the autoactivation of the molecule. Following binding of Zn^{2+}, factor XII undergoes a conformational change that induces an enzymatic activity to the molecule only when bound to a negatively charged surface. Activation of the zymogen by kallikrein, trypsin, or plasmin results in an enzyme with decreased coagulant activity. Two major forms of activated factor XII exist: αXIIa form which is a M_r 80,000 protein composed of two chains associated through disulfide bonds and the βXIIa form (or HFf form) which results from cleavage at Arg^{334}, Arg^{343}, and Arg^{353} and is composed of fragments of M_r 28,000 to M_r 30,000 that has no surface binding capabilities but it's still capable in activating prekallikrein.

Factor XIII (Plasma Transglutaminase)   α-Thrombin activates factor XIII following cleavage at Arg37 in the α subunit. Following cleavage of the α chain an active site sulfhydryl is exposed and factor XIIIa will catalyze the formation of amide bonds between glutamine and lysyl side chain residues of the α and γ chains of fibrin through intermediate steps of acylation and deacylation resulting in homopolymers. Fibrin was found to have six lysyl glutamyl cross links per molecule.

Fibrinogen   Following cleavage of the Aα chain at Arg^{16} and the Bβ chain at Arg^{14} two fibrinopeptides (A and B) are released from the fibrinogen molecule. As a consequence of these cleavages there is formation of the insoluble fibrin plug. The release of fibrinopeptide A generates a fibrin molecule exposing a polymerization site (amino acid residues 17–20) on the Aα chain. These regions will bind to complimentary regions on the D domain of fibrin to form the protofibrils. Subsequent cleavage by α-thrombin on the Bβ chain and release of fibrinopeptide B exposes additional polymerization sites and promotes lateral growth of the fibrin network. Plasmin dissolves the fibrin clot following several sequential cleavages giving rise to several soluble degradation products.

High molecular weight kininogen (Fitzgerald factor)
Cleavage of high molecular weight kininogen by kallikrein results in the liberation of the bradykinin

domain and the rearrangement of the cysteine protease inhibitor domain which is opposite to the prekallikrein binding region. Proteolysis of high molecular weight kininogen allows the molecule to perform its cell antiadhesive activity.

Prekallikrein (Fletcher Factor)   Prekallikrein is activated to kallikrein by factor XIIa on a surface in the presence of high molecular weight kininogen or by factor βXIIa in a fluid phase. Activation occurs following cleavage at Arg^{371} and results in the formation of a 371 amino acids heavy chain which is associated to the light chain via a disulfide bond. Thus, activation of the molecule will not result in a change in molecular weight. In the absence of factor XII, prekallikrein will not be activated on an artificial surface.

Protein C   Protein C is converted to its active form, activated protein C (APC), by cleavage of the heavy chain at Arg^{169} (Figure 2). This cleavage is catalyzed by the α-thrombin-thrombomodulin complex. It has been also shown that the meizothrombin-thrombomodulin and the factor Xa-thrombomodulin complex can also activate protein C to generate APC.

Protein S   No proteolysis is necessary for activation. However, both α-thrombin and factor Xa abolish the anticoagulant activity of protein S by proteolysis at specific amino acid bonds located within the 29 amino acid long α-thrombin sensitive loop. This loop is located between the γ-carboxyglutamate and the first EGF domain and corresponds to exon IV of the protein S gene. α-Thrombin converts protein S into a two-chain inactive molecule following cleavage at Arg^{49} and Arg^{70} whereas factor Xa inactivates protein S following single cleavage at Arg^{60}. The latter cleavage is abolished in the presence of factor Va.

Thrombomodulin   No proteolysis is necessary for activation.

Tissue Factor (TF)   No proteolysis is necessary for activation.

Tissue Factor Pathway Inhibitor (TFPI)   No proteolysis is necessary for activation. However, more than 70 % of plasma TFPI circulates in various forms which are been truncated at a variety of positions at the COOH-terminal end.

Biological Activity

Antithrombin III (AT-III)   The general mechanism of inhibition of proteases by serpin's is a covalent entrapment of the active site of the enzyme by the serpin. In the absence of sulfated polysaccharides (heparin/heparan sulfate) the coagulation proteases interact slowly with the reactive site bond of AT-III producing a stable complex. The initial binding of the protease/inhibitor complex is facilitated by the presentation of a pseodosubstrate by the inhibitor. In the presence of sulfated polysaccharides the reaction is considerably accel-

erated because of a polysaccharide induced conformational change of the reactive site of the protease inhibitor. Upon interaction at this site with the enzyme the inhibitor changes its conformation which prevents the protease from performing the cleavage at the presented site resulting in a stable intermediate complex. The enzyme may be released from the complex when it is able to escape the trap upon cleavage of the peptide bond, this will result in an inactive altered serpin and free enzyme. The inhibition thrombin by AT-III has been used as a model to study the mechanism of serpins.

The importance of AT-III in vivo as an anticoagulant protein is indicated by the thromboembolic complications occurring in AT-III deficient patients. AT-III has been shown to inhibit the enzyme activity of thrombin, factor Xa, factor IXa, factor VIIa, factor XIa and factor XIIa [38]. The most relevant targets of AT-III in vivo are probably thrombin, factor Xa and factor IXa as indicated by circulating complexes of these enzymes with AT-III. Factor Xa is relatively protected against inactivation by AT-III when bound to cofactor (factor Va) on a phospholipid surface [39]. Upon dissociation from the cofactor factor Xa will be inactivated by AT-III in the circulation. Factor VIIa activity and its inactivation by AT-III are only significant when the protease is bound to TF. TF "opens" the active site of factor VIIa and this alteration of the active site probably sensitizes factor VIIa to inactivation by the active site directed AT-III. Factor VIIa is unique among the proteases of blood coagulation in that it becomes only sensitive for inactivation when bound to its cofactor. This probably provides the mechanism by which traces of free factor VIIa may circulate in blood [40].

Factor II (Prothrombin)   The single chain zymogen (which circulates in plasma at a concentration of 1.4 μM) is activated during coagulation to the potent serine protease α-thrombin, by the enzymatic complex termed prothrombinase, which is composed of the enzyme factor Xa, associated to its cofactor factor Va, on a membrane-surface in the presence of Ca^{2+} ions. α-Thrombin is responsible for the cleavage of fibrinogen to fibrin leading to the ultimate step in coagulation, the formation of the fibrin clot. Prothrombin deficiency is manifested as an autosomal recessive trait and is observed rarely. Several prothrombin deficiency states which are associated with hemorrhagic manifestations have been reported in the literature. Recently, a polymorphism corresponding to a G→A substitution in the 3' untranslated region of the prothrombin gene (nucleotide 20210) was associated with an increased level of prothrombin procoagulant activity. Individuals with this mutation have a two-to five-fold increase in the risk of venous thrombosis. This mutation is very common among the Caucasian population.

Factor V (Proaccelerin)   Factor V is the precursor of factor Va which is the required cofactor of prothrombinase which in turn is responsible for the efficient catalysis of the activation of prothrombin to α-thrombin. Incorporation of factor Va into prothrombinase increases the rate at which prothrombin is converted to α-thrombin by 300,000-fold relative to the rate of the reaction produced by factor Xa alone. A total deficiency in factor V is lethal in mice whereas, in humans individuals who are deficient in plasma factor V (< 2% factor V clotting activity) have been described. Thus, platelet factor V is extremely important for normal hemostasis.

Dahlbäck et al. [41] observed that plasma from an individual with venous thrombosis had an abnormal response to APC in a modified aPTT (activated Partial Thromboplastin Time) assay [41]. When APC is introduced into normal plasma which has been preincubated with an aPTT reagent a prolongation of clotting time which is proportional to the concentration of APC used has been observed. In plasma from some patients, higher concentrations of APC are required to obtain similar prolongation of clotting time as seen with normal plasma. This condition was called "APC-resistance". The molecular defect in individuals with APC-resistance was identified at the genetic level by Bertina et al. [42] who showed that individuals with APC-resistance have a mutation in the factor V gene (a G to A substitution at nucleotide 1691). This mutation results in an $Arg^{506} \rightarrow Gln$ mutation in the factor V molecule. This mutation is present in up to 60% of patients with venous thrombosis and in 3-5% of normal individuals. As a consequence of the mutation the abnormal molecule, called factor V^{LEIDEN}, does not possess the APC-cleavage site at Arg^{506}. When isolated from the plasma of patients homozygous for the $Arg^{506} \rightarrow Gln$ mutation, factor Va^{LEIDEN} is inactivated by APC at a rate slower than that observed for normal factor Va [43]. However, inactivation still proceeds as a consequence of cleavage at Arg^{306}. Recently it has been proposed that low concentrations of heparin could accelerate cleavage at Arg^{306} and inactivation of factor V by APC. These data together with the fact that at elevated APC concentrations, inactivation of membrane-bound factor V by APC may occur faster that activation of the procofactor (α-thrombin) would suggest an increased antithrombotic effect of APC in the presence of low doses of heparin.

Interestingly, and in contrast to the observations made with factor Va^{LEIDEN}, the APC-catalyzed rate of inactivation of factor V^{LEIDEN} is comparable to that observed with normal factor V. These results are in agreement with the data demonstrating the initial membrane-dependent cleavage site in normal factor V which leads to inactivation, occurs at Arg^{306}. The APC-catalyzed inactivation of membrane-bound factor V^{LEIDEN} also occurs through initial cleavage at Arg^{306} at a comparable rate to normal factor V, indicating that inactivation of factor V does not necessarily require cleavage at Arg^{506}.

Regulation by proteolysis for activation /inactivation of the cofactor plays a preeminent role in clinical assays for the definition of one individual's genetic status. Thus, the quality of the reagents for the assay must be always determined. For example, during execution of the APC-resistance assay, plasma is incubated with reagent that will activate factor V for a given period of time. Then APC is added together with Ca^{2+}. The clotting time of the sample is measured and compared to the clotting time in the absence of APC. In plasma from normal individuals the prolongation of clotting time is correlated with the inactivation of factor Va by APC. In plasma from individuals homozygous for the $Arg^{506} \rightarrow Gln$ mutation, a lesser prolongation of the clotting time is observed in the presence of APC. Since factor Va, not factor factor V is procoagulant active, all assays that rely on the in situ activation of factor V to support the clotting reaction are highly sensitive to the status of factor V activation in the plasma sample. Furthermore, since factor V inactivation occurs by an initial cleavage at Arg^{306} while factor Va inactivation results from sequential cleavages at Arg^{506} followed by Arg^{306} the factor V activation status can and will compromise the APC-resistance assay if the procofactor is not properly and timely activated. As a consequence a patient with thrombotic disorders who is suspected to be homozygous for the factor V^{LEIDEN} mutation and who's plasma factor V was not properly activated to factor Va because of a defective assay reagent will have a normal sensitivity to APC when compared to normal plasma assayed under similar condition since inactivation will only involve cleavage at Arg^{306} and will escape detection.

Factor VII   Factor VII is the precursor of factor VIIa which is the initiator of blood clotting. Factor VIIa binds to the endothelial cell receptor tissue factor which is only exposed following injury of the vasculature or cytokine stimulation, to form the extrinsic tenase complex which is responsible for the initiation of blood coagulation. This complex is primarily responsible for the cleavage/activation of factor X and factor IX and for the production of small quantities of active enzyme (i.e. factor Xa and factor IXa).

Factor VIII (Antihemophilic Factor)   Factor VIII is the precursor of factor VIIIa which is the required cofactor of the intrinsic tenase complex which in turn is responsible for the efficient catalysis of the activation of factor X to factor Xa. Incorporation of factor VIIIa into the intrinsic tenase complex increases the rate at which factor X is converted to factor Xa by approximately 200,000-fold relative to the rate of the reaction produced by factor IXa alone. Over 200 point mutations of the factor VIII gene have been reported in the literature. The importance of the corresponding phenotype varies, however, all individuals affected (all males) have bleeding problems. An examination of the second order rate constants of the activation of factor X by the factor VIIa/TF complex (extrinsic tenase) when compared to the second order rate constant of the activation of factor X by the factor IXa/factor VIIIa complex (intrinsic tenase) demonstrates that the latter complex is approximately 50 times more efficient in producing factor Xa than the former. Thus, the preferred pathway for factor Xa formation in plasma where low (physiological) levels of factor VIIa are present, is the pathway that utilizes the intrinsic tenase complex. Thus, from a physiological point of view factor VIIIa and factor IXa are clearly identified as essential for normal hemostasis.

The importance of the correct regulation of factor Va cofactor activity is very much illustrated by the "beneficial" effect of factor V^{LEIDEN} in some individuals with severe hemophilia A. Different pathology (phenotype) has been reported in several patients with hemophilia carrying the same mutation(s). The reported mutations normally result in a phenotype which is characterized as severe bleeding disorders. However, while some hemophiliacs are classified clinically as severe hemophiliacs, several other individuals carrying the same mutation(s) were classified as mild/moderate hemophiliacs and only had minor bleeding problems. This classification is based on the level of measurable factor VIII activity in a one-stage clotting assay using factor VIII deficient plasma. The net effect of a factor VIII deficiency is a reduced α-thrombin formation rate, whereas in the case of an abnormal factor V molecule that would be resistant to inactivation by APC (i.e. factor V^{LEIDEN}) the net effect would be an increase in α-thrombin formation. The patients classified as severe hemophiliacs were carrying normal factor V whereas the other were heterozygous for the factor V^{LEIDEN} mutation. Since factor Va^{LEIDEN} is inactivated by APC with a rate slower than normal factor Va, hemophiliac patients who posses the factor V^{LEIDEN} gene may have a milder bleeding syndrome than hemophiliac patients with normal factor V because of increase rate in prothrombin activation. This hypothesis was recently confirmed *in vitro*, and the data demonstrated an increase in α-thrombin generation in individuals with hemophilia A which is proportional on the plasma level of factor V^{LEIDEN}. Thus, it appears that if an individual possesses both, a factor VIII mutation that normally results in severe bleeding and factor V^{LEIDEN} the extended lifetime of factor Va^{LEIDEN} is able to partially compensate for an abnormal or absent factor VIII molecule resulting in sufficient α-thrombin generation to provide hemostasis.

Factor IX (Christmas Factor)   Factor IX is the precursor of factor IXa which is the required enzyme for the intrinsic tenase complex. Factor IXa binds to its cofactor factor VIIIa on a membrane surface in the presence of divalent metal ions, to form the intrinsic tenase complex which is responsible for efficient activation of factor X during blood coagulation.

Factor X (Stuart Factor)   Factor X is the precursor of factor Xa which is the required enzyme for the prothrombinase complex. Factor Xa binds to its cofactor factor Va on a membrane surface in the presence of divalent metal ions, to form the prothrombinase complex which is responsible for efficient activation of prothrombin during blood coagulation.

Factor XI   Recent research has attempted to uncover why factor XI deficiency is biologically important for normal coagulation, whereas its presumed activator, factor XII, is unnecessary for clotting in vivo. An alternate means of activating factor XI, independent of factor XII and the early contact pathway, is implied by the clinical data. Recent data demonstrated activation of factor XI by α-thrombin in a purified system. The rate of this reaction was greatly accelerated by negatively charged glycosaminoglycans such as dextran sulfate [44]. α-Thrombin activated factor XIa at femto- to picomolar levels in the absence of factor XII or negatively charged surfaces has been detected during the initiation phase of thrombin generation using sensitive assays [45]. Such observations support a role for thrombin-activated factor XI in thrombin activation and coagulation.

Factor XII (Hageman factor)   Deficiency of factor XII is not associated with any bleeding tendency in vivo. Cleavage of factor XII by kallikrein is central to the initiation of the intrinsic pathway of the blood coagulation cascade. Surface bound αXIIa activates factor XI to factor XIa. Secondary cleavage of αXIIa by kallikrein results βXIIa which in turn will activate kallikrein, factor VII and the complement cascade.

Factor XIII (Plasma Transglutaminase)   Factor XIII is the precursor form of the glutaminyl-peptide γ-glutamyl transferase, called factor XIIIa (fibrinoligase, plasma transglutaminase, and fibrin stabilizing factor). This is the last of the zymogens of the blood coagulation process to become activated and is the only enzyme in the system that is not a serine protease. α-Thrombin will activate factor XIII to factor XIIIa following cleavage at Arg^{36} of the NH_2-terminal portion of the α chain and exposure of the active site sulfhydryl. Factor XIIIa will catalyze the conversion of soluble fibrin to an insoluble fibrin clot by crosslinking the lysine and glutamine side chains of fibrin to form homopolymers.

Fibrinogen   Fibrinogen is a multi-functional molecule possessing diverse biological activities within coagulation and hemostasis. Fibrinogen also participates in extravascular inflammatory responses. Following addition of α-thrombin to fibrinogen two different fibrinopeptide release rates will result in two separate changes in fibrinogen. Briefly, α-thrombin removes fibrinopeptide A from the fibrinogen molecule to form soluble monomers of fibrin which will polymerize to yield a double stranded protofibril. Removal of fibrinopeptide B from the protofibril by α-thrombin will result in the aggregation of the protofibrils into fibers. Finally, the γ-chains of two adjacent monomers will be cross-linked by their respective D regions in the presence of factor XIIIa, resulting in a cross-linked clot.

Albeit the greatest attention upon the transition of fibrinogen to fibrin and the polymerization of the fibrin derivative substantial work has also focused in the past on the capacity of fibrinogen to mediate platelet aggregation through the binding of the protein to the platelet glycoprotein receptor GPIIb/IIIa. This phenomenon which is induced by a wide variety of physiologic and pharmacological stimuli establishes a role of potential significance for the molecule in the primary phase of hemostasis.

High molecular weight kininogen (Fitzgerald factor)
Deficiency of high molecular kininogen is not associated with any bleeding tendency in vivo. The most important function of the kininogens is the release of bradykinin upon their activation. Bradykinin operates as an anti-thrombotic/profibrinolytic agent. Further, bradykinin is a potent stimulator of endothelial cell prostacyclin synthesis, an inhibitor of platelet function. Bradykinin also prevents subedothelial cell-dependent smooth muscle proliferation.

Prekallikrein (Fletcher Factor) Deficiency of prekallikrein is not associated with any bleeding tendency in vivo. Following activation the light chain of α-kallikrein reacts with circulating plasma protease inhibitors (α_2-macroglobulin, and C1 inhibitor). The interaction with C1 inhibitor results in the loss of proteolytic and amidolytic activity of the protein.

Protein C Protein C is the precursor of activated protein C (APC) which is the required enzyme for the normal inactivation of factor Va and the arrest of α-thrombin production by the prothrombinase complex. α-Thrombin bound to its endothelial cell surface receptor thrombomodulin will cleave protein C at Arg^{12} in the heavy chain to generate APC.

Protein S Protein S is an important coagulation inhibitor as evidenced by the thrombotic tendency of protein S deficient individuals. However the mechanism of anticoagulant action of protein S remains controversial. It has been reported that: a) protein S functions as a cofactor of APC; b) protein S eliminates the protection of factor Va by factor Xa against inactivation by APC; c) protein S inhibits prothrombinase (factor Xa/factor Va) and intrinsic tenase (factor IXa/factor VIIIa) activity in a APC-independent manner. The enhancement of the activity of APC by protein S ranges from 2 to 10-fold, and is rather dependent on the experimental conditions under which cofactor activity is assessed. Addition of protein S to APC results in an increase in the rate of the membrane dependent cleavages by APC in factor Va (Arg^{306}) and factor VIIIa (Arg^{336}).
The independent inhibitory activity of protein S was first hypothesized to be the result of interactions of protein S with factor Xa, factor Va or factor VIIIa. Protein S, however, inhibits all phospholipid dependent reactions. The APC-independent inhibitory action of protein S on the intrinsic tenase complex and on the prothrombinase complex is reported to correlate with the apparent phospholipid binding properties of the various protein S preparations. Based on these observations it was hypothesized that protein S inhibits these reactions by competing for the procoagulant phospholipid available in the reaction. A very significant inhibitory effect of protein S on thrombin generation initiated by TF in the presence

of quiescent platelets is observed. These data demonstrate that the inhibitory effect of protein S which is only expressed when a limited number of phospholipid binding sites for the procoagulant enzyme complexes are present is potentiated in the presence of the cell surfaces usually available in vivo for clot formation and is caused by competition for negatively charged phospholipid. Whether the APC-independent effect of protein S on the procoagulant reaction is important in vivo remains a question.

Thrombomodulin Thrombomodulin is the required endothelial cell surface cofactor for the activation of protein C to APC by α-thrombin. Thrombomodulin forms a 1:1 stoichiometric complex with α-thrombin. The activation of protein C to APC is accelerated approximately by 1000-fold when compared to the activation rate of protein C by α-thrombin alone. Further, binding of α-thrombin to thrombomodulin completely alters the procoagulant activity of the enzyme. Following binding to thrombomodulin, thrombin no longer triggers platelet aggregation or clots fibrinogen. The α-thrombin -thrombomodulin complex also no longer activates factor V, or inactivates protein S.

Tissue Factor (TF) Tissue factor (TF) is the membrane-bound glycoprotein that is exposed at the site of the vascular injury and is responsible for the initiation of the coagulation process. TF is expressed by a variety of vascular and extravascular cells. Normally the endothelial cells cover the luminal side of the blood vessels and form a barrier between the flowing blood and the underlying cells. Upon exposure to the blood flow TF will bind factor VIIa. The catalytic efficiency of the enzymatic complex which initiates coagulation by activating factor X and factor IX increases by three orders of magnitude as compared with the catalytic efficiency of the enzyme, factor VIIa alone.

Tissue Factor Pathway Inhibitor (TFPI) The in vivo role of TFPI in coagulation is still under investigation since no TFPI deficient individuals are yet found. However, it has been demonstrated that complete deficiency of TFPI in mice is incompatible with birth and survival [46]. TFPI complexes with the limited quantities of factor Xa formed initially by the extrinsic tenase, and factor Xa-TFPI subsequently inhibits the extrinsic tenase via formation of a quaternary complex with TF and factor VIIa. Factor Xa-dependent inhibition of the extrinsic tenase by TFPI down-regulates further generation of factors Xa and IXa. Available factor Xa that is produced by the extrinsic tenase is therefore limited by these factors. TFPI will inactivate factor Xa by reversible binding of the second Kunitz domain to the active site of the enzyme. The following step is the inhibition of the catalytic activity of TF/factor VIIa complexes by formation of the quaternary complex TF/factor VIIa/TFPI/factor Xa. This complex formation depends on the binding of the first Kunitz domain of TFPI to the active site of factor VIIa. The quaternary complex is stable and can be dissociated by EDTA. The membrane interaction of the Gla-domain of factor Xa in complex with TFPI seems of major importance since Gla-domain-less factor Xa does not function as a cofactor for TFPI in the inhibition of the TF/factor VIIa complex. The physiological function of TFPI seems therefore to stem primarily from its ability to regulate TF-dependent coagulation and not by its ability to inhibit FXa directly. *In vitro* studies it have shown that TFPI exerts a potent inhibition of TF/factor VIIa activity in flow models using purified human coagulation factors. The direct activation of factor X by TF/factor VIIa is blocked in time by TFPI in the absence of factor IX and factor VIII. Activation of traces of factor IX by TF/factor VIIa before inactivation by TFPI allows continuation of factor X activation by the factor IXa/factor VIIIa activity in the presence of TFPI. While the direct activation of factor Xa by TF/factor VIIa is rapidly inhibited by TFPI, the traces of factor IXa formed by the transient TF/factor VIIa activity results in sufficient factor Xa generation to sustain the hemostatic reaction.

Role in Vascular Biology

Physiological Function

While the importance of the TF pathway in coagulation is clear, a role for contact initiation is questionable. Deficiencies in factors VII, VIII, IX and X and V are invariably associated with bleeding disorders; however, hemorrhagic tendencies resulting from deficiency of prekallikrein, factor XII or high molecular weight kininogen are unknown. While the role of factor XI is suggested by several studies, the bleeding disorders associated with factor XI deficiency (Hemophilia C) are variable in their severity and frequency. Rapaport et al. [47] have shown a major and minor form of factor XI deficiency. Levels of factor XI below 20% are considered most commonly associated with significant hemorrhage upon surgical challenge.

The recent years an earlier theory of coagulation has resurfaced. This theory of coagulation has been first proposed by Paul Morawitz [1] in 1905 who postulated that the key proteins required for normal blood clotting and physiological hemostasis are thromboplastin, prothrombin and fibrinogen. The material called thromboplastin in the early days is the equivalent of the tissue factor/phospholipid used in the present days. TF (the integral membrane glycoprotein) is now accepted as the in vivo initiator of thrombin generation in the revised theory of coagulation. While the majority of proteins involved in blood coagulation circulate as inactive zymogens that require proteolytic activation in order to function, approximately 1% of the circulating factor VII molecules are active (factor VIIa) and possess an active site which is not inhibited by circulating stoichiometric protease inhibitors. Following injury to the vasculature and subsequent exposure of TF, the circulating factor VIIa molecules can bind to the exposed TF forming the extrinsic tenase complex (TF/factor VIIa) and initiate the blood coagulation process. The TF pathway proceeds by assembly of three distinct complexes. The first is the extrinsic tenase (factor VIIa and the membrane bound cofactor TF). This complex activates a fraction of the circulating zymogen factors X and IX to their respective active forms, factors Xa and IXa. but only transiently at relatively low levels as the result of stringent down-regulation of the factor VIIa/TF complex by TFPI and AT-III. This initial amount of factor Xa provides limited amounts of α-thrombin, adequate to activate the cofactors [48]. Factor IXa assembles with factor VIIIa to form the intrinsic tenase complex, which produces additional factor Xa. Free factor Xa assembles with factor Va into the prothrombinase complex on the cell surface, which is the activator of prothrombin to α-thrombin. This delayed burst of factor Xa results in a burst of prothrombinase, which in turn creates the burst of thrombin activity required for clotting and adequate hemostasis. Thrombin-activated factor XIa serves to augment this burst of factor Xa by increasing the quantity of factor IXa-factor VIIa complex. As a result there is measurable increase in the final amount of thrombin generated when factor XI participates in the reaction when compared with reactions without factor XI. The extrinsic tenase activity alone is attenuated by AT-III and tissue factor pathway inhibitor (TFPI), while the intrinsic tenase is largely under the control of AT-III. Thus, factor Xa generated by the extrinsic tenase is insufficient to maintain an ongoing hemostatic response. Under these conditions, the intrinsic tenase complex provides the additional factor Xa required to maintain thrombin generation. This is the reason that individuals which lack factor VIII (hemophilia A) and factor IX (hemophilia B) bleed profusely albeit the presence of normal concentrations of circulating factor VII. This theory of coagulation is supported by observations concerning the severity of hemophilias. In factor IX or factor VIII deficiency (hemophilias A and B), the delayed burst of thrombin is not observed, because only factor Xa supplied directly by factor VIIa/TF supports thrombin generation. Thus, this low level of thrombin is not adequate to ensure hemostasis. Studies of the platelet-fibrin plug in hemophilias A and B indicate that the platelet plug formed initially in response to injury, ruptures subsequently due to defective fibrous transformation and stabilization within the plug. This theory that supports an active involvement of factor XIa in blood clotting taken alone does not adequately explain, however, why factor XI deficiency (hemophilia C) exhibits the type of infrequent and variable hemorrhagic symptoms which accompany this defect. However recent observations have demonstrated that in factor XI deficient individuals, when very low TF concentrations are used to initiated the coagulation reaction, platelet activation, factor V activation and fibrin formation is significantly delayed [49]. Thus, factor XI deficiency will result in delayed hemostasis only at low concentrations of TF.

α-Thrombin will finally cleave fibrinogen and activate the circulating platelets which together will form the hemostatic plug. α-Thrombin participates in its own down-regulation by binding to the endothelial cell receptor thrombomodulin, and initiating the protein C pathway, which in turn leads to the formation of APC. APC is required for efficient neutralization of factor Va and VIIIa cofactor activities which result in the inactivation of the prothrombin and factor Xa activating complexes [50]. This inactivation can only occur in the presence of the appropriate membrane surface. Thus, while following α-thrombin activation, factor VIIIa at physiological concentration (0.7 nM), is rapidly and spontaneously inactivated by dissociation of the A2 domain from the rest of the cofactor, APC is required for down-regulation of α-thrombin formation by prothrombinase. APC down-regulates the prothrombinase complex by cleaving specific peptide bonds on the heavy chain of factor Va which also results in the dissociation of the A2 domain of factor Va from the rest of the molecule. Dissociation of the A2 domain of both cofactors impairs their ability to interact with the other protein components of prothrombinase and intrinsic tenase and results in the arrest of α-thrombin formation. Interestingly, it has been demonstrated that effective down-regulation of α-thrombin generation by the protein C pathway in combination with TFPI and AT-III occurs because APC prevents the coexistence of the factor Va heavy and light chains [51].

Pathology, Clinical Relevance and Therapeutic Implications

Most of the pathology associated with the coagulation factors has been described above. However efficient hemostasis requires Ca^{2+} ions and the presence of a negatively charged membrane surface.

Common to each vitamin K-dependent zymogens relevant to coagulation is a highly-conserved NH_2-terminal "Gla-domain," which binds multiple calcium ions and contains 9-12 γ-carboxyglutamic acid residues. At the COOH-terminus is the serine protease domain, which is largely homologous to that of trypsin and chymotrypsin, with insertions which alter the macromolecular substrate specificity of each enzyme. Between the NH2- and COOH-terminal domains is a region of the protein which varies in structure among the zymogens, containing either epidermal growth factor-like domains or kringle domains.

The binding of calcium to a number of sites in the Gla-domain stabilizes the structure in this region of the protein, which is disordered in the absence of calcium. The importance of the Gla-domain to the function of these coagulation zymogens is exemplified in warfarin therapy, which blocks the vitamin K-dependent carboxylation of glutamate residues. Cooperativity is observed for the binding of calcium to prothrombin, and the calcium-bound configuration exhibits a greater affinity for negatively-charged phospholipid than the apoform. Despite the remarkable complexity of this cooperative calcium effect, no regulatory role for calcium has been elucidated in coagulation. Investigations in porcine carotid arteries demonstrated that the growing thrombus formed in response to vascular injury is a neo-tissue, with packed platelets inaccessible to the flow of oxygened blood outside the aggregate. Within this structure, anoxia develops and interstitial calcium levels fall as a result of platelet membrane depolarization, a result which was not observed when blood clotted *in vitro*. Administration of hirudin a very potent inhibitor of α-thrombin, aided clot dissolution and reversed the calcium effects, providing the first in vivo evidence of calcium modulation within a reversibly-formed thrombus. Along with the data for cooperative calcium binding by the vitamin K-dependent proteins, these observations suggest a mechanism wherein complex-dependent coagulation may be effectively ablated by a modest reduction in the extracellular calcium concentration. These observations allows for the possibility of a regulatory role for calcium in coagulation.

In quiescent cells, distinct activities maintain lipid asymmetry with the negatively charged phospholipids (phosphatidyl serine (PS) and phosphatidyl inositol (PI)) almost exclusively located in the inner membrane. Neutral choline lipids (i.e., phosphatidyl choline (PC)) are moved negligibly or very slowly. Activation of the cell causes internal calcium release resulting in a redistribution of the internalized negatively charged lipids, the extent of which appears to depend upon the level and type of activation. With platelets, strong phospholipid redistribution is observed upon treatment with calcium ionophores (such as A23817), diamide, and a combination of thrombin and collagen, while weaker effects are detected following treatment with thrombin or collagen alone; ADP and epinephrine have little effect. In platelets, differences among the activators have been reported to correlate with the level of intrinsic tenase and prothrombinase activity observed. Formation of procoagulant microvesicles is also observed following extended activation of platelets, but PS accumulation can occur significantly in advance of microvesiculation.

Defects in platelet phospholipid reorganization have been noted in a bleeding disorder known as Scott syndrome. Scott syndrome is characterized by a failure to accumulate PS at the platelet surface. Poor binding of the cofactors results, leading to impaired tenase and prothrombinase activity. This disease has been detected in a French family, and was found to be hereditary in nature. While it is evident that the defect affects Ca^{2+}-dependent lipid scrambling, the exact molecular cause of the disease is unknown. Since the defect can be demonstrated in erythrocytes as well as platelets, a common genetic origin in various cell types is suggested. Whereas calcium ionophores induce loss of lipid asymmetry on red cells leukocytes and endothelium, α-thrombin and other platelet activators do not universally cause loss of lipid asymmetry. However, erythrocyte procoagulant lipid activity has been noted in reversible red cell sickling and diabetes. High glucose buffers also lead to a loss of phospholipid asymmetry in erythrocytes and apoptosis in endothelium. Apoptosis is accompanied by a loss of lipid asymmetry in a variety of cell types, including lymphocytes, vascular smooth muscle cells and endothelium.

Michael Kalafatis, Cornelis van't Veer and Kenneth G. Mann

References

1. Morawitz P (1905) Ergebn Physiol 4:307-423
2. Davie EW and Ratnoff OD (1964) Science 145:1310-1312
3. Lundblad RL et al (1976) Methods Enzymol 45:156-176
4. Tracy PB et al (1982) Blood 60:59-63
5. Mann KG et al (1981) Biochem 20:28-33
6. Nesheim ME and al (1984) J Biol Chem 259:3187-3196
7. Kisiel W and Davie EW (1975) Biochem 14:4928-4934
8. Vehar G and Davie E, (1980) Biochem 19:401-410
9. Vehar G (1984) Nature 312:337-342
10. Thompson AR (1986) Blood 67:565-572
11. Fujikawa K et al (1972) Biochem 11:4882-4891
12. Bouma BN and Griffin JH (1977) J Biol Chem 252:64327-66437
13. Scheraga HA and Laskowski M, Jr (1957) Adv Protein Chem 12:1-5
14. Kisiel W and Davie EW (1981) Methods Enzymol 80:320-332
15. Scharfstein J, et al (1978) J Exp Med 148:207-222
16. Tsiang M et al (1992) J Biol Chem 267:6164-6170
17. Drake TA et al (1989) J Cell Biol 109:389-395
18. Broze GJJr (1995) Ann Rev Med 46:103-112
19. MacGillivray RTA and Davie EW (1984) Biochem 23:1626-1634
20. Jenny RJ et al (1987) Proc Natl Acad Sci USA 84:4846-4850
21. Cripe LD et al (1992) Biochem 31:3777-3785
22. O'Hara PJ et al (1987) Proc Natl Acad Sci USA 84:5158-5162
23. Gitschier J et al (1984) Nature 312:326-330
24. Kurachi K and Davie EW (1982) Proc Natl Acad Sci USA 79:6461-6464
25. Leytus SP et al (1984) Proc Natl Acad Sci USA 81:3699-3702
26. Fujikawa K et al (1986) Biochem 25:2417-2424
27. Ichinose A and Davie EW (1988) Proc Natl Acad Sci USA 85:5829-5833
28. Chung DW et al (1990) Adv Exp Med Biol 28:139-148
29. Long GL et al (1984) Proc Natl Acad Sci USA 81:5653-5656
30. Schmidel DK et al (1990) Biochem 29:7845-7852
31. Owen WG and Esmon CT, (1981) J Biol Chem 256:5532-5535
32. Mackman N et al (1989) Biochem 28:1755-1762
33. Heldebrant CM et al (1973) J Biol Chem 248:7149-7163
34. Suzuki K et al (1982) J Biol Chem 257:6556-6564
35. Kalafatis M et al (1994) J Biol Chem 269:31869-31880
36. Fay PJ and Smudzin TM (1992) J Biol Chem 267:13246-13250
37. Lollar P and Parker C (1990) J Biol Chem 265:1688-1692
38. Olson ST and Shore JD (1982) J Biol Chem 257:14891-14895
39. Marciniak E (1973) Br J Haematol 24:391-400
40. Morrissey JH et al (1993) Blood 81:734-744
41. Dahlbäck B et al (1993) PNAS 90:1004-1008
42. Bertina RM, et al (1994) Nature 369:64-67
43. Kalafatis M et al (1995) J Biol Chem 270:4053-4057
44. Gailani D and Broze GJ, Jr (1991) Science 253:909-912
45. von dem Borne PA et al (1994) Thromb Haemost 72:397-402
46. Huang Z et al (1993) J Biol Chem 268:26950-26955
47. Rapaport SI et al (1961) Blood 18:149-165
48. Pieters J et al (1989) Blood 74:1021-1024
49. Cawthern KM et al (1998) Blood 91:4581-4592
50. Esmon CT (1987) Science 235:1348-1352
51. van 't Veer C et al (1997) J Biol Chem 272:7983-7994

Collagen

Definition *Extracellular matrix protein that is a constituent of basement membranes. A large number of collagen subtypes have been described ranging from I to XVIII.*

See: →Extracellular matrix

Collagenase

Definition *Proteolytic enzymes that degrade collagen. Important in cell invasion, tissue remodeling or angiogenesis. Also identified as matrix metalloproteinases 1, 8, or 18.*

See: →Matrix metalloproteinases

Colony-Stimulating Factors

Definition *Cytokines that regulate the hematopoietic cell development and function. A number of CSF have been characterized including monocyte/macrophage colony stimulating factor (M-CSF) and granulocyte macrophage colony stimulating factor (GM-CSF).*

See: →Cytokines in vascular biology and disease; →Megakaryocytes

Complement

Definition *Enzymatic cascade that mediates a number of biological functions including host defense against infection, initiation of an inflammatory response or processing and clearance of immune complexes.*

See: →Complement system

Complement S-Protein

See: →Vitronectin/vitronectin receptors

Complement System
(Interaction of Vascular Cells with)

Introduction The complement system mediates a number of biological functions that participate in host defense against infection, initiation of the inflammatory reaction, processing and clearance of immune complexes and regulation of the immune response [1, 2]. Pathogens, altered host cells and immune complexes trigger complement activation, resulting in the production of biologically active complements fragments. Except for lysis and for the non-cytolytic cellular responses elicited by the terminal C5b-9 sequence, most of the biological effects derived from complement activation depend on ligand-receptor interactions between complement proteins and specific receptors on cells. Inadequate regulation or extensive complement activation may alter the physiological functions of normal cells and contribute to pathology.

Characteristics   The human complement system comprises 23 plasma components and regulatory proteins that represent 5 % of the plasma protein content. Upon activation, complement components interact within distinct and finely regulated functional units (Figure 1). The classical and the alternative pathways of activation both form specific enzymatic complexes termed C3 convertases that cleave C3 and generate the major cleavage fragment C3b. A single amplification pathway exists that augments C3 cleavage once initial C3b has been generated and covalently linked to complement activating surfaces. A common effector sequence comprising C3b and components C5 to C9 generates the opsonizing, vasoactive, leukocyte-attracting, immune regulatory and cytolytic activities of complement. The classical pathway of activation comprises the C1 complex formed by one molecule of C1q, two molecules of C1r and two molecules of C1s, the components C2 and C4 and the regulatory proteins C1 inhibitor (C1-inh), C4-binding protein (C4BP) and I. The alternative and amplification pathways involve the components C3, B and D, and the regulatory proteins P, H and I. When not engaged in the assembly of a membrane-bound C5b-9 complex, the components C5-C9 bind the serum protein S, also termed vitronectin to form a fluid phase cytolytically-inactive SC5b-9 complex. Several membrane-associated proteins widely distributed on human cells, i.e. DAF, MCP and CD59 regulate the cleavage of C3 and the formation of the C5b-9 complex. Phagocytic cells and lymphocytes express specific receptors for complement proteins and their fragments, which upon interaction with their ligands, elicit the various cellular responses that initiate inflammatory process.

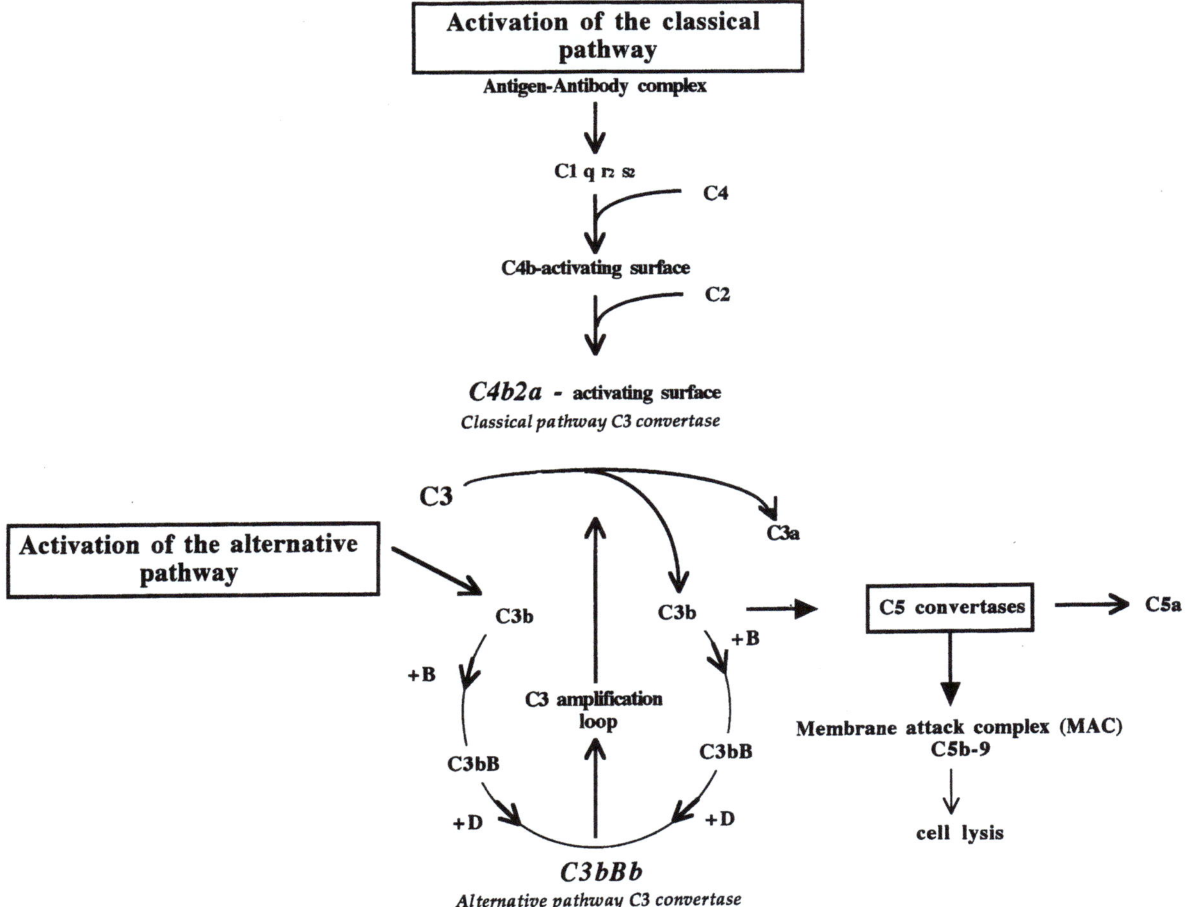

Figure 1. Activation of the human complement system

Generation of classical pathway C3 convertase Classical pathway activation by immune complexes is initiated by the binding of the globular heads of C1q to the CH2 domain of complexed IgG or IgM [3, 4]. Of the various immunoglobulin isotypes, only IgM and IgG3, 1 and 2 bind C1q to trigger complement activation. Binding of C1q to the immunoglobulin results in a conformational change in the distal portion of C1q that renders the C1r and C1s molecules within the C1 complex inaccessible to the C1 inhibitor. C1r and C1s are normally associated with C1q in a pentameric C1q, C1r2, C1s2 calcium-dependent complex. Bacteria, viruses and virus-infected cells, products of membrane and tissue damage directly bind C1q and may activate the classical pathway in the absence of antibody [5]. Activation of C1 results in expression of a proteolytic activity by C1s. The substrates for the active C1s serine protease are C4 and C2. C4 is first cleaved into a small anaphylatoxin fragment, C4a, and a large fragment, C4b, that expresses a labile binding site which mediates covalent attachment of the molecule to cell membranes or to the constant regions of immunoglobulins. Surface-fixed C4b binds C2 which is, in turn, cleaved by C1s into a small fragment C2b that is released in the fluid phase and a large fragment C2a that remains associated with C4b on the activator. The C4b,2a complex is the classical pathway C3 convertase in which the C2a fragment carries the proteolytic site for cleavage of C3. The next step in the classical pathway activation process is the cleavage of C3 by the immune-complex-bound C3 convertase, which results in the release of the anaphylatoxin C3a and the generation of C3b. As it is the case for C4b, nascent C3b transiently expresses a reactive site that allows the formation of a covalent bond with hydroxyl or amino groups on immune complexes and/or on bystander surfaces [6]. Binding of multiple C3b molecules to the target surface will then change the specificity of the C3 convertase C4b,2a to a C5 convertase denoted [C4b,2a(C3b)n] and trigger the activation of the terminal sequence C5-C9.

Generation of the alternative pathway C3 convertase
The alternative pathway represents a natural system for resistance to infection in the non-immunized host [7]. Activation by the alternative pathway displays unique features. First, antibody is not required, although it can facilitate the activation process. Second, activation proceeds both in the fluid phase and on cell surfaces. Activation of the alternative pathway is initiated by the low rate interaction in normal plasma of a "C3b like" form of C3 and factors B, D and P, resulting in the formation of an initial C3 convertase. Upon generation in the fluid phase, C3b randomly attaches to bystander surfaces following exposure of the reactive site of the molecule. Binding of C3b to an activating surface of the alternative pathway is followed by the formation of a bimolecular complex with factor B, cleavage of B by factor D, and assembly of the alternative pathway amplification convertase C3b,Bb on the target surface. Once formed, the amplification convertase C3b,Bb cleaves C3

generating molecules of C3b that bind to the activating surface. The binding of multiple C3b molecules in the vicinity of the C3b,Bb enzyme changes the specificity of the alternative pathway C3 convertase to a C5 convertase [C3b, Bb(C3b)n] and initiates activation of the terminal sequence C5-C9.

Assembly of the C5b-9 membrane attack complex
Formation of the C5b-9 complex (membrane attack complex, MAC) on target cells is initiated by the cleavage of C5 into C5a and C5b by cell-bound C5 convertases. Generated C5b is rapidly released in an inactive form in the fluid phase unless it forms a stable bimolecular C5b6 complex with C6. The C5b6 complex can reversibly bind to cell membranes. With the addition of C7-C9, it lyses unsensitized bystander cells, a process known as "reactive lysis". Binding of one molecule of C7 to C5b,6 creates a trimolecular complex which inserts firmly into the lipid bilayer of the target cell membrane. Assembly of the C5b-9 complex is further completed by the binding of one molecule of C8 and several molecules of C9. The interaction of the terminal complement components C5-C9 on the surface of susceptible cells results in the formation of protein-lined hydrophilic transmembrane channels which cause osmotic lysis of the cells [8].

Regulation

Activation of the classical and alternative pathways is subjected to fine regulatory processes, i.e. the spontaneous and rapid dissociation of enzymatic complexes and the regulatory activity of several plasma and cell-associated proteins.

Formation of the C1 complex is regulated by C1 inhibitor (C1-inh) which belongs to the family of human serum protease inhibitors (serpins). C1-inh inhibits auto-activation of C1r and C1s into native C1 [9]. The half-life of the C4b,2a enzyme is limited by the intrinsic decay of the C2a subunit and by the decay-accelerating activity of C4BP in plasma and of decay-accelerating factor (DAF, CD55) that is expressed on the membrane of a large variety of autologous cells. The formation of the classical pathway C3 convertase is also limited by the cleavage and inactivation of C4b by factor I in the presence of the cofactors C4BP or membrane cofactor protein (MCP, CD46). Cleavage of C4b generates C4c and C4d.

Regulation of the activity of the alternative pathway does not depend on the initial binding of B to C3b which is unaffected by the nature of the surface to which C3b becomes covalently attached. Discrimination between "activating" and "non-activating" surfaces of the alternative pathway depends on the relative capacity of factor H to bind to C3b, that prevents formation of the C3b,Bb enzyme and allows for inactivation of C3b by factor I. Formation of the C3b,Bb convertase is inhibited on "non-activating" surfaces by the preferential binding of factor H to C3b as compared with factor B. Sialic acid and sulfated mucopolysaccharides (e.g. heparan sulphate) modulate the ability of cell surfaces to activate the alternative pathway by enhancing the interaction between surface-bound C3b and the regulatory protein H [10]. Several membrane proteins inhibit formation of the classical and/or alternative pathway C3 convertases on the cell surface thus contributing to the protection of host cells from damage by autologous complement. They include decay-accelerating factor (DAF, CD55), the C3b receptor (CR1, CD35) and membrane cofactor protein (MCP, CD46) which accelerate the dissociation of C3 convertases and serve as cofactors for proteolytic cleavage of C3b by factor I [11, 12]. Inactivation of C3b by factor I generates the iC3b, C3dg and C3d fragments that remain covalently linked to the target surface of complement activation but are unable to perpetuate complement activation. Each of the C3 cleavage products C3b, iC3b, C3dg and C3d may react with one or several of four types of C3 receptors CR1 (CD35), CR2 (CD21), CR3 (CD11b/CD18) and CR4 (CD11c/CD18) expressed on phagocytes and lymphocytes [11]. The critical importance of C3 in the complement cascade is evident from: (1) its position at the convergence of the classical and alternative pathways; (2) its role in the activation and amplification of the alternative pathway; (3) the fact that C3 is a major step for regulation of complement activity; (4) the multiple biological activities associated with the cleavage products of the C3 molecule.

Cytolytic damage by the membrane attack complex (MAC) is modulated by a plasma protein termed S-protein or vitronectin. The S-protein blocks the action of C5b-9 by binding to the newly-formed C5b-9 complex rendering it unable to attach to the surface of target cells. Nucleated cells which are relatively resistant to lysis possess the ability to repair complement-mediated damage by both endocytosis and exocytosis of the C5b-9 channels. MAC formation is also inhibited on cells by two membrane proteins, CD59 and C8 binding protein (C8bp) that inhibits homologous complement-mediated lysis and thus acts as homologous restriction factor (HRF) [13]. CD59 that is anchored to the membrane via a glycosyl phosphatidyl inositol linkage, binds to neo-epitopes on C8 and C9 that are exposed during MAC assembly [14].

Molecular Interactions

Complement-mediated processing of immune complexes   The transient occurrence of circulating immune complexes is now regarded as a physiological phenomenon. The role of complement in handling immune complexes in the circulation and at a site of a tissue lesion is decisive in that it may favor the removal of immune aggregates or mediate a clinically overt inflammatory process [15]. Activation of the classical or the alternative pathway by immune complexes results in the covalent binding of C4b and C3b to the antigen and/or to the immunoglobulin, that further allow an optimal interaction of the immune complexes with cellular complement receptors. Complement activation is also involved in maintaining immune complexes in solution and in the transport of C3-bearing complexes to phago-

cytes in the spleen and/or the liver. The interaction of classical pathway proteins with immune complexes during the formation of the complexes prevents immune precipitation [16], whereas incubation of preformed complexes with serum (i.e. with complement proteins) leads to their disaggregation and solubilization through an alternative pathway-mediated mechanism [17]. Once immune complexes have been opsonized with complement, their clearance from the circulation depends on the interaction with C3b receptors (CR1) on erythrocytes and on the functional state of the reticuloendothelial system. Erythrocyte CR1 is the major source of CR1 in whole blood. CR1 endows the cells with the capacity for repeated uptake and release of C3b-bearing immune complexes which allows the erythrocyte to transport the complexes and deliver them through the portal circulation to the liver [18]. In the liver, the complexes are stripped from the cells and transferred to Kupffer cells. An impaired clearance of immune complexes in the liver has been found in hypocomplementemic monkeys in whom complexes appear widely distributed throughout the body.

Receptors for complement proteins   Complement activation results in the generation of diffusible biologically-active peptides and the deposition of complement fragments on target surfaces of complement activation. Both diffusible anaphylatoxins and target-bound activation fragments may interact with receptors on effector cells to trigger specific cellular responses [11]. In addition, the insertion of C5b-9 into the lipid bilayer of cell membranes may directly induce lysis of target cells. Although nucleated cells are relatively resistant to membrane attack by complement, sublytic amounts of C5b-9, C5b67 and C5b678 mediate signal transduction in various cell types [19]. These include the mobilization of Ca^{2+} from intracellular stores, stimulation of arachidonic acid metabolism, release of prostaglandins and leukotrienes and the production of oxygen metabolites by leukocytes that induces activation of endothelial cells.

Activation of complement results in the rapid generation of the anaphylatoxins C3a, C4a and C5a through cleavage of the α-chains of C3, C4 and C5. The peptides diffuse into plasma and extracellular fluids where their binding to specific receptors mediate a number of biological functions. Thus, the anaphylatoxins cause smooth muscle contraction and increased vascular permeability. C5a is more active than C3a, and C4a has little activity. In plasma, the C-terminal arginines of C3a and C4a are removed physiologically by carboxypeptidase N, converting the anaphylatoxins in stable biologically inactive C3a-desArg and C4a-desArg derivatives. C3a binds to receptors expressed on mast cells, basophils, monocytes, smooth muscle cells, lymphocytes and platelets. A receptor for C3a on macrophages has recently been cloned [20]. C3a has also been reported to suppress immunological functions such as polyclonal and antigen specific humoral immune responses. In addition to being the most potent anaphylatoxin, C5a at nanomolar concentration is chemoattractant for leukocytes causing directed migration of neutrophils, eosinophils, basophils and monocytes against the gradient of concentration that occurs when the peptide diffuses away from the site of complement activation. C5a may have systemic effects due to its relative resistance to cleavage by carboxypeptidase N and by the retention of significant biological activity of C5a-desArg. The C5a receptor belongs to the rhodopsin superfamily and is coupled to a G protein [21]. Triggering of the C5a receptor on specific target cells (i.e. neutrophils, mastocytes, basophils, monocytes) results in receptor phosphorylation and activation of intracellular signal transduction pathways. The recent cloning of C5a receptor allowed the demonstration that its expression extends to non-myeloid cells such as lung vascular smooth muscles and endothelial cells [22].

The C3b receptor (CR1, CD35) is a single chain polymorphic glycoprotein (Table 1) [23]. The most common allotype has a molecular weight of 250 kDa. CR1 functions as a high-affinity receptor for C3b that is covalently bound to particles, cells or immune complexes. C3b is thus presented as a multivalent ligand for the receptor. CR1 also binds with lower affinity the surface-fixed fragments iC3b and C4b. CR1 is expressed on erythrocytes, neutrophils, monocytes, eosinophils, mastocytes, B lymphocytes, a subset of T lymphocytes and thymocytes, follicular dendritic cells, Kupffer cells and glomerular podocytes. Soluble and membrane-associated CR1 efficiently regulate alternative pathway and classical pathway activation by impairing the formation of the C3 convertases and promoting cleavage of C3b and C4b by factor I. CR1 is a potent anti-inflammatory agent since the administration of soluble recombinant CR1 prevents tissue damage induced by complement activation in experimental models of myocardial reperfusion injury and xenograft rejection [24]. CR1 expresses additional functions that differ with the cell type that carries the receptor (see below).

CR3 (the iC3b receptor, CD11b/CD18) is a member of the β_2-integrin family, which also includes the leukocyte adhesion molecules LFA-1 (CD11a/CD18) and p150,95 (CD11c/CD18)[25]. All three molecules are heterodimers sharing a common β_2 subunit associated with a specific α-chain. The α and β chains of CR3 are of respective Mr 165 and 95 kDa. CR3 may bind multiple ligands. The divalent cations Ca^{++} and Mg^{++}are essential for the stabilization and function of the α/β complex since both subunits contribute to ligand binding. CR3 is present in an inactive state on circulating leukocytes. Acquisition of an active state when cells are triggered by inflammatory mediators, reinforces the binding reaction of iC3b to CR3 and allows for expression by CR3 of binding sites for other ligands including coagulation factor X, fibrinogen, intercellular adhesion molecule (ICAM-1, CD54), and betaglycans. The p150/95 (CD11c/CD18) protein may also be considered as a C3 complement receptor (CR4), since it binds iC3b and C3dg. The cellular expression of CR3 is restricted to monocytes, macrophages, neutrophils, eosinophils, basophils and natural killer (NK) cells. In resting monocytes and granulocytes, 90 % of CR3 is stored as intracellular pools localized in peroxydase-negative granules. In response to a variety of stimuli including chemoattractants, leucotrienes, cytokines, the intracellular pools of CR3 translocate and fuse with the plasma membrane. Most stimuli that upregulate membrane expression of CR3 also induce qualitative changes allowing CR3-mediated biological functions. In addition to phagocytic function, CD11b/CD18 is essential for spreading, adhesion and transendothelium migration of leukocytes during the inflammatory process.

The CR2 receptor (CD21), a 145 kDa type I transmembrane glycoprotein, is the cellular receptor for the iC3b,C3dg, C3d cleavage fragments of C3 and for Epstein Barr virus (EBV) [23]. More recently, CD21 was also shown to bind CD23 that serves as the low-affinity receptor for IgE [26]. CD21 is only expressed on B lymphocytes, approximately 50 % of peripheral blood T lym-

Table 1.   Cellular receptors for target-bound C3 activation fragments

Receptor	Molecular Weight (kD)	Ligands	Cellular Distribution
CR1 (CD35)	160-250	C3b (C4b, iC3b)	erythrocytes, monocytes/macrophages, neutrophils, eosinophils, basophils, B lymphocytes, subset of T lymphocytes, NK cells, follicular dendritic cells, Kupffer cells, glomerular podocytes
CR2 (CD21)	145	iC3b, C3dg, C3d CD23, EBV	B lymphocytes, subset of T lymphocytes and thymocytes, follicular dendritic cells
CR3 (CD11b, CD18)	α chain: 165 β chain:* 95	iC3b fibrinogen, ICAM-1 factor X	monocytes/macrophages, neutrophils, eosinophils, basophils, follicular dendritic cells, Kupffer cells
CR4 (CD11c, CD18)	α chain: 150 β chain:* 95	iC3b, C3dg	macrophages, neutrophils, Kupffer cells

* CR3 and CR4 share common β chain of β2 integrins

phocytes, a subpopulation of immature thymocytes and on follicular dendritic cells. CR2 plays an important role in antigen-induced B cell proliferation. The interaction of CR2 with CD23 contributes to the regulation of IgE production, germinal center B-cell survival, and B cell-presentation of soluble antigen to T cells.

The biological relevance of cellular receptors for C1q is as yet poorly understood. Lymphocytes, monocytes, neutrophils, platelets express a C1q receptor which binds to C1q that remains fixed to classical pathways activators following the dissociation of macromolecular C1 by C1inh. Human umbilical vein endothelial cells express receptors for both the collagen-like and the globular domain of C1q, which may have implications for the role of C1q in vascular inflammatory and thrombotic lesions [27].

Cells and Cellular Interactions Complement mediates a number of important host defense reactions including the recognition of pathogens and altered host cells, the production of an acute inflammatory response, opsonization, phagocytosis and cytolysis which facilitate the elimination of pathogens. Endothelial cells normally provide a barrier to the egress of proteins and cells from blood vessels, maintain an anticoagulant environment intravascularly and remain not adherent for leukocytes under physiological conditions. Generation of anaphylatoxins modify vascular permeability, attract inflammatory cells and modify their adhesion properties to the endothelium. Uncontrolled complement activation may occur following the entrapment of bacterial antigens in chronic inflammatory sites, during immune complexes-associated diseases or, e.g., following intravascular complement activation during hemodialysis. In such pathological situations complement may directly induce vascular inflammatory injury by the generation of anaphylatoxins and the membrane attack complex. Complement may also induce vascular injury in an indirect fashion by generating chemotactic peptides which activate leukocytes inducing their adherence to target tissues and the secondary release of inflammatory mediators and toxic oxygen products.

Endothelial cells and complement

C5a causes the rapid expression of P-selectin, the secretion of von Willebrand factor and adhesion of human neutrophils to human umbilical vein endothelial cells, indicating that C5a behaves as an important inflammatory mediator for early adhesive interactions between neutrophils and endothelium [28]. Formation of the membrane attack complex directly influences the integrity of the vascular endothelium. The sequential association of C5b with complement components C6-C9 establishes pores on endothelial cell surfaces. The pores when present in sufficient numbers, mediate cell lysis. In sublytic amounts, the membrane attack complex leads to modification of coagulant status and to activation of endothelial cells. Thus, the assembly of C5b-9 triggers von Willebrand factor production and induces membrane cell vesiculation, leading to the

expression of a prothrombinase complex [29]. By inducing upregulation of P selectin and the synthesis of the chemokines IL-8 and MCP-1 [30], C5b-9 promotes the adhesion of phagocytes to endothelial cells and amplifies the recruitment of leukocytes at the site of inflammation. The expression of the complement regulatory proteins DAF, MCP and CD59 by endothelial cells limits the insertion of autologous MAC [31]. The inhibitory capacity of these membrane-associated regulatory components has been extensively documented in discordant xenograft models. The use of organs from transgenic pigs expressing human DAF and CD59 revealed striking protection from complement-mediated injury of xenogenic endothelial cells [32]. Finally, the synthesis of the alternative pathway components B and C3 by endothelial cells *in vitro* has been shown to be enhanced by the pro-inflammatory cytokine IL-1, suggesting that complement protein secretion may take an active part in the local deposition of C3 fragments on endothelial cells [33].

Leukocytes and complement Neutrophils and monocytes express receptors for anaphylatoxins and receptors for C3 fragments. Neutrophil stimulation by C5a results in respiratory burst, enhances neutrophil turnover of arachidonic acid and stimulates cellular production of 5-HETE and leukotriene B4 [34]. These lipids are neutrophil chemotactic factors. C5a causes degranulation of leukocytes, inducing the release of lysosomal enzymes from neutrophils and the release of ECP and EPO by eosinophils which may mediate changes in the integrity of blood vessels. For example, the release of proteases from neutrophils induces rapid cleavage of heparan sulfate from the surface of endothelial cells that may enhance their capacity to activate the alternative pathway. Neutrophil activation by C5a increases the membrane expression of CR1 and CR3 which are stored in secondary granules, allowing for a better attachment of the cells to targets opsonized with C3b and iC3b [35,36]. C5a and C5a-desArg also enhance the adhesiveness of neutrophils to foreign surfaces and endothelial cells and reversibly aggregate the cells *in vitro* and *in vivo* [37]. The latter effect is secondary to enhanced expression of the adhesion-promoting molecule CR3. CR3 mediates the intravascular aggregation of leukocytes which causes leukopenia in individuals undergoing hemodialysis with membranes that activate complement. This mechanism is probably essential for the pathogenesis of pulmonary endothelial damage in the adult respiratory distress syndrome (ARDS). Recent data indicate that oxygen-derived free radicals released from sequestrated neutrophils play a major role in endothelial cell damage. Triggering of C5a receptor leads to production of IL-1 by monocytes and IL-8 chemokine by eosinophils, pro-inflammatory cytokines which, in turn, can activate endothelial cells.

The interaction of CR1 and CR3 with complement-opsonized targets will cause particles to adhere to monocytes and neutrophils [38]. The main function of

CR1 on these cells is to enhance phagocytosis of IgG-coated particles and mediate the internalization of small ligands bearing C3b. The binding of fibronectin or laminin to C3b-coated particles confers the ability to ingest the opsonized targets in the absence of antibody on both monocytes and C5a-stimulated neutrophils. The enhancing effect of connective tissue proteins on phagocytosis may be particularly relevant to the pathogenesis of vascular lesions where extracellular matrices become exposed. The interaction of bound C3b with CR1 has also been shown to induce enzyme release by neutrophils, trigger the oxidative metabolism and activate the arachidonic acid pathway in neutrophils and monocytes. The expression of CR1 as that of CR3 is increased by IL-1 and chemoattractants and decreased by IFNγ. CR3 is probably the most important receptor for phagocytosis of opsonized bacteria. Triggering of C3 receptors on monocytes with polymeric C3b and iC3b has also been reported to result in intracellular accumulation and release of IL-1. As mentioned earlier, CR3 is a member of the b2 integrin family which is involved in adhesion events. By enhancing the expression of CR3 and by activating the receptor, C5a and other chemoattractants (PAF acether or leukotrienes) increase the adhesiveness of the cells to vascular endothelium and pathogens. It should be mentioned here that activation of endothelial cells with cytokines such as IL-1 upregulates the expression of ICAM-1, which serves as an additional ligand for CR3 that may facilitate the adhesion of monocytes and granulocytes to the vascular endothelium.

Additional Features   Complex interactions occur between complement and proteins of the coagulation and fibrinolytic systems. The serine esterase enzymes and the serine protease inhibitors of the systems are structurally related. *In vitro* plasmin generation may directly act on C1 to activate the complement system or inactivate C1inh, thereby releasing C1 from inhibition. Plasmin and kallikrein also cleave C3 and C5 *in vitro*, although the cleavage efficiency is low compared to that normally obtained with physiological convertases, raising the question of the physiological relevance of this interaction.

Clinical Relevance   There are several mechanisms by which antigen-antibody complexes may form and/or deposit in the vascular wall: (1) local interaction between antigen and pre-formed antibody (as is the case in the Arthus reaction); (2) deposition of circulating immune complexes; and (3) the interaction of circulating antibody with an antigen *in situ*, whether the antigen is a constitutive antigen of the vascular wall or whether it is an exogenous "planted" antigen. The deposition or *in situ* formation of immune complexes and the subsequent activation of complement mediate immunologically-induced vascular injury [39]. A critical function of complement with regard to the pathogenesis of vasculitis is its ability to prevent immune complex precipitation and to prepare optimally solubilized complexes for their intravascular transport to the sites of removal in the reticulo endothelial system. Thus, inherited deficiencies of proteins of the classical pathway are associated with an increased incidence of autoimmune and immune complex-mediated diseases [40,15]. Patients with homozygous C4 or C2 deficiencies have a high incidence of systemic lupus erythematosus.

Systemic small vessel vasculitides, polyarteritis nodosa and related syndromes, pulmonary endothelial cell injury in the adult respiratory distress syndrome and vascular hyperacute rejection of transplants are examples of immunologically-mediated vascular damage where complement is involved. For diagnostic purposes, antibodies to complement proteins and to neoantigens expressed by activation products of complement may be used to detect and characterize deposits of complement in pathological tissues by indirect immunofluorescence. The availability of monoclonal and polyclonal reagents to C5b9 has allowed for the detection of terminal complexes in the skin, vessels and kidney lesions in SLE, in vasculitis, and in ischaemic areas of myocardial infarction, bullous pemphigus and synovial tissues in rheumatoid arthritis. Deposits of C3 and of C5b-9 have been found in association with immune complexes along the glomerular capillary walls and in the mesangium in a variety of glomerular diseases (Figure 2).

Because the complement system has both protective and autoaggressive potential, therapeutic modulation of complement is difficult to design and conduct unless it may be targeted in the future to the sites of complement activation. At the present time, no therapeutic intervention specifically aimed at down-regulating complement is used in vascular diseases where complement activation is involved. Yet, clinical trials of recombinant CR1, the most promising inhibitor of complement in the fluid phase, have recently been initiated.

*Elizabeth Fischer, Véronique Frémeaux-Bacchi
and Michel D. Kazatchkine*

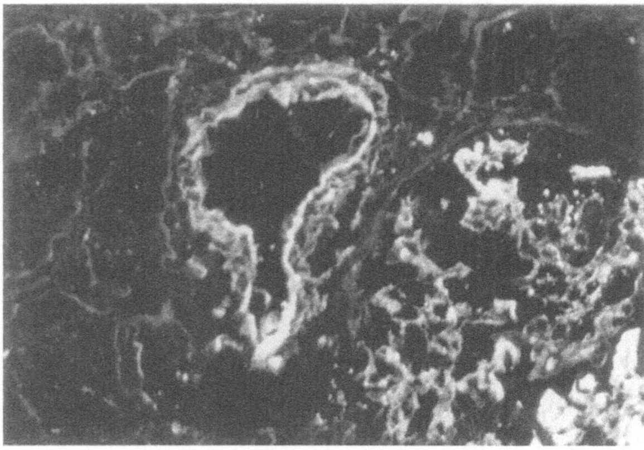

Figure 2. Nephritis during systemic lupus erythematosus. Staining of adjacent glomerulus and artery with anti-C5b-9 monoclonal antibody. x 450

References

1. Ross GD, Medof ME (1985) Adv Immunol 37:217-267
2. Müller-Eberhard HJ (1988) Ann Rev Biochem 57:321-347
3. Cooper NR (1985) Adv Immunol 37:151-207
4. Schumaker VN et al (1987) Ann Rev Immunol 5:21-42
5. Sim RB, Reid KBM (1991) Immunol Today12:307-311
6. Pangburn MK et al (1981) J Exp Med 154:856-861
7. Kazatchkine MD, Nydegger UE (1982) Prog Allergy 30:193-234
8. Bhakdi S, Tranum-Jensen J (1982) J Cell Biol 94:755-759
9. Ziccardi RJ (1983) Springer Semin Immunopathol 6:213-230
10. Kazatchkine MD et al (1979) J Immunol 122:75-89
11. Fearon DT, Wong WW (1983) Ann Rev Immunol 1:243-27
12. Hourcade D et al (1989) Adv Immunol 45:381-416
13. Sugita Y, Masuho Y (1995) Immunotechnology, 1:157-168
14. Okada N et al (1989) Int Immunol 1:205-208
15. Nydegger UE, Kazatchkine MD (1986) Prog Allergy 39:361-392
16. Schifferli JA et al (1982) Clin Exp Immunol 45:555-562
17. Fujita T et al (1981) J Exp Med 154:1743-1751
18. Cornacoff JB et al (1983) J Clin Invest 71:236-247
19. Nicholson-Weller A, Halperin JA (1993) Immunol Res 12:244-257
20. Crass T et al (1995) Eur J Immunol 26:1944-1950
21. Wetsel RA (1995) Curr Opin Immunol 7:48-53
22. Haviland DL et al (1995) J Immunol 154:1861-1869
23. Ahearn JM, Fearon DT (1989) Adv Immunol 46:183-219
24. Kalli KR et al (1994) Springer Semin Immunopathol 15:417-431
25. Arnaoult MA (1990) Blood 75:1037-1050
26. Aubry J et al (1992) Nature 358: 505-508
27. Peerschke EIB et al (1996) J Immunol 157:4154-4158
28. Foreman KE et al (1996) Inflammation 20:1-9
29. Hamilton KK et al (1990) J Biol Chem 265:3809-3814
30. Kilgore KS et al (1996) Amer J Pathol 149:953-961
31. Brooismans RA et al (1992) Eur J Immunol 22:3135-3140
32. Rosengard AM et al (1995) Transplantation 59:1325-1333
33. Dauchel H et al (1990) Eur J Immunol 20:1669-1675
34. Vogt W (1986) Complement 3:177-188
35. Fearon DT, Collins LA (1983) J Immunol 130:370-375
36. Yancey KB et al (1985) J Immunol 135:465-470
37. Craddock PR et al (1977) J Clin Invest 60:260-264
38. Brown EJ (1992) Infectious Agents and Diseases 1:63-70
39. Kazatchkine MD et al (1989) In: Camillieri JP, Berry CL, Fiessinger JN, Bariety J (eds) Diseases of the arterial wall. Springer, Berlin Heidelberg New York, pp 391-422
40. Morgan BP, Walport MJ (1991) Immunol Today 7:95-306

Contractility

Definition   *Contractility, the shortening of cells through the sliding action of actin-myosin and Ca²⁺ in smooth muscle cells. Extracellular factors of vasoconstriction are angiotensin II and endothelin.*

See: →Smooth muscle cells

COX-1

Definition   *Cyclooxygenase-1*

See: →Cyclooxygenase; →Bleeding disorders; →Prostacyclin; →Prostaglandins; →Thromboxanes

COX-2

Definition   *Cyclooxygenase-2*

See: →Cyclooxygenase; →Bleeding disorders; →Prostacyclin; →Prostaglandins; →Thromboxanes

CPAE

Definition   *Calf pulmonary artery endothelial cell*

See: →Angiogenin; →von Willebrand Factor

CR1, 2, 3 and 4

Definition   *Complement receptor 1, 2, 3 and 4*

See: →Complement system (interaction of vascular cells with)

CSF

Definition   *Colony-stimulating factor*

See: →Colony-stimulating factors; →Atherosclerosis

Cyclic Nucleotides

Definition   *Cyclic adenine monophosphate (cAMP) is involved in vascular smooth muscle relaxation. Cyclic guanosine monophosphate (cGMP) binds to type I (smooth muscle cells) or type II G-kinases (epithelial cells). Production of cyclic nucleotides is under the control of cyclases and phosphodiesterases.*

See: →Platelet stimulus-response coupling; →Vasomotor tone regulation, molecular mechanisms of

Cyclooxygenase (Cox)

Definition
Same as PGH synthase-1 (Cox-1) and PGH synthase-2 (Cox-2), enzyme involved in the generation of prostaglandins

See: →Platelet stimulus-response coupling; →Prostacyclin; →Prostaglandins

Cytokines in Vascular Biology and Disease

Introduction   Interactions between immune mediators and vascular cells regulate basic functions of the immune system such as lymphocyte migration, regional immunity, and inflammatory responses. Immune mediators also modulate cardiovascular functions controlling both systemic and regional hemodynamics. In this review, we explore current evidence for a role of cytokines in the development of cardiovascular diseases. In particular, we will focus on how cytokines regulate cell surface receptors on vascular cells and modulate the functions of vascular endothelial and smooth muscle cells, with emphasis on two basic vascular activities: vascular tone and hemostasis. We also discuss the involvement of cytokines in thrombosis and the stability of atherosclerotic plaques. Finally, the crucial role of signal transduction pathways in mediating the function of cytokines will be examined.

Characteristics (of atherosclerotic vascular disease)
A chronic pathologic process in the intima of large arteries, atherosclerosis starts when cholesterol-bearing low density lipoproteins (LDL) infiltrate the arterial intima in such amounts that the normal elimination mechanisms are exceeded [1, 2]. In the intima, LDL adheres to proteoglycans of the extracellular matrix and is oxidized by free oxygen radicals and enzymes produced by vascular endothelial cells and macrophages [3]. Thies leads to the inflammatory proliferative disease of atherosclerosis.

The atherosclerotic plaque is the site of a local immune activation. In its fibrous cap, approximately 20 % of the cells are T lymphocytes [3] and nearly half of these show signs of activation [5, 6]. Cytokines produced by T cells, macrophages and smooth muscle form a local network that plays a pivotal role in the development of cardiovascular diseases. The expression of cytokines is tightly controlled in the producing cells, and one of the most important regulatory steps is the control of gene transcription. Several lines of evidence indicate that transcription factors are crucial in determining changes in cell function, growth and differentiation, leading to characteristic patterns of gene expression during inflammation. The final stage of plaque development is characterized by the appearance of fissures, thrombi, and actual plaque rupture [7–9].

Vascular contractility plays an important role in the pathophysiology of atherosclerosis. Vascular contractility is determined by interaction between actin and myosin filaments in smooth muscle cells. Some of the filament-forming proteins such as α-SM-actin are specific for smooth muscle cells but the filaments themselves are not as well developed as in striated muscle. The contractile process is dependent on energy provided by ATP derived from mitochondrial respiration. It is regulated by several different soluble factors including α-adrenergic agonists from local nerve endings and from the blood, circulating angiotensin, and endothelium-derived factors such as endothelin and nitric oxide (NO).

Regulation

Molecular Interactions
Cytokines in fatty streak formation and antigen presentation   Oxidized LDL is a ligand for the macrophage scavenger receptors (MSR), which bind proteins and particles with clustered negative charges [10]. Receptor-ligand complexes are internalized and degraded intracellularly, and fragments of the ligands can associate with MHC molecules and may be presented as T cell antigens [11, 12]. LDL cholesterol entering the macrophage via MSR accumulates in cytoplasmic droplets; the increasing intracellular cholesterol accumulation gradually transforms the macrophage into a foam cell, which is the hallmark of atherosclerosis [13]. Foam cells form fatty streaks, i.e. lesions of the arterial intima which predispose to atherosclerosis. Two macrophage surface receptors, CD36 (a class B scavenger receptor) and the SR-A macrophage scavenger receptor (a class A scavenger receptor), have been identified as major receptors that bind and internalize OxLDL [3]. Recently, the importance of MSR in the formation of fatty streaks has been further addressed by targeted disruption of the SR-A gene in mice [14]. Deletion of SR-A results in a reduction in the size of atherosclerotic lesions in animals deficient in apolipoprotein E. Expression of both class A and class B MSR in monocyte/macrophages is dependent on the differentiation state as well as exposure to soluble mediators (cytokines and growth factors).

Macrophage colony-stimulating factor (M-CSF) is a differentiating factor for monocyte/macrophage development as well as a regulator of lipoprotein uptake. M-CSF, which is produced by plaque cells, increases SR-A expression on the mRNA level and enhances foam cell formation [15–17]. Thus, M-CSF can be expected to have profound effects on atherosclerosis. Its importance was recently confirmed by the observation that *op/op* mice, which carry a defective M-CSF gene, only form diminutive fatty streak lesions when fed an atherogenic diet, even when bred onto the highly atherosclerosis-prone, apolipoprotein E-knockout background [18, 19]. In addition to M-CSF, an S-100-like chemotactic cytokine has also been shown to control scavenger receptor expression [20].

Proinflammatory cytokines and endotoxins may exert effects on MSR opposite to those of M-CSF. Both interferon-γ (IFN-γ) and tumor necrosis factor-α (TNF-α) downregulate SR-A expression and inhibit foam cell formation [21–23]. In the case of IFN-γ, modulation is accomplished by destabilization of cytoplasmic mRNA, whereas TNF-α inhibits transcription of the SR-A gene [22, 24]. Endotoxin has similar effects on SR-A expression as TNF-α and appears to act by inducing autocrine TNF-α secretion that in turn inhibits SR-A gene transcription [22]. The importance of proinflammatory cytokines for SR-A expression is supported by the *in vivo* observation that areas of atherosclerotic plaques with inflammatory activity and T cell activation show reduced expression of SR proteins [23]. This suggests that macrophage-T cell interactions may be decisive for the development of the fatty streak.

Apart from its role in scavenging oxLDL leading to forming fatty streak, macrophages are professional antigen presenting cells that may initiate immune reactions in atherosclerotic plaques. This has been addressed in a recent review [25]. Finally, recent studies suggest that the vascular endothelium not only plays an important role in the cytokine-regulated recruitment of leukocytes from blood to tissues, but also performs a series of other functions relating to immunity, inflammation, hemostasis, and permeability, which can all be modulated by cytokines.

Cytokines in the transition into atherosclerosis RT-PCR and immunohistochemical analyses have shown that the T cell cytokine, IFN-γ is produced in the plaque [5, 23]. Furthermore, smooth muscle cells, endothelial cells, and macrophages all express high levels of MHC class II proteins (HLA-DR), suggestive of cellular responses to IFN-γ [26]. In addition to the specific T cell cytokines, proinflammatory and macrophage-stimulating cytokines as well as chemokines are also produced in the atherosclerotic plaque. TNF-α and IL-1 are both found in plaques and can be produced both by endothelial cells (preferentially IL-1), smooth muscle cells, and macrophages [27–29]. They may be important as regulators both of smooth muscle cell growth and the activation of endothelial cells, macrophages, and T cells. Finally, chemokines such as monocyte chemoattractant protein-1 (MCP-1) [30, 31] and IL-1 [32] could be important for the recruitment of inflammatory and immunocompetent cells into the plaque. Thus, cytokines produced by inflammatory cells in the atherosclerotic plaque form a network that may direct the progression of the disease. Recent studies of disease development in murine models of atherosclerosis support a role for immune factors. Thus, immunosuppressive treatment aggravates fatty streak formation in C57BL/6 and immunodeficient MHC class I knockout mice of the C57BL/6 background develops significantly larger fatty streaks than wild type C57BL/6 mice [33]. These observations imply that immune factors are anti-atherogenic. They are also supported by observations in rabbits implying an anti-atherogenic role of immune factors [34] and an anti-fatty streak effect of treatment with the cytokine, LIF (leukemia inhibitory factor) [35].

The results of similar studies in atherosclerosis-prone ApoE-knockout mice are in apparent contrast to these findings. Double-knockouts lacking both ApoE and RAG-1 (recombinase-activating gene-1) exhibit a reduction of lesion size by approximately 40% compared to the ApoE single knockout [36]. Similarly, double-knockouts lacking ApoE and the interferon-γ receptor show a lesion reduction by 60% [37]. Finally, transplant atherosclerosis induced by allgrafting of blood vessels was found to be dependent on CD4+ T cells, MHC class II, and B cells [38]. To interpret these findings, we must bear in mind that the immune system is a complex defense system that contains counterbalancing activities. For instance, activation of proinflammatory Th1 responses is counteracted by the Th2 and T suppressor cells, and many of the transcriptional effects of inflammatory cytokines such as TNF-α and IFN-γ are counteracted by TGF-β and IL-10. Therefore, not only the extent but also the type of immune activation is likely to be decisive for the immunopathogenesis of a disease. In atherosclerosis of ApoE-knockout mice, recent data show that the balance between Th1 and Th2 responses is determined by serum cholesterol levels [39]. Lack of IL-10 increases fatty streak formation in S578L/6 mice [194] while treatement of apo E-/- or LDL receptor-/- mice with immunomodulating antibodies reduce atherosclerosis [195-196]. More information will therefore be needed to understand the conditions promoting proatherogenic and antiatherogenic immune responses.

Role of transcription factors in cardiovascular diseases Among transcription factors defined to date, NF-κB (nuclear factor-κB), AP-1 and STATs (signal transducers and activators of transcription) have been intensively studied. NF-κB is a homo- or heterodimeric combination of Rel-related proteins. Members of this family are p65 (RelA), RelB, p50 (NF-κB1), p52 (NF-κB2) and c-Rel [40-45]. They all share a conserved domain, the Rel homology domain, that participates in DNA binding and dimerization. In most cell types, NF-κB resides in the cytoplasm in a latent form that is associated with the inhibitor IκB. Three major IκB molecules, IκBα, IκBβ and IκBε, have been identified [46–49], which retain NF-κB in the cytoplasm by masking its nuclear localization signal domains. In response to multiple stimuli, including cytokines, phorbol esters, bacterial lipopolysaccharide (LPS), and viral products, the inhibitor is proteolyzed and NF-κB translocates into the nucleus, where it activates its target genes [50-53]. The NF-κB family of transcription factors plays a crucial role in the expression of a large array of genes essential to the immune response in the atherosclerotic lesion, including the genes encoding IL-1β, TNF-α, tissue factor, VCAM-1 (vascular cell adhesion molecule-1), ICAM-1 (intercellular adhesion molecule-1), iNOS (inducible NO synthase), MCP-1and M-CSF (for reviews, see [54, 55]). Human vascular smooth muscle cells express predominantly RelA, p50 and p52 precursor (p100) as well as its inhibitors including at least IκBα and IκBβ [56, 57]. Activation of NF-κB in vascular smooth muscle cells involves two phases. In the absence of stress or cytokine stimuli, vascular smooth muscle cells display a constitutive NF-κB activation, which seems to be composed of a p50 homodimer [56]. This constitutive NF-κB signal is serum-dependent, and does not initiate transcription of inflammatory genes. The constitutive activation of NF-κB may be needed for driving cell proliferation (our unpublished data and [56]). Upon cytokine stimulation, an inducible activation of NF-κB will be triggered, leading to strong translocation of a p65/p50 heterodimer into the nucleus. This inducible signal exerts a critical regulatory role in the transcription of inflammatory genes such as the iNOS gene in smooth muscle cells and macrophages [57-59]. Interestingly, intimal smooth muscle cells, a population of cell isolated from the intima of injured rat arteries, show a different NF-κB activation than normal medial smooth muscle cells. Compared with medial smooth muscle cells, intimal smooth muscle cells have a high constitutive NF-κB activity and a hyperinducible activation of NF-κB ([58], and our unpublished data). This special feature is in agreement with a high ability for iNOS transcription in the intimal smooth muscle cells, and may be the reason why inflammatory genes are expressed preferentially by intimal smooth muscle cells in injured vessels and in atherosclerotic lesions.

Although the importance of NF-κB in cytokine expression has been well established *in vitro*, the work to unravel the role of NF-κB in the development of atherosclerosis has just started. Brand and colleagues have recently reported that NF-κB is activated in macrophages, smooth muscle cells and endothelial cells in human atherosclerotic lesions, but not detected in vessel free of atherosclerosis [60]. Using double immunostaining, they have further shown that activation of NF-κB correlates with expression of tissue factor and ICAM-1, two NF-κB dependent inflammatory genes. Bourcier et al have also found that activated NF-κB is preferentially found in smooth muscle cells of human atherosclerotic lesions [57]. These findings suggest that NF-κB is important in the regulation of cytokine networks during the development of atherosclerosis.

Because of its pivotal role in the regulation of cytokine production, the NF-κB signal transduction pathway is an appealing target for therapeutic intervention in cardiovascular disorders. Antioxidants such as PDTC (pyrrolidone derivative of dithiocarbamate) and NAC (*N*-acetyl cysteine) are effective inhibitors of NF-κB activation *in vitro* [61, 62]. In an experimental septic model, pretreatment of rats with PDTC compromises the activation of NF-κB, and prevents expresssion of inducible NO synthase in lung tissue [63]. Although PDTC has to be applied before endotoxin treatment and the toxic effects as well as the possibility for long term application of the substance are uncertain, prevention of inducible NF-κB activation is a promising area for future research. In agreement with this study, Böhrer and his colleagues injected mice intravenously with a plasmid construct encoding IκBα, and achieved a 20–35% transfection efficiency for monocytes/ macrophages and endothelial cells *in vivo*. They report that pretransfection of mice with IκBα attenuated NF-κB activation and tissue factor expression in renal tissue and significantly increased survival rate following LPS treatment [64]. No high-selective, non-toxic inhibitor of NF-κB is yet available, but Morishita et al have developed an alternative strategy, using a DNA fragment containing the consensus for the NF-κB binding site to block NF-κB driven transcription. By infusion of the decoy DNA into rat coronary arteries, they successfully prevented NF-κB activation and inhibited myocardial infarction [65, 66]. The therapeutic benefit seen with antioxidants and DNA constructs against NF-κB activation in diseased state such as myocardial infarction and septic shock, together with the rapidly increasing knowledge of the signalling processes involved, will certainly encourage the search for more selective and effective low molecular drugs in the near future.

Obviously, research on the role of NF-κB in inflammatory and immune cardiovascular disorders is still at an early stage. Many questions remain to be answered. For instance, two studies recently reported that oxidized LDL, generally regarded as a major pathogenic initiator of atherosclerosis, inhibits NF-κB activation in macrophages and smooth muscle cells [67, 68], raising the question as to which stimuli and mechanisms are responsible for the regulation of NF-κB signalling in atherosclerosis.

STATs comprise another family of transcription factors. To date, seven STAT genes have been identified [69]. As originally defined, activation of STATs involves ligand-dependent activation of a particular class of receptor-associated tyrosine kinases, the JAK proteins, which phosphorylate themselves and receptor components, creating recruitment sites for STATs. The STATs are phosphorylated, dissociate from the receptor-JAK complex, and then translocate to the nucleus where they participate in transcriptional gene activation. The JAK-STAT mediated cytokine-response has been more precisely defined than the NF-κB transduction pathway. One example is the interferon-induced signalling pathway. During the response to interferon-α (IFN-α), JAK1 and Tyk2 tyrosine kinases are activated, leading to activation of STAT1 and STAT2. Activated STAT1 and STAT2 assemble together with ISGF3 (interferon-stimulated gene factor 3), and translocate into the nucleus where they bind to interferon-stimulated response elements in the promoter of IFN-α stimulated genes. The response to IFN-γ is mediated by activation of JAK1 and JAK2 associated with IFN-γ receptors, leading to phosphorylation of STAT1. The activated STAT1 forms homodimers, the so-called γ-activated factors, translocate into nucleus, and bind the γ-activated site (GAS) element in the promoters of IFN-γ induced genes [70, 71]. The JAK-STAT pathway is crucial for cytokine-mediated gene responses and a central determinator of their specificities (reviewed in [69, 72]). Therefore, this transcription pathway attracts considerable attention in the study of inflammatory and immune diseases.

It is now evident that infiltration of T cells occurs in the initial stage of atherosclerosis, and also happens in the development of transplant arteriosclerosis [5, 25, 73, 74]. Studies of cytokine expression suggest a Th1 predominant pattern in these conditions ([75] and unpublished results). However, the molecular mechanisms responsible for regulating the differentiation of Th1 and Th2 cells are less well defined.

Several lines of evidence have suggested that STATs may participate in the Th1 development. In STAT4 knockout mice, IL-12 function is impaired, resulting in multiple deficiencies including impaired induction of IFN-γ, reduced proliferation and cytolytic function of natural killer cells as well as hampered defective Th1 differentiation [76-79], indicating that STAT4 is required for mediating IL-12 regulated functions. Furthermore, exposure of murine CD4+ T cells to IFN-γ or human CD4+ cells to IFN-α maintained expression of the IL-12 receptor β2 chain, which is specifically expressed by Th1 cells but not Th2 cells [80-82]. Since the function of IFN-γ and IFN-α both depend on the activation of STATs, these data again suggest that differential regulation of STATs may determine T cell differentiation, resulting in a specific cytokine pattern in a local environment.

Although it is well recognized that transcription pathways have a central role during the development of cardiovascular diseases, the crosstalk between different signalling pathways in determining the response of cells to stimuli and ultimately controlling cytokine networks is unclear. Future studies should gain a more detailed insight into the various mechanisms that activate the different signalling pathways.

Cells and Cellular Interactions

Fatty streak formation and antigen presentation One of the most important, cytokine-regulated endothelial functions is the capacity of these cells to activate immunocompetent T cells by presenting foreign antigens. Endothelial cells cannot present antigens under baseline conditions but stimulation with the immune-regulating cytokine, IFN-γ, renders the endothelium capable of antigen presentation. This activity was discovered by Hirschberg, Thorsby and their colleagues, who showed that cultures of umbilical vein endothelial cells can activate allogeneic T lymphocytes in mixed cultures [83]. Pober et al. unveiled the cytokine-dependent mechanism that endows antigen-presenting capacity to the endothelium. IFN-γ induces transcriptional expression of MHC class II molecules, which can bind fragments of internalized peptide antigens in an endosomal compartment. The MHC-oligopeptide fragments are transported to the cell surface, where they can be recognized by antigen-specific CD4+ T lymphocytes [84-87]. Although IFN-γ alone is required for induction of MHC class II expression by endothelial cells, several other cytokines modulate the IFN-γ induced MHC expression, including TNF-α , IL-1, and IL-3 [84, 88].

Antigen presentation to CD4+ T cells depends not only on MHC expression but also on the ability of the antigen-presenting cell to internalize and process antigen, secrete IL-1, and express costimulatory adhesion molecules such as ICAM-1, LFA-3, and B7. The IFN-γ activated endothelial cell must therefore also be able to carry out these functions [84] . Similar and in part identical molecular interactions operate in endothelial antigen presentation via the synthetic route to HLA class I-restricted CD8+ T lymphocytes [89].

In vivo, endothelial cells of the microvasculature are often seen to express MHC class II molecules [84] and such expression can also be observed in large vessel endothelium under pathological conditions such as chronic rejection of organ transplants [90, 91]. Due to its great surface area, the endothelium should have a huge potential for antigen presentation and probably represents an important amplification loop for immune activation.

An interesting functional difference exists between endothelial cells, which respond to IFN-γ by MHC class II expression and antigen-presenting capacity, on the one hand, and fibroblasts and smooth muscle cells, which also respond to IFN-γ by MHC class II expression but which cannot *de novo* activate resting T cells. This difference could be due to differences in the capacity to express costimulatory molecules and/or antigen processing capacity [92, 93].

Cytokines in the transition into atherosclerosis Cell culture studies have shown that proinflammatory and immune-regulatory cytokines modify proliferative responses in vascular cells. IL-1 has a weak, growth-promoting effect on smooth muscle cells by upregulating PDGF receptors and increasing autocrine PDGF production [94]. This is, however, counteracted by its induction of the growth-inhibitory prostaglandin, PGE1 [94]. *In vivo*, it is likely that cellular growth and differentiation is regulated in a complex fashion by integration of signals derived from cytokines, growth factors, and autocoids [25, 95, 96].

Interferons are potent growth inhibitors for cultured smooth muscle cells [93, 97-100]. This is due to a direct inhibitory effect on growth factor-induced progression through the first part of the G1 phase of the cell cycle [98, 101]. IFN-γ also inhibits α-actin and collagen production by smooth muscle cells [97, 102], which could contribute to its drastic effects on arterial scar formation (see below). The inhibitory effect of IFN-γ on smooth muscle cell replication is, however, dependent on the simultaneous presence of growth factors in the extracellular milieu. Thus, IFN-γ may upregulate PDGF receptors, leading to a paradoxically increased growth factor sensitivity in a serum-free environment. The divergent effects of IFN-γ in growth factor-poor vs. growth factor-rich cell culture environments might reflect a differential regulation of vascular cell proliferation by IFN-γ producing T and NK cells depending on the local availability of growth factors in different stages of inflammation *in vivo*.

Animal experiments using the rat carotid artery injury model have revealed that IFN-γ is an *in vivo* modulator of smooth muscle cell proliferation and tissue accumulation in the arterial intima. Injection of recombinant IFN-γ inhibits smooth muscle cell proliferation and reduces the size of intimal hyperplastic lesions [103, 104]. Administration of IFN-γ during the first week after injury causes persistent growth inhibition and reduces lesion size even 10 weeks later [103]. This suggests that interferon-mediated growth inhibition during the early phase of the response may determine the final outcome of the lesion.

Removal of IFN-γ producing T lymphocytes using cytolytic antibodies, in contrast, increases smooth muscle cell proliferation and lesion formation in the injured artery [104]. Similarly, restenotic lesions become significantly larger in T cell deficient, homozygous rnu/rnu rats compared to T cell competent, heterozygous rnu/+ littermates [104]. Together, the results from genetically T cell defective, T cell depleted, and IFN-γ treated rats suggest that the Th1 response inhibits vascular and other connective tissue repair processes.

It appears likely that growth factors govern the formation of the fibrous cap but the precise molecular mechanisms are not fully known [2, 95]. It is, however, clear that plaque smooth muscle cells express receptors for PDGF and FGF and that plaque macrophages and endothelial cells produce mRNA for PDGF [105]. Since endothelial dysfunction and damage is often observed at this stage, it is possible that microthrombi formed on denuded plaque surfaces release PDGF that stimulates smooth muscle cell immigration and the formation of the cap.

Apoptosis of smooth muscle cells has recently been demonstrated in human atherosclerotic plaques, indicating that death of vascular smooth muscle cells also influences the final composition and tensile strength of the plaque [106-108].

Among the cytokines existing in atherosclerotic lesions, TNF induces apoptosis by a mechanism that is at least partly clarified. This process is dependent on the binding of TNF to TNF receptor-1 which is associated with a "death domain" in the cytoplasmic region [109, 110]. Other cytokines including IL-1, IL-2 and IFN-γ can induce apoptosis directly as well as indirectly through induction of TNF from target cells, including smooth muscle cells. TNF and some other cytokines are also able to induce nitric oxide production, which depresses smooth muscle cell function and can induce apoptosis [111-113]. As growth hormone is able to inhibit the production of proinflammatory cytokines in many cell types, it may also play an important role in the regulation of apoptosis induced by these cytokines. Whether a specific cytokine inhibits or suppresses apoptosis depends on their effects on cell death regulatory genes such as bcl-2 and iap family members, Fas receptor, and others. Finally, the intracellular pathways of cytokine receptor-mediated control of apoptosis have begun to be unravelled, implicating specific intracellular receptor domains and protein kinases in the regulation of apoptosis [114-116].

Smooth muscle cells produce the extracellular matrix of the vessel wall and therefore have a high capacity to synthesize structural proteins such as collagens, elastin, basement membrane components, and core proteins of proteoglycans. This is reflected in the ultrastructure, which is dominated by endoplasmic profiles together with contractile filaments [117]. In fact, the proportion between these two components can be used to determine the phenotypic state of the smooth muscle cell [118]. Thus, contractile filaments dominate in contractile smooth muscle cells of the media, which regulate vascular tone, while endoplasmic reticular structures are abundant in "synthetic" smooth muscle cells found in the intima and in cell culture systems.

Agents such as PDGF, which induce smooth muscle cell proliferation, also affect differentiation and matrix formation [119, 120]. Vice versa, agents that degrade the extracellular matrix modulate smooth muscle cell phenotype. This implies that these phenomena are linked and that smooth muscle cell growth and differentiation could be regulated by cell-matrix interactions. Studies on cytokine effects on smooth muscle cells support this notion.

The extracellular matrix of the vessel wall is controlled by inflammatory cytokines. Interferon-γ is a potent inhibitor of collagen synthesis [102] and also inhibits production of α-actin, the major component of contractile filaments, and DNA synthesis in smooth muscle cells [97]. IL-1 and TNF may exert important control of the extracellular matrix by inducing metalloproteinases that degrade matrix components [121, 122]. Proteolytic degradation of the matrix reduces adhesive interactions between smooth muscle cells and their microenvironment. This, in turn, causes the smooth muscle cells to dedifferentiate and proliferate in response to growth factors [123, 124].

The antiinflammatory cytokine, TGF-β, exerts effects on smooth muscle cells that are opposite to those of the proinflammatory ones. Thus, TGF-β stimulates collagen [125, 126] and α-actin production [127, 128] and induces fibrotic hyperplasia when transfected into the arterial wall [129]. Consequently, antibodies to TGF-β inhibit neointimal hyperplasia after mechanical injury [130, 131]. TGF-β also modulates growth of smooth muscle cells but its effects are complex, dependent on the phenotypic state, and mediated via expression of other growth factors and their receptors [132, 133].

Plaque complications It has been suggested that macrophage activation causes the appearance of fissures, thrombi, and actual plaque rupture by secreting proteases that degrade the extracellular matrix [134-137]. Mast cells, which are also present at sites of plaque rupture [138, 139], can exert similar functions [140]. Interestingly, leukocyte elastase and mast cell chymase activate TGF-β [141]; this may counteract tissue destruction by inducing a fibrotic antiinflammatory response. Proinflammatory cytokines (TNF-α, IFN-γ) could play a role in this process by virtue of their macrophage-activating and matrix-inhibiting activities [142]. In addition, such cytokines (TNF-α, IL-1, IFN-γ) enhance PAI-1 (plasminogen activator inhibitor-1) and downregulate tPA (tissue plasminogen activator) expression by endothelial cells (see above); this would be expected to tip the balance towards thrombus formation on the arterial surface. Under normal conditions, the endothelial cells constitute a non-thrombogenic surface due to their production of antithrombotic factors such as prostacyclin [143], heparin-like molecules [144], nitric oxide [145, 146], and an activator of the fibrinolytic system, tissue-type plasminogen activator (t-PA) [147]. However, endothelial cells are also able to produce procoagulant substances, e.g. von Willebrand factor [148], tissue factor [149] and inhibitors of the fibrinolytic system, i.e. plasminogen activator inhibitor-1 (PAI-1) [150] and plasminogen activator inhibitor-2 (PAI-2) [151].

TNF-a and IL-1 have been shown to increase the expression of PAI-1 [152-154]. The effect of TNF-α on PAI-1 expression has been confirmed *in vivo* in healthy volunteers [155,156]. TNF-α also increases the expression of u-PA, urokinase plasminogen activator, [157,158] and PAI-2 in cultured endothelial cells [159]. These effects are modulated by IFN-γ: the TNF-α-induced expression of u-PA and PAI-1 is antagonized by IFN-γ [160-162], while the TNF-α-induced expression of PAI-2 is synergistically increased by IFN-γ [160]. The net effect of TNF-α and IFN-γ is antifibrinolytic in cultured endothelial cells due to increased expression of PAI-1 and PAI-2. Increased expression of PAI-1, t-PA and u-PA has been demonstrated in smooth muscle cells and macrophages of atherosclerotic arteries. The expression of PAI-1 exceeds that of the plasminogen activators, suggesting that fibrinolysis is inhibited [163-167].

Cytokine-induced expression of plasminogen activators and their inhibitors may have a role in the development of the plaque as well as in the induction of thrombotic complications. t-PA and u-PA could induce degradation of the extracellular matrix by activating matrix metalloproteinases (MMP). These enzymes degrade extracellular matrix and can be activated by plasmin. Their activation is inhibited by plasmin inhibitors, tissue inhibitors of metalloproteinases, and PA inhibitors [168-170] . PAI-1 may limit MMP activation and thus protect a developing plaque from uncontrolled matrix degradation. In advanced atherosclerosis, increased expression of MMPs and matrix degrading activity has been detected in vulnerable regions of the plaque [136,171] and it is possible that cytokine-induced MMPs may promote destabilization and rupture of the plaque.

Studies in mice with targeted gene inactivation of t-PA, u-PA, PAI-1, the urokinase receptor (u-PAR), and plasminogen revealed that vascular injury-induced neointima formation is reduced in mice lacking u-PA-mediated plasmin proteolysis, unaltered in t-PA- or u-PAR-deficient mice, and that PAI-1 plays an inhibitory role in vascular wound healing and arterial neointima formation after injury [172,173]. The plasminogen system could therefore exert several effects in advanced atherosclerotic lesions. By inhibiting MMP activation, it may protect the plaque against rupture. However, if the plaque is already destabilized, PAI-1 may prevent thrombosis from occurring on the plaque surface. The role of proinflammatory cytokines in these processes is intriguing and deserves further clarification.

Thrombosis on atherosclerotic plaques is thought to be the cause of myocardial infarction and stroke [174]. Tissue factor (TF) is the cellular receptor for coagulation factor VIIa and is generally viewed as the primary physiological initiator of blood coagulation. In atherosclerotic plaques, TF has been found in smooth muscle cells, macrophages, endothelial cells, and in the lipid-rich necrotic core [175-177], implicating a role for TF in mediating thrombosis associated with atherosclerosis. Induction of the TF gene in human monocytic cells and endothelial cells exposed to bacterial lipopolysaccharide or cytokines is mediated by a distal enhancer (-227 to -172 bp) containing two AP-1 sites and a κB site. TNF-α induces a procoagulative state of endothelial cells by stimulating their production of tissue factor. Additionally, procoagulant activity of atherosclerotic lesions is also induced by a T cell-derived cytokine, macrophage procoagulant-inducing factor (MTIF).

Regulation of vascular tone Proinflammatory cytokines (TNF, IL-1, IFN-γ) modulate smooth muscle cell contractility on several levels of this regulatory system. The most long-lasting effect on smooth muscle cell contractility is caused by direct interference with the production of contractile filaments. For example, IFN-γ down-regulates expression of the α-SM-actin gene [97]. A more rapid effect on vascular contractility is accomplished by cytokine regulation of NO production (Reviews: [178, 179]). As mentioned, NO is normally released by endothelial cells. It diffuses over to smooth muscle cells, where it nitrosylates the heme group of guanylyl cyclase, resulting in an activation of the enzyme to produce cGMP. The elevated cGMP level activates the myosin kinase cascade, resulting in dissociation of myosin from actin and a relaxation of the smooth muscle cells. The artery is normally under modest NO-dependent vasodilation, and systemic administration of NO synthase inhibitors results in increased vascular tone and elevated blood pressure [178].

Endothelial NO synthesis is accomplished by the enzyme, endothelial NO synthase (eNOS or NOS-3). It is constitutively expressed as a protein but requires activation by Ca^{++}/calmodulin [180]. This occurs after stimulation of the endothelial cells by bradykinin, acetylcholin and other stimuli. Since NO synthase inhibitors increase vascular tone under baseline conditions, the normal vessel wall is probably in a state of NO-dependent, partial relaxation [178, 181].

Smooth muscle cells do not normally produce NO but can be stimulated to do so by proinflammatory cytokines. Both TNF-α, IL-1, lipopolysaccharide (LPS), and IFN-γ induce the production of large amounts of NO in cultured smooth muscle cells [182-185]. The cytokine-inducible NO synthase isoform, iNOS or NOS-2, binds calmodulin with high affinity immediately after translation and is therefore independent of Ca^{++} for its activity [186, 187]. This results in a high capacity for NO synthesis, which lasts until the enzyme protein is degraded [187]. The iNOS expressed by smooth muscle cells appears to be identical to the one expressed by cytokine-activated macrophages and is highly conserved between species [188].

The dichotomy of inflammatory signalling is also reflected in the regulation of iNOS. While transcription of this gene is induced by IL-1, TNF-α and IFN-γ, it is inhibited by IL-4. Furthermore, TGF-β downregulates iNOS by suppressing transcription as well as reducing mRNA and protein stability [189, 190].

NO produced by cytokine-stimulated smooth muscle cells acts as an auto- and paracrine mediator to control contractility and metabolism of the cells. Moderate cytokine stimulation induces sufficient NO to nitrosylate smooth muscle cell heme proteins, including guanylyl cyclase [191]. At higher levels of cytokine stimulation, the output of NO is sufficient to react with iron atoms in the iron-thiol groups of many intracellular enzymes, which therefore lose their activity [191]. This causes inhibition of mitochondrial respiration, ATP deficiency and a switch to anaerobic glycolysis in the cell [184, 191]. The combined effects of guanylyl cyclase activation and reduced energy levels are likely to explain the vasodilation of inflammation [184].

In vivo, the iNOS gene is activated in smooth muscle cells during the response to vascular injury. De-endothelializing balloon catheter injury to the rat carotid artery causes rapid expression of iNOS in the underlying smooth muscle cell layer [192]. iNOS expression is maintained in the smooth muscle cells for more than a week, probably due to cytokine stimulation. Enzyme expression is, however, induced so rapidly after deendothelialization (hours) that it is unlikely to be caused by *de novo* production of IFN-γ, TNF-α and/or IL-1 from infiltrating leukocytes. Instead, there may be release of extracellularly deposited cytokines or activation of transcription through cytokine-independent mechanisms. iNOS activity is likely to be important by dilating vessels during inflammation, modulating platelet deposition and thrombus formation on the injured vessel, and controlling smooth muscle cell proliferation during the response to injury [145, 192, 193].

Clinical Relevance   see under the different subsections

Göran Hanson and Zhong-qun Yan

References

1. Steinberg D et al (1989) N Engl J Med 320:915-24
2. Ross R (1993) Nature 362:801-9
3. Steinberg D (1997) J Biol Chem 272:20963-6
4. Jonasson L et al (1986) Arteriosclerosis 6:131-8
5. Hansson GK et al (1989) Am J Pathol 135:169-75
6. Stemme S et al (1992) Immunology 36:233-42
7. Kristensen SD et al (1997) Am J Cardiol 80:5E-9E
8. Davies MJ (1997) N Engl J Med 336:1312-4
9. Falk E (1992) Circulation: 3:30-42
10. Doi TK et al (1993) J Biol Chem 268:2126-33
11. Abraham R et al (1995) J Immunol 154:1-8
12. Nicoletti A et al (1998) Eur J Immunol, in press
13. Brown MS, Goldstein JL (1983) Annu Rev Biochem 52:223-61
14. Suzuki H Y et al (1997) Nature 386:292-6
15. Li H et al (1995) J Clin Invest 95:122-33
16. Clinton SK et al (1992) Am J Pathol 140:301-16
17. Antonov AS et al (1997) J Clin Invest 99:2867-76
18. Smith JD et al (1995) Proc Natl Acad Sci USA 92:8264-8
19. Qiao JH et al (1997) Am J Pathol 150:1687-99
20. Lau W et al (1995) J Clin Invest 95:1957-65
21. Fong LG et al (1990) J Biol Chem 265:11751-60
22. van Lenten BJ, Fogelman AM (1992) J Immunol 148:112-6
23. Geng YJ et al (1995) Arterioscler Thromb Vasc Biol 15:1995-2002
24. Dufva M et al (1995) J Lipid Res 36:2282-90
25. Hansson GK (1997) Curr Opin Lipidol 8:301-11
26. Jonasson L et al (1985) J Clin Invest 76:125-31
27. DeGraba TJ (1997) Neurology: S15-9
28. Poston RN et al (1992) Am J Pathol 140:665-73
29. Wilcox JN et al (1994) J Atheroscler Thromb: S10-3
30. Nelken NA et al (1991) J Clin Invest 88:1121-7
31. Ylä-Herttuala S et al (1991) Proc Natl Acad Sci USA 88:5252-6
32. Koch AE et al (1993) Am J Pathol 142:1423-31
33. Fyfe AI et al (1994) J Clin Invest 94:2516-20
34. Roselaar SE et al (1995) J Clin Invest 96:1389-94
35. Moran CS et al (1994) Arterioscler Thromb 14:1356-63
36. Dansky HM et al (1997) Proc Natl Acad Sci USA 94:4642-6
37. Gupta S et al (1997) J Clin Invest 99:2752-61
38. Shi C et al (1996) Proc Natl Acad Sci USA 93:4051-6
39. Zhou X, Hansson GK (1998) J Clin Invest 101:1717-1725
40. Ghosh S e al (1990) Cell 62:1019-29
41. Ruben SM et al (1991) Science 251:1490-3
42. Schmid RM et al (1991) Nature 352:733-6
43. Ryseck RP et al (1992) Mol Cell Biol 12:674-84
44. Bours V PR et al (1992) Mol Cell Biol 12:685-95
45. Verma IM et al (1995) Genes Dev 9:2723-35
46. Haskill S et al (1991) Cell 65:1281-9
47. Thompson JE et al (1995) Cell 80:573-82
48. Whiteside ST et al (1997) Embo J 16:1413-26
49. Siebenlist U (1997) Biochim Biophys Acta 1332: R7-R13
50. Beg AA et al (1992) Genes Dev 6:1899-913
51. Henkel T et al (1993) Nature 365:182-5
52. Traenckner EB et al (1994) Embo Journal 13:5433-41
53. Scherer DC et al (1995) Proc Natl Acad Sci USA 92:11259-63
54. Baldwin AJ (1996) Ann Rev Immunol 14:649-83
55. Barnes PJ, Larin M (1997) New Engl J Med 336:1066-1071
56. Lawrence R et al (1994) J Biol Chem 269:28913-8
57. Bourcier T et al (1997) J Biol Chem 272:15817-24
58. Yan ZQ, Hansson GK (1998) Circ Res 82:21-29
59. Spink J et al (1995) J Biol Chem 270:29541-7
60. Brand K et al (1996) J Clin Invest 97:1715-22
61. Schreck R et al (1991) Embo J 10:2247-58
62. Schreck R et al (1992) J Exp Med 175:1181-94
63. Liu SF et al (1997) J Immunol 159:3976-83
64. Böhrer H et al (1997) J Clin Invest 100:972-85
65. Morishita R et al (1997) Nat Med 3:894-9
66. Sawa Y et al (1997) Circulation: 96(9 Suppl): II-II2804
67. Ohlsson BG et al (1996) J Clin Invest 98:78-89
68. Ares MP et al (1995) Arterioscler Thromb Vasc Biol 15:1584-90
69. Darnell JJ (1997) Science 277:1630-5
70. Wong LH et al (1997) J Biol Chem 272:28779-85
71. Pellegrini S, Dusanter FI (1997) Eur J Biochem 248:615-33
72. Silvennoinen O et al (1997) Apmis 105 no 7:497-509
73. Zhou X et al (1996) Am J Pathol 149:359-66
74. Hansson GK et al (1991) Arterioscl Thromb 11:745-50
75. Stemme S et al (1995) Proc Natl Acad Sci USA 92:3893-7
76. Thierfelder WE et al (1996) Nature 382:171-4
77. Kaplan MH et al (1996) Nature 382:174-7
78. Jacobson NG et al (1995) J Exp Med 181:1755-62
79. Cho SS et al (1996) J Immunol 157:4781-9
80. Rogge L et al (1997) J Exp Med 185:825-31
81. Szabo SJ et al (1997) J Exp Med 185:817-24
82. Groux H et al (1997) J Immunol 158:5627-31
83. Hirschberg H et al (1980) J Exp Med: 249s-255s
84. Pober JS, Cotran RS (1990) Physiol Rev 70:427-51
85. Epperson DE, Pober JS (1994) J Immunol 153:5402-12
86. St Louis JD et al (1993) J Exp Med 178:1597-605
87. Murray AG et al (1994) Immunity 1:57-63

88. Korpelainen EI et al (1995) Blood 86:176-82
89. Kaneko K et al (1994) Clin Exp Immunol 98:264-9
90. Marelli BF et al (1996) J Exp Med 183:1603-12
91. Seino K et al (1995) Int Immunol 7:1331-7
92. Fabry Z et al (1990) J Neuroimmunol 28:63-71
93. Stemme S et al (1990) Immunology 69:243-9
94. Libby P et al (1988) J Clin Invest 81:487-98
95. Libby P, Hansson GK(1991) Lab Invest 64:5-15
96. Hansson GK et al (1989) Arteriosclerosis 9:567-78
97. Hansson GK et al (1989) J Exp Med 170:1595-608
98. Hansson GK et al (1988) Circ Res 63:712-9
99. Warner SJ et al (1989) J Clin Invest 83:1174-82
100. Palmer H, Libby P (1992) Lab Invest 66:715-21
101. Bennett MR et al (1994) Circ Res 74:525-36
102. Amento EP et al (1991) Arterioscler Thromb 11:1223-30
103. Hansson GK, Holm J (1991) Circulation 84:1266-72
104. Hansson GK et al (1991) Proc Natl Acad Sci USA 88: 10530-4
105. Wilcox JN (1993) Am J Cardiol 72:88E-95E
106. Isner JM et al (1995) Circulation 91:2703-11
107. Geng YJ, Libby P (1995) Am J Pathol 147:251-66
108. Bennett MR et al (1995) J Clin Invest 95:2266-74
109. Tartaglia LA et al (1993) Cell 74:845-53
110. Smith CA et al (1994) Cell 76:959-62
111. Geng YJ et al (1996) Arterioscler Thromb Vasc Biol 16:19-27
112. Fukuo K et al (1996) Hypertension: 823-6
113. Chin YE et al (1997) Mol Cell Biol 17:5328-37
114. Reed JC (1997) Nature 387:773-6
115. Nagata S (1997) Cell 88:355-65
116. Salvesen GS, Dixit VM (1997) Cell 91:443-6
117. Chen YH et al (1997) Circulation 95:1169-75
118. Thyberg J et al (1995) Cell Tissue Res 281:421-433
119. Reidy MA (1994) Ann NY Acad Sci 714:225-30
120. Thyberg J (1996) Int Rev Cytol 169:183-265
121. Galis ZS et al (1995) Proc Natl Acad Sci USA 92:402-6
122. Galis ZS et al (1995) Ann NY Acad Sci 748:501-7
123. Krettek A et al (1997) Arterioscler Thromb Vasc Biol 17:2897-903
124. Snow AD et al (1990) Am J Pathol 137:313-30
125. Lawrence R et al (1994) J Biol Chem 269:9603-9
126. Davidson JM et al (1993) J Cell Physiol 155:149-56
127. Shi Y et al (1996) Arterioscler Thromb Vasc Biol 16:1298-305
128. Desmouliere A et al (1993) J Cell Biol 122:103-11
129. Nabel EG et al (1993) Proc Natl Acad Sci U S A 90:10759-63
130. Rasmussen LM et al (1995) Am J Pathol 147:1041-8
131. Wolf YG et al (1994) J Clin Invest 93:1172-8
132. Stouffer GA, Owens GK (1994) J Clin Invest 93:2048-55
133. Majack RA et al (1990) J Cell Biol 111:239-47
134. Farb A et al (1996) Circulation 93:1354-63
135. Libby et al (1996) Curr Opin Lipidol 7:330-5
136. Shah PK et al (1995) Circulation 92:1565-9
137. Moreno PR et al (1994) Circulation 90:775-8
138. Kaartinen M et al (1996) Circulation 94:2787-92
139. Kovanen PT et al (1995) Circulation 92:1084-8
140. Saarinen J et al (1994) J Biol Chem 269:18134-40
141. Taipale J et al (1995) J Biol Chem 270:4689-96
142. Libby P et al (1997) Ann NY Acad Sci 811:134-42
143. Moncada S et al (1978) Nature 273:767-8
144. Marcum JA et al (1984) J Clin Invest 74:341-50
145. Yan ZQ et al (1996) Circ Res 79:38-44
146. Radomski MW et al (1993) Cardiovasc Res 27:1380-2
147. van Hinsbergh VWM (1988) Haemostasis 18:307-27
148. Jaffe EA et al (1974) Proc Natl Acad Sci USA 71:1906-9
149. Lyberg T et al (1983) Br J Haematol 53:85-95
150. Philips M et al (1984) Biochim Biophys Acta 802:99-110
151. Schleef RR et al (1988) J Cell Physiol 134:269-74
152. Yamashita M, Yamashita M (1997) Thromb Res 87:165-70
153. Emeis JJ et al (1995) Blood 85:115-20
154. Samad F et al (1996) J Clin Invest 97:37-46
155. van der Poll T et al (1991) J Exp Med 174:729-32
156. van Hinsbergh VWM et al (1990) Blood 76:2284-9
157. van Hinsbergh VWM et al (1990) Blood 75:1991-8
158. Niedbala MJ, Stein M (1991) Biomed Biochim Acta 50:427-36
159. Zoellner H et al (1993) Thromb Haemost 69:135-40
160. Arnman V et al (1995) Thromb Res 77:431-40
161. Gallicchio M et al (1996) J Immunol 157:2610-7
162. Niedbala MJ, Picarella MS (1992) Blood 79:678-87
163. Schneiderman J et al (1992) Proc Natl Acad Sci USA 89:6998-7002
164. Raghunath PN et al (1995) Arterioscler Thromb Vasc Biol 15:1432-43
165. Padro T JJ et al (1995) Arterioscler Thromb Vasc Biol 15:893-902
166. Lupu F et al (1995) Arterioscler Thromb Vasc Biol 15:1444-55
167. Reidy MA et al (1996) Circ Res 78:405-14
168. Gomis RF et al (1997) Nature 389:77-81
169. Olson MW et al (1997) J Biol Chem 272:29975-83
170. Carmeliet P et al (1997) Ann NY Acad Sci 811:191-206
171. Knox JB et al (1997) Circulation 95:205-12
172. Carmeliet P et al (1997) Circulation 96:3180-91
173. Carmeliet P et al (1997) Circ Res 81:829-39
174. Fuster V et al (1992) N Engl J Med 326:242-50
175. Wilcox JN et al (1989) Proc Natl Acad Sci USA 86:2839-43
176. Thiruvikraman SV et al (1996) Lab Invest 75:451-61
177. Annex BH et al (1995) Circulation 91:619-22
178. Moncada S, Higgs EA (1995) FASEB J 9:1319-30
179. Moncada S, Higgs EA (1991) E J Clin Invest 21:361-74
180. Moncada S et al (1991) Pharmacological Reviews 43:109-42
181. Chu A et al (1991) J Clin Invest 87:1964-8
182. Beasley D et al (1991) J Clin Invest 87:602-8
183. Busse R, Mülsch A (1990) Febs Letters 275:87-90
184. Geng YJ et al (1992) Circ Res 71:1268-76
185. Wood KS et al (1990) Biochem Biophys Res Comm 170:80-8
186. Cho HJ et al (1992) J Exp Med 176:599-604
187. Xie QW et al (1992) Science 256:225-8
188. Geng YJ et al (1994) Biochim Biophys Acta 1218:421-4
189. Perrella MA et al (1994) J Biol Chem 269:14595-600
190. Vodovotz Y et al (1993) J Exp Med 178:605-13
191. Geng YJ et al (1994) Exp Cell Res 214:418-28
192. Hansson GK et al (1994) J Exp Med 180:733-8
193. Yan ZQ, Hansson GK (1998) Circ Res 82:21-29
194. Mallat Z et al (1999) Circ Res 85:17-24
195. Mach F, Schönbeck U, Sukhova GK, Atkinson E, Libby P (1998) Nature 394:200-3
196. Nicoletti A, Kareri S, Caligiari G,Bariety J, Hansson GK (1998) J Clin Invest 102:910-918

DAF

Definition *Decay-accelerating factor*

See: →Complement system (interaction of vascular cells with)

DAG

Definition   *Diacylglycerol*

See: →Lipid mediators; →Angiogenin; →Platelet stimulus-response coupling; →Vasomotor tone regulation, molecular mechanisms of

Diabetes mellitus

Definition   *Disease due to abnormal glucose metabolism leading to hyperglycemia. Two types are recognized which include diabetes mellitus type I (juvenile, insulin-dependent diabetes mellitus) and type II (non-insulin-dependent diabetes mellitus).*

See: →Blood cells, interaction with vascular cells

Diabetic Microangiopathy

Definition   *Proliferation of retinal blood vessels accompanied by leakage and hemorrhage leading to the destruction of the retina.*

See: →Endothelial cells

DIP

Definition   *Degranulation-inhibiting protein*

See: →Angiogenin

DMS

Definition   *Demarcation membrane system*

See: →Megakaryocytes

DNA

Definition   *Desoxyribonucleic acid*

DTS

Definition   *Dense tubular system*

See: →Platelet stimulus-response coupling

ECGS

Definition   *Endothelial cell growth supplement*

See: →Endothelial cells

ECM

Definition   *Extracellular matrix*
See: →Extracellular matrix

EDHF

Definition   *Endothelial-derived hyperpolarizing factor*

See: →Thrombosis

EDRF

Definition   *Endothelial cell-derived relaxing factor*

See: →Nitric Oxide

EDS

Definition   *Ehlers-Danlos syndrom*

See: →Bleeding disorders

EGF

Definition   *Epidermal growth factor*

See: →Coagulation factors; →Matrix metalloproteinases; →Procoagulant activities; →Signal transduction mechanisms in vascular biology; →Smooth muscle cells; →Transforming growth factor β

ELAM

Definition   *Endothelial cell leukocyte adhesion molecule*

See: →Selectins; →Atherosclerosis; →Blood cells, interaction with vascular cells

Endostatin

Definition   *Fragment of collagen XVIII with antiangiogenic activity*

See: →Angiogenesis inhibitors

Endothelial Cell-Derived Relaxing Factors (EDRF)

Definition   *Same as nitric oxyde (NO).*

See: →Nitric oxyde

Endothelial Cells

Definition   *Mesoderm-derived cells that constitute the inner lining of blood vessels in contact with blood. These cells play an important role in phenomenon such as angiogenesis, atherosclerosis, blood pressure regulation and leukocyte trafficking. The most frequently used endothelial cells for in vitro culture are human umbilical vein endothelial cells (HUVECs), human foreskin microvascular endothelial cells, bovine capillary endothelial cells*

(ACE, adrenal gland; BCE, brain), bovine aortic endothelial cells (BAE; FBAE, fetal), or mouse lung microvascular endothelial cells (LE II).

See also: →Angiogenesis; →Angiogenesis inhibitors; →FGF-1 and -2; →Ontogeny of the vascular system; →Vascular endothelial growth factor; →Vasomotor tone regulation, molecular mechanisms of

Introduction The adult vasculature consists of large arteries which progressively branch out into small vessels of various caliber, like a fractal and are terminated by pre-capillary arterioles. Capillaries irrigate tissues and are connected to post-capillary venules that associate together to form larger veins.

Tissues that require an increase in oxygen and nutrients release endothelial cell modulators towards pre-existing vessels which induce the sprouting of new capillaries. Over the past two decades, major efforts have been made to isolate and cultivate endothelial cells *in vitro*. It is generally admitted that growth and differentiation of cultured cells *in vitro* mimic the events leading to sprouting of endothelial cells and the formation of new blood vessels. Cultured endothelial cells proliferate, migrate, differentiate and express genes required for maintaining an anti-thrombotic surface as well as regulating the vasotonus or the trafficking of circulating cells when the appropriate growth factors and extracellular matrix components are supplied. It is obvious that the signals leading to the conversion of the quiescent phenotype to a growth factor-responsive phenotype are complex and cannot be fully understood by only an *in vitro* approach. Nevertheless, endothelial cell culture systems have provided a valuable tool for the study of blood vessel cell interactions.

Structure More than 10^{12} endothelial cells border the surface of blood vessels which is estimated at 1000 m^2 in adults [1]. Resting endothelial cells are heterogenous and differ in size, morphology and physiological functions depending on the vessel caliber and the organ. Endothelial cells from large vessels are generated from angioblasts derived from the mesenchyme by the process of vasculogenesis. In the adult, they form a typical cobblestone monolayer and control blood pressure by acting on vasoconstriction or vasodilatation. In contrast to microvascular endothelial cells, these cells are not involved in neovascularization, in the blood-tissue exchange of nutriments and oxygen or in the removal of waste components [2,3]. The interaction with the extracellular matrix contributes to vessel morphology. For example, endothelial cell derived from large vessels and cultured on a extracellular matrix synthesized by the kidney-derived epithelial MDCK cell line become fenestrated.

Microvascular endothelial cells are derived from hemangioblasts, which are common precursors for both endothelial and hematopoietic cells [4]. They are devoted to blood-tissue exchange and therefore exhibit variable ability in maintaining capillary permeability. They are classified into three different types on the basis of their morphology: continuous, fenestrated or discontinuous [5]. In continuous capillaries, the endothelial cytoplasm is continuous and the luminal and abluminal plasma membranes are only separated by sparse tight junctions. This provides a strong barrier as observed in lungs, high endothelial veinules or the central nervous system. Fenestrae are specialized plasma membrane microdomains appearing as pores of 60 nm containing a diaphragm. They are found mainly in endocrine glands, the choroid plexus, the gastrointestinal tract and kidney glomeruli. In discontinuous endothelium or sinusoidal cells, the clustered holes in the plasma membrane are larger (80–200 nm) and do not contain diaphragms. These sinusoidal cells are specialized in the exchange of large particles as it occurs in the liver or in the exchange of blood cells as it occurs in bone marrow or the spleen.

Regulation of Cell Function

Cell to Cell Interactions During the past years a considerable body of evidence has established the roles of several master genes for blood vessel formation during embryogenesis. Gene knock-out studies have demonstrated that VEGF [6-9] and VEGF-R2 [10,11] are required for angioblast proliferation during vasculogenesis and angiogenesis and VEGF-R1 for vessel assembly [12]. The Tie-2/angiopoietin-1-angiopoietin 2 system seems to be required later for maturation, branching and organization of large and small vessels [13]. It has been demonstrated that angiopoietin-1 activation of Tie-2 leads to the recruitment of peri-endothelial supporter cells and further to the production of extracellular matrix through TGF-β activation. This contributes to the maturation of blood vessels. On the contrary, angiopoietin-2 activation of Tie-2 has opposite effects. The first evidence of the existence of a molecular distinction between arteries and veins has been provided recently [14] by gene knock-out studies of ephrin-B2 (that specifies arteries) and its receptor Eph-B4 (that specifies veins).

Although the primary capillary plexus is remodeled many times during embryogenesis, expression of most of the genes implicated in vasculogenesis is reduced soon after birth and endothelial cell proliferation ceases. In fact, endothelial cells are among those exhibiting the lowest replication rate in the adult body with only 0.01% of the cells entering the cell cycle at any time.

Molecular Interactions and Signaling Mechanisms The elucidation of the mechanisms of action in embryogenesis of the two endothelial cell-specific regulatory systems, the members of the VEGF/VEGF-Rs and the angiopoïetins/Tie-2 families prompted several studies to determine their relative expression in normal and pathological tissues [15, 16].

VEGF is expressed in adult tissues in the vicinity of fenestrated endothelia in the choroid plexus and the kidney, suggesting a role in maintaining permeability [17]. Its expression is upregulated by hypoxia, hormones, hypoglycemia and advanced glycation end products in angiogenic tissues and it seems to be the only angiogenic factor expressed in uncontrolled angiogenesis such as diabetic retinopathy [18]. VEGF also contributes to endothelial cell survival. This has been demonstrated by conditional expression of VEGF in tumors. Downregulation of VEGF induces shedding of the endothelial cells of neovessels [19]. VEGF-R1 is expressed in large- and small-vessel endothelial cells where it mediates NO-dependent vasodilatation [20], permeability and possibly endothelium survival. VEGF-R2 is expressed in several organs which do not exhibit any sign of proliferation such as the kidney or the retina. VEGF-R3 expression is restricted to the lymphatic endothelium [21]. Angiopoietin-1 is widely expressed in adult tissues whereas the expression of its antagonist angiopoietin-2 is readily detectable in ovaries, placenta and the uterus, which are the three predominant sites of vascular remodeling in adults [22]. Tie-2 and Tie-1 are also widely expressed in the adult endothelium and are involved in the control of vessel integrity.

Several genes which are not expressed in the quiescent endothelium are expressed in the angiogenic endothelium. For example, the integrins $\alpha_v\beta_3$ and $\alpha_v\beta_5$ [23], the PEX fragment of the metalloprotease MMP 2 [24] and the homeobox gene Hox D3 [25].

Monoclonal antibodies against proliferating endothelial cells have allowed the identification of genes specifically expressed by angiogenic endothelial cells (Figure 1) such as endoglin [26], endosialin [27], type VIII collagen [28] and a splice variant of fibronectin [29].

Culture of Endothelial Cells   The isolation of endothelial cells presents three main difficulties: the organ of origin, the purity of the cell culture and the stability of the phenotype in culture.

Depending on the goal of the study, it is important to obtain endothelial cells cultured from the appropriate organ. For instance, if the aim of a study is to identify gene regulation in diabetic retinopathy or in skin ulceration, it would be preferable to work with endothelial cells cultured from the retina or the skin. Depending on whether the study is related to endothelium relaxing factors or to the mechanisms of action of angiogenic factors, endothelial cells cultured from large arteries should be used in the first case, and endothelial cells from microvessels in the second.

Human umbilical vein endothelial cells (HUVECs) are the most commonly used endothelial cells. They are easy to obtain by veinous infusion of collagenase and then flushing out a pure preparation of endothelial cells [30]. The isolation of endothelial cells from large vessels such as aorta does not require proteolytic treatment but a gentle scraping of the intima with a rubber policeman. By contrast, the obtention of microvessel derived endothelial cells is a major challenge. The enzymatic treatment depends on the resistance of the junctions between the capillaries and the neighboring cells or the extracellular matrix. For instance, skin endothelial cells require a long treatment to remove epithelial cells whereas retinal capillaries are easily dissociated from the neural retina. Once capillary fragments have been dissociated, they must be separated from any other cells that are present before seeding in culture dishes. Several strategies are used such as mesh filtering (for example, retinal capillaries are retained on a 40 μm mesh filter and selectively captured in a 60 μm mesh filter), gradient centrifugation, flow cytometry or magnetic separation, assuming that an antibody directed against an endothelial marker is available. The capillary cells are then seeded on culture dishes coated with various substrates. Bovine gelatin (0.2 % diluted in PBS) is most commonly used but pure preparations of type I or type IV collagen, fibronectin, laminin or extracellular matrix deposited by other cultured cells (more often corneal endothelial cells and ideally endothelial cells from the same organ) are also used, although they are more expensive. The ability of endothelial cells to bind specifically to lectins such as Ulex Europaeus type I is useful to remove the contaminating cells by washing. Bovine or mouse endothelial cells are cultured in medium containing 10 % calf serum whereas human cells generally require 20 % fetal calf serum. Growth advantage for endothelial cells is achieved by adding ECGS (Endothelial Cell Growth Supplement, which is an almost crude brain extract containing FGF1, FGF2 and VEGF) or purified growth factors and inhibitors of contaminating cells, for instance a high concentration of heparin for smooth muscle cells. Once the colonies of endothelial cells have started to emerge, they are identified by their cobblestone morphology and separated from the other cells. A simple technique consists in sucking-up the endothelial cell colonies with a pipette tip under the control of a microscope. Cloning rings or flow cytometry can also be used.

The endothelial origin of the cells in culture must be ascertained by the detection of endothelial specific markers such as von Willebrand factor, CD31, CD34, angiotensin converting enzyme or AcLDL uptake. However, neither the appearance of a cobblestone morphology which can transiently change (after FGF-2 addition for example), nor the presence of a single marker is

ENDOTHELIAL PHENOTYPES

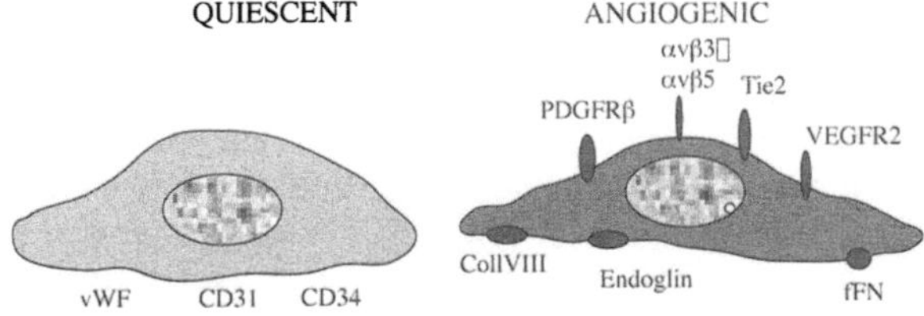

Figure 1. *Markers of the endothelial cell phenotype.* Quiescent endothelial cells can be visualized by immunohistochemistry for Von Willebrand factor, CD 31 and CD 34. Angiogenic endothelial cells express in addition to these constitutive markers VEGF-R2, Tie-2, PDGF-Rβ, integrins αvβ3 and αvβ5, as well as collagen VIII, endoglin and a fragment of fetal fibronectin.

sufficient to prove the endothelial origin of a culture. Although pioneer publications described VEGF as an selective endothelial growth factor, the detection of VEGF-Rs would be misleading for the identification of endothelial cells. They are in fact expressed in several non-endothelial cultured cells such as corneal endothelium, lens epithelium, retinal pigment epithelium or monocytes. When pure cultures are obtained, the stability of their phenotype must be carefully examined.

The measure of the expression of several genes such as thrombospondin [31] or PDGF-Rβ [32] may distinguish between proliferating and non-proliferating endothelial cells. In our laboratory we have selected, from retina-derived capillaries, clones of non-angiogenic cells which do not differentiate in 3D cultures in the presence of VEGF (Figure 2) or clones of angiogenic cells which do differentiate. Although both strains differentiate upon FGF-2 addition, VEGF induces an increase in bcl-2 expression and prevents TNF-α-dependent apoptosis only in the angiogenic cells.

It has also been recently reported that endothelial precursors can be mobilized from the bone marrow and localize in angiogenic foci [33]. It seems that angiogenic territories can induce the activation of the chemokine receptor CXCR4 through its SDF-1 ligand and stimulate the mobilization and circulation of endothelial precursors presenting the CD34 and VEGF-R2 markers at their surface [34]. These cells further colonize angiogenic sites,whatever their origin, inflammatory or tumoral, where they proliferate and differentiate by the mechanism of vasculogenesis (Figure 3).

Role in Vascular Biology

Physiological Function   Physiological angiogenesis is always transient whereas pathological angiogenesis is uncontrolled. If the increase of tissue mass observed during corpus luteum maturation or re-growth following organ removal are angiogenesis dependent, it is not clear whether the signals are similar to those leading to wound healing, for instance. In experimental angiogenesis, growth factors act in an inflammatory context created by the surgical traumatism. VEGF and angiopoietin-2 are up-regulated and angiopoietin-1 is down-regulated, which contributes to the loosening of matrix contacts and induce the switch from the quiescent phenotype to the activated one. This allows access of VEGF to VEGF-R2 and VEGF-R1 and therefore the switch from the activated phenotype to an angiogenic phenotype. The switch of the quiescent to activated phenotype does not require growth factors. The use of anti-idiotypic antibodies to VEGF which behave like circulating agonists of VEGF-R2 demonstrated that systemic activation of VEGF-R2 is sufficient to promote angiogenesis, provided that endothelial cells had received a phenotype-converting signal ([35]; Sordello, unpublished results) such as surgical traumatism (corneal pocket assay), estrogen (tumor take of estrogeno-dependent cancer xenografts) or androgen bolus (prostate re-growth after castration).

However, it is not yet understood why angiogenic endothelial cells switch their angiogenic phenotype to a quiescent phenotype and stop growing in physiological conditions and fail to do so in the case of diabetic retinopathy or cancer metastasis, for instance.

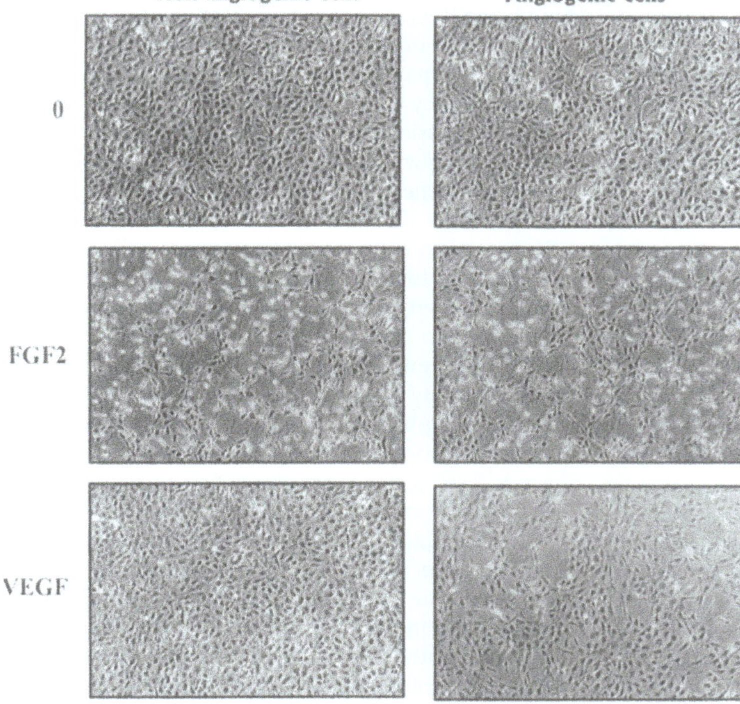

Figure 2. *Angiogenesis in vitro.* Angiogenic and non-angiogenic endothelial cells, cultured from retinal capillaries, were seeded on collagen gels in the presence of FGF-2 or VEGF. VEGF induces tube formation only on angiogenic cells, whereas FGF-2 induces differentiation in both cell types.

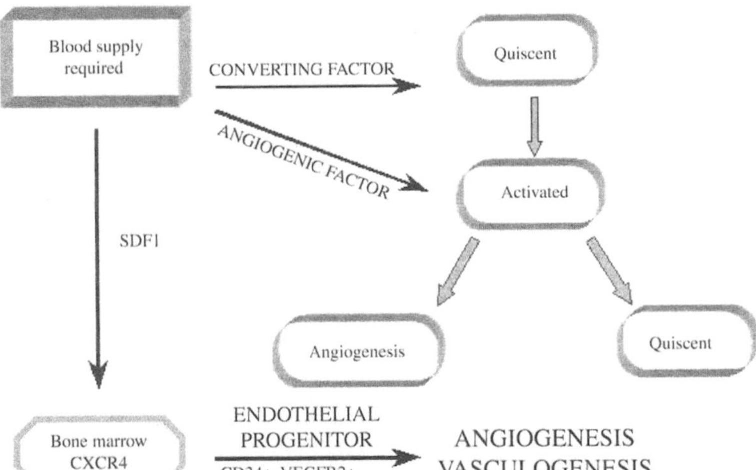

Figure 3. Schematic representation of the different events occuring during angiogenesis.

Pathology Several pathologies involve uncontrolled proliferation of endothelial cells. We shall focus our description on ocular neovascularization and cancer.

Diabetic microangiopathy induces capillary occlusion leading to retinal ischemia. Despite the absence of new vessel formation in the normal eye, diabetic retinopathy is complicated by retinal, papillary and iris neovascular proliferation that can cause intravitreal haemorrhages, retinal detachment and neovascular glaucoma. Another hallmark is the increase of retinal vessel permeability. Isaac Michaelson [36] hypothesized that a soluble factor might induce retinal vessel growth in response to ischemia. Over the last few decades several different activities were designated as Factor X until the discovery of VEGF. VEGF is thought to be a good candidate for triggering ocular neovascularization since (1) it is an autocrine factor for vascular endothelial and pigment epithelial retinal cells, (2) its expression is up-regulated by hypoxia, (3) it induces vascular permeability and (4) its glycosylation or its glycation enhance its angiogenic activity. Analysis of pathological samples has confirmed this hypothesis. VEGF immunoreactivity has also been detected in the endothelium of choriocapillaries as well as in intravascular leukocytes and migrating retinal pigment epithelium, whereas it is not detected in normal retinas. VEGF mRNA has been identified as the sole angiogenic factor expressed in fibrovascular membranes collected from diabetic patients [18]. The levels of VEGF in the vitreous of these patients are significantly elevated compared to the levels of VEGF in the vitreous of patients with retinal detachment. Age-Related Macular Degeneration (AMD) is the leading cause of blindness, this disease is characterized by malformations of choroid vessels and atrophy of the retinal pigment epithelium. Immunohistochemistry studies of neovascular membranes taken from patients with this disease have revealed that VEGF is up-regulated, as well as VEGFR1, and its concentration is increased in the vitreous body. VEGF is not always up-regulated in ocular

pathologies associated with angiogenesis, but rather with stress. For instance, tears contain high concentrations of VEGF after refractive keratotomy.

The demonstration that VEGF mRNA is expressed in cancer cells, whereas the VEGF protein is also accumulated in endothelial cells located in the vicinity of cancer cells [37] has paved the way to numerous studies confirming the essential role of VEGF in tumoral angiogenesis [17, 38]. VEGF is usually overexpressed in the hypoxic periphery of necrotic areas, whereas VEGFR1 and VEGFR2 are overexpressed in contiguous endothelial cells. Counting endothelial cells has been proposed as an independent factor reflecting the metastatic potential of prostate and breast cancer [39]. Indeed, a good correlation between vascularity and VEGF expression has been found, and VEGF seems to represent a useful prognostic marker. VEGF immunoreactivity is increased in the plasma of cancer patients [40] and its decrease might indicate the efficacy of the chemotherapy employed [41].

VEGF expression has also been found in several pathologies in which angiogenesis is a pro-eminent feature such as rheumatoid arthritis, bullous pemphigoid and psoriasis. In neonatal hemangiomas VEGF and VEGFR2 are co-expressed in the stromal cells. This is the first example to date in which VEGF confers a growth advantage *in vitro* and *in vivo* to a benign tumor [42].

Clinical Relevance and Therapeutic Implications The genetic stability of endothelial cells is an advantage for drug targeting because it is unlikely that such cells will acquire mutations and become resistant to chemotherapy. This concept was developed in 1971 by Judah Folkman [43] and has recently been confirmed by the same group [44]. Nude mice bearing tumors were treated with conventional chemotherapy. After three cycles of treatment some clones became resistant and the treatment became inefficient. Conversely, nude mice treated with the potent anti-angiogenic agent endostatin responded to each cycle and even ceased to relapse after a various number of cures. However, experimental angiogenic assays have already ascribed an angiogenic activity to so many factors that it would be illusive to attempt to control angiogenesis by interfering with only one angiogenic pathway.

So far many drugs have been used as anti-angiogenic agents. The most promising ones seem to be triggered on the VEGF/VEGF-Rs and integrin $\alpha_v\beta_3$ systems [45]. Use of monoclonal antibodies against VEGF gave the first demonstration that tumor growth is angiogenesis-dependent [46] and also anti-VEGFR2 antibodies are sufficient to halt invasion and angiogenesis of tumors, even when VEGF is up-regulated [47]. Other anti-angiogenic agents naturally occuring by proteolytic cleavage of larger proteins such as angiostatin [48] and endostatin [49] are also very promising.

Helen Hutchings, Nathalie Ortéga,
Jean-Francois Tournier and Jean Plouet

References

1. Jaffe EA et al (1973) J Clin Invest 52:2745-2756
2. Augustin HG et al (1994) BioEssays 16:901-906
3. Risau W (1995) Faseb J 9:926-933
4. Eichmann A et al (1997) Proc Natl Acad Sci 94:5141-5146
5. Bennet et al (1959) Am J Physiol 196:381-390
6. Ferrara N et al (1989) Biochem Biophys Res Commun 161:851-858
7. Plouët J et al (1989) EMBO J 8:3801-3806
8. Ferrara N et al (1996) Nature 380:439-442
9. Carmeliet P et al (1996) Nature 380:435-439
10. Millauer B et al (1993) Cell 72:835-846
11. Shalaby F et al (1995) Nature 376:62-66
12. Fong GH et al (1995) Nature 376:66-70
13. Sato TN et al (1995) Nature 376:70-74
14. Wang HU et al (1998) Cell 93:741-753
15. Hanahan D et al (1996) Cell 86:353-364
16. Hanahan D (1997) Science 277:48-50
17. Ferrara N et al (1997) Endocr Rev 18:4-25
18. Malecaze F et al (1994) Arch Ophthalmol 112:1476-1482
19. Benjamin LE (1997) Proc Natl Acad Sci 94:8761-8766
20. Malavaud B et al (1997) Cardiovasc Res 36:276-281
21. Korpelainen EI et al (1998) Curr Biol 10:159-164
22. Maisonpierre PC et al (1997) Science 277:55-60
23. Friedlander M et al (1995) Science 270:1500-1502
24. Brooks PC et al (1998) Cell 92:391-400
25. Boudreau N et al (1997) J Cell Biol 139:257-264
26. Burrows FJ et al (1995) Cancer Res 1:1623-1634
27. Rettig WJ et al (1992) Proc Natl Acad Sci 89:10832-10836
28. Paulus W et al (1991) Br J Cancer 63:367-371
29. Carnemolla B et al (1989) J Cell Biol 108:1139-1148
30. Jaffe EA et al (1987) Hum Pathol 18:234-239
31. RayChaudhury A et al (1994) J Cell Science 107:39-46
32. Battegay EJ et al (1994) J Cell Biol 125:917-928
33. Asahara T et al (1997) Science 275:964-967
34. Shi Q et al (1998) Blood 92:362-367
35. Ortéga N et al (1997) Am J Path 151:1215-1224
36. Michaelson I (1948) Trans Ophtalmol Soc UK 68:137-180
37. Dvorak HF et al (1991) J Exp Med 174:1275-1278
38. Dvorak HF et al (1995) Am J Pathol 146:1029-1039
39. Weidner N et al (1993) Am J Pathol 143:401-409
40. Gasparini G et al (1997) J Natl Cancer Inst 89:139-147
41. Dirix LY et al (1997) Br J Cancer 76:238-243
42. Bérard M et al (1997) Am J Pathol 150:1315-1326
43. Folkman J (1971) N Engl J Med 285:1182-1186
44. Boehm T et al (1997) Nature 390:404-407
45. Brooks PC et al (1994) Cell 79:1157-1164
46. Kim KJ et al (1993) Nature 362:841-844
47. Skobe M et al (1997) Nature Med 3:1222-1227
48. O'Reilly MS et al (1994) Cell 79 315-328
49. O'Reilly MS et al (1997) Cell 88:277-285

eNOS

Definition   *Endothelial nitric oxide synthase*

See: →Cytokines in vascular biology and disease; →Nitric oxide

EOMA

Definition   *Endothelioma*

See: →Angiogenesis inhibitors

EPA

Definition   *Eicosapenta(e)noic acid*

See: →Prostacyclin; →Prostaglandins; →Thromboxanes

EPA

Definition   *Erythroid-potentiating activity*

See: →Tissue inhibitors of metalloproteinases

EPCR

Definition   *Endothelial protein C receptor*

See: →Thrombin

Eph

Definition   *Ephrin*

See: →Ephrins

Ephrins

Definition   *Molecules involved in the development of the nervous system. These molecules also play an important role in vascular development since they specify veins (Eph-B4) and arteries (Eph-B2).*

See: →Angiogenesis; →Endothelial cells

Epibolin

See: →Vitronectin/vitronectin receptors

EPO

Definition   *Erythropoietin*
See: →Megakaryocytes

Epoprostenol

See: →Prostacyclin

ERK

Definition   *Extracellular signal-regulated kinase*

See: →Mitogen-associated kinases; →Signal transduction mechanisms in vascular biology; →Smooth muscle cells

Erythrocytes/Red Blood Cells

Definition   *Non-nucleated cells that carry hemoglobin*

See: →Blood cells, interaction with vascular cells

ESIMS

Definition   *Electrospray ionisation mass spectrometry*

See: →Fibrin/fibrinogen

ET-1

Definition   *Endothelin-1*

See: →Smooth muscle cells

Extracellular Matrix (Basement Membranes)

Synonym: Basal lamina

Definition   *Matrix found outside the cells composed of proteins or proteoglycans such as collagen, laminin, fibronectin, decorin, fibrin or proteoheparan sulfates. The subendothelial extracellular matrix is named basement membrane.*

See also: →Fibrinolytic, hemostatic and matrix metalloproteinases, role of; →Metalloproteinases; →Vitronectin/ vitronectin receptors; →von Willebrand Factor

Introduction   The thin extracellular matrix underlying the endothelium is termed the basement membrane [1]. Basement membranes are also found underlying epithelial cells and surrounding fat, muscle and Schwann cells. The amount and type of the components present in the basement membrane varies depending on the tissue and on the stage of development [2]. Although only visible at the level of the electron microscope in normal tissue, these extracellular matrices have important functions

Table 1. Functions of Basement Membrane

Separate endothelium from underlying stroma
Barrier to the passage of macromolecules and cells
Storage depot for growth factors and cytokines
Filter for nutrients and wastes
Maintain differentiated endothelium

based on both their structure and biological activity (Table 1).

In particular, the basement membrane separates endothelium from the underlying stroma, acts as a barrier to the passage of macromolecules and cells, stores important cytokines and growth factors involved in repair, and maintains the differentiated phenotype of the endothelium [3, 4]. Given these diverse and important biological functions, it is not surprising that acquired and genetic diseases of the basement membrane can be life threatening.

Structure   Basement membranes are composed of many interacting molecules. These molecules can in some cases bind to themselves, to each other, and to the endothelial cells. The most abundant and ubiquitous components include the glycoproteins laminin and entactin, collagen IV, and perlecan (heparan sulfate proteoglycan) [5] (Table 2).

These molecules interact via specific and multiple sites to form the highly organized structure of the basement membrane. A number of growth factors and cytokines are also present in the basement membrane and some have been shown to interact with collagen IV and perlecan. It should be noted that the amount and type of these components varies depending on the tissue location of the basement membrane and the developmental stage. Thus, multiple interactions form the elastic vascular basement membrane which is able to stretch to accommodate blood flow.

Laminin (Mr=800,000) is composed of three chains, designated α, β and γ which form a cruciform-like structure held together by disulfide bonds. Several isomeric chains (5 α, 3 β and 2 γ) have been identified and found to form at least 11 different laminins based on their composition including laminin-1 composed of $\alpha 1$, $\beta 1$ and $\gamma 1$, laminin-2 composed of $\alpha 2$, $\beta 1$ and $\gamma 1$, laminin-3 composed of $\alpha 1$, $\beta 2$ and $\gamma 1$, laminin-4 composed of $\alpha 2$, $\beta 2$ and $\gamma 1$, etc [6]. The chains are products of separate genes and no alternative splicing has yet been reported. These chains preferentially appear in tissue-specific locations and some basement membranes may contain more than one type of laminin molecule. The laminin variant(s) found in the endothelial basement membrane has not yet been identified and may represent a new form. Entactin (Mr=158,000) is a relatively small dumbbell-shaped, highly protease-sensitive, monomeric glycoprotein with globular domains at both ends which binds to laminin in the cross region [1]. Collagen IV (Mr=450,000) is a trimeric, non fibrillar collagen composed of two identical $\alpha 1$(IV) chains and one different

Table 2. Abundant Basement Membrane Components

Component	Size Mr	Structure	Function
laminin	800,000	α1β1γ1 trimers α2β1γ1 trimers etc	adhesion, migration, proliferation, differentiation
collagen IV	450,000	α1(IV)2α2(IV) α2(IV)2α2(IV) etc	adhesion
entactin	150,000	monomeric	adhesion
perlecan	800,000	core protein with three GAG chains	adhesion, filtration

Other components which are variable include SPARC, collagen VII, amyloid P, agrin, bFGF, TGF beta, EGF, PDGF, plasminogen activator, gelatinase A and B (reviewed in [14]).

α2(IV) chain [7]. The trimers interact via their uncleaved terminal domains to assemble into a flexible network. Studies suggest that the trimers assemble at their carboxy termini and then form an extended lattice via their amino terminus binding to three other collagen IV molecules. Disulfide bonds and covalent crosslinks stabilize the structure. At least six different collagen α(IV) chains have been described. Perlecan (Mr=800,000) has three heparan sulfate side chains attached to a large (Mr=200,000) protein core which has a number of regions homologous to other molecules including laminin, immunoglobulin superfamily, and LDL receptor domains [1, 2, 8].

The interactions of these components have been studied using fragments and intact molecules with standard solid and liquid phase assays as well as with rotary shadowing electron microscopy [9]. Laminin binds to collagen IV via its short arms at the amino terminus, to entactin near the cross region and to perlecan at several sites with a major site being the carboxy terminus. Collagen IV has a least three binding sites for laminin and two for perlecan.

Regulation of Cell Function Despite its small size and amorphous appearance, the basement membrane has multiple important functions (Table 1). It separates two different tissue types: the endothelium and stroma. The basement membrane also serves to regulate the passage of macromolecules and cells due to the high negative charge of the proteoglycans. This function is important in the kidney and in vessels for regulating waste and nutrients. Proteoglycans and other basement membrane components bind many growth factors and cytokines and, thus, the basement membrane has been described as a storage depot [2]. Growth factors are generally small in size and are highly diffusable. Growth factors bind to the basement membrane in an inactive form and are released when needed to influence cell proliferation, migration, differentiation, and synthesis and remodelling of the basement membrane.

Basement membranes are also important in maintaining the differentiated state of the endothelium [3, 10, 11]. When plated on a basement membrane matrix, many epithelial and endothelial cells attach and differentiate. For example, neuronal cells extend long processes. Endothelial cells form capillary-like structures with a lumen within 18 hours [10, 11] (Figure 1). This morphological differentiation mimics many of the steps in vessel formation where the cells attach, migrate and form tubes. Protein synthesis including collagen synthesis

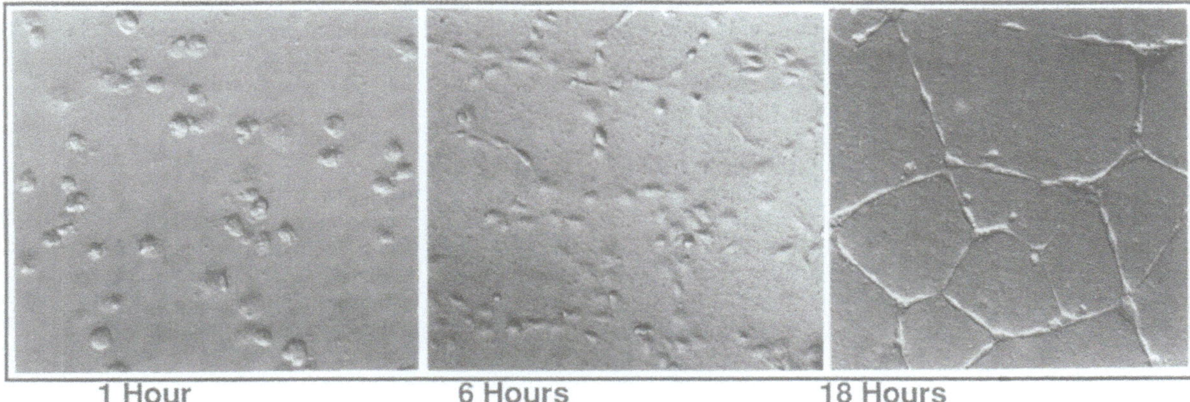

Figure 1. Morphological differentiation of endothelial cells on basement membrane. Human umbilical vein endothelial cells were plated on basement membrane and photographed at various times.

and protease activity are required during normal vessel formation and in this *in vitro* assay. It should be noted that cells that differentiate in this assay do not proliferate. This rapid *in vitro* differentiation has been used as an assay to screen for angiogenic and antiangiogenic compounds [12] (Table 3). Many of the known angiogenic factors, such as bFGF and HGF, are active in this assay and in more recognized assays. In addition, inhibitors of matrix synthesis or turnover block tube formation and angiogenesis in several assays.

Basement membrane components including laminin, collagen IV, entactin and perlecan promote cell adhesion. Laminin can also promote cell migration, differentiation, neurite outgrowth and proliferation. Multiple active sites on laminin have been identified using fragments and synthetic peptides. The RGD (arg-gly-asp) sequence which is present in laminin, collagen IV and entactin among others is known to be important in cell adhesion and spreading [13]. This sequence promotes endothelial cell spreading and inhibits angiogenesis and tumor growth. The YIGSR (tyr-ile-gly-ser-arg) sequence on the β1 chain also inhibits angiogenesis and tumor growth but does not promote cell spreading. The IKVAV (ile-lys-val-ala-val) sequence on the α1 chain promotes angiogenesis and tumor growth [14]. It is not clear why different sites on the molecule should have opposing activities. It is probable that some sites are not available in the basement membrane matrix at the same time due to molecular interactions and protein conformation and, thus, the functional sites are likely regulated.

Cell to Cell Interactions

The basement membrane appears to promote cell-cell interactions. When endothelial cells attach to basement membrane *in vitro*, they migrate towards each other and then form tube-like structures within 18 hours [11] (Figure 1). When these cells make similar contact on plastic dishes, no tubes are observed. Other extracellular matrices, such as collagen I, do not induce tube formation unless PMA is added. Here it should be noted that the tubes that form are inside out with the basement membrane deposited in the lumen. The mechanism for the cell-cell interactions is not known but is very important in vessel formation and integrity. During angiogenesis, the vessel is not functional until the basement membrane is deposited suggesting this important role *in vivo*.

Molecular Interactions

Many molecular interactions of variable affinities determine the final structure of the basement membrane. All of the major components (Table 2) interact with each other via multiple sites. These components also bind other basement membrane molecules [2] (Table 4). The major components, laminin and collagen IV, both exist in multiple isoforms and each can form huge networks by self assembly [1]. These networks can then bind together via direct interactions or via entactin and perlecan. The interactions between laminin, collagen IV, entactin and perlecan are important in forming the structure of the basement membrane. Given that multiple isoforms of laminin and collagen IV exist in different basement membranes and that basement membranes have different functions and in the case of the vessel wall must also be able to stretch, it is likely that the exact assembly mechanism and interactive sites may vary in different tissues. Due to the limited amount of the material available for study, this has not yet been shown. Thus, the assembly of basement membrane is highly complex and dependent on strong and weak interactions as well as on interactions with adjacent cells that produce the matrix components.

The interactions of the growth factors and cytokines with other components of the basement membrane may not be important to the structure of the basement membrane but are likely to be important to its function [2]. These factors are stored in the basement membrane matrix and are released due to proteolytic processing.

Table 3. Angiogenic Factors Active in *in vitro* Tube Assay and *in vivo*

Factor	Activity *in vitro* in Tube assay	Activity *in vivo*
laminin peptide IKVAV	stimulates	stimulates in sponge implant
laminin peptide YIGSR	inhibits	inhibits in CAM
laminin peptide RGD	inhibits	inhibits tumor growth
bFGF	stimulates	stimulates in many assays
α interferon	stimulates	inhibits in some assays
γ interferon	inhibits	not tested
HGF	stimulates	stimulates in matrigel plug
haptoglobin	stimulates	stimulates in sponge implant
estrogen	stimulates	stimulates in matrigel plug
inhibitors of collagen synthesis	inhibits	inhibits in CAM
stimulators of collagen synthesis	stimulates	stimulates wound repair
TIMPs	inhibits	inhibits tumor growth
PAI-1	inhibits	not done
thymosin alpha 1	stimulates	stimulates in matrigel plug
thymosin beta 4	stimulates	stimulates in matrigel plug

bFGF, basic fibroblast growth factor; CAM, chick chorioallantoic membrane; HGF, hepatocyte growth factor; TIMP, tissue inhibitor of metalloproteinases; PAI-1, plasminogen activator inhibitor

Table 4. Molecules Interacting with the Major Basement Membrane Components

Component	Binding Factors
laminin	laminin, collagen IV, entactin, perlecan
collagen IV	collagen IV, laminin, perlecan, BMP-2, BMP-7, TGF-β
entactin	laminin, collagen IV
perlecan	laminin, collagen IV, TGF-β, FGFs 1-9, VEGF, PDGF,GM-CSF, HB-EGF, HGF, IP-10, PF-4, IL-3, IL-4, Il-8, etc

Once released, these factors can regulate cell behavior. Thus, a quick response can be obtained without the need for synthesis of new factors. In this way, the basement membrane matrix can be thought to play a major role in growth factor signalling.

Interactions of the basement membrane with endothelial cell surfaces are also important and likely involve several receptor-ligands. The major cell surface matrix protein receptors are members of the integrin family which are αβ heterodimers. Integrins α1β1 and α1β2 bind collagen whereas α6β1 and α6β4 bind to laminin. In addition to the β1 containing integrins, αvβ3 and αvβ5 have been found on endothelial cells and are expressed during the formation of new vessels [15]. Integrin αvβ3 is important in angiogenesis induced by bFGF whereas αvβ5 is implicated in angiogenesis induced by VEGF. Thus, depending on the pathway of induction different integrin receptors may be involved. Multiple receptor interactions with each matrix molecule occur due to the high number of cell binding sites on these molecules. Perlecan and entactin also bind to integrins but these interactions are likely to be less abundant and weaker. The RGD sequence has been reported on fibronectin to bind to integrin α5β1 but it is not known if this is the case for the RGD in the basement membrane proteins and endothelial cells. In addition, the other cellular receptors for endothelial cells have not yet been defined. Non-integrin receptors have been described for laminin and collagen IV but their role in endothelial cell behavior is not known.

Signalling Mechanisms   Endothelial cells respond to a diverse array of signals when interacting with the basement membrane components and stored growth factors and cytokines. It should also be considered that the response may involve migration, proliferation, growth, etc. depending on whether the tissue is undergoing development, repair, or angiogenesis. The response to growth factors, such as bFGF or VEGF, involves migration and proliferation whereas the response to laminin involves adhesion and differentiation. Thus, the signalling response is dependent on the stimulus. Much is known about the response to the growth factors and receptor tyrosine kinase signalling has been described in vascular development [16]. Some of these receptor

tyrosine kinases are expressed at high levels in endothelial cells. For example, some growth factors, such as VEGF, which binds specifically to endothelial cells, binds to several of these receptor tyrosine kinases. In addition, kinases have been shown *in vitro* to be important for tube formation on collagen and on the basement membrane matrix [17]. For example, PMA increases tube formation on collagen and on the basement membrane matrix and a protein kinase C inhibitor blocks tube formation. The basement membrane controls endothelial cell sensitivity to soluble growth factors by binding integrin receptors and thereby activating a chemical signalling cascade that rapidly integrates with growth factor-induced signalling mechanisms [18]. It has been proposed that the integrin and growth factor receptors are activated via the cytoskeleton and focal adhesion complex. Recently it has been reported in endothelial cells that there are molecular connections between integrins, the cytoskeleton, and nuclear scaffolds suggesting a pathway for mechanical signal transfer from the basement membrane [19]. The matrix may also regulate sensitivity to growth factors, in part, by controlling cellular levels of phosphotidylinositol bis-phosphate (PIP$_2$), the substrate for phospholipase C-gamma [20]. Thus, matrix may increase PIP$_2$ by clustering cell surface integrin receptors and locally activating PIP kinase. In summary, some of the signalling mechanisms are beginning to be studied but much more needs to be done in this area given the number of stimuli and variety of responses.

Additional Features   The formation of new blood vessels or angiogenesis is envisioned as a four step process: 1) breakdown of the basement membrane, 2) migration of endothelial cells, 3) proliferation and tube formation, and 4) synthesis of the basement membrane [21]. Thus, the timing and balance of breakdown and synthesis of the basement membrane are important in new vessel formation [22]. The growth factors released during the breakdown of the basement membrane contribute to new vessel formation. The proteases and endogenous protease inhibitors that control its breakdown have been defined and include metalloproteinases and TIMPs [23]. The importance of the basement membrane in regulating angiogenesis has been shown using inhibitors of collagen IV synthesis which block angiogenesis [24]. Thus, changes in the basement membrane regulate vessel structure and repair.

Angiogenesis has received a lot of attention lately because of its well defined role in promoting tumor progression and metastasis [25]. Much is now known about the process and the factors which stimulate and inhibit it. Clinical trials for certain cancers have begun with compounds that block angiogenesis [26, 27]. Some of the targeted compounds affect basement membrane synthesis and degradation [28]. For example, batimastat and marimastat inhibit matrix metalloproteinases involved in basement membrane breakdown. These inhibitors work well in blocking tumor growth in animal models and are in clinical trials with humans. Other therapeutics are aimed at the integrins involved in vessel formation.

Kaposi's sarcoma is a complex cancer frequently found in AIDS patients which involves an increased number of vessels [29]. Hepatocyte growth factor, a highly angiogenic molecule, is found in the conditioned medium of cells from Kaposi's sarcoma patients. Hepatocyte growth factor appears to function by promoting angiogenesis and by inhibiting the expression of thrombospondin-1, an extracellular matrix molecule known to inhibit angiogenesis [30]. Certain inhibitors of angiogenesis may also be useful in treating this cancer as well.

Role in Vascular Biology

Pathology   A variety of diseases affect the glomerular basement membrane. This basement membrane is composed of factors made by the epithelial and endothelial cells. Autoimmunity to the glomerular basement membrane is found in Goodpasture's syndrome and in Alport's disease. In both of these diseases, there is an absence in the fusion of the glomerular and endothelial basement membrane. In Goodpasture's syndrome, the autoantibodies recognize the noncollagenous domain of the $\alpha3(IV)$ chain. In Alport's disease, there is a mutation in the $\alpha5(IV)$ chain [27]. Other inflammatory diseases have also been identified which affect the kidney.

In diabetes, there is a thickening of the basement membrane particularly in areas of high filtration/flow. This can lead to considerable neuropathy and nephropathy and circulatory problems. There appears to be a reduced amount of proteoglycan and more laminin and collagen IV in the basement membrane from diabetics [31]. Since there is no increased expression in the mRNA for these components [32], it is possible that there may be a reduced degradation of the basement membrane in diabetics. The proteoglycan is responsible for the filtration function. The loss of this function may stimulate a mild compensatory increase in the synthesis of the basement membrane components. It has also been reported that increased glycosylation of basement membrane components in diabetics may affect their molecular interactions and biological activity [33]. Thus, there may be multiple changes in the amount and composition of basement membrane in diabetics.

Hynda K. Kleinman, Katherine
M. Malinda and M. Lourdes Ponce

References

1. Timpl R, Brown JC (1996) BioEssays 18:123-132
2. Tiapole J, Keski-Oja J (1997) FASEB J 11:51-59
3. Grant DS et al (1990) NY Acad Sci 588:61-72
4. Sage EH, Vernon RB (1994) J Hypertension S145-S152
5. Yurchenco PD, Schittny JC (1990) FASEB J 4:1577-1590
6. Bergeson RE et al (1994) Matrix Biol 14:209-211
7. Yurchenco PD, Ruben GC (1987) Biochemistry 23: 1839-1850
8. Noonan D, Hassell JR (1993) In: Rohrbach DH, Timpl R (eds) Molecular and Cellular Aspects of Basement Membrane. Academic Press, pp 189-210
9. Laurie GW (1986) J Mol Biol 189:206-216
10. Schnaper HW et al (1993) Kidney Int 43:20-25
11. Grant DS, Kleinman HK (1997) In: Goldberg ID, Rosen EM (eds) Regulation of Angiogenesis Birkhäuser, pp 317-333 (Experientia Supplementum, Vol 79)
12. Cockerill GW et al (1995) Int Rev Cytol 159:113-160
13. Yamada KM (1991) J Biol Chem 266:12809-12812
14. Powell SK, Kleinman HK (1997) Int J Biochem Cell Biol 29:401-414
15. Brooks PC (1996) Eur J Cancer 32A:2423-2429
16. Merenmies J et al (1997) Cell Growth Differentiation 8:3-10
17. Kinsella J et al (1995) Exp Cell Res 199:56-62
18. Plopper GE et al (1995) Mol Biol Cell 6:1349-1365
19. Maniotis AJ et al (1997) Proc Natl Acad Sci USA 4:849-854
20. McNamee HP et al (1996) Exp Cell Res 224:116-122
21. Folkman J, Haudenschild C (1980) Nature 299:551-556
22. Liotta LA et al (1990) Sem Cancer Biol 1:99-106
23. Ingber DE, Folkman J (1989) Cell 58:803-805
24. Haralobopoulos GC et al (1994) Lab Invest 71:575-582
25. Folkman J (1996) Scientific American 2-5
26. Ellis LM, Fidler IJ (1996) Eur J Cancer 32A:2451-2460
27. Barina M (1997) Science 275:482-484
28. Talbot DC, Brown PD (1996) Eur J Cancer 32A:2528-2533
29. Abrahamson DR (1991) Seminars Nephrology 11:375-389
30. Noel A et al (1993) Adv Molec Cell Biol 6:271-292
31. Abrahamson DR (1986) J Pathol 149:257-278
32. Kolbe M et al (1990) Conn Tiss Res 25:77-85
33. Rohrbach DH, Murrah VA (1993) In: Rohrbach DH, Timpl R (eds) Molecular and Cellular Aspects of Basement Membrane. Academic Press, pp 385-419

FA

Definition   *Fanconi's anemia, Fanconi's syndrome*

See: →Bleeding disorders

FAD

Definition   *Flavin adenine dinucleotide*

See: →Nitric oxide

FAK

Definition   *Focal adhesion kinase*

See: →Focal adhesion kinase; →Signal transduction mechanisms in vascular biology; →Platelet stimulus-response coupling

FBAE

Definition   *Fetal bovine aortic endothelial cell*

See: →Endothelial cells

FGF

Definition   *Fibroblast growth factors*

See: →Fibroblast growth factors; →FGF receptors

FGF Receptors

Synonym: FGF receptor 1 is also known as FGFR1, flg (fms-like gene) [1], cek1 (chicken embryo kinase 1) [2], and N-sam [3]. FGF receptor 2 is also known as FGFR2, bek (bacterially expressed kinase) [4], cek3 [2], K-sam [5], and keratinocyte growth factor receptor [6]. FGF receptor 3 is also known as FGFR3, cek2 [2], and flg-2 [7]. FGF receptor 4 is also known as FGFR4.

Definition   *Receptors that bind FGF molecules at the cell surface. Four main tyrosine-kinase containing receptor types are described. Each type may have subtypes generated by alternate splicing.*

See: →FGF receptors; →FGF-1 and -2; →Signal transduction mechanisms in vascular biology

Introduction   The FGF receptors are four related tyrosine kinases that transmit signals upon binding the members of the FGF family of growth factors. The four FGF receptors all share a similar structure: an extracellular portion containing three immunoglobulin-like domains and a stretch of acidic amino acids, a transmembrane region, and an intracellular portion consisting of a long juxtamembrane domain and a split tyrosine kinase domain [8-11]. Each FGF receptor can be expressed in a variety of forms as a result of alternate splicing of the mRNA.

An important splicing event for FGFR1, 2, and 3 involves the second half of the third immunoglobulin-like domain. Two alternate exons (generally called IIIb and IIIc) can be spliced at this site, generating membrane bound receptors that have different ligand binding affinities [12–17]. All receptor isoforms generated by this splicing event bind FGF-1 with high affinity. Receptor isoforms incorporating the IIIc exon bind FGF-2 with high affinity, whereas those incorporating the IIIb exon bind FGF-2 with much lower affinity [18].

For FGFR1, a third possible exon (IIIa) can occur in the second half of the third immunoglobulin-like domain. Transcripts including this exon encode a truncated secreted receptor that binds FGF-2 [19]. Secreted receptors can also be generated by proteolytic degradation of the receptor near its transmembrane region [20]. Secreted receptor molecules have been found in blood and the basement membrane of endothelial cells [21, 22]; it is not clear if these molecules are generated by alternate splicing or proteolysis. The secreted receptors may act as carriers of FGFs or may modulate FGF activity by competing with cell-surface receptors.

Alternative splicing in the tyrosine kinase region of FGFR1 can generate isoforms with inactive kinase domains [23]. Overexpression of these receptor isoforms diminishes the activity of FGFR1 isoforms with the intact kinase domains [24]. Another splicing event can generate receptor isoforms missing the first of the three immunoglobulin-like domains [25-28]. Other described splicing events generate receptor isoforms that lack the stretch of acidic amino acids [6], contain a two amino acid insert in the intracellular juxtamembrane region, or have alternate C-terminal tails [5, 28]. The functional significance of these last splicing events has not been determined. Thus, a combination of splicing events can produce FGF receptors with a variety of affinities for FGFs and a variety of activities.

Characteristics

Molecular Weight   The isoforms of FGFR1 and FGFR2 with three immunoglobulin-like domains have predicted molecular weights of 89,437 and 89,750, respectively. When expressed in NIH 3T3 cells, the mature proteins have molecular weights of 150,000 and 135,000, respectively [9]. The increased apparent molecular weight is due to glycosylation as inhibition of N-linked glycosylation with tunicamycin gives receptor proteins of lower molecular weight [9]. The difference in size of the mature receptors is consistent with the presence of an additional glycosylation site in FGFR1 compared to FGFR2. When the isoform of FGFR1 with three immunoglobulin-like domains is expressed in L6 cells, receptors with molecular weights of 145,000 and 125,000 are detected [29]. The 125,000 form is a precursor of the 145,000 mature protein that differs in the extent of glycosylation. Mature FGFR3 expressed in COS cells has a molecular weight of 125,000, but inhibition of glycosylation gives a protein of 97,000 [10]. The primary translation product of FGFR4 mRNA is 88,000. After glycosylation, FGFR4 molecules with molecular weight of 95,000 and 110-115,000 are obtained [11].

Isoforms of FGFR1 and FGFR2 lacking the first immunoglobulin-like domain have molecular weights approximately 30,000 less than the isoforms containing all three immunoglobulin-like domains due to the absence of approximately 10 kDa of amino acid sequence and two glycosylation sites [23, 24, 27, 30]. Alternative splicing of the FGFR3 and FGFR4 mRNAs in this region would eliminate the same amount of amino acid coding sequence but only one glycosylation site.

Domains   The four FGF receptors all share a similar structure [8-11]. The extracellular portion contains three immunoglobulin-like domains. A stretch of 4 to 8 consecutive acidic amino acids is located between the first and second immunoglobulin-like domains. The first immunoglobulin-like domain and/or the acidic box may be eliminated in some isoforms of the receptors as a result of alternate splicing of the mRNA. The hydrophobic transmembrane domain contains 21 amino acids. The intracellular portion consists of a long juxtamembrane domain, a split tyrosine kinase domain, and the C-terminal tail. The tyrosine kinase domain is split by a 14 amino acid insert.

Other regions of the extracellular portion of the FGF receptors that might have functional importance have been identified. These include a proposed heparin-binding domain within the second immunoglobulin-like domain [31], and a region between the first and second immunoglobulin-like domains containing sequences that bear homology to the cadherin cell adhesion recognition sequence [32], to N-CAM, and to the neural adhesion molecule L1 [33].

Binding Sites and Affinity   The interactions of the FGF receptors with members of the FGF family of growth factors are complicated. Most members of the FGF family can bind to several different FGF receptors. Each FGF receptor can bind multiple members of the FGF family with differing affinities [18]. Alternate splicing of the mRNA can generate isoforms of one receptor that bind different FGFs. Thus, there is incredible redundancy in the FGF system, where a large family of growth factors interacts with overlapping specificity with a variety of alternately spliced receptors.

The ligand binding site is contained in the second and third immunoglobulin-like domains of the receptors. Alternate splicing in the C-terminal half of the third immunoglobulin-like domain of FGFR1, FGFR2, and FGFR3 generates isoforms that differ in their affinity for members of the FGF family, demonstrating that this portion of the receptor determines ligand specificity [12, 13, 15, 16, 28]. However, replacement of the C-terminal half of the third immunoglobulin-like domain of FGFR1 with the IIIb exon of FGFR2 is not sufficient to confer FGFR2-IIIb-like binding to FGFR1, suggesting that other portions of the receptor also contribute to ligand binding [34]. Replacement of the second immunoglobulin-like domain of the hybrid molecule with the homologous sequences from FGFR2 conferred FGFR2-IIIb-like ligand specificity, demonstrating that portions of the second immunoglobulin domain contribute to ligand interactions [34]. This is supported by experiments in which the second or third immunoglobulin-like domains of FGFR2-IIIb were fused to the immunoglobulin heavy chain Fc domain [35]. Hybrid molecules carrying the second immunoglobulin-like domain of FGFR2-IIIb could bind FGF-1 with high affinity but not FGF-7, whereas those carrying the third immunoglobulin-like domain bound FGF-7 with high affinity but not FGF-1. Experiments with proteolytic fragments and site directed mutations of FGFR1-IIIc and FGFR2-IIIb molecules show that a fragment containing the second immunoglobulin-like domain and a portion of the N-terminal half of the third immunoglobulin-like domain was sufficient to bind FGF-1, FGF-2, and FGF-7 but lost the ability to distinguish between these molecules [36]. Binding of FGF-7 was restricted by inclusion of additional C-terminal residues. These results suggest that the second immunoglobulin-like domain and a portion of the N-terminal half of the third immunoglobulin-like domain are sufficient to bind FGF ligands and that the ability to distinguish between ligands is conferred by the third immunoglobulin-like domain.

An alternate splicing event that includes or excludes the first immunoglobulin-like domain has no effect on the specificity of the receptor for the different FGFs but may alter the affinity of the receptor for a particular ligand.

The FGFR1 isoform missing the first immunoglobulin-like domain has a higher affinity for FGF-1 than the isoform containing all three immunoglobulin-like domains [37].

All receptor isoforms bind FGF-1 with high affinity [18]. The FGFR1 isoforms containing exon IIIc encoded sequences also bind FGF-2 and FGF-4 with high affinity and FGF-5 and FGF-6 with 10-fold lower affinity [9, 13, 18, 27, 38, 39]. The FGFR1 isoforms containing exon IIIb encoded sequences bind FGF-2 , FGF-3, and FGF-4 with lower affinity than FGF-1 [13, 18, 40]. The IIIc isoform of FGFR2 binds FGF-2 and FGF-4 with high affinity and FGF-5, FGF-6, FGF-8, and FGF-9 with 5 to 10-fold lower affinity [38, 41-43]. The IIIb isoform of FGFR2 binds FGF-7 with high affinity, FGF-3 with about 10-fold lower affinity, and FGF-2 with even lower affinity [15, 40, 44]. The IIIc isoform of FGFR3 also binds FGF-9 with high affinity and FGF-2, FGF-4, and FGF-8 with lower affinity [42, 43, 45, 46]. The IIIb isoform of FGFR3 binds FGF-9 with lower affinity than FGF-1 [47]. FGFR4 binds FGF-2, FGF-4, FGF-5, and FGF-6 with about 10-fold lower affinity than FGF-1 [39, 48, 49]. FGF-8 and FGF-9 also bind FGFR4 [43, 47].

Additional Features   Affinity of FGF receptors for various members of the FGF family is also modulated by heparan sulfate proteoglycans. Cells expressing FGF receptors but devoid of heparan sulfates do not respond to FGF-1 or FGF-2 in the absence of heparin or heparan sulfate [50, 51]. Initial experiments suggested that this requirement for heparin in biological responses to FGF reflected a requirement for heparin in the binding of FGFs to their receptors [50, 52]. However, titration of the binding of radiolabelled FGF-2 to FGFR1 expressed on cells lacking heparan sulfates revealed that FGF-2 bound to receptors in the absence of added heparin. Addition of heparin or heparan sulfates increased the affinity of FGF-2 for its receptor by 5-fold [53]. Similar results were obtained when binding of unlabelled FGF-2 to the purified extracellular domain of FGFR1 was measured by differential calorimetry [54]. FGF-2 also bound to its receptor in the absence of heparin in this system, but addition of heparin increased binding affinity 10-fold. In addition, measurement of the binding of unlabelled FGF-1 to the extracellular domain of FGFR2 by differential calorimetry showed that heparin was not necessary for FGF-1 binding and did not affect the affinity of the interaction [55]. Heparin or heparan sulfates increase the affinity of FGF-2 for its receptors by decreasing the dissociation rate of the FGF-2 - receptor complex [53, 56, 57]. These results suggest that trimolecular complexes of FGF-2, receptor, and heparan sulfate are formed and that these complexes are more stable than complexes of FGF-2 and receptor alone.

A recent model for the role of heparan sulfate in potentiating FGF bioactivity is that the glycosaminoglycan causes growth factor dimerization and the dimerized growth factors each bind to a receptor molecule, causing receptor dimerization [55]. This model is based on the observation that heparin can cause dimerization of FGF-1 and FGF-2 *in vitro* [46, 55]. In addition, when FGF receptor was expressed in mutant CHO cells lacking heparan sulfates, cross-linked dimers of the receptor could not be obtained after addition of FGF-1 and cross-linking reagent. In contrast, if FGF receptors were expressed in wild-type CHO cells expressing heparan sulfates or if exogenous heparin was added to the mutant CHO cells expressing receptors, cross-linked dimers of the receptor were observed. Analysis of the crystal packing of FGF-2 has identified faces that may interact strongly enough to give dimers [58]. The faces of FGF-2 that interact with heparin are in the correct position to promote interaction between the dimerization motifs when two FGF-2 molecules bind to the same heparin molecule. This observation suggests that FGF-2 may dimerize and cause dimerization of receptors in the absence of heparin, but heparin potentiates the dimerization. However, the crystal structure of FGF-2 in complex with the extracellular portion of FGFR1 shows that the FGF-2 molecules do not directly interact in the complex and sit on opposite sides of a receptor dimer [168]. The crystal structure shows that heparin interacts with both receptor and ligand and most likely acts to stabilize receptor dimers.

Structure

Sequence and Size   The isoform of FGFR1 with three immunoglobulin domains has 822 amino acids. FGFR2 has 821 amino acids. FGFR3 has 806 amino acids. FGFR4 has 802 amino acids.

Homologies   The FGF receptors are highly homologous to each other. The highest homology is between FGFR1 and FGFR2 with 72% amino acid identity. The lowest homology is between FGFR1 and FGFR4 with 55% amino acid identity. FGFR3 shares 61 to 66% identity with the other FGF receptors. Highest homology among the receptors is in the tyrosine kinase domains. Lowest homology is in the kinase insert, the first immunoglobulin-like domain, and the region between the third immunoglobulin-like domain and the transmembrane domain [59].

Conformation   The conformation of the extracellular portion of the FGF receptors has been modeled on the structure of immunoglobulins [60]. Recently the crystal structure of the extracellular domain has been solved [168]. The immunoglobuline-like domains adopt a β-barrel structure similar to the structure of telokin. FGF-2 interacts extensively with both the second and third immunoglobulin-like domains. The tyrosine kinase domain of FGFR1 has been crystallized and analyzed by X-ray diffraction [61]. The tyrosine kinase domain is bilobate with an overall structure very similar to the tyrosine kinase domain of the insulin receptor. ATP binding is coordinated mainly by residues in the N-terminal lobe and substrate peptide binding and catalysis are performed by residues in the C-terminal lobe. The

major difference from the insulin receptor tyrosine kinase is in the autoactivation loop of the C-terminal lobe. In insulin receptor, the activation loop blocks the ATP binding site and the binding site for substrate peptides. Upon phosphorylation, the conformation of the activation loop is altered, allowing access to the ATP and substrate binding sites. In unphosphorylated FGFR1, the ATP binding site is accessible, but the binding site for substrate peptides is blocked by the autoactivation loop. The autoactivation loop of FGFR1 makes relatively few interactions that would stabilize it in an inhibiting conformation, suggesting that autoinhibition of FGFR1 tyrosine kinase is weaker than the autoinhibition of insulin receptor tyrosine kinase.

Additional Features   As with other tyrosine kinase receptors, ligand binding to the FGF receptors results in receptor dimerization and transphosphorylation [62]. Seven tyrosine residues in the cytoplasmic domain of FGFR1 have been identified as sites of autophosphorylation [63]. Phosphorylation of one of these tyrosines regulates binding of phospholipase Cγ1 to the receptor [64]. Phospholipase Cγ1 binding to activated FGFR1 results in its phosphorylation and activation [65]. However, replacement of the tyrosine in the phospholipase Cγ1 binding site with phenylalanine abrogates phospholipase Cγ1 binding, phosphorylation, and activation, but has little effect on a number of biological responses to FGFs [66-69]. Indeed, only two of the phosphorylated tyrosines are required for FGF receptor signaling [63], and these are located in the activation loop of the tyrosine kinase domain [61]. Activated FGF receptors also phosphorylate SHC and a novel membrane-anchored adapter protein, FRS2. These proteins, in turn, bind to and phosphorylate grb-2 and SOS, leading to activation of the ras signaling pathway [70, 71]. Other signaling molecules that are phosphorylated as a result of activation of the FGF receptor are 80K-H, src, and cortactin [72, 73]. The role of these molecules in FGF signaling are not clear, but signaling requires long-term activation of the receptor and the pattern of phosphorylated proteins changes throughout this process [74, 75].

Other FGF receptors may utilize different signaling pathways. Activation of FGFR4 induced a much weaker phosphorylation of phospholipase Cγ, SHC, and ERK proteins than that induced by activation of FGFR1 [39, 76]. In addition, FGFR4 was unable to induce phosphorylation of a characteristic 80 kDa FGFR1 substrate protein in L6 rat myoblasts [39]. In contrast, a phosphorylated 85 kDa protein could be co-immunoprecipitated with activated FGFR4 but not FGFR1 in L6 cells [76]. Thus, FGFR4 may interact with different signaling molecules than FGFR1.

Gene

Gene Structure   The FGFR3 gene has been the most completely characterized [77, 78]. It consists of 19 exons and 18 introns spanning 16.5 kb. The FGFR1 gene has at least 18 exons and 17 introns [12, 59]. The exon/intron boundaries are conserved between FGFR3 and FGFR1 genes. Interestingly, the FGFR1, FGFR2, and FGFR3 genes all have an intron interrupting 5' non-coding sequence of the mRNA [77, 79, 80]. The FGFR2 and FGFR4 genes have been partially sequenced in the region encoding the third immunoglobulin domain. The structure of the FGFR2 gene in this region is very similar to the structure of the FGFR1 gene [12]. The FGFR4 gene has introns in positions similar to the FGFR1 and FGFR2 genes, but the introns are smaller. In addition, the FGFR4 gene lacks the alternatively spliced IIIb exon and, therefore, has only one exon encoding the second half of the third immunoglobulin domain [48].

Chromosomal Localization   The FGFR1 gene is located on human chromosome 8 in the region p11.2-p12 [1] and mouse chromosome 8. The FGFR2 gene has been localized on human chromosome 10 in the region q26 [81] and mouse chromosome 7 [82]. The FGFR3 gene is located on the short arm of human chromosome 4 in the region p16.3 [83] and on mouse chromosome 5 [82]. The FGFR4 gene is located near the terminus of the long arm of human chromosome 5 (q33-qter) [84] and on mouse chromosome 13 [82].

Gene Expression   FGF receptors are expressed in a variety of tissues throughout development and in the adult organism. In general, isoforms of the receptors containing the IIIb exon are expressed in epithelial tissues whereas those expressing the IIIc exon are expressed in mesenchymal tissues [16, 17, 85]. FGFR1 is widely expressed in the early mouse and chicken embryo, primarily in mesenchymal tissues [86, 87]. FGFR1 is expressed at high levels in human fetal brain, skin, growth plates of developing bones, and calvarial bones [11]. FGFR2 is also highly expressed in early mouse and chicken embryos, primarily in epithelial tissues [86, 87]. FGFR2 is expressed at high level in human fetal brain, lung, and skin [11]. Expression of FGFR3 and FGFR4 is more restricted. FGFR3 is expressed in early mouse and chicken embryos in cartilage, intestine, lung, and brain [86, 88]. FGFR3 is expressed at high level in human fetal brain, lung, intestine, kidney, skin, growth plates of developing bones, and calvarial bones [11]. FGFR4 is expressed in early mouse embryo in skeletal muscle, liver, lung, pancreas, and andrenal cortex [89]. FGFR4 is expressed at high level in human fetal adrenal, lung, and pancreas [11]. In organs where multiple FGF receptor family members are expressed, the pattern of expression of each of the receptors is unique. For example, in developing long bones, FGFR3 is expressed in the cartilage model and FGFR2 is expressed in the perichondrium [86-88]. As the cartilage is replaced by bone, FGFR3 expression is reduced in the hypertrophied cartilage, FGFR2 is expressed in the periostium, and FGFR1 is expressed in osteoblasts [86-88]. In the lung, FGFR3 is expressed in the smooth muscle surrounding the air sacs and bronchioles, FGFR1 is expressed in the epithelial cells lining the bronchi and in the surrounding mesenchyme, FGFR2 is expressed in the epithelial cells lin-

ing the air sacs, and FGFR4 is expressed in the mesenchyme and epithelium of the air sacs [86, 89]. Thus, the different receptors are expressed in a cell-type specific manner.

Gene Regulation   Regulation of FGF receptor gene expression is likely to be quite complex as each of the genes is expressed in a cell and time-dependent manner during development. FGFR1, FGFR2, and FGFR3 genes, like other growth factor receptor genes, have promoters that lack TATA box and CAAT box sequences. In the mouse FGFR1 promoter, there are four consensus sequences for SP1 binding sites and binding sites for Oct-1, AP1, and AP2 transcription factors [80]. Mouse FGFR2 promoter has two SP1 binding sites [79]. The human FGFR3 promoter has seven SP1, three AP2, two Krox 24, one IgHC.4, and one Zeste binding site [77]. The contribution of these sites to gene expression has not been elucidated.

In endothelial cells, interleukin 1 or interferon γ treatment decreases FGF receptor levels [90, 91]. In contrast, FGFR1 expression is increased when endothelial cells are induced to form tubes in culture [1]. Expression of FGFR1 mRNA and protein is increased in fibroblasts treated with platelet-derived growth factor BB [92]. Treatment of cells with FGF ligands has been reported to have opposite effects on FGF receptor expression depending on the cell type, causing increased expression in a pancretatic cell line and mammary carcinoma cells and decreased expression in NIH 3T3 cells [80, 93, 94].

FGF receptor expression is regulated during differentiation of various tissues. For example, FGFR1 mRNA is down-regulated and FGF receptor proteins are lost as myoblasts differentiate into myocytes [95, 96]. Stimulation of FGF receptor activity by addition of FGF-1, FGF-2, or FGF-4 prevents myoblasts from differentiating *in vitro* [97]. Moreover, constitutive expression of FGFR1 in myoblasts prevents them from differentiating into muscle *in vivo* [98].

Processing and Fate   FGF receptors are heavily glycosylated [99]. The glycosyl groups contain N-acetyl glucosamine residues that may be important for receptor function or stability. When the isoform of FGFR1 with three immunoglobulin domains is expressed in L6 myoblasts, two glycosylated receptor molecules with molecular weights of 125,000 and 145,000 are detected [29]. The lower molecular weight molecule is a precursor to the higher molecular weight one. As only the high molecular weight molecule can bind extracellular FGF, the lower molecular weight form may be an intracellular form carrying high-mannose glycosyl groups rather than mature complex glyosyl groups.

Another processing step may be the association of a divalent cation with the extracellular region of the receptor. The sequence of acidic amino acids that lies between the first and second immunoglobulin-like domains of FGFR1 has been shown to bind copper and calcium ions [100]. Binding of copper ions to this region increases the affinity of the receptor for heparin.

Once FGF receptors bind their ligands, they are internalized [101]. Internalization of FGFR1 is attenuated by mutation of tyrosine 766 to phenylalanine [101]. As phosphorylation of tyrosine 766 by activated receptors causes binding and activation of phospholipase Cγ1 [66, 67], phospholipase C may play a critical role in the receptor internalization process. When the isoform of FGFR1 containing 2 immunoglobulin-like domains is internalized, it is translocated to a compartment near the nucleus, where it appears to continue to signal [29, 102]. Eventually, the internalized FGF receptors are degraded, leading to a down-regulation of cell surface receptors [103, 104].

Biological Activity   The FGF receptors transduce signals to the cell after binding members of the FGF family of growth factors. One major role of FGF receptors *in vivo* is the regulation of bone development. A variety of human genetic diseases that result in skeletal malformations have been shown to be the result of mutations in FGF receptor genes. Achondroplasia, the most common inherited form of dwarfism is the result of a mutation in the transmembrane sequence of FGFR3 [105, 106]. A neonatal lethal skeletal disorder, thanatophoric dysplasia, is caused by mutations in either the tyrosine kinase domain or the extracellular region of FGFR3 [107]. Some cases of Pfeiffer syndrome, in which the bones of the head fuse prematurely causing craniofacial abnormalities, have been shown to be the result of a specific pro to arg substitution in the first half of the third immunoglobulin-like domain of FGFR1 [108]. Other, phenotypically similar, cases of Pfeiffer syndrome have been shown to be the result of mutations in FGFR2 [109]. In addition, similar cranio-facial syndromes, Crouzon, Jackson-Weiss, and Apert, are also caused by mutations in FGFR2 [109-111]. These are all dominant mutations. These abnormalities correlate with what is known about FGF receptor expression during bone development. FGFR3 is expressed in cartilage, while FGFR1 is expressed in osteoblasts and FGFR2 is expressed in the perichondrium [87, 88]. The phenotype in the FGFR1 mutation does not seem to be the result of inactivation of the receptor as a knockout of the mouse FGFR1 gene is lethal shortly after the blastocyst stage in the homozygote but the heterozygote shows no phenotypic changes [112, 113]. Knockout of the FGFR3 gene in mice causes an increase in long-bone length [114]. On the basis of the genetics of the diseases, the mutations causing these skeletal disorders are expected to cause constitutive activation of the receptor. Indeed, recent reports show that a mutation in the transmembrane domain of FGFR3 that causes achondroplasia, mutation in the tyrosine kinase domain of FGFR3 that causes thanatophoric dysplasia, and a cys to tyr mutation in the third immunoglobulin-like domain of FGFR2 that causes Crouzon syndrome all result in constitutive activation of the receptors [115-120]. The mutation in Crouzon syndrome creates a free cysteine residue and appears to promote receptor activation by creating disulfide cross links between receptor molecules result-

ing in receptor dimerization [115, 118]. Many of the other mutations in the extracellular region of the FGF receptors responsible for craniofacial syndromes also create or eliminate cysteine residues or are near cysteine residues where they may disrupt disulfide bonding. These mutations may also activate receptors by creating covalent receptor dimers. Interestingly, although the FGF receptors are also expressed in a variety of other tissues during development [87, 88], these mutations seem to affect only skeletal development in humans.

Role in Vascular Biology

Physiological Function   One of the major roles proposed for the FGFs *in vivo* is in the induction of new blood vessel growth or angiogenesis [121]. Angiogenesis occurs physiologically in the development of the vascular system during embryonic, fetal, and adolescent growth, during the normal cycling of the female reproductive system, and during the healing of wounds. Angiogenesis also contributes to several pathologies either directly, as in diabetic retinopathy, or indirectly by supporting the growth of pathologic tissues, as in rheumatoid arthritis and tumor growth. Neovascularization occurs from capillaries and is initiated when the capillary endothelial cells penetrate through their basement membranes, migrate toward the source of angiogenic inducer, and proliferate, forming new cords of endothelial cells that eventually develop into capillaries [122]. The FGFs have effects on cultured endothelial cells that are consistent with a role in this process. FGF-2 induces an invasive phenotype in cultured endothelial cells [123], enabling them to penetrate basement membranes *in vitro* [124]. The ability to penetrate the basement membrane is dependent on the increased production of the proteolytic enzymes plasminogen activator and collagenase in response to FGF-2 [124-126]. The induction of these enzymes by FGF-2 is dependent on the presence of FGF receptors [69, 127]. In addition, both FGF-1 and FGF-2 are chemotactic for endothelial cells [125, 128], suggesting that these factors support the directed growth of capillaries during angiogenesis. Finally, FGF-1 and FGF-2 stimulate endothelial cell proliferation [129]. Thus, FGF-1 and FGF-2 have properties expected for angiogenic factors and, indeed, induce angiogenesis *in vivo* in a number of model systems [130-135].

However, the roles of the FGFs and their receptors in physiological angiogenesis *in vivo* have been difficult to sort out, not only because of the overlapping biological properties of the members of the FGF family, but also because similar biological effects are also induced by unrelated growth factors. For example, FGFR1 was first identified as a gene that is upregulated in endothelial cells forming tubes *in vitro* [1], so it might be expected to have a critical role in angiogenesis *in vivo*. However, knockout of the FGFR1 gene is lethal at embryonic stages but has no effect on early vascular development [112, 113]. Similarly, no vascular defects are noted in patients with cranial-facial syndromes resulting from activating mutations in the FGF receptors. Although FGFR1 is expressed on some cultured endothelial cells and in aortic endothelial cells *in vivo* [12, 87, 136, 137], expression of FGFR2 has been detected in other endothelial cells [136]. Thus, it is possible that vascular development is mediated through other FGF receptors in FGFR1 knockout mice or that FGFR1 mediates angiogenesis at specific sites at later stages of development.

In several studies, administration of neutralizing antibodies to FGF-2 has had no effect on tumor growth in animals suggesting that these antibodies had no effect on tumor angiogenesis [138, 139]. Furthermore, knockout of the FGF-2 gene has no effect on vascular development or tumor growth [140, 141](C. Basilico, personal communication). It is possible that the role of FGF-2 is filled by other members of the FGF family in these systems.

However, it appears the FGF-2 - FGF receptor system is important in some instances of angiogenesis. Collagen sponges were implanted subcutaneously in rats, and the effect of antibodies to FGF-2 on the subsequent formation of granulation tissue in these sponges was investigated [142]. When a pellet that slowly released neutralizing antibody to FGF-2 was included in the sponge, vascularization of the sponge and granulation tissue formation were inhibited. In addition, treatment of animals with antibodies to FGF-2 gives a small reduction in growth of gliomas [143-145], and treatment of melanomas with anti-sense FGF-2 or FGFR1 constructs inhibits melanoma growth [146, 147]. Where studied, these treatments seem to exert their affects, at least in part, through the inhibition of angiogenesis. Thus, FGFs may have a role in the vascularization of certain types of tumors or may contribute to angiogenesis in wound healing.

Pathology   FGF and its receptors are proposed to have two roles in vascular pathologies. First, FGF-2 is important for the regrowth of the endothelium in injured vessels [148, 149]. Second, FGFs seem to have a role in the initiation of neointimal thickening after injury in a rat model. The formation of the neointima after injury is partly due to effects of FGF-2 on stimulation of smooth muscle cell migration from the media and partly to its effects on smooth muscle cell proliferation [149, 150]. The expression of FGFR1, the predominant FGF receptor expressed by vascular smooth muscle cells, is upregulated in these cells after injury [149, 151-153]. Addition of neutralizing antibodies to FGF-2 inhibits both smooth muscle cell migration from the media and smooth muscle cell proliferation [150, 154]. Similarly, addition of saporin conjugated to FGF-2, which targets the toxin to cells expressing FGF receptors [155], reduces neointimal thickening [151, 153, 156, 157]. Thus, FGFs and their receptors may contribute to the initial thickening of the neointima after injury. However, the role of FGFs in established lesions may be different, as the regrowth of the neointima after balloon catheterization is not inhibited by antibodies to FGF-2 [158].

Clinical Relevance and Therapeutic Implications   As noted above, toxins conjugated to FGF are being used experimentally to target smooth muscle cells expressing FGF receptors in the neointima [151, 153, 156, 157]. This approach may be limited, however, as other cells expressing FGF receptors will also be targeted by the toxin conjugate.

As the FGFs are potent angiogenic factors, the FGF - FGF receptor system is also being used experimentally to increase collateral blood flow in areas of ischemia. In dog, rat, and rabbit models, administration of FGFs has been shown to increase capillary density, increase collateral blood flow, and reduce infarct size [159-165]. Increased vascular growth seems to be limited to the area of ischemia, and the added FGFs have no apparent affects on mature vessels in other vascular beds [166, 167]. These experiments may lead to an effective angiogenic therapy for patients whose ischemia is not amenable to other treatments.

David Moscatelli

References

1. Ruta M et al (1988) Oncogene 3:9-15
2. Pasquale EB (1990) Proc Natl Acad Sci USA 87:5812-5816
3. Hattori Y et al (1992) Cancer Res 52:3367-3371
4. Kornbluth S et al (1988) Mol Cell Biol 8:5541-5544
5. Hattori Y et al (1990) Proc Natl Acad Sci USA 87:5983-5987
6. Miki T et al (1991) Science 251:72-75
7. Avivi A et al (1991) Oncogene 6:1089-1092
8. Lee PL et al (1989) Science 245:57-60
9. Dionne CA et al (1990) EMBO J 9:2685-2692
10. Keegan K et al (1991) Proc Natl Acad Sci USA 88:1095-1099
11. Partanen J et al (1991) EMBO J 10:1347-1354
12. Johnson DE et al (1991) Mol Cell Biol 11:4627-4624
13. Werner S et al (1992) Mol Cell Biol 12:82-88
14. Dell KR, Williams LT (1992) J Biol Chem 267:21225-21229
15. Miki T et al (1992) Proc Natl Acad Sci USA 89:246-250
16. Chellaiah AT et al (1994) J Biol Chem 269:11620-11627
17. Avivi A et al (1993) FEBS Lett 330:249-252
18. Ornitz DM et al (1996) J Biol Chem 271:15292-15297
19. Duan D-SR et al (1992) J Biol Chem 267:16076-16080
20. Levi E et al (1996) Proc Natl Acad Sci USA 93:7069-7074
21. Hanneken A et al (1994) Proc Natl Acad Sci USA 91:9170-9174
22. Hanneken A et al (1995) J Cell Biol 128:1221-1228
23. Hou J et al (1991) Science 251:665-668
24. Shi E et al (1993) Mol Cell Biol 13:3907-3918
25. Johnson DE et al (1990) Mol Cell Biol 10:4728-4736
26. Reid HH et al (1990) Proc Natl Acad Sci USA 87:1596-1600
27. Mansukhani A et al (1990) Proc Natl Acad Sci USA 87:4378-4382
28. Champion-Arnaud P et al (1991) Oncogene 6:979-987
29. Prudovsky I et al (1996) J Biol Chem 271:14198-14205
30. Bernard O et al (1991) Proc Natl Acad Sci USA 88:7625-7629
31. Kan M et al (1993) Science 259:1918-1921
32. Byers S et al (1992) Dev Biol 152:411-414
33. Mason I (1994) Current Biology 4:1158-1161
34. Zimmer Y et al (1993) J Biol Chem 268:7899-7903
35. Cheon H et al (1994) Proc Natl Acad Sci USA 91:989-993
36. Wang F et al (1995) J Biol Chem 270:10222-10230
37. Wang F et al (1995) J Biol Chem 270:10231-10235

38. Clements DA et al (1993) Oncogene 8:1311-1316
39. Wang J-K et al (1994) Mol Cell Biol 14:181-188
40. Mathieu M et al (1995) J Biol Chem 270:24197-24203
41. Mansukhani A et al (1992) Proc Natl Acad Sci USA 89:3305-3309
42. Hecht D et al (1995) Growth Factors 12:223-233
43. Blunt A et al (1997) J Biol Chem 272:3733-3738
44. Yayon A et al (1992) EMBO J 11:1885-1890
45. Lin H et al (1997) FEBS Lett 411:389-392
46. Ornitz DM, Leder P (1992) J Biol Chem 267:16305-16311
47. Santos-Ocampo S et al (1996) J Biol Chem 271:1726-1731
48. Vainikka S et al (1992) EMBO J 11:4273-4280
49. Ron D et al (1993) J Biol Chem 268:5388-5394
50. Rapraeger AC et al (1991) Science 252:1705-1708
51. Ornitz DM et al (1992) Mol Cell Biol 12:240-247
52. Yayon A et al (1991) Cell 64:841-848
53. Roghani M et al (1994) J Biol Chem 269:3976-3984
54. Pantoliano MW et al (1994) Biochemistry 33:10229-10248
55. Spivak-Kroizman T et al (1994) Cell 79:1015-1024
56. Moscatelli D (1992) J Biol Chem 267:25803-25809
57. Nugent MA, Edelman ER (1992) Biochemistry 31:8876-8883
58. Venkataraman G et al (1996) Proc Natl Acad Sci USA 93:845-850
59. Johnson DE, Williams LT (1993) Adv Cancer Res 60:1-41
60. Xu J et al (1992) J Biol Chem 267:17792-17803
61. Mohammadi M et al (1996) Cell 86:577-587
62. Bellot F et al (1991) EMBO J 10:2849-2854
63. Mohammadi M et al (1996) Mol Cell Biol 16:977-989
64. Mohammadi M et al (1991) Mol Cell Biol 11:5068-5078
65. Burgess WH et al (1990) Mol Cell Biol 10:4770-4777
66. Mohammadi M et al (1992) Nature 358:681-684
67. Peters KG et al (1992) Nature 358:678-681
68. Spivak-Kroizman T et al (1994) J Biol Chem 269:14419-14423
69. Roghani M et al (1996) J Biol Chem 271:31154-31159
70. Kouhara H et al (1997) Cell 89:693-702
71. Wang J et al (1996) Oncogene 13:721-729
72. Zhan X et al (1994) J Biol Chem 269:20221-20224
73. Goh K et al (1996) J Biol Chem 271:5832-5838
74. Zhan X et al (1993) J Biol Chem 268:9611-9620
75. Flaumenhaft R et al (1989) J Cell Physiol 140:75-81
76. Vainikka S et al (1994) J Biol Chem 269:18320-18326
77. Perez-Castro A et al (1997) Genomics 41:10-16
78. Wuchner C et al (1997) Hum Genet 100:215-219
79. Avivi A et al (1992) Oncogene 7:1957-1962
80. Saito H et al (1992) Biochem Biophys Res Commun 183:688-693
81. Mattei M et al (1991) Hum Genet 87:84-86
82. Avraham K et al (1994) Genomics 21:656-658
83. Thompson L et al (1991) Genomics 11:1133-1142
84. Armstrong E et al (1992) Genes Chromosomes Cancer 4:94-98
85. Orr-Urtreger A et al (1993) Dev Biol 158:475-486
86. Patstone G et al (1993) Dev Biol 155:107-123
87. Peters KG et al (1992) Development 114:233-243
88. Peters K et al (1993) Dev Biol 155:423-430
89. Stark KL et al (1991) Development 113:641-651
90. Cozzolino F et al (1990) Proc Natl Acad Sci USA 87:6487-6491
91. Friesel R et al (1987) J Cell Biol 104:689-696
92. Landgren E et al (1996) Exp Cell Res 223:405-411
93. Estival A et al (1996) J Biol Chem 271:5663-5670
94. Saito H et al (1991) Biochem Biophys Res Commun 174:136-141
95. Olwin BB, Hauschka SD (1988) J Cell Biol 107:761-769
96. Moore JW et al (1991) Development 111:741-748
97. Olwin BB, Rapraeger A (1992) J Cell Biol 118:631-639
98. Itoh N et al (1996) Development 122:291-300
99. Feige J-J, Baird A (1988) J Biol Chem 263:14023-14029
100. Patstone G, Maher P (1996) J Biol Chem 271:3343-3346
101. Sorokin A et al (1994) J Biol Chem 269:17056-17061
102. Prudovsky I et al (1994) J Biol Chem 269:31720-31724
103. Moscatelli D, Devesly P (1990) Growth Factors 3:25-33
104. Moscatelli D (1988) J Cell Biol 107:753-759
105. Rousseau F et al (1994) Nature 371:252-254
106. Shiang R et al (1994) Cell 78:335-342
107. Tavormina PL et al (1995) Nature Genet 9:321-328
108. Muenke M et al (1994) Nature Genetics 8:269-274
109. Wilkie AOM et al (1995) Curr Biol 5:500-507
110. Reardon W et al (1994) Nature Genetics 8:98-103
111. Jabs EW et al (1994) Nature Genetics 8:275-279
112. Deng C-X et al (1994) Genes & Dev 8:3045-3057
113. Yamaguchi TP et al (1994) Genes & Dev 8:3032-3044
114. Deng C et al (1996) Cell 84:911-921
115. Neilson KM, Friesel RE (1995) J Biol Chem 270:26037-26040
116. Webster MK, Donoghue DJ (1996) EMBO J 15:520-527
117. Webster M et al (1996) Mol Cell Biol 16:4081-4087
118. Mangasarian K et al (1997) J Cell Physiol 172:117-125
119. Li Y et al (1997) Oncogene 14:1397-1406
120. Neilson K, Friesel R (1996) J Biol Chem 271:25049-25057
121. Folkman J, Klagsbrun M (1987) Science 235:442-447
122. Ausprunk DH, Folkman J (1977) Microvasc Res 14:53-65
123. Montesano R et al (1986) Proc Natl Acad Sci USA 83:7297-7301
124. Mignatti P et al (1989) J Cell Biol 108:671-682
125. Moscatelli D et al (1986) Proc Natl Acad Sci USA 83:2091-2095
126. Presta M et al (1986) Mol Cell Biol 6:4060-4066
127. Besser D et al (1995) Cell Growth Diff 6:1009-1017
128. Terranova VP et al (1985) J Cell Biol 101:2330-2334
129. Gospodarowicz D et al (1987) Endocrine Reviews 8:95-114
130. Shing Y et al (1985) J Cell Biochem 29:275-287
131. Thomas KA et al (1985) Proc Natl Acad Sci USA 82:6409-6413
132. Lobb RR et al (1985) Biochemistry 19:4969-4973
133. Hayek A et al (1987) Biochem Biophys Res Commun 147:876-880
134. Thompson JA et al (1988) Science 241:1349-1352
135. Thompson JA et al (1989) Proc Natl Acad Sci USA 86:7928-7932
136. Bastaki M et al (1997) Arterioscler Thromb Vasc Biol 17:454-464
137. Liaw L, Schwartz S (1993) Arterioscler Thromb 13:985-993
138. Matsuzaki K et al (1989) Proc Natl Acad Sci USA 86:9911-9915
139. Dennis P, Rifkin D (1990) J Cell Physiol 144:84-98
140. Zhou M et al (1998) Nat Med 4:201-207
141. Ortega S et al (1998) Proc Natl Acad Sci USA 95:5672-5677
142. Broadley KN et al (1989) Lab Invest 61:571-575
143. Gross J et al (1993) J Natl Cancer Inst 85:121-131
144. Takahashi J et al (1991) FEBS Lett 288:65-71
145. Stan A et al (1995) J Neurosurg 82:1044-1052
146. Becker D et al (1992) Oncogene 7:2303-2313
147. Wang Y, Becker D (1997) Nat Med 3:887-893

148. Lindner V et al (1990) J Clin Invest 85:2004-2008
149. Lindner V, Reidy MA (1993) Circ Res 73:589-595
150. Jackson CL, Reidy MA (1993) Am J Pathol 143:1024-1031
151. Casscells W et al (19hager2) Proc Natl Acad Sci USA 89:7159-7163
152. Xin X et al (1994) Biochem Biophys Res Commun 204:557-564
153. Farb A et al (1997) Circ Res 80:542-550
154. Lindner V et al (1992) J Clin Invest 90:2044-2049
155. Lappi DA et al (1991) J Cell Physiol 147:17-26
156. Lindner V et al (1991) Circ Res 68:106-113
157. Mattar S et al (1996) J Surg Res 60:339-344
158. Koyama H, Reidy M (1997) Circ Res 80:408-417
159. Yanagisawa-Miwa A et al (1992) Science 257:1401-1403
160. Baffour R et al (1992) J Vasc Surg 16:181-191
161. Lazarous D et al (1995) Circulation 91:145-153
162. Unger E et al (1994) Am J Physiol 266:H1588-H1595
163. Uchida Y et al (1995) Am Heart J 130:1182-1188
164. Rosengart T et al (1997) J Vasc Surg 26:302-312
165. Yang H et al (1996) Circ Res 79:62-69
166. Jacot J et al (1996) J Anat 188:349-354
167. Shou M et al (1997) J Am Coll Cardiol 29:1102-1106
168. Plotnikov AN et al (1999) Cell 98:641-650

FGF-1 (Fibroblast Growth Factor-1), FGF-2 (Fibroblast Growth Factor-2)

Synonym: FGF-1: acidic Fibroblast Growth Factor, Heparin-Binding Growth Factor-I; FGF-2: Basic Fibroblast Growth Factor, Heparin-Binding Growth Factor II

Introduction   Fibroblast Growth Factor-1 and Fibroblast Growth Factor-2 belong to an increasing family of growth factors, comprising to date 18 members [1-4]. FGF-1 and FGF-2 are the oldest members of this growth factor family. These factors have been shown to stimulate endothelial cell-growth and migration in *in vitro* as well as in *in vitro* angiogenesis and have long been considered to be the principal tumor angiogenic factors.

Mature FGF-1 exists as an 18 kDa molecular form. Truncated FGF-1 forms and a smaller-splicing variant of 6 kDa have also been described [3]. FGF-2 exists in 4 molecular forms comprising three high molecular weight (21.5, 22 and 24 kDa, HMW FGF-2) and one 18 kDa (18 kDa FGF-2) forms [5,6]. The HMW FGF-2 forms are initiated at three CUG codons whereas 18 kDa FGF-2 is initiated at AUG. HMW FGF-2 is mainly nucleus-associated, whereas 18 kDa is cytoplasmic or found at the cell surface or the extracellular matrix.

Angiogenesis involves different steps that include migration, proliferation and tubulogenesis. These steps are integrated in space and time by an orderly sequence of extracellular and intracellular molecular events. FGF-1 and FGF-2 are implicated in the control of migration, proliferation and tubulogenesis [1, 3].

A number of angiogenic inhibitors has been discovered that are able to antagonize FGF-1 or FGF-2 activity, among them Platelet Factor-4 (PF-4) [7], Angiostatin [8-10], Endostatin [11] and the 16 kDa human Prolactin Fragment (16 kDa hPRL) [12]. These factors may inhibit FGF activity at the FGF receptor level or may alter downstream signaling.

In this review, we will focus on some recent observations that relate to the roles and mechanisms of action of FGF-1 and FGF-2 and their implications for vascular biology. We will also discuss the different molecular forms of FGF-2 and what we know about their biological activities and mechanisms of action. We will finally evaluate the contributions of FGF-1 or FGF-2 to vascular biology.

Characteristics

Molecular Weight   Full-length FGF-1: 18 kDa; truncated FGF-1 forms: 17 kDa, 16.5 kDa; alternately-spliced FGF-1: 6 kDa; High Molecular Weight FGF-2: 3 forms of 21.5, 22, and 24 kDa.; Standard FGF-2: 1 form of 18 kDa.

Domains   FGF-1 and FGF-2 have been isolated initially as a 18 kDa molecule from several tissues and cell types including cells of mesenchymal origin and several tumor cell lines [1,3]. The N-terminus of FGF-1 is acethylated and contains a nuclear localization sequence [3]. FGF-1 has three cystein residues and contains a receptor and heparin-binding domains [3]. FGF-1 contains also a copper-binding site. Copper may induce FGF-1 dimer formation. FGF-2 forms of higher molecular weight have been detected in guinea pig brain, placenta, endothelial cells and several tumor cells [1]. These forms are initiated at CUG codons that lie upstream from the initial 18 kDa FGF-2-initiating AUG. Several domains may be important for FGF-2 function. Residues 13-30 and 106-129 are believed to represent the receptor binding sites [13, 14]. The inverse RGD sequences PDGR and EDGR are possibly involved in the modulation of mitogenicity [15]. Two potential phosphorylation sites exist: one at serine 64 and the other at threonine 112. Serine 64 and threonine 112 can be phosphorylated by protein kinase A and protein kinase C, respectively [16]. The cellular kinases responsible for FGF-2 phosphorylation may be localized both in the nucleus and at the cell surface [17, 18]. FGF-2 contains four cysteines; however, there are no intramolecular disulfide bonds [19]. The unique feature of the HMW FGF-2 forms, which distinguishes them from 18 kDa FGF-2, is the amino-terminal extension. In the largest form, this sequence contains 9 Gly-Arg repeats. At least six of the arginines in these Gly-Arg motifs are methylated [20, 21]. Neither the exact number nor the functional significance of the methylated arginines are known, but they may be involved in nuclear transport or retention. The forms initiated using the CUG codons [22, 22.5, and 24 kDa) are predominantly localized in the nucleus, whereas the AUG-initiated form (18 kDa) is localized primarily in the cytoplasm [1]. This may depend, however, upon the specific cells examined and the levels of FGF-2 expressed.

Binding Sites and Affinity   Extracellularly, FGF-1 and FGF-2 exert their biological effects by interacting with specific cell surface receptors. Four major tyrosine kinase receptor families have been identified that include FGFR1 (flg), FGFR2 (bek), FGFR3, FGFR4 [1, 22-

27]. The overall structure of FGFR1 to R4 comprises a conserved tyrosine kinase domain, a transmembrane domain and an extracellular ligand binding domain. A number of spliced variants for the extracellular ligand binding domain has been described. These may contain 2 or 3 immunoglobulin (IgG)-like loops.

FGF / FGF Receptor Interactions   The ligand specificities of the different FGF receptors and their association with FGF-2 is determined by alternate splicing of exon III that encodes the C-terminal half of the third IgG-like domain. This region is encoded by exon IIIa, IIIb or IIIc in FGF-R1-3 [23-26]. Alternate splicing of the exons IIIb and IIIc dramatically affects the binding specificity for FGFs. FGF-2 binds preferentially to the IIIc FGFR splicing form [23]. FGFR4 is devoid of exon IIIb and FGF-2 therefore also binds preferentially to this receptor subtype. Furthermore, IgG loops I and III, the inter-loop sequence between loop I and II, and the acidic domain of FGF-R1 also modulate receptor-ligand interactions [26-27]. Alternately-spliced loop I lowers the binding affinity of FGF by interacting with the other loops. Furthermore, the inter-loop sequence between loop I and II except the acidic domain is rather inhibitory since deletion of this sequence increases FGF binding. Conversely, the acidic domain between IgG loop I and II seems to be important for FGF/FGF receptor interactions because the deletion of this domain decreases FGF-2 binding and activity.

The number of binding sites for cells in culture varies between 2000 and 80,000 sites per cell. For example, endothelial cells express 5,000- 70,000 / cell and vascular smooth muscle cells 2,000-20,000 / cell. The affinities of FGF-1 FGF-2 to high-affinity cell surface receptors are variable ranging from 10 pM to 100 pM K_d in culture cells. The approximate respective affinities for vascular cells in culture are: microvascular endothelial cells 20-50 pM K_d , human umbilical vein endothelial cells 20 K_d, and vascular smooth muscle cells 20-50 K_d.

Heparan sulfate / FGF Interactions   Heparan sulfate proteoglycans are involved in the regulation of FGF-1 or FGF-2 activity. If everyone agrees about the importance of heparan sulfate/FGF interactions, the significance of this interaction is still a matter of controversy. Several investigators have reported that heparan sulfates are absolutely necessary for FGF signaling because of inducing FGF-2 dimerization necessary for FGF receptor activation [28-34]. This was based on experiments that used mutant CHO cells which lack heparan sulfates or the lymphoid cell line BaF3, which do not synthesize heparan sulfates or NIH 3T3 cells treated with chlorate that blocks sulfation of heparan sulfates [28-33]. Mutant CHO cells, or BaF3 when transfected with FGF-R1, failed to bind or respond to FGF-2 in the absence of heparin and NIH 3T3 cells treated with clorate no longer responded to FGF-2. This is also supported by a recent nuclear magnetic retenance (NMR) study indicating that FGF-2 is active in a dimeric state [34]. However, Moscatelli and co-workers reported different results [35,

36]. 32 D cells, which also lack heparan sulfates and which are transfected with FGFR1 bind FGF-2 in the absence of exogenous heparin and also induces c-fos activation [35]. Heparin does increase FGF-2 binding to its receptor threefold. In addition, Moscatelli [36] demonstrated that heparin stabilizes the FGF-2/receptor complex, thereby decreasing the rate of dissociation of FGF-2 from its receptor. Using X-ray crystallography, Venkataraman et al., [37] recently showed convincing evidence that FGF-2 self-associate in the absence of heparin and that heparin may associate to self-associated FGF-2 additionally stabilizing FGF-2 oligomeric complexes. Taking into account the latter observations, this would indicate that heparin is not absolutely required for FGF-2 dimerization and only stabilizes FGF-2 dimers.

What is the nature of the heparan sulfate proteoglycans involved in FGF binding? N-Syndecan, perlecan, glypican 1 and 3 are among the candidate proteoglycans involved in FGF-2 binding [38-42]. N-Syndecan is abundant in neonatal brain and binds FGF-2 but not FGF-1 or any other growth factors [38]. On the other hand, Aviezer et al. have reported that perlecan but not syndecan promotes binding of FGF-2 to FGFRs, mitogenesis and angiogenesis [39]. Furthermore, glypican 1 and 3 also bind FGF-2 [40, 41]. Thus, a number of proteoglycans are candidates for FGF-2 binding. Aviezer et al [42] provided strong arguments for the crucial role of perlecan in the regulation of FGF-2 binding and biological activity. They reported that the specific reduction of perlecan expression levels by transfection of NIH 3T3 cells with perlecan antisense cDNA dramatically impaired FGF-2 binding and biological activity. These data should be validated in endothelial cells to clearly establish the nature of the proteoglycans that are involved in angiogenesis.

Heparan sulfates might have another role in FGF/FGF receptor interactions. Richard et al [43] have demonstrated that heparan sulfates will participate with FGF-2 and FGFR1 in the attachment of circulating cells to adherent cells. 32 D cells transfected with FGFR1 attach in the presence of extracellular FGF-2 to monolayers of CHO, Balb/C, NIH 3T3 and bovine capillary cells. Protein synthesis is not required for this interaction. The physiological significance of this finding remains to be established.

The binding of FGF-2 to heparan sulfate proteoglycans has another interesting consequence. Tumova and Bame [44] recently reported that FGF-2 protected HSPG from degradation by heparanases. This has been demonstrated for free glycosaminoglycans in solution where the heparan sulfates remain attached to the core protein. FGF-2 inhibits heparanase action by binding heparan sulfate chains at a sequence close to the cleavage site but different to it, since cleaved heparan sulfate chains are still able to bind FGF-2.

Two other heparan sulfate /FGF-2/ FGFR interactions are of significance. Firstly, internalization of FGF-2 not only involves FGF receptors, but is also mediated via heparan sulfates [45]. Secondly, heparin may activate FGFR4 by associating directly with this receptor type

[46]. If true, this may be the first example of a molecule that is not a peptide but a glycosaminoglycan, able to activate a growth factor receptor.

FGFs display interactions with other proteoglycan types. Kinsella et al. [47] reported that biglycan is induced by FGF-2 or wounding of endothelial cell monolayers *in vitro*. Biglycan is localized at the tips and edges of lamellopodia in migrating cells. Release of endogenous FGF-2 seems to be responsible for altered biglycan expression during endothelial cell migration. The number of binding sites for FGF-2 to heparan sulfate proteoglycans is approximately 1,000,000 / cell for different cell types in culture.

Additional Features   Activation of FGFR induces receptor autophosphorylation and the association with downstream substrates. The elucidation of the crystal structure of FGFR1 may provide clues into the mechanisms of receptor activation (48). The activation loop is structured in such a way that the substrate binding site is blocked by Arg 661 and Pro 663 leaving the ATP binding site accessible. This implies an auto-inhibitory mechanism that may be overcome by conformational changes induced by ligand binding. This will allow the transphosphorylation event to occur. The FGF-dependent signaling pathways are in the process of being identified and fall into four groups: (1) PLC-g activation, (2) activation of the ras/raf signaling cascade, (3) association with tyrosine kinase such as the src kinase and phosphorylation of cortactin, (4) signaling at an intracellular (nuclear) site. FGFR1 may associate directly with PLC-γ and stimulate phosphatidyl inositol (PI) hydrolysis [49, 50]. The biological significance of this association is not yet clear. Although internalization of FGFR1 is inhibited by mutation of the Tyr-766 autophosphorylaton site, this will not alter mitogenesis, plasminogen activator expression or neurite outgrowth in PC12 cells [49-51]. Similarly, PLC-g activation after FGF-R1 occupancy is not required for mesoderm induction in Xenopus [52]. FGFR1 activates the ras/raf signaling pathway. Of the other six additional autophosphorylation sites (Y-463, Y-583, Y-585, Y-653, Y-654 and Y-730) described, autophosphorylation on tyrosine 653 and 654 is important for activation of tyrosine kinase activity of FGFR1 [53]. FGFR1 is able to phosphorylate Shc and an unidentified Grb2-associated phosphoprotein of 90 kDa (pp90). Binding of the Grb2/Sos complex to phosphorylated Shc and pp90 represents the key link between FGFR1 and the ras/raf signaling pathway. Among the other tyrosine kinases that may associate with the activated FGFR1 is c-src. C-src phosphorylates cortactin in response to FGFR1 activation [54]. FGFR1 dependent c-src activity and phosphorylation is drastically impaired in senescent endothelial cells [55]. Activation of the MAP-kinase pathway by FGF-1 is correlated with cell proliferation whereas activation of the src kinase is correlated with migration [56].

A body of observation suggests that FGF-1 and FGF-2 may be translocated to the nucleus and signal at an intracellular site [57-66]. This is based on the following observations. Nuclear translocation of FGF-1 and FGF-2 has been reported to occur at the entry and during S phase of the cell cycle [57-59]. Maximal DNA synthesis requires an exposure to FGF-1 for at least 12 hrs [59]. Not only FGFR1 has been shown to be translocated to the nuclear membrane [59-61] but also be associated with the nuclear matrix [62]. Nuclear membrane and matrix-associated FGFR1 may undergo receptor activation and autophosphorylation [59-61]. Furthermore, intracellular FGFR1 has been described to exist in an unglycosylated form [63]. FGF-1 contains a nuclear targeting motive. If this motive is deleted, FGF-1 still activates FGFR1 but will no longer sustain cell growth [64]. If FGF-1 is fused with diphtheria toxin and added to toxin-resistant cells, which lack functional FGF receptors, the chimera is internalized and stimulates DNA synthesis [65]. Furthermore, when the nuclear targeting sequence of FGF-1 is injected into cells, it will stimulate DNA synthesis on its own [66]. Finally, it appears that FGFR1α which exhibits three IgG loops but not the two IgG-loop FGFR1β form is translocated to the nucleus upon FGF-1 stimulation [67]. The implications of these data are twofold i.e. (1) FGFs may signal at an intracellular site to achieve DNA synthesis, (2) the mechanisms of intracellular signaling might involve classical FGF receptors such as FGFR1 or may depend upon the activation of novel factors not related to any known FGF receptor type. It is not clear whether 18 kDa FGF-2 also might be able to signal at an intracellular site as this factor does not exhibit the nuclear localization sequence of FGF-1. The model of nuclear accumulation of exogenous FGF and signaling at an intracellular site has been challenged recently by molecular modeling and deletion mutagenesis [68]. According to the reported data, the presumed nuclear localization sequence of FGF-1 indirectly stabilizes the major receptor-binding domain. Mutants also exhibit an increase in heparin-dependency, loss of receptor binding and mitogenic activity. These results are of major concern for the validity of an intracellular action mechanism for exogenous FGFs.

Translation of FGF-2   Translation of FGF-2 forms occurs with use of internal ribosomal entry sites (IRES) [69]. Vagner et al. [70] performed an interesting study about the translational regulation of HMW FGF-2 and 18 kDa FGF-2. HMW FGF-2 was produced in transformed, heat-shocked or stressed cells whereas the 18 kDa was exclusively produced by normal cells. CUG initiation was dependent on cis-elements within the 5' region of the FGF-2 mRNA. In addition, several proteins, including a specific protein of 60 kDa (p60), were bound to the 5' region of the FGF-2 mRNA. Thus, translation of the HMW FGF-2 forms is dependent on specific trans-acting factors.

The role of eukaryotic initiation factor 4E (eIF4E) has recently been investigated [71]. Cells overexpressing eIF4E produce large amounts of FGF-2, in particular the largest CUG-initiated form. In addition, breast carcinoma cells that express increased amounts of eIFE4 exhibit also FGF-2 translational products. This is also observed by *in vitro* transcription/translation assays where the addition of eIF4E preferentially increases the largest CUG-initiated form and the AUG-initiated form.

Structure

Sequence and Size   FGF-1 has been identified as a 154 amino acid protein with truncated forms of 140 and 134 amino acids. The shorter forms of FGF-1 originate from proteolytic cleavage and disulfide rearrangement. FGF-2 was first identified as a 146 amino acid protein isolated from the pituitary [72]. When FGF-2 cDNAs were cloned [73, 74], an AUG codon was found in the proper context to initiate translation of a protein of 155 amino acids, and no in-frame AUG codons were found upstream. Therefore, translation was predicted to initiate at this AUG codon. However, FGF-2 molecules, both longer and shorter than that predicted from the cDNA sequence, were found in guinea pig brain, rat brain, liver, human placenta, prostate and several types of cultured cells [75-82]. The shorter forms are derived from the 155 amino acid FGF-2 by proteolytic degradation [76]. The origin of higher molecular weight forms (196, 201, and 210 amino acids) was elucidated by *in vitro* transcription/translation analysis that revealed that CUG codons, 5' to the AUG codon used for the translation initiation of the 155 amino acid form, were used as initiation codons for the larger species [5,6]. Alternative translation occurs by internal ribosomal entry sites in the FGF-2 mRNA [69]. When the FGF-2 cDNA is expressed in cells, the AUG- and three CUG-initiated forms migrate on SDS-PAGE gels with molecular weights of 18, 22, 22.5, and 24 kDa, respectively. Figure 1 depicts the structure of FGF-1 and FGF-2 as outlined above.

Homologies   FGF-1 and FGF-2 have homologies of 30–50 % with the other FGF family members.

Conformation   The three-dimensional structure has been established for FGF-1 and 18 kDa FGF-2 by X-ray crystallography [83, 84]. FGF-2 contains 12 anti-parallel β-sheets organized into a trigonal pyramidal structure. The high-resolution structure revealed by nuclear magnetic resonance (NMR) of recombinant FGF-2 also showed some interesting features [85]. FGF-2 consists of 11 anti-parallel β-strands. These β-strands are arranged in three groups of three or four connected strands. Two strands of each group form a β-sheet barrel of six anti-parallel β-strands. Furthermore, α-helix structure is present between residues 131-136 that is part of the heparin-binding site. This is a very important finding because the existence of this helix structure, that has not been detected by X-ray crystallography, may have important consequences for the understanding of the FGF-2 / heparin interactions.

Gene

Gene Structure   The FGF-1 is encoded by a single copy gene and contains three exons and two introns. [3]. The FGF-2 gene has a length of 56 kb and contains three exons and two introns each of 16 kb [74]. This gene is transcribed into five mRNA species of 7-6.5, 3.7, 2.2, 1.8 and 1.1 kb. Although the coding sequences are identical, the length of the mRNAs differ because of variable numbers of polyadenylation sites. Five GC boxes are found within the FGF-2 promoter and possible SP1 and AP1 binding sites.

Chromosomal Localization   The human FGF-1 and FGF-2 gene are localized on chromosome 5 (5q31-33) and 4 (4q25) respectively.

Gene Expression   see below

Gene Regulation   Several factors regulate the expression of endogenous FGF-1 and FGF-2. FGF-1 is mainly expressed in neural tissue and neuroendocrine tissue. Furthermore, FGF-1 expression has been detected in the prostate, in Langerhans Islets, in the heart, in the kidney and in a number of normal and transformed cell lines. FGF-2 is expressed in almost all organs and in a large number of cells in culture. Pharbol esters enhances FGF-2 mRNA and protein expression in human umbilical vein and bovine adrenal cells [86, 87], and forskolin induces the expression of all FGF-2 isoforms in these cells. FGF-2 accumulates preferentially in the cytosol after pharbol ester stimulation, whereas forskolin induces nuclear accumulation. Interleukin-1 and TGF-β regulate FGF-2 expression [88, 89]. p53 may regulate FGF-2 expression as well, because co-transfection of p53 and FGF-2 cDNAs into human glioblastoma or hepatocellular carcinomas has revealed that the FGF-2 promoter is responsive to p53 [90]. Whereas wild type p53 represses FGF-2 expression, mutant p53 enhances FGF-2 expression.

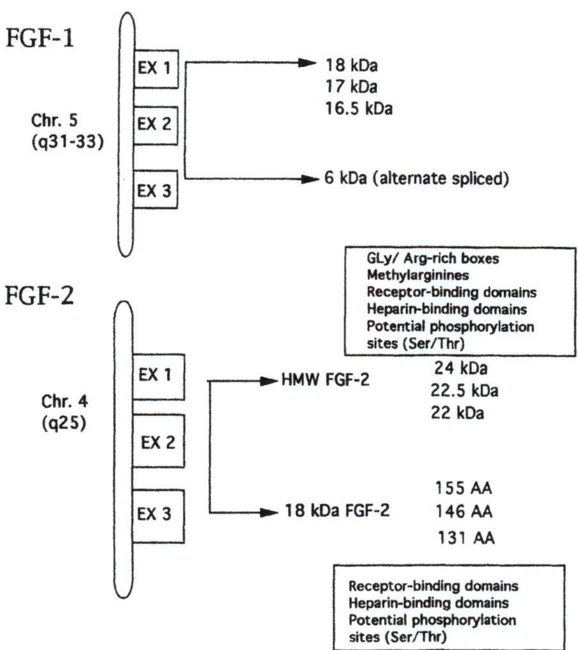

Figure 1. The different FGF-1 and FGF-2 forms

Processing and Fate

FGF-1   FGF-1 was found to be internalized in cultured cells and detected in the nucleus [3]. Nuclear accumulation correlated with progression in the S-phase [3]. A putative nuclear localization sequence (NYKKPKL) has been identified in the structure of FGF-1. Deletion of this sequence was shown to inhibit cell growth but did not affect receptor autophosphorylation [3]. A translocation to the nucleus of FGFR1 was observed. When diphtheria toxin-insensitive cells that lack FGF receptors were exposed to FGF-1 fused to diphtheria toxin, stimulation of DNA synthesis was observed [65]. As mentioned above, these experiments indicate that nuclear accumulation is important for the growth-promoting activity of FGF-1.

FGF-1 lacks a classical signal sequence. A number of arguments support the view that an alternative transport mechanism is involved in FGF-1 release. Jackson et al. [91] showed that heat-shock induces FGF-1 release in NIH 3T3 cells transfected with a cDNA encoding FGF-1. The released complex was identified by Western Blotting after ammonium sulfate precipitation. Furthermore, cysteine residues are involved in the release of FGF-1 in response to temperature [91]. This has been assessed by mutagenesis of Cys 30 Cys 97 and Cys 131 that have been converted into serines. These mutants are not released into the medium in response to heat-shock or brefeldin A treatment. FGF-1 may bind phospholipids [92]. The phospholipid binding domain lies between amino acid 114 and 131 but is not involved in FGF-1 export [92]. Serum starvation also induces FGF-1 release resulting in the appearance of high molecular mass products in the medium, that are converted to standard FGF-1 by reducing agents [93]. These data suggest that FGF-1 is released by a mechanism involving heat-shock proteins, stress and disulfide bonds. However this mechanism seems not to be involved in FGF-2 release. Shi et al. [94] demonstrated that only FGF-1 but not FGF-2 is released by heat-shock. Furthermore, FGF-1 but not FGF-2 contains a cytosolic retention domain that is absolutely required for FGF-1 export. Recently, a molecule, synaptotagmin-1 has been implicated in FGF-1 export [95]. It has been demonstrated that a fragment of synaptotagmin-1 (p40) is associate with FGF-1 and released. Antisense synaptotagmin-1 gene expression was able to inhibit FGF-1 release. Furthermore, a calcium-binding protein S100A13 is associated together with FGF-1 and synaptotagmin-1 in a ternary complex [96]. These data significantly contribute to our understanding of FGF-1 release.

Endogenous FGF-2 forms have distinct intracellular or extracellular distributions   It has been clearly shown by a number of investigators that 18 kDa or HMW FGF-2 segregate differently into distinct cellular and extracellular compartments [1]. 18 kDa FGF-2 is preferentially found in the cytoplasm, at the cell membrane whereas HMW FGF-2 is nuclear. How is this differential distribution governed and what its the biological significance? What is the nature of the FGF-2 domains involved?

18 kDa FGF-2 is found in the cytoplasm but is also released from the cell. FGF-2 release is divided into two different events i.e. (1) release of 18 kDa FGF-2 from the inside to the extracellular compartment (export), (2) release of cell surface or extracellular matrix bound 18 kDa FGF-2 into the extracellular environment. FGF-2 export involves two distinct mechanisms, one is passive by injury or cell death, the other is active by a novel export mechanism. The latter is supported by a number of experiments. Mignatti et al. [97] demonstrated that migration of an isolated single cell expressing FGF-2 can be inhibited by FGF-2-neutralizing antibodies. This indicates that FGF-2 is exported into the medium by a living cell and that other than injury-related mechanisms are at work. Furthermore, drugs that interfere with the ER/golgi pathway do not influence FGF-2 release. FGF-2 is exported from COS-1 cells by an alternative energy-dependent, non-ER/Golgi pathway [98]. COS-1 cells transfected with all four FGF-2 isoforms only release 18 kDa FGF-2. Export is not inhibited by brefeldin A. However, insertion of the transmembrane domain of vesicular stomatitis virus into the 18 kDa FGF-2 cDNA blocked export. What is the significance of FGF-2 release? FGF-2 may be implicated in the angiogenic switch. It has been reported that FGF-2 is released by fibrosarcoma cells at the onset of the angiogenic switch [99]. However, these data need to be ascertained and their validity tested in other carcinomas that are known to undergo multistage carcinogenesis and the angiogenic switch.

Extracellularily, endogenous 18 kDa may be ribosylated [100]. Jones and Baird [100] recently reported that 18 kDa FGF-2 is ribosylated at the surface of cells expressing an arginine-specific   glycophosphatidylinositol-anchored mono-ADP-ribosyltransferase. The ribosylation is inhibited by excess NAD, heparin and the putative receptor binding domain peptide analogue of FGF-2.

HMW FGF-2 is preferentially targeted into the nucleus. This has been demonstrated for several cell types including endothelial cells [1]. In addition, the N-terminus of HMW FGF-2 is required for nuclear accumulation. HMW FGF-2 forms are post-translationally modified which includes the presence of methylated arginines [101,102]. When extracts from ^{35}S-methionine-labelled cells are immunoprecipitated with anti-FGF-2 antibodies and analyzed by SDS-polyacrylamide electrophoresis, an increase of 1- to 2-kDa in the apparent molecular mass is observed [102]. This post-translational modification is inhibited by methionine starvation and by the methyltransferase inhibitors 5'-deoxy-5'-methylthioadenosine (MTA) and 3-deaza-adenosine. Inhibition of the methylation-dependent modification results in a significant decrease in the nuclear accumulation of HMW FGF-2. These results indicates that methylation is involved in the intracellular distribution of HMW FGF-2. The exact mechanism of nuclear accumulation remains to be established. In addition, endogenous FGF-2 is bound to ribosomes. Klein et al. [103] showed that 18 kDa or HMW FGF-2 are bound to ribosomal proteins or rRNA, since

RNAse or puromycin treatment did not affect binding of endogenous FGF-2 to ribosomes. Binding had a stoichiometry of 1:1 and remained unaffected by high salt treatment. This indicates a strong association with ribosomes.

Endogenous FGF-2 forms induce different cellular phenotypes   In a series of systematic studies it has been demonstrated that FGF-2 isoforms yield distinct phenotypes when expressed in cells [104]. Cells expressing 18 kDa FGF-2 showed an increase in migration whereas cells expressing only HMW FGF-2 migrate as control cells. Cells expressing HMW FGF-2 only exhibit increased proliferative behaviour. These conclusions received overwhelming support by the studies conducted by Vagner et al. [70], and Joy et al [105]. Vagner et al. [70] analyzed a number of transformed cell lines including endothelial cells of human origin. In all cell lines, increased growth properties and transformation were always associated with HMW FGF-2 expression but not with 18 kDa expression. Finally, Joy et al. [105], demonstrated that nuclear accumulation of FGF-2 is associated with proliferation of human astrocytes and glioma cells. Cell proliferation was attenuated by 5'-deoxy-5'methylthioadenosine that acts intracellularily.

The induction of migration by endogenous 18 kDa FGF-2 correlates with an induction of β_1 integrin expression. 18 kDa FGF-2 but not HMW FGF-2 regulates β_1 integrin expression [106]. 18 kDa FGF-2 induces cell migration but not HMW FGF-2. Migration induced by 18 kDa FGF-2 is correlated with a modulation of integrin expression. Cells expressing 18 kDa FGF-2 exhibit increased cell surface levels of $\alpha_5\beta_1$ integrin, whereas cells expressing only HMW FGF-2 expressed $\alpha_5\beta_1$ levels similar to parental cells. Immunoprecipitation of biosynthetically labeled cells indicated that expression of 18 kDa FGF-2 increases the biosynthesis and the rate of maturation of α_5. Northern Blot analysis demonstrated that 18 kDa FGF-2 increases the level of α_5 mRNA but did not affect the β_1-subunit transcription levels. Experiments using luciferase reporter gene activity revealed increased α_5-promoter activity indicating that the enhanced α5-transcript level is due to modulation of the transcription rate. This indicates that 18 kDa FGF-2 regulates integrin-expressing at the transcriptional level for the α-subunit and at the level of processing for the α- and the β-subunits.

Endogenous FGF-2 may also stimulate FGF receptor levels. Using AR42J cells, Estival et al. [107] demonstrated an increase in high affinity and low affinity receptors in cells expressing HMW or 18 kDa FGF-2. Although both FGF-2 isoforms regulate FGF receptor levels, only HMW FGF-2 but not 18 kDa FGF-2 stimulate FGFR1 levels by increasing FGFR1 mRNA stability.

Endogenous FGF-2 forms are probably acting by distinct signaling mechanisms   Supertransfection with a cDNA encoding the dominant negative FGF receptors affects the phenotype in cells expressing 18 kDa FGF-2 but not in cells expressing exclusively HMW FGF-2 [104]. This indicates that the events triggered by 18 kDa FGF-2 are dependent on FGF receptor cell surface activation, whereas HMW FGF-2 signals intracellularily.

What is the nature of the molecules with which endogenous FGF-2 might associate? The presence of nuclear FGF receptors have been reported by several groups [54-66]. As indicated above, it has been reported that FGFR1 is found in the nucleus, has kinase activity and undergoes autophosphorylation. More importantly, FGFR1 is found in the nucleoplasm and the nuclear matrix. This finding is very surprising; one would expect that a nuclear matrix associated FGFR1 would have its transmembrane domain deleted while retaining the tyrosine kinase domain. Such a splicing variant has never been described for any transmembrane tyrosine kinase receptor. Thus, it is difficult to invoke nuclear matrix-associated FGFR1 activation for endogenous FGF-2 signaling. A different view has been put forward by others. Using the two-hybrid system, it has been shown that intracellular FGF-2 associates with molecules that are not classical FGF receptors (H. Prats, personal communications). The nature of these molecules and their roles in nuclear accumulation and biological activity is now being investigated.

Other evidences for potential direct activation of gene transcription or kinase activity include the effect of FGF-2 on PGK gene transcription and on CKII kinase activity. Nakanishi et al [108] showed that *in vitro* PGK gene transcription is regulated by 18 kDa FGF-2. It has been reported that FGF-2 stimulates de-phosphorylation of nucleolin by CKII in an *in vitro* system [109]. This results from direct binding of FGF-2 to the regulatory β-subunit of Casein kinase2 and direct stimulation of CK activity toward nucleolin. The validity of the latter two observations remains to be clarified in relation to endogenous FGF-2.

Biological Activity

Stimulation   FGF-1 and FGF-2 stimulates a number of biological processes that include proliferation, migration, protease production, hormone production or cell differentiation (see under the different subsections).

Inhibition   Inhibitory effects have also been reported. FGF-2 inhibits the proliferation of Ewing sarcoma cells and mammary carcinoma cells (MCF 7) [110]. The mechanisms of this inhibitory effects are at present unclear, but may involve the stimulation by FGF-2 of inhibitory molecules such as TGF-β.

Role in Vascular Biology

Physiological Function   FGF-2 induces endothelial cell proliferation, migration, and angiogenesis *in vitro* [1]. FGF-2 regulates the expression of several molecules thought to mediate critical steps during angiogenesis. These include interstitial collagenase, urokinase-type plasminogen activator (uPA), plasminogen activator inhibitor (PAI-1), uPA receptor, and β_1 integrins (For reviews see [111-113]. These molecules may be involved in the invasive phenotype displayed by endothelial cells during angiogenesis. Angiogenesis induced by FGF-2 also involves $\alpha_v\beta_3$ integrin because antibodies directed against this integrin subtype block angiogenesis *in vitro* and *in vivo* [114]. In addition, the extracellular matrix provides tensional signals to the FGF-2-activated endothelial cells to allow capillary cord formation [115]. Flamme and Risau [116] performed an interesting study to determine how endothelial and hematopoietic cell lineages emerge. They showed that FGF-2 induces differentiation of both endothelial cells and hematopoietic cells from dissociated quail epiblasts *in vitro*. In long-term cultures, the induced endothelial cells give rise to vascular structures. Based on this study, FGF-2 may be important for embryonic vascular development. The opposing effects of TGF-β on FGF-2 activity in BAE cells has been extensively investigated [117]. TGF-β inhibits FGF-2 induced cell migration and protease production. FGF-2 stimulates uPA expression, which, in turn, activates latent TGF-β. Activated TGF-β stimulates PAI-1 synthesis which inhibits uPA, shutting down subsequent TGF-β formation. This creates a loop regulating both TGF-β activation and FGF-2 activity. In addition, TGF-β is a biphasic regulator of FGF-2-induced angiogenesis [118, 119]. At low concentrations, TGF-β stimulates FGF-2 action *in vitro*, whereas at a high concentrations, it inhibits FGF-2 action. Several inhibitors of angiogenesis have been described. These include heparin [120], heparinase [121], platelet factor-4 [7, 122], suramin [123, 124], angiostatic steroids [125], thalidomide [126] and angiostatin [8]. Among the heparinases, only heparinase I and III inhibit FGF-2-induced angiogenesis *in vitro* and *in vivo* [121]. Tissue inhibitor of metalloproteinase-2 (TIMP-2) inhibits FGF-2-induced human microvascular endothelial cell proliferation [127]. Interferon-α and -β downregulate FGF-2 expression in human renal, bladder, prostate, colon and breast carcinomas [128]. This might account for the benefit observed after interferon treatment in these vascularized neoplasms. However, interferon-a and interleukin-2 in combination stimulate endothelial cell growth and *in vivo* angiogenesis [129]. Like FGF-2, FGF-1 stimulates angiogenesis *in vitro* and *in vivo* [3,130]. However, FGF-1 is not significantly expressed in endothelial cells, which suggests a paracrine mechanism of action. Smooth muscle cells express FGF-1, which may play a role in the induction of neovascularization within the atherosclerotic lesion [130, 131]. It is also possible that other members of the FGF family, if expressed at appropriate levels in specific sites, will be angiogenic as they can bind to the same receptors as FGF-1 and -2. The fate of FGF-2 applied to the vessel wall was examined by Edelman et al. [132] who characterized intravenously injected ^{125}I-labeled FGF-2 vs. controlled perivascular released growth factor. Whereas intravenously administered FGF-2 was rapidly cleared from the circulation, FGF-2 from slow release polymers was delivered to the extravascular space without transendothelial transport for longer periods of time. The deposition of FGF-2 delivered by the slow release system was 40 times greater than by intravenous administration. Thus, an intact endothelium is not required for FGF-2 to reach the sub-endothelium, and this passage is not mediated by transendothelial transport. Systemic administration of FGF-2 in rabbits lowers the blood pressure [133]. This hypotensive action is due to the induction of endothelial cell-relaxing factor (EDRF, NO) synthesis and/or adenosine triphosphate-sensitive potassium ion channels. Therefore, FGF-2 may play a role in the regulation of blood pressure and may be of therapeutic use in the treatment of hypertension. FGF-2 improves myocardial function in chronically ischemic porcine hearts [134]. Periadventitial administration of FGF-2 in a gradual coronary occlusion model resulted in an improvement of coronary blood flow and a reduction in the infarction size. Furthermore, intracoronary injection of FGF-2 improved cardiac systolic function and reduced infarction size in a canine experimental myocardial infarct model [135].

Pathology

Tumor Angiogenesis With the discovery of VEGF, FGF-1 or FGF-2 were no longer regarded by a number of investigators as playing a significant role in physiological or pathological angiogenesis [136]. VEGF exhibits several advantages over FGF-1 or FGF-2. It is released with high efficiency from cells and its action is more restricted than that of FGF-1 or FGF-2 which acts upon many cells. In addition, the gene knock-out of VEGF and VEGF receptors provides compelling evidence for a major role of this growth factor in embryonic angiogenesis [136]. Furthermore, VEGF expression also correlates with angiogenesis in several tumors [136]. FGF-1 or FGF-2 on the other hand is released from most cell-types with low efficiency. However, there are a number of reasons that implicates FGFs at least in pathological angiogenesis. Many tumors express high levels of FGF-2, which may be released by necrosis. Furthermore, several studies indicate that FGF-2 may be released by a non-lysis-dependent mechanism. Endothelium may also be induced to produce its own FGF-2, as a secondary autocrine or intracrine cytokine, and may thereby control angiogenesis. In addition, FGF-2 and VEGF may act synergistically as it has been demonstrated in *in vitro* angiogenic assays [137, 138]. Recent observations also suggest an interplay between FGF-2 and VEGF providing a rational for the involvement of both factors in angiogenesis. Plouet and co-workers have shown in corneal endothelial cells that the effect of endogenous 189 amino acid VEGF is mediated via the liberation of FGF-2 from the matrix [139]. VEGF competes with FGF-2 bound to ECM, thus allowing its passage into the medium. Along this line, Mandriota et al have reported that VEGF induces FGF-2 protein in capillary endothelial cells [140]. This constitutes a secondary cytokine loop. Most interestingly, Mignatti and co-workers [141] undertook a systematic analysis about the regulation of VEGF by exogenous and endogenous FGF-2. They could show that endogenous FGF-2 increases VEGF expression. Only the expression of the 165 VEGF form was stimulated by FGF-2. Endogenous HMW FGF-2 was more efficient in the induction of VEGF than 18 kDa FGF-2. In addition, during FGF-2-induced corneal angiogenesis *in vivo*, high levels of VEGF expression are detected in growing capillaries. These results provide a new conceptual framework about the role of FGF-2 in angiogenesis.

It is important to validate these observations by a FGF-1 or FGF-2 gene knock-out. Inactivation of the FGF-2 gene did not show vascular development abnormalities in the embryo or adult [142-144]. FGF-2 knock-out animals showed a decrease in the migration of cortical neurons, hypotension and delay in wound-healing [142-144]. This does not rule out an involvement of FGF-2 in embryonic angiogenesis or tumor angiogenesis since functional redundancy may occur to rescue the phenotype. Double knock-outs including other FGF members or crosses of FGF-2 knock-out mice with mouse strain deficient for other molecules may reveal an angiogenic phenotype. Exciting times lie ahead for FGF biologists once these mice are made available to the scientific community. FGF-2-dependent tumor angiogenesis may also involve the release of a FGF-2 binding protein (BP) [145]. This has been shown for several tumors including squamous cell carcinoma (SCC) and [145]. Once released, BP will mobilize FGF-2 bound to heparan sulfates and allow its association with endothelial cell FGF receptors.

Tumor angiogenesis may be regulated not only at the FGF level but also at the receptor level. Arbeit et al., [146] have analyzed the patterns of expression of the FGF/FGFR system in multistage carcinogenesis using transgenic mice expressing the early region of high-risk papilloma virus type 16 under the control of human-keratinocyte-14 enhancer/promoter. While FGF-1 is upregulated in dysplasia, FGF-2 is constitutively expressed at all stages. FGFR1 was upregulated in well-differentiated squamous cancers and co-localized with angiogenic capillaries in the dermis underlying dysplastic lesions and within papillary fronds of invasive cancers. Furthermore, FGFR1 is expressed within the tumor cells in moderate to poorly differentiated malignant squamous cell carcinoma. These data indicate that the FGF/FGF receptor system might play a significant role in multistage carcinogenesis and tumor angiogenesis in the epidermis. Thus, tumor angiogenesis might also be controlled at a receptor level. Figure 2 depicts the different putative mechanisms described herein.

Atherosclerosis FGF-2 stimulates smooth muscle cell proliferation [147]. An elegant series of experiments on the role of FGF-2 in neointimal cell proliferation and atherogenesis were performed by Reidy and co-workers [148, 149], who demonstrated that the infusion of neutralizing antibodies to FGF-2 after balloon injury of the rat aorta inhibits neointimal cell proliferation. By *in situ* hybridization, FGF-2 mRNA was detectable at the wound edge of the endothelial cell layer and in migrating or proliferating smooth muscle cells. Expression of FGF-2 mRNA and FGFR1 mRNA was observed in replicating endothelial and smooth muscle cells. In agreement with these results,

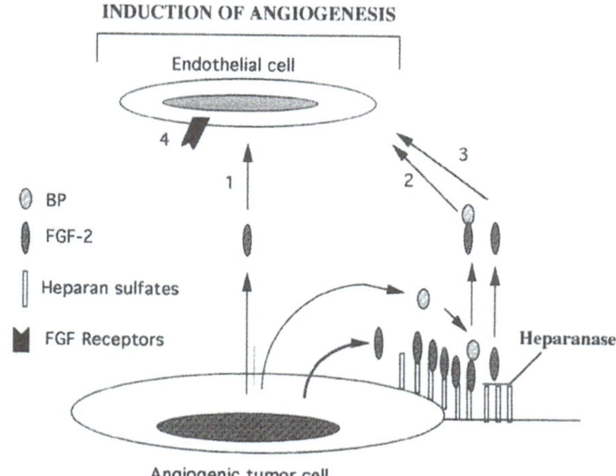

Figure 2. Putative mechanisms of action of FGF-2 in tumor angiogenesis. 1, Release of FGF-2 and direct activation of endothelial cell FGF receptors; 2, Release of FGF-2 from the extracellular matrix by BP which will transfer FGF-2 to endothelial cells; 3, Release of FGF-2 from the extracellular matrix by heparanase or proteases, released FGF-2 will stimulate endothelial cell FGF receptors; 4, upregulation FGF receptors on endothelial cells.

Casscells et al. [150] observed that FGFRs were upregulated in smooth muscle cells after vessel injury. Upregulation of FGFR expression renders smooth muscle cells susceptible to the lethal effects of FGF-2 coupled to saporin. Furthermore, FGF-2 and FGFR1 mRNAs are upregulated in human atherosclerotic arteries and increased mRNA expression is specifically associated with neovascularization of the atheromatous lesion [151]. Thus, in injured arteries, the FGF-2 ligand/receptor system may be involved in neointimal formation. In support of this hypothesis, FGF-2 was found to be released after vessel injury [152]. Brogi et al. [153] showed that all cells of arteries contain FGF-1 and FGF-2. However, FGF-1 mRNA was detected in only one of five control arteries tested, whereas all five atheromatous arteries contained FGF-1 mRNA. FGF-2 mRNA was expressed in both control and atheromatous arteries. Immunolocalization revealed abundant FGF-2 in control arteries but little in plaque. FGF-1 immunoreactivity was absent in control arteries but was high in atheroma-containing arteries. All arterial cells and arteries contained FGFR1. Only smooth muscle cells and control vessels had FGFR2 mRNA, although endothelial cells and some arteries contained FGFR4 mRNA. These data suggest that FGF-1, but not FGF-2, may be important in atherogenesis. However, FGF-2 may play a role in the early stages of formation of the atherosclerotic lesion, whereas FGF-1 is active at a later stage.

Clinical Relevance and Therapeutic Implications Clinical data also support a role of FGF-2 in tumor angiogenesis [154, 155]. Cerebrospinal fluid of children and adults with brain tumors contain an angiogenic activity identical to FGF-2, that correlates with the extent of tumor microvessel formation [154]. Furthermore, FGF-2 levels in urine samples from a large number of patients with a wide variety of solid tumors, lymphomas or leukemias correlate with the severity of the disease [155]. Patients with aggressive and metastatic tumors have high FGF-2 levels, whereas patients with less aggressive tumors had low FGF-2 levels.

FGF-2 also seems to be implicated in the formation of the collateral circulation after ischemia [156, 157]. Preclinical and clinical studies are underway to evaluated the validity of FGf-based strategies (protein or DNA) in the treatment of coronary ischemia or peripheral arterial occlusion.

Andreas Bikfalvi

References

1. Bikfalvi A et al (1997) Endocr Rev 18:26-45
2. Smallwood PM et al (1996) Proc Natl Acad Sci USA 93:9850-9857
3. Friesel RE, Maciag T (1995) FASEB J 9:915-925
4. Hu MCT et al (1998)Mol Cell Biol 18:6063-74
5. Florkiewicz RZ, Sommer A (1989) Proc Natl Acad Sci USA 1989 86:3978-3981
6. Prats H et al (1989) Proc Natl Acad Sci USA 86:1836-1840
7. Maione TE et al (1990) Science 247:77-79
8. O'Reilly MS et al (1994) Cell 79:315-328
9. O'Reilly MS et al (1996) Nat Med 2:689-692
10. Hanahan D, Folkman J (1996) Cell 86:353-364
11. O'Reilly M.S. et al (1997) Cell 88:277-285
12. Ferrara N. et al (1991) Endocrinology 129:896-900
13. Baird A et al (1998) Proc Natl Acad Sci USA 85:2324-2328
14. Yayon A et al (1993) Proc Natl Acad Sci USA 90, 10643-10647
15. Presta M et al (1991) J Cell Physiol 149:512-524
16. Feige JJ, Baird A (1989) Proc Natl Acad Sci USA 86:3174-3178
17. Vilgrain I, Baird A (1991) Mol Endocrinol 5:1003-1012
18. Vilgrain I et al (1993) FEBS Lett 331:228-232
19. Thompson SA (1992) J Biol Chem 267:2269-2273
20. Sommer AD et al (1989) Biochem Biophys Res Commun 74:969-976
21. Burgess W et al (1991) Cell Regul 2:87-93
22. Jaye M et al (1992) Biochem Biophys Acta 1135:185-199
23. Ornitz DM et al (1996) J Biol Chem 1996 271:15292-15297
24. Werner S et al (1992) Mol Cell Biol 12:82-88
25. Wang F et al (1995) J Biol Chem 270:1022-1030
26. Zimmer Y et al (1993) J Biol Chem 268:7899-7903
27. Wang F et al (1995) J Biol Chem 270:10231-10235
28. Yayon A et al (1991) Cell 64:841-848
29. Ornitz DM et al (1992) J Biol Chem 267:16305-16311
30. Ornitz DM et al (1992) Mol Cell Biol 12:240-247
31. Rapraeger AC et al (1991) Science 252:1705-1708
32. Olwin BB, Rapraeger AC (1992) J Cell Biol 118:631-639
33. Rapraeger AC et al (1994) Methods Enzymol 245:219-240
34. Moy FJ et al (19969 Biochemistry 35:13552-13561
35. Rohgani M et al (1994) J Biol Chem 269:22156-22162
36. Moscatelli D (1993) J Biol Chem 267:25803-25809
37. Venkataraman G et al (1996) Proc Natl Acad Sci USA 93:845-850
38. Chernousov MA, Carey DJ (1993) J Biol Chem 268:16810-16814
39. Aviezer D et al (1994) Cell 79:1005-1013
40. Steinfeld R et al (19969 J Cell Biol 133:405-416
41. Song HH et al J (1997) Biol Chem 272:7574-7577
42. Aviezer D et al (1997) Mol Cell Biol 17:1938-1946
43. Richard C et al J Biol Chem 1996 270:24188-24196
44. Tumova S, Bame KJ (1997) J Biol Chem 272:9078-9085
45. Roghani M, Moscatelli D (1992) J Biol Chem 267:22156-22162
46. Gao G, and Goldfarb M (1995) EMBO J 14:2183-2190
47. Kinsella MG et al (1997) J Biol Chem 272:318-325
48. Mohammadi M et al (1996) Cell 86:577-587
49. Mohammadi M et al (1992) Nature 358:681-684
50. Peters K et al (1992) Nature 358:678-681
51. Spivak-Kroizman T et al (1994) J Biol Chem 269:14419-14423
52. Muslin AJ et al (1994) Mol Cell Biol 14:3006-3012
53. Mohammadi M et al (1996) Mol Cell Biol 16:977-989
54. Zhan X et al (1994) J Biol Chem 269:20221-20224
55. Garfinkel S et al (1996) J Cell Biol 134:783-791
56. LaVallee TM et al (1998) J Cell Biol 141:1647-58
57. Bouche G (1987) Proc Natl Acad Sci USA 84:6770-6774
58. Baldin V (1990) EMBO J 9:1511-1517
59. Zhan X (1993) J Biol Chem 268:9612-9619
60. Prudovsky I (1996) J Biol Chem 269:31720-31724
61. Maher PA (1996) J Cell Biol 134:529-536
62. Stachowiak MK (1996) Mol Cell Biol 7:1299-1317
63. Maher PA (1996) J Cell Physiol 169:380-390
64. Friedman S (1994) Biochem Biophys Res Commun 198:1203-1208
65. Wiedlocha A et al (1994) Cell 76:1039-1051
66. Lin YZ et al (1996) J Biol Chem 271:5305-5308
67. Prudovsky IA et al (1996) J Biol Chem 271:14198-14205
68. Luo Y et al (1997) J Biol Chem 271:26876-26883
69. Vagner S et al (1995) Mol Cell Biol 15:35-44
70. Vagner S et al (1997) Cell Biol 135:1391-1402
71. Kevil C et al (1995) Oncogene 11:2339-2348
72. Bohlen P et al (1984) Proc Natl Acad Sci USA 81:5364-5368
73. Abraham JA et al (1986) Science 233:545-548
74. Abraham JA et al (1986) EMBO J 5:2523-2528
75. Gospodarowicz D et al (1986) Endocrinology 118:82-90
76. Klagsbrun M et al (1987) Proc Natl Acad Sci USA 84:1839-1843
77. Moscatelli D et al (1987) Proc Natl Acad Sci USA 84:5778-5782
78. Sommer A et al (1987) Biochem Biophys Res Commun 144:543-550
79. Story MT et al (1987) Biochem Biophys Res Commun 142:702-709
80. Presta M et al (1988) Biochem Biophys Res Commun 155:1161-1172
81. Presta M et al (1989) Biochem Biophys Res Commun 164:1182-1189
82. Brigstock DR et al (1990) Growth Factors 4:45-52
83. Ericksson AE et al (1991) Proc Natl Acad Sci USA 88:3441-3445
84. Zhu X et al (1992) Science 251:90-93
85. Moy FJ et al (19969 Biochemistry 35:13552-13561
86. Bikfalvi A et al (1989) J Cell Physiol 144:151-158
87. Stachowiak MK et al (1990) J Cell Biol 12:203-223
88. Gaye G, Winkles JA (1991) Proc Natl Acad Sci USA 88:296-300
89. Partovaara L et al (1993) Growth Factors 9:81-86
90. Ueba T et al (1994) Proc Natl Acad Sci USA 91:9009-9013
91. Jackson A et al (1995) J Biol Chem 270:33-36
92. Tarantini F et al (1995) J Biol Chem 270:29039-29042

93. Shin JT et al (1996) Biochim Biophys Acta 1312:27-38
94. Shi J et al (1998) J Biol Chem 272:1142-1147
95. La Vallee TM et al (1998) J Biol Chem 273:22217-22223
96. Carreira CM et al J (1998) Biol Chem 273:22224-22231
97. Mignatti P et al (1992) Cell Physiol 151:81-93
98. Florkiewicz R et al (1995) J Cell Physiol 162:388-399
99. Kandel J et al (1991) Cell 66:1095-1104
100. Jones EM et al (1997) Biochem J 323:173-177
101. Burgess W et al (1991) Cell Regul 2:87-93
102. Pintucci G et al (1996) Mol Biol Cell 7:1249-1258
103. Klein S et al (1996) Growth Factors 13:219-228
104. Bikfalvi A et al (1995) J Cell Biol 129:233-243
105. Joy A et al (1997) Oncogene 14:171-183
106. Klein S et al (1996) J Biol Chem 271:22583-22590
107. Estival A et al (1996) J Biol Chem 271:5663-5670
108. Nakanishi Y et al (1992) Proc Natl Acad Sci USA 89:5216-5220
109. Bonnet H et al (1996) J Biol Chem 271:24781-24787
110. Schweigerer L et al (1987) J Clin Invest 80:1516-1520
111. Montesano R (1992) Eur J Clin Invest 22:504-515
112. Mignatti P, Rifkin DB (1993) Phys Rev 73:161-195
113. Klein S et al (1993) Mol Biol Cell 4:973-982
114. Brooks PC et al (1994) Science 264:569-571
115. Ingber D (1991) J Cell Biochem 47:236-241
116. Flamme I, Risau W (1992) Development 116:435-439
117. Rifkin DB et al (1993) Throm Haemost 70:177-179
118. Pepper MS et al (1990) J Cell Biol 111:743-755
119. Pepper MS et al (1993) Exp Cell Res 204:356-363
120. Moscatelli D (1987) J Cell Physiol 131:123-130
121. Sasisekharan R et al (1994) Proc Natl Acad Sci USA 91:1524-1528
122. Perollet C et al (1998) Blood 91:3289-99
123. Moscatelli D, Quarto N (1989) J Cell Biol 109:2519-2527
124. Takano S et al (1994) Cancer Res 54:2654-2660
125. Blei F et al (1994) J Cell Physiol 155:568-578
126. D'Amato RJ et al (1994) Proc Natl Acad Sci USA 91:4082-4085
127. Murphy A et al (1993) J Cell Physiol 157:351-358
128. Singh RK et al (1995) Proc Natl Acad Sci USA 92:4562-4566
129. Cozzolino F et al (1993) J Clin Invest 91:2504-2512
130. Maciag T (1990) Important Adv Oncol 85-98
131. Nabel EG et al (1993) Nature 362:844-846
132. Edelman ER et al (1993) Proc Natl Acad Sci USA 90:1513-1517
133. Cuevas P et al (1991) Science 254:1208-1210
134. Harada K et al (1991) J Clin Invest 94:623-630
135. Yanagisawa-Miwa A et al (1994) Science 257:1401-1403
136. Ferrara N, Davis-Smyth T (1997) Endocr Rev 18:1-25
137. Goto F et al (1993) Lab Invest 69:508-517
138. Pepper MS et al (1996) Curr Top Microbiol Immunol 213:31-67
139. Jonca F et al (1997) Biol Chem 272:24203-9
140. Mandriota SJ, Pepper MS (1997) J Cell Sci 110:2293-302
141. Seghezzi G et al (1998) J Cell Biol 141:1659-73
142. Dono R et al (1998) EMBO J 17:4213-25
143. Ortega S et al (1998) Proc Natl Acad Sci USA 95:5672-7
144. Zhou M et al (1998) Nat Med 4:201-7
145. Czubayko F et al (1997) Nature Med 3:1137-1140
146. Arbeit JM et al (1996) Oncogene 13:1847-1857
147. Schwarz SM, Liau L (1993) J Cardiovasc Pharmacol 1:31-49
148. Lindner V et al (1992) J Clin Invest 90:2044-2049
149. Lindner V, Reidy MA (1993) Circ Res 73:589-595
150. Casscells W et al (1992) Proc Natl Acad Sci USA 89:7159-7163
151. Hughes SE et al (1993) Cardiovasc Res 27:1214-1219
152. Villaschi S, Nicosia R (1993) Am J Pathol 143:181-190
153. Brogi F et al (1993) J Clin Invest 92:2408-2418
154. Li VW et al (1994) Lancet 344:82-86
155. Nguyen M et al (1994) J Natl Cancer Inst 86:356-361
156. Arras M et al (1998) J Clin Invest 101:40-50
157. Arras M et al (1998) Nat Biotechnol 16:159-162

Fibrin/Fibrinogen

Definition *Fibrinogen is a soluble plasma protein composed of three pairs of non-identical polypeptide chains. Fibrinogen is converted into fibrin by thrombin. This is the final reaction of the blood coagulation cascade.*

See also: →Fibrinolytic, hemostatic and matrix metalloproteinases, role of; →Thrombin; →Thrombosis

Introduction The identification of fibrinogen as a protein and its isolation from plasma goes back to the latter half of the past century (for review see [1]). A key biological relevance of fibrinogen is its participation as the structural element of a blood clot in physiological hemostasis and in the pathophysiology of thrombosis. In order to understand both processes better, we need to know more about the functional states of this molecule under different physiological and pathophysiological conditions. Although fibrinogen offers many opportunities for therapeutic intervention, there is still difficulty intervening in diseases as diverse as thrombosis, malignancy and renal disease, to name a few. Despite the amount of work published, this molecule still offers new aspects of involvement with diseases and cellular interaction that continue our renewed interest and have not finished surprising us. For early work on different areas of research on fibrinogen we refer the reader to previous extensive reviews [1–4].

In this chapter we will focus on general structural-function aspects and recent studies on the cellular and molecular interactions of fibrinogen that have appeared in the past five years.

In an extensive review on fibrinogen, fibrin and their roles in hemostasis and thrombosis, Birger Blombäck appropriately characterized fibrinogen as a "watery, dimeric molecule with many constituents and forms" [1]. Fibrinogen, synthesized in the liver, is a glycoprotein present in blood plasma of all vertebrates. The concentration of the protein in normal human plasma is about 3g per liter (~ 9 µM). Since fibrinogen is an acute-phase protein, its level is increased significantly after infection or injury and in certain disease states. Original work by Meade and collaborators [5] and subsequent studies by others, have shown that fibrinogen is an important primary cardiovascular risk factor. More recent studies indicate that fibrinogen level is associated with increased mortality in patients with coronary heart disease [6].

Fibrinogen is a soluble plasma protein composed of three pairs of non-identical polypeptide chains called Aα (610 residues), Bβ (461 residues) and γ (411 residues). Since fibrinogen is a dimer, the formula for the molecule can be given by $(Aα, Bβ, γ)_2$. The six chains are held together by 29 inter- and intrachain disulfide bonds, the protein does not have any free sulfhydryl groups. Three of the 29 bridges, the so-called symmetrical disulfides (between two Aα-chains at residue AαCys28 and the two γ-chains at γCys8 and γCys9, respectively) play an important role in maintaining fibrinogen as a dimeric structure. Several studies have suggested that, in addition to the symmetrical disulfides, either hydrophobic forces [7] and/or additional non-symmetric disulfide bridges [8] contribute to the strong association of fibrinogen half-molecules.

The conversion of fibrinogen to fibrin is perhaps the most important reaction in the coagulation cascade. Thrombin, a trypsin-like enzyme with highly restricted specificity, cleaves two small acidic peptides, fibrinopeptide A (FPA, Aα 1-16) and fibrinopeptide B (FPB, Bβ 1-14) from the NH_2-terminal ends of the Aα- and Bβ-chain, respectively. Once activated by thrombin, the fibrinogen molecule is converted into an insoluble fibrin gel or meshwork which can be further stabilized by introduction of covalent bonds by plasma transglutaminase (Factor XIIIa). The latter enzyme catalyzes Ca^{2+}-dependent acyl transferase reactions which result in ε-(γ-glutamyl)lysine bonds. Initially these cross-links are formed between COOH-terminal regions of the γ-chain on adjacent fibrin molecules. Subsequently, these same bonds also link fibrin α-chains. Of the three, only the β-chain of fibrin is totally inert to cross-linking by Factor XIIIa.

Characteristics

Molecular Weight   Until very recently, molecular weight estimations of fibrinogen and its component chains were based on sedimentation analysis, size exclusion chromatography as well as gel electrophoresis. From primary amino acid sequence analysis, the average size of the Aα-, Bβ- and γ-chain should be about 66, 52 and 46 kDa, respectively. However, two of the three chains contain a single biantennary oligosaccharide (on BβAsn364 [9] and γAsn52 [10, 11]) composed of mannose, galactose, glucosamine and sialic acid. Taking into account the structure of the sugar side-chain and assuming complete sialylation, the predicted mass for the Bβ- and γ-chains would be about 54 and 48.4 kDa. From these considerations, and mindful of the fact that fibrinogen is dimeric, summation of chain mass leads one to assign a value of about 340 kDa for the full size, undegraded molecule. The latter accounts for about 70 % of blood plasma fibrinogen. Smaller forms of fibrinogen, approximately 305 and 270 kDa, can also be identified in human blood. The latter, comprising about 25 % and 5 %, respectively, of blood fibrinogen differ from the major form in that they either lack one, or both,

COOH-terminal portions of fibrinogen Aα-chain [12]. In recent studies using electrospray ionization mass spectrometry (ESI MS), Brennan and colleagues not only determined the mean masses of the dominant isoforms of the three chains, but have also used the technique to deduce the nature of a mutation (BβArg14→Cys) in a patient diagnosed with dysfibrinogenemia (also see below) who had experienced severe thrombotic complications [13, 14]. ESI MS data on chains – isolated by reverse-phase high pressure liquid chromatography of dithiothreitol-reduced plasma fibrinogen obtained from the blood of six normal subjects – gave mean molecular weights of $66,196 \pm 64$ Da for the Aα-, $54,200 \pm 12$ for the Bβ- and $48,370 \pm 10$ for the γ-chain. Fibrin chains, analyzed using a similar approach, showed, as expected, no change for the γ-chain but lower average mass values for both the desFPA α-chain (64,650 Da) and desFPB β-chain (52,658 Da). These results are in good agreement with an expected decrease of 1519 Da for the loss of the non-phosphorylated form of FPA and 1535 Da for the loss of FPB [13].

Domains

Disulfide-bonded structure   Initial disulfide-bonded structure studies on fibrinogen primary structure were carried out by Blombäck and his collaborators using cyanogen bromide (CNBr) fragments of the molecule. Cleavage of fibrinogen with CNBr results in some 30 fragments, the largest of which is often referred to as the "NH_2-terminal disulfide knot" or N-DSK. The latter fragment represents about 16 per cent of the mass of fibrinogen, contains 11 of the 29 disulfide bridges in the molecule, and, like fibrinogen itself, N-DSK is a dimer whose formula is given by $(Aα 1-51, Bβ 1-118, γ 1-78)_2$. N-DSK (58 kDa) contains the NH_2-terminal portions of all three fibrinogen chains with microheterogeneity being associated with both the Aα and γ-chains [10, 15, 16]. The N-DSK domain is crucial in the fibrinogen-to-fibrin transition (see below). In addition to N-DSK, four other disulfide-containing CNBr fragments may be obtained from fibrinogen. In contrast to N-DSK, these other fragments are monomeric. One such fragment has been named Hi2-DSK (28 kDa). The latter is derived from the middle portion of the Aα-chain (Aα 241-476), has a single intra-chain disulfide bridge (AαCys442-Aα Cys472), contains unique non-identical repeat units and includes two Gln residues (328 and 366) which are believed to be major amine acceptor sites in Factor XIIIa-mediated cross-linking [17]. Both N-DSK and Hi2-DSK are hydrophilic whereas the three remaining disulfide-containing structures are hydrophobic and have been termed Ho1-DSK (42.5 kDa, six disulfides), Ho2-DSK (7 kDa, one disulfide) and Ho3-DSK (7 kDa, one disulfide), respectively. N-DSK and the other four monomeric CNBr fragments contain all the disulfides found in fibrinogen [18].

Plasmin-generated domains   Proteolytic cleavage of fibrinogen with plasmin results in a number of fragments, and heterogeneity of some of these is more restricted when digestion buffers contain calcium in the range 2-10mM [19]. Under such conditions, fibrinogen is progressively cleaved to transient degradation products called Fragment X (Fg-X, 225-333 kDa) and Fragment Y (Fg-Y, 150-170 kDa) as well as terminal core products called Fragment D (Fg-D, 93 kDa) and Fragment E (Fg-E, 50 kDa). A very similar pattern of degradation is obtained when non-cross-linked fibrin is used as substrate. When FXIIIa-stabilized fibrin (see above) is cleaved with plasmin, a different set of terminal core fragments is obtained. Instead of Fg-D, plasmin-cleaved cross-linked fibrin yields a dimerized (Fg-DD or D-dimer, 186 kDa) Fragment D species [20]. The latter are derived from different fibrin molecules and are joined together by the ε-(γ-glutamyl)lysine bonds at the COOH-terminal portion of the γ-chain of each fragment. Different Fg-E species have been identified in plasmin digests of cross-linked fibrin [21]. Despite the fact that the Fg-E species derived from fibrinogen and fibrin display significant heterogeneity, most satisfy the formula (Aα 20-78, Bβ 54-120, γ 1-53)$_2$ and, therefore, are structurally related to N-DSK. The involvement of the Fg-E (or N-DSK) and Fg-D domains in the fibrinogen-to-fibrin transition will be discussed below.

Binding Sites   Domains on fibrinogen which interact with a number of different cells and proteins have been identified. Fibrinogen and fibrin serve as adhesion molecules for a variety of cells including platelets, endothelial cells and leucocytes. The crucial binding of fibrin(ogen) to platelets through the membrane glycoprotein receptor GPIIb/IIIa, a member of the integrin family also referred to as integrin $\alpha_{IIb}\beta_3$, will be described later in this chapter. Fibrinogen not only has the ability to interact with eukaryotic cells/cell fragments but one of its important functions also includes its capacity to agglutinate (clump) certain prokaryotic cells. Pathogenic staphylococci bind to the COOH-terminal part of fibrinogen γ-chain [22]. In these and related studies it was established that about 2,000 fibrinogen molecules bind per cell (K$_d\sim$ 10^{-8} M) and that a pentadecapeptide corresponding to γ397-411 can block this interaction. As will be discussed below, this region of fibrinogen γ-chain can not only clump bacteria but also contains the essential recognition sequence that mediates platelet aggregation [23].

A number of proteins interact with fibrinogen and perhaps the most important reactions are those involving thrombin, the enzyme responsible for the fibrinogen-to-fibrin transition, and plasminogen, the pro-enzyme form of the fibrinolytic enzyme plasmin. Since the early structural characterization of fibrinopeptides from various species, much has been learned about the mechanism of interaction of thrombin with fibrinogen. Studies by Scheraga and colleagues suggested binding of thrombin to the Aα 7-16 part of fibrinogen, with AαPhe9 being in close proximity to the cleavage site at AαArg16-Gly17 [24]. In addition to the substrate site, thrombin binds to other regions of the E domain. Both Bβ 15-42

and Aα 27-50, and possibly also a segment(s) of the γ-chain in the E domain, have been implicated. Measurements of the strength of thrombin binding to fibrinogen have not been reported due to the fact that such interactions lead to fibrin formation. The dissociation constant ($\sim$2 x 10^{-6} M) for the thrombin-fibrinogen complex has been estimated from kinetic data on the rate of release of FPA. In addition to the substrate site(s), thrombin binds to fibrin at two classes of non-substrate sites. Mosesson and co-workers have recently re-examined these and found that the low-affinity site probably involves more than one peptide sequence in the E domain [25]. In these studies evidence was also presented that the anionic COOH-terminal γ' sequence (see below) contains the high-affinity (K$_d\sim$3 x 10^{-7} M) thrombin binding site.

Plasminogen (92 kDa) binds to fibrinogen, however, binding is more avid to urea-denatured or surface-bound fibrinogen. Presumably these treatments induce conformational changes in fibrinogen which expose interaction sites. Plasminogen binds specifically to fibrin through its lysine-binding site. The binding of plasminogen to a fibrin clot is markedly increased in the presence of tissue plasminogen activator (t-PA, 70 kDa), the physiologically most important enzyme which converts plasminogen to plasmin. The latter has a very high affinity for fibrin and, in the presence of plasminogen, the dissociation constant for the t-PA-fibrin complex is 1.4 x 10^{-7} M. The formation of a ternary complex between fibrin, plasminogen and t-PA leads to the acceleration of plasminogen activation by several orders of magnitude [26]. Two possible, fibrin-related, sites involved in complex formation have been characterized, one in the Aα-(Aα 148-160) and the other in fibrinogen γ-chain (γ 311-379) [27, 28].

Other important interactions of fibrin(ogen) include those with fibronectin, α_2-antiplasmin and Factor XIII. Fibronectin binds to fibrin(ogen) and covalently cross-linked (by Factor XIIIa) fibrin-fibronectin complexes have been identified in blood clots. The latter are formed between lysyl residues in the COOH-terminal region of fibrinogen Aα-chain and glutaminyl residues in fibronectin [29]. It is believed that such complexes may be important in wound healing in that they could enhance the adhesion and migration of certain cells (e.g., fibroblasts and endothelial cells). Similarly to fibronectin, α_2-antiplasmin is also covalently cross-linked to fibrin α-chain by Factor XIIIa [30]. The precise location of the donor lysine is unknown, but glutamine in the second position of α_2-antiplasmin serves as acceptor. Following activation of plasminogen in plasma, α_2-antiplasmin is the principal inhibitor of the active enzyme. Cross-linking to fibrin may be one way by which α_2-antiplasmin controls fibrinolysis [26]. Recombinant human placental Factor XIIIa-subunit binds native fully hydrated fibrin gel columns. It was further demonstrated that binding required prior activation with thrombin and that it could be inhibited by both Hi2-DSK (Aα 241-476) and a monoclonal antibody (anti Aα 389-402) directed to a segment of this peptide [31]. Recent studies by Mosesson's group have shown that the anionic COOH-terminal γ' (see below) sequence not only forms the non-substrate high-affinity binding site for thrombin but also binds directly to the b-subunit of the zymogen form of Factor XIII [32]. More recent data suggest that 2 heterodimeric fibrinogen molecules, containing one normal length γ-chain and one γ'-chain (see below), bind one molecule of Factor XIII and that a synthetic peptide corresponding to the extension found at the COOH-terminus of the γ'-chain inhibits binding of the zymogen to the γ'-chain. It has been suggested that the negatively-charged residues in this extension mediate the interaction of heterodimeric fibrinogen molecules with the positive segment in the Factor XIII β-subunit sushi domain [33].

Heparin binding domains (HBDs) are known to be present on a number of adhesive glycoproteins and these regions are involved in cell-cell and cell-matrix interactions. Studies have now shown that the new NH_2-terminus of fibrin β-chain (Bβ 15-42) also contains an HBD [35]. It is believed that this domain is inaccessible in intact fibrinogen but becomes available following release of FPB by thrombin. Support for this hypothesis was obtained by measuring the relative strength of the complex formed between heparin and intact fibrinogen (K_d = 1.8 x 10^{-5} M), N-DSK (K_d =8 x 10^{-7} M) and thrombin-digested N-DSK (K_d=3 x 10^{-7} M). Although it has yet to be identified *in vivo*, a complex between heparin, thrombin and fibrin monomer would have a significant affect on regulation of hemostasis and thrombosis [34]. In addition to binding cells and other proteins, fibrinogen is also known to contain three high-affinity binding sites for calcium [35]. The latter ion is an essential cofactor for fibrin polymerization, promotes Factor XIIIa cross-linking, increases the heat stability of the molecule and, when bound, protects fibrinogen from more extensive degradation with plasmin. A calcium-binding site is located in fibrinogen Fragment D (K_d~10^{-5} M) and recent crystallographic studies with a recombinant fragment corresponding to γVal143-Val411 (see below) confirm that this segment of Fragment D contains a single calcium-binding site formed by side-chain carboxyl groups of γAsp318 and γAsp 320 and main-chain carbonyl oxygens of γPhe322 and γGly324 [36].

Undoubtedly the most important functional binding sites in fibrinogen are those which are involved in the fibrinogen-to-fibrin transition. The thrombin-catalyzed release of FPA and FPB from the NH_2-terminal portions of fibrinogen results in the formation of fibrin monomer. The latter molecules spontaneously begin to polymerize and, once the process is complete, a firm gel results. As mentioned above, fibrin can be further stabilized by Factor XIIIa-mediated cross-linking of adjacent monomers. The molecular interactions involved in fibrin polymerization have been studied for a number of years, the main goal being the identification of binding sites in the different chemically or proteolytically-derived domains identified above. Affinity chromatography studies of Heene and Matthias strengthened the

notion that fibrin polymerization results from the interaction of complementary binding sites on adjacent monomer molecules [37]. Subsequent studies showed that one such site (or sites) was located on the Fg-D domain, was also present on intact fibrinogen and that its complementary site (or sites) was located on the N-DSK domain [38, 39]. The N-DSK site is normally hidden and becomes functional only after activation with thrombin or similar coagulant enzymes. A defective N-DSK site in congenitally abnormal fibrinogen Detroit (see below), resulting from the mutation AαArg19→Ser, was believed to be responsible for impaired hemostasis in several members of a Detroit family that were homozygous for this trait [40]. Laudano and Doolittle showed that one of the sites in the N-DSK domain was in a peptide segment of the Aα-chain which is contiguous with FPA. This conclusion was based on the observation that Gly-Pro-Arg (Aα 17-19), or longer peptides beginning with this same sequence, bind fibrinogen and can inhibit fibrin monomer polymerization [41]. Studies by Shainoff and Dardick using a pro-coagulant from the copperhead snake venom also suggested that another binding site in this same domain may be contained in peptide segment Bβ 1-42 [42]. The coagulant-exposed sites are sometimes referred to as "knobs" and their complementary sites in the D domain have been called "holes". In more recent photoaffinity label studies, Doolittle and colleagues, have localized the ever-present, coagulant-independent, "hole" polymerization surface to γ337-379, specifically the region around γTyr363 [43, 44]. Ongoing crystal structure studies (see below), using some of the fragments just discussed, should provide greater details on how fibrinogen is converted to fibrin. Additionally, further work should clarify the exact role of the COOH-terminal part of fibrinogen Aα-chain in the so-called "second-stage" of fibrin polymerization [45].

Structure

Sequence and Size   The predominant form of vertebrate fibrinogens is a complex, disulfide-linked dimeric protein of approximately 340,000 Mr, and is composed of pairs of three different polypeptide chains, designated Aα, Bβ and γ [2, 3, 46]. The primary structure of each of the three chains has been determined by amino acid and nucleic acid sequence analyses ([47-62], reviewed in [2] and [3]). The mature Aα-, Bβ- and γ-chain polypeptides of human fibrinogen contain 610, 461 and 411 amino acids, respectively.

Homologies

Evolution and Homologies   It has long been known that the three non-identical chains that compose the vertebrate fibrinogens are descended from a common ancestor [63]. The reader is encouraged to consult several excellent reviews for a more detailed analysis of the evolution of the vertebrate fibrinogens [3, 63-68]. It has been proposed that the original molecule was either a homotrimer or a dimer thereof. Amino acid and cDNA sequence homology suggest that the genes encoding the Aα-, Bβ- and γ-chain polypeptides arose by gene duplication and subsequent evolutionary divergence of an ancestral gene [49, 51, 56, 69-75]. It is hypothesized that this ancestral gene duplicated to form the α-chain gene and a pre β-γ-chain gene approximately 1 billion years ago. The pre β-γ-chain gene then duplicated approximately 500 million years ago to yield the individual β- and γ-chain genes. It has been hypothesized that the three genes share a common regulatory mechanism, possibly mediated via similar sequence motifs in the immediate 5' flanking regions of each gene and their interaction with common transcriptional regulatory molecules [2, 3]. Significant regions of homology among the three fibrinogen genes have been identified, localized in sequences immediately upstream from the sites of fibrinogen mRNA transcription initiation [49, 50, 53, 55, 56, 72, 76-80]. These homologous regions represent *cis*-acting regulatory elements, with the consensus sequences of such elements being necessary for physiological induction of gene expression.

Structural similarities   Significant similarity between the fibrinogen Bβ- or γ-chain genes and portions of heterologous genes has been described. Thus, stretches of nucleotide sequence identified as "fibrinogen-like" are found in other proteins, including the gene encoding scabrous of the developing Drosophila eye, fibroleukin, and the tenascin family of extracellular matrix proteins. Scabrous, a secreted dimeric glycoprotein related in part to fibrinogen and tenascins [81], directs identical precursor cells to adopt different fates [82] through direct cell contact [81]. Fibroleukin, formerly identified as protein pT49 from cytotoxic T-lymphocytes [73, 83], may play a role in physiologic lymphocyte functions at mucosal sites [84]. The tenascin/cytoactin family of extracellular matrix proteins influence neuron-glia interactions [85] and play a transient role in early matrix organization and wound healing [86-89].

Functional Similarities   Three-dimensional studies on defined fragments of fibrinogen are revealing new insights about both fibrin formation and its dissolution [63, 65, 90]. These studies are also showing exactly what structural modifications have accompanied changes in function for the various domains, and may soon reveal the subtleties of how this large, complex glycoprotein is transformed into a fibrin clot [65]. Fibrinogen Aα-, Bβ- and γ-polypeptide chains are held together by 29 pairs of intra- and inter-chain disulfide bond pairs, with the cystein residues in mammalian fibrinogen highly conserved. The γ-chain is involved in almost every known function ascribed to fibrin(ogen) in primary and secondary hemostasis, except susceptibility to thrombin cleavage [46]. Thrombin cleavage releases short peptides, designated FPA and FPB, from the aminotermini of the Aα- and Bβ-chains, respectively. While the primary structure of the fibrinopeptides varies considerably [91, 92], the thrombin Arg-Gly cleavage site is high-

ly conserved [3, 63-65, 92]. The newly exposed N-termini of the α- and β-chains are important in the lateral association of fibrin monomers in fibrin gel formation [1, 66, 93-94]. Furthermore, the primary structure of the β-chain neo-N-terminus is highly conserved [92], likely reflecting its role in fibrin polymerization as well as conservation of function of the β15-42 heparin binding domain in support of cell-fibrin(ogen) interactions [34, 95-98]. The domains of the γ-chain involved in both Ca^{2+} binding and polymerization of fibrin monomers into the fibrin gel are also functionally conserved [99-101]. The carboxyterminal 12 amino acids of the γ-chain (γ400-411) are involved in fibrinogen-mediated support of platelet aggregation [102] via the $\alpha_{IIb}\beta_3$ receptor [103]. In addition, the factor XIIIa-mediated cross-linking sites Q-398 and K-406 on the γ-chain, which are essential for covalent bond formation during stabilization of the fibrin gel [104], are structurally and functionally conserved as well [105, 106]. Interestingly, a monoclonal antibody (MAb), H9B7, specific for C-terminal residues γ406-411 of human fibrinogen [107] cross-reacts strongly with reduced and denatured γ-chains of plasma fibrinogens from bovine, ferret, and guinea-pig, indicating that, in addition to the primary structure and functional capacity for platelet aggregation, the antigenic structure at the region of the platelet binding site is also conserved [92].

Conformation   Hall and Slayter, using shadowing techniques, were first to show by electron micrography that fibrinogen was a trinodular or three-globule structure [108]. Since that time others, using either shadowing or negative staining techniques, have supported this trinodular model. The rod-like molecule is about 450Å long. The diameters of the central E domain and the two identical distal D domains measure about 50Å and 65Å, respectively. The central E domain is connected to the two distal D domains by coiled helices composed of all three fibrinogen chains [109]. These coiled-coils measure about 150Å in length and are less than 15Å thick. Later work by Weisel and collaborators provided more detailed information on the morphology of fibrinogen. Studying crystals and microcrystals prepared from protease-modified fibrinogen, these investigators developed a three-dimensional low resolution model for the molecule. This so-called "heptad" model consists of a linear arrangement of seven domains. In this model, each distal domain of Hall and Slayter has been subdivided into two globular domains. The latter are believed to be composed of the folded COOH-termini of the Bβ- and γ-chains, respectively. Furthermore, a small plasmin-sensitive domain is situated in the interdomainal coiled-coil connectors, approximately half the distance between the central and each distal domain [110]. Recently a three-dimensional molecular image of human fibrinogen was obtained using atomic force microscopy (AFM). Since these experiments were performed on a hydrophobic surface under aqueous conditions, the inter-domain coiled-coil regions were not observed in the fully hydrated molecules. Fibrinogen monomers were seen as overlapping trinodular ellipsoids. Dimers and trimers, linked through adjacent D globular domains, were also visualized and these showed an increased affinity for the hydrophobic surface [111].

As already mentioned, the COOH-terminus of fibrinogen γ-chain contains a segment which is involved in fibrin polymerization. In addition, this region of the chain contains other important structural features including the Factor XIIIa cross-linking site, a platelet receptor recognition site, a ligand for the clumping receptor on pathogenic staphylococci and a calcium-binding site. Detailed crystal structure of this region should therefore provide very important information in our understanding of the fibrinogen-to-fibrin transition and blood-clot formation. Several years ago, a clever approach was used to map, by x-ray diffraction analysis, the structure of a γ-chain segment containing the platelet and bacteria recognition site as well as the donor and acceptor cross-linking residues. Crystals were made from a chimeric protein composed of egg-white lysozyme whose COOH-terminal end contained fibrinogen γ 398-411 [112]. It was shown that the lysozyme structure was relatively unaffected by the fibrinogen segment. However, the γ-chain segment of the chimer displayed a turn distinct from the type II β turn observed by NMR studies of a synthetic peptide only slightly smaller than γ 398-411. Since there were both intra- and intermolecular hydrogen bond interactions, it is possible that the parent lysozyme affected the folding of the fibrinogen peptide extension [112].

More recently, laboratories at the University of California in San Diego and the University of Washington in Seattle have reported on the crystal structures of fibrinogen fragment D and fibrin fragment D-dimer [90, 113] as well as a recombinant 30 kDa COOH-terminal fragment of human fibrinogen γ-chain [36]. The two types of fragments D have been co-crystallized with small peptides that mimic the thrombin-exposed fibrin polymerization domains, the so-called "knobs" (see above). Different crystals were obtained with Fragment D alone or with a peptide (Gly-Pro-Arg-Pro-amide) which serves as a surrogate for the new thrombin-exposed site at the NH_2-terminal end of α-chain of fibrin. In fact, the conformation of Fragment D must be altered substantially upon reaction with this peptide since crystals of the pure fragment shattered when added to a solution containing the peptide [90]. Crystals prepared with D-dimer and both Gly-Pro-Arg-Pro-amide and Gly-His-Arg-Pro-amide (thrombin-exposed site at the NH_2-terminal end of β-chain of fibrin) show each ligand in its complementary "hole" or "polymerization pocket" in the COOH-terminal portion of γ- and β-chain, respectively [113]. The Seattle group has made crystals with a yeast (*Pichia pastoris*)-derived 30 kDa COOH-terminal fragment (rFbg γC30) corresponding to γVal143-Val411. The latter contains the "hole/polymerization pocket" and all other functional sites identified above, including the one involved in calcium binding. The three-dimensional structure of this

recombinant fragment indicates that it contains a single calcium-binding site as well as a deep binding pocket that includes γTyr363 [36]. Crystals prepared with this same fragment complexed with peptide Gly-Pro-Arg-Pro show only subtle conformational changes [114]. These results are at odds with the apparent large structural change observed when the disulfide-linked three-chain Fragment D binds peptide Gly-Pro-Arg-Pro-amide [90]. It is hoped that future studies with larger fragments – including those with the yet-to-be crystallized D-dimer/Fragment E complex – will provide greater insight, not only into the three-dimensional structure of fibrinogen, but also as to how fibrin forms.

Additional Features   The covalent structure of the three fibrinogen chains was completed by 1979 [10, 15, 17, 18, 115, 116]. As already mentioned, fibrinogen is a "...dimeric molecule with many constituents and forms". There are a number of sites on fibrinogen which can exist as one or more variants and these can be either on one or both sides of the dimeric molecule. Regional variants can be classified as non-inherited and inherited. The reader can refer to an excellent source book on the subject of structural and/or functional inherited variants (dysfibrinogenemias) [117] as well as a review on the most common non-inherited regional variants [12]. Two cases of dysfibrinogenemia have already been mentioned in the present review, Fibrinogen Detroit I (AαArg19$\rightarrow$Ser) and Christchurch II (BβArg14$\rightarrow$Cys). The former homozygous variant was associated with a severe bleeding disorder, the latter (heterozygous) with severe thrombosis.

Many years ago it was shown that a small fraction ($\leq$15%) of human plasma fibrinogen eluted differently from the major fibrinogen-containing peak on DEAE-cellulose chromatography. Subsequent analyses revealed that the minor fraction was heterodimeric, containing both the normal (γA, 411 residues) and slightly larger (γ', 427 residues) γ-chains [118, 119]. The γ'-chain arises by alternative mRNA splicing where the last four residues of γA are replaced by a 20 amino acid residue stretch which includes two sulfated tyrosine residues. As already discussed, γ'-chains bind both thrombin and the β-subunit of factor XIII. Other studies have shown that, compared to γA/γA fibrinogen, clots made with the heterodimeric (γA/γ') species are more resistant to fibrinolysis [120].

DNA sequence analysis confirmed and extended the structural studies principally carried out in separate laboratories headed by B. Blombäck, A. Henschen and R.F. Doolittle. However, these later studies also led to the discovery that blood of vertebrates contains a small population of fibrinogen molecules whose size is significantly larger than the more abundant form [71, 121]. This "new" fibrinogen, containing normal size Bβ- and γ-chains and two copies of extended Aα-chain (α_E, ~ 110 kDa), has been termed Fib$_{420}$. The COOH-terminal domain of the αE isoform is made up of a 236 amino acid residue extension that is glycosylated and is highly homologous to the COOH-terminal regions of fibrinogen Bβ- and γ-chains. One of every 100 fibrinogen molecules in normal adult blood is Fib420 and very recent results indicate that umbilical cord blood is at least three times more enriched in the larger size species [122]. Ongoing studies are aimed at understanding the functional importance of Fib_{420}.

Gene

Gene Structure   Each of the three chains of fibrinogen is encoded by separate but closely-linked genes [123, 124], which are clustered on a 50 kb region of the distal third of the long arm of chromosome four [123]. The allelic gene for each chain is present as a single-copy in the genome and their arrangement at the human fibrinogen locus is such that the Aα- and γ-chain genes are aligned in the same direction and are transcribed toward the Bβ-chain gene, which is transcribed from the opposite DNA strand in the direction of the Aα- and γ-chain genes [123] (see Figure 1). The Aα-, Bβ- and γ-chain genes contain six, eight, and ten exons, respectively [2, 23, 52, 53, 55, 56, 69-72, 124, 125].

Gene Expression

Post-transcriptional mRNA processing   The fibrinogen Aα-, Bβ- and γ-chain genes are single-copy and each primary transcript is produced from a single transcription initiation event; however, multiple mRNAs exist for each gene [2]. These mRNAs are produced by a combination of alternative polyadenylation site selection as well as, in the case for the Aα- and γ-chain gene primary transcripts, alternative RNA processing that produces two distinct isoform variants of the Aα- and γ-chain polypeptides differing in their carboxyterminal sequences [2, 48, 51, 55-58, 68, 70, 71, 107, 119, 126-134]. The predominant form of the γ-chain mRNA results from the splicing of the ten exons encoding the γ-chain gene {also designated γA or γMet-412(rat) and 5γo or γVal-411(human)}; the minor retained in the mature AαE transcripts. The translated sequence of AαE confers homology to the Bβ- and γ-chains' carboxytermini, a characteristic not seen in the shorter, predominant Aα-chain [71]. Interestingly, the mature mRNA for the major form of the Aα-chain polypeptide encodes an additional 15 amino acids which are not found in the Aα-chain of circulating fibrinogen [61, 135]. It is appears that the proteolytic cleavage of the Aα-chain removing the carboxyl-terminal 15 amino acids is a normal and specific processing event occurring during the maturation of the nascent polypeptide [136], although the mechanism for and biological significance of this post-translational processing event have yet to be determined. Expression of human γ' and AαE is apparently less efficient, since plasma fibrinogen composed of these subunits accounts only for approximately ten and two percent in contrast to their counterparts, γ- and Aα-chains, respectively [70, 71, 107, 127, 130, 131, 134, 137]. However, approximately 30% of the mature g-chain transcripts in rat hepatocytes are specific for γ' [2]; the reason for this species difference in relative abundance of hepatic fibrinogen γ' mRNAs remains unknown.

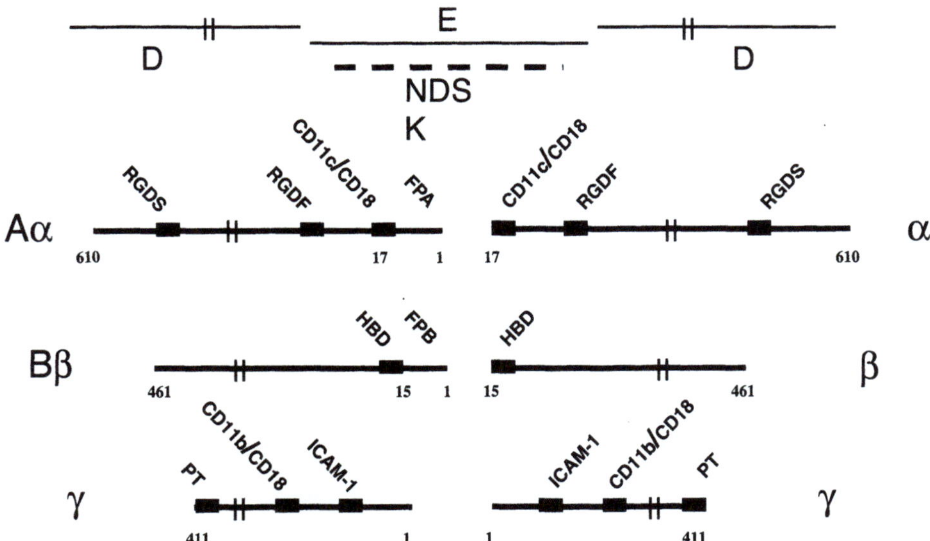

Figure 1. *Schematic representation of the human FBG locus and regulatory regions of the Aα, Bβ and γ chain genes.* Fibrinogen Aα, Bβ and γ chain genes are single copy, and each primary transcript is produced from a single transcription initiation event. The FBG locus is found on chromosome 4q23-q32. The genes are drawn approximately to scale (kb bar); however, the scale of the 5' flanking region of each gene is expanded as noted by the scale bar, which is numbered in bp from +1 to -1102/-1400/-3000. The direction of transcription of each gene is indicated by the arrow. The symbol X at the 5' end of both the β and γ chain genes represents the hypothetical location of putative regulatory regions for the coordinated expression of the fibrinogen genes (31,32,56). The map of the fibrinogen locus and regulatory elements was reprinted with permission from [80].

Functional attributes of the isoform variants of plasma fibrinogen The functional implications of the γ- and γ'-chains are only beginning to be understood. Although the products of the same genes [138], differences in plasma and platelet fibrinogen molecular weight and immunological reactivity have been demonstrated [107, 130, 139]. Because the alternative splicing event disrupts the carboxyterminus of the coding sequence of the γ-chain polypeptide, the participation of the fibrinogen γ'-chain in support of platelet aggregation is either reduced [107, 133, 140] or lost [137]. In contrast, the involvement of both γ and γ' fibrinogens in fibrin polymerization and crosslinking are not affected [133, 141]. In addition, the carboxyterminal sequences of the human and rat γ'-chains are more acidic than their respective shorter-length counterparts. A unique function for γ' fibrinogen has recently been identified. Siebenlist and colleagues [32] have shown that fibrinogen containing γ'-chains have a physiologically significant affinity for the factor XIII B subunits, which are found in plasma but not platelet factor XIII, suggesting that through this interaction fibrinogen serves as a carrier for the plasma zymogen in circulating blood. Whether other functional attributes can be assigned to the γ' sequences remains uncertain; however, given the evolutionary conservation among γ-chains, it is likely that the alternative splicing pattern would have been selected against had there not been some selective pressure to maintain it. This is particularly striking when one considers the number of sites essential to the function of fibrinogen in hemostasis situated in close proximity to the carboxyterminal site of γ-chain variation (Figure 2). It has been proposed that the Aα-chain RGD sites (Figure 3)

mediate γ'-fibrinogen support of platelet aggregation. Earlier reports demonstrated that plasma γ'-fibrinogen is 50 % less effective in supporting ADP-induced platelet aggregation because it binds less well to platelets than γ-fibrinogen [140, 141]. However, through molecular biological and biochemical techniques, it has been shown conclusively that the disruption of the platelet recognition domain by alternative splicing produces a fibrinogenγ'-chain that is incapable of binding to the $\alpha_{IIb}\beta_3$ receptor in support of platelet aggregation under physiologic conditions [23, 137, 142, 143].

The functional significance of the AαE fibrinogen, termed fibrinogen-420 due to its apparent molecular weight in SDS-polyacrylamide gels, is less well defined [68, 70, 71, 121, 122]. However, several lines of evidence suggest that the AαE subunit, alone or incorporated into fibrinogen, is more stable than the predominant Aα-chain [121]. Fibrinogen-420 is cleaved by thrombin and polymerizes into a fibrin gel. Furthermore, the fibrinogen AαE chain is glycosylated; whereas, the common Aa-chain in the major form of plasma fibrinogen is nonglycosylated [70, 144]. Fibrinogen-420 levels are increased by mediators of the APR, but the relative abundance of the AαE to the Aα product remains the same. However, it appears that expression of the AαE chain containing fibrinogen is higher in new-borns than adults, suggesting a role for fibrinogen-420 in development [122].

Origins of plasma and megakaryocyte fibrinogen
Platelet biosynthesis occurs in the bone marrow progenitor cells, the megakaryocyte. Megakaryocytes are

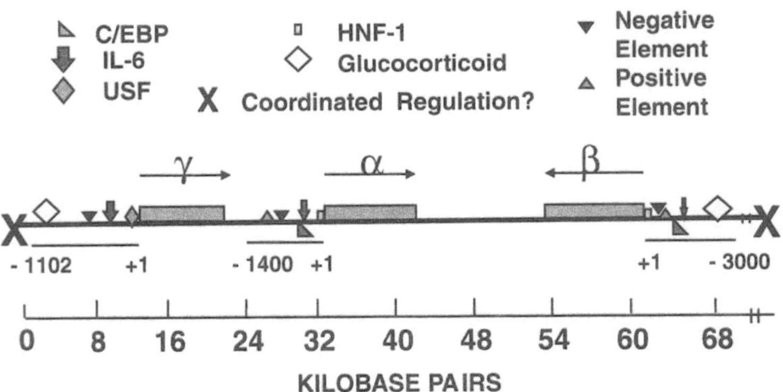

Figure 2. *Alternative mRNA processing of the human fibrinogen γ chain gene.* The upper bar represents the exon (open) and intron (striped) organization of the 3' end of the γ chain gene. The nucleotide numbers above the bar indicate exon/intron junctions. The amino acid sequence of the major form of the γ chain is represented by the broken line directly under the open bar, indicating the intron I gap in coding sequence of the gene. During RNA splicing, intron I is removed and exons IX and X are joined by RNA ligation to generate the mature γ chain mRNA. The amino acids important in mediating support of platelet aggregation are shown (γ400-His His Leu Gly Gly Ala Lys Gln Ala Gly Val-411). The residues involved in factor XIIIα mediated crosslinking of the γ chains are denoted with a dot (•) above the residue. The γ' chain of fibrinogen is generated when the last intron (intron I) is not spliced out, but is retained as an exon, resulting in continuation of the open reading frame which encodes an additional 20 amino acids. This alternative splicing event disrupts the coding region for the platelet recognition domain such that γ'-fibrinogen is incapable of supporting platelet aggregation, whereas, fibrin polymerization and γ-γ crosslinking (•) are not affected. This schematic figure of the human fibrinogen γ chain gene alternative splicing was reprinted with permission from [107].

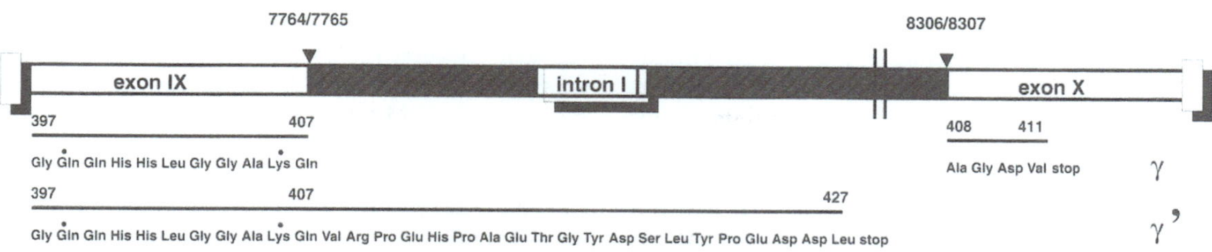

Figure 3. *Diagram of fibrin(ogen) chemical and enzymatic cleavage fragments and cell recognition domains.* The N-terminal plasmin cleavage fragment E and C-terminal fragments D are represented by the lines above the diagram. The N-terminal disulfide knot (NDSK), which represents the minimum sequence of the central domain after CNBr cleavage, is shown by the dashed line above the schematic. The left-half of the dimeric fibrin(ogen) molecule represents fibrinogen with intact fibrinopeptides A and B (FPA and FPB); the right half represents thrombin-cleaved fibrin N-termini at positions α17 and β15. The following receptor-cell binding recognition domains as represented by boxes are shown: CD11c/CD18, Aα17-19; RGDF, Aα95-98; RGDS, Aα572-575; intercellular adhesion molecule-1 (ICAM-1), γ117-133; CD11b/CD18, γ190-202; and platelet recognition domain (PT), γ400-411. The heparin binding domain (HBD) is shown at β15-42. NDSK is composed of Aα1-51, Bβ1-118, γ1-78; fibrin-NDSK- Aα17-51, Bβ15-118, γ1-78; Fragment E3- Aα20-78, Bβ54-133, γ1-53; and Fragments D1A- Aα105-206, Bβ134-461, γ63-406. This schematic figure of fibrinogen cleavage fragments and cell binding domains was reprinted with permission from [98].

large, fully-differentiated cells which undergo endomitoses during development, resulting in polyploidy (as high as 64N) and expansion of cytoplasmic volume [145]. As megakaryocytes mature, coagulation proteins required for the specialized hemostatic functions of platelets are synthesized in the cytoplasm and stored in their α-granules [146]. Platelet α-granules have been shown to contain, along with other proteins involved in hemostasis, a pool of pre-formed fibrinogen representing roughly 3 % to 10 % of total platelet protein. Platelets are then produced by fragmentation of the megakaryocyte cytoplasm and are shed into the circulation [145, 147-150]. Stored α-granule fibrinogen is released from platelets upon activation by a number of aggregating agents, including thrombin, collagen, ADP, prostaglandins and epinephrine, which are either released or are present at the site of vascular injury [150, 151].

Megakaryocytes, like platelets and other cells, are capable of taking up and storing circulating molecules [152] and thus, the mere presence of certain proteins in cells does not necessarily indicate *de novo* synthesis. However, synthesis of platelet factor 4, β-thromboglobulin and platelet-derived growth factor (PDGF) is limited to cells of the megakaryocyte/platelet lineage [145, 153, 154]. Although platelets are anucleate cells and are not capable of transcribing new mRNA molecules, they contain remnants of the protein synthetic apparatus of megakaryocytes [145, 153, 155]. Thus, the presence of these proteins in platelets implies that their synthesis indeed occurs in megakaryocytes. In addition, von Willebrand Factor (vWF), coagulation factor V, and thrombospondin have been shown to be synthesized by megakaryocytes, although each is known to be produced in other tissues [145, 148, 156-158]. The megakaryocyte has been considered the site of Aα-, Bβ- and/or γ-chain gene expression [128, 159, 160] and fibrinogen biosynthesis as well [145, 161, 162]. Megakaryocytes fixed immediately *ex vivo* were shown to express only γ-chain mRNA, and only in the most immature megakaryocytes [159], suggesting that megakaryocyte expression of the γ-chain gene is developmentally regulated. Furthermore, megakaryocytes from patients with high-grade T-cell lymphoma were shown to express Aα-fibrinogen mRNA [163]. However, the origin of megakaryocyte and platelet fibrinogen is now thought to be due to $\alpha_{IIb}\beta_3$-mediated endocytosis of plasma fibrinogen and its subsequent storage in α-granules [152, 158, 164-169]. Endocytosis of plasma fibrinogen into megakaryocyte α-granules requires the presence of two intact platelet recognition domains defined by γ-chain residues 400-411 [165], as does fibrinogen support of platelet aggregation [23, 137, 143]. In summary, the data suggest that the origin of megakaryocyte and platelet fibrinogen may involve both endocytosis of plasma fibrinogen, as well as endogenous biosynthesis in megakaryocyte precursor cells at defined stages of maturation, or possibly as a result of altered cell metabolism and gene expression during disease processes.

Gene Regulation

Liver-specific and ubiquitous expression of the fibrinogen genes   It has been shown that the expression of the separate mRNAs for the three fibrinogen chains is highly coordinated [72, 170], and this is reflected in the fact that, at least in hepatocytes, the relative proportion of each mRNA species is held nearly equal [2]. However, in the absence of serum, the *in vitro* expression of the Aα-chain of chick fibrinogen is uncoordinated with expression of the β and γ-chains genes [171], suggesting that activation of serum response signal transduction pathways exert control over fibrinogen gene expression. Under normal, i.e., basal, conditions the mRNAs of the three fibrinogen genes are constitutively expressed in the liver [2, 76, 172, 173-176]. In addition, the γ-chain gene is ubiquitously expressed in hepatocytes and a variety of extrahepatic epithelial cells *in vitro* [47, 176] and tissues *in vivo* [128, 129]. The expression of the γ-chain gene, but not the Aα- and Bβ-chain genes, in extrahepatic tissues is likely due to activation of the γ-fibrinogen gene by the ubiquitous transcription factors Sp1, a CCAAT-binding factor, and the upstream stimulatory factor (USF), formerly known as the adenovirus major late transcription factor (MLTF), that interact and stimulate constitutive/basal transcription from the γ-chain promoter [47, 49, 177]. In contrast, the expression of the fibrinogen Aα- and Bβ-chain genes is due to activation of predominantly liver-specific transcription factors including hepatocyte-specific nuclear factor-1 (HNF-1) [50, 173-176, 178] (Figure 1).

Fibrinogen and systemic inflammation While fibrinogen is a principal factor in the maintenance of hemostasis, it is also an important component of homeostasis, as it is one of several hepatic proteins whose plasma levels are upregulated during a systemic inflammatory response [2, 124, 78, 79, 179-182]. The mechanism by which the host responds to disturbances in homeostasis due to infection, tissue injury, neoplastic growth or immunological disorders is termed the acute phase response (APR) [78, 179-181, 183, 184]. The APR is characterized by a series of local and systemic reactions that result in activation of a variety of cell types to produce the pro-inflammatory cytokines, interleukin-1 (IL-1), IL-6, and tumor necrosis factor-α. These cytokines in turn can act on distant tissues and cells, resulting in fever, production of glucocorticoids, stimulation of proliferation by cells of the immune system, and changes in synthesis of plasma proteins produced by the liver. The proteins produced in the liver which can be altered during the APR are collectively known as the acute phase proteins or reactants. The APR is conserved across species with fibrinogen being universally upregulated 2 to 10-fold in all species examined including rats, mice, humans, rabbits and ferrets [184, 185]. Simultaneously, serum albumin concentrations decrease by ~50% during an APR. Increased levels of circulating fibrinogen serve to restore homeostasis by providing a provisional matrix to promote wound healing and tissue remodeling [186], while decreases in albumin serve to compensate for the increase in concentrations of circulating acute phase proteins [179, 181, 183]. The upregulation of the fibrinogen genes during systemic inflammation is mediated by IL-6 and the presence of type II-IL-6 response elements on the 5' flanking regions of all three fibrinogen genes has been confirmed [49, 76, 77, 79, 187]. When hepatocytes are incubated in the presence of IL-6, all three fibrinogen mRNAs increase simultaneously and to the same extent at the level of gene transcription, leading to an increase in the amount of fibrinogen secreted [78, 80, 180, 182]. Furthermore, as is the case for the fibrinogen genes, glucocorticoids can act synergistically with IL-6 in upregulation of class II acute phase proteins [76, 77, 80, 179-181, 183, 188, 189]. A glucocorticoid response region has been identified between -2900 and -1500 on the 5' flanking region of the Bβ-chain gene [77], and recently a functional glucocorticoid response element was found at bases -1116 to -1102 on the 5' flanking region of the human fibrinogen γ-chain gene [188] (Figure 1).

Extrahepatic gene expression and production of fibrinogen Several extrahepatic sites of fibrinogen synthesis have been identified, suggesting that this protein may function independently of hemostasis in cellular adhesive interactions or in the maintenance of structural integrity of these tissues. As discussed above, the fibrinogen Aα-, Bβ- and γ-chains are all expressed by hepatocytes [179]; however, γ-chain gene expression has been demonstrated in extrahepatic tissues *in vivo* including brain, lung, and marrow [128, 129, 159]. Furthermore, *in vitro* studies have indicated that several non-hepatic epithelial cells synthesize and secrete fibrinogen. Fibrinogen synthesis was demonstrated in human cervical carcinoma cells, but not primary cervical epithelial cells cultured under basal conditions [190]. Upon estrogen stimulation of ovarian granulosa cells, Bβ- and γ-chain polypeptides were secreted [191], and expression of Aα-, but not Bβ- or γ-chain mRNA was demonstrated in rat kidneys [175]. Epithelial cells from human intestine (Caco-2) respond to IL-6 induction by a modest increase in synthesis and secretion of fibrinogen [192]. In addition, the synthesis and secretion of fully-assembled fibrinogen by a lung alveolar epithelial cell line, A549, has been recently demonstrated [80]. Although little constitutive fibrinogen expression occurs in the lung cells, the fibrinogen genes are transcriptionally upregulated 5 to 10-fold after induction with dexamethasone and IL-6 [80]. This increased expression of fibrinogen genes in lung epithelial cells is consistent with the induction observed in hepatic fibrinogen gene expression in response to pro-inflammatory cytokines.

Functional significance of lung epithelial cell derived fibrinogen Previous studies have identified fibrillar strands within the provisional matrix of cutaneous and vascular wounds as fibrin [193]. During coagulation, additional adhesive glycoproteins from plasma become incorporated into the fibrin clot by covalent cross-linking providing, in addition to the hemostatic plug, a scaffold for cell migration and proliferation, a reservoir for growth factors, proteases, and protease inhibitors, and a substrate for induction and modulation of cell function [193-195]. Once re-epithelialization is complete or vascular integrity re-established, fibrin is dissolved through the action of plasmin. The synthesis and basolateral secretion of fibrinogen by lung epithelial cells in response to IL-6 and dexamethasone [80, 196] or during *Pneumocystis carinii* infection [185] suggests that fibrinogen synthesized at the site of injured tissues or derived from plasma due to increased vascular permeability may incorporate into the provisional matrix. Recently, it was shown that lung epithelial cell derived fibrinogen assembles into a pre-established, mature extracellular matrix independently of conversion to fibrin and colocalizes with other fibrillar matrix proteins, including fibronectin, laminin, and collagen type IV [197]. Furthermore, basic fibroblast growth factor (bFGF/FGF-2) binds specifically and saturably to fibrinogen and fibrin with high affinity [198], implicating fibrin(ogen) as a molecule essential to the maintenance of not only hemostasis, but homeostasis as well by exerting biological effects locally at sites of tissue injury.

The current understanding is that the ubiquitous expression of the γ-fibrinogen gene under basal conditions in extrahepatic tissue is not coordinated with expression of the Aα- and Bβ-chain gene, but that in response to pro-inflammatory mediators, the coordinated upregulation of the Aα-, Bβ- and γ-chain genes occurs. However, the mechanisms controlling the coordinated upregulation of the fibrinogen genes during a systemic (hepatic) inflammatory response are not fully understood, although recent studies have shown that fibrinogen gene expression in response to IL-6 induction involves several different transcription factors [49, 50, 76, 77, 79, 173-176, 178, 180, 187-189]. Intensive effort has been applied to studying the upstream regions of the fibrinogen Bβ-chain gene for sequences that are responsible for regulation of fibrinogen expression. The reasons for this are two-fold: 1) translation of the nascent Bβ-chain of fibrinogen is the rate limiting step in fibrinogen biosynthesis and assembly [199-201], and 2) the organization of the fibrinogen genes is unusual for a multi-subunit protein [72, 123] (Figure 1). These studies indicate that IL-6 and glucocorticoid response elements are important; however, the direct contribution of Bβ-chain 5' flanking regions to coordinating Aα- and γ-chain gene expression has not been demonstrated. Presently, the data suggest that the low, basal level of γ-chain mRNA expression is driven by its ubiquitously activated promoter in epithelial cells of diverse tissues [80, 128, 129, 159, 196]. Under basal conditions, the transcription factor HNF-1 exhibits a limited tissue distribution with high levels of HNF-1 mRNA in liver and kidney, but comparatively little in lung [202]. However, after induction by IL-6 and glucocorticoids, not only is the g-chain gene upregulated in lung epithelial cells, but the Aα- and Bβ-chain genes are transcriptionally activated as well [80]. Together, the data suggest that cis-acting regulatory regions of the γ-chain may function in conjunction with cell-type specific trans-acting factors that are activated by a pro-inflammatory response pathway, resulting in the upregulation of all three fibrinogen genes in extrahepatic tissues. Moreover, tissue-specific transcription requires a battery of transcription factors that are transcriptionally regulated themselves [203]. The inducible expression of fibrinogen Aα-, Bβ- and γ-chain mRNAs and secretion of intact fibrinogen polypeptides in response to pro-inflammatory mediators by lung epithelium provides evidence that extrahepatic epithelial cells contribute to changes in acute phase proteins during an inflammatory response.

Downregulation of fibrinogen gene expression   Constant exposure to IL-6 is required for the continual increase in expression of the fibrinogen mRNAs [204]. The half-life for each fibrinogen mRNA species is approximately 8 hours following stimulation with IL-6, and the decline in levels of all three fibrinogen transcripts is tightly coordinated. Inhibitors of transcription (actino-mycin-D) or translation (cycloheximide) significantly diminish the decay of mRNA following a maximal induction of fibrinogen mRNA expression by IL-6 [204]. Furthermore, the addition of cycloheximide to hepatocytes increases fibrinogen mRNA levels, but only if the cells have been stimulated with IL-6. Taken together, these results provide evidence that transcription and translation are required for fibrinogen mRNA degradation in hepatocytes and suggest that the turnover of fibrinogen mRNAs is stringently coordinated, involving specific regulatory molecules yet to be characterized [204].

In addition to the cis-acting response elements and trans-acting factors responsible for positive regulation of the fibrinogen genes, regions of the 5' flanking gene sequences have been identified as negative regulators of transcription (Figure 1). To date, how these cis-acting elements control fibrinogen gene expression have yet to be elucidated. However, paradoxically, the pro-inflammatory cytokine, IL-1β, which induces expression of class I acute phase proteins, has no effect [184] or an inhibitory effect [205] on hepatocyte synthesis of fibrinogen. In addition, transforming growth factor-β (TGF-β) induces a decrease in the basal level of fibrinogen mRNAs in cultured hepatocytes [206]. Furthermore, TGF-β efficiently antagonizes the IL-6 induction of fibrinogen mRNA at late (12-48 h) but not early (6 h) times after IL-6 treatment. This effect is apparently mediated by posttranscriptional mechanism(s). These findings, together with previously reported data on the inhibitory effect of TGF-β on acute-phase genes (e.g., ApoA1 and albumin), suggest a role for TGF-β in the regulation of liver-specific gene expression. The early stimulatory and late inhibitory effect exerted by IL-6 and TGF-β, respectively, on fibrinogen mRNA levels may play a role in the regulatory mechanism(s) of clot formation in a variety of pathologic conditions [206]. In summary, the control of fibrinogen expression is subjected to coordinated constitutive and inducible liver-specific regulatory mechanisms, as well as inducible ubiquitous regulation in a variety of epithelial cells from diverse tissues, including liver and lung. The mechanisms of fibrinogen gene expression during inflammation, and the biological role of fibrinogen and fibrin in normal and pathologic diseases processes, will remain an area of intensive and exciting research investigations.

Biological Activity   Fibrinogen participates in a number of biological and physiological activities including platelet aggregation, clot formation and wound healing. Different moieties of fibrin(ogen) bind to vascular and circulating cells. These interactions can be in part related to their role in hemostasis and in part to other physiological functions of fibrinogen. Non-hemostatic functions of fibrinogen involve fibrin(ogen)-cell interactions including cell migration, chemotaxis, immunosuppression, mitogenesis and inflammation.

Platelet aggregation   Fibrinogen binds to platelet receptor GPIIb/IIIa, after platelet activation by appropriate agonists (e.g. ADP), in the presence of calcium ions, bridging one platelet to the other in aggregates [207-209]. This represents a fundamental cell-cell adhesion-mediated integrin process in normal hemostasis [210]. In Glanzmann's thrombastenia, a hereditary bleeding disorder, a genetic defect in GPIIb/IIIa receptor prevents normal fibrinogen binding to the platelet membrane [211]. Fibrinogen carries multiple RGD binding sites specific for the platelet integrin GPIIb/IIIa (also referred to as integrin $\alpha_{IIb}\beta_3$). Two RGD sequences are located on each Aα-chain (95-97 and 572-574). The γ-chain COOH-terminal dodecapeptide (400-411), competes with the RGD peptide for binding to $\alpha_{IIb}\beta_3$ [212-214] and binds to a discrete sequence in the platelet receptor [215]. Further studies with recombinant γ- and γ'-chains demonstrated that, to support platelet aggregation. the sequence γ 408-411 was necessary and had to be carboxyterminal [216]. However, studies with natural fibrinogen variants and mutant forms of recombinant fibrinogen have emphasized the importance of the dodecapeptide sequence in the fibrinogen γ-chain carboxyterminus for fibrinogen binding and platelet aggregation [217, 218]. In the resting platelet in the circulation, the integrin $\alpha_{IIb}\beta_3$ is maintained in an inactive conformation with a low affinity for fibrinogen. During platelet activation, the affinity of $\alpha_{IIb}\beta_3$ for fibrinogen increases greatly by a process termed inside-out signaling [219]. Recent studies have shown that resting platelets bound to surface immobilized fibrinogen leads to adhesion via $\alpha_{IIb}\beta_3$ [220], while binding to soluble or plasma fibrinogen leads to platelet aggregation, requiring their activation [142, 221-223]. A few recent publications have extensively reviewed this area [224-227]. While the complete signal transduction pathway has not been elucidated, it seems that a G-protein coupled pathway ultimately leads to activation of $\alpha_{IIb}\beta_3$ integrin complex resulting in conformational alterations needed for fibrinogen binding [219, 228-230]. The platelet receptor $\alpha_{IIb}\beta_3$ also binds fibronectin, vitronectin, von Willebrand Factor (vWF) and thrombospondin mediating adhesion and spreading [212, 231]. The $\alpha_v\beta_3$ vitronectin receptor, on a variety of cell types including platelets, endothelial cells (EC) and smooth muscle cells (SMC) also binds fibrinogen. A monoclonal antibody specific for the $\alpha_{IIb}\beta_3$ receptor (7E3) has been shown to recognize the $\alpha_v\beta_3$ "vitronectin" receptor as well [224]: Fab fragments of this antibody inhibit both fibrinogen and vitronectin binding to $\alpha_{IIb}\beta_3$ [232] The same antibody, 7E3, blocks CD11b/CD18 (Mac-1, $\alpha_M\beta_2$) dependent adhesion to fibrin(ogen) and ICAM-1 [233]. Therefore 7E3 not only blocks $\alpha_{IIb}\beta_3$ mediated platelet aggregation but also CD11b/CD18-mediated monocyte adhesion and binding of $\alpha_v\beta_3$ receptor(s) to fibrinogen. Collectively, interference of these binding events may contribute to the effectiveness of therapeutic infusion of 7E3, such as that achieved in the recent EPIC trial (Evaluation of IIb/IIIa Platelet Receptor Antagonist 7E3 in Preventing Ischemic Complications Trial) [232, 233].

Clot retraction   Integrin $\alpha_{IIb}\beta_3$ is essential for platelet-dependent clot retraction, that is absent or diminished to different extents in patients with Glanzmann's thrombastenia [234]. Clot-retraction mediated by nucleated cells is predominantly supported by integrin $\alpha_v\beta_3$ [235, 236]. Although fibroblasts also mediate clot-retraction [237] and express $\alpha_v\beta_3$, direct evidence supporting this mechanism is not available.

Role in Vascular Biology   Fibrin(ogen) deposits are common in cell injury and inflammation, where they represent a major component of the extracellular matrix. In recent years a large number of studies have shown a multitude of effects of fibrin(ogen) derived fragments on different cell types. These phenomena might have a role in a number of physiological and pathophysiological processes such as wound healing, angiogenesis, placenta development, inflammation, atherosclerosis, malignancy and other vascular and renal disease.

Fibrinogen and endothelial cells (EC)   Fibrinogen binds to EC through several distinct receptors [238] including integrin ($\alpha_v\beta_3$ to Aα 572-575 and $\alpha_5\beta_1$) [103, 239-241] and non-integrin binding sites, such as intercellular adhesion molecule 1, (ICAM-1) [242], heparin-binding site (HBS) [98] and others [243, 244]. Fibrin(ogen) and fibrin(ogen)-derived fragments induce a number of biological activities of endothelial cells such as adhesion, migration, spreading [238, 245] and mitogenesis [96, 246]. Stimulation of EC cells by fibrin, on their apical or basal surface, elicits different biological responses; cross-linked fibrin overlay disrupts the organization of the monolayer [247], induces the release of tissue plasminogen activator, prostacyclin [248] von Willebrand factor [97], and induces ICAM-1 and IL-8 expression [249, 250]. Fibrin II matrix (desFPA-FPB fibrin) interacts with the basolateral face of EC supporting their adhesion, growth, and spreading [96]. In a recent study it has been shown that fibrin binding to EC in monolayers is mediated, at least in part, by a heparin-dependent mechanism [98]. A heparin-dependent interaction may serve to facilitate fibrin interactions with integrins or other cellular receptors. Recent studies suggest that binding of the Bβ 15-42 domain, of conformationally altered fibrinogen, to EC involves cell-surface heparan sulfate proteoglycans [251]. A diagram of fibrinogen-cell recognition domains is shown in Figure 3 [98].

Although fibrin is not associated with developmental angiogenesis, it is present in repair angiogenesis, in cancer, vascular, rheumatoid and other diseases. Both fibrin [252, 253] and fibrin(ogen) degradation products [254, 255] have been shown to stimulate angiogenesis in different *in vitro* models. In particular, the requirement for fibrin II (desFPAdesFPB fibrin) [256] and for a specific fibrin gel network have indicated preferred molecular and biophysical characteristics of fibrin gels favoring vessel growth. For instance, fibrin gels with high rigidity, measured as decreased turbidity by absorbance, and thinner fibers, stimulate capillary growth as opposed to porous gels with thicker fibers, that stimulate migration instead [257]. Two integrin receptors, mediating distinct angiogenic pathways, have been identified as $\alpha_v\beta_3$ and $\alpha_v\beta_5$ [258]. Integrin $\alpha_v\beta_3$ binds fibrinogen to EC [103]; addition of antibodies to $\alpha_v\beta_3$ *in vitro*, inhibit EC adhesion and proliferation while promoting capillary formation [256, 259]. Recent studies in progress indicate that the VE-cadherin receptor, which specifically binds the amino-terminal sequence of fibrin II β-chain (β 15-42), might be involved in angiogenesis [243]. A diagram of fibrin(ogen) chemical and enzymatic cleavage fragments and cell recognition domain is shown in Figure 3. For recent reviews on comprehensive aspects of angiogenesis see [222, 260-262], for adhesion molecules in vascular biology see "Cell Adhesion in Vascular Biology", (Series Editors, MH Ginsberg, ZM Ruggeri, AP Varki, J Clin Invest, Vol 98, Issue 8-12, 1996; Vol 99, Issue 1-11; Vol 100, Issue 3, 1997).

Fibrinogen and leukocytes   Fibrin is formed at sites of injury, infection and both acute and chronic inflammation. Both macrophages and neutrophils have specific binding and adhesion receptors for fibrin(ogen) [263-266]. CD11b/CD18 (Mac-1 or $\alpha_M\beta_2$) is a member of a family of leukocyte integrins, and an inducible high-affinity receptor for fibrinogen, on both macrophages and neutrophils [266-269]. Binding of CD11b/CD18 to fibrinogen is inhibited by the synthetic peptide sequence γ190-202 [270]. In addition to fibrinogen, CD11b/CD18 binds to several cellular or soluble ligands such as ICAM-1 [271], factor X [272], complement protein fragment C3bi [266], heparin [273], oligodeoxynucleotides [274] and others [275]. Moreover, CD11b/CD18 on neutrophils binds to fibrinogen that is already bound to platelets by $\alpha_{IIb}\beta_3$, acting as a cell-cell bridging molecule [276].

Stimulation of neutrophils with tumor necrosis factor promotes their attachment to fibrin(ogen) coated surfaces, by binding of their integrin CD11c/CD18 ($\alpha_X\beta_2$, p150/95) receptor to fibrin amino-terminal α-chain [277]. Intercellular adhesion molecule-1 (ICAM-1), member of the immunoglobulin superfamily, binds leukocytes to fibrinogen [278] at γ 117-133 [279], enhancing their adhesion to endothelial cells [242].

Circulating monocytes are among the earliest cells recruited to sites of vessel injury and have the potential to interact with other vascular cells by secreting growth factors and cytokines. Fibrin increases the expression and release of IL-1β [280], and fibrin(ogen) degradation products upregulates both IL-1β and IL-6 in peripheral blood mononuclear cells *in vitro* [281]. In addition, fibrin(ogen) derived fragments, acting as soluble or cell-bound chemoattractants, stimulate cells to emigrate from the vasculature towards sites of injury, infection and inflammation [278, 282]. Peptide Bβ 1-42 is a potent chemoattractant for neutrophils, and this property is shared by other fragments of the amino-terminus of the Bβ-chain of fibrin(ogen), namely fibrinopeptide B (FPB, Bβ 1-14) and fibrin Bβ 15-42 [283]. These fragments are chemotactic for both neutrophils and mesenchymal cells [283].

Mononuclear phagocytes, including circulating monocytes and tissue macrophages (e.g. Kupffer cells), are responsible for the clearance and degradation of circulating fibrin and contribute to the resolution of vascular and extravascular fibrin deposits [284, 285]. Fibrin(ogen) is internalized and degraded by macrophages [286, 287] possibly contributing to local fibrinolysis and wound healing.

Fibrinogen and other vascular and non-vascular cells
Fibrin provides a provisional matrix for neovascularization during wound healing, and also serves as a scaffold to stimulate growth and migration of other cells, such as fibroblasts, which synthesize new matrix proteins to ensure adequate healing. Fibroblasts adhere to fibrinogen coated surfaces by RGD-dependent and independent mechanisms, involving a_v integrins [288] including $\alpha_v\beta_3$ [289] and ICAM-1 [288], but migrate more effectively on cross-linked fibrin matrix [290]. Confluent fibroblasts in culture bind fibrinogen in the fragment E domain [291]. Both fibrinogen and fibrin stimulate migration of smooth muscle cells (SMC) [292].

Low molecular weight FDPs elicit release of mitogens from EC that stimulate growth of both EC and mesenchymal cells [246]. In addition to being mitogenic for EC and fibroblasts, fibrinogen (fragment D) increases the production of early progenitors in long-term human bone-marrow cultures and stimulates growth of T- and B-lymphoma derived cell lines with an RGD-independent mechanism [293]. Moreover, fibrinogen has a direct effect on the proliferation of purified early human hematopoietic progenitors [294] (Figure 3).

Synthesis of Fg in cultured cells, other than normal or transformed hepatocytes, has been recently reported [190, 295]. The authors suggest this could represent a cytokine-induced response to injury and inflammation, by cells of epithelial origin, in a variety of pathophysiological processes. Moreover, lung epithelial cells assemble the secreted Fg in a matrix, that could also be part of the response to inflammation [296]. Transcripts for the mRNA of the Aα, Bβ and γ fibrinogen chains were found in syncitial trophoblast in normal placenta at term [297]. Whether this might represent a source for the presence of fibrin(ogen) deposits in human placenta [298] remains to be established. These studies are discussed further under "Gene Regulation".

Pathology   Detection of fibrin in tissue by Immunohistochemistry
Formation of fibrin deposits occurs in vascular and extravascular spaces in tissues during many systemic diseases that result from a breakdown of host defence mechanisms. This is true of inflammatory diseases in general and more specifically in renal disease, neoplasia and atherosclerosis. Therefore, a cause and effect relationship has frequently been suggested between fibrin deposition and the pathogenesis of these disorders. This information has been derived from routine histochemical stains for "fibrin", namely Lendrum and PTAH (Phosphotungstic Acid Hematoxylin), from immunofluorescence with polyclonal antibodies, and from electron microscopy studies. The recent development of specific monoclonal anti-

bodies that recognized epitopes exposed in the process of clotting and/or fibrin(ogen)lysis, has allowed the characterization of the molecular nature of fibrin deposits directly in tissues [298, 299].

Atherosclerosis The association between thrombosis and atherosclerosis can be dated back to Rokitansky in the past century [300]. A number of immunochemical and imunohistochemical studies showed both different pattern of distribution and presence of different fibrin(ogen)-related antigens in atherosclerotic plaques (reviewed in [301]). Specific fibrin(ogen)-related antigens were quantified in normal and atherosclerotic human arteries, and in thrombi, [302, 303]. Pathological thrombi consist mainly of fibrin II (desFPA, desFPB fibrin). Intact fibrinogen was predominant in normal vessels with a progression to mainly fibrinogen and fibrin I (desFPAfibrin) in fatty and fibrous plaques and mostly fibrin II in complicated plaques. Similarly, the identification of the distribution of fibrinogen, fibrin II and fragment D/D-dimer, showed that increased fibrin formation and degradation in atherosclerotic plaques was associated with more advanced lesions. Moreover, the pattern of distribution of the different molecular forms of fibrin(ogen), suggested that the presence of fibrin in the vessel wall might be derived from different sources such as mural thrombus and microfibrin deposits formed around macrophages and smooth muscle cells. Recent data supported the hypothesis that both macrophages and SMC, expressing tissue factor, might initiate intravessel fibrin formation. Subsequent studies on the identification of fibrin deposits in atherosclerotic plaques confirmed and extended previous work [304] by showing the presence of fibrin(ogen) complexes crosslinked by tissue transglutaminase. Fibrin(ogen) binds to most cells present in atherosclerotic lesions *in vitro*, although their direct binding of fibrinogen *in vivo* or *ex vivo* has not yet been specifically demonstrated.

Malignancy An association between blood coagulation and malignancy has been supported by numerous clinical and experimental studies [305-307]. However, although the amount of fibrin deposited has not been shown to correlate with the degree of malignancy, the pattern and distribution of fibrin in tumor stroma seem to be constant for a given type of tumor [308]. A number of immunohistochemical studies on the distribution of different molecular forms of fibrin(ogen) in human tumors have been performed, using the same monoclonal antibodies used for atherosclerotic plaques [299]. Extracellular fibrin deposits were only detected along the margins of tumor cell nests in SCCL (small cell carcinoma of lung), renal cell carcinoma and malignant melanoma [309]. Fibrinogen, rather than fibrin, was abundant in the stroma of breast and colon carcinomas (reviewed in [309]). Whether this is due to an interaction of fibrinogen with other proteins of the extracellular matrix or whether it might be attributed to loss of antigenicity of the antibodies used due to degradation of fibrin(ogen) by proteolytic enzymes other than plas-

min remains to be determined [310,311]. A recent study has shown that fibrin II is mainly adjacent to tumor associated macrophages in lung adenocarcinoma [307].

Renal disease Accumulation of fibrin seems to play a major role in the pathogenesis of experimental and human glomerular disease occurring by different mechanisms, and there is a broad correlation between the amount of fibrin deposited and the histologic damage [312, 313]. Immunohistochemistry with MoAbs to fibrin(ogen)-related antigens in microangiopathy, primary glomerulonephritis and lupus nephritis showed that fibrin formation and lysis occurred at different levels of the renal vasculature in both systemic disorders of coagulation and in many primary glomerulopathies of diverse etiology [314]. Fibrinolysis was greater in glomerular capillaries than in larger vessels and in microangiopathies than in other renal disease. These data suggested that damage to renal endothelium by several mechanisms might involve the formation and lysis of fibrin deposits in kidney disease of different origin [314]. More recently, renal biopsies selected from 24 patients with lupus nephritis were examined. In this group, it was found that the presence of fibrin and fragment D/D dimer deposits in glomeruli correlated with the histologic activity index [315].

Placenta development The presence of fibrin in placental villi is still a process of unknown etiology although a few hypothesis have been suggested [316]. Placenta fibrin increases with gestational age and recent immunohistochemical studies with MoAbs to fibrin(ogen) showed that different molecular forms of fibrin(ogen) can be detected [298]. Fibrinogen, fibrin and fibrin(ogen) degradation products are present in perivillous and intervillous deposits in chorionic villi, anchoring villi and along the chorionic and basal plate [298]. Fibrinogen and fibrin were also detected associated with macrophages and Hofbauer cells. Fibrin and fragment D/D-dimer were detected along trophoblast basement membrane [298]. Those data seemed to indicate that maternal, rather than fetal fibrinogen is involved in the formation of fibrin. In another study, fibrin (detected with fibrin-specific MoAb/T2G1) was found associated with villous epithelial denudation. This suggested that fibrin formation participates in the repair of the denuded syncitial trophoblast layer [317]. Earlier studies showed that Factor XIII-deficient patients could not complete their pregnancy without replacement therapy [318]. This suggested that formation of cross-linked fibrin is a necessary process for normal progression and completion of pregnancy.

Clinical Relevance and Therapeutic Implications
Human fibrinogen deficiency (afibrinogenemia) and transgene models Congenital afibrinogenemia is a rare disorder inherited as an autosomal recessive or intermediate trait [319, 320]. Synthesis of fibrinogen by liver cells is markedly reduced or absent, but the genetic defect is still unknown [321]. Its clinical manifestations vary in degree of severity among patients. However,

excessive bleeding of every type, such as from the umbilical cord, epistaxis, after venipuncture and surgical intervention, hematoma following minor trauma and spleen rupture, are common [322]. Defective and delayed wound healing has been noted [320]. Coagulation tests are extremely prolonged and plasma is unclottable while platelet aggregation is partially compensated [323] by binding of von Willebrand Factor to GPIIb/IIIa, in place of fibrinogen [324]. Fibrinogen replacement treatment has to be delivered to afibrinogenemic women during the entire pregnancy for a successful outcome [325, 326]. Patients are treated with replacement therapy [327]. Fibrinogen-deficient mice, born from crossing heterozygous Aα+/- mice, show no evidence of fetal loss and blood samples fail to clot or to support ADP-induced platelet aggregation *in vitro* [328]. Most newborns, displaying signs of bleeding, control the loss of blood and clear the affected tissue, surviving the neonatal period (90%). Hepatic and renal hematomas are encapsulated by fibroblasts that do no migrate into it, and the lesions resolve as thick fibrotic scars. Pregnancy of fibrinogen null mice results in fatal uterine bleeding early in gestation [328].

In combined fibrinogen (Fg)/plasminogen (Plg) deficiency, the absence of fibrinogen alleviates the diverse pathologies associated with tissue damage due to vascular and extravascular fibrin deposits that remain undegraded in Plg deficient mice [329]. Skin incisions, performed in the single and combined fibrinogen/plasminogen deficiency, show that wound healing is severely impaired in Plg deficient mice, while Fg and Fg/Plg deficient mice show a healing time similar to controls. However, their scabs are less stable than in control animals and bleed occasionally in the first few days postsurgery [329]. Similarly, healing of corneal epithelium is impaired in plasminogen-deficient mice and restored in fibrinogen- or fibrinogen/plasminogen-deficient mice [330].

Disfibrinogenemia A vast number of congenital abnormalities in fibrinogen structure and/or function, with normal antigen levels, have been described. The reader is encouraged to consult specialized reviews for this large body of knowledge [1, 117, 331-333].

Plasma fibrinogen concentration The plasma concentration of fibrinogen is about 9 μM and its half-life is 3-5 days [334, 335]. It is also distributed in interstitial fluid and lymph; the concentration in thoracic duct lymph is 30-50% of that in plasma [336]. Fibrinogen synthesis can be stimulated both *in vivo* and *in vitro* by a number of factor such as 1) defibrinogenating agents, 2) hormones and their analogues, 3) during the "acute phase reaction" induced by trauma or injury i.e., infection, neoplasia, inflammation, pregnancy and 4) by injection of biological agents such as thrombin, endotoxin, growth hormones and prostaglandins [337]. In tissue culture, fibrinogen synthesis can be stimulated by dexamethasone [338], suppressed by the estrogen 17-beta-estradiol [339] and downregulated by Vitamin E with

EPA (eicosapentanoic acid) or by oleic acid and palmitic acid alone [340]. Fibrin degradation products E has been reported to induce monocytes and fibroblasts to secrete IL-6 and stimulate hepatocytes to synthesize fibrinogen [341]. In recent years numerous epidemiological studies have shown that increased fibrinogen level is an independent risk factor for coronary, peripheral and cerebral vascular disease [342-354]. Specific therapies for the modulation of fibrinogen levels are not yet available [346, 347].

Degradation of fibrin(ogen) by non-serine proteases
Whether fibrin is formed in the normal hemostatic process or as part of pathological processes, its fate is to be subsequently degraded by proteolytic enzymes. The main enzyme that degrade fibrin(ogen) *in vivo* is plasmin. However, other enzymes can degrade fibrin(ogen) both *in vivo* and *in vitro* to different extents. Endogenous leukocyte proteases [348,349] such as elastase and cathepsin-G [350, 351] can partially degrade fibrin(ogen). Human neutrophil elastase (HNE) activity has been demonstrated *in vivo* by increased circulating levels of Aα 1-21 [352]. This Aα-chain amino-terminal peptide is specifically released by HNE in patients with congenital deficiency of α_1-proteinase inhibitor, the main physiological inhibitor of HNE [352]. Human neutrophil elastase cleaves fragment D-dimer at γ 305-306, γ 347-348 and γ 357-358 [353].

Snake venom enzymes that exhibit specificity for the β-chain of fibrin are serine proteinases [354]. A second class of snake venom enzymes that preferentially degrade the Aα-chain of fibrinogen and also the α- and β-chain of fibrin are zinc metalloproteases [354]. An endopeptidase from Puff adder venom (*Bitis arietans*) cleaves the γ-chain cross-linking site thereby cleaving Fragment D-dimer into a D-like monomer [355]. Similar activity has been shown for a proteinase from *Aeromonas hydrophila* [356]. Enzymes from two different leeches [357, 358] seem to share similar properties. A number of enzymes from the matrix metalloproteinase family (MMP) degrade fibrin(ogen) to a different extent [310, 359]. MMP-3 (stromelysin 1) degrades fibrinogen (Fg) and lyses cross-linked (XL-Fb) clots completely, degrading the D-dimer fragment to a D-like monomer by hydrolysis of γ Gly 404-Ala 405 peptide bond in the cross-linking region (γ 398-406). MMP-1 (collagenase) and MMP-2 (gelatinase A) can only partially degrade Fg and MMP-2 has only a limited capacity to degrade XL-Fb [310]. Similarly to MMP-3, MMP-7 (matrilysin) completely solubilizes XL-Fb clots [359]. However, Fragment D-dimer obtained after MMP-7 degradation of XL-Fb is similar to Fragment D-dimer from plasmin degradation. In contrast, Fragment D-like-monomer from MMP-3 degradation of both Fg and XL-Fb is similar to Fragment D from plasmin degradation of Fg. The diverse pattern of proteolysis of Fg and XL-Fb by different classes of MMPs might be relevant in the degradation of the extracellular matrix in inflammation, wound healing, atherosclerosis and other pathophysiological processes. Very recently, a tryptase from mouse mast cells has been shown to possess anticoagulant activity by degrading all three fibrinogen chains, both in purified system and in normal mouse plasma [360].

Acknowledgements: We would like to thank all our collaborators and coworkers who have contributed to our work over the years. We apologize to all authors whose work has not been cited, since for space reasons we were not able to refer in detail to the numerous contributions in this very fertile area. Many grateful thanks to Dr. Ellinor Peerschke for critically reviewing the biological, pathophysiological and clinical parts of the manuscript. PJS-H wishes to thank David S. Arvan and Dr. Gayule Guadiz for their excellent literature reviews, which formed the basis for the introductions of their research masters and doctoral dissertations, respectively. P.J. Simpson-Haidaris has published under the name of P.J. Haidaris. Portions of this work were supported by grants HL50615, HL30616, HL49610, AI07362, AI23302 (National Institutes of Health, Bethesda, MD) and by the Bugher Foundation, The Retirement Research Foundation and the Lindsley F. Kimball Research Institute at NYBC.

Alessandra Bini, Patricia Haidaris
and Bohdan J. Kudryk

References

1. Blombäck B (1996) Thromb Res 83:1-75
2. Crabtree GR (1987) In: Stamatoyannopoulos G et al (eds) The Molecular Basis of Blood Diseases. Saunders, Philadelphia, pp 631-661
3. Doolittle RF (1984) Annu Rev Biochem 53:195-229
4. Scheraga HA, MJ Laskowski (1957) Adv Prot Chem 12:1-131
5. Meade TW et al (1980) Lancet 1:1050-1054
6. Benderly M et al (1996) Arterioscler Thromb Vasc Biol 16:351-356
7. Procyk R et al (1992) Biochemistry 31:2273-2278
8. Zhang J-Z, Redman CM (1994) J Biol Chem 269:652-658
9. Watt KWK et al (1979) Biochemistry 18:68-76
10. Blombäck B et al (1973) J Biol Chem 248:5806-5820
11. Gati WP, Straub PW (1978) J Biol Chem 253:1315-1321
12. Henschen AH (1993) Thromb Haemost 70:42-47
13. Brennan SO (1997) Thromb Haemost 78:1055-1058
14. Brennan SO et al (1997) Thromb Haemost 78:1484-1487
15. Blombäck B et al (1972) J Biol Chem 247:1496-1512
16. Hessel B et al (1979) Eur J Biochem 98:521-534
17. Doolittle RF et al (1979) Nature 280:464-468
18. Gårdlund B et al (1977) Eur J Biochem 77:595-610
19. Nieuwenhuizen W, Haverkate F (1983) Ann NY Acad Sci 408:92-96
20. Pizzo SV et al (1973) J Biol Chem 248:4584-4590
21. Olexa SA et al (1981) Biochemistry 20:6139-6145
22. Strong DD et al (1982) Biochemistry 21:1414-1420
23. Farrell DH et al (1992) Proc Natl Acad Sci USA 89:10729-10732
24. Meinwald YC et al (1980) Biochemistry 19:3820-3825
25. Meh DA et al (1996) J Biol Chem 271:23121-23125
26. Collen D, Lijnen HR (1987) In: Stamatoyannopoulos G et al (eds) The Molecular Basis of Blood Diseases. Saunders, Philadelphia, PA, pp 662-688
27. Voskuilen M et al (1987) J Biol Chem 262:5944-5946
28. Yonekawa O et al (1992) Biochem J 283:187-191
29. Mosher DF (1975) J Biol Chem 250:6614-6621
30. Sakata Y, Aoki N (1982) J Clin Invest 69:536-542
31. Procyk R et al (1993) Thromb Res 71:127-138
32. Siebenlist KR et al (1996) Biochemistry 35:10448-10453
33. Farrell DH et al (1997) Blood 90(Suppl 1):465a (Abstract 2064)
34. Odrlijin TM et al (1996) Blood 88:2050-2061
35. Marguerie G et al (1977) Biochim Biophys Acta 490:94-103
36. Yee VC et al (1997) Structure 5:125-138
37. Heene DL, Matthias FR (1973) Thromb Res 2:137-154
38. Kudryk B et al (1973) Thromb Res 2:297-304
39. Kudryk B et al (1974) J Biol Chem 249:3322-3325
40. Kudryk B et al (1976) Thromb Res 9:25-36
41. Laudano AP, Doolittle RF (1980) Biochemistry 19:1013-1019
42. Shainoff JR, Dardik BN (1979) Science 204:200-202
43. Shimizu A et al (1992) Proc Natl Acad Sci USA 89:2888-2892
44. Yamazumi K, Doolittle RF (1992) Proc Natl Acad Sci USA 89:2893-2896
45. Gorkun OV et al (1994) Biochemistry 33:6986-6997
46. Hantgan RR et al (1994) In: Colman RW et al (eds) Haemostasis and Thrombosis: Basic Principles and Clinical Practice. Lippincott, Philadelphia, pp 277-300
47. Chodosh LA et al (1987) Science 238:684-688
48. Morgan JG et al (1987) Nucleic Acids Res 15:2774-2776
49. Mizuguchi J et al (1995) J Biol Chem 270:28350-28356
50. Hu CHet al (1995) J Biol Chem 270:28342-28349
51. Fornace AJ Jr et al (1984) J Biol Chem 259:12826-12830
52. Crabtree GR, JA Kant (1982) Cell 31:159-166
53. Crabtree GR et al (1993) Ann NY Acad Sci 408:457-468
54. Crabtree GR, JA Kant (1981) J Biol Chem 256:9718-9723
55. Chung DW et al (1990) Adv Exp Med Biol 281:39-48
56. Rixon MW et al (1985) Biochemistry 24:2077-2086
57. Chung DW, EW Davie (1984) Biochemistry 23:4232-4236
58. Chung DW et al (1983) Ann N Y Acad Sci 408:449-456
59. Chung DW et al (1983) Biochemistry 22:3250-3256
60. Chung DW et al (1983) Biochemistry 22:3244-3250
61. Rixon MW et al (1983) Biochemistry 22:3237-3244
62. Chung DW et al (1981) Proc Natl Acad Sci USA 78:1466-1470
63. Doolittle RF et al (1997) CIBA Found Symp 212:4-17; discussion 17-23
64. Doolittle RF (1993) Thromb Haemost 70:24-28
65. Doolittle RF (1996) FASEB J 10:1464-1470
66. Henschen A (1983) Thromb Res Suppl:27-39
67. Medved LV (1990) Blood Coag Fibrinolysis 1:439-442
68. Fu Y et al (1995) Genomics 30:71-76
69. Crabtree GR et al (1985) J Mol Biol 185:1-19
70. Weissbach L, G Grieninger (1990) Proc Natl Acad Sci USA 87:5198-5202
71. Fu Y et al (1992) Biochemistry 31:11968-11972
72. Fowlkes DM et al 1984 Proc Natl Acad Sci USA 81:2313-2316
73. Koyama T et al (1987) Proc Natl Acad Sci USA 84:1609-1613
74. Doolittle RF et al (1976) Biochim Biophys Acta 453:439-452
75. Takagi T, RF Doolittle (1975) Biochim Biophys Acta 386:617-622
76. Dalmon J et al (1993) Mol Cell Biol 13:1183-1193
77. Huber P et al (1990) J Biol Chem 265:5695-5701
78. Evans E et al (1987) J Biol Chem 262:10850-10854
79. Zhang Z et al (1995) J Biol Chem 270:24287-24291
80. Simpson-Haidaris PJ (1997) Blood 89:873-882
81. Lee EC et al (1996) Mol Cell Biol 16:1179-1188

82. Baker NE et al (1990) Science 250:1370-1377
83. Ruegg C, R Pytela (1995) Gene 160:257-262
84. Marazzi S et al (1998) J Immunol 161:138-147
85. Jones FS et al (1988) Proc Natl Acad Sci USA 85:2186-2190
86. Clark RA et al (1997) J Cell Biol 137:755-765
87. Vrucinic-Filipi N, R Chiquet-Ehrismann (1993) Symp Soc Exp Biol 47:155-162
88. Van Eyken P et al (1990) J Hepatol 11:43-52
89. Copertino DW et al (1995) Proc Natl Acad Sci USA 92:2131-2135
90. Everse SJ et al (1995) Protein Sci 4:1013-1016
91. Dayhoff MO (1972) National Biomedical Research Foundations 5:D90-D96
92. Courtney MA et al (1994) Blood Coag Fibrinolysis 5:487-496
93. Pandya BV et al (1985) J Biol Chem 260:2994-3000
94. Budzynski AZ (1986) Crit Rev Oncol Hematol 6:97-146
95. Bunce LA et al (1992) J Clin Invest 89:842-850
96. Sporn LA et al (1995) Blood 86:1802-1810
97. Ribes JA et al (1989) J Clin Invest 84:435-442
98. Odrljin TM et al (1996) Arterioscler Thromb Vasc Biol 16:1544-1551
99. Boyer MH et al (1972) Blood 39:382-387
100. Endres GF, HA Scheraga (1972) Arch Biochem Biophys 153:266-278
101. Mihalyi E (1988) Biochemistry 27:967-976
102. Kloczewiak M et al (1982) Biochem Biophys Res Commun 107:181-187
103. Cheresh DA et al (1989) Cell 58:945-953
104. Pisano JJ et al (1968) Science 160:892-893
105. Chen R, RF Doolittle (1971) Biochemistry 10:4487-4491
106. Simpson-Haidaris PJ et al (1995) Gene 167:273-278
107. Haidaris PJ et al (1989) Blood 74:743-750
108. Hall CE, Slayter HS (1959) J Biophys Biochem Cytol 5:11-17
109. Doolittle RF et al (1978) J Mol Biol 120:311-325
110. Weisel JW et al (1985) Science 230:1388-1391
111. Marchant RE et al (1997) Thromb Haemost 77:1048-1051
112. Donahue JP et al (1994) Proc Natl Acad Sci USA 91:12178-12182
113. Everse SJ et al (1997) Blood 90(Suppl 1):256a-257a (Abstract 1128)
114. Pratt KP et al (1997) Proc Natl Acad Sci USA 94:7176-7181
115. Lottspeich F, Henschen A (1977) Hoppe-Seyler's Z Physiol Chem 358:935-938
116. Henschen A, Lottspeich F (1977) J Biophys Biochem Cytol 358:1643-1646
117. Ebert RF (1994) Index of Variant Human Fibrinogens CRC Press Boca Raton FL
118. Mosesson MW et al (1972) J Biol Chem 247:5223-5227
119. Wolfenstein-Todel C, Mosesson MW (1980) Proc Natl Acad Sci USA 77:5069-5073
120. Falls LA, Farrell DH (1997) J Biol Chem 272:14251-14256
121. Fu Y, Grieninger G (1994) Proc Natl Acad Sci USA 91:2625-2628
122. Grieninger G et al (1997) Blood 90:2609-2614
123. Kant JA et al (1985) Proc Natl Acad Sci USA 82:2344-2348
124. Kant JA, GR Crabtree (1983) J Biol Chem 258:4666-4667
125. Fornace AJ Jr et al (1984) Science 224:161-164
126. Francis CW, MW Mosesson (1989) Thromb Haemost 62:813-814
127. Francis CW et al (1988) Proc Natl Acad Sci USA 85:3358-3362
128. Haidaris PJ, MA Courtney (1990) Blood Coag Fibrinolysis 1:433-437
129. Haidaris PJ, MA Courtney (1992) Blood 79:1218-1224
130. Homandberg GA et al (1985) Thromb Res 39:203-209
131. Homandberg GA et al (1985) Thromb Res 39:263-269
132. Legrele CD et al (1982) Biochem Biophys Res Commun 105:521-529
133. Mosesson MW et al (1984) Blood 63:990-995
134. Wolfenstein-Todel C, MW Mosesson (1981) Biochemistry 20:6146-6149
135. Kant JA et al (1983) Proc Natl Acad Sci USA 80:3953-3957
136. Farrell DH et al (1993) J Biol Chem 268:10351-10355
137. Lawrence SO et al (1993) Blood 82:2406-2413
138. Doolittle RF et al (1974) Science 185:368-370
139. Francis CW et al (1984) Thromb Haemost 51:84-88
140. Peerschke EI et al (1986) Blood 67:385-390
141. Haidaris PJ et al (1989) Blood 74:2437-2444
142. Zaidi TN et al (1996) Blood 88:2967-2972
143. Farrell DH, P Thiagarajan (1994) J Biol Chem 269:226-231
144. Grieninger G et al (1984) Biochemistry 23:5888-5892
145. Breton-Gorius J, W Vainchenker (1986) Semin Hematol 23:43-67
146. Kieffer N et al (1987) Eur J Biochem 164:189-195
147. Martin JF et al (1986) In: Levine RF et al (eds) Megakaryocyte Development and Function, Vol 215. Liss, New York, p 405
148. Schick BP, PK Schick (1986) Semin Hematol 23:68-87
149. Smith MP et al (1997) Baillieres Clin Haematol 10:125-148
150. Schick PK, BP Schick (1986) Prog Clin Biol Res 215:265-279
151. Furie B, BC Furie (1992) N Engl J Med 326:800-806
152. Handagama PJ et al (1989) J Clin Invest 84:73-82
153. Ryo R et al (1983) J Cell Biol 96:515-520
154. Schick BP, RD Thornton (1993) Leukemia 7:1955-1959
155. Adachi M et al (1991) Exp Hematol 19:923-927
156. Sporn LA et al (1985) J Clin Invest 76:1102-1106
157. Harrison P, EM Cramer (1993) Blood Rev 7:52-62
158. Cramer EM et al (1989) Blood 73:1123-1129
159. Courtney MA et al (1991) Blood 77:560-568
160. Uzan G et al (1986) Biochem Biophys Res Commun 140:543-549
161. Belloc F et al (1985) Thromb Res 38:341-351
162. Leven RM et al (1985) Blood 65:501-504
163. Podolak-Dawidziak M et al (1995) Br J Haematol 91:362-366
164. Harrison P et al (1989) J Clin Invest 84:1320-1324
165. Handagama PJ et al (1995) Blood 85:1790-1795
166. Louache F et al (1991) Blood 77:311-316
167. Cramer EM et al (1990) Prog Clin Biol Res 356:131-142
168. Handagama PJ et al (1990) Prog Clin Biol Res 356:119-130
169. Handagama PJ et al (1990) Am J Pathol 137:1393-1399
170. Crabtree GR, JA Kant (1982) J Biol Chem 257:7277-7279
171. Plant PW, G Grieninger (1986) J Biol Chem 261:2331-2336
172. Grieninger G et al (1983) Ann N Y Acad Sci 408:469-489
173. Courtois G et al (1988) Proc Natl Acad Sci USA 85:7937-7941
174. Courtois G et al (1987) Science 238:688-692
175. Baumhueter S et al (1990) Genes Dev 4:372-379
176. Baumhueter S et al (1989) Ann N Y Acad Sci 557:272-278, discussion 279
177. Morgan JG et al (1988) Mol Cell Biol 8:2628-2637
178. Baumhueter S et al (1988) EMBO J 7:2485-2493
179. Dowton SB, HR Colten (1988) Semin Hematol 25:84-90

180. Otto JM et al (1987) J Cell Biol 105:1067-1072
181. Weber J (1993) Biologic Therapy Cancer Updates 3:1-9
182. Fuller GM et al (1985) J Cell Biol 101:1481-1486
183. Castell JV et al (1990) Hepatology 12:1179-1186
184. Amrani DL (1990) Blood Coag Fibrinolysis 1:443-446
185. Simpson-Haidaris PJ et al (1998) Infect Immun: in press
186. Donaldson DJ et al (1989) J Cell Sci 94:101-108
187. Anderson GM et al (1993) J Biol Chem 268:22650-22655
188. Asselta R et al (1998) Thromb Haemost 79:1144-1150
189. Princen HM et al (1984) Biochem J 220:631-637
190. Lee SY et al (1996) Thromb Haemost 75:466-470
191. Parrott JA et al (1993) Endocrinology 133:1645-1649
192. Molmenti EP et al (1993) J Biol Chem 268:14116-14124
193. Clark RAF (1996) In: Clark RAF (ed) The molecular and cellular biology of wound repair. Plenum Press, New York, p 617
194. Clark RA et al (1985) J Invest Dermatol 84:378-383
195. Clark RA et al (1982) J Invest Dermatol 79:264-269
196. Guadiz G et al (1997) Am J Respir Cell Mol Biol 17:60-69
197. Guadiz G et al (1997) Blood 90:2644-2653
198. Sahni A et al (1998) J Biol Chem 273:7554-7559
199. Roy SN et al (1991) J Biol Chem 266:4758-4763
200. Roy SN et al (1990) J Biol Chem 265:6389-6393
201. Yu S et al (1984) J Biol Chem 259:10574-10581
202. Xanthopoulos KG et al (1991) Proc Natl Acad Sci USA 88:3807-3811
203. Xanthopoulos KG, J Mirkovitch (1993) Eur J Biochem 216:353-360
204. Nesbitt JE, GM Fuller (1991) Biochim Biophys Acta 1089:88-94
205. Conti P et al (1995) Mol Cell Biochem 142:171-178
206. Hassan JH et al (1992) Thromb Haemost 67:478-483
207. Marguerie GA et al (1980) J Biol Chem 255:154-
208. Marguerie GA et al (1979) J Biol Chem 254:5357-
209. Mustard JF et al (1978) Blood 52:453
210. Shattil SJ (1995) Thromb Haemost 74:149-155
211. Coller BS (1980) Blood 55:169-178
212. Ginsberg MH et al (1995) Thromb Haemost 74:352-359
213. Hawiger J (1987) Hum Pathol 18:111-122
214. Hynes RO (1991) Thromb Haemost 66:40-43
215. D'Souza SE et al (1991) Nature 350:66-68
216. Hettasch JM et al (1992) Thromb Haemost 68:701-706
217. Farrell DH (1993) Proc Natl Acad Sci 89:10729-10732
218. Kirschbaum NE et al (1992) Blood 79:2643-2648
219. Ginsberg MH et al (1992) Curr Opin Cell Biol 4:766-771
220. Gartner TK et al (1993) Thromb Res 71:47-60
221. Moskowitz KA et al (1998) Thromb Haemost 79:824-831
222. Pepper MS (1997) Arterioscler Thromb Vasc Biol 17:605-619
223. Savage B et al (1995) J Biol Chem 270:28812-28817
224. Coller BS (1997) J Clin Invest 99:1467-1471
225. Du X, H Ginsberg (1997) Thromb Haemost 78:96-100
226. Ruggeri ZM (1997) Thromb Haemost 78:611-616
227. Shattil SJ, MH Ginsberg (1997) J Clin Invest 100:1-5
228. Brass LF et al (1997) Thromb Haemost 78:581-589
229. Clark EA, JS Brugge (1995) Science 268:233-238
230. Shattil SJ, LF Brass (1987) J Biol Chem 262:992-1000
231. Plow EF et al (1985) Proc Natl Acad Sci 82:8057-8061
232. Coller BS (1997) Throm Haemost 78:730-735
233. Simon DI et al (1997) Arterioscler Thromb Vasc Biol 17:528-535
234. George JN et al (1990) Blood 75:1383-1395
235. Katagiri Y et al (1995) J Biol Chem 270:1785-1790
236. Yi-Ping C et al (1995) Blood 86:2606-2615
237. Niewiaroski S et al (1972) Proc Soc Exp Biol Med 140:1199
238. Dejana E et al (1988) Haemostasis 18:262-270
239. Cheresh DA (1987) Proc Natl Acad Sci USA 84:6471-6475
240. Suehiro et al (1997) J Biol Chem 272:5360-5366
241. Thiagarajan P et al (1996) Biochemistry 35:4169-4175
242. Languino LR et al (1993) Cell 73:1423-1434
243. Bach TL et al (1997) Blood 90 (Suppl 1):290a
244. Erban JK, DD Wagner (1992) J Biol Chem 267:2451-2458
245. Francis CW et al (1993) Blood Cells 19:291-307
246. Lorenzet R et al (1992) Thromb Haemostas 68:357-363
247. Kadish JL et al (1979) Tissue Cell 11:99-108
248. Kaplan KL et al (1988) Arteriosclerosis 9:43-49
249. Qi J et al (1997) Blood 90:3595-3602
250. Qi J et al (1997) J Immunol 158:1880-1886
251. Simpson-Haidaris PJ et al (1997) Blood (Suppl 1 Part 1):464a
252. Dvorak HF et al (1987) Lab Invest 57:673-686
253. Zimrin AB et al (1995) Biochem Biophys Res Commun 213:630-638
254. Stirk CM et al (1993) Atherosclerosis 103:159-169
255. Thompson WD et al (1985) J Pathol 145:27-37
256. Chalupowicz DG et al (1995) J Cell Biol 130:207-215
257. Nehls V, R Herrmann (1996) Microvasc Res 51:347-364
258. Friedlander M et al (1995) Science 270:1500-1502
259. Gamble JR et al (1993) J Cell Biol 121:931-943
260. Folkman J (1995) Nature Medicine 1:27-30
261. Hanahan D, J Folkman (1996) Cell 86:353-364
262. Van Hinsbergh VWM et al (1997) In: Goldberg ID, Rosen EM (eds) Regulation of angiogenesis. Birkhäuser, Basel, pp 391-411
263. Altieri DC et al (1986) J Clin Invest 78:968-976
264. Colvin RB, HJ Dvorak (1975) J Exp Med 142:1377-1390
265. Sherman LA, J Lee (1977) J Exp Med 145:76-85
266. Wright SD et al (1988) Proc Natl Acad Sci USA 85:7734-7738
267. Altieri DC et al (1988) J Cell Biol 107:1893-1900
268. Gustafson EJ et al (1989) J Cell Biol 109:377-387
269. Trezzini C et al (1988) Biochem Biophys Res Commun 156:477-484
270. Altieri DC et al (1993) J Biol Chem 268:1847-1853
271. Diamond MS et al (1991) Cell 65:961-971
272. Altieri DC, TS Edgington (1988) J Biol Chem 263:7007-7015
273. Diamond MS et al (1995) J Cell Biol 130:1473-1482
274. Benimetskaya L et al (1997) Nature Medicine 3:414-420
275. Plow EF, L Zhang (1997) J Clin Invest 99:1145-1146
276. Weber C, TA Springer (1997) J Clin Invest 100:2085-2093
277. Loike JD et al (1991) Proc Natl Acad Sci USA 88:1044-1048
278. Diamond MS, TA Springer (1993) J Cell Biol 120:545-556
279. Altieri DC et al (1995) J Biol Chem 270:696-699
280. Perez RL, J Roman (1995) J Immunol 154:1879-1887
281. Robson SC et al (1994) Brit J Haematol 86:322-326
282. Senior RM et al (1986) Brit J Clin Invest 77:1014-1019
283. Skogen WF et al (1988) Blood 71:1475-1479
284. Barnhart MI, DC Cress (1967) Adv Exp Med Biol 1:492-502
285. Sherman LA S et al (1975) J Lab Clin Med 86:100-111
286. Gonda SR, JR Shainoff (1982) Proc Natl Acad Sci USA 79:4565-4569
287. Simon DI et al (1993) Blood 82:2414-2422
288. Farrell DH, A Al-Mondhiry (1997) Biochemistry 36:1123-1128
289. Gailit J et al (1997) Exp Cell Res 232:118-126

290. Colvin RB et al (1979) Lab Invest 41:464-473
291. Dejana E et al (1984) Eur J Biochem 139:657-662
292. Naito M et al (1992) Atherosclerosis 96:227-234
293. Hatzfeld JA et al (1982) Proc Natl Acad Sci USA 79:6280-6284
294. Yi-Qing Z et al (1993) Blood 82:800-806
295. Haidaris PJS (1997) Blood 89:873-882
296. Guadiz G et al (1997) Blood 90:2644-2653
297. Galanakis D et al (1996) Thromb Res 81:263-269
298. Bini A et al (1989) Lab Invest 60:814-821
299. Bini A (1987) Thesis. Columbia University College of Physicians and Surgeons
300. Rokitansky C (1852) A manual of Pathologic Anatomy. The Sidhenam Society London
301. Bini A, BJ Kudryk (1992) Ann NY Acad Sci 667:112-126
302. Bini A et al (1987) Blood 69:1038-1045
303. Bini A et al (1989) Arteriosclerosis 9:109-121
304. Smith EB et al (1990) Arteriosclerosis 10:263-275
305. Rickles FR, RL Edwards (1983) Blood 62:14-31
306. Sack GH et al (1977) Medicine 56:1-37
307. Shoji M et al (1998) Am J Pathol 152:399-411
308. Dvorak HF (1986) New Engl J Med 315:1650-1659
309. Zacharski LR et al (1992) Fibrinolysis 6 Suppl:1:39-42
310. Bini A et al (1996) Biochemistry 35:13056-13063
311. Stetler-Stevenson WG et al (1993) Annu Rev Cell Biol 9:541-573
312. Hancock W, R Atkins (1985) Seminars in Nephrology 5:69-77
313. Vassalli P, RT Mc Cluskey (1964) Ann NY Acad Sci 11:1052-1062
314. Kudryk BJ et al (1991) In: Goldstein J (ed) Biotechnology of Blood. Butterworth, Stoneham MA
315. D'Agati V et al (1990) J Am Soc Nephrol 7:559a
316. Faulk WP (1989) Am J Reprod Immunol 19:132-135
317. Nelson MD et al (1990) Am J Pathol 136:855-865
318. Duckert F (1972) Ann NY Acad Sci 202:190-199
319. Al-Mondhiry H, WC Ehmann (1994) Am J Hematol 46:343-347
320. Bithell TC (1993) In: Lee GR et al (eds) Wintrobe's Clinical Hematology, Vol 2. Lea and Febiger, Philadelphia, pp 1422-1472
321. Uzan G et al (1984) Biochem Biophys Res Commun 120:376
322. Ehmann WC, H Al-Mondhiry (1994) Am J Med 96:92-94
323. Inceman S et al (1966) J Lab Clin Med 68:21-32
324. De Marco L et al (1986) J Clin Invest 77:1272-1277
325. Grech H et al (1991) Br J Hematol 78:571
326. Inamoto Y, T Terao (1985) Am J Obstet Gynecol 153:803
327. Ness PM, HA Perkins (1979) JAMA 241:1690
328. Suh TT et al (1995) Genes Dev 9:2020-2033
329. Bugge TH et al (1995) Genes Dev 9:794-807
330. Kao WW-Y et al (1998) Invest Ophthalmol Vis Sci 39:502-508
331. Haverkate F, M Samama (1995) Thromb Haemost 73:151-161
332. Martinez J (1997) Curr Opin Hematol 4:357-365
333. McDonagh J et al (1993) In: Colman, RW et al (eds) Hemostasis and thrombosis. Basic principles and clinical practice. Lippincott, Philadelphia, pp 314-334
334. Collen D et al (1972) Brit J Haematol 22:681-700
335. Rausen AA et al (1961) Blood 18:710-716
336. Barnhart MI et al (1970) Thromb Diath Haemost 39 (Suppl):143-159
337. Aronsen KF et al (1972) Scand J Clin Lab Invest Suppl 124:127-136
338. Amrani DL et al (1983) Biochim Biophys Acta 743:394-400
339. Wangh LJ et al (1983) J Biol Chem 258:4599-4605
340. Kiserud CE et al (1995) Thromb Res 80:75-83
341. Wang Y, GM Fuller (1991) Biochim Biophys Res Commun 175:562-567
342. Ernst E, KL Resch (1993) Ann Intern Med 118:956-963
343. Kannel WB et al (1990) Am Heart J 120:672-676
344. Kannel WB et al (1987) JAMA 258:1183-1186
345. Meade TW et al (1986) Lancet:533-537
346. Ernst E (1992) Clin Pharm 11:968-970
347. Handley DA, TE Hughes (1997) Thromb Res 87:1-36
348. Bilezikian SB, HL Nossel (1977) Blood 50:21-28
349. Plow EF, TS Edgington (1975) J Clin Invest 56:30-38
350. Gramse M et al (1978) J Clin Invest 61:1027-1033
351. Plow EF (1980) Biochim Biophys Acta 630:47-56
352. Weitz JI et al (1996) J Clin Invest 78:155-162
353. Mumford RA et al (1991) Ann NY Acad Sci 624:167-178
354. Guan AL et al (1991) Arch Biochem Biophys 289:197-207
355. Purves L et al (1987) Biochemistry 26:4640-4646
356. Loewy AR et al (1993) J Biol Chem 268:9071-9078
357. Budzynski AZ (1991) Blood Coag Fibrinolysis 2:149-152
358. Zavalova LL et al (1993) Thromb Res 71:241-244
359. Bini A et al (1997) Blood 90:465a
360. Huang C et al (1997) J Biol Chem 272:31885-31893

Fibrinolytic, Hemostatic and Matrix Metalloproteinases, role of

Synonym: Fibrinolytic, hemostatic and matrix metalloproteinases: *in vivo* role as deduced from targeted manipulation of the mouse genome.

Introduction Targeted manipulation of the mouse genome has made it possible to unravel the *in vivo* role of candidate genes in an unprecedented manner. Whereas initial studies suggested a redundant role of the plasminogen (Plg) system during embryonic development, subsequent analysis has revealed its essential role in numerous (patho)-biological processes after birth. Conversely, gene-inactivation studies of coagulation proteinases demonstrated their unexpected role during embryonic development, and confirmed their well-known role in adult hemostasis. More recently, the matrix metalloproteinases have been shown to play a pleiotropic role in a variety of biological processes *in vivo*. Frequently, the study of the *in vivo* role of proteinases required the development and use of specialized murine models of hemostasis and thrombosis, arterial neointima formation after vascular wound healing and allograft transplantation, atherosclerosis, aneurysm formation, myocardial ischemia, angiogenesis, tumor growth, metastasis, renal, pulmonary, gastrointestinal and skin inflammation, infection, and brain function. Future studies promise to further unravel the implication of these proteinases in biological processes, relevant for the understanding of clinical pathology.

Characteristics

The coagulation system Initiation of the plasma coagulation system is triggered by tissue factor (TF), which functions as a cellular receptor and cofactor for activation of the serine proteinase factor VII (FVII) to factor VIIa (FVIIa) [1-4]. Activation of factor VII can be performed by FVIIa (autoactivation), factor IX, factor X, factor XII, thrombin or hepsin [2]. The TF·FVIIa complex activates factor X either directly, or indirectly via activation of factor IX, resulting in the activation of prothrombin to thrombin and in the conversion of fibrinogen to fibrin. Factor X can, however, also become activated by FVIIa bound to an as yet undefined cellular binding site, as well as by other processes involving Mac-1, glycoprotein-C and tumor procoagulant. Because of an efficient FXa-dependent feedback inhibition of TF·FVIIa by the Kunitz domain-type inhibitor tissue factor pathway inhibitor (TFPI) (which is synthetized primarily by and bound to the endothelium and its surrounding extracellular matrix in microvessels), coagulation is initiated but rapidly shut off. Therefore, sustained progression of the coagulation has been proposed to depend on a positive feedback stimulation by thrombin and factor Xa, which activate factor XI, factor VIII and factor V (the latter two serving as membrane-bound receptors/cofactors for the proteolytic enzymes factor IXa and factor Xa, respectively) [5]. Apart from antithrombin III, anticoagulation results from interaction of thrombin with thrombomodulin, activating thereby protein C which, together with protein S, inactivates factor Va and factor VIIIa in a negative feedback loop [6,7]. The coagulation system may be involved in functions beyond hemostasis including cellular migration and proliferation, immune response, metastasis and brain function [1-3, 8, 9]. Increasing evidence suggests that tissue factor participates in processes beyond initiation of fibrin formation: (i) it is a member of the immunoglobulin superfamily and expressed as an immediate early gene during inflammation and immune challenge [1, 2]; (ii) tissue factor is expressed in embryonic tissues including the brain, the visceral endoderm cells in the yolk sac, the heart, the kidneys, and at later stages, in the smooth muscle cells of larger blood vessels [10, 11]; (iii) it is involved in cellular activation, in intracellular calcium transients, and its intracellular domain interacts with the cytoskeleton and mediates signalling during metastasis [2, 12-18]; (iv) tissue factor has been implicated in adhesion and in cell-cell contact during cytokinesis, possibly through association with the cytoskeleton [19]; (v) it promotes migration of smooth muscle cells *in vitro* and during arterial stenosis *in vivo*, and this chemotactic action is blocked by TFPI *in vitro* but not by thrombin inhibitors ([20, 21] and A. Clowes, personal communication); (vi) tissue factor may participate in angiogenesis during wound healing and cancer [22-28], TF-specific antibodies block corneal angiogenesis (N. Mackman, personal communication), and FVIIa induces release of vascular endothelial growth factor (VEGF) via interaction with TF [28]. Its precise role and relevance in these processes *in vivo* remains, however, largely unknown.

The plasminogen system   The plasminogen system is composed of an inactive proenzyme plasminogen (Plg) that can be converted to plasmin by either of two plasminogen activators (PA), tissue-type PA (t-PA) or urokinase-type PA (u-PA) [29, 30]. This system is controlled at the level of plasminogen activators by plasminogen activator inhibitors (PAIs), of which PAI-1 is believed to be physiologically the most important [31-33], and at the level of plasmin by a2-antiplasmin [29]. Due to its fibrin-specificity, t-PA is primarily involved in clot dissolution, although it has also been invoked in ovulation, bone remodelling and brain function [29, 30]. Cellular receptors for t-PA and Plg have been identified which might localize plasmin proteolysis to the cell surface [34, 35]. u-PA also binds a cellular receptor, the urokinase receptor (u-PAR), and has been implicated in pericellular proteolysis during cell migration and tissue remodelling in a variety of normal and pathological processes including angiogenesis, atherosclerosis and restenosis [36, 37]. u-PAR binds to vitronectin [38], whereas PAI-1 controls recognition of vitronectin by u-PAR or the $\alpha_v\beta_3$-integrin receptor, suggesting a role in coordinating cell adhesion and migration [39-41]. In addition, growing evidence implicates u-PAR as a trophic factor in morphogenic processes, possibly via intracellular signalling [41]. It is presently unclear whether or in which conditions binding of u-PA to u-PAR is required *in vivo*. Plasmin is able to degrade fibrin and extracellular matrix proteins directly or, indirectly, via activation of other proteinases (such as the metalloproteinases) [42, 43]. Plasmin can also activate or liberate growth factors from the extracellular matrix including latent transforming growth factor (TGFβ₁), basic fibroblast growth factor (bFGF) and VEGF [42, 44]. Cell specific clearance of plasminogen activators or of complexes with their inhibitors by low-density lipoprotein receptor-related protein (LRP) or gp330 may modulate pericellular plasmin proteolysis [45].

Matrix metalloproteinases   Matrix metalloproteinases (MMPs) constitute a rapidly expanding family of proteinases able to degrade most extracellular matrix components [46-49]. In the mouse, MMP-13 (collagenase 3) is the primary interstitial collagenase, whereas MMP-2 (gelatinase-A) and MMP-9 (gelatinase-B) degrade collagen type IV, V, VII and X, elastin and denatured collagens. The stromelysins-1 and -2 (MMP-3 and MMP-10) and matrilysin (MMP-7) break down the proteoglycan core proteins, laminin, fibronectin, elastin, gelatin and non-helical collagens, while the macrophage metalloelastase (MMP-12) primarily degrades insoluble elastin in addition to collagen IV, fibronectin, laminin, entactin and proteoglycans. As many as 20 MMP family members have been identified to date, among them the membrane-type metalloproteinases (MT1- to MT4-MMP) [50-55]. Control of MMP activity is mediated by tissue inhibitors of MMPs (TIMPs) in a tissue- and substrate-specific manner [56]. Since most MMPs are secreted as zymogens, they require extracellular activation. u-PA-generated plasmin is a likely pathological activator of several zymogen MMPs [57]. The activity of MMPs may not be restricted to degradation of extracellular matrix proteins, but may also involve the activation or processing of cytokines or growth factors [58], similar to the role of other metalloproteinases in shedding of membrane-bound receptors [59], in liberation of Fas ligand [60], or in activation of tumor necrosis factor (TNF)-α [61].

Clinical Relevance

Hemostasis: vessel fragility versus clot formation

Embryonic hemostasis   Following initial differentiation of stem cells into endothelial cells and their assembly into endothelial cell-lined channels (vasculogenesis), the embryonic vasculature further develops via sprouting of new channels from pre-existing vessels (angiogenesis), a process which is recapitulated in adulthood during tissue neovascularization [62, 63]. Once the endothelial cells are assembled into vascular channels, they become surrounded by smooth muscle cells (large vessels) or pericytes (small vessels), that affect maturation of the blood vessels not only by providing the fragile primitive blood vessels structural support but also by controlling endothelial cell proliferation and differentiation, vascular permeability and tone [63-66]. In addition, mural cells influence remodelling and branching of endothelial tubes, and protect them against environmental stress of oxygen fluctuations. Although vascular smooth muscle cells/pericytes hvae been implicated in the pathogenesis of vasculopathies during adulthood (atherosclerosis, restenosis and diabetic retinopathy), their role during vascular development has only recently been studied [63]. Surprisingly, targeted gene inactivation of some coagulation factors has revealed their possible implication in blood vessel development.

Targeted inactivation of the tissue factor (TF) gene resulted in increased fragility of endothelial cell-lined channels in the yolk sac in 80 to 100% of the mutant embryos [67-70]. At a time when the blood pressure increased during embryogenesis (day 9 of gestation), the immature TF deficient blood vessels ruptured, formed micro-aneurysms and 'blood lakes', and failed to sustain proper circulation between the yolk sac and the embryo [68]. Since these vitello-embryonic vessels are essential for transferring maternally derived nutrients from the yolk sac to the rapidly growing embryo, the embryo wasted and died due to generalized necrosis. In advanced stages of deterioration, blood leaked into the extracoelomic cavity. Similar observations were made when TF deficient embryos were cultured *in vitro*, suggesting that the observed vascular defects in the yolk sac were not merely due to a possible defect in feto-maternal exchange [68].

Visceral endoderm cell function appeared normal, suggesting that these cells were not responsible for the vascular defects. In contrast, defective development and/or recruitment of periendothelial mesenchymal cells (primitive smooth muscle cell or pericytes) appeared to be a likely cause of the vascular fragility. These cells surround the endothelium in yolk sac vessels, form a primitive "muscular" wall and provide structural support by their close physical association and their increasing production of extracellular matrix proteins. Microscopic and ultrastructural analysis revealed that deficiency of tissue factor resulted in a 75% reduction of the number of mural cells, and a diminished amount of extracellular matrix [68]. Immunocytochemical analysis further revealed a reduced level of smooth muscle a-actin staining in these cells, suggesting impaired differentiation. Although our analysis did not reveal overt endothelial defects, we cannot exclude subtle abnormalities since their growth, differentiation and survival is largely determined by the presence of pericytes [65, 71]. An unresolved issue is whether the absence of TF reduces VEGF expression in these embryos [28, 72], which could affect endothelial [73] as well as periendothelial [74-76] cell function. Recent studies indicate that the TF deficient lethality was rescued by low levels (~1%) of expression of a human TF minigene, suggesting that only minimal amounts are required for hemostasis during embryonic development [77].

In contrast to the severe tissue factor deficient embryonic lethality, deficiency of FVII did not compromise embryonic development and only caused fatal bleeding after birth [78]. Variations in genetic background did not explain the different phenotypes since deficiency of TF, when generated in a similar mixed C57Bl/6J x 129/SvJ background as mice lacking FVII, still caused 85% embryonic lethality [70]. At present, the mechanism of action of embryonic TF remains largely speculative. Although tissue factor was easily detectable at 9.5 days of gestation, *FVII* mRNA levels in the yolk sac were minimal, and FVII plasma levels in wild type embryos at 11.5 days of gestation were only ~0.2% of those present in adult wild type mice. Minimal transfer of maternal FVIIa was detectable but only at supraphysiological maternal FVIIa plasma levels [78]. It remains unknown whether such minimal transfer of maternal FVIIa might rescue development of FVII deficient embryos, by triggering generation of downstream coagulation factors. The requirement of such coagulation factors for development might then also explain why deficiencies of factor V [79], prothrombin [80, 81] and of its protease activated receptor PAR-1 [82] also resulted in embryonic lethality at 9-10 days. However, the greater penetrance of the lethality in embryos lacking TF than in embryos lacking factor V [79], prothrombin [80, 81] or PAR-1 [82] may suggest a more essential role for TF, possibly independent of FVII.

An unresolved issue is whether these coagulation factors affect development by generation of fibrin or, instead, (or perhaps in addition) by their morphogenic activities. Indeed, fibrin deposition in the yolk sac of wild type embryos was not detectable, as would be expected in case TF-deficient embryos died due to a fibrin-dependent hemostatic defect. Furthermore, intracardial injection of a TF·FVIIa inhibitor (rNAPc2, which blocks TF·FVIIa in a factor Xa-dependent manner, but does not inhibit factor Xa activity by itself) in early stage embryos failed to induce bleeding at levels that consistently induced bleeding in postnatal mice by impairment of platelet/fibrin clot formation, suggesting, perhaps, that early embryonic hemostasis may be less dependent on fibrin formation than anticipated. The abnormal embryonic development in embryos lacking factor V [79], prothrombin [80, 81] or PAR-1 [82] in contrast to the normal development of fibrinogen deficient embryos [83], the failure of large doses of thrombin to induce fibrin formation in early stage embryos [78], and the normal development of platelet deficient embryos [84] may underscore the notion that hemostasis in murine embryos does not depend on fibrin formation and platelet activation to the same extent as in adult mice.

Alternatively, factor Xa and thrombin may affect proliferation and migration of vascular cells directly [8, 9, 85, 86], or indirectly via release of platelet-derived growth factor-B [87]. TF may also be directly mitogenic or chemotactic for smooth muscle cells, independently of thrombin generation [20, 21]. Although the cytosolic tail of TF has been implicated in interactions with the cytoskeleton [18, 19] and in intracellular activation [14, 15], morphogenic properties of TF appear (thus far) to depend also on FVIIa [13, 18, 20]. Whether TF would act independently of FVII (as has been suggested during cellular migration and adhesion; T. Luther, personal communication] or, perhaps, via interaction with another ligand (apolipoprotein B100 [88]; plasminogen [89]], remains to be determined.

More than half (60%) of the embryos, expressing a mutant TFPI without the first Kunitz domain (which impairs interaction with factor VIIa but not with factor Xa], die around a similar time as the TF deficient embryos because of impaired vascular integrity and bleeding, with remarkable phenocopy of the TF deficient vascular defects [90]. The remainder of the TFPI deficient embryos develop until birth but suffer fatal consumptive coagulopathy around birth. However, in contrast to the marked intravascular thrombosis in TFPI deficient neonates, surprisingly little fibrin was observed in 9.5 day old TFPI deficient embryos. It is at present undetermined whether bleeding in early TFPI deficient embryos was due to exhausted fibrin formation, or due to abnormal vascular integrity, for example caused by depletion of morphogenic coagulation factors (factor Xa or thrombin].

Other coagulation factors might also appear to be involved in morphogenic processes during early embryogenesis, possibly in blood vessel formation. Deficiency of factor V resulted in embryonic lethality in approximately half of the homozygously deficient embryos, possibly due to vascular defects in the yolk sac [79].

Thrombin has been implicated in processes beyond hemostasis. Indeed, it is mitogenic for fibroblasts and vascular smooth muscle cells, chemotactic for monocytes and activates endothelial cells [9]. Loss of prothrombin resulted in embryonic death around days 9 and 10 of gestation in 50% of homozygous deficient embryos [80, 81]. Whereas in one study, prothrombin deficient embryos appeared to die because of vascular defects (similar to those found in the tissue factor deficient embryos] [80], in another study, affected embryos suffered fatal bleeding from apparently normal yolk sac vessels [81]. By *in situ* hybridization, prothrombin mRNA was detectable within the visceral endoderm of the yolk sac by 9.5 days of gestation, but prothrombin activity was undetectable in blood isolated from 18.5 day old prothrombin deficient embryos [81]. Despite these differences, thrombin-mediated proteolysis appears to be crucial for vascular development. Many of the cell signalling activities of thrombin appear to be mediated by the currently identified thrombin receptors, PAR-1 and PAR-3 [91]. Expression studies have suggested that PAR-1 participates in inflammatory, proliferative or reparative responses such as restenosis, atherosclerosis, neovascularization and tumorigenesis [9]. In addition, *in situ* analysis indicated that this receptor is expressed during early embryogenesis in the visceral endoderm of the yolk sac (S. Coughlin, personal communication], in the developing heart and blood vessels, in the brain and in several epithelial tissues [10]. Targeting of PAR-1 resulted in a block of embryonic development in approximately 50% of the homozygous deficient embryos around a similar developmental stage as in TF deficient embryos, presumably resulting from abnormal yolk sac vascular development [82, 92]. Thrombomodulin deficient embryos also die during early gestation but the precise cause of lethality remains unclear [93]. Notably, fibrin deposits were not observed. Generation of mice with targeted mutations of thrombomodulin may help to resolve its mechanism of action. Mice expressing a mutant thrombomodulin without cytosolic tail or without lectin-like domain develop, however, normally [E. Conway, P. Carmeliet, and D. Collen, unpublished observations].

In summary, hemostasis in the early embryo (e.g. around 9 days of gestation] may be less dependent on fibrin formation and platelet function (thrombocytopenic embryos develop normally and bleed only postnatally [84]] than anticipated. Later during embryogenesis, probably around midgestation, when the embryo produces larger quantities or a more complete set of coagulation factors, hemostasis appears to become more typically dependent on fibrin formation as during adulthood. Indeed, embryos expressing a mutant factor V Leiden (D. Ginsburg, personal communication], TFPI [90] or lacking protein C [94] progressively develop fibrin deposits before birth (beyond 12 days of gestation]. It is therefore not surprising that postnatal bleeding occurs in (the surviving fraction of] mice deficient of factor V [79], factor VII [78], factor VIII [95], factor IX [96, 97], prothrombin [80, 81] and fibrinogen [83] due to defective clot formation following trauma of normally developed blood vessels [see below].

Adult hemostasis: bleeding   Deficiency of tissue factor, PAR-1or factor V (in approximately 50% of the embryos] resulted in bleeding due vascular defects ("vascular bleeding"). This section describes the bleeding, resulting from defects in clot formation in the absence of vascular defects ("hemostatic bleeding"). Inactivation of the intrinsic pathway by deletion of factor XI [96, 97] had little effect as mutant animals showed normal survival, fertility and fecundity, but no signs of bleeding. A small fraction of tissue factor deficient mice developed till term and died because of fatal bleeding [67-70]. Factor VII deficient mice died due to massive intraabdominal bleeding within the first three days postnatally whereas the remainder of the factor VII deficient neonates died due to intracranial bleeding within two to three weeks [78]. Deficiency of factor VII in patients results in severe bleeding when plasma levels of factor VII are below 2% of normal plasma levels [98]. However, lack of early postnatal death in patients with reduced factor VII plasma levels may relate to the fact that factor VII was not completely absent [98] in contrast to the factor VII deficient mice which completely lack this factor. Factor V deficient mice appeared to suffer a severe bleeding phenotype (resulting in early postnatal death), suggesting critical hemostatic functions of thrombin activation beyond fibrin generation. The more severe phenotype of factor V deficiency in mice than in humans is consistent with the detection of residual factor V activities in most patients. Deficiency of factor VIII (hemophilia A) in patients predisposes to spontaneous and trauma-induced bleeding into joints and soft tissues [99]. Mice deficient in factor VIII (hemophilia A) or in factor IX (hemophilia B) suffered life-threatening bleeding in association with tail injury but did not appear to bleed spontaneously [95-97]. The hemophilic mice constitute useful models for studying the immune response that limits recombinant factor VIII substitution in hemophilic patients as well as for the testing of possible gene therapy strategies [100]. The surviving fraction of prothrombin deficient mice all died within the first days after birth due to massive hemorrhage [80, 81]. Surprisingly, PAR-1 deficient mice that developed properly, did not reveal signs of bleeding, suggesting that other related thrombin receptors may play a significant role in platelet activation and hemostasis. Deficiency of fibrinogen resulted in overt intra-abdominal, subcutaneous, joint and/or periumbilical bleeding in the neonatal period [83]. These are the common sites of spontaneous bleeding events in humans with acquired or congenital coagulation disorders [101]. The bleeding manifestations in adult fibrinogen deficient mice (e.g. hemoperitoneum, epistaxis, hepatic, renal, intraintestinal, intrathoracic and soft tissue hematomas) are generally comparable to those observed in the rare human congenital disorder afibrinogenemia and probably resulted from coincidental mechanical trauma [101]. Although the afibrinogenemic murine blood was totally unclottable and platelets failed to aggregate, bleeding was not consistently life-threatening. Possibly, bleeding was controlled by the residual thrombin generation and platelet activation. Whether other platelet receptors beyond the GPIIb/IIIa receptor or other ligands than fibrinogen (including vitronectin, fibronectin or von Willebrand

factor) might rescue deficient platelet interactions remains to be determined.

Hemostasis involves platelet deposition and coagulation to stabilize hemostatic plugs. Failure to stabilize the clot, e.g. as a result of hyperfibrinolytic activity might result in delayed rebleeding. A hemorrhagic tendency has indeed been observed in patients with increased plasma t-PA or reduced plasma α_2-antiplasmin or PAI-1 activity levels [102, 103]. Delayed rebleeding might also explain the hemorrhagic tendency in transgenic mice, expressing high levels of plasma u-PA and in transgenic mice overexpressing GM-CSF, in which increased production of u-PA by peritoneal macrophages occurs [104]. Contrary to patients with low or absent plasma PAI-1 levels, PAI-1 deficient mice did not reveal spontaneous or delayed rebleeding, even after trauma [105]. Lower plasma PAI-1 levels and the occurrence of alternative PAIs in murine plasma (unpublished data) might explain the less pronounced hyperfibrinolytic phenotype and the species-specific difference in the control of plasmin proteolysis.

Adult hemostasis: thrombosis   Deficient fibrinolytic activity, e.g. resulting from increased plasma PAI-1 levels or reduced plasma t-PA or plasminogen levels might participate in the development of thrombotic events [31-33, 103]. Fibrin surveillance in the different knock-out mice was analyzed under basal conditions and after challenge. In unstressed conditions, u-PA deficient mice developed occasional minor fibrin deposits in liver and intestines and excessive fibrin deposition in chronic non-healing skin ulcerations, whereas in t-PA deficient mice, no spontaneous fibrin deposits were observed [106, 107]. Mice with a single deficiency of plasminogen (Plg) or a combined deficiency of t-PA and u-PA, however, revealed extensive intravascular and extravascular fibrin deposits in several organs [106-109] [and unpublished observations]. Interestingly, mice with a combined deficiency of t-PA and u-PAR did not display such excessive fibrin deposits, suggesting that sufficient plasmin proteolysis can occur in the absence of u-PA binding to u-PAR [110, 111]. Loss of both plasminogen activators or of plasminogen severely affected general health and caused a multi-organ dysfunction syndrome characterized by dyspnea, anemia, sterility, cachexia and premature death [106, 107].

After traumatic or inflammatory challenge, mice with a single deficiency of t-PA or u-PA were significantly more susceptible to venous thrombosis, for example following local injection of proinflammatory endotoxin in the footpad [106]. Significant fibrin and matrix deposition was present in Plg- deficient mice following skin wounds [112] or during experimental glomerulonephritis [113]. Similar to Plg-deficient patients, Plg-deficient mice also suffered increased and prolonged arterial thrombosis, but only after injury [114] or after myocardial ischemia (unpublished observations). That arterial thrombosis only develops in Plg-deficient mice after injury may relate to the fact that mice, in contrast to men, do not normally develop vasculopathies, which provide highly thrombogenic surfaces, such as for example on ruptured atherosclerotic plaques. Deficiency of PAI-1 protected mice against arterial thrombosis following perivascular application of thrombogenic ferric chloride [115].

The increased thrombotic susceptibility of t-PA deficient and of combined t-PA:u-PA or Plg-deficient mice can be explained by their significantly reduced rate of spontaneous lysis of ^{125}I-fibrin-labelled pulmonary plasma clots [106, 107]. On the contrary, PAI-1-deficient mice were virtually protected against development of venous thrombosis following injection of endotoxin, consistent with their ability to lyse these plasma clots at a significantly higher rate than wild type mice [105, 116]. The increased susceptibility of u-PA-deficient mice to thrombosis associated with inflammation or injury, might be due to their impaired macrophage function. Indeed, thioglycollate-stimulated macrophages (which are known to express cell-associated u-PA) isolated from u-PA deficient mice, lacked plasminogen-dependent breakdown of ^{125}I-labelled fibrin (fibrinolysis) or of ^{3}H-labelled subendothelial matrix (mostly collagenolysis), whereas macrophages from t-PA deficient or PAI-1 deficient mice did not [106, 107]. Intravenous injection of adenoviruses, expressing a recombinant PAI-1 resistant human *t-PA* (*rt-PA*) gene, in t-PA-deficient mice increased plasma rt-PA levels 100 to 1000-fold above normal and restored their impaired thrombolytic potential in a dose-related way [117]. Conversely, adenovirus-mediated transfer of recombinant human PAI-1 in PAI-1 deficient mice resulted in 100- to 1000-fold increased plasma PAI-1 levels above normal and efficiently reduced the increased thrombolytic potential of PAI-1 deficient mice [unpublished observations].

Lipoprotein(α) contains the lipid and protein components of low-density lipoprotein plus apolipoprotein(α) [118]. Extensive homology of apolipoprotein(α) to plasminogen has prompted the proposal that apolipoprotein(α) forms a link between thrombosis and atherosclerosis, but *in vitro* studies have not yielded conclusive evidence. Transgenic mice overexpressing apolipoprotein(α) displayed reduced thrombolytic potential but only after administration of pharmacological doses of recombinant t-PA, suggesting a mild hypofibrinolytic condition [119]. Studies using transgenic mice overexpressing lipoprotein(α) extended these findings and revealed that spontaneous lysis of ^{125}I-fibrin labelled pulmonary plasma clots (thus not lysis induced by exogenous administration of recombinant t-PA) was also reduced [unpublished observations].

Deficiencies of anti-coagulant factors in humans predispose to thrombosis. Deficiency of TFPI [90] or protein C [94] in mice resulted in postnatal lethality due to disseminated intravascular coagulation with secondary bleeding due to exhaustion of coagulation factors. Heterozygous thrombomodulin-deficient mice were viable and did not appear to develop spontaneous thrombosis [93], possibly indicating that the mice need to be challenged either genetically (by cross-breeding them with other thrombosis-prone transgenic mice) or

physiologically (by administration of proinflammatory reagents, injury etc). However, transgenic mice with a mutated thrombomodulin^{Q387P} gene (which reduces interaction with protein C) survived embryonic development and revealed an increased spontaneous thrombotic incidence [120]. The recently generated mutant factor V mice, engineered to have a similar activated protein C-resistance phenotype as in humans, might be valuable to examine the role of this anticoagulant protein *in vivo* [D.Ginsburg, personal communication].

Neointima formation   Vascular interventions for the treatment of atherothrombosis induce "restenosis" of the vessel within three to six months in 30 to 50 % of treated patients [121, 122]. The risk and costs associated with reinterventions represent a significant medical problem, mandating a better understanding at the molecular level of this process. Arterial stenosis may result from remodelling of the vessel wall (such as occurs predominantly after balloon angioplasty) and/or from accumulation of cells and extracellular matrix in the intimal layer (such as occurs predominantly after intraluminal stent application) [123, 124]. Several candidate molecules have been identified based on correlative expression studies, but their *in vivo* role has frequently remained obscure. Surprisingly, the genetic basis of the susceptibility to arterial stenosis has only been limitedly studied. Although the availability of transgenic mice offers a novel opportunity to study the role of candidate genes in this process, the lack of feasible and appropriate mouse models of arterial stenosis has limited such progress. We and others have developed models of arterial injury in the mouse and, although they may not represent an ideal model of human restenosis, they make it possible to study the biological role and mechanism of candidate genes, and to assemble a molecular analysis of the underlying mechanisms [125, 126].

Proteinases such as the plasminogen and the metalloproteinase system participate in the proliferation and migration of smooth muscle cells, and in the matrix remodelling during arterial wound healing [47, 127-129]. PAI-1 is expressed by uninjured vascular smooth muscle cells [32, 130]. u-PA, t-PA and (to a lesser degree) PAI-1 activity in the vessel wall are significantly increased after injury, coincident with the time of smooth muscle cell proliferation and migration [131-135]. Of the MMPs, only MMP-2 appears to be expressed in the quiescent smooth muscle cells, whereas expression of MMP-3, MMP-7, MMP-9, MMP-12 and MMP-13 is induced in injured, transplanted or atherosclerotic arteries [136-142].

Two experimental models of arterial injury were used, one based on the application of an electric current [125] and the other on an intraluminal guidewire [134, 135]. The electric current injury model differs from mechanical injury models in that it induces a more severe injury across the vessel wall resulting in necrosis of all smooth muscle cells. This necessitates wound healing to initiate from the adjacent uninjured borders and to progress into the central necrotic region, allowing to quantitate the migration of smooth muscle cells. Microscopic and morphometric analysis revealed that the rate and degree of neointima formation and the neointimal cell accumulation after injury were similar in wild type, t-PA deficient and u-PAR deficient arteries [134, 143]. However, neointima formation in PAI-1 deficient arteries occurred earlier after injury [135]. In contrast, both the degree and the rate of arterial neointima formation in u-PA-deficient, Plg-deficient and combined t-PA:u-PA-deficient arteries was significantly reduced until 4 to 6 weeks after injury [114, 134]. Infiltration by leukocytes and myofibroblasts into the media and adventitia, and the associated adventitial remodelling were also significantly reduced in Plg- deficient mice, whereas intravascular thrombosis (albeit transient) was more frequent and extensive in mice lacking u-PA or Plg [114]. Similar genotypic differences were obtained after mechanical injury [134, 135], which more closely mimics the balloon-angioplasty injury in patients.

Evaluation of the mechanisms responsible for these genotype-specific differences in neointima formation revealed that proliferation of medial and neointimal smooth muscle cells was only marginally different between the genotypes [114, 134, 135, 143]. Impaired migration of smooth muscle cells is a likely cause of reduced neointima formation in mice lacking u-PA-mediated plasmin proteolysis since smooth muscle cells migrated over a shorter distance from the uninjured border into the central injured region in Plg-deficient than in wild type arteries [114, 134]. In addition, migration of u-PA deficient smooth muscle cells, but not of t-PA-deficient or u-PAR- deficient smooth muscle cells, cultured in the presence of serum, was impaired after scrape wounding [134]. Notably, when smooth muscle cells were cultured without serum, u-PA was essential for migration induced by basic fibroblast growth factor (bFGF), whereas t-PA was required for migration induced by platelet-derived growth factor (PDGF)-BB [144]. The requirement of u-PA is consistent with the more than 100-fold increased expression levels of u-PA mRNA, immunoreactivity and zymographic activity by migrating smooth muscle and inflammatory cells. Although our results demonstrate that migration of smooth muscle cells requires plasmin proteolysis, it is possible that PAI-1 may also influence cellular migration by affecting vitronectin-dependent cell adhesion through interaction with the avß3-integrin receptor [39]. However, vitronectin and PAI-1 poorly colocalized in the healing arteries [135].

That u-PAR deficient arteries developed a similar degree of neointima was not due to lack of u-PAR expression in wild type arteries as revealed by the expression of functional u-PAR by smooth muscle cells *in vitro* and *in vivo* [143]. Instead, immunogold labelling of u-PA in injured arteries revealed that u-PA was present on the cell surface of wild type smooth muscle cells and accumulated in the pericellular milieu (associated with extracellular matrix components such as collagen fibers) around u-PAR deficient cells [143]. In fact, u-PA accumulated to slightly increased levels in the pericellular milieu. Degradation of ^{125}I-labelled fibrin or activation of proMMP-9 and proMMP-13 was similar for wild type and u-PAR deficient cells. Taken together, these data suggest that sufficient pericellular plasmin proteolysis is present in the absence of binding of u-PA to its cellular receptor. Possibly, the role of u-PAR in biological processes may depend on its topographical and temporo-spatial expression pattern. Somewhat surprisingly, no genotypic differences were observed in reendothelialization [114, 134, 135, 143] suggesting a cell-type specific requirement of plasmin proteolysis for cellular migration.

More recently, the role of MMPs was investigated. Only low levels of MMP-2 were detected in a quiescent artery. In contrast, following injury, significantly induced expression levels of MMP-2, MMP-3, MMP-9, MMP-12 and MMP-13 were observed across the entire injured vessel wall [145]. Similar levels of proMMP-2 and active MMP-2 were observed in arterial extracts of wild type and Plg-deficient mice, confirming that activation of proMMP-2 is not dependent on plasmin [see also below]. In contrast, significantly lower levels of active MMP-9 were present in Plg- deficient than in wild type arteries [145]. Since MMP-9 is primarily expressed by leukocytes, which are involved in the healing of the injured arteries, the lower active MMP-9 levels may contribute to the impaired medial and adventitial remodelling and to the reduced neointima formation.

The involvement of plasmin proteolysis in neointima formation was supported by intravenous injection of a replication-defective adenovirus expressing human PAI-1 in PAI-1 deficient mice [135]. This resulted in preferential infection of hepatocytes and in more than 100- to 1000-fold increased plasma PAI-1 levels. Although the injured arterial segment was not infected, PAI-1 immunnoreactivity was detected in the developing neointima, presumably due to deposition of plasma PAI-1. This resulted in a similar degree of inhibition of neointima formation as observed in u-PA deficient mice without noticeable toxic liver necrosis or intravascular thrombosis. Proteinase-inhibitors have been suggested as anti-restenosis drugs, and recent studies indicate that seeding of retrovirally transduced smooth muscle cells, expressing high levels of PAI-1, inhibits balloon angioplasty induced arterial stenosis.[A. Clowes, personal communication]. In addition, use of a viral PAI-1 like serpin (SERP-1) reduces lesion formation in cholesterol-fed injured rabbits [146].

Atherosclerosis Atherosclerotic lesions initially consist of subendothelial accumulations of foamy macrophages (fatty streaks) which subsequently develop into fibro-proliferative lesions by infiltration of myofibroblasts and accumulation of extracellular matrix [147]. A fibrous cap rich in smooth muscle cells and extracellular matrix overlies a central necrotic core containing dying cells, calcifications and cholesterol crystals. As long as these lesions do not critically limit blood flow, they may grow insidiously. However, clinical syndromes of myocardial or peripheral tissue ischemia due to occluding thrombosis are frequently triggered by rupture of unstable plaques, and constitute the primary cause of cardiovascular morbidity and mortality in Western societies [148, 149]. In addition, the atherosclerotic wall may become thinner due to media necrosis, ultimately resulting in aneurysm formation and fatal bleeding. Aneurysm formation is a major cause of mortality in elderly patients, responsible for more than 15,000 deaths annually in the United States alone [150-152]. The pathogenetic mechanisms of atherosclerotic aneurysm remain, however, largely undefined.

Epidemiologic, genetic and molecular evidence suggests that impaired fibrinolysis resulting from increased PAI-1 or reduced t-PA expression, or from inhibition of plasminogen activation, may contribute to the development and/or progression of atherosclerosis [153-155]. It has been proposed that this results from increased fibrin and matrix deposition, which could promote plaque growth directly via incorporation of its mass, or indirectly via its mitogenic/chemotactic properties for smooth muscle cells, fibroblasts, endothelial cells and leukocytes. Indeed, PAI-1 plasma levels are elevated in patients with ischemic heart disease, angina pectoris and recurrent myocardial infarction [156]. Recent genetic analyses revealed a link between polymorphisms in the PAI-1 promoter and the susceptibility of atherothrombosis [154]. Adipocytes may significantly contribute to the increased plasma PAI-1 levels in obese patients prone to ischemic heart disease [157]. A possible role for increased plasmin or matrix metallo-proteolysis in atherosclerosis is, however, suggested by the enhanced expression of t-PA, u-PA and several MMPs in plaques [136-141, 158, 159]. Proteolysis might indeed participate in neovascularization and rupture of plaques, or in ulceration and rupture of aneurysms [129, 148, 149]. A causative role of the Plg and/or MMP system in these processes has, however, not been conclusively demonstrated.

Therefore, atherosclerosis was studied in mice deficient in apolipoprotein E (apoE) [160], singly or combined deficient in t-PA, u-PA, Plg or PAI-1 [161]. No differences in the size or the predilection site of early fatty streaks and more advanced plaques were observed between mice with a single deficiency of apoE or with a combined deficiency of apoE and t-PA, or of apoE and u-PA, suggesting that plasmin is not essential for subendothelial infiltration by macrophages. Conversely, deficiency of apoE and Plg resulted in accelerated atherosclerosis [161]. Possibly, this discrepancy relates to differences in genetic background or to the type of diet. Whether Plg deficiency affected matrix deposition, cell accumulation or cell function in plaques remains unknown. Further, it remains to be determined whether the poor general health of the Plg-deficient mice with their associated generalized state of increased inflammatory stress, as well as their significantly lower levels of high density lipoproteins may have contributed to the accelerated atherosclerosis [161]. Similar studies in mice lacking both apoE and fibrinogen indicated that lesions of all stages developed to a similar extent and with an indistinguishable histological appearance as in mice lacking only apoE [162]. This was somewhat surprising as fibrin has been detected in plaques and suggested to accelerate lesion development through its adhesive, chemotactic or mitogenic properties for inflammatory, smooth muscle and endothelial cells. Perhaps these findings may relate to a neglible contribution of fibrinogen to lesion development in this extremely lipid-driven model of atherosclerosis which is devoid of plaque rupture-induced thrombosis. As suggested by epidemiological studies [163], it is, however, also possible that the role of fibrinogen as a risk factor may become more apparent in transgenic mouse models of atherosclerosis when its plasma levels are increased instead of being reduced.

Mice with a combined deficiency of apoE and PAI-1 developed normal fatty streak lesions but, subsequently, revealed a transient delayed progression to fibroproliferative plaques (unpublished observations). Whether the increased plasmin proteolytic balance in these mice might prevent matrix accumulation and, consequently, delay plaque progression, or whether more abundant plasmin increased activation of latent TGF-β_1 with its pleiotropic role on smooth muscle cell function and matrix accumulation, remains to be determined.

Significant genotypic differences were observed in the integrity of the atherosclerotic aortic wall. Indeed, destruction of the media with resultant erosion, transmedial ulceration, necrosis of medial smooth muscle cells, aneurysmal dilatation and rupture of the vessel wall were more frequent and severe in mice lacking apoE or apoE:t-PA than in mice lacking apoE:u-PA [57]. At the ultrastructural level, elastin fibers were eroded, fragmented and completely degraded, whereas collagen bundles and glycoprotein-rich matrix were disorganized and scattered in apoE-deficient and in apoE:t-PA-deficient mice, but not in apoE:u-PA-deficient mice, which were virtually completely protected. Mac3 immunostaining and ultrastructural analysis revealed that macrophages were absent in the media of uninvolved arteries, that they were only able to infiltrate into the media of atherosclerotic arteries after they degraded the elastin fibers, and that media destruction progressed in an intima-to-adventitial gradient [57]. Plaque macrophages (and especially those infiltrating into the media) expressed abundant amounts of *u-PA* mRNA, antigen and activity at the base of the plaque, similar as in patients [158, 159]. In contrast, t-PA and PAI-1 were confined to the more apical regions within the plaque. Thus, a dramatic increase of free u-PA activity (which is minimal in quiescent arteries) was generated by the infiltrating plaque macrophages. Since plasmin by itself is unable to degrade insoluble elastin or fibrillar collagen, it most likely activated other matrix proteinases. Because of their increased expression in human atherosclerotic plaques and aneurysms, matrix metalloproteinases (MMPs) constituted likely candidates. Macrophages in murine atherosclerotic plaques abundantly expressed MMP-3,

MMP-9, MMP-12 and MMP-13 [57]. Furthermore, cultured peritoneal macrophages derived from wild type mice, or from mice deficient in t-PA, MMP-3, MMP-7 and MMP-9 degraded 3H-elastin in a plasminogen-dependent manner, whereas u-PA-deficient or MMP-12-deficient macrophages were unable to. In addition, wild type and t-PA-deficient but not u-PA-deficient cultured macrophages activated secreted proMMP-3, proMMP-9, proMMP-12 and proMMP-13 but only in the presence of plasminogen, indicating that u-PA-generated plasmin was responsible for activation of these proMMPs. These plasmin-activatable metalloproteinases colocalized with u-PA in plaque macrophages, suggesting that plasmin is a likely activator of proMMPs *in vivo*. Another possible mechanism of action of plasmin is that it mediates the degradation of glycoproteins in the stroma of the aortic wall, thereby exposing the highly insoluble elastin to elastases and facilitating elastolysis *in vivo* [164]. Our unpublished data that u-PA-deficient macrophages degraded 3H-fucose labelled glycoproteins in a subendothelial matrix less efficiently than wild type macrophages supports such a role. Taken together, these results implicate an important role of u-PA in the structural integrity of the atherosclerotic vessel wall, likely via triggering activation of matrix metalloproteinases, and suggest that increased u-PA levels are a risk factor for progression of aneurysm formation. Direct proof whether and which MMPs are involved in media destruction and aneurysm formation has to await a similar analysis in mice that are combined deficient of apoE and each of these MMPs.

Graft arterial disease. The obstruction of the lumen due to intimal thickening (resulting from accumulation of smooth muscle cells, leukocytes and extracellular matrix, and leading to tissue ischemia) significantly limits the success of organ transplantation in a majority of patients. A mouse model of transplant arteriosclerosis has been developed, that mimics in many ways the accelerated arteriosclerosis in coronary arteries of transplanted cardiac allografts in man [142,165]. In this model, host-derived leukocytes adhere to and infiltrate beneath the endothelium and form a predominantly leukocyte-rich neointima within 15 days after transplantation. Subsequently, leukocytes infiltrate into the media where they activate (through production of cytokines and growth factors) the quiescent smooth muscle cells to proliferate and migrate into the intima. The role of proteinases in this process remains largely undefined. Therefore, carotid arteries from B.10A(2R) wild type mice were transplanted in C57Bl6:129 Plg-deficient mice. Such analysis revealed that neointima formation within 15 days was not significantly different among genotypes, but that significantly more leukocytes infiltrated into the media of the allografts transplanted in wild type than in Plg-deficient recipients [142]. In addition, adventitial infiltration by leukocytes and accumulation of myofibroblasts was markedly greater in allografts transplanted in wild type than in Plg-deficient mice. Within

45 days after transplantation, the neointima was, however, much larger and contained 10-fold more α-actin positive smooth muscle cells in allografts transplanted in wild type than in Plg-deficient mice. In addition, media necrosis and fragmentation of the elastic laminae were more severe in transplants in wild type than in Plg-deficient mice [142]. Graft thrombosis was, however, much more frequent and extensive in Plg-deficient mice. Expression of u-PA, MMP-2, MMP-3, MMP-9, MMP-12 and MMP-13, and to a smaller extent of t-PA, were significantly increased within 15 days after transplantation, when cells actively migrate. Notably, the reverse transplantation of Plg-deficient donor arteries into wild type recipients did not affect the development of the neointima, indicating that circulating plasminogen is essential. To summarise, it appears that plasmin is not essential for leukocytes to adhere and to infiltrate beneath the endothelium (as also suggested by the similar size of atherosclerotic plaques in mice lacking t-PA or u-PA, see above). However, plasmin mediates lysis of arterial thrombi, and is required for leukocytes to fragment the elastic laminae and to infiltrate into the media. Similar to the atherosclerotic aorta (see above), destruction of the medial stroma was conditional on prior elastic lamina degradation by macrophages. Since plasmin is unable to degrade elastin, collagen and other matrix components in the media, it presumably activates other matrix degrading proteinases, likely of the MMP family. Once present in the media, leukocytes activate medial smooth muscle cells to proliferate and emigrate into the intima, a process which is also mediated by plasmin (as also exemplified for neointimal smooth muscle cell accumulation after arterial injury; see above).

Myocardial ischemia Acute myocardial infarction due to occlusion of coronary arteries is a leading cause of morbidity and mortality in Western societies. It depresses cardiac performance, and may induce arrhythmias, infarct expansion, ventricular wall rupture, and aneurysm formation [148, 149]. Although proteinases have been implicated in cardiac remodelling [166-172], and in growth and remodelling of collateral vessels [173], today however, surprisingly little is known about their *in vivo* relevance.

Recently, a mouse model of chronic myocardial infarction was used to evaluate the role of the plasminogen system in cardiac healing. Initial studies reveal that the plasminogen system is importantly involved in this process (in collaboration with M. Daemen and J. Smits, Maastricht, the Netherlands). Indeed, following ligation of the left anterior descending coronary artery, wild type or t-PA-deficient mice heal their ischemic myocardium within two weeks via scar formation, i.e. the ischemic myocardium becomes infiltrated by leukocytes, endothelial cells and fibroblasts with resultant deposition of collagen. In a fraction of these mice, rupture of the ischemic myocardium occurs shortly after infarction due to excessive u-PA-generated plasmin proteolysis by infiltrating wound cells. In sharp contrast, mice lacking u-PA or Plg are protected against ventricular wall rupture, but fail to heal the ischemic myocardium which remains largely devoid of infiltrating leukocytes, endothelial cells and fibroblasts. Mural thrombosis in the ventricular cavity occurs, however, more frequently in Plg- and u-PA-deficient mice than in wild type mice. How these morphologic observations correlate with expression of fibrinolytic or matrix metalloproteinase enzymes and whether cardiac function is affected differently in the various genotypes remains to be determined. Nevertheless, the data show that u-PA-generated plasmin proteolysis is required for healing, but needs to be carefully balanced to avoid tissue destruction and ventricular wall rupture. Studies are underway to investigate the role of the various MMPs in the respective knockout mice.

Angiogenesis Migration of endothelial cells involves proteolysis of the extracellular matrix. Quiescent endothelial cells constitutively express t-PA, MMP-2 and minimal MMP-1[29, 30, 48]. Net proteolysis is, however, prevented by coincident expression of PAI-1, TIMP-1 and TIMP-2 [32, 130]. In contrast, when endothelial cells migrate, they significantly upregulate u-PA, u-PAR, and, to a lesser extent, t-PA at the leading edge of migration [36, 48, 131, 134, 174]. Although PAI-1 is also increased, its expression at different locations and times allows a net increase in fibrinolytic activity [133, 135, 175]. A variety of cytokines and growth factors with angiogenic activity modulate the expression of these proteinases. VEGF-A and bFGF (synergistically) induce expression of u-PA, t-PA and u-PAR [42, 176, 177], whereas TNF-α and interleukin (IL)-1 upregulate expression of MMP-1, MMP-3 and MMP-9. bFGF and VEGF are counteracted in a negative feedback by TGF-β_1, which downregulates u-PA and induces PAI-1 [42].

Immunoneutralization or chemical inhibition of PAs and MMPs reduce endothelial cell migration *in vitro* [36, 48]. Surprisingly, mice deficient in u-PA and/or t-PA [134], PAI-1 [105, 116, 135], u-PAR [110, 143], Plg [108, 114] or α_2-antiplasmin (unpublished observations) develop normally without overt vascular anomalies. Whether this relates to insufficient expression, to redundancy or to compensation of these proteinases during vascular development, or alternatively, to the fact that embryonic vessels have a poorly developed basement matrix [178] (rendering the need for proteinases less significant) remains to be determined.

Formation of new blood vessels during adulthood resumes during reproduction, wound healing, chronic inflammation, tissue ischemia, or cancer. A consistent observation in PA-deficient mice is their impaired wound healing, in part due to reduced migration of inflammatory cells, fibroblasts, smooth muscle cells, and keratinocytes. Surprisingly, migration of endothelial cells or formation of new blood vessels are not (much) affected during skin [112] or corneal [179] wound healing in Plg-deficient mice, during reendothelialization alongside denuded vessels in mice deficient in t-PA, u-

PA, u-PAR, PAI-1 and Plg [57, 114, 135, 143], during primary tumor formation [180] or during dissemination of Lewis lung carcinoma [181], although possible defects were not always quantified. Thus, despite a reduced proteolytic potential, endothelial cells were still able to migrate, suggesting compensation by other proteinases. In addition, the increased and persistent deposition of fibrin, which by itself as well as through its proteolytic derivatives, can recruit endothelial and inflammatory cells (which produce angiogenic factors) may have provided a strong stimulus for neoavascularization [63, 182-185]. In contrast, hemangioblastoma formation after Polyoma middle-T retroviral infection was dependent on generation of u-PA-mediated plasmin [186]. In summary, these data suggest that migration of endothelial cells alongside a denuded vessel does not require u-PA-generated plasmin, whereas invasion of endothelial cells through an anatomic barrier of extracellular matrix may (Polyoma tumor model) or may not (cornea, skin healing) require plasmin proteolysis. Whether these differences relate to the composition and/or thickness of the extracellular matrix, or to the expression pattern of proteinases by endothelial cells during these different conditions remains to be determined.

The plasminogen system may be implicated in the inhibition of angiogenesis through generation of inhibitors, such as angiostatin (containing the first four kringles) or the kringle 5 domain of plasminogen [187-190]. Notably, plasmin and several MMPs are able to generate angiostatin from plasminogen, further underscoring the close interaction between both proteinase systems [191, 192]. Although such mechanisms may be therapeutically relevant for tumor dormancy in the adult [187, 193], it was somewhat surprising that plasminogen deficient and control mice developed a similar number of metastatic foci with a comparable degree of vascularization, regardless whether the primary Lewis lung tumor was resected or not.

Mice deficient in MMP-3 [194], MMP-7 [195], MMP-7 [196], MMP-11 (P. Basset, personal communication), MMP-12 [197] or TIMP-1 [198, 199] did not exhibit abnormal angiogenesis during development or pathology (to the extent studied). In contrast, vascularization of tumors implanted in gelatinase-A deficient mice was reduced [196], the mechanisms of which remain undefined. Furthermore, mice deficient in gelatinase-B (MMP-9) exhibited an abnormal pattern of growth plate vascularization during postnatal skeletal growth [200]. Endochondral bone formation is characterized by capillary invasion, degradation of the hypertrophic cartilage, and apoptosis of hypertrophic chondrocytes. Lack of gelatinase-B reduced vascularization and chondrocyte apoptosis, thereby inducing accumulation of hypertrophic cartilage. Interestingly, transplantation of wild type bone marrow cells rescued vascularization and ossification of gelatinase-B deficient growth plates. It remains to be determined whether gelatinase-B facilitates capillary ingrowth by 'clearing the path' through matrix degradation, or, perhaps, by generating angiogenic activators or inactivating angiogenic inhibitors (gelatinase-B is able to proteolytically process IL-1β and substance P [201]). Whether it liberates growth factors sequestered within the matrix, similar to stromelysin, interstitial collagenase and plasmin which degrade perlecan and release bound bFGF, remains to be determined [202].

Tissue factor, the initiator of the coagulation system, has been implicated in angiogenesis during wound healing in mice and during cancer and metastasis in mice and patients [22-28]. In fact, the role of TF in dissemination of tumor cells depended on both the interaction of factor VIIa with the extracellular TF domain and on its short cytosolic domain, suggesting intracellular signalling [13, 18, 26]. Furthermore, TF-specific antibodies block corneal angiogenesis (N. Mackman, personal communication), and FVIIa induces release of vascular endothelial growth factor (VEGF) via interaction with TF [28]. In contrast, growth and vascularization of embryonic stem cell-derived tumors were similar for wild type and TF deficient embryonic stem cells when subcutaneously inoculated in nude mice [70]. However, since the angiogenic activity of TF in human tumors has been related to its expression by endothelial and smooth muscle cells, the lack of any effect of the TF genotype of the embryonic tumor cells on tumor growth and angiogenesis may not be surprising, as both endothelial and vascular smooth muscle cells invading the embryonic stem cell-derived tumors were derived from the wild type hosts and thus could express TF in both wild type and TF deficient tumors.

Tumor growth and dissemination   Pericellular plasmin proteolysis has been proposed to play a role in tumor invasion and metastasis by facilitating the migration of malignant cells through anatomical barriers via degradation of extracellular matrix constituents. Increased expression of u-PA, u-PAR and PAI-1 by tumor cells or by the surrounding stroma has indeed been observed [203-206]. In addition, use of antisense mRNA, or administration of natural or synthetic serine proteinase inhibitors, u-PAR antagonists, or of anticatalytic PA antibodies reduced, whereas genetically engineered overexpression of u-PA increased tumor dissemination [207, 208]. Furthermore, profibrinolytic system components have been used as markers for poor prognosis of certain tumors in man.

Studies in which tumor cells were inoculated in heterotopic sites in transgenic mice generally indicated a positive effect of host-derived u-PA-generated plasmin proteolysis on primary tumor growth and dissemination. For example, deficiency of host plasminogen reduced the growth of a primary Lewis lung carcinoma (3LL), delayed its dissemination to regional lymph nodes, and prolonged the survival after primary tumor resection, although, overall, the effects were rather small and transient [181]. Thus, although plasmin-mediated proteolysis contributes to the morbidity and mortality of Lewis lung carcinoma in mice, sufficient proteolytic activity is generated in plasminogen-deficient mice for efficient tumor development and metastasis. Growth and dissemination of orthotopic skin tumors were studied *in situ* by application of 7,12-dimethylbenz(α)anthracene and croton oil [180]. Although cellular blue nevi were induced in both wild type and in u-PA deficient mice, radial and vertical progression of these lesions was reduced in u-PA-deficient mice. Nevertheless, more than 95% of cellular blue nevi invaded the underlying tissues in both genotypes. As these tumors produce t-PA, invasion of melanocytic lesions may depend also (or more) on t-PA than on u-PA activity. Notably, however, progression to melanomas was not observed in u-PA-deficient mice. Whether this trophic effect of u-PA relates to its proteolytic activity (for example by liberating growth factors sequestered within the matrix, or by affecting matrix-induced cell growth), or instead (or in addition), on some non-proteolytic morphogenic action of u-PA [41] remains to be determined. To summarise, u-PA generally exerts a positive effect on tumorigenesis, although its mode of action (e.g. on the growth, dissemination or progression of the tumors) may differ according to the models studied.

Based on its ability to block u-PA proteolysis, PAI-1 would be anticipated to impair tumorigenesis. This is indeed observed in a number of transgenic tumor models. Overexpression of human PAI-1 in transgenic mice failed to affect primary tumor growth, nor did it alter the fibrin deposition around the tumors, but significantly reduced the number of pulmonary metastases [209]. Since PAI-1 was localized to the endothelial lining of small vessels in the primary tumor and in the lungs of 3LL-bearing PAI-1 transgenic mice, reduced lodging of the 3LL tumor cells to the pulmonary vessels might explain inhibition of metastatic spread. In another study, deficiency of host PAI-1 or overexpression of murine PAI-1 in transgenic mice did not affect primary tumor growth, nor did these genetic manipulations alter the number of pulmonary metastases after intravenous inoculation of tumor cells [210]. The reason for these different findings remains to be unravelled.

The role of PAI-1 in tumor growth and metastasis remains, however, controversial as epidemiologic studies indicate that PAI-1 is a marker of poor prognosis for survival of patients suffering from a variety of different cancers [211]. Although a role for PAI-1 in cell adhesion (promoting cellular migration [39, 40]), or in stabilization of the extracellular matrix surrounding sprouting neovessels [212] (a prerequisite for tumor growth [213]) has been suggested, the molecular basis of the protumorigenic activity of PAI-1 has remained elusive. Therefore, malignant keratinocytes, plated on collagen disks, were transplanted in a transplantation chamber onto the dorsal muscle fascia of wild type or PAI-1-deficient hosts, and their growth and vascularization examined [214]. When transplanted into wild type hosts, endothelial and stromal host cells migrated upwards into the collagen gel (presumably in response to production of angiogenic factors by tumor cells), nourishing the malignant keratinocytes which subsequently colonized the entire collagen gel and invaded into the host stroma. In contrast, endothelial migration and tumor invasion were markedly reduced to absent in PAI-1 deficient hosts. As the tumor cells produced similar amounts of t-PA, u-PA and PAI-1, the genotypic effect on tumor invasion appeared to be attributable to PAI-1, produced by host mesenchymal cells and sprouting endothelial cells. Further support for a role of host-derived PAI-1 was derived from the observation that adenoviral PAI-1 gene transfer into PAI-1 deficient hosts restored the invasive behaviour of the tumor cells. Whether PAI-1 facilitates endothelial cell migration via interaction with integrins [40, 41], or whether it prevents excessive degradation of extracellular matrix and provides structural support for the growing neovessels [212], remains to be determined.

MMPs have been implicated in the destruction of basement membrane components and connective tissue during tumor cell invasion and dissemination [215-217]. Although MMPs are usually expressed in the stroma of many human neoplasms, matrilysin (MMP-7) is primarily produced by (epithelial) tumor cells. Incidence and size of intestinal tumors in the *Min* (*m*ultiple *i*ntestinal *neo*plasia) mouse (a model of human familial adenomatous polyposis) were significantly suppressed in mice lacking MMP-7 [195]. Surprisingly, MMP-7 was immunolocalized to the lumenal surface of dysplastic glands rather than to the basement membrane or extracellular matrix, suggesting that its substrates may not be restricted to the extracellular matrix. Whether matrilysin locally perturbs tumor growth by activating lumenal or membrane-bound cytokines or growth factors remains to be determined.

Stromelysin-3 (MMP-10), expressed by fibroblasts around malignant cells, has been associated with poor clinical outcome in human carcinomas [218]. Deficiency of stromelysin-3 decreased the number and size of benign tumors and of malignant carcinomas induced by 7,12-dimethylbenz(α)anthracene [219]. In contrast to wild type fibroblasts, stromelysin-3 deficient fibroblasts were unable to promote implantation of malignant epithelial cells in nude mice. Possibly, this paracrine action of stromelysin-3 may rely on the release or activation of extracellular matrix-associated growth factors.

Gelatinase-A (MMP-2) binds to a membrane-type (MT)-MMP [220] and to the $\alpha_v\beta_3$ integrin [221], present on both tumor or endothelial cells. Growth and vascularization of B16-BL6 melanomas or Lewis lung carcinomas were reduced in gelatinase-A-deficient mice [196]. In addition, formation of granulation tissue was also reduced as compared to wild type mice, suggesting a significant role for host-derived gelatinase-A in tumor invasion and angiogenesis.

Stromelysin-1 (MMP-3) has been implicated in basement membrane destruction. Expression of autoactivating stromelysin-1 isoforms in transgenic mice caused epithelial cell proliferation and differentation in mammary glands of virgin females, resulting in precocious alveolar maturation, and branching [222]. Possibly, this

morphogenic action of MMP-3 relates to the generation or activation of a growth stimulator, the destruction of a growth inhibitor, or the release of growth factors sequestered within the basement membrane. During pregnancy and lactation, MMP-3 transgene expression resulted in degradation of the basement membrane (nidogen) and epithelial apoptosis [222], a process that could be rescued by concommittant overexpression of TIMP-1 in these mice [223]. That loss of the basement membrane structure induces epithelial apoptosis extends previous *in vitro* data that proteolytic fragments of basement membrane not only fail to support differentiation of mammary epithelium, but also act in a dominant-negative manner to inhibit extracellular matrix-mediated differentiation [224]. MMP-3 overexpression has been reported to increase [225] or reduce [226] the incidence but not the invasion of tumors, the precise mechanism of which remains to be elucidated.

TIMP-1 has been implicated as a tumor suppressor. TIMP-1 deficient embryonic stem cells or oncogene-transformed fibroblasts were generally more tumorigenic than wild type cells [199, 227]. Conversely, overexpression of TIMP-1 suppressed the development and vascularization of hepatocellular carcinomas [228] and inhibited intradermal growth of T-lymphoma cells [229]. Potential mechanisms of action of TIMP-1 include an effect on extracellular matrix-directed gene expression or apoptosis, or on the bioavailability of growth factors sequestered within the matrix [230, 231].

Chronic inflammatory disorders

Kidney Plasminogen activators have been implicated in renal biology [30]. u-PA is released in the urine by the epithelial cells lining the straight proximal and distal tubules, whereas t-PA is produced by glomerular cells and by epithelial cells lining the distal part of the collecting ducts [232]. Impaired fibrinolysis, resulting from reduced u-PA or increased PAI-1 activity, has been implicated in the deposition of fibrin and of extracellular matrix components in chronic renal inflammatory disorders in patients [233] and in endotoxin-treated [234] or in MRL lpr/lpr mice (which spontaneously develop autoimmune lupus nephritis) [235]. Electron microscopy of t-PA:u-PA- deficient mice revealed fibrin deposition in both the intravascular and extravascular glomerular compartments (unpublished observations). The role of plasmin was further analyzed using an experimental model of glomerulonephritis induced by glomerular membrane-specific antibodies [113]. Mice deficient of Plg or of t-PA:u-PA suffered severe glomerulonephritis, characterized by increased accumulation of fibrin and infiltration of macrophages resulting in renal failure. Presumably, fibrin provides a strong stimulus for macrophage recruitment (which can occur thus in the absence of plasmin proteolysis), resulting in an intense inflammatory reaction with associated necrosis of the glomerular tuft and hypocellularity of the affected glomeruli. Mice deficient in t-PA showed significantly increased renal failure, glomerular hypercellularity, glomerular macrophage infiltration and fibrin deposition, although not sufficient to cause glomerular necrosis [113]. In contrast, mice deficient in u-PA alone did not show significant differences in histological indices of disease, although they did show a trend towards reduced glomerular macrophage infiltration and increased renal impairment, perhaps related to the high expression of u-PA in the renal tubular epithelial cells. Whereas the reduced glomerular macrophage infiltration in u-PA-deficient mice suggests that u-PA participates in macrophage invasion, the latter can still occur in the absence of plasmin proteolysis, suggesting that plasmin assists but is not essential for macrophage infiltration. Loss of u-PAR did not alter the glomerulonephritis.

Lung   Alveoli are proficient at clearing extravascular fibrin due to a net fibronolytic balance. However, during adult respiratory distress syndrome, idiopathic pulmonary fibrosis, sarcoidosis, and bronchopulmonary dysplasia (diseases that frequently progress to pulmonary fibrosis with respiratory failure), fibrinolytic activity is reduced (decreased u-PA and increased PAI-1 or α_2-antiplasmin levels) resulting in fibrin deposition and formation of hyaline membranes [236]. Administration of bleomycin (which induces fibrotic pneumonitis) or exposure to hyperoxia (which causes respiratory distress in premature babies or adults) induced excessive deposition of fibrin and collagen-rich matrix, caused formation of hyaline membranes, and shortened the survival of PAI-1 overexpressing but not of PAI-1-deficient or α_2-macroglobulin-deficient mice [237, 238]. Notably, fibrin deposition is known to impair gas exchange, to inactivate surfactant, and to stimulate the proliferation of collagen-synthesizing fibroblasts. In a more recent analysis, intratracheal bleomycin instillation induced lung fibrosis with reduced leukocyte infiltration into the lungs of u-PA-deficient and Plg-deficient mice, whereas wild type mice suffered widespread pulmonary hemorrhages (V. Ploplis, personal communication). t-PA-deficient mice exhibited an intermediate phenotype, whereas u-PAR-deficient mice were not affected. Initial analysis suggests that reduced macrophage recruitment to the lungs may be related to decreased activation of matrix metalloproteinases.
Pulmonary emphysema, characterized by destructive enlargement of terminal airspaces, significantly contributes to cigarette smoking-induced chronic obstructive pulmonary disease (COPD) [239]. Although lung-specific overexpression of a human collagenase transgene induced emphysema [240], it is not certain whether expression of the transgene during growth or development interfered with normal elastic fiber assembly, perhaps through destruction of the elastic fiber microfibrillar scaffold. Development of emphysema in patients with functional deficiency of α_1-antiproteinase (resulting in unopposed neutrophil elastase activity) or in animals after intratracheal instillation of elastases suggested that destruction of elastic fibers is a central feature in the pathogenesis of emphysema [241]. However,

macrophages and not neutrophils, are the predominant defense cells in the inflamed lung. To determine which proteinases are responsible for lung destruction characteristic of pulmonary emphysema, macrophage elastase- (MMP-12) deficient mice, which mice exhibit reduced macrophage invasion through reconstituted basement membranes *in vitro* and *in vivo* [197], were subjected to cigarette smoke [242]. Macrophage infiltration in MMP-12-deficient lungs was reduced as compared to wild type mice and, remarkably, MMP-12-deficient mice did not develop emphysema in response to chronic exposure to cigarette smoke [242, 243]. Smoke-exposed MMP-12-deficient mice that received monthly intratracheal instillations of monocyte chemoattractant protein-1 (MCP-1) showed significant recruitment of alveolar macrophages (indicating that monocytes can be recruited to the lung in the absence of MMP-12), but they failed to develop airspace enlargement. These data may suggest that MMP-12, produced by resident macrophages in the lung, produce a chemotactic factor for monocytes/macrophages (possibly elastin degradation products; Shapiro S., personal communication), which subsequently mediate elastolysis and pulmonary destruction.

Gastro-intestinal tract   Mice with deficiency of α_2-macroglobulin suffered increased mortality upon experimental induction of acute pancreatitis, possibly because of uncontrolled proteolysis [244]. Plasminogen- deficient mice displayed fibrin-rich gastric ulcerations when infected by the pathogen *Helicobacter Pylori* [108]. Mice deficient for the major lysosomal aspartic proteinase cathepsin D developed normally during the first two weeks after birth, stopped thriving by the third week, and died in a state of anorexia after 4 weeks [245]. Cachexia appears to result from atrophy of the intestinal mucosa, presumably due to a defective developmental switch to the adult-type epithelial phenotype, allowing penetration of gut bacteria and development of endotoxin shock. Whether cathepsin D controls (in)activation of signalling proteins by limited proteolysis in the endosomal and/or lysosomal compartment remains to be determined.

Skin, joints, and cornea   Migration of keratinocytes involves expression of proteinases at the leading edge of their migration front [246]. Using a model of skin wound healing, plasminogen deficient mice exhibited delayed and impaired closure of skin wounds [112]. Notably, keratinocyte migration appeared to be reduced but, surprisingly, the granulation tissue appeared normal except for a more abundant presence of fibrin(ogen) and fibronectin at the wound edges. In fact, Plg-deficient mice, like their wild type controls, had abundant infiltration of macrophages, neutrophilic granulocytes and fibroblast-like cells, and pronounced neovascularization, consistent with the above described hypothesis that fibrin provides a strong stimulus for wound cells. However, wound healing in combined plasminogen and fibrinogen-deficient mice was not impaired, indicating that fibrin mediates, to a large degree, the effects of plasminogen deficiency [247]. Similar findings were observed during healing of surgically induced corneal epithelial defects [179] or of spontaneously occurring conjunctival lesions (ligneous conjunctivitis) [248]. In contrast to the complete healing of corneal defects in wild type mice, corneal healing in plasminogen-deficient mice was impaired and complicated by severe and persistent inflammatory responses, the formation of retrocorneal fibrin deposits, corneal cloudiness caused by scar-tissue formation, and often stromal neovascularization. Conjunctival lesions in Plg-deficient mice exhibited epithelial disruption with associated hypertrophy, disorganization, reduplication, inflammatory infiltrates, and stromal vascularization consistent with chronic, recurrent ulceration and attempted re-epithelialization. Possibly, repetitive injury due to scratching or abrasion of foreign bodies against the conjunctival epithelium may have initiated the development of ligneous lesions. Loss of fibrinogen in Plg-deficient mice restored corneal and conjunctival healing, indicating that healing is related to fibrinogen. Taken together, the plasminogen system appears to play a significant role in the tissue remodelling during wound healing, in part mediated by its role in fibrin surveillance. This notion is supported by the observation that fibrinogen-deficient mice had an unusual wound healing response in which the migrating and proliferating cells (primarily fibroblasts) form a thick layer *encapsulating* but not *infiltrating* hematomas [83]. It is thus possible that fibrin provides a critical initial matrix for the movement of cells into sites of injury.

Different degrees of wound healing responses have been reported, depending on the environmental conditions and infectious challenges. The most significant phenotype occurred in u-PA-deficient mice after infection with botryomycosis [249]. In contrast to their wild type littermates, housed in the same environmental conditions, u-PA-deficient mice developed a suppurative infection of the skin characterized by the presence of abscesses and granulomas, containing large numbers of polymorphonuclear leukocytes and histiocytes that were surrounded by a capsule of fibrous connective tissue. Such destructive tissue remodelling is indeed more severe than observed in combined t-PA:u-PA-deficient or Plg-deficient mice, indicating that the phenotypes observed in these knock-out mice is importantly determined by the infectious or inflammatory challenge.

Although MMPs are likely involved in similar processes, to date, little is known about their *in vivo* role as deduced from the gene targeting studies. In the skin, dermal fibroblasts and epidermal keratinocytes synthetize tissue collagenase, especially at the leading edge of a wound. Overexpression of collagenase in the skin caused hyperkeratosis, acanthosis, and epidermal hyperplasia [250]. At the ultrastructural level, intercellular spaces were widely opened with cell separation and adhesion limited to the desmosomes and reduced intercellular granular material [250]. Furthermore, the transgenic skin exhibited greater susceptibility to 7,12-dimethyl-benz

[α]anthracene-induced tumorigenesis. MMP-3 has been implicated in the loss of cartilage in rheumatoid arthritis and osteoarthritis. Surprisingly, deficiency of MMP-3 did not prevent nor reduce the cartilage destruction associated with collagen-induced arthritis, suggesting redundancy or compensation by other proteinases [194].

Infection   The expression of proteinases [in particular of the u-PA:u-PAR system) is thought to be critical for the ability of leukocytes to degrade matrix proteins and to traverse tissue planes during recruitment to inflammatory sites. u-PA has, however, also been implicated in the modulation of cytokine and growth factor expression. It is required for TNF-α expression by mononuclear phagocytes, for activation of latent TGF-β_1 [42], and may also be involved in the release of interleukin-1 [IL-1) [251]. In addition, serine proteinase inhibitors reduce IL-2 expression [252]. u-PA-deficient mice were unable to mount an adequate pulmonary inflammatory response to a challenge with the non-lethal 52D *cryptococcus neoformans* pathogen [253]. They were unable to recruit sufficient mononuclear phagocytes, neutrophils and lymphocytes, did not contain the infection to the lung and could not eliminate the organism, which disseminated widely and ultimately infected the brain, leading to death. This pattern of wide dissemination and death with strain 52D has only been seen in profoundly immunoincompetent mice. Whereas u-PA and u-PAR may promote recruitment of monocytes and neutrophils by enhancing the degradation of matrix components, u-PA may also play a role in lymphocyte recruitment by modulating the cytokine network. Thus, the absence of u-PA may result in inadequate signalling via IL-1 or IL-2, significant modulators of lymphocyte cell function.

Treatment of patients with TNF-α frequently induces transient thrombocytopenia. Platelet consumption and trapping within organs was significantly decreased in u-PA-deficient but not in t-PA-deficient mice, consistent with a reduced activation of u-PA-deficient platelets *in vitro* [D. Belin, personal communication). Another interesting observation is that α_2-macroglobulin- deficient mice were significantly more resistant to lethal doses of endotoxin. Thus, the role of the plasminogen system in the inflammatory response may extend beyond the proteolytic activities required to allow for the movement of cells, and may participate in the orchestration of cytokine networks which serve to intensify the inflammatory response.

A number of invasive bacteria can interact with the host plasminogen system by expressing endogenous plasminogen activators and by binding plasminogen directly through bacterial cell-surface receptors, allowing them thereby to utilize the plasminogen activators of the host for activation [254]. Once bound on the bacterial surface, plasmin is protected from inhibition by α_2-antiplasmin, mediates degradation of the extracellular matrix and penetration through the endothelium. Binding of Plg on the spyrochete *Borrelia burgdorferi* occurs as the tick feeds on the host's blood, and plasminogen activators are derived from the host blood meal. Studies in Plg-deficient mice indicate that Plg was required for efficient dissemination of *B. burgdorferi* within the tick and for enhancement of spirochetemia in mice, but was not critical for transmission and infection [255]. Thus, bacteria can use vertebrate proteinases to disseminate in an invertebrate vector. More recent studies reveal a similar requirement of host-derived Plg by other bacteria [e.g. Yersinia Pestis] for their dissemination [T. Bugge, personal communication]. Bacterial strains expressing a plasminogen activator (*pla+*) escaped elimination by the host immune system and were almost a million-fold more pathogenic than *pla-* strains (not expressing such plasminogen activator) in wild type mice but not in Plg-deficient hosts.

Neutrophil elastase [NE) is a potent serine proteinase exhibiting antibacterial activity. NE deficient mice were more susceptible than wild type mice to sepsis and death following intraperitoneal infection with Gram-negative [*Klebsiella pneumoniae* and *Escheria coli*) but not with Gram positive [*Staphylococcus aureas*) bacteria [256]. NE deficient neutrophils migrated normally to sites of infection. However, ultrastructural analysis revealed that internalized bacteria were in various stages of degradation within phagosomes in wild type but not in NE-deficient neutrophils, indicating that NE is required for maximal intracellular killing of Gram negative bacteria.

Brain function Evidence has been provided that the plasminogen system might be involved in brain function. Fibrinolytic system components are expressed in specialized areas of the brain during development or in adulthood following different forms of brain activity [257]. In addition, *in vitro* studies with cultured neurons revealed that these cells are able to produce and respond to plasminogen activators [258]. Restricted and temporal specific expression of t-PA in the nervous system during development has also been observed in transgenic mice expressing the *LacZ* marker gene driven by various t-PA promotor constructs [259, 260]. Ectopic and over-expression of murine u-PA in the brain [e.g. in the hippocampus and in the limbic system) was associated with impaired learning of tasks in transgenic mice, reduced food intake, body weight and size, and increased longevity [261].

cAMP-dependent *de novo* synthesis of proteins including t-PA has been proposed to participate in long term potentiation [LTP]. Deficiency of t-PA [but not of u-PA) abolished the late phase of LTP in both the Schaffer collateral and mossy fiber pathways of the hippocampus after induction by electrical stimulation or by treatment with dopamine agonists or cAMP analogues [262]. Somewhat surprisingly, t-PA deficiency did not affect hippocampus-related learning tasks including spatial memory [Barnes circular maze and Morris water maze tests), exploration in a novel environment, and context conditioning [262]. t-PA deficiency significantly impaired, however, active avoidance learning and slightly affected the acquisition learning [Morris water maze test). Since t-PA is expressed in certain nuclei of the limbic system, this impairment might be due to abnormal coping of t-PA-deficient mice with stress. Another study reported that t-PA-deficient mice completely lacked conventional, homosynaptic late LTP at the Schaffer collateral-CA1 pyramidal cell synapses and exhibited a different form of [heterosynaptic) LTP that not only required glutaminergic but also GABA-dependent transmission [263]. This heterosynaptic form of potentiation provided t-PA-deficient mice with an output of CA1 neurons similar to that seen in wild type mice during conventional late LTP. Compensation of conventional LTP by a GABA-dependent potentiation could explain the relative lack of hippocampal-related learning defects. In summary, these data suggest that t-PA plays a significant role in the late phase of LTP as a downstream target of cAMP.

Another remarkable observation is that t-PA- and Plg-deficient mice were resistant to neuronal degeneration after intrahippocampal injection of excitotoxin glutamate agonists [264, 265]. Surprisingly, mice deficient for both plasminogen and fibrinogen were also resistant, indicating that plasmin acts on substrates other than fibrin during experimental neuronal degeneration [266]. Intracerebral injection of excitotoxins in wild type mice depolarizes neurons and induces influx of calcium, rapid release of intracellularly stored t-PA, and *de novo* production of t-PA. Increased t-PA levels [released by neurons and also by microglial cells, the non-neuronal macrophage-like cells that are transformed from a resting to an activated state upon neuronal injury) activate plasminogen (locally produced by neurons) to plasmin, which degrades laminin in the extracellular matrix between neurons and glia [267]. Presumably, the loss of cell attachment sites by plasmin renders the neurons more susceptible to death [anokis) upon excitotoxin stimulation. The lack of neuronal degeneration in excitotoxin-injected t-PA-deficient mice appears to be due, at least in part, to a failure of microglial cell activation. As microglial activation was reduced in t-PA-deficient mice but not in Plg-deficient mice or in wild type mice, plasminogen but not t-PA lies downstream of microglial activation, suggesting some other activity or function for t-PA aside from plasminogen cleavage during microglial activation [266]. Since the experimental paradigm of excitotoxin injection mimicks both the development and the type of cell death that occurs in many neuropathologies (ischemic stroke, Alzheimer's disease, amyotrophic lateral sclerosis), these data suggest that t-PA activity might contribute to pathologies associated with accelerated neuronal degeneration such as Alzheimer's disease. Recent data indicate that t-PA-deficient mice are also protected against ischemic neuronal cell death [268]. As t-PA (administered during thrombolytic therapy) may reach the neurons at risk through the disrupted blood-brain barrier, a therapy that could be devised at degrading fibrin but not laminin would minimize any possible deleterious effect of local plasmin production.

t-PA has been implicated in the migration of granule neurons in the developing cerebellum [269]. Interestingly, there were two to three times more granule neurons in the molecular layer of the developing cerebellum of neonatal t-PA-deficient mice, suggesting that the absence of t-PA leads to retardation in granule neuron migration. Such retardation was not observed in wild type or u-PA-deficient mice [Seeds and Haffke, personal communication]. Transsection or crush of peripheral motor nerves leads to a retrograde reaction in the neuronal cell bodies accompanied by the activation of glial cells in the vicinity of the damaged neurons. These microglia extend processes into the synaptic clefts thus stripping synapses from the motor neuron cell bodies, a process that was proposed to be mediated by plasminogen activators based on the induced expression of t-PA, u-PA and PAI-1 [270]. Initial studies suggest that stripping of the synapses from the motorneuron cell bodies still occurs in t-PA-deficient mice [Reddington et al, personal communication]. Further studies are required to determine the precise involvement of the plasminogen system in the tissue remodelling accompanying neuronal injury.

Peter Carmeliet and Désiré Collen

References

1. Edgington TS et al (1991) Thromb Haemost 66:67-79
2. Camerer E et al (1996) Thromb Res 81:1-41
3. Petersen LC et al (1995) Thromb Res 79:1-47
4. Davie EW(1995) Thromb Haemost 74:1-6
5. Broze GJ, Jr (1995) Annu Rev Med 46:103-112
6. Esmon CT (1992) Arterioscler Thromb 12:135-145
7. Perry DJ (1994) Blood Rev 8:37-55
8. Altieri DC (1995) Faseb J 9):860-865
9. Coughlin SR(1994) Semin Hematol 31:270-277
10. Soifer SJ et al (1994) Am J Pathol 144:60-69
11. Luther T (1996) Am J Pathol 149:101-113
12. Mueller BM et al (1992) Proc. Natl Acad Sci USA 89:11832-11836
13. Mueller BM, Ruf W (1998) J Clin Invest 101:1372-1378
14. Pendurthi UR et al (1997) Proc Natl Acad Sci USA 94:12598-12603
15. Poulsen LK et al (1998) J Biol Chem 273):6228-6232
16. Zioncheck TF et al (1992) J Biol Chem 1992;267:3561-3564
17. Masuda M et al (1996) Eur J Immunol 26:2529-2532
18. Ott I et al (1998) J Cell Biol 140:1241-1253
19. Müller M et al (1998) Am. J Pathol (submitted)
20. Sato Y et al (1997) Thromb Haemost 78:1138-1141
21. Sato Y et al (1996) Thromb Haemost 75:389-392
22. Zhang Y et al (1994) J Clin Invest 94:1320-1327
23. Zhang Y et al (1996) J Clin Invest 97:2213-2224
24. Contrino J et al (1996) Nat Med 2:209-215
25. Koomagi R, Volm M (1998) Int J Cancer 79:19-22
26. Ruf W, Mueller BM (1996) Curr Opin Hematol 3:379-384
27. Shoji M et al (1998) Am J Pathol 152:399-411
28. Ollivier V et al (1998) Blood 91:2698-2703
29. Collen D, Lijnen HR (1991) Blood 78:3114-3124
30. Vassalli JD et al (1991) J Clin Invest 88:1067-1072
31. Wiman B (1995) Thromb Haemost 74:71-76
32. Schneiderman J, Loskutoff DJ (1991) Trends Cardiovasc Med 1:99-102
33. Lawrence DA, Ginsburg D (1995) In: Roberts HR, High KA (eds) Molecular Biology of Thrombosis and Hemostasis. Marcel Dekker, New York, pp 517-543
34. Hajjar KA (1995) Thromb Haemost 74:294-301
35. Plow EF et al (1995) Faseb J 9:939-945
36. Vassalli JD (1994) Fibrinolysis 8(Suppl 1):172-181
37. Blasi F et al (1994) Fibrinolysis 8(Suppl 1):182-188
38. Wei Y et al (1994) J Biol Chem 269:32380-32388
39. Stefansson S, Lawrence DA (1996) Nature 383:441-443
40. Deng G et al (1996) J Cell Biol 134:1563-1571
41. Chapman HA (1997) Current Opinion in Cell Biology 9:714-724
42. Saksela O, Rifkin D (1988) Annu Rev Cell Biol 4:93-126
43. Chapman HA, Jr., Stone OL (1984) Biochem J 222:721-728
44. Ferrara N (1993) Trends Cardiovasc Med 6:244-250
45. Andreasen PA et al (1994) FEBS Lett 338:239-245
46. Murphy G (1995) Acta Orthop Scand (suppl 256) 66:55-60
47. Dollery CM et al (1995)Circ Res 77:863-868
48. Mignatti P, Rifkin DB (1996) Enzyme Protein 49:117-137
49. Matrisian LM (1990) Trends Genet 6:121-125
50. Jenkins GM et al (1998) Circulation 97:82-90
51. Sato H et al (1997) Thromb Haemost 78:497-500
52. Apte SS et al (1997) J Biol Chem 272:25511-25517
53. Mattei MG et al (1997) Genomics 40:168-169
54. Llano E et al (1997) Biochemistry 36:15101-15108
55. Sedlacek R et al (1998) Immunobiology 198:408-423
56. Gomez DE et al (1997) Eur J Cell Biol 74:111-122
57. Carmeliet P et al (1997) Nature Genetics 17:439-446
58. Suzuki M et al (1997) J Biol Chem 272:31730-31737
59. Preece G et al (1996) J Biol Chem 271:11634-11640
60. Kayagaki N et al (1996) J Exp Med 182:1777-1783
61. Black RA et al (1997) Nature 385:729-733
62. Risau W (1997) Nature 386:671-674
63. Carmeliet P, Collen D (1998) Kidney Int 53:1519-1549
64. Nehls V, Drenckhahn D (1993) Histochemistry 99:1-12
65. Beck L, D'Amore P (1997) FASEB J 11:365-373
66. D'Amore PA, Smith SR (1993) Growth Factors 8:61-75
67. Bugge TH et al (1996) Proc Natl Acad Sci USA 93:6258-6263
68. Carmeliet P et al (1996) Nature 383:73-75
69. Toomey JR et al (1996) Blood 88:1583-1587
70. Toomey JR et al (1997) Proc Natl Acad Sci USA 94:6922-6926
71. Hirschi KK, d'Amore PA (1996) Cardiovasc Res 32:687-698
72. Clauss M et al (1990) J Exp Med 172:1535-1545
73. Ferrara N, Davis-Smyth T (1997) Endocrine Reviews 18:4-25
74. Grosskreutz CL et al (1996) Invest Opthalmol Vis Sci 37:5470-5475
75. Takagi H et al (1996) Diabetes 45:1016-1023
76. Nomura M et al (1995) J Biol Chem 270:28316-28324
77. Parry GC et al (1998) J Clin Invest 101:560-569
78. Rosen E et al (1997) Nature 390:290-294
79. Cui J et al (1996) Nature 384:66-68
80. Xue J et al (1998) Proc Natl Acad Sci USA 95:7603-7606
81. Sun WY et al (1998) Proc Natl Acad Sci USA 95:7597-7602
82. Connolly AJ et al (1996) Nature 381:516-519
83. Suh TT et al (1995) Genes Dev 9:2020-2033
84. Shivdasani RA et al (1995) Cell 81:695-704
85. Gasic GP et al (1992) Proc Natl Acad Sci USA 89:2317-2320
86. Feng NK et al (1996) J Clin Invest 98:1493-1501
87. Ross R (1986) Cell 46:155-169
88. Ettelaie C et al (1996) Arterioscler Thromb Vasc Biol 16:639-645

89. Fan Z et al (1998) Blood 91:1987-1998
90. Huang ZF et al (1997) Blood 90:944-951
91. Ishihara H et al (1997) Nature 386:502-506
92. Darrow AL et al (1996) Thromb Haemost 76:860-866
93. Healy AM et al (1995) Proc Natl Acad Sci USA 92:850-854
94. Jalbert L et al (1998) J Clin Invest (under revision)
95. Bi L et al (1995) Nat Genet 10:119-121
96. Kung SH et al (1998) Blood 91:784-790
97. Wang L et al (1997) Proc Natl Acad Sci USA 94:11563-11566
98. Tuddenham EG et al (1995) Thromb Haemost 74:313-321
99. Hoyer LW (1996) N Engl J Med 330:38-47
100. Snyder RO et al (1997) Nat Genet 16:270-276
101. al Mondhiry H, Ehmann WC (1994) Am J Hematol 46:343-347
102. Fay WP et al (1992) N Engl J Med 327:1729-1733
103. Aoki N (1989) Blood Rev 3:11-17
104. Elliott MJ et al (1992) J Immunol 149:3678-3681
105. Carmeliet P et al (1993) J Clin Invest 92:2756-2760
106. Carmeliet P et al (1994) Nature 368:419-424
107. Carmeliet P et al (1995) Ann NY Acad Sci 748:367-381
108. Ploplis VA et al (1995) Circulation 92:2585-2593
109. Bugge TH et al (1995) Genes Dev 9:794-807
110. Dewerchin M et al (1996) J Clin Invest 97:870-878
111. Bugge TH et al (1996) Proc Natl Acad Sci USA 93:5899-5904
112. Romer J et al (1996) Nat Med 2:287-292
113. Kitching AR et al (1997) J Exp Med 5:963-968
114. Carmeliet P et al (1997) J Clin Invest 99:200-208
115. Farrehi PM et al (1998) Circulation 97:1002-1008
116. Carmeliet P et al (1993) J Clin Invest 92:2746-2755
117. Carmeliet P et al (1997) Blood 90:1527-1534
118. Liu AC, Lawn RM (1994) Trends Cardiovasc Med 4:40-44
119. Palabrica TM et al (1995) [published erratum appears in Nat Med 1995 Jun;1:598] Nat Med 1:256-259
120. Weiler-Guettler H et al (1998) J Clin Invest 101:1983-1991
121. Libby P et al (1992) Circulation 86(Suppl III):III47-52
122. Clowes AW, Reidy MA (1991) J Vasc Surg 13:885-891
123. Schwartz SM et al (1995) Thromb Haemost 74:541-551
124. Kakuta T et al (1994) Circulation 89:2809-2815
125. Carmeliet P et al (1997) Am J Pathol 150:761-777
126. Carmeliet P et al (1998) Cardiovasc Res 39:8-33
127. van Leeuwen RT (1996) Fibrinolysis 10:59-74
128. Carmeliet P, Collen D (1996) Sem Thromb Hemost 22:525-542
129. Celentano DC, Frishman WH (1997) J Clin Pharmacol 37:991-1000
130. Simpson AJ et al (1991) J Clin Pathol 44:139-143
131. Clowes AW et al (1990) Circ Res 67:61-67
132. Reidy MA et al (1996) Circ Res 78:405-414
133. Fearns C et al (1996) In: van Hinsbergh VWM (ed) Advances in vascular biology. Gordon and Breach Publishers, Camberwell, Victoria, Australia, pp 207-227
134. Carmeliet P et al (1997) Circ Res 81:829-839
135. Carmeliet P et al (1997) Circulation 96:3180-3191
136. Newman KM et al (1994) J Vasc Surg 20:814-820
137. Galis ZS et al (1995) Proc Natl Acad Sci USA 92:402-406
138. Galis ZS et al (1994) Circ Res 75:181-189
139. Halpert I et al (1996) Proc Natl Acad Sci USA 93:9748-9753
140. Irizarry E et al (1993) J Surg Res 54:571-574
141. Sakalihasan N et al (1996) J Vasc Surg 24:127-133
142. Moons L et al (1998) J Clin Invest (in press)
143. Carmeliet P et al (1998) J Cell Biol 140:233-245
144. Herbert JM et al (1997) J Biol Chem 272:23585-23591
145. Lijnen R et al (1998) Arterioscl Thromb Vasc Biol 18:1035-1045
146. Lucas A et al (1996) Circulation 94:2890-2900
147. Ross R (1993) Nature 362:2844-2850
148. Fuster V 1994) Circulation 90:2126-2146
149. Libby P (1995) Circulation 91:2844-2850
150. Belkin M et al (1994) Curr Opin Cardiol 9:581-590
151. Ernst CB (1993) N Engl J Med 328:1167-1172
152. Halloran BG, Baxter BT (1995) Semin Vasc Surg 8:85-92
153. Juhan-Vague I, Collen D (1992) Ann Epidemiol 2:427-438
154. Hamsten A, Eriksson P (1994) Fibrinolysis 8(Suppl 1):253-262
155. Schneiderman J et al (1992) Proc Natl Acad Sci USA 89:6998-7002
156. Hamsten A et al (1987) Lancet 2:3-9
157. Loskutoff DJ, Samad F (1998) Arterioscl Thromb Vasc Biol 18:1-6
158. Schneiderman J et al (1995) J Clin Invest 96:639-645
159. Lupu F et al (1995) Arterioscler Thromb Vasc Biol 15:1444-1455
160. Plump AS et al (1992) Cell 71:343-353et al (1990)
161. Xiao Q et al (1997) Proc Natl Acad Sci USA 94:10335-10340
162. Xiao Q et al (1998) J Clin Invest 101:1184-1194
163. Ernst E, Resch KL (1993) Ann Intern Med 118:956-963
164. Chapman HA, Jr et al (1984) J Clin Invest 73:806-815
165. Shi C et al (1994) Circ Res 75:199-207
166. Spinale FG et al (1998) Circ Res 82:482-495
167. Tyagi SC (1998) J Cell Biochem 68:403-410
168. Funck RC et al (1997) Adv Exp Med Biol 432:35-44
169. Robert V et al (1997) Lab Invest 76:729-738
170. Booz GW, Baker KM (1995) Cardiovasc Res 30:537-543
171. Cleutjens JP et al (1995) Am J Pathol 147:325-338
172. Brilla CG, Rupp H (1994) Cardiologia 39(Suppl 1):389-393
173. Schaper W, Ito WD (1996) Circ Res 79:911-919
174. Bacharach E et al (1992) Proc Natl Acad Sci USA 89:10686-10690
175. Pepper MS et al (1992) J Cell Physiol 153:129-139
176. Pepper MS et al (1991) Biochem Biophys Res Comm 181:902-906
177. Pepper MS et al (1992) Biochem Biophys Res Comm 189:824-831
178. Wilting J, Christ B (1996) Naturwissenschaften 83:153-164
179. Kao WW et al (1998) Invest Opthalmol Vis Sci 39:502-508
180. Shapiro RL et al (1996) Cancer Res 56:3597-3604
181. Bugge TH et al (1997) Blood 90:4522-4531
182. Dvorak HF et al (1995) Am J Pathol 146:1029-1039
183. Thompson WD et al (1993) Blood Coagul Fibrinolysis 4:113-115
184. Nagy JA et al (1995) Cancer Res 55:360-368
185. Nagy JA et al (1995) Cancer Res 55:369-375
186. Sabapathy KT et al (1997) J Cell Biol 137:953-963
187. O'Reilly MS et al (1994) Cell 79:315-328
188. Cao Y et al (1996) J Biol Chem 271:29461-19467
189. Cao Y et al (1997) J Biol Chem 272:22924-22928
190. O'Reilly MS et al (1996) Nature Medicine 2:689-692
191. Dong Z et al (1997) Cell 88:801-810
192. Stathakis P et al (1997) J Biol Chem 272:20641-20645
193. Holmgren L et al (1995) Nature Med 1:149-153
194. Mudgett JS et al (1998) Arthritis Rheum 41:110-121
195. Wilson C et al (1997) Proc Natl Acad Sci USA 94:1402-1407
196. Itoh T et al (1998) Cancer Res 58:1048-1051

197. Shipley JM et al (1996) Proc Natl Acad Sci USA 93:3942-3946
198. Nothnick WB et al (1997) Biol Reprod 56:1181-1188
199. Soloway PD et al (1996) Oncogene 13:2307-2314
200. Vu TH et al (1998) Cell 93:411-422
201. Vu TH, Werb Z (1998) In: Parks WC, Mecham RP (eds) Matrix metalloproteinases. Academic Press, San Diego
202. Whitelock JM et al (1996) J Biol Chem 271:10079-10086
203. Xing RH, Rabbani SA (1996) Int J Cancer 67:423-429
204. Ossowski L (1992) Cancer Res 52:6754-6760
205. Dano K et al (1994) Fibrinolysis 8:189-203
206. Schmitt M et al (1995) J Obstet Gynacol 21:151-165
207. Ossowski L et al (1991) Cancer Res 51:274-281
208. Fazioli F, Blasi F (1994) Trends Pharmacol Sci 15:25-29
209. Donati MD (1995) Thrombosis Haemostasis 74:278-281
210. Eitzman DT et al (1996) Blood 87:4718-4722
211. Pappot H et al (1995) Biol Chem Hoppe Seyler 376:259-267
212. Montesano R et al (1990) Cell 62:435-445
213. Folkman J (1995) Nature Med 1:27-31
214. Bajou K et al (1998) (submitted)
215. MacDougall JR, Matrisian LM (1995) Cancer Metastasis Rev 14:351-362
216. Ray JM, Stetler-Stevenson WG (1994) Eur Resp J 7:2062-2072
217. Seiki M (1996) Curr Top Microbiol Immunol 213:23-32
218. Rouyer N et al (1994) Invasion Metastasis 14:946-953
219. Masson R et al (1998) J Cell Biol 140:1535-1541
220. Sato H et al (1994) Nature 370:61-65
221. Brooks PC et al (1996) Cell 85:683-693
222. Sympson CJ et al (1994) J Cell Biol 125:681-693
223. Alexander CM et al (1996) J Cell Biol 135:1669-1677
224. Streuli CH et al (1995) J Cell Biol 129:591-603
225. Sympson CJ et al (1995) Sem Cancer Biol 3:159-163
226. Witty JP et al (1995) Mol Biol Cell 6:1287-1303
227. Alexander CM, Werb Z (1992) J Cell Biol 118:727-739
228. Martin DC et al (1996) Oncogene 13:569-576
229. Kruger A et al (1997) Blood 90:1993-2000
230. Boudreau N et al (1995) Science 267:891-893
231. Boudreau N et al (1995) Trends Cell Biol 5:1-5
232. Sappino AP et al (1991) J Clin Invest 87:962-970
233. Tomooka S et al (1992) Kidney Int 42:1462-1469
234. Yamamoto K, Loskutoff DJ (1996) J Clin Invest 97:2440-2451
235. Yamamoto K, Loskutoff DJ (1997) Am J Pathol 151:725-734
236. Bertozzi P et al (1990) N Engl J Med 322:890-897
237. Eitzman DT et al (1996) J Clin Invest 97:232-237
238. Barazzone C et al (1996) J Clin Invest 98:2666-2673
239. MacKenzie TD et al (1994) N Engl J Med 330:975-980
240. D'Armiento J et al (1992) Cell 71:955-961
241. Shapiro SD (1994) Am J Respir Crit Care Med 150:S160-S164
242. Hautamaki RD et al (1997) Science 277:2002-2004
243. Shapiro SD (1997) Matrix Biology 15:527-533
244. Umans L et al (1995) J Biol Chem 270:19778-19785
245. Saftig P et al (1995) EMBO J 14:3599-3608
246. Kaharri VM et al (1997) Exp Dermatol 6:199-213
247. Bugge TH et al (1996) Cell 87:709-719
248. Drew AF et al (1998) Blood 91:1616-1624
249. Shapiro RL et al (1997) Am J Pathol 150:359-369
250. D'Armiento J et al (1995) Mol Cell Biol 15:5732-5739
251. Matsushima K et al (1986) J Immunol 136:2883-2891
252. Auberger P et al (1993) Immunology 70:547-550
253. Gyetko MR et al (1996) J Clin Invest 97:1818-1826
254. Boyle MDP, Lottenberger R (1997) Thromb Haemost 77:1-10
255. Coleman JL et al (1997) Cell 89:1111-1119
256. Bellaaouaj A et al (1998) Nature Medicine 4:615-618
257. Sappino AP et al (1993) J Clin Invest 92:679-685
258. Krystosek A, Seeds NW (1981) Science 213:1532-1534
259. Theuring F et al (1995) Fibrinolysis 9:277-287
260. Carroll PM et al
261. Meiri N et al (1994) Proc Natl Acad Sci USA 91:3196-3200
262. Huang YY et al (1996) Proc Natl Acad Sci USA 93:8699-8704
263. Frey U et al (1996) J Neurosci 16:2057-2063
264. Tsirka SE (1997) J Mol Med 75:341-347
265. Tsirka SE et al (1997) J Neurosci 17:543-552
266. Tsirka TE et al (1997) Proc Natl Acad Sci USA 94:9779-9781
267. Chen ZL, Strickland S (1997) Cell 91:917-825
268. Wang YF et al (1998) Nat Med 4:228-231
269. Seeds NW et al (1990) In: Lander JM (ed) Molecular aspects of development and aging of the nervous system. Plenum Press, New York
270. Nakajima K et al (1996) J Neurochem 66:2500-2505

Fibrinopeptides

Definition *Small acidic peptides that are cleaved by thrombin from the N-terminal chain of the Aα and Bβ chain of fibrinogen. There are two fibrinopeptides, fibrinopeptide A (FPA) and fibrinopeptide B (FPB).*

See: →Fibrin/fibrinogen; →Thrombin

Fibroblast Growth Factors

Definition *Family of growth factors that act on a number of mesoderm-derived and mesenchymal cells. 18 family members are described to date.*

See: →FGF-1 and -2; →FGF receptors; →Signal transduction mechanisms in vascular biology; →Atherosclerosis; →Cytokines in vascular biology and disease; →Extracellular matrix

Fibronectin

Definition *Extracellular matrix molecule that contains RGD motifs. Fibronectin binds to $\alpha 4\beta 1$, $\alpha 5\beta 1$ or $\alpha v\beta 3$ integrin cell surface receptors.*

See: →Extracellular matrix; →Angiogenesis inhibitors

FLK-1

Definition *Fetal liver kinase-1*

See: →Vascular endothelial growth factor receptors; →Angiogenesis; →Signal transduction mechanisms in vascular biology

Flt-1

Definition   *Fms-like tyrosine kinase-1*

See: →Vascular endothelial growth factor receptors

FLT4

Definition   *Fms-like tyrosine kinase-4*

See: →Vascular endothelial growth factor receptors; →Tyrosine kinase receptors for factors of the VEGF family

FMN

Definition   *Flavin mononucleotide*

See: →Nitric Oxide

Focal Adhesion Kinase (FAK)

Definition   *Kinase found at sites of focal cell adhesion sites that play a role in the assembly of actin stress fibers.*

See: →Signal transduction mechanisms in vascular biology; →Platelet stimulus-response coupling

FPA

Definition   *Fibrinopeptide A*

See: →Fibrinopeptides; →Fibrin/fibrinogen

FPB

Definition   *Fibrinopeptide B*

See: →Fibrinopeptides; →Fibrin/fibrinogen

FSH

Definition   *Follicle-stimulating hormone*

See: →Hormonal regulation of vascular cell function in angiogenesis

G-Protein-Coupled Receptors

Definition   *Receptors with seven transmembrane-spanning regions that are coupled to heterotrimeric GTP-binding proteins*

See: →Signal transduction mechanisms in vascular biology; →Vasomotor tone regulation, molecular mechanisms of; →Platelet stimulus-response coupling

G-Proteins-GTPase

Definition   *Small molecular weight proteins that through their capacity to hydrolyses GTP act as cellular switches. Involved in signal transduction.*

See: →Signal transduction mechanisms in vascular biology; →Vasomotor tone regulation, molecular mechanisms of; →Platelet stimulus-response coupling

GAG

Definition   *Glycosaminoglycan*

See: →Glycosaminoglycans; →FGF-1, FGF-2; →FGF receptors

GAP

Definition   *GTPase-activating protein*

See: →Signal transduction mechanisms in vascular biology

GDP

Definition   *Guanosine diphosphate*

See: →Platelet stimulus-response coupling

Glanzmann's Thrombasthenia

Definition   *Thrombopathy due to abnormal glycoprotein IIbIIIa on platelets.*

See: →Bleeding disorders; →Vascular integrins; →Platelet stimulus-response coupling

Glycosaminoglycans

Definition   *Polysaccharides composed of disaccharide units (usually uronic acid and hexosamine) containing sulfate. Constituent of proteoglycans.*

See: →FGF receptors; →FGF-1 and -2

GM-CSF

Definition   *Granulocyte-macrophage colony stimulating factor*

See: →Colony-stimulating factors; →Atherosclerosis; →Megakaryocytes

GnRH

Definition   *Gonadotropin-releasing hormone*

See: →Hormonal regulation of vascular cell function in angiogenesis

GPIb

Definition   *Glycoprotein Ib*

See: →Bleeding disorders; →Thrombin; →von Willebrand factor

GPIIb/IIIa

Definition   *Glycoprotein IIb/IIIa*

See: →Bleeding disorders; →Fibrin/fibrinogen; →Megakaryocytes

GP-IX

Definition   *Glycoprotein IX*

See: →Bleeding disorders; →von Willebrand factor

Graft Arterial Disease

Definition   *Obstruction of arterial lumen due to intimal thickening after organ transplantation*

See: →Fibrinolytic, hemostatic and matrix metalloproteinases, role of

Grb2

Definition   *Growth factor receptor bound-2*

See: →FGF-1, FGF-2; →Signal transduction mechanisms in vascular biology

GT

Definition   *Glanzmann's thrombasthenia*

See: →Glanzmann's thrombasthenia

GTP

Definition   *Guanosine triphosphate*

See: →Signal transduction mechanisms in vascular biology

HARP

Definition   *Heparin-affin regulatory protein*

See: →Heparin-affin regulatory protein

HB

Definition   *Hemoglobin*

HB

Definition   *Heparin binding*

HB-GAM

Definition   *Heparin-binding growth-associated molecule*

See: →Heparin-affin regulatory protein

HBD

Definition   *Heparin-binding domain*

See: →Fibrin/fibrinogen

HDL

Definition   *High density lipoprotein*

See: →Lipoproteins

Heparan Sulfates

Definition   *Glycosaminoglycan composed of disaccharide units similar to heparin. Heparan sulfates are more uniform and shorter than heparin. Heparan sulfates are connected to a protein core to form proteoheparan sulfates and are widespread at animal cell surfaces and matrices.*

See: →FGF-1 and -2; →FGF receptors; →Extracellular matrix

Heparin

Definition   *Glycosaminoglycan produced in mast cells of many tissues and composed of disaccharide units. This disaccharide units are D-glucuronique acid/L-iduronic acid and N-acethyl D-Glucosamine and have two or three sulfates per repeat unit.*

See: →FGF-1 and -2; →FGF receptors; →Thrombosis; →Heparin-affin regulatory peptide

Heparin-Affin Regulatory Peptide, HARP

Synonym: Pleiotrophin (PTN), Heparin-binding growth associated molecule (HB-GAM), Osteoblast specific factor (OSF-1)

Definition *Heparin-binding growth factor of 18 kDa size. It belongs to the midkine/pleiotropin family and is angiogenic and highly expressed in several type of tumors.*

Introduction   The development of vascular tissues constitutes an important process during the organogenesis and contributes to the maintenance of normal tissues. Two mechanisms including vasculogenesis and angiogenesis are involved in the formation of new blood vessels. It is currently accepted that vasculogenesis which constitutes the differentiation of endothelial cell precursors from mesoderm and their subsequent organisation into capillary-like tubes only occurs during early embryogenesis. Angiogenesis is defined as the formation of new capillaries from pre-existing capillaries taking place both during development and perinatal life. In addition to its key role during tissue formation, angiogenesis displays important functions in tissue homeostasis [1]. Therefore, during adulthood, angiogenesis is tightly regulated and occurs in the uterus during the estrous cycle, in the follicule during its development, in the placenta and the mammary glands during pregnancy. Angiogenesis is also involved in wound healing, and in inflammatory and proliferative processes such as proliferative retinopathy, rheumatoid arthritis and tumour development. The significant interest in tumor angiogenesis derives from the concept developed by Folkman in which primary tumours must recruit endothelial cells from the surrounding stroma to constitute its own capillaries network leading to tumor growth and metastasis [2].

Given the significance of angiogenesis in several physiological and pathophysiological processes, many studies have been devoted to the characterisation and purification of regulatory peptides that can either stimulate or inhibit angiogenesis. Furthermore, studies of these molecules strongly suggest that multiple mechanisms are involved in the regulation of angiogenesis. Among those that modulate either positively or negatively the activation of endothelial cells, are polypeptides produced by normal and tumor cells involved in the control of cellular proliferation and differentiation. These cytokines are secreted by cells and can be trapped in the extracellular matrix acting directly or indirectly on endothelial cells at very low concentrations. During the past three decades, an important number of these regulatory peptides have been isolated and characterised. Among these polypeptides, heparin-affin regulatory peptide (HARP) also named Pleiotrophin (PTN) [3] or heparin-binding growth associated molecule (HB-GAM) [4] constitutes one of these molecules involved in the mechanisms that controlled both cell proliferation and differentiation. HARP belongs to a family of heparin-binding growth factors which includes midkine (MK) [5] also called retinoic acid-d heparin-binding protein (RI-HB) [6].

Initially, HARP was isolated from perinatal rat brain and from adult bovine brain as a molecule that induces neurite outgrowth [4] suggesting that this polypeptide is involved in the neuronal physiology [7]. Other studies have shown that this molecule was also present in non-neuronal tissues including heart [8], uterus [9], cartilage [10] and bone extracts [11] suggesting that the biological activities of HARP is not restricted to the neurite outgrowth as initially reported [4].

The aim of this article is to summarise recent findings on the biology of HARP, with special focus on its role in angiogenesis according to recent advances in this field.

Characteristics

Molecular Weight   The apparent molecular weight of HARP deduced by SDS-PAGE is 18 kDa. It does not reflect the amino acid composition of the mature polypeptide since mass spectroscopy studies of the purified bovine HARP revealed a mass unit of 15,291 [8]. The aberrant mobility of HARP on SDS-PAGE, as compared to its amino acids sequence, is likely due to a high number of basic amino acids present in the molecule.

Domains   The tertiary structure of the mature protein has not yet been established; however, the arrangement of disulfide bonds of the mature protein led Hulmes et al. to propose a two domains structure, N- and C-terminal [12]. Each domain consists of a cluster of basic residues (see Table 1), disulfide bridges and one heparin-binding site. Furthermore, using circular dichroism analysis, Fabri et al. have suggested that these two domains are predominantly formed with β-sheet and random coil (63% and 23%, respectively) [13]. As shown in Table 1, angiogenic heparin-binding molecules including FGF-1, FGF-2, VEGF, HIV tat-1 and tat-2, HGF, HB-EGF, angiogenin, HARP and MK, have a cluster of basic residues. Sequence alignments of these basic domains show that HARP basic domain matches with a consensus sequence described by Albini et al. involved in heparin-binding and angiogenic activity [14].

Binding Sites and Affinity   HARP presents a strong affinity toward heparin. This property was used intensively during the purification procedures using heparin-affinity chromatography (elution at 0.9 M NaCl). Recently, Kinnunen et al. have investigated the interaction of HARP with heparin using *in vitro* competition binding assay of tritiated heparin as tracer [15]. According to these authors, a minimum of 10 monosacharide residues is required for the interaction between HARP and heparin. Furthermore, the 2-O-sulfated iduronic acid units of heparin are important in this interaction with HARP whereas glucosamine N-sulfate and 6-O-sulfate groups seems to be implicated to a lesser degree [15]. Heparan sulfate chains from proteoglycans located at the cell surface or in the extracellular matrix also present a strong binding capacity for HARP. Recent studies have described interactions between HARP and two cell surface heparan sulfate proteoglycans (HSPG), syndecan-1 and syndecan-3.

Table 1. Motifs of angiogenic heparin-binding molecules.

PROTEIN	Basic rich domain
CONSENSUS SEQUENCE	b-........BBbBb.B..BK
HARP (1-14)	1 GKKEKPEKKVKKSD
HIV2-tat (70-92)*	70 K-GLGICYERKGRRRRTPKKTKTH
HB-EGF (85-114)*	85 ATPNKEEHGKRKKKGKGLGKKRDPCLRKYK
HARP (108-132)	108 KLTKPKPQAESKKKKKEGKKQEKML
FGF-2 (116-141)*	116 RSRKYTSWYVALKRTGQYKLGSKTGPGQP
HGF (25-50)*	25 IAIPYAEGQRKRRNTIHEFKKSAKTT
VEGF (145-170)*	145 RGKGKGPKRKRKKSRYKSWSVPCGP
HIV1-tat (41-65)*	41 KGLGISYGRKKRRQRRRPPQGNQAH
MK (106-122)	106 PKTKAKAKAKKGKG-KD
MK (1-11)	1 KKKDKVKKGGP
Angiogenin (24-50)	24 RYCESIMRRRGLTSPCKDINTFIN
FGF-1 (15-42)*	15 KFNLPPGNYKKPKLLYCSNGGHFLRILP
FGF-1 (115-140)*	115KKHAEKNWFVGLKKNGSCKRGPRTHYGQK

Alignments of basic domains of angiogenic heparin-binding molecules. Amino acid is indicated using the single amino acid code and basic residue are shown in bold type." - " represents a gap. Shading residues are those matching the consensus sequence corresponding to the sequence (b.......BBbBb.B..BK) involved in heparin binding and angiogenic activity. (*) From Albini et al. [14]. B is arginine or lysine and b represents a basic residue.

Syndecan-3, also named N-syndecan, was firstly demonstrated to interact *in vitro* with HARP with an apparent K_d of 0.8 nM. This HSPG is implicated in the neurite outgrowth activity of HARP since anti-N-syndecan antibodies can inhibit this activity [16]. Using N18 neuroblastoma cells that overexpressed N-syndecan, Kinnunen et al. have recently shown that N-syndecan was distributed to the growth cones and the filipodia of the neurites in the presence of HARP [17]. In these cells, binding of HARP to N-syndecan induces neurite outgrowth through the cortactin-src kinase signalling pathway. However, this interaction between HARP and N-Syndecan is competed by FGF-2 (K_d=0.6 nM) and suggests that the same or closely-localised binding structure is involved in the binding of the two growth factors to N-syndecan. This result raises the question of a possible regulatory mechanism for neurite-promoting outgrowth or other biological activities through competition with other heparin-binding growth factors, as has been described for FGF-2 [18, 19]. Recently, syndecan-1 was reported to bind HARP and MK proteins [20] and was found to be co-distributed with these two growth factors during the development of the mouse embryo. In a similar way, the major basement membrane HSPG perlecan was reported as a binding molecule for HARP [21] and the syndecan-4, also named ryudocan, was shown to bind MK [22]. A recent report of Maeda et al. has described the binding of

HARP to 6B4, another proteoglycan also named phosphacan which represents the major chondroitin sulfate proteoglycan present in brain and which is also an extracellular variant of a receptor-like protein tyrosine phosphatase b (RPTPb) [23]. Binding studies of HARP to phosphacan revealed two sites with low (K_D=3 nM) and high (K_D=0.25 nM) affinity. As shown in this study, anti-6B4 proteoglycan antibodies were able to suppress the HARP-induced neurite outgrowth of cultured neurones *in vitro*. This last finding displays considerable importance with regard to the neurite outgrowth activity of HARP since it was recently demonstrated than phosphacan and RPTPb can interact heterophilically with the adhesion molecules N-CAM, Ng-CAM [24] and tenascin [25] which are cell surface or extracellular matrix proteins involved in regulation of both neuronal adhesion and neurite extension. The role of RPTPb in the neurite outgrowth and in the migration of embryonic rat cortical neurons induced by HARP has been newly confirmed [26]. Using *in vitro* cell system, HARP-induced neuronal migration has been strongly suppressed using antibodies raised against the extracellular domain of RPTPb, soluble RPTPb, protein phosphatase inhibitors or chondroitin sulfate C. More recently, we have shown that dermatan sulfate binds HARP with high affinity (K_d=51 nM) and modulates its mitogenic activity [27].

In the "quest" for HARP high affinity binding sites and associated signalisation pathway involved in mitogenic activity, Kuo et al. have reported binding sites (K_D=0.6 nM) in cell lines such as NIH 3T3, NRK, A431, HepG2, NB41A13 and PC12 [28]. In a similar way, Li and Deuel have detected a 195-200 kDa protein that is phosphorylated on both tyrosine and serine residues during HARP treatment of NIH 3T3 and NB41A3 cells [29]. Although no evidence has been presented showing interaction between HARP and these 195-200 kDa proteins, it is tempting to speculate that these polypeptides are directly involved in the HARP binding. It is noteworthy that all these studies have been performed using preparation of HARP that displayed very little or no mitogenic activity for the cell lines investigated. Very recently, Wellstein 's group [30] studied the signal transduction pathways leading to the mitogenic activity of recombinant HARP using bovine epithelial lens (BEL) cells. As we have previously described, these cells are very responsive to HARP with regard to its mitogenic activity [31]. In this system, Souttou et al. have demonstrated that HARP could transduce a mitogenic signal via MAP kinase and PI-3 kinase pathways. The authors have also demonstrated phosphorylation of tyrosine residues of 190 and 215 kDa proteins after HARP stimulation [30]. The apparent molecular mass of these two phosphotyrosine proteins are consistent with those described by Li and Deuel [29].

Structure

Sequence and Size

The primary structure of HARP deduced from the cDNA is a 168 amino acid polypeptide which contains a highly hydrophobic N-terminal sequence of 32 amino acids corresponding to its signal peptide sequence. The originally purified mature peptide from bovine brain and from uterus tissues corresponds to a 132 amino acid polypeptide which is cleaved downstream of this signal peptide (N-terminal sequence; NH_2-GKKEKP-) [3]. However the N-terminal sequence of the mitogenically active human recombinant HARP purified from eucaryotic expression system [32] or from conditioned media of naturally producing BEL cells (unpublished results) and NIH 3T3 cells (P. Böhlen, personal communication) begins three amino acids upstream of the predicted point of cleavage (i.e. NH_2-AEAGKKE-). The isolation of this NH_2-extended form with three additional amino acids raised an interesting question with regard to the processing of the molecule during its biosynthesis and in regards to its mitogenic activity (see biological activity). The NH_2-extended molecular form could result from an unusual cleavage site which has also been observed for MK [33].

Sequence of the mature HARP presents 24 % of basic residues that form two clusters at the N- and C-terminal regions of the molecule [34]. It also contains 10 cystein residues, but whether or not all these cysteins residues are engaged in disulfide bonds remains controversial. Hampton et al. have reported four free cystein residues [8]. In contrast, other investigators have reported that all cysteins are engaged in disulfide bonds formation [13, 28, 35]. The primary structure of HARP also contains three potential nuclear targeting sequences based on the consensus sequence K-R/K-X-R/K described by Chelsky et al. [36]. Despite the presence of these targeting sequences, localisation of HARP in the nucleus of cells has not yet been confirmed. However, Take et al. have demonstrated by ligand blotting experiments the binding of HARP to nucleolin [37]. Mature forms of HARP do not seem to be glycosylated.

Homologies

HARP shared approximately 50 % amino acid identity with Midkine (MK)/Retinoic-acid Induced-heparin binding (RI-HB) protein. The MK protein was originally identified as a product of a retinoic-acid inducible gene during midgestation period of mouse embryogenesis [5]. Its chicken homologue, the RI-HB protein was purified from chicken embryo tissues [38]. Both proteins have a strong affinity for heparin, promote neurite outgrowth and present mitogenic activity [6, 39]. According to these structural and biological data, HARP and MK/RI-HB constitute a new family of heparin-binding proteins [40, 41]. Homologies between HARP and MK toward their structural relationship and their biological activities is also supported by purification of a NH_2-extended form of MK protein from bovine follicular fluid which is mitogenic for bovine aortic smooth muscle cells [42].

Conformation

No data are actually available for this molecule.

Gene

Gene Structure

The HARP gene including rat, human and mouse has been described as a single copy ($\geq$ 42 kb), arranged at least in five exons [43-45]. The open reading frame (ORF) is encoded on 4 exons with the N-terminal signal peptide and the first 7 amino-acids of the mature protein contained in the first exon. A 5'-untranslated region (5'-UTR) exon (U1) has been reported by several groups [43, 44]. An additional 5'-UTR region (U2) has been found in the human and mouse gene [46, 47]. However, identified transcription start points are mainly localised in U1 exon. Genomic analysis is in agreement with the size of the cDNAs obtained from molecular cloning and consistent with a 1650 nt transcript.

Analysis of the human HARP promoter revealed no TATA box but the presence of a CAAT box [44, 48]. Different consensus sequences for transcription factor binding sites (MyoD, AP1, AP2, CBP, NFkB, CArG), retinoic acid responsive elements and serum response elements have also been identified. The 3'-untranslated sequence showed 3 repeats of the ATTTA motif. This sequence which is implicated in mRNA stability has been found in the 3' region of several oncogenes. The 5'- and 3'-untranslated regions of human HARP gene are highly homologous to the antisense cDNAs of heat shock protein 70 and of ribosomal protein L7, respectively [43]. The post-transcriptional regulation of human gene expression by these two regions remains to be clarified.

Chromosomal Localization   The human HARP gene is localised on chromosome 7, band q 33-34 [45, 48] and rodent genes have been mapped on chromosome 4 and 6 for mouse and rat, respectively [48, 49].

Gene Expression   HARP has been initially isolated in 1989 from early postnatal rat brain [4] and from bovine uterus [9]. As detailed in Vanderwinden et al. using *in situ* hybridisation, HARP is developmentally expressed in many neuroectodermal and mesodermal tissues but not in the endoderm, ectoderm and trophoblasts [50]. The pattern of expression suggests functions in neurogenesis, cell migration, secondary organogenesis induction and mesoderm-epithelial interactions. HARP is differentially expressed but generally down-regulated after birth suggesting a temporally regulation. However, HARP is expressed in various cancer cell lines and was proposed to play a role in an autocrine loop in mammary tumor progression [51].

Nervous system   HARP is mostly expressed in the early postnatal period in developing rat cerebral cortex [50, 52, 53] but persists during adulthood [54, 55]. The molecular layer of the rat cerebellar cortex was strongly labelled during the postnatal days 1 to 8, associated with radial fibres which could be from Golgi epithelial cells [54]. These cells seem to play a crucial role for the migration of granular cells during differentiation of the cerebellar cortex. HARP could mediate adhesion phenomena by binding to heparan-like molecules associated with neuronal membranes. In mouse, HARP mRNA are strongly expressed in neopallial cortex and midbrain [20] and could be involved in neural-glial interactions during mouse development and in mature brain [56]. Indeed, mRNA are expressed during the perinatal period in neurones as well as glial cells [57]. In adult brain, HARP is mainly expressed in hippocampal and cortical neurones [50, 55, 57]. HARP mRNA and protein are also detected in the peripheral nervous system. Mouse and chicken dorsal root ganglia as well as neuronal projections, strongly express HARP [20, 58] suggesting a putative role in the growth of sensory afferent fibres to and within the spinal cord. It is also expressed in the developing fibre tracts of the rat peripheral nervous system and, similarly to cerebellar cortex, may guide axonal process of brain neurones and promote formation of neural connections via extracellular component binding [52].

Organs and tissues undergoing epithelial-mesenchymal interactions   HARP protein is expressed in the extracellular component of established mammary cell lines like MDA-MB 231 strain [59] and is also found in the mammary gland in normal breast tissue [59-61]. Isolated from bovine uterus [9], HARP is therefore present in the urogenital system. The molecule has been identified in bovine follicular fluid [42] and pig uterine luminal fluid [62]. Cells isolated from endometrial epithelium and stroma express HARP mRNA and its expression has been observed in the myometrium of adult rat uterus [50]. In addition to myometrium, HARP mRNA have also been detected in endothelial cells and the protein has been associated to myometrium including endothelial and epithelial cells [77]. The relevance of these results are discussed in the "role in vascular biology" section. In the male reproductive system, HARP has also been found associated with prostate *in vivo* [83] and also expressed in cell culture in few epithelial cell lines originated from the prostate [64]. HARP protein is associated with the digestive and respiratory system, the sense organs, hair, whiskers, facial processes and limb buds [20].

Muscle   In the differentiating neuromuscular system, HARP has been found associated with the surface of muscles of developing rat limb [65].

Others   HARP is also expressed in cardiovascular and skeletal system and in bovine epiphysis and nasal cartilage from new-born calves [10].

Gene Regulation   HARP regulation is still poorly documented. In FGF-2-treated BALB/c 3T3 cells, J. Merenmies described a down-regulation of HARP mRNA [66]. In serum-starved NIH3T3 cells stimulated by the Platelet-derived growth factor or by FGF-2, HARP mRNA was increased [67] and a positive retinoic regulation was observed in mouse and human teratocarcinoma cells, P9 and NT2/D1 respectively [41]. However, retinoic acid failed to induce HARP gene expression in NIH3T3 cells [67]. HARP mRNA expression can also be modulated by a psychoactive component of cannabis [68] and down-regulated in osteoblast-like cells by vitamin D3 [69]. It seems that HARP may also be considered as a marker of neuronal injury [70]. Indeed, we have shown an increase of HARP mRNA expression in the central nervous system following traumatic and ischemic brain injury (in preparation). In another study, in the lesioned rat brain, an upregulation of HARP immunoreactivity occurred 5 days after lesion, with staining of the cytoplasm of astrocytes along the margin of the wound and the endothelial cells in the microvessels (in preparation). Moreover, the molecule is found in senile plaques in Alzheimer's disease and Down's syndrome [71]. HARP is also expressed in primary neuroblastomas with favourable prognosis but absent in primary aggressive neuroblastomas and cell lines [72].

HARP mRNA expression observed in the myometrium of adult rat uterus has been shown to be increased during the gestation period . More recently, such a variation of mRNA expression was also observed in our laboratory with maximal level on early oestrous and diestrous [77]. These results suggest a putative hormonal regulation of HARP expression in the uterus which is also supported by the stimulation of HARP expression by progesterone in ovariectomized animals. Hormonal regulation of HARP mRNA by either dihydrotestosterone, testosterone and estrogen has also been demonstrated in prostate epithelial cell line PNT-1A [64].

Biological Activity   Since the first study reporting the purification and the characterisation of HARP, several reports have seriously questioned the biological activity of this heparin-binding molecule. As shown in Table 2, the mitogenic activity of HARP is still largely controversial and the biological properties of this molecule have been a matter of intriguing controversies between several laboratories [8, 32, 34, 73]. Initially described as a molecule that induced neurite outgrowth of neuronal cells [4], other groups reported that HARP displayed mitogenic activity on a wide variety of cell types (see Table 2). In contrast, several other investigators reported a lack of mitogenic effect using their purified HARP preparation, claiming, without clear demonstration, that the growth promoting activity of these HARP could be due to the presence of FGFs or other related growth factors contaminating HARP preparations [8, 73]. Nevertheless, we found that, using an enzyme immunoassay [74], we were not able to detect FGF-1 or FGF-2 in our mitogenically active form of HARP isolated from adult bovine brain [75]. More recently, Wellstein's group has showed that anti-FGF 2 neutralising antibodies failed to block the proliferation of HARP-stimulated BEL cells demonstrating that, in this case, FGF-2 is not involved in the stimulation induced by HARP [30].

In order to clarify these controversial results, mitogenic activity of recombinant HARP produced in eucaryotic [3, 32, 73] as well as procaryotic [35, 76] expression systems has been investigated. As presented in Table 2, only recombinant forms of HARP produced in mammalian expression systems displayed cellular growth activity. The failure to obtain mitogenically active HARP may be explained by differences in protein translation and maturation between these different systems. As we reported in 1994 [32] and as confirmed by Wellstein's group in 1997 [30], it seems clear now that distinct molecular forms of HARP exist and could interfere differentially in the mitogenic activity [63]. This emerging possibility is in agreement with the interesting findings of Szabat [65] that showed that the short form of HARP (NH_2-GKKEKP-) strongly inhibits the proliferation of FGF-stimulated mesenchymal cells in limb buds. Similar effects have also been observed in epithelial cells [65].

Activities in neuronal and neuro-muscular systems   The first observation on the neurite outgrowth activity of HARP was reported by Rauvala [4]. Subsequent studies have established that neurite-promoting activity of HARP occurs using rat embryon cerebral neurones in a concentration ranging from 40 to 200 ng / 0.2 cm² / well [73] only

Table 2.   Mitogenic activity of HARP

Origin of the purified protein	NH_2-seq.	Mitogenicity	Tested cells	References
Native protein				
Bovine brain	n.d.	–	Endothelial cells, fibroblasts Balb c3T3	[31]
	n.d.	+	Bovine brain capillary cells	[72]
	n.d.	–	NIH 3T3	[84]
Bovine uterus	GKK	+	NIH 3T3	[9]
	n.d.	+	NRK fibroblast cells	[3]
Bovine follicular fluid	GKK	+	Bovine aortic smooth muscle cells	[39]
Murine bone	GKK	+	MC3T3.E1 (Osteoblast-like cells)	[85]
Human embryonic kidney cells	GKK	+	HUVEC, Endothelial fetal bovine heart cells	[56]
Human MDA-MB231 breast cancer cells	xKK	+	SW13 (human epithelial cells)	[56]
Recombinant protein				
COS7 cells (bovine and human cDNA)	n.d.	+	NRK fibroblast cells	[3]
NIH 3T3 cells (human cDNA)	AEA	+	Bovine brain capillary cells	[29]
	AEA	+	BEL (Bovine epithelial lens)	[28]
SW13 cells (human cDNA)	n.d.	+	BEL (Bovine epithelial lens)	[27]
SW13 and 293 cells (human cDNA)	n.d.	+	SW13 (human epithelial cells), NRK (rat fibroblast cells)	[56]
E. coli cells (murine cDNA)	GKK	–	ACE (bovine adrenal endothelial cells)	[73]
E. Coli cells (human cDNA)	n.d.	–	Bovine aortic endothelial cells, NIH 3T3	[38]
Baculovirus-infected SF9 cells (human cDNA)	n.d.	–	SW13 (human epithelial cells), NRK (rat fibroblast cells)	[56]
Baculovirus-infected SF9 cells (rat cDNA)	GKK	–	Balb/C 3T3 fibroblast, CHO, N18 neuroblastoma cells, PC12 cells, C6 glioma cells	[70]

Native and recombinant proteins have been purified from culture media except for reference [3] in which both media and lysates have been tested. (n.d., non determinated; x, used in the NH_2-terminal sequence, corresponds to a lack in the sequencing reaction).

when it is bound to the substratum. Under these conditions, neurite-outgrowth promoting effects were observed in the neurones obtained from 17-19 days-old embryos whereas slight induction was observed using cells obtained from 20-21-days-old embryos or using PC12 cells [73]. In rat brain, expression of HARP protein and mRNA is relatively low during the embryonic phase, increases to maximal levels during the early postnatal growth phase, which is characterised by extensive outgrowth of axons and dendrites, and then decreases during maturation. These data suggest that HARP may contribute to events involved in brain differentiation and organisation.

Synapse formation occurs during the perinatal stage when HARP expression peaks and is also expressed in developing muscle *in vivo*. As the molecule is present at the surface of muscle cells before the arrival of axons, HARP could favor growth cones to reach their targets and initiate synapse formation [65]. Neuromuscular junction is characterised by a high concentration in heparan sulfate proteoglycans localised in the extracellular matrix of skeletal muscles. Moreover, HARP, which binds with high affinity to heparin and HSPG, is concentrated and colocalised with acetylcholine receptors (AChR) clusters. The fact that HARP could be implicated in postsynaptic induction has also been demonstrated *in vitro* [21]. Indeed, HARP-coated beads, added to cultured Xenopus muscle cells, induces AChR clustering [21]. If muscle cells were treated with anti-HARP antibodies before seeding with spinal cord neurones, a suppression of nerve-induced clustering process was observed [21].

Role in epithelium/mesenchyme interactions   During fetal development and organogenesis, HARP could be implicated in epithelium/mesenchyme interaction. Using a polyclonal anti-HARP antibody, Mitsiadis et al. have studied its distribution during mouse embryonic development [20]. The expression of the protein in many organs undergoing morphogenesis was concomitant with epithelial differentiation. A specific localisation in basement membranes and epithelial cell surfaces of developing organs was observed. In these experiments, HARP, MK, FGFs and syndecan-1 were frequently detected colocalised in embryonic mesenchyme under active cell division. However, HARP released from HARP-coated beads in limb and jaw mesenchyme, did not stimulate cell proliferation. In aiming to clarify whether HARP is involved in cell proliferation or in cell differentiation, the developing limb model has been used [65]. Several arguments support a role of HARP in proliferation arrest: 1) the pattern of HARP expression is gradually distributed from proximal to distal limbs and is associated with the surface of growth arrest cells; 2) Exogenous HARP strongly inhibits proliferation of mesenchymal cells in limb buds, as well as epithelial cells (but to a lesser extent). This inhibitory activity of proliferation could be explained by a competition between FGF-2 and HARP in the binding to the heparan sulfate chains that deprive FGF of its essential carbohydrate

sites required for its mitogenic activity. So HARP expression seems to be linked to proliferation arrest and myogenic differentiation.

Role in genital tract   In genital tract, MK and HARP were isolated from the bovine follicular fluid and were reported to be weakly mitogenic for bovine aortic smooth muscle cells [42]. HARP was purified from uterine luminal flushings from non pregnant adult pigs and 3 forms (25 kDa, 18 kDa and 14 kDa) were characterised by Western blotting experiments using an affinity-purified antiserum raised against the N-terminal 14 amino acids. The lower form seems C-terminally truncated [62]. Since several proteins that function in the maintenance of early pig pregnancy are also produced during the oestrus cycle, HARP could also be implicated in this function. In agreement with these data, HARP expression during estrous cycle in rat uterus was studied in our laboratory. We demonstrate that HARP expression peaks at early estrus and diestrus and is up-regulated by progesterone which was the most important steroid hormone during the early events of pregnancy [77].

Role in Vascular Biology
Physiological Function   HARP, purified from adult bovine brain, was first described as a mitogenic factor for endothelial cells [75, 78] and its role in angiogenesis is now supported by several recent reports.

Human HARP from transfected human embryonal kidney cell line 293 was found mitogenic for fetal bovine heart (FBHE) and human umbilical vein (HUVEC) endothelial cells . We have also demonstrated angiogenic activity of human recombinant HARP produced from transfected NIH 3T3. Using an *in vitro* assay described by Montesano et al., [79], exogenous HARP (50 ng/ml), like FGF2 (30 ng/ml), was found to induce migration of aortic bovine endothelial cells in collagen gels and to form pseudo-capillary structures [32]. Using HARP-targeting ribozyme, it was shown that angiogenesis induced by HARP can be suppressed [80]. Therefore, in melanoma cells that constituvely express a high level of the molecule, Czubayko et al. showed that the reduction of HARP did not affect cell growth *in vitro*, but tumor growth and angiogenesis (evaluated as the number of blood vessels) were concomitantly decreased with HARP expression in athymic mice [81]. The metastatic spread of the tumors was prevented while apoptosis in the tumour was increased. In addition to the stimulation of HUVEC growth by HARP, Choudhuri et al. confirmed the *in vivo* role of the molecule [82]. Using HARP-overexpressing MCF-7 breast carcinoma cells and xenograft experiments, the protein was found to enhance tumour growth, endothelial proliferation and vascular density. In this study, a strong angiogenic response was also demonstrated using the rabbit corneal assay [82].

HARP localisation also corroborates with its role in angiogenesis. Indeed, HARP (mRNA and protein) are expressed in endothelial cells from human mammary glands [60], rat endometrium [77] and human prostate [83]. We also showed that HARP is secreted in the conditioned medium from bovine adrenal capillary endothelial cells in culture (unpublished results). The presence of HARP in endothelial cells is also suspected since intravenous heparin injection allows HARP detection in the plasma [84]. Moreover, chemically-synthesised C-domain of HARP (43 amino acids) which could form a compact structure, enhanced plasminogen activator activity and decreased plasminogen inhibitor (PAI-1) levels in bovine aortic endothelial cells [85].

More recently [86], HARP gene expression was shown to be upregulated in macrophages, astrocytes and endothelial cells in areas of developing neovasculature after focal cerebral ischemia in adult rat [86].

Pathology   HARP has not yet been directly linked to pathology. Studies on knockout animal models or measures of HARP level in pathological tissue or fluids are still needed for establishing links to pathologies. However, a few reports describe HARP expression in pathologies like Alzheimer's disease (see "gene regulation" section) but the importance of the molecule has not yet been established.

Clinical Relevance and Therapeutic Implications   Isolated from conditioned medium of highly tumorigenic human breast cancer line MDA-MB-231 [51], HARP has been suspected as a growth factor involved in the development of human breast cancer. In this context, using RNAse protection assay and Northern blot analysis, additional studies support this hypothesis by showing that HARP was expressed in the majority of the breast cancer [59]. However, the precise pattern of HARP expression in normal tissue has not been clearly established. More recently, comparison of HARP expression in breast cancer tissues versus normal tissues have been performed using RT-PCR [61]. This study shows that expression of HARP was equivalent in both normal and malignant human breast tissues in contradiction with another study in which it was stated that HARP mRNA was not detected in normal human mammary tissue [59]. However, this report did not mention the origin of the normal tissues and the number of the biopsies analysed. In addition, both studies were performed using transcripts isolated from human resected tissues composed of a wide variety of cell types. No information was provided on the cellular distribution of HARP mRNA. In order to work out this controversy and analyse the potential role of this growth factor in mammary gland, we have investigated the cellular localisation of HARP mRNA as well as HARP polypeptide in the normal human mammary gland taken from areas opposite to invasive adenocarcinomas (n=11) or from reduction mammoplasties (n=6) [60] as well as in invasive adenocarcinomas (n=7) (unpublished data). We found that HARP protein and its mRNA were present in both normal and malignant human mammary gland and localised both in myoepithelial cells and in endothelial cells. In agreement with the results reported by Garver, these observations suggest that the direct role of HARP in breast cancer progression appears less clear than previously suggested by Fang [59]. In our group, we have also analysed the

involvement of HARP molecule in human prostate cancer, since, as already mentioned, HARP expression is regulated by androgen [64]. In preliminary experiments, we performed RNase protection assay on biopsies from 4 normal human prostates and on resected tissues from 7 cases of benign prostate hyperplasia and 6 cases of prostate cancer. HARP transcripts were always detected in the samples, but no correlation was found between the level of HARP mRNA expression and the type of investigated tissue. However, in order to study the role of HARP in the prostate, we investigated the cellular localisation of HARP mRNA by *in situ* hybridation on frozen sections of human prostate tissue. HARP mRNA was localised to the fibro-muscular stroma in both normal tissue, benign hyperplasia and prostate cancer. No HARP mRNA was detected in epithelial cells. In contrast, while no staining corresponding to HARP protein was observed in normal epithelial cells, strong staining was observed in prostate cancer epithelium. This result suggests that HARP may act in a paracrine manner, secreted by stromal mesenchymal cells and stimulating cancer epithelial cells. Morever, HARP induced the formation of colonies in soft agar assay using SW13 or NRK cells [59].

Experimental proof that HARP can act as a tumor and angiogenic growth factor is supplied from *in vivo* studies which indicate that injecting athymic nude mice with cells that overexpress HARP causes the development of highly vascular tumors [87-89]. In addition, involvement of HARP in tumor development has been investigated by targeting HARP mRNA with ribozyme that inactivate HARP by generating dominant negative mutants. Using melanoma cells that expressed a high level of HARP, Wellstein's group showed that a decrease of HARP expression reduced tumor angiogenesis and, subsequently, tumor growth and its metastatic capacity. A similar observation was made using a mutant cDNA that encoded a truncated form of HARP which heterodimerized with endogenous HARP acting as a dominant negative effector of HARP [87]. These authors established that expression of the mutant gene product prevented transformation of cells like MDA-MB 231 cells that expressed endogenously HARP. These studies indicated that the maintenance of transformed phenotype of the highly malignant human breast cancer cell line MDA-MB 231 was dependent on the constitutive expression of HARP. Furthermore, the ability to suppress cell transformation with a dominant negative suggests a key role of HARP in the regulation of neoplastic transformation [81, 87].

Further studies on HARP molecule including molecular structure, receptor characterisation, biosynthesis, regulation of its expression and transduction pathways as well as the involvement of this molecule in angiogenesis remain to be performed. Further studies in these fields may open up new strategies in the development of drugs for clinical applications in which HARP is involved.

José Courty, Pierre Emmanuel Hilhiet, Jean Delbé,
Danièle Caruelle and Denis Barritault

References

1. Pepper M (1997) Vasc Med 1:253-266
2. Folkman J (1972) Ann Surg 175:409-416
3. Li YS et al (1990) Science 250:1690-1694
4. Rauvala H (1989) EMBO J 8:2933-2941
5. Kadomatsu K et al (1988) Biochem Biophys Res Commun 151:1312-1318
6. Raulais D et al (1991) Biochem Biophys Res Commun 174:708-715
7. Merenmies J et al (1990) J Biol Chem 265:16721-16724
8. Hampton BS et al (1992) Mol Biol Cell 3:85-93
9. Milner PG et al (1989) Biochem Biophys Res Commun 165:1096-1103
10. Neame PJ et al (1993) J Orthopaed Res 11:479-491
11. Gieffers C et al (1993) Eur J Cell Biol 62:352-361
12. Hulmes JD et al (1993) Biochem Biophys Res Commun 192:738-746
13. Fabri L et al (1992) Biochem Int 28:1-9
14. Albini A et al (1996) Oncogene 12:289-297
15. Kinnunen T et al (1996) J Biol Chem 271:2243-2258
16. Raulo E et al (1994) J Biol Chem 269:12999-13004
17. Kinnunen T et al (1998) J Biol Chem 273:10702-10708
18. Walker A et al (1994) J Biol Chem 269:931-935
19. Rapraeger AC et al (1991) Science 252:1705-1708
20. Mitsiadis TA et al (1995) Development 121:37-51
21. Peng HB et al (1995) J Neurosci 15:3027-3038
22. Kojima T et al (1996) J Biol Chem 271:5914-5920
23. Maeda N et al (1996) J Biol Chem 271:21447-21452
24. Milev P (1994) J Cell Biol 127:1703-1715
25. Grumet M et al (1994) J Biol Chem 269:12142-12146
26. Maeda N et al (1998) J Cell Biol 142:203-216
27. Vacherot F et al (1999) J Biol Chem 274:7741-7747
28. Kuo MD et al (1992) Biochem Biophys Res Commun 182:188-194
29. Li YS et al (1993) Biochem Biophys Res Commun 195:1089-1095
30. Souttou B et al (1997) J Biol Chem 272:19588-19593
31. Delbé J et al (1995) J Cell Physiol 164:47-54
32. Laaroubi K et al (1994) Growth Factors 10:89-98
33. Muramatsu H et al (1993) Dev Biol 159:392-402
34. Böhlen P et al (1991) Growth Factors 4:97-107
35. Seddon AP et al (1994) Prot Express Purif 5:14-21
36. Chelsky D et al (1989) Mol Cell Biol 9:2487-2492
37. Take M et al (1994) J Biochem 116:1063-1068
38. Vigny M et al (1989) Eur J Biochem 186:733-740
39. Muramatsu H et al (1991) Biochem Biophys Res Commun 177:652-658
40. Laaroubi K et al (1995) Progr in Growth Factor Res 6:25-34
41. Kretschmer PJ et al (1991) Growth Factors 5:99-114
42. Ohyama Y et al (1994) Mol Cell Endocrinol 105:203-208
43. Lai S et al (1992) Biochem Biophys Res Commun 187:1113-1122
44. Kretschmer PJ et al (1993) Biochem Biophys Res Commun 192: 420-429
45. Milner PG et al (1992) Biochemistry 31: 12023-12028
46. Katoh K et al (1992) DNA & Cell Biol 11:735-743
47. Lai S et al (1995) Gene 153:301-302
48. Li YS et al (1992) J Biol Chem 267:26011-26016
49. Hornum L et al (1996) Mamm Genome 7:923-924
50. Vanderwinden JM et al (1992) Anat & Embryology 186:387-406
51. Wellstein A et al (1992) J Biol Chem 267: 2582-2587
52. Rauvala H et al (1994) Dev Brain Res 79:157-176
53. Matsumoto K et al (1994) Neurosci Lett 178:216-220
54. Wewetzer K et al (1995) Brain Res 693:31-38
55. Bloch B et al (1992) Dev Brain Res 70:267-278
56. Silos-Santiago I et al (1996) J Neurobiol 31:283-296
57. Wanaka A et al (1993) Dev Brain Res 72:133-144
58. Nolo R et al (1996) Eur J Neurosci 8:1658-1665
59. Fang W et al (1992) J Biol Chem 267:25889-25897
60. Ledoux D et al (1997) J Histochem Cytochem 45:1239-1245
61. Garver RI, Jr. et al (1994) Cancer 74:1584-1590
62. Brigstock DR et al (1996) J Endocrinol 148:103-111
63. Zhang L et al (1995) J Cell Sci 108:323-331
64. Vacherot F et al (1995) In Vitro Cell & Dev Biol 31:647-648
65. Szabat E et al (1996) Dev Biol 178:77-89
66. Merenmies J (1992) FEBS Lett 307:297-300
67. Li YS et al (1992) Biochem Biophys Res Commun 184:427-432
68. Mailleux P et al (1994) Neurosci Lett 175:25-27
69. Tamura M et al (1994) Endocrinology 3:21-24
70. Takeda A et al (1995) Neuroscience 68:57-64
71. Wisniewski T et al (1996) Neuroreport 7:667-671
72. Nakagawara A et al (1995) Cancer Res 55:1792-1797
73. Raulo E et al (1992) J Biol Chem 267:11408-11416
74. Caruelle D et al (1988) Anal. Biochem. 173:328-339
75. Courty J et al (1991) Biochem Biophys Res Commun 180:145-151
76. Takamatsu H et al (1992) Biochem Biophys Res Commun 185:224-230
77. Milhiet PE et al (1998) J Endocrinol 158:389-399
78. Böhlen P et al (1988) J Cell Biochem (Suppl) 12A:221
79. Montesano R et al (1986) Proc Natl Acad Sci USA 83:7297-7301
80. Czubayko F et al (1995) Breast Cancer Res & Treatment 36:157-168
81. Czubayko F et al (1996) Proc Natl Acad Sci USA:14753-14758
82. Choudhuri R et al (1997) Cancer Res 57:1814-1819
83. Vacherot F et al (1999) Prostate 38:126-136
84. Novotny WF et al (1993) Arteriosclerosis & Thrombosis 13:1798-1805
85. Kojima S et al (1995) Biochem Biophys Res Commun 206:468-473
86. Yeh HJ et al (1998) J Neurosci 18:3699-3707
87. Zhang N et al (1997) J Biol Chem 272:16733-16736
88. Czubayko F et al (1994) J Biol Chem 269:21358-21363
89. Chauhan AK et al (1993) Proc Natl Acad Sci USA 90:679-682

Heparin-Induced Thrombocytopenia (HIP)

Definition *Thrombocytopenia due to an immunological mechanism involving heparin-dependent antibodies*

See: →Platelet stimulus-response coupling

Hepatocyte Growth Factor (HGF)

Synonym: Scatter factor

Definition *Mitogen for hepatocytes, epithelial cells and endothelial cells. HGF induces dissociation/motility of epithelial cells. HGF is also involved in the maturation of* *hematopoietic progenitors, liver regeneration, kidney tubule formation and mammary gland development.*

Introduction   HGF/Scatter Factor owes its double name to the separate discovery of the abilities to induce growth of hepatocytes and dissociation/motility of epithelial cells ("scattering") [1-3]. More recently it became clear that the specific genetic program activated by HGF regulates complex morphogenetic processes including extracellular matrix invasion, cell migration, cell polarization and tubulogenesis and angiogenesis (reviewed in [4]). HGF is the prototype of a family of soluble factors, including at least one other member, originally isolated with the name of macrophage stimulatory protein (MSP), which regulates chemotaxis and phagocytosis of macrophages (reviewed in [4]). HGF is secreted by several mesenchymal-derived cells, while the specific receptor is expressed on a variety of target cells. Although HGF is normally found in serum, it is thought to work predominantly in a paracrine mode. *In vitro* HGF is a powerful mitogen and motogen of hepatocytes, other epithelial cells [1,5] and endothelial cells [6,7], regulates the maturation of hematopoietic progenitors [8], and enhances the synthesis of enzymes involved in the degradation of extracellular matrix [9]. Furthermore, HGF induces epithelial and endothelial cell-growth in a three-dimensional gel to form respectively tubular and capillary-like structures [6,10,11]. *In vivo,* HGF is a mediator of kidney and liver regeneration (reviewed in [4]), promotes the migration of myoblast during embryogenesis [12], coordinates the formation of kidney tubules from metanephric tissue [13] and the development of the mammary gland during pregnancy and lactation [11].
These features indicate that HGF plays a relevant role in differentiation and vascularization during organogenesis.

Characteristics
Molecular Weight   90 kD (the molecular weight can be modified by glycosilation from 87 kD to 101 kD)

Domains   HGF is a disulphide-linked heterodimers of a a subunit of 55-65 kD and a b subunit of 32-36 kD [2,3] . The a subunit contains four typical N-terminal kringle domains, similar to those observed in plasminogen. The exact function of kringles is presently unclear; they comprise a cysteine-rich double looped structure, acting as protein-protein interaction motif. A short sequence folded in a hairpin loop has been identified that comprises the receptor binding site. The a subunit is also expected to contain the heparin binding site, although its sequence is still undefined.
The b subunit is closely related to the catalytic domain of serine proteases; however, the serine residue of the active site is substituted with a tyrosine. Therefore, HGF, although sharing with plasminogen structural homology and activation mechanism, has lost protease activity during evolution.

Binding Sites and Affinity   Both mitogenic and motogenic responses to HGF are mediated through a high affinity receptor (K_d=50-200 pM), encoded by *Met* gene

(reviewed in [4]). Furthermore, HGF binds cell surface with a low affinity interacting with heparansulphates [3] and these binding sites seem to be relevant in biological responses [14,15]. MET, with SEA [16], an orphan receptor, and RON, the receptor of macrophage stimulating protein [17], belongs to a distinct subfamily of single-pass cell membrane tyrosine kinase receptor. *Met* gene encodes a 170 kD precursor that is glycosilated to form a 190 kD species, and then proteolytically cleaved into two distinct disulfide linked subunits. The activating protease is not known; the specific sequence of the cleavage site, however, suggests that they should belong to the "furin" family [18]. The two subunits of these receptors (a: 50 kD, and b: 145 kD) are both necessary for the biological activity, although the minimal binding site for the factor has not been mapped yet. The extracellular domains do not contain obvious protein patterns, but comprise a number of cysteines, well aligned in their relative positions. It has been identified as an eight-cysteine-motif, called MRS (for MET Related Sequence), which interestingly is also observed in a family of putative receptors similar to MET: the SEX protein family [19].

Additional Features   The intracellular domain of MET receptor includes a well-conserved tyrosine kinase catalytic domain, flanked by distinctive juxtamembrane and C-tail sequences. In MET, the phosphorylated tyrosine residues in position 1234-1235 have a positive regulatory-effect on the enzyme activity. Upon ligand stimulation, two receptors are probably induced to dimerize and to activate each other by transphosphorylating these regulatory tyrosines [4].
The tyrosine kinase activity of MET is negatively regulated by the phosphorylation of a serine residue in the juxtamembrane sequence [20]. The C-tail domain of MET receptor is of importance for biological activity. It includes two tyrosine residues, that, when phosphorylated, together form a specific docking site for multiple signal transducers and adaptors, such as GRB-2, phosphatidylinositol 3-kinase, phospholipase C γ, STAT3 and Shc. If these docking tyrosine residues are mutated, signal transduction does not ensue [21].
Most of the transducers of MET bind the phosphotyrosine of the receptor via SH2 domains. Among them, GRB-2 has the highest affinity for tyrosine[1356] and a pivotal role, being the adaptor molecule for SOS-1 and the initiator of the RAS transduction pathway, leading to mitogen-activated protein kinase activation. This pathway is essential for the biological activity of the receptor, although insufficient to explain the full specific effects of MET activation [22]. Phosphatidylinositol 3-kinase, RAC and BAG-1 are other molecules associated with the cytoplasmic domain of the HGF-receptor [23, 24] and are involved in the control of cytoskeleton and in the anti-apoptotic effects triggered by HGF.

Structure
Sequence and Size   Gene Bank accession code: M29145
Size of human HGF mRNA: 2756 bp

```
   1  gggctcagag ccgactggct cttttaggca ctgactccga acaggattct ttcacccagg
  61  catctcctcc agagggatcc gccagcccgt ccagcagcac catgtgggtg accaaactcc
 121  tgccagccct gctgctgcag catgtcctcc tgcatctcct cctgctcccc atcgccatcc
 181  cctatgcaga gggacaaagg aaaagaagaa atacaattca tgaattcaaa aaatcagcaa
 241  agactaccct aatcaaaata gatccagcac tgaagataaa aaccaaaaaa gtgaatactg
 301  cagaccaatg tgctaataga tgtactagga ataaaggact tccattcact tgcaaggctt
 361  ttgttttttga taaagcaaga aaacaatgcc tctggttccc cttcaatagc atgtcaagtg
 421  gagtgaaaaa agaatttggc catgaatttg acctctatga aaacaaagac tacattagaa
 481  actgcatcat tggtaaagga cgcagctaca agggaacagt atctatcact aagagtggca
 541  tcaaatgtca gccctggagt tccatgatac cacacgaaca cagcttttttg ccttcgagct
 601  atcggggtaa agacctacag gaaaactact gtcgaaatcc tcgaggggaa gaaggggggac
 661  cctggtgttt cacaagcaat ccagaggtac gctacgaagt ctgtgacatt cctcagtgtt
 721  cagaagttga atgcatgacc tgcaatgggg agagttatcg aggtctcatg gatcatacag
 781  aatcaggcaa gatttgtcag cgctgggatc atcagacacc acaccggcac aaattcttgc
 841  ctgaaagata tcccgacaag ggctttgatg ataattattg ccgcaatccc gatggccagc
 901  cgaggccatg gtgctatact cttgaccctc acacccgctg ggagtactgt gcaattaaaa
 961  catgcgctga caatactatg aatgacactg atgttccttt ggaaacaact gaatgcatcc
1021  aaggtcaagg agaaggctac aggggcactg tcaataccat ttggaatgga attccatgtc
1081  agcgttggga ttctcagtat cctcacgagc atgacatgac tcctgaaaat ttcaagtgca
1141  aggacctacg agaaaattac tgccgaaatc cagatgggtc tgaatcaccc tggtgttttta
1201  ccactgatcc aaacatccga gttggctact gctcccaaat tccaaactgt gatatgtcac
1261  atggacaaga ttgttatcgt gggaatggca aaaattatat gggcaactta tcccaaacaa
1381  tctgggaacc agatgcaagt aagctgaatg agaattactg ccgaaatcca gatgatgatg
1441  ctcatggacc ctggtgctac acgggaaatc cactcattcc ttgggattat tgccctattt
1501  ctcgttgtga aggtgatacc acacctacaa tagtcaattt agaccatccc gtaatatctt
1561  gtgccaaaac gaaacaattg cgagttgtaa atgggattcc aacacgaaca aacataggat
1621  ggatggttag ttttgagatac agaaataaac atatctgcgg aggatcattg ataaaggaga
1681  gttgggttct tactgcacga cagtgtttcc cttctcgaga cttgaaagat tatgaagctt
1741  ggcttggaat tcatgatgtc cacggaagag gagatgagaa atgcaaacag gttctcaatg
1801  tttcccagct ggtatatggc cctgaaggat cagatctggt tttaatgaag cttgccaggc
1861  ctgctgtcct ggatgatttt gttagtacga ttgatttacc taattatgga tgcacaattc
1921  ctgaaaagac cagttgcagt gtttatggct ggggctacac tggattgatc aactatgatg
1981  gcctattacg agtggcacat ctctatataa tgggaaatga gaaatgcagc cagcatcatc
2041  gagggaaggt gactctgaat gagtctgaaa tatgtgctgg ggctgaaaag attggatcag
2101  gaccatgtga ggggggattat ggtggcccac ttgtttgtga gcaacataaa atgagaatgg
2161  ttcttggtgt cattgttcct ggtcgtggat gtgccattcc aaatcgtcct ggtattttttg
2221  tccgagtagc atattatgca aaatggatac acaaaattat tttaacatat aaggtaccac
2281  agtcatagct gaagtaagtg tgtctgaagc acccaccaat acaactgtct tttacatgaa
2341  gatttcagag aatgtggaat ttaaaatgtc acttacaaca atcctaagac aactactgga
2401  gagtcatgtt tgttgaaatt ctcattaatg tttatgggtg ttttctgttg ttttgtttgt
2461  cagtgttatt ttgtcaatgt tgaagtgaat taaggtacat gcaagtgtaa taacatatct
2521  cctgaagata cttgaatgga ttaaaaaaac acacaggtat atttgctgga tgataa
```

```
MWVTKLLPALLLQHVLLHLLLLPIAIPYAEGQRKRRNTIHEFKKSAKTTLIKIDPALKI
KTKKVNTADQCANRCTRNKGLPFTCKAFVFDKARKQCLWFPFNSMSSGVKKEFGHE
FDLYENKDYIRNCIIGKGRSYKGTVSITKSGIKCQPWSSMIPHEHSFLPSSYRGKDLQE
NYCRNPRGEEGGPWCFTSNPEVRYEVCDIPQCSEVECMTCNGESYRGLMDHTESGKI
CQRWDHQTPHRHKFLPERYPDKGFDDNYCRNPDGQPRPWCYTLDPHTRWEYCAIKT
CADNTMNDTDVPLETTECIQGQGEGYRGTVNTIWNGIPCQRWDSQYPHEHDMTPEN
FKCKDLRENYCRNPDGSESPWCFTTDPNIRVGYCSQIPNCDMSHGQCYRGNGKNY
MGNLSQTRSGLTCSMWDKNMEDLHRHIFWEPDASKLNENYCRNPDDDAHGPWCYT
GNPLIPWDYCPISRCEGDTTPTIVNLDHPVISCAKTKQLRVVNGIPTRTNIWMVSLRY
RNKHICGGSLIKESWVLTARQCFPSRDLKDYEAWLGIHDVHGRGDEKCKQVLNYDG
LLRVAHLYIMGNEKCSQHHRGKVTLNESEICAGAEKIGSGPCEGDYGGPLVCEQHK
MRMVLGVIVPGRGCAIPNRPGIFVRVAYYAKWIHKIILTYKVPQS
```

Homologies   Phylogeny of the serine proteinase domains and analysis of intron-exon boundaries and kringle sequences indicate that HGF, MSP, plasminogen and apolipoprotein have evolved from a common ancestral gene that consisted of an N-terminal domain corresponding to plasminogen activation peptide, 3 copies of the kringle domain, and a serine proteinease domain.

Conformation

The X-ray analysis of HGF crystal has not yet been done. A model of tertiary structure of b subunit (the serine protease-like domain) is based on the crystal structures of some proteases [25]. In the globular structure of this subunit, two aminoacids (Glutamine^{534} and Tyrosine^{673}) replace Histidine^{57} and Serine^{195} of the catalytic triade of serine proteases. Valine^{692} of HGF replaces Serine^{214} of proteases, which participates to the catalytic pocket. A distinct pattern of charged and hydrophobic residues in the helix-strand-helix motif characterizes the b subunit and could be important for receptor interaction. Three-dimensional models of kringle and proteinase domains suggest a mechanism for the formation of a non-covalent HGF homodimer, that may be responsible for receptor activation [26].

Additional Features   The identification of the minimal biologically active sequences of HGF has been done by the use of mutants. A first HGF variant consists of the first two NH$_2$ terminus kringles and is the naturally occurring product of an alternatively spliced HGF mRNA [27]. This protein fails to stimulate cell proliferation and blocks the mitogenic activity of HGF, but maintains motogenic activity [28]. A HGF variant, truncated after the first kringle domain , also blocks mitogenic activity of HGF [25], but at high concentration it promotes receptor phosphorylation, and the presence of heparin converts it to partial mitogenic agonists [15]. Single or double mutations in the protease-like domain of the b subunit result in variants that are completely defective for mitogenic activity, yet exhibit binding affinities similar to HGF [25].

Gene

Gene Structure   The human HGF-encoding gene is composed of 18 exons and 17 introns, spans about 70 Kb. The first exon contains the 5'-untranslated region and the signal peptide. The next ten exons encode the a subunit. Each kringle domain of this subunit is encoded by two exons observed in plasminogen. The twelfth exon contains the short spacer between the a and b subunits and the remaining six exons comprise the b subunit [29].

Chromosomal Localization   The human gene encoding HGF is located on chromosome 7 (q11.2-q21.1).

Gene Expression   In adult tissues, HGF is widely expressed in mesenchymal cells (reviewed in [4]). During gastrulation, HGF is selectively expressed in the endoderm and in the mesoderm. At stage E13, it contributes to the development of epithelial organs and at E18 is widely distributed in developing epithelia, limb buds and neural tissue [30]. The complete deletion of the gene causes lethality at E15 with hepatic anomalies [31]. During embryogenesis, HGF seems to be relevant in the development of myocytes and muscles. Chimaeric HGF homozygous mice survived for a few days after birth and show an underdevelopment of upper body muscles [31], due to the deficient migration of the myoblast precursors . In wild mice the migration of these cells depends on the HGF secreted by stromal cells. HGF also dissociates epithelial cells to allow the migration of myogenic precursors. HGF and its receptor are also expressed transiently in premyocardium, but not in heart progenitor cells. The expression persists through the first looping stage when the heart morphology begins to elaborate [32].

Gene Regulation   The promoter region of mouse HGF gene contains a non canonical TATA box (ATAAA). In the regulatory region there are a number of putative regulatory elements, such as four interleukin-6 (IL-6) response elements, two potential binding sites for NF-IL-6, a transforming growth factor β inhibitory element, a cAMP response element, two estrogen response elements, a potential vitamin D response element, two liver-specific transcription factor binding sites, and a B cell and macrophage-specific transcriptional factor binding site [33]. A cell-type-specific transcriptional repressor in the promoter region has been identified. This region is the binding site for a nuclear protein present in epithelial cells, but not in mesenchymal cells and is evidently responsible for the suppression of gene expression in this cell type [34]. More recently, it has been demonstrated that p53, but not mutated proteins, activates the promoter [35]. Several cytokines, steroids, growth factors up-regulate HGF gene expression and its synthesis and release. They include basic fibroblast growth factor (bFGF), interleukin-1 and 6, tumor necrosis factor, estradiol and 1,25 dihydroxyvitamin D3 (reviewed in [36]). Increase of HGF expression is also observed after tissue (kidney and liver) injury (reviewed in [4]).

Processing and Fate   HGF is secreted by mesenchymal cells as a biologically inactive single-chain precursor (92 kD) and is cleaved (Arginine^{494}-Valine^{495}) to mature heterodimer in the extracellular environment by serine proteases [37, 38]. Plasminogen activator urokinase irreversibly binds HGF forming a stochiometric 1:1 complex in which the precursor becomes activated. Another HGF activating enzyme has been identified and is similar to coagulation factor XII [39]. This molecule has a molecular weight of 34 kD and consists of two chains held together by a disulfide bond. This HGF activator derives from a zymogen which is activated by thrombin by the cleavage of the bond between Arginine^{407} and Isoleucine^{408} [38].

Biological Activity   Subnanomolar concentrations of HGF promotes the proliferation of several epithelial cells, melanocytes and endothelial cells. In addition to its mitogenic activity, HGF stimulates cellular motility which can be manifested as scattering of cells that grow in tight clusters or as enhanced movement across membranes in a Boyden chamber (reviewed in [4]). HGF is also cytotoxic for some tumour cells (reviewed in [4]). HGF is able to protect epithelial cells from apoptosis induced in several experimental conditions ([24]; reviewed in [4]).

Primary hepatocytes derived from transgenic mice expressing constitutively activate HGF receptor become immortalized (reviewed in [4]). They are protected by apoptosis induced by anti-FAS antibodies. The effect of HGF on cell movement is complex. Firstly, it reorganizes the cytoskeleton and dissociates cell-cell junctions. However, the full motogen phenotype requires RNA transcription and synthesis of proteinases that dissolve extracellular matrix and permit invasion [9]. Following their dissociation, the cells start to proliferate as if a survival/growth program had been concurrently activated. The current hypothesis underlines that HGF may be important in local stromal-epithelial/endothelial communication. Notably, HGF receptor is expressed on the basolateral surface where HGF secreted by stromal cells accumulates (reviewed in [4]).

Complex morphogenetic processes have been reconstituted *in vitro* by growing epithelial cells (derived from kidney tubules, colon, prostate, lung mammary gland) in collagen gel [10]. In these conditions, most cells form spheroids and become susceptible to apoptosis. The addition of HGF promotes the growth and cell sprouting to form branch tubules. HGF also exerts an autocrine/paracrine stimulation of macrophages and osteoclasts. For instance murine macrophages stimulated by HGF release platelet-activating factor [40] and osteoclasts proliferate [41]. HGF stimulates growth and differentiation of erythroid precursors [8].

Role in Vascular Biology

Physiological Function   By an autocrine/paracrine mechanism, HGF activates in endothelial cells a genetic program related to angiogenesis and inflammation. Endothelium expresses and produces HGF, but the level of expression is not uniform. It is expressed in sinusoidal endothelial cells and is up-regulated after liver injury [42, 43]. HGF is also strongly localized to the endothelial cells lining the villous vasculature and the vasculosyncytial membrane [44]. *In vitro* endothelial cells produce spontaneously HGF, which is decreased by high concentrations of glucose [45] and increased by heparin [46]. HGF has been shown to stimulate *in vitro* HGF receptor activation on endothelial cells, their motility (chemotaxis) and growth [6, 47, 48]. *In vitro* it also promotes angiogenesis and the repair of wounded endothelial monolayer [6, 49]. *In vivo*, when injected in rabbit cornea, HGF promotes neovascularization without inflammation [6]. The angiogenic response to HGF has also been observed in an experimental model system of implanted reconstituted membrane [7, 14]. In this system, the full biological response is elicited by wild type HGF, but not by truncated molecules and requires the receptor kinase activation [14]. The naturally occurring variant consisting of the NH$_2$-terminal hairpin and the first kringle domain stimulates *in vitro* angiogenesis. A similar truncated form with a second kringle is devoid of this activity [50]. The expression of HGF and HGF receptor in developing chick embryo is consistent with their role in angiogenesis rather than in vasculogenesis

[51]. HGF exerts *in vitro* proliferative effects on endothelial cells in an additive manner with FGF [52]. It has been recently reported that HGF stimulates the production of VEGF-A which potentiates the angiogenic activity of HGF [14,53]. Macrophages challenged with HGF release platelet activating factor, a powerful motogen of endothelial cells, which potentiates the *in vivo* angiogenic effect of HGF [40].

HGF is also involved in the regulation of adhesion molecules on an endothelial surface. It up-regulates CD44, E-selectin, $\alpha_2\beta_1$ integrin, but is ineffective on expression of ICAM-1 and VCAM [6, 54, 55], suggesting a selected role of HGF in cell transmigration.

Pathology   The angiogenic effect of HGF is relevant in vascularization of some solid tumours and in metastasis dissemination. HGF plays a role in angiogenesis associated to Kaposi's sarcoma [56], glial tumours [57], endometrial carcinomas [58] and pleural mesotheliomas [59]. HGF transgenic mice develop a broad array of histologically distinct tumours of both mesenchymal and epithelial origin [60]. HGF increases the adhesion of cancer cells to the endothelium and facilitates their migration to extravascular tissues [54,55].

There is also evidence that HGF is involved in vascular remodeling and angiogenesis in liver cirrhosis [43], in inflammatory arthritis [61] and in a model of myocardial ischemia and reperfusion [62]. It has been recently demonstrated that HGF serum levels correlate with the severity of hypertension [63].

Federico Bussolino and Paolo Comoglio

References

1. Stoker M et al (1987) Nature 327:238-242
2. Nakamura T et al (1989) Nature 342:440-443
3. Naldini L et al (1991) EMBO J 10:2867-2878
4. Tamagnone L et al (1997) Cytokine & Growth Factor Rev (in press)
5. Nakamura T et al (1986) Proc Natl Acad Sci USA 86:6489-6493
6. Bussolino F et al (1992) J Cell Biol 119:629-641
7. Grant DS et al (1993) Proc Natl Acad Sci USA 90:1937-1941
8. Galimi F et al (1994) J Cell Biol 127:1743-1754
9. Pepper et al (1992) J Biol Chem 267:20493-20496
10. Montesano R et al (1991) Cell 67:901-908
11. Soriano JV et al (1995) J Cell Sci 108:413-430
12. Maina F et al (1996) Cell 87:531-542
13. Sonnenberg E et al (1993) J Cell Biol 123:223-235
14. Silvagno F et al (1995) Arterioscler Thomb Vasc Biol 15:1857-1865
15. Schwall RH et al (1996) J Cell Biol 133:709-718
16. Huff JL et al (1993) Proc Natl Acad Sci USA 90:6140-6144
17. Gaudino G et al (1994) EMBO J 13:3524-3532
18. Mark MR (1992) J Biol Chem 267:26166-26171
19. Maestrini E et al (1996) Proc Natl Acad Sci USA 93:674-678
20. Gandino L (1994) J Biol Chem 269:1815-1820
21. Ponzetto C et al (1994) Cell 77:261-271
22. Ponzetto C et al (1996) J Biol Chem 271:14119-14123
23. Ridley et al (1995) Mol Cell Biol 15:1110-1122
24. Bardelli A et al (1996) EMBO J 15:6205-6212
25. Lokker NA et al (1992) EMBO J 11:2503-2510
26. Donate LE et al (1994) Protein Sci 3:2378-2394
27. Chan AM (1991) Science 254:1382-1385
28. Hartmann G et al (1992) Proc Natl Acad Sci USA 89:11574-11578
29. Seki T et al (1991) Gene 1012:213-219
30. Andermarcher E et al (1996) Develop Genetics 18:254-266
31. Brand-Saberi R et al (1996) Develop Biol 179:303-308
32. Rappolee DA et al (1996) Circ Res 78:1028-1036
33. Liu Y et al (1994) J Biol Chem 269:4152-4160
34. Liu Y et al (1994) Mol Cell Biol 14:7046-7058
35. Metcalfe AM et al (1997) Nucleic Acids Res 25:983-986
36. Zargenar R (1995) EXS 74:33-49
37. Naldini L et al (1992) EMBO J 13:4825-4833
38. Shimomura T et al (1993) J Biol Chem 268:22927-22932
39. Miyazawa K et al (1993) J Biol Chem 268:10024-10028
40. Camussi et al (1997) J Immunol 158:1302-1309
41. Grano M et al (1996) Proc Natl Acad Sci USA 93:7644-7648
42. Maher JJ (1993) J Clin Invest 91:2244-2252
43. Yamaguchi K et al (1996) Scand J Gastroenterol 31:921-927
44. Kilby MD et al (1996) Growth Factors 13:133-139
45. Morishita R et al (1997) Diabetes 46:138-142
46. Matsumoto K et al (1996) Biochem Biophys Res Commun 227:455-461
47. Rubin JS et al (1991) Proc Natl Acad Sci USA 88:415-419
48. Yo Y et al (1998) Kidney Int 53:50-58
49. Sato Y et al (1993) Exp Cell Res 204:223-229
50. Montesano R et al (1998) Cell Growth Differ 9:355-365
51. Thery C et al (1995) Develop Genetics 17:90-101
52. Nakamura Y et al (1996) J Hypert 14:1067-1072
53. Van Belle E et al (1998) Circulation 97: 381-390
54. Hiscox S, Jiang WG (1997) Biochim Biophys Res Commun 233:1-5
55. Kawakami-Kimura N et al (1997) Br J Cancer 75:47-53
56. Naidu YM et al (1994) Proc Natl Acad Sci USA 91:5281-5285
57. Rosen EM et al (1996) Int J Cancer 67:248-255
58. Wagatsuma S et al (1998) Cancer 82:520-530
59. Tolnay E et al (1998) Cancer Res Clin Oncol 124:291-296
60. Takayama H et al (1997) Proc Natl Acad Sci USA 94:701-706
61. Koch AE et al (1997) Arthritis Rheum 39:1566-1575
62. Ono K et al (1997) Circulation 95:2552-2558
63. Nakamura S et al (1998) Biochem Biophys Res Commun 242:238-243

HGF

Definition   *Hepatocyte growth factor*

See: →Hepatocyte growth factor

HHT

Definition   *Hereditary hemorrhagic telangiectasia*

See: →Bleeding disorders

HIF

Definition   *Hypoxia-inducible factor*

See: →Angiogenesis

HIP

Definition *Heparin-induced thrombocytopenia*

See: →Heparin-induced thrombocytopenia

HMG COA

Definition *3-hydroxyl-3-methylglutaryl Coenzyme A*

See: →Lipoproteins

HNF-1

Definition *Hepatocyte-specific nuclear factor-1*

See: →Fibrin/fibrinogen

HO-1

Definition *Heme oxygenase-1*

See: →Smooth muscle cells

Hormonal Regulation of Vascular Cell Function in Angiogenesis

Introduction Angiogenesis, the sprouting of new vessels from the existing vasculature, is crucial for a wide variety of physiologic processes such as embryonic development, normal growth and differentiation and reproductive functions. It also occurs in pathological conditions in response to injury, tumorigenesis and diabetes mellitus. Physiological hormonally-controlled angiogenesis is fundamental for the cyclic growth of the ovarian corpus luteum and of the endometrium, as well as for the changes of the reproductive organs during pregnancy.

This review discusses the current state of knowledge regarding physiologic angiogenic processes and their regulation in the female reproductive tissues, emphasizing the role of vascular endothelial growth factor (VEGF) as a major angiogenic inducer. It also examines briefly the production of angiogenic factors in the male reproductive tract and in endocrine tissues. In addition, implications of this subject of research for the regulation of fertility are discussed. Tumorigenesis in the reproductive tract is not discussed in the present review.

A large body of evidence has established important roles for vascular endothelial growth factor (VEGF) and its receptors in physiological angiogenesis, as well as in pathological neovascularization (see reviews [1,2]). VEGF, also known as vascular permeability factor (VPF), stimulates endothelial cell growth and angiogenesis and is a potent inducer of microvascular hyperpermeability. Other biological activities of VEGF include induction of protease expression, such as the serine proteases urokinase and tissue-type plasminogen activator (PA) or the

metalloprotease interstitial collagenase (MMP1), consistent with a pro-degradative environment that facilitates migration and sprouting of endothelial cells (see review [2]). Molecular cloning of the complementary DNA for this growth factor revealed that alternative messenger RNA (mRNA) splicing results in the generation of 121-,165-, 189- and 206 aminoacid-encoding mRNA forms. VEGF is up-regulated by factors such as hypoxia, elevated cAMP concentrations, and by multiple growth factors and cytokines such as epidermal growth factor (EGF), TGFβ, and $IL_1β$. Two high-affinity endothelial cell tyrosine kinase VEGF receptors, Flt-1 (VEGFR-1) and KDR/Flk-1 (VEGFR-2), are known to be expressed by quiescent and/or proliferating endothelial cells and transduce the cellular signal. Recently, several additional members of the VEGF/VPF family have been discovered: VEGF-B, VEGF-C and placenta growth factor (PIGF) [3-5]. VEGF-C exerts its effects via VEGFR2 (KDR/flk-1) and VEGFR3 (flt-4) and promotes lymphatic endothelial proliferation. PIGF binds to Flt-1, whereas the cellular binding components for VEGF-B remain to be elucidated.

Characteristics

Female reproductive organs The female reproductive organs, ovary, uterus and placenta, exhibit dynamic and periodic growth in response to ovulation and gestation; in these tissues, angiogenesis occurs as a physiological process [6]. The female reproductive system provides a unique model for studying regulation of angiogenesis during growth and differentiation of normal adult tissues.

Ovary Angiogenesis is a significant component of the cyclic development and differentiation of the corpus luteum (CL). During follicular growth, the theca interna becomes highly vascularized. Following ovulation, the thecal microvessels invade the ruptured follicle and form a rich capillary network within the developing CL [7-9]. These changes suggest the local release of angiogenic factors.

Endometrium The human endometrium undergoes a complex process of cell proliferation and differentiation, including growth of the vascular elements, for the preparation of a receptive endometrium. A complex subepithelial capillary plexus develops during the follicular phase and throughout the secretory phase; spiral arteries appear after ovulation; angiogenesis and changes in vascular permeability throughout the menstrual cycle promote the transformation from a thin, dense endometrium into the thick, highly oedematous secretory endometrium. Angiogenesis is also required to support endometrial regeneration after the shedding of the uterine surface in the absence of implantation. These changes are tightly regulated, and the central role of VEGF as a paracrine factor controlling this angiogenic process has been established.

Placenta Normal development and function of the placenta require invasion of the maternal decidua by trophoblastic cells, followed by abundant and organized vascular growth. Extensive angiogenesis is required to establish the vascular structures that are necessary for the efficient transplacental transport of oxygen and nutrients from the mother to the fœtus. The human placenta is a rich source of angiogenic growth factors and expresses their receptors.

Mammary gland   Vascularization of the mammary gland comprises the larger arteries which pass along the larger ducts to break up into a dense capillary network located on the external surface of the basal lamina of secretory portions (alveoli). The mammary tissue undergoes a cyclic remodeling through the menstrual cycle and through pregnancy and lactation; these changes are associated to increased blood flow and edema of breast connective tissue.

Male reproductive organs   Normal growth and differentiation of somatic cells of the testis, Leydig and Sertoli cells are dependent on hypophyseal gonadotropins, luteinizing hormone (LH) and follicle-stimulating hormone (FSH), respectively. The action of these gonadotrophins on the somatic cells also results in the establishment of the appropriate microenvironment for the process of spermatogenesis in the seminiferous tubule. Spermatozoa development is stimulated by testosterone under the control of LH, and by Sertoli cells under FSH regulation. Once formed, the testis secrete testosterone which stimulates the formation of the male reproductive tract. A developed microcirculation of the testis is necessary for its function and therefore for the development of the entire male reproductive tract.

Regulation

Molecular Interactions

Ovary   The first evidence supporting the hypothesis that VEGF may be a physiologic regulator of angiogenesis was provided by *in situ* hybridization studies on the rat ovary [10]. Several studies have examined VEGF expression and binding sites in the cycling ovary. In mouse ovary, VEGF induction occurs in a spatial and temporal fashion, the expression being restricted to areas that acquire new capillary networks as the theca interna layers and the corpus luteum [11]. VEGF is also expressed in the CL of ewes and primates during neovascularization [12, 13], and in primary cultures of bovine and human ovarian granulosa cells [14, 15]. One recent study on the normal cycling human ovary reports that VEGF is predominantly localized in the thecal cell layer at the more advanced stages of follicle development and in smaller amounts in the granulosa cell layer; subsequent to ovulation it is localized in granulosa and thecal lutein cells within the highly vascularized CL [16] (Figure 1 a, b); no expression of VEGF is found in the atretic follicles or in degenerating corpora lutea [16]. Ovarian granulosa cells also express other VEGF isotypes, i.e. VEGF-B and VEGF-C, under differential hormonal regulation [17].

The presence of VEGF receptors in responsive cells from intact tissues was analyzed by examining binding of biologically active, iodinated VEGF 165 to rat ovarian sections [18]. In developing follicles, specific, high-affinity binding sites are absent in the avascular granulosa cells, while intense binding is associated with vessels in the theca. Intense binding for VEGF is present in the thecal and granulosa layers of the mature CL, correlating with VEGF expression in the vascularized CL.

Based on these studies, it has been suggested that VEGF production by ovarian cells plays a role in the modifications of vascular permeability and the angiogenic process which are associated with ovulation and corpus luteum formation. Ovulation is accompanied by the rapid swelling of the follicle and by a large increase in the permeability of capillaries that surround the follicles, beginning a few hours after the gonadotropin surge. The pre-ovulatory increase in rat follicular vascular permeability and edema is closely associated with a sustained increase in VEGF expression [19], strongly suggesting a role of VEGF/VPF in this process. In the primate, the onset of the midcycle surge of LH signals the beginning of luteinization of the pre-ovulatory follicle; during the formation of the CL, high levels of VEGF mRNA are expressed after the LH surge and correlate with the pattern of vascular changes [13]. Treatment with a GnRH antagonist also reduces the concentration of mRNA VEGF in monkey corpora lutea [13], indicating that VEGF expression in the CL is under the control of gonadotropins. In addition, VEGF gene expression is induced in cultured bovine granulosa cells by luteotrophic hormone (LH), a known activator of adenylate cyclase, and the stimulation by LH can be mimicked by forskolin that directly activates adenylate cyclase [15]. An increase in gonadotropin secretion also contributes to revascularization of the immature rat ovary after autotransplantation by up-regulating the VEGF gene expression [20]. In addition to gonadotropins, hypoxia stimulates VEGF expression in tumors and in several types of cultured cells, including ovarian granulosa cells; other factors (i.e., cytokines and hormones, including steroid hormones) may also play a role; however, the role of ovarian steroids in angiogenesis is difficult to establish, due to the fact that ovarian cells themselves produce estradiol and progesterone. Recently, direct evidence for the involvement of VEGF as mediator of angiogenesis was established [21]. Treatment with truncated soluble Flt-1 receptors, which inhibit VEGF bioactivity, resulted in virtually complete suppression of CL angiogenesis and development, as well as progesterone release in CL, in a rat model of hormonally induced ovulation.

The mechanisms involved in the modifications of vascular permeability in the ovary are unknown. Endothelial fenestration is induced by VEGF [22]. At the ultrastructural level in ovarian tumoral cells, VEGF binding is localized on the abluminal plasma membrane and vesiculovacuolar organelles that provide an important pathway for extravasation of circulating macromolecules [23], which may result in an extracellular matrix

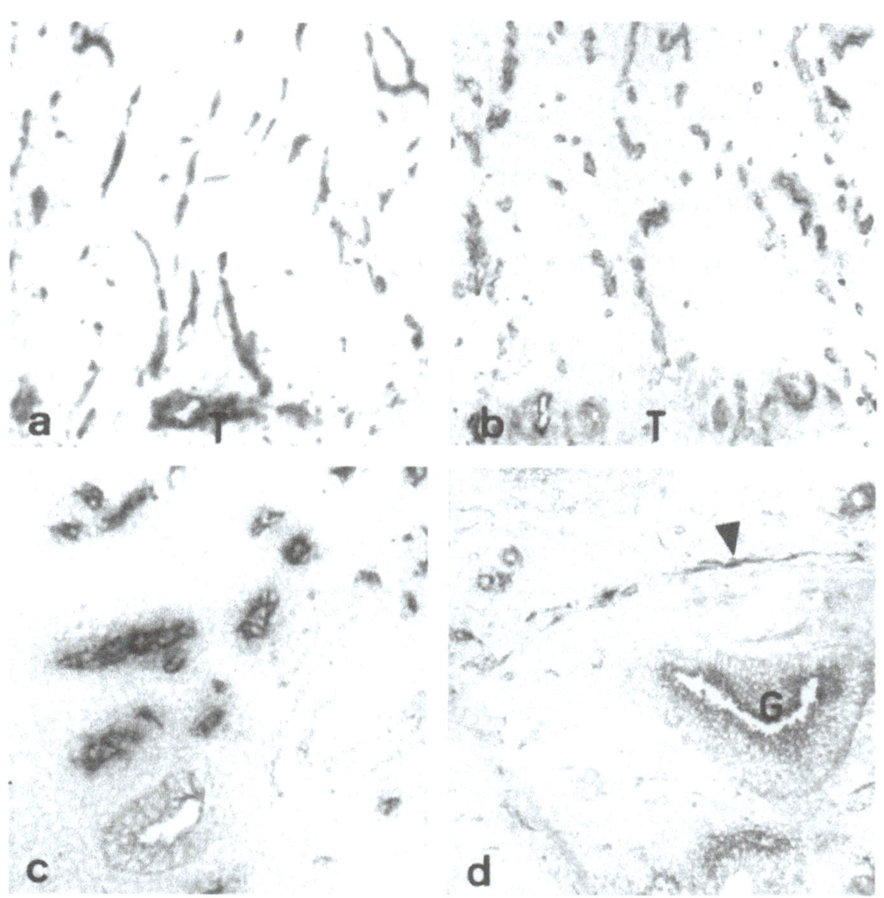

Figure 1. Immunocytochemical detection of VEGF in ovary and endometrium. b, VEGF in sow ovary; note the labeling on capillaries of corpus luteum. T, thecal lutein cells. d, VEGF in human endometrium; note the labeling on glands (G) and on narrow capillaries which have not formed a lumen (arrow). a, c. Immunostaining of endothelial cells with anti-von Willebrand Factor in sections from porcine ovary (a) and human endometrium (c).

which favor angiogenesis. In addition to its role in the ovary, VEGF may increase vascular permeability within the fallopian tube, and modulate tubal luminal secretions [16].

Endometrium

In vivo VEGF expression   Several studies performed on human and mammalian endometrium have reported the expression of VEGF mRNA by Northern blot or *in situ* hybridization, and the presence of VEGF protein by immunocytochemistry [24-28]. VEGF mRNA is present throughout the menstrual cycle, with an increase in the late proliferative and luteal phases. VEGF protein is expressed by glandular epithelium and cells within the stroma; it is also present on vascular endothelium in both the proliferative and secretory endometrium, in association with narrow capillaries, endothelial strands which have not yet formed a lumen and a few spiral arterioles [28] (Figure 1 c, d). The distribution of VEGF on capillaries may correspond to fixation to both high affinity VEGF receptors and to plasma membrane proteoglycans [28], and supports a paracrine role for VEGF in angiogenesis within the cyclic endometrium. VEGF also appears to be involved in the induction of vascular hyperpermeability at implantation and in the angiogenic process that follows in the rabbit [29]. The role of

estradiol as an inducer of angiogenesis has also been demonstrated in a murine model [30].

In vitro regulation of VEGF by ovarian steroids   The expression of VEGF mRNA and protein is modulated by ovarian steroids in isolated endometrial stromal or epithelial cells [27-28,31]. The addition of physiological concentrations of estradiol (E2) to epithelial and stromal cells in culture significantly increases VEGF expression over control values; this effect persists after 12 days of E2 in stromal cells ([28]). Major VEGF transcripts most likely correspond to two of the four transcripts generated by alternative splicing and translated into VEGF 125 and 165 isoforms. The addition of E2 plus progesterone, or medroxyprogesterone acetate, does not significantly increase VEGF expression over E2 stimulation alone [27,28].

Mechanism of regulation   The mechanisms by which β-estradiol influences VEGF expression is not clear. It may be due to a transient increase in steady state VEGF mRNA. Estradiol binds to specific nuclear receptors [32] that regulate gene transcription through consensus sequences (hormone responsive elements) present in the promoter region of the hormone responsive gene [33]. In the human endometrium, the estrogen receptor (ER) is expressed in glandular and stromal cells and is hormonally regulated in these cell types during the menstrual cycle [34]. It is also expressed at low levels in smooth muscle cells of spiral arteries during the luteal phase and during pregnancy [35]. In addition, another estrogen receptor (named ERβ) recently discovered [36] is present in the vasculature; its function remains to be elucidated. Another possible regulatory mechanism of gene expression could occur at a post-transcriptional level by modifying the stability of mRNA [37]. Estradiol has been shown to increase (albumin) or decrease (vitellogenin) the stability of mRNA. Stabilization of VEGF mRNA by estradiol has not been described.

Placenta   A number of laboratories have examined the distribution of VEGF protein, its mRNA and receptors within early gestational human placenta and decidua. Within a few days after implantation, VEGF expression is first detected in the giant cells of the trophoblast, placental macrophages (Hofbauer cells) and decidual cells [38-40], suggesting an important role of this factor in the induction of vascular growth of the decidua and placenta. VEGF is still expressed in the villous trophoblast shell during the first and second trimester of gestation and in the invading extravillous trophoblast, while the syncytiotrophoblast expresses no VEGF (or only a low level) [40].
KDR was found in association with endothelial cells of the placenta and the decidua, while Flt-1 was mainly detected in cells around the villous trophoblast [41-42], suggesting that the growth, differenciation and migration of trophoblast is mediated through the spatial and temporal regulation of the flt-1 receptor [41]. Recently sFlt-1, a potent antagonist of VEGF, was shown to be present in the mouse placenta, associated with an increased expression as gestation progresses, suggesting a novel mechanism of regulation of angiogenesis by alternative splicing of Flt-1 premRNA.
VEGF 165 stimulates the release of nitric oxide (NO) by the activation of constitutive NO synthase and the incorporation of H^3 thymidine in a transformed human trophoblastic cell line [67]. Although few functional studies have been performed, the presence of VEGF and its receptors in placental tissues throughout gestation strongly suggest that VEGF plays an important role in the development and maintenance of the uteroplacental unit during pregnancy.
Significant defects in the vasculature of several organs including the placenta are observed in VEGF gene knockout mice [43,44]. Little is known of the significance and function of other members of the VEGF family, which includes an additional 145 aa isoform of VEGF expressed by placental cells [45], VEGF-B, and VEGF-C, and placenta growth factor (PlGF) [5], in the placental vascularization.
During development in the mouse and rat embryos, the VEGF mRNA is expressed in several organs, including heart, vertebral column, kidney and brain [38,39,46]. Although not covered by this review, VEGF and its receptors have an important function both in the differentiation of the endothelial lineage and in the neovas-

cularization of developing organs, acting in a paracrine fashion [38,39,46].

Mammary gland   Little information is currently available concerning the expression and regulation of VEGF in normal human breast tissue, in contrast to breast tumors where it has been established that its concentration is markedly increased and correlates with microvessel density and disease-free survival [47,48]. Myoepithelial cell-derived bFGF may be an important paracrine factor controlling cyclic epithelial cell survival and growth in the normal human breast [49]. Furthermore, thrombospondin is expressed in the normal mammary gland at times coincident with vascular regression (review in [50]).

Male reproductive organs   In a survey of human tissues by *in situ* hybridization, high levels of VEGF/VPF expression were found in normal prostatic glands. This finding prompted further studies of VEGF/VPF expression in the human male genital tract.
Human testicular tissue including Leydig and Sertoli cells as well as seminiferous tubules express VEGF, and both types of VEGF receptors Flt-1 and KDR are present on perivascular cells, suggesting the role of VEGF as a paracrine mitogenic and angiogenic factor responsible for the development of testicular vascularization and for the maintenance of the permeability of testicular microvasculature [51].
The epithelium of prostatic and seminal vesicles strongly express VEGF, as found by *in situ* hybridization and immunocytochemistry. VPF is also present in substantial quantities in seminal plasma, exceeding the high levels reported in malignant effusions [52]; high levels of VPF are found in post-vasectomy ejaculates indicating that the prostate and seminal vesicles, rather than the testis or epididymis, are the major source of VPF in this fluid [52]. Preliminary data suggest that the secretion of VEGF by both epithelial cells from human adult prostate and fibroblasts from fetal prostate is under androgenic regulation ([53]; Sordello and Plouet, personal communication). The function of VPF in the male genital tract is unclear, but its strong expression in the prostate and seminal vesicles and its secretion at substantial levels into the seminal fluid argue for an important role in determining the fluid and protein composition of semen and thus an effect on sperm motility or survival. Regardless of the precise mechanisms, estrogens appear to be essential for male fertility. Hess et al [54] have shown that transgenic mice lacking the a form of the estrogen receptor have impaired sperm production due to a defect in resorbtion of fluid in the efferent ductules of the testis, as well as subsequent atrophy of the testis providing evidence of a physiological role for estrogens in the male reproductive organs.

Additional Features
Other angiogenic factors in the female reproductive organs   In physiologic angiogenesis which occurs in the female reproductive system in response to ovulation

or gestation, growth of new capillaries is tightly controlled by an interplay of growth regulatory proteins which can either stimulate or inhibit blood vessel growth.

Basic fibroblast growth factor (bFGF) was the first of these angiogenic factors to be identified, and shown to stimulate endothelial cell proliferation *in vitro,* to induce angiogenesis *in vivo,* and to be frequently present at sites of capillary growth [55]. However, it is not clear whether bFGF is in fact necessary for inducing physiological angiogenesis *in vivo.* The enzyme platelet-derived endothelial cell growth factor (thymidine phosphorylase (PD-ECGF/TP) is another angiogenic factor produced by the endometrium and regulated by ovarian steroids and cytokines during the endometrial cycle [56]; the precise role of additional angiogenic factors remains to be elucidated.

Recently, a novel family of endothelium-specific receptor tyrosine kinases, the Tie family consisting of two receptors Tie-1 and Tie-2 [57], has been shown to be essential for the development of embryonic vasculature; one of the ligands for Tie-2 is angiopoietin-1 [58]; Tie-2 is expressed in the quiescent vasculature, as well as in the ovarian and endometrial capillaries during hormone-stimulated angiogenesis [59], suggesting a dual function, in both vascular growth and maintenance.

Thrombospondin-1 (TSP-1, a 450 kDa extracellular matrix glycoprotein) has recently been identified as a negative regulator of the angiogenic response *in vivo* and *in vitro* and is found in a variety of normal adult tissues, endometrium, ovary and mammary gland. TSP-1 is indeed temporally regulated during the endometrial cycle; its expression is restricted to the secretory phase and to the basement membranes of small capillaries of the functional endometrium, consistent with its role as a suppresser of vessel remodeling [50,60].

Other endocrine glands   Contrasting to the ovary, where development is accompanied by an active cycle of angiogenesis during luteal formation and by vascular degeneration during luteal regression, in other endocrine glands the endothelium is normally quiescent and the developed vasculature provides a pathway for the specific secretory products. Particularly high levels of VEGF binding sites are present in the adrenal cortex, which is extremely well vascularized, with a ratio of parenchymal cells to capillary endothelial cells approaching 1/1 [61]. In the rat adenohypophysis, intense VEGF mRNA and VEGF binding are present in the pars distalis and pars nervosa correlating with the rich microvascular bed; in contrast, a very low VEGF mRNA hybridization signal or low VEGF binding is detectable in the poorly vascularized pars intermedia (reviewed in [1]). High levels of VEGF binding sites are also present in the pancreas [62]. Thyroid-stimulating hormone (TSH) does not appear to stimulate VEGF in normal thyroid cells, in contrast to cell lines derived from thyroid cancer [63]. The precise role of VEGF in the normal physiology in these glands has yet to be defined.

Clinical Relevance   Overexpression of VEGF/VPF has also been observed in non-malignant pathological ovarian situations characterized by hypervascularization, angiogenesis and the generation of new stroma. For example, VEGF up-regulation is implicated in the development of ovarian hyperstimulation syndrome, characterized by massive transudation of protein-rich fluid from the vasculature of the peritoneal cavity during gonadotropin-induced ovulation [64]. Angiogenesis is important in the pathogenesis of endometriosis, a condition characterized by ectopic endometrium implants in the peritoneal cavity. Elevation of VEGF in the peritoneal fluid of patients with endometriosis has been reported [27]. Immunohistochemistry indicates that activated peritoneal fluid macrophages as well as macrophages within the ectopic endometrium are the main source of VEGF in this pathology. In addition, VEGF secreted by peritoneal fluid macrophages is increased in response to ovarian steroids, and the macrophages express the VEGF receptors flt and KDR, suggesting autocrine regulation [65].

In vivo data show that several reproductive processes can be controlled through the inhibition of angiogenesis. Angiogenesis inhibition by chronic administration of the angiogenic inhibitor fumagillin (or AGM-1470, secreted by certain fungi) to non-pregnant cycling females results in inhibition of endometrial maturation and corpora lutea formation. Furthermore, chronic administration of this drug to pregnant mice results in complete failure of embryonic growth due to interference with decidualization, placental and yolk sac formation and embryonic vascular development [66]. The above mentioned studies suggest that VEGF play a role in the development and differentiation of the ovarian corpus luteum and endometrium through its angiogenic properties. A defective production in the ovary or endometrium could be a pathogenic factor in infertility.

Martine Perrot-Applanat

References

1. Ferrara N et al (1992) Endocr Rev 13:18-32
2. Ferrara N, Davis-Smyth T (1997) Endocr Rev 18:4-25
3. Maglione D et al (1991) Proc Natl Acad Sci USA 88:9267-9271
4. Joukov V et al (1996) EMBO J 15:290-298
5. Park JE et al (1994) J Biol Chem 269:25646-25664
6. Reynolds LP, Killilea SD, Redmer DA (1992) Faseb J 6:886-892
7. Bassett DL (1943) Am J Anat 73:251-291
8. Clark JG (1990) Johns Hopkins Hosp Rep 9:593
9. Koos RD (1993) In: Adashi EY, Leung PCK (eds) The ovary. Raven Press, New York, pp 433-453
10. Philips HS et al (1990) Endocrinology 127:965-967
11. Shweiki D et al (1993) J Clin Invest 91:2235-2243
12. Redmer DA et al (1996) J Reprod Fertil 108:157-165
13. Ravindranath N et al (1992) Endocrinology 131:254-260
14. Kamat BR et al (1995) Am J Path 146:157-165
15. Garrido C et al (1993) Growth Factors 8:109-117
16. Gordon JD et al (1996) J Clin Endocrinol Metab 81:353-359
17. Laitinen M et al (1997) Endocrinology 138:4748-4756
18. Jakeman LB et al (1992) J Clin Invest 89:244

19. Koos RD (1995) Biol Reprod 52:1426-1435
20. Dissen GA et al (1994) Endocrinology 134:1146-1154
21. Ferrara N et al (1998) Nature Medecine 4:336-340
22. Roberts WG, Palade GE (1995) J Cell Sci 108:2369-2379
23. Qu-Hong et al (1995) J Histochem Cytochem 43:381-389
24. Charnock-Jones DS et al (1993) Biol Reprod 48:1120-1128
25. Cullinan-Bove K, Koos RD (1993) Endocrinology 133:829-837
26. Torry DS et al (1996) Fertil Steril 66:72-80
27. Shifren JL et al (1996) J Clin Endocrinol Metab 81:3112-3118
28. Bausero P et al (1998) Angiogenesis 2:167-182
29. Das SK et al (1997) Biol Reprod 56:1390-1399
30. Morales DE et al (1995) Circulation 91:755-763
31. Zhang L et al (1995) J Cell Sci 108:323-331
32. Evans RM (1988) Science 240:889-895
33. Beato M et al (1996) Ann NY Acad Sci 784:93-123
34. Garcia E et al (1988) J Clin Endocrinol Metab 67:80-88
35. Perrot-Applanat M et al (1994) J Clin Endocrinol Metab 78:216-224
36. Mosselman S et al (1996) FEBS Letters 392:49-53
37. Carter BZ, Malter JS (1991) Lab Invest 65:610-621
38. Jakeman LB et al (1993) Endocrinology 133:848-859
39. Breier G et al (1992) Development 114:521-532
40. Ahmed A et al (1995) Growth factors 12:235-243
41. Clark DE et al (1996) Human Reprod 11:1090-1098
42. Charnock-Jones DS et al (1994) Biol Reprod 51:524-530
43. Ferrara N et al (1996) Nature 380:439-442
44. Carmeliet P et al (1996) Nature 380:435-439
45. Poltorak Z et al (1997) J Biol Chem 272:7151-7158
46. Breier G, Risau W (1996) Trends in Cell Biology 6:454-456
47. Obermair A et al (1997) Int J Cancer 74:455-458
48. Zhang HAT et al (1995) J Natl Cancer Institute 87:213-219
49. Gomm JJ et al (1997) Exp Cell Res 234:165-173
50. Iruela-Arispe ML, Dvorak HF (1997) Thrombosis and Haemostasis 78:672-677
51. Ergun S et al (1997) Mol Cell Endocrinol 131:9-20
52. Brown LF et al (1995) J Urology 154:576-579
53. Greenberg PD (1997) Proc 79 th Annual Meeting of the Endocrine Society. Minneapolis (Abstract)
54. Hess RA et al (1997) Nature 390:509-512
55. Schweigerer L et al (1987) Nature 325:257-259
56. Zhang L et al (1997) Endocrinology 138:4921-30
57. Sato TN et al (1993) Proc Natl Acad Sci USA 90:9355-9358
58. Suri C et al (1996) Cell 87:1171-1180
59. Wong AL et al (1997) Circ Res 81:567-574
60. Iruela-Arispe ML et al (1996) J Clin Invest 97:403-412
61. Hamaj M, Harrison TS (1984) In: Abramson DI, Dobrin PD (eds) Blood vessels and lymphatics in organs systems. Academic Press, New York, p 280
62. Christifori G et al (1995) Mol Endocrinol 9:1760-1770
63. Soh EY et al (1996) Surgery 120:944-947
64. McClure N et al (1994) Lancet 344:235-269
65. McLaren J et al (1996) J Clin Invest 98:482-489
66. Klauber N et al (1997) Nature Medicine 3:443-446
67. He Y et al (1999) Mol Endocrinol 13:537-545
68. Ahmed A et al (1997) Lab Invest 76:779-791

HSPG

Definition   *Heparan sulfate proteoglycan*

See: →FGF-1, FGF-2; →FGF receptors; →Heparin-affin regulatory protein

HUVE, HUVEC

Definition   *Human umbilical vein endothelial cell*

See: →Angiogenin; Endothelial cells; →Signal transduction mechanisms in vascular biology

ICAM-1

Definition   *Intercellular adhesion molecule-1*

See:→ Intercellular adhesion molecule-1 and 2; →Blood cells, interaction with vascular cells

IDL

Definition   *Intermediate density lipoprotein*

See: →Lipoproteins

IFN

Definition   *Interferon*

See: →Interferons; →Cytokines in vascular biology and disease

IGF-1

Definition   *Insulin-like growth factor-1*

See: →Atherosclerosis; →Thrombosis

IL-1, 2 etc.

Definition   *Interleukin-1, -2 etc.*

See: →Interleukins

INF

Definition   *Interferon*

See: →Interferons; →Cytokines in vascular biology and disease

Inflammatory Cells

Definition   *White blood cells that include monocytes/ macrophages, polynucleated leukocytes and lymphocytes.*

See: →Cytokines in vascular biology and disease; →Blood cells, interaction with vascular cells; →Atherosclerosis; →Complement system

iNOS

Definition   *Inducible nitric oxide synthase*

See: →Angiogenesis inhibitors; →Cytokines in vascular biology and disease; →Nitric oxide

Integrins

Definition
Heterodimeric (one α and one β subunit) cell surface receptors for extracellular matrix molecules. They are connected to the cytoskeleton and are involved in inside-out and outside-in signalling.

See: →Vascular integrins; →Platelets; →Angiogenesis; →Vitronectin/vitronectin receptors; →Extracellular matrix; →Fibrin/fibrinogen; →Thrombin; →von Willebrand Factor

Intercellular Adhesion Molecule-1 and 2 (ICAM-1 and ICAM-2)

Definition
ICAM-1 (CD54) ICAM-2 (CD102)

See: →Blood cells, interaction with vascular cells

Interferons (INF)

Definition
Cytokines with multiple regulatory functions produced by different cell types during inflammation. Three types are described: INF α (leukocyte interferon), INF β (fibroblast interferon) and INF γ (immune interferon).

See: →Cytokines in vascular biology and disease

Interleukins

Definition   *Cytokines that mediate communication between leukocytes. They have also a role in hematopoiesis, atherosclerosis/thrombosis and angiogenesis.*

See: →Cytokines in vascular biology and disease; →Megakaryocytes; →Fibrin/fibrinogen

IP3

Definition
Inositol 3-phosphate

See: →Phospholipases; →Vasomotor tone regulation, molecular mechanisms of

ISGF3

Definition   *Interferon-stimulated gene factor-3*

See: →Cytokines in vascular biology and disease

ITAM

Definition   *Immunoreceptor tyrosine-based activation motif*

See: →Platelet stimulus-response coupling

JAK

Definition   *Janus kinase*

See: →Cytokines in vascular biology and disease; →Signal transduction mechanisms in vascular biology

JNK

Definition   *c-jun N-terminal kinase*

See: →Mitogen-associated kinases; →Signal transduction mechanisms in vascular biology

kD

Definition   *Kilodalton*

kDa

Definition   *Kilodalton*

KDR

Definition   *Kinase domain region*

See: →Angiogenesis; →Tyrosine kinase receptors for factors of the VEGF family; →Vascular endothelial growth factor family

L-NA

Definition   *NG-nitro-L-arginine*

See: →Nitric oxide

L-NAME

Definition   *NG-nitro-L-arginine methyl ester*

See: →Angiogenesis; →Angiogenesis inhibitors; →Nitric oxide

L-NIO

Definition   *N-iminoethyl-L-ornithine*

See: →Nitric oxide

L-NMMA

Definition   *NG-monomethyl-L-arginine*

See: →Nitric oxide

LAD

Definition   *Leucocyte adhesion deficiency*

See: →Blood cells, interaction with vascular cells

Laminin

Definition   *Trimeric extracellular matrix molecule that binds integrin family members such as αvβ3. Main component of the basal membrane.*

See: →Extracellular matrix

LAP

Definition   *Latency-associated protein*

See: →Thrombospondins; →Transforming growth factor β

LCAT

Definition   *Lecithin cholesterol acyl transferase*

See: →Lipoproteins

LDL

Definition   *Low density lipoprotein*

See: →Cytokines in vascular biology and disease; →Lipoproteins

Leukocyte Adhesion Molecules

Definition   *Three integrin of the β2 subtype are expressed in leukocytes and are involved in adhesion to extracellular matrix or cells. CD11a/CD18 (LFA1) is expressed in lymphocytes, CD11b/CD18 (Mac1) is expressed in monocytes and CD11c/CD18 (P150/95) is expressed in granulocytes.*

See: →Blood cells, interaction with vascular cells

LFA1, 2 etc.

Definition   *Leukocyte function antigens-1, -2 etc.*

See: →Blood cells, interaction with vascular cells

LH

Definition   *Luteinizing hormone*

See: →Hormonal regulation of vascular cell function in angiogenesis

LIBS

Definition   *Ligand-induced binding site*

See: →Vascular integrins

LIF

Definition   *Leukemia inhibitor factor*

See: →Cytokines in vascular biology and disease; →Megakaryocytes

Lipid Mediators

Definition   *Lipids involved in cell signalling such as phosphatidyl inositol phosphate (PIP$_2$), or diacylglycerol (DAG)*

See: →Signal transduction mechanisms in vascular biology; →Platelet stimulus-response coupling; →Prostaglandins; →Prostacyclin; →Thromboxanes; →Vasomotor tone regulation, molecular mechanisms of; →Thrombin

Lipoproteins

Definition   *Lipid-protein complexes that circulate in the blood. These include chylomicrons, low density lipoproteins (LDL), oxydized-LDL, high density lipoproteins (HDL), very few density lipoproteins (VLDL), Lipoprotein(a), apolipoprotein (a) and apolipoprotein A.*

See: →Fibrinolytic, hemostatic and matrix metalloproteinases, role of; →Thrombosis; →Atherosclerosis; →Cytokines in vascular biology and disease; →Thrombin

Introduction   Lipids play a central role in human metabolism, participating in such diverse functions as the maintenance of cellular integrity, the storage of energy, the provision of metabolic intermediates and the transmission and transduction of signals.
We have the capacity to synthesise them in abundance but most of us consume enough or more than enough in our diet to satisfy our daily requirements. Each day we ingest 100-150 g of triglyceride with 0.5-1.0 g of cholesterol and require to transmit them from sites of absorption in the gut to the liver and tissue for utilisation and storage. Such hydrophobic substances are dispersed through the aqueous environment of the plasma by the surrounding of the insoluble globules of dietary cholesteryl ester and triglyceride with a shell of amphipathic proteins (apolipoproteins) and phospholipids, hydrophilic on one face and hydrophobic on the other. These lipid-protein complexes (lipoproteins) are called chylomicrons. In the fasting state, when intestinal lipoprotein production is minimal, smaller, triglyceride rich lipoproteins (very-low density

lipoproteins – VLDL) continue to be made and secreted by the liver. Intravascular remodelling results in their conversion to species of reduced size and lipid content which express a wide spectrum of physical properties (intermediate density lipoprotein – IDL and low density lipoprotein – LDL).

All these lipoproteins deliver lipids to the peripheral tissue. The transport of cholesterol from peripheral tissues back to the liver, the main organ of cholesterol utilisation corresponds to the so-called "reverse cholesterol transport" and is assumed by high density lipoproteins (HDL). Lipoproteins constitute a complex poly-dispersed system and are classified according to different physical and biochemical criteria.

Characteristics

Molecular Weight   Lipoproteins may be isolated by molecular size chromatography. A six percent agarose gel is the most appropriate available medium for separating the major lipoproteins in the plasma. Chylomicrons and VLDL are excluded from the gel and appear in the void volume while LDL and HDL penetrate and are retarded by the matrix mesh. Recently, a simple and rapid method has been developed for lipoprotein analysis using high performance gel permeation chromatography which permits direct measurement of the cholesterol, phospholipids or triglycerides content of each lipoprotein class present in small amounts (10-20 µl) of untreated serum (Figure 1).

Domains   The apolipoprotein component of lipoproteins are their major antigenic determinants and confer upon them their metabolic properties. Alaupovic and his colleagues [3] developed the nomenclature system of apolipoproteins which pertains their classification today.

According to their concept, the plasma lipoproteins can be viewed as a system of discrete species of families distinguishable from each other on the basis of their apolipoprotein composition. Lipoproteins containing apolipoprotein B correspond to the LpB family and are associated with chylomicrons, VLDL, IDL and LDL and lipoproteins containing apolipoprotein A-I correspond to the LpA family and are associated with HDL. However, particles can contain other apolipoproteins such as C-I, C-II, C-III, E, A-II, A-IV, J, etc... Therefore LpA and LpB families are constituted of a mixture of simple particles containing only apo A-I or apo B (LpA-I and LpB respectively) or additional apo(s) leading to the formation of complex particles such as LpB:C-III, LpB:E, LpB:C-III:E, LpA-I:A-II, LpA-I:A-IV, etc.).

Binding Sites and Affinity   The LDL or apo B/E receptor originally described by Goldstein and Brown [5] is now known to influence not only the metabolism of LDL but also that of a number of other apolipoprotein B containing particles in the plasma, including VLDL and IDL. This membrane protein appears on the cytoplasmic membranes of cells in response to their need for cholesterol. It recognises specific positively charged domains on apolipoprotein B and E and is responsible for the facilitated endocytosis of particles which contain these proteins.

The affinity of E containing lipoproteins for the receptor is higher by an order of magnitude than that of those containing only apolipoprotein B. The receptor appears to be responsible for about one half of LDL catabolism in normal individuals. The activity of the receptor pathway in cultured cells is closely regulated by intracellular sterol levels. When cells are replete with cholesterol, receptor activity on their membranes falls.

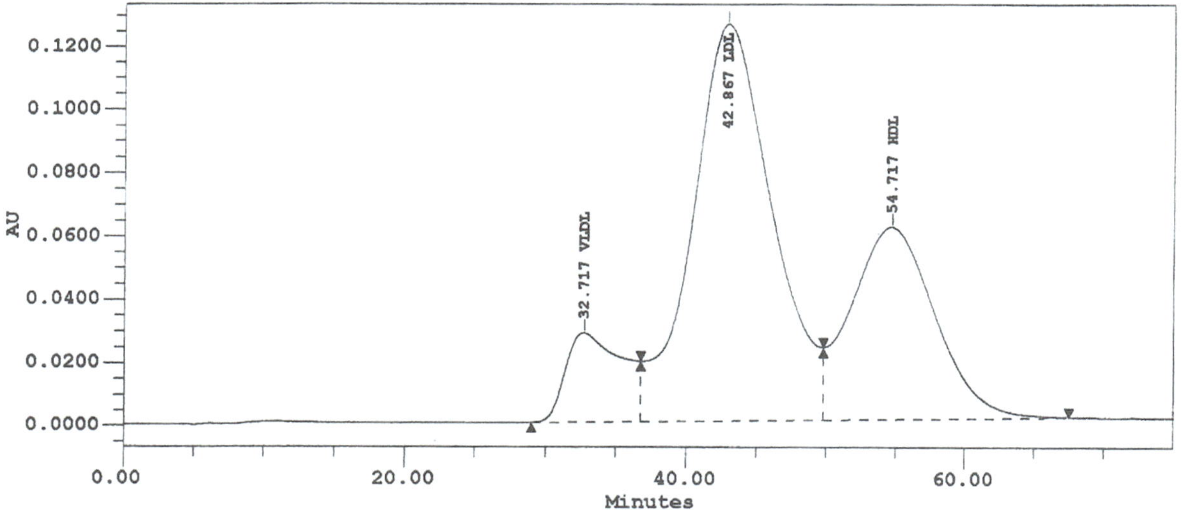

Figure 1. High performance gel chromatography (Superose 6B) of human serum. 10 µl of serum were injected into the Superose 6B gel chromatography column and cholesterol concentration was continuously monitored in the effluent of the column.

The receptor appears to occupy a pivotal role in sterol metabolism since it is not only an indicator of intracellular sterol needs but also by virtue of its quantitative importance in lipoprotein catabolism regulates the level of extracellular sterol pool. Two thirds of the LDL particles are metabolised by LDL receptors in liver and extrahepatic cells, with the liver accounting for 75 % of the total expression of LDL receptors *in vivo*. LDL receptor are effectively regulated. Hepatic LDL receptors are suppressed whenever the liver content of cholesterol increases or its demand for cholesterol is reduced. Conversely, the number of LDL receptors increases when hepatic cholesterol synthesis is blocked by drugs that inhibit hydroxy methyl-glutaryl Co enzyme A reductase (compactin, mevilonin, simvastatin, pravastatin, etc...), and when bile acid binding resins are given. All of the changes in receptor activity alter the rate of uptake of LDL by the liver and cause reciprocal changes in plasma LDL level. In familial hypercholesterolemia several groups of mutations occur naturally in the structural gene for the LDL receptor which disrupt its normal function and lead to severe hypercholesterolemia.

Chylomicron remnants are cleared rapidly from the plasma and are taken up by the liver and other tissues. The clearance of chylomicron remnants by the liver is mediated by apo E, and genetic and other evidence indicates the presence of a chylomicron remnant receptor that is distinct from the LDL receptor. Herz *et al* [6] cloned a cell surface protein that was designated as the LDL receptor related protein (LRP) because of its sequence homology with the LDL receptor. The LRP has been shown to bind and internalise apo E enriched lipoproteins. Based on the structural similarities between the LDL receptor and LRP, and the ability of the LRP to interact with the remnant particles, it has been postulated and then demonstrated that the LRP is involved in the hepatic clearance of chylomicron remnants.

Additional Features

Hydrated density Gofman and his associates [1] proceeded to characterise the spectral distribution of human plasma lipoproteins using analytical ultracentrifugation. Their detection of absorption minima at solvent densities of 1.006, 1.019, 1.063, 1.125 and 1.210 kg/l laid the foundation for the classification of lipoproteins into species of very low density lipoprotein (VLDL, d < 1.006 kg/l), intermediate density lipoprotein (IDL, d=1.006– 1.019 kg/l), low density lipoprotein (LDL, d=1.019–1.063 kg/l) and high density lipoprotein (HDL2, d=1.063–1.125 ; HDL3, d=1.125–1.210 kg/l) and led to the development of the sequential preparation flotation procedure [2] which is widely employed today to isolate the major plasma lipoprotein fractions. Table 1 describes some of the properties of the five major plasma lipoprotein density classes.

Electrophoretic mobility The lipoprotein fractions carry an electrical charge and migrate in an electric field, permitting their resolution into four major classes (Table 1, Figure 2). Chylomicrons become trapped in the interstices of the support medium and therefore remain at the origin of the electrophoretogram. VLDL migrates with the α_2 (or pre-β) globulin fraction, LDL with the β-globulins, and HDL with the α_1-globulin.

A variety of support media, including paper, agarose, starch gel, agarose polyacrylamide have been employed for lipoprotein electrophoresis.

Processing and Fate

Metabolism of dietary lipids Each day we ingest about 1 g of cholesterol and about 100 g of triglycerides. Under the action of digestive enzymes in the intestinal lumen triglycerides absorption is virtually quantitative, while only 50 % of the cholesterol is taken up, the remainder being lost in the faeces. Within the enterocyte, absorbed cholesterol and fatty acids are rapidly reconstituted to cholesterol-ester and triglycerides and packaged into

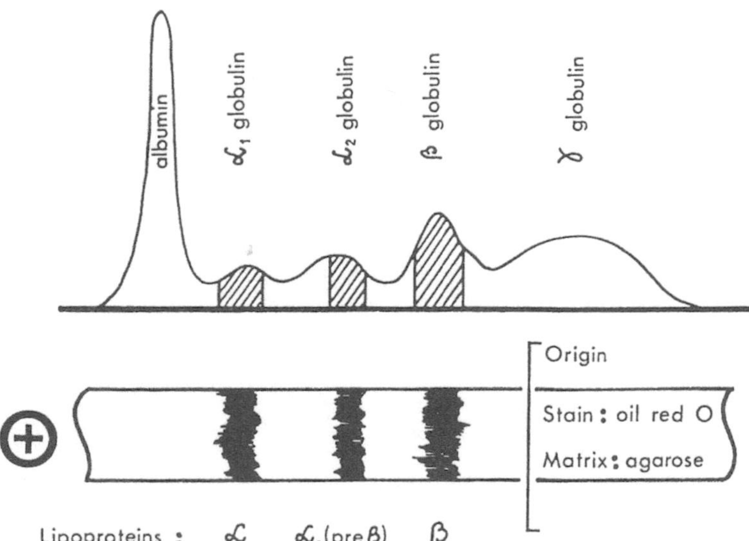

Figure 2. Electrophoresis of serum lipoproteins

Table 1. Classification of the human plasma lipoproteins

Lipoprotein	Buoyant density (kg/l)	Flotation rate at 26 °C and at a background density of		Mean particle mass (10^6 daltons)	Electrophoretic mobility	Apolipoprotein content A B C D E	Mean diameter (nm)
		1.063 kg/l	1.20 kg/l				
Chylomicrons	< 0.95	< 400	–	1000	origin	1 40 47 trace 12	100–1000
Very Low Density Lipoproteins	0.95–1.006	20–400	–	4.5–100	α_2(pre-β)	trace 54 38 trace 8	30–70
Intermediate Density Lipoproteins	1.006–1.019	12–20	–	4.0	α_2-β	–	25
Low Density Lipoproteins	1.019–1.063	0–12	–	3.0	β	1.4 90 5 trace 3.5	20
High Density Lipoproteins Subfraction 2	1.063–1.125	–	0–3.5	0.4	α_1	82 3.7 8.9 2.9 2.1	10
High density lipoproteins	1.125–1.21	–	3.5–9	0.2	α_1	95 Nil 2.8 2.2 0.4	8

layered chylomicron particles containing about 1% apolipoprotein, mainly apo B48, but chylomicrons also contain other apo(s) such as apo A-I, A-II, A-IV also elaborated in the gut and secreted with chylomicrons, while additional apo(s) are acquired from interstitial fluid HDL. The latter including apo C(s) and apo E. A number of rapid changes take place in the chylomicron particle when it enters the plasma. Within the capillary beds of skeletal muscle and adipose tissues it is exposed to the lipolytic action of endothelium bound lipoprotein lipase which hydrolyses its triglycerides core, releasing free fatty acids and mono- and diglycerides for energy production or storage.

Sequential delipidation steps result in the generation of a so-called "chylomicron remnant". This particle is structurally distinct from its parent. It acquires lipid, particularly cholesterol ester, from HDL via the agency of cholesteryl ester transfer protein (CETP). This is exchanged for triglyceride which is transferred in the opposite direction. In addition to these alterations in its hydrophobic core, the surface coat of the remnant shows substantial differences from that of its parent. All of the apo B initially present on the particle is retained, but the relative amounts of the other proteins are altered.

Apo A-I and apo A-II are transferred to HDL. The apolipoproteins C content initially rises as a result of transfer from HDL. This is of particular significance as far as apo C-II is concerned since the latter is an obligatory cofactor for lipoprotein lipase, increasing the particle's affinity for the enzyme. The role of apo C-III on the chylomicron surface is less clear although it may be responsible for regulating lipolysis of the particle or delaying its hepatic clearance.

As the chylomicron progresses down the lipolytic cascade the apolipoproteins C are transferred gradually back into the reservoir within HDL. Apolipoprotein E, which is acquired by the particle simultaneously with apolipoproteins C is not involved in lipolysis, but thereafter plays a role in triggering hepatic assimilation of the remnant.

The liver is responsible for the efficient and rapid clearance of cholesteryl ester rich chylomicron remnants from the circulation. Hepatic paranchymal cells contain, on their surfaces, receptors which recognise and bind to apo E of remnants. The metabolism of chylomicrons is depicted schematically in the general model of lipoprotein metabolism shown in Figure 3A.

Hepatic lipid metabolism   The liver, although it has the ability to make the cholesterol and triglyceride that it needs, prefers to utilise preformed components that are available either from the diet as outlined above or from adipose tissue by the fatty acid/albumin transport mechanism. Under conditions in which their availability is limited, the liver can generate all the fatty acid it requires from small molecular weight precursors. Similarly a cholesterol deficit can be met by endogenous synthesis, following activation of the enzyme 3-hydroxy-3-methylglutaryl Coenzyme A reductase (HMG CoA reductase).

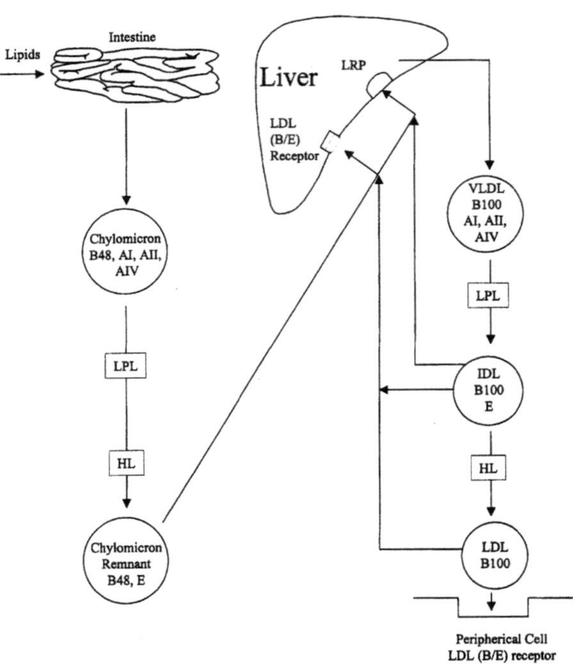

Figure 3A. Chylomicron, VLDL and LDL metabolism. LPL, lipoprotein lipase; HL, hepatic lipase; LRP, LDL receptor related protein.

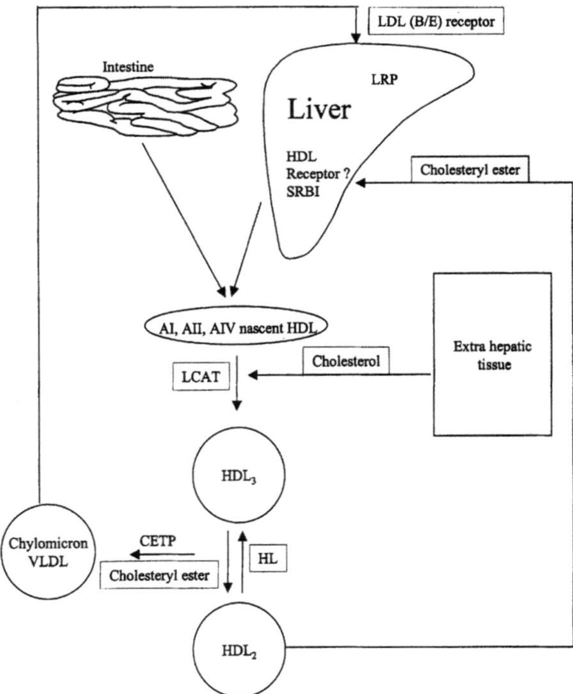

Figure 3B. Reverse cholesterol transport. LCAT, lecithin cholesteryl acyl transferase; CETP, cholesteryl ester transfer protein; HL, hepatic lipase.

Lipids synthesised in the liver have a number of fates. First, a significant proportion of cholesterol and triglyceride is secreted, in the form of VLDL, the major vehicle for plasma triglyceride transport in the post-absorptive state. Secondly, lipid surplus to requirements may be stored (temporarily) as cytoplasmic oil droplets within the hepatocytes. And, thirdly, the liver has the unique ability to eliminate cholesterol into bile, either unchanging or following oxidation to bile acids.

The post-absorptive state

Metabolism of apolipoprotein B containing lipoproteins: Although chylomicrons are secreted by the gut following a meal, in a recent study by Schneeman *et al* [4] that examined the effects of an ordinary test meal in healthy men, the rise in triglyceride-rich lipoproteins (TGRL) was attributed to an 80 % increase in the number of VLDL particles of hepatic origin. As this accumulation of VLDL particles was correlated with reduced efficiency of chylomicron removal, the increase in VLDL was ascribed to a preferential clearance of chylomicrons by lipoprotein lipase. The large VLDL is the main competitor of chylomicrons rather than the small VLDL particles.

Triglyceride-rich VLDL enter a metabolic cascade. Similar to that described for chylomicrons, apolipoprotein B100 is integral to the particle and essential for its normal hepatic secretion. Radiolabelling of apo B100 of VLDL (density less than 1.006 Kg/l) indicated that VLDL are transformed to intermediate density lipoprotein (IDL) which are converted to low-density lipoprotein (LDL). In this process, the particle's core is hydrolysed by lipoprotein lipase and hepatic lipase.

Lipoprotein lipase acts primarily on triglyceride-rich lipoproteins of dietary (chylomicrons) and hepatic origin (VLDL). LPL activity is essential for the clearance of chylomicrons and the conversion of VLDL1 into VLDL2, and VLDL2 into IDL. Hepatic lipase functions at a lower level in the delipidation cascade, facilitating the conversion of IDL into large LDL. It may also act on smaller VLDL particles and have a role in promoting interconversions between individual LDL subclasses. Again, in parallel with chylomicron metabolism, under the action of CETP, cholesteryl esters are acquired from HDL, by exchange, while surface coat apo C's are transferred in the opposite direction. LDL therefore represents a "remnant" of VLDL catabolism in which the triglyceride core is virtually eliminated and apo B remains the sole apolipoprotein component. In most subjects, the rate of synthesis of apolipoprotein B into VLDL exceeds that into LDL. Not all VLDL particles are therefore destined to complete the cascade conversion to LDL. Using procedures which permit subfractionation of the VLDL spectrum the evidence which emerged indicated that lipolysis of large VLDL generated remnants, (small VLDL/IDL flotation interval), the majority of which were removed directly from the plasma without appearing in LDL. The latter seemed to come from rapid and quantitative transformation of small VLDL particles which has been secreted directly by the liver.

HDL metabolism and reverse cholesterol transport (Figure 3B): HDL lie in the density interval 1.063 – 1.210 Kg/l and are the smallest of the lipoproteins and constitute an heterogeneous mixture of particles endowed with a diversity of metabolic properties. The HDL fraction not only is synthesised *de novo* by intestine and liver but also acquires components from lipoproteins in the circulation and from peripheral tissues (Figure 3). HDL represents an amalgam of diverse components which come together following:

- direct secretion by the liver and intestine,
- transfer from other lipoproteins,
- transfer from peripheral tissues.

The major HDL proteins, apolipoproteins A-I and A-II are elaborated in precursor form in the liver and intestine. Apo A-I forms three types of stable structure with lipids: small lipid-poor complexes; flattened discoidal particles containing only polar lipids (phospholipid and cholesterol), and spheroidal particles containing both polar and non polar lipids.

Prebeta 1 HDL, "lipid-poor apo A-I" Plasma HDL normally contains 2-5 % of small particles distinguished by a slow (prebeta-) electrophoretic mobility from the bulk of alpha-migrating HDL. The apparent molecular mass of these particles was 60-70 Kda in different studies, with a calculated diameter of 5-6 nm. The protein moiety contains only apo A-I. Lipid content of 10-40 % have been reported. Higher concentrations of prebeta-1 HDL are present in large vessel lymph. Lecithin and sphingomyelin are present at almost equal molar concentrations in prebeta-1 HDL. Prebeta-1 HDL were completely lost from HDL density range by ultracentrifugal flotation. Apolipoprotein A-I presents a unique organisation in prebeta-1 HDL.

Prebeta 2 HDL, discoidal HDL: An HDL fraction (prebeta-2 HDL) present in plasma to the extent of 2-5 % of particles contains three apolipoproteins A-I per particle as the only protein. It is rich in lecithin and contains a smaller proportion of sphingomyelin and cholesterol. Prebeta 1 and prebeta 2 differ mainly by their proportion of phospholipids particularly lecithin. Cholesteryl ester is not detected in prebeta 2. Sectional election microscopy indicates that discoidal HDL are made of a single lipid bilayer probably stabilised by protein at the periphery. Discoidal HDL are present at increased concentration in large vessel lymph.

Spherical HDL: Most HDL in plasma are present as spherical particles (9-12 nm diameter) with alpha-migration in agarose electrophoresis. Three size subclasses predominate (HDL3, HDL2a and HDL2b) in order of increasing size and lipid content. Most HDL contain both apo A-I and apo A-II, the second major HDL protein. The proportion of total HDL containing only apo A-I (LpA-I) was 11-45 % in different studies. HDL also contain minor apolipoproteins (A-IV, E, C-I, etc...).

Other proteins involved in HDL metabolism: Lecithin: cholesterol acyltransferase (LCAT): LCAT reacts with discoidal and spherical HDL, transferring the 2-acyl group of lecithin or phosphatidyl ethalonamine to the free hydroxyl residue of fatty alcohols, particularly cholesterol, to generate cholesteryl esters (retained in HDL) and lysolecithin. A high concentration of LCAT (relative to apo A-I) is found in a prebeta-migrating high molecular weight complex (prebeta-3 HDL) which is probably a complex between discoidal (prebeta-2) HDL and LCAT, with the addition, possibly transient, of other catalytic proteins such as CETP.

Cholesteryl ester and phospholipid transfer proteins (CETP, PLTP): CETP catalyses an exchange of neutral lipids, particularly triglycerides and cholesteryl esters, between all the major lipoprotein classes. The cDNA of PLTP, predicts a protein sequence with 25% homology with CETP but highly purified PLTP, devoid of CETP activity, converts HDL3 of particle diameter 8.7 nm into main populations, one of larger (10.7 nm) and one of smaller (7-8 nm) particle size.

Lipoprotein lipase and hepatic lipase: It has long been known that the action of lipoprotein lipase on triglyceride-rich lipoproteins produces HDL2-like particles. Under appropriate conditions, HDL2 is converted to HDL3 species by the action of hepatic lipase.

Proteins active in the hepatic uptake of cell-derived cholesterol: Three different mechanisms have been described for the clearance of LCAT-derived cholesteryl esters by hepatocytes. LDL receptor is active on hepatocytes in the endocytosis of LDL and VLDL remnant lipoproteins. Large apo E-rich HDL also have a high affinity for the LDL receptor although the concentration of such particles in human plasma is very small. Retroendocytosis of HDL by rat hepatocytes has been reported. A disproportional uptake of cholesteryl ester from HDL not involving lipoprotein apo E or the LDL receptor has been described. This selective uptake of cholesteryl ester may be mediated by a docking receptor for HDL cholesteryl esters, called scavenger receptor SR-BI [7].

Molecular mechanisms in reverse cholesterol transport (Cholesterol transfer from the cell surface): The peripheral tissues involved in reverse cholesterol transport transfer cholesterol only in its unesterified form. Much of the largest part of cholesterol from labelled peripheral cells is initially transferred to HDL even though this fraction contains only about one quarter of total plasma free cholesterol. Cholesterol efflux could be based on diffusion and could be receptor-dependent or receptor independent. Cholesterol transfer between synthetic lipid vesicles in evidently diffusional. As the aqueous solubility of cholesterol, although low, is certainly finite, diffusion must also contribute to total cholesterol efflux from cell membranes. The real question is, does diffusion account for the whole of cellular efflux, or only part? The most recent data suggests that efflux from nucleated cells reflects the existence of multiple plasma membrane cholesterol pool. If description from the membrane is rate-limiting, the complex kinetics of efflux can still be explained in terms of diffusion, if plasma membrane micro-domains with different desorption ratio are defined by different local lipid compositions. Intracellular lipid-binding protein sterol carrier protein-2 (SCP-2) to cell membrane preparations significantly increase the fast component of sterol efflux from cell membranes.

Overall the evidence that sterol efflux can be modulated by intracellular lipid transfer protein appears convincing. In the second hypothesis, cholesterol efflux follows from protein C kinase-mediated signalling, initiated by the high affinity binding of HDL to a cell surface HDL binding protein. Cholesterol is transferred to the cell surface from a pool of new synthesis in the endoplasmic reticulum.

While there is no doubt that signalling intermediates are present *in vitro* after HDL binding, the significance of the observation to normal physiology is not yet fully established. The phenomenon has been found mainly in lipid-loaded cells such as adipocytes and foam cells. Peripheral cells such as fibroblasts, when not cholesterol-loaded, showed little or no efflux via this pathway. The contribution of signal-mediated efflux to reverse cholesterol transport in unloaded cells, which represent the vast majority of cells *in vivo*, is probably therefore quite small.

Neither hypothesis satisfactorily explains the specificity for HDL (particularly lipid-poor LpA-I only, prebeta-1 HDL) as cholesterol acceptor now reported in a number of different studies. Many questions arise from recent experimental data. How good is the evidence that lipid-poor (prebeta-1) HDL is the preferred acceptor of cell-derived cholesterol? Does HDL-mediated efflux require HDL binding to the cell surface? And is HDL-mediated efflux linked to a particular membrane cholesterol pool?

Role in Vascular Biology

Pathology

Lipoprotein disease and classification of the hyperlipemias The Fredrickson's approach [8] remains the cornerstone for the practical management of patients with plasma lipid disorders. This classification is based on quantification of plasma lipoprotein fractions. Table 2 shows the phenotype and the genotype of hyperlipoproteinemia and the associated genetic disorder.

Atherosclerosis The vast majority of patients suffering from coronary heart disease presents lipoprotein disorders such as: high LDL cholesterol, high IDL cholesterol, hypertriglyceridemia associated with low HDL cholesterol levels. The pathogenic role of LDL is clearly demonstrated.

Plasmatic LDL cross the vascular endothelial barrier and are submitted in the arterial intima to the oxidative action of endothelial cells, macrophages and vascular smooth muscle cells. The resulting oxidised LDL are recognised by "scavenger receptors" expressed on macro-

Table 2. Hyperlipoproteinemia : Phenotype and Genotype

Phenotype Synonym		Clinical Presentation	Laboratory findings	Associated genetic disorder
Type I	Primary hyper-chylomicronemia Burger-Grutz disease	Childhood disease Eruptive papular xanthomatosis Hepatosplenomegaly Pancreatitis Lipimia retinalis	Triglyceride ↑↑ Cholesterol Fasting chylomicronemia Post heparin lipase deficiency	Familial lipoprotein lipase deficiency
Type IIa	Hypercholesterolemia	Present in childhood but usually manifest in adult life by premature vascular disease. Tendinous xanthomatis and family history may be present.	Triglyceride → Cholesterol ↑ LDL increased	Polygenic hyper-cholestrolemia Familial hyper-cholesterolemia LDL receptor deficiency
Type IIb	Combined hyperlipidemia	Present in childhood but usually manifest in adult life by premature vascular disease. Tendinous xanthomatosis and family history may be present.	Triglyceride ↑ Cholesterol ↑↑ LDL and VLDL increased	Familial combined hyperlipidemia Familial hyper-cholesterolemia
Type III	Familial dysbetalipoproteinemia	Planar palmar xanthomata	Triglyceride ↑ Cholesterol ↑	Apolipoprotein E_2 homozygosity
	Remnant removal disease	Tuberous and tendinous xanthomata	Chylomicron and VLDL remnants (and IDL) increased	
	Floating beta disease	Premature central and peripheral vascular disease		
Type IV	Endogenous hypertriglyceridemia	Premature vascular disease. Occasional eruptive xanthomatosis – hepatosplenomegaly and pancreatitis	Triglyceride ↑↑ Cholesterol → VLDL increased	Familial hyper-triglyceridemia Familial combined hyperlipidemia
Type V	Mixed hyperlipoproteinemia	Eruptive papular xanthomatosis Hepatosplenomegaly Pancreatitis Lipimia retinalis	Triglyceride ↑↑ Cholesterol ↑ Fasting chylomicronemia VLDL increased LDL and HDL decreased	Apolipoprotein CII deficiency Familial combined hyperlipidemia Familial hyper-triglyceridemia

phages and vascular smooth muscle cells, that internalise these modified LDL. The overloading of these cells with oxidised LDL leads to their transformation into foam cells which constitute the fatty streak, the initial lesion of atheroma. Then, oxidised LDL induces smooth-muscle cells proliferation, cytotoxicity and participates to the constitution of the mature athenoma.
It is now clearly demonstrated that reduction of the LDL cholesterol level significantly decreases the risk of coronary heart disease.

Patrick Duriez and Jean-Charles Fruchart

References

1. Gofman JW et al (1942) J Biol Chem 179:973-978
2. Hanel RJ et al (1955) J Clin Invest 341345-1356
3. Alaupovic P, La Ricerva (1982) Clin Lab 12:3-21
4. Schneeman BO et al (1993) Proc Natl Acad Sci (USA) 90: 2069-2073
5. Goldstein JL, Brown MS (1977) Ann Rev Biochem 46:897-930
6. Herz J (1988) EMBO J 7:4119-4127
7. Acton S et al (1996) Science 271:518-520
8. Friedrickson DS et al (1967) N Engl J Med 276:34-44, 94-103, 148-156, 215-225, 273-281

LPL

Definition   *Lipoprotein lipase*

See: →Lipoproteins

LPS

Definition   *Lipopolysaccharide*

See: →Cytokines in vascular biology and disease; →Nitric oxide; →Thromboxanes

LRP

Definition   *Low density lipoprotein receptor-related protein*

See: →Lipoproteins

LTBP

Definition   *Latent TGF-β binding protein*

See: →Transforming growth factor-β

LTP

Definition   *Long term potentiation*

See: →Fibrinolytic, hemostatic and matrix metalloproteinases, role of

Lymphocytes

Definition   *White-blood cells that produce antibodies (B cells) or have cytotoxic or immuno-modulatory functions (T cells).*

See: →Blood cells, interaction with vascular cells; →Cytokines in vascular biology and disease

M-CSF

Definition   *Macrophage colony stimulating factor*

See: →Colony-stimulating factors; →Cytokines in vascular biology and disease

MAC

Definition   *Membrane-attack complex*

See: →Complement system (interaction of vascular cells with)

Macrophage Scavenger Receptor (MSR)

Definition   *Receptor expressed in macrophages that binds proteins and particles with clustered negative charge*

See: →Blood cells, interaction with vascular cells

Macrophages/Monocytes

Definition   *Inflammatory cells that are involved in phagocytosis*

See: →Blood cells, interaction with vascular cells; →Atherosclerosis

MAPK

Definition   *Mitogen activated protein kinase*

See: →Signal transduction mechanisms in vascular biology

Matrix Metalloproteinases (MMPs)

Synonym: Matrix metalloproteinases (MMPs) or matrixins are Ca-containing Zn-endopeptidases which constitute a family belonging to the super family of zinc metalloproteases.

Definition   *Proteinases that degrade matrix molecules comprising at least 14 members. They are numbered MMP1,2 etc. Besides soluble MMPs, membrane-anchored MMPs (MT-MMP) have also been described.*

See: →Fibrinolytic, hemostatic and matrix metalloproteinases, role of; →Smooth muscle cells; →Angiogenesis inhibitors; →Cytokines in vascular biology and disease

Introduction   The MMP gene family consists of at least 14 structurally related members. They can be defined according to several common characteristics:

- the catalytic mechanism depends on zinc at the active center,
- the proteinases are secreted as zymogens,
- the zymogens can be activated by proteinases or by organomercurials,
- activation is accompanied or followed by a loss of Mr of about 10,000,
- the cDNA sequences all show homology to that of interstitial collagenase (MMP-1),
- the enzymes cleave one or more components of the extracellular matrix (ECM),
- activity is inhibited by specific tissular inhibitors (tissue inhibitors of metalloproteinases, TIMPs).

MMPs can be divided into subgroups of collagenases, gelatinases, stromelysins, and membrane-type MMPs (MT-MMPs) according to their substrate specificity and primary structure (Tables 1 and 4).

Table 1. The Matrix Metalloproteinase Family

MMP sub-groups	Main name	Alternative names	MMP no.
Collagenases	Collagenase-1 (EC 2.4.24.7)	Interstitial collagenase, fibroblast collagenase, vertebrate collagenase	MMP-1
	Collagenase-2 (EC 3.4.24.34)	Neutrophil collagenase	MMP-8
	Collagenase-3	–	MMP-13
Gelatinases	Gelatinase A (EC 3.4.24.24)	72-kDa type IV collagenase	MMP-2
	Gelatinase B (EC 3.4.24.35)	92-kDa type IV collagenase	MMP-9
Stromelysins	Stromelysin-1 (EC 3.4.24.17)	Procollagenase activator, proteoglycanase	MMP-3
	Stromelysin-2 (EC 3.4.24.22)	–	MMP-10
	Stromelysin-3	–	MMP-13
	Matrilysin (EC 3.4.24.23)	Putative mealloproteinase-1(PUMP-1), uterine metalloproteinase, matrin	MMP-7
	Metalloelastase (EC 3.4.24.65)	–	MMP-12
Membrane-type matrix metalloproteinases (MT-MMPs)	MT1-MMP	–	MMP-14
	MT2-MMP	–	MMP-15
	MT3-MMP	–	MMP-16
	MT4-MMP	–	MMP-17

Controlled proteolysis by MMPs plays a main role in detachment and migration of cell, as well as in tissue remodelling in several physiological situations, e.g. developmental tissue morphogenesis, tissue repair, and angiogenesis. MMPs also play a pathogenetic role in excessive degradation of ECM macromolecules , e.g. in rheumatoid arthritis, osteoarthritis, pulmonary diseases, tumor invasion and metastasis, as well as in athero-arteriosclerosis.

Among members of the MMP family, mainly interstitial collagenase (MMP-1), gelatinases A (MMP-2) and B (MMP-9), stromelysin-1 (MMP-3), matrilysin (MMP-7), metalloelastase (MMP-12) and MT1-MMP (MMP-14) have been implicated in vascular tissue injuries. Several different types of cells are able to produce these MMPs: smooth muscle cells, macrophages, and endothelial cells.

Characteristics

Molecular Weight The molecular weights of MMPs involved in vascular pathologies range between 28,000 and 92,000 (Table 2). Particularly, MMP-1 has a predicted molecular weight of 51,929 [1]. MMP-2 and MMP-9 have predicted molecular weights of 72,000 [2] and 78,426 [3] respectively. MMP-3 is synthesized as a preproenzyme form with a calculated size of 53,977 [4]. The zymogen form of MMP-7 has a molecular weight of 28,000 [5] while that of MMP-12 is 54,000 [6]. The predicted molecular weight of the MMP-14 proenzyme is 66,000 [7].

Domains A pre- or signal domain, a latency or pro-domain, a catalytic domain, and a COOH-terminal domain constitute the prototype of MMP domain structure [8]. The signal domain comprises a region of amino acids that cells cleave prior to MMP secretion. The propeptide domain contains a conserved motif PRCGXPD which maintains MMPs in a latent state. The catalytic domain presents a highly conserved zinc-binding site consisting of the HEXGHXXGXXH sequence. Structural integrity of the zinc-binding site is maintained by a strictly conserved "Met-turn" [9]. A hinge region followed by the hemopexin/vitronectin-like domain constitutes the C-terminal domain of MMP. MMP-1 , MMP-3 and MMP-12 contain these different domains [10]. On the contrary, MMP-7 contains only the pre-, pro- and catalytic domains. Additionally, MMP-2 [2] and MMP-9 [3] both contain fibronectin type II-like inserts within the catalytic domain while MMP-9 further exhibits a domain homologous to α_2(V) chain of type V collagen. MMP-14, as a member of the membrane-type MMP sub-group, presents an additional sequence of hydrophobic residues at its C-terminal end, which constitutes the transmembrane domain responsible for cell membrane localization of this MMP; MMP-14 also contains an 11-residue insert [7] located between the propeptide and the catalytic domain, with a RXKR recognition motif for the Golgi-associated proteinase, furin, previously identified in MMP-13 maturation [11].

Binding Sites and Affinity Hemopexin-like domain, present in all MMPs except MMP-7, is generally considered to play a role in matrix binding and specificity. For instance, truncated MMP-1, in which the C-terminal domain has been removed, fails to cleave triple helical collagen; however, such a removal has no effect on the substrate specificity of MMP-3 [12]. As indicated above, both MMP-2 and MMP-9, also designated gelatinase A and gelatinase B respectively, possess additional fibronectin type II modules which contribute to their gelatin affinity [13].

We and others have shown that MMP-2 bound to the cell surface of many cell types [14, 15, 16, 17]. It has been shown that MMP-14 acts as a cell surface TIMP-2 receptor, this complex in turn acting as a receptor for the latent MMP-2, by binding to the carboxyl-end of the enzyme [18, 19]. Recently, Brooks and co-workers suggested that MMP-2 could interact with the integrin $\alpha_V\beta_3$ [20]. MMP-1 [21, 22] and MMP-9 [23, 24] have been also found to be associated on the surface of cells from various origins.

Active MMPs exhibit high affinity for tissue inhibitors of metalloproteinases, TIMPs; interestingly TIMP-1 and TIMP-2 form non-covalent complexes with latent MMP-9 and MMP-2 respectively [10] [Hornebeck and Emonard, this book]. Furthermore, TIMP-4 has been recently shown to also associate with proMMP-2 [25].

Additional Features   Main physical properties of MMPs involved in vascular pathology are summarized in Table 2.

Structure

Sequence and Size   The nucleotide sequence of MMP-1 cDNA codes for a 469 amino acid-long preprocollagenase with a hydrophobic signal peptide of 19 amino acids and possesses two potential N-glycosylation sites, Asn[120] and Asn[143] [1].The cDNA for MMP-2 codes for a mature proenzyme of 631 amino acids with two potential N-linked glycosylation sites, Asn[546] and Asn[613], and a leader sequence of 29 amino acids; however, secreted MMP-2 does not contain N-linked oligosaccharides [2]. The MMP-3 cDNA codes for a 477 amino acid-long preprostromelysin with a hydrophobic signal peptide of 17 amino acids; the protein sequence contains two putative N-glycosylation sites, Asn[120] and Asn[398] [4]. The MMP-7 cDNA has coding capacity for a preproenzyme of 267 amino acids and secreted proMMP-7 contains 250 residues [5]. The cDNA for MMP-9 encodes for a 707-amino acid precursor including a 19 amino acid-long signal peptide; three potential N-linked glycosylation sites are found in the predicted amino acid sequence of MMP-9 at positions 38, 120, and 127 [3]. The deduced MMP-12 sequence comprises 472 amino acids with a signal peptide of 18 residues and three potential sites for N-linked glycosylation; however, secreted MMP-12 does not appear to display significant glycosylation based on the similar SDS-PAGE migration of recombinant MMP-12 derived from bacteria (nonglycosylated) and native enzyme [6]. The MMP-14 cDNA has a large open reading frame which encodes a protein of 582 amino acids with a 20-residue signal peptide [7].

Homologies   As discussed above (§ Characteristics: Domains), all MMPs exhibit two highly conserved sequences: one is PRCGXPD, localized in the propeptide domain, the other being HEXGHXXGXXH present in the active site domain, which shares very high homology with the sequence HELTHA found in the bacterial metalloendopeptidase thermolysin [28]. The C-terminal domain which is present in all MMPs except MMP-7 shows some similarity to the hemopexin family of proteins. This hemopexin domain of four short tandem repeats also occurs twice in vitronectin [8]. Sang and Douglas [26] have recently studied the homology relationship for each member of the MMP family (Table 3). The active site domain of both MMP-2 and MMP-9 exhibits a 175 residue insertion consisting of three 58 or 59 amino acid head to tail repeats, which shares significant homology with the collagen binding domain of fibronectin [2, 3, 29]. In addition, MMP-9 contains a proline-rich insertion of 53 amino acids in the C-terminus of this domain which presents homology with the α_2 chain of type V collagen [3].

Table 2.   Physical Properties of MMPs Involved in Vascular Diseases

MMP	Zymogen form			Active form		
	Mr (Da)	Number of Residues	Isoelectric point[a]	Mr (Da)	Number of Residues	Isoelectric point[a]
MMP-1[1]	52,000 57,000[b]	450	6.89	42,000 47,000[b]	369	6.63
MMP-2[2]	72,000	631	5.04	66,000	551	4.88
MMP-3[4]	57,000 60,000[b]	460	5.68	45,000	376	5.36
MMP-7[5]	28,000	250	7.78	19,000	173	9.63
MMP-9[3]	92,000[b]	688	5.81	84,000[b]	615	5.41
MMP-12[6]	54,000	454	9.01	45,000	371	9.01
MMP-14[27]	63,000	562	8.23	54,000	471	6.65

[a] from ref. [26];   [b] minor glycosylated form

Table 3. Percent Homology of MMPs Involved in Vascular Diseases[a]

Identity	Similarity						
	MMP-1	MMP-2	MMP-3	MMP-7	MMP-9	MMP-12	MMP-14
MMP-1	100	63	71	62	56	65	56
MMP-2	44	100	63	63	66	62	59
MMP-3	54	45	100	65	41	70	55
MMP-7	46	50	48	100	40	65	57
MMP-9	39	50	59	62	100	55	55
MMP-12	50	43	53	47	33	100	57
MMP-14	40	41	39	40	35	41	100

[a] from ref. [26]; Separated by the "100 percent" homologies, the lower left values are the identity percentages and the upper right values are the similarity percentages.

Conformation   The secondary structure of the catalytic domain of the inhibitor-free catalytic domain of human recombinant MMP-1 has been recently elucidated by multidimensional NMR [30]. The overall structure of MMP-1 is composed of a β-sheet consisting of five β-strands in a mixed parallel and anti-parallel arrangement and three α helices. Crystallographic study of the catalytic domain of MMP-1 complexed with an inhibitor revealed that, in addition to the catalytic zinc, there is a second zinc ion and a calcium ion which play a major role in stabilizing the tertiary structure of the molecule [31].

The secondary structure of the catalytic domain of human recombinant short-form MMP-3 complexed with an inhibitor have also been determined by NMR spectroscopy [32]. Three helices have been delineated and a four-stranded β-sheet with three parallel and one antiparallel strands have been identified; this study also demonstrates that the six histidine residues (151, 166, 179, 201, 205, and 211), invariant in the MMP family, form two clusters: one ligates the catalytic zinc (His^{201}, His^{205}, and His^{211}), and the other ligates a structural zinc (His^{151}, His^{166}, and His^{179}).

The X-ray crystal structure of the catalytic domain of recombinant MMP-7 complexed with different inhibitors has been determined at a maximum resolution ranging from 1.9 to 2.4 Å [33], as previously done for the catalytic domain of MMP-1 [31]. MMP-7 shares a remarkable similarity in secondary structure with MMP-1 and MMP-3. Overall, its structure consists of three α helices interspersed among a five-stranded β-sheet. Browner et al [33] have determined that four metal ions were bound to the MMP-7 molecule, a catalytic zinc ion, a structural zinc ion, and two calcium ions. The catalytic zinc is complexed with the three histidine residues in the HEXGHXXGXXH region. The structural zinc is ligated in a similar fashion via three histidine residues upstream of the HEXGHXXGXXH sequence. Binding of the calcium ions appears to stabilize the β-sheet structure, and the active site of MMP-7 lies in the cleft between the β-sheet and the central helix.

The crystal structure of the hemopexin-like C-terminal domain of MMP-2 reveals that it consists of a four-bladed β-propeller protein [34, 35]; the four blades are arranged around a channel-like opening in which a Ca^{2+} ion and a Na^{2+}-Cl ion pair are bound.

Additional Features   The structures for the catalytic domains of several MMPs have recently become available including MMP-1 [31, 36], MMP-3 [37, 38], and MMP-7 [33]. These structures revealed that S'_1 subsite is the most well defined pocket in these MMPs and consists of a hydrophobic pocket which varies greatly in its depth. Mutational analyses of the S'_1 pocket [39] revealed that residue 214 (numbering according [33]) which lies at the bottom of the S'_1 pocket is critical in determining its shape. For MMP-1 and MMP-7, which have an Arg^{214} and a Tyr^{214}, respectively, these residues points into the S'_1 pocket, thus forming a shallow pocket. MMP-3 has a Leu^{214} residue which points away from the pocket, thus allowing a deep S'_1 pocket to exist. MMP-12 also contains a Leu residue at identical position, suggesting that it may also exhibit a deep S'_1 pocket [40]. These S'_1 pocket "types" have been corroborated by substrate specificity studies using libraries of synthetic peptides [41]. MMP-1 and MMP-7, which have shallow S'_1 pockets, prefer small hydrophobic amino acids at the P'_1 position. In contrast, MMP-3 and MMP-12, which have deep S'_1 pockets, can accommodate large and small P'_1 amino acids with similar efficiency.

Gene

Gene Structure   The MMP-1 gene is constituted of 10 exons and spans 8.2 kb; the highly conserved zinc-binding site is encoded by exon 5 [42]. Similarly, MMP-3 gene consists of 10 exons and spans 13 kb [43]. The MMP-12 gene spans 13 kb and contains 10 exons and 9 introns; exon sizes range from 92 bp (exon 9) to 247 bp (exon 2), and introns range from 100 bp (intron 3) to 2,600 bp (intron 8) [44]. Size of the 10 exons and intron-exon borders are highly conserved among the MMPs. The MMP-7 gene of approximately 9.65 kb, is composed of 6 exons [45]. As compared with other MMP genes, those encoding MMP-2 [46] and MMP-9 [47] are considerably larger (26-27kb) and consist of 13 exons, all intron locations of the MMP-2 gene coinciding with intron locations in the MMP-9 gene [47]. The precise arrangement of exons is conserved in the different members of the MMP gene family: the pre-domain and a portion of the pro-domain are contained in exon 1, exon 2 encodes the remainder of the pro-domain and the N-terminal portion of the catalytic domain, the catalytic domain is spread over exons 2-5, with the zinc binding region located in exon 5 , and the C-terminal domain arises from exons 6-10. MMP-7, which lacks the C-terminal hemopexin-like domain, has a unique sixth exon which encodes the final 9 amino acids of the coding sequence and the 3'-untranslated sequences [45]. MMP-2 and MMP-9 have three additional exons encoding internal repeats which resemble the collagen-binding domains of fibronectin [46, 47]. The sequence coding for a unique 48-residue segment in MMP-9, homologous to $\alpha_2(V)$ chain of type V collagen, is not present in a separate exon but is contained in exon 9 which also codes for sequences with homology to the other MMPS [47].

Chromosomal Localization   The chromosomal localization of these MMPs has been determined (Genome Database: http://www.gdb.org). A cluster containing several MMP genes has been identified in the long arm of chromosome 11. Genes coding for MMP-1, MMP-3, MMP-12 and MMP-7 have been mapped respectively in this order, from centromere to telomere, in the locus 11q22.2-22.3 [44, 48]. The MMP-2 gene maps to 16q13 [46], while the gene coding for MMP-9 maps to chromosome 20 at 20q12-13 [49]. Gene for MMP-14 is assigned to 14q12.2 [50].

Gene Expression   A single mRNA species of 2.5 kb codes for MMP-1 [1]. Northern analysis reveals a single mRNA band of 3.1 kb which hybridizes to the MMP-2 cDNA [2]. MMP-3 is encoded within a single 2.3 kb mRNA [4]. The MMP-7 cDNA probe detects a 1.2 kb mRNA by Northern blot analysis [5]. The mRNA encoding MMP-9 has a size of 2.3 kb [3]. MMP-12 and MMP-14 transcripts have sizes of 1.8 and 4.5 kb respectively [6, 27].

Gene Regulation   The basal gene expression of most metalloproteinases (MMP-1, MMP-3, MMP-7, MMP-9, MMP-12) is low but their mRNAs can be induced by a variety of cytokines and growth factors, such as tumor necrosis factor-α (TNF-α), interleukin-1 (IL-1), IL-6, epidermal growth factor (EGF), platelet-derived growth factor (PDGF), or fibroblast growth factor (FGF) [51], but also by oncogen products or tumor promoters (12-O-tetradecanoylphorbol-13-acetate, TPA). A canonical TATA box has been described in these MMPs. The 5'-flanking regulatory regions of these inducible genes contain an AP-1 *cis*-regulatory element (or TRE for "TPA-responsive element") in the proximal promoter approximately at position -70 [44, 45, 52]. The extracellular stimuli mentioned above induce the activation of nuclear AP-1 transcription factor complex that binds to the AP-1 site and initiates transcription. AP-1 dimers are composed of members of Jun and Fos gene families [53]. The expression of *c-fos/c-jun* complex results from activation of three classes of mitogen activated protein kinases (MAPKs), extracellular stimulus regulated kinase (ERK), stress activated protein kinase/Jun N-terminal kinases (SAPK/JNKs), and p38 [53]. The promoter regions of these MMPs contain another *cis*-element, a PEA3 binding site, which number, position, and orientation vary considerably among proteinases. The PEA3 motif serves as a binding site for the products of the ETS gene family [54]. Both AP-1 and PEA3 act synergistically and confer responsiveness to tumor promoters and a variety of cytokines, growth factors, and oncogenes [55, 56]. Induction of metalloproteinase gene expression can also be negatively regulated. Transforming growth factor-β (TGF-β) has been reported to block the induction of expression of several MMPs by binding to an inhibitory element of the promoter, the TGF-β-inhibitory element (TIE). TIE has been shown in MMP-1 [52], MMP-3 [57], and MMP-7 [45]. Other potential *cis*-acting elements have been described. Thus, the promoter region of MMP-7 contains different additional sites with similarity to consensus binding sequences for factors such as glucocorticoids (GATA-1), interferon-γ, C/EBP, and NF-IL6 [58]. The MMP-9 promoter contains binding sites for NF-κB and SP1, which are not present in the MMP-1 or MMP-3 promoters [59]. The MMP-12 promoter contains potential *cis*-acting elements including LBP-1 and TRF, of which functional significance is however uncertain [44].

On the contrary, MMP-2 is constitutively expressed. Unlike other MMP promoters, the MMP-2 promoter lacks the TRE/AP-1-binding site [60], the upstream TIE and the PEA3 transcription elements [61]. It contains a noncanonical TATA box and two SP1 sites [62]. Additionally, a binding site for p53, a tumor suppressor and a transcription factor, has recently been identified in the promoter of the gene encoding MMP-2 [63].

Additional Features   Different agents modulate MMP expression by vascular smooth muscle cells (SMCs). For instance, basic FGF [64] and IL-4 [65] both induce the expression of MMP-1 by SMCs, which is not expressed by these cells in basal culture conditions; however, basic FGF has no effect on MMP-2, MMP-3 or MMP-9 production [64]. TNF-α induces a rapid activation of the c-ets-1 gene, which codes a transcription factor known to regulate several MMPs involved in matrix degradation during SMC migration [66]. Platelet-derived growth factor also stimulates MMP-1 synthesis by these cells [67]. Lipid hydroperoxides such as linoleic acid hydroperoxide have been shown to stimulate MMP-1 synthesis by human aortic intimal SMCs [68]. Recently, 17β-oestradiol, which possesses anti-atherosclerotic properties, has been demonstrated to increase MMP-2 expression in human coronary artery and umbilical artery vascular SMCs [69]. Heparin appears to have an opposite effect, by inhibiting the expression of MMP-1, MMP-3 and MMP-9 [70].

Angiogenic factors, such as acid and basic FGFs, VEGF, and EGF, induce the expression of MMP-1 mRNA by vascular endothelial cells [71]. TNF-α or IL-1α both enhance MMP-9 expression by these cells. It has recently been shown that α-thrombin up-regulates MMP-1 and MMP-3 production by artery endothelial cells [72] and that platelet factor 4 inhibits this stimulatory effect [73].

Processing and Fate   All MMPs are synthesized as pre-proenzymes, and most of them are secreted from cells as proenzymes or zymogens [74]. The latency of MMPs is dependent on "cysteine switch" formed by interaction of a conserved cysteine in the propeptide with the zinc in the highly conserved catalytic site, blocking the access of the catalytic site to substrate [75, 76, 77]. The activation processes encompass three different mechanisms: i) stepwise activation in the extracellular space, ii) intracellular activation and iii) activation on the cell surface.

Stepwise activation in the extracellular space: the zymogen of most MMPs reviewed (MMP-1, -3, -7, -9, and -12) can be activated by proteinases and by non-proteolytic compounds such as SH reactive agents and denaturants [74]. Proteinases first hydrolyze the proteinase susceptible "bait" region located in the middle of the propeptide. This cleavage induces conformational changes in the propeptide and allows the final activation site to be cleaved by a second proteolytic step. This latter reaction is generally catalyzed by a MMP, except the first proteinase activator. For instance, treatment of proMMP-1 with plasmin or plasma kallikrein gives intermediates which can be fully activated by MMP-3 [78]. In a similar fashion, MMP-7 activates proMMP-1 [79, 80]. Pro MMP-2 is readily activated by a mercurial compound but is resistant to activation by many endopeptidases [81]. MMP-1 [80] and MMP-7 [80, 82] have been shown to partially activate proMMP-2; however action of MMP-7 remains controverted [79]. Significantly, thrombin has been demonstrated to activate proMMP-2, even when the proenzyme is bound to TIMP-2 [83, 84]. ProMMP-3, which appears to be resistant to proteolytic activation by other MMPs, is activated by serine proteinases such as neutrophil elastase, plasma kallikrein, and plasmin [85]. ProMMP-7 is partially activated by plasmin and neutrophil elastase, while MMP-3 activates proMMP-7 to its full activity in a single-step mechanism [79]. ProMMP-9 can be activated by serine proteinases such as plasmin [86] or tissue kallikrein [87] but also by other MMPs including MMP-3 [88] and MMP-2 [89].

Intracellular activation: the propeptide domain of MT1-MMP contains a sequence RXKR which has been previously described in stromelysin-3 (MMP-11). Pei and Weiss [90] first found that this sequence was recognized by a Golgi-associated subtilisin-like proteinase, furin, thus allowing intracellular activation of proMMP-11. More recently, they demonstrated that this intracellular process also occurred with latent MT1-MMP [91]. Subsequently, Sato et al [92] reported that MT1-MMP expressed in *Escherichia coli* was activated by furin *in vitro*. However, a recent study suggests that latent MT1-MMP could be transported to the plasma membrane and activated by plasmin extracellularly [93]. Recent studies also suggest that proMMP-2 could also be activated intracellularly [94, 95].

Cell surface activation of proMMP-2:   the cell surface localization of proMMP-2 activation is considered to be important for pericellular degradation of extracellular matrix during cell migration. Plasma-membrane dependent activation of proMMP-2 has been demonstrated in human vascular endothelial cells [96]. Expression of MMP-14 correlated with proMMP-2 activation has recently been been found in rabbit neointimal development [97]. We have previously shown that TIMP-2 bound to the surface of tumor cells [98]. Strongin et al [18] proposed that TIMP-2 contributes to proMMP-2 activation by binding to activated MT1-MMP in the plasma membrane. A recent study suggests that TIMP-2 binds to the zinc catalytic site of MT1-MMP [19]. This bimolecular complex then binds proMMP-2 to form a trimolecular complex which allows the presentation of proMMP-2 to a neighbouring TIMP-2-free active MT1-MMP [99]. This second MT1-MMP molecule cleaves the Asn^{37}-Leu^{38} bond of proMMP-2 to generate an intermediate MMP-2 species [100]. The mechanism by which such an intermediate form is processed to the mature MMP-2 species remains controverted. Sato at al [101] suggested that cell-surface binding concentrates the MMP-2 intermediate form locally to allow autoproteolytic processing to the fully active form. On the other hand, recent data strongly suggest the participation of the urokinase-plasmin system in this second step of proMMP-2 activation [86, 102]. Heparan sulfate [99] and integrin $\alpha_V\beta_3$ [103] could also participate in proMMP-2 activation by MT1-MMP at the cell surface.

We have previously shown that once bound to elastin fibers, proMMP-9 was resistant to proteolytic activation but remained sensitive to an organomercurial compound [104]. These data suggested that, *in vivo*, non-proteolytic processing of proMMP-9 could occur, such as activation by reactive oxygen species produced by macrophages-derived foam cells [105]. On the other hand, proMMP-2 bound to elastin was activated by autolysis [104], as previously described for proMMP-2 bound to heparin [106]. The proteolytic activity of MMPs is inhibited by non-specific inhibitors, e.g. α_2-macroglobulin and α_1-antiprotease [107], as well as specific inhibitors, TIMPs [Hornebeck and Emonard, this book].

Biological Activity   Proteolysis of the extracellular matrix (ECM) appears to be the major function exerted by MMPs. Table 4 summarizes the ECM subtrates cleaved by MMPs, almost *in vitro*. Fibulin-2 has an affinity for the platelet $\alpha_{IIb}\beta_3$ [108]. This macromolecule might have importance during vascular injuries since it is abundant in small and large vessel walls [109], and is degraded by MMP-7 [110]. Vitronectin can bind to plasminogen activator inhibitor type 1, plasminogen, complement complex, and antithrombin III in the circulation, and to collagen and elastin in extracellular matrix [111]; this ECM component, which is a substrate for cellular adhesion and stimulates motility of SMCs [112] is degraded by several MMPs including MMPs-2, -3, -7, and -9 [113].

Table 4. ECM Substrate Specificity

MMP	ECM substrate specificity
MMP-1	Fibrillar collagens (III>I), II, VII, VIII, X, entactin
MMP-2	Gelatin, collagens I, IV, V, VII, X, XI, fibronectin, laminin, elastin, SPARC, aggrecan, decorin, entactin, vitronectin, large tenascin-C
MMP-3	Collagens II, III, IV, V, IX, X, XI, gelatin, fibronectin, laminin, entactin, SPARC, aggrecan, perlecan, decorin, vitronectin, large tenascin-C
MMP-7	Same as MMP-3, elastin, fibulin-2, small tenascin-C
MMP-9	Gelatin, collagens IV, V, XIV, elastin, entactin, SPARC, aggrecan, vitronectin
MMP-12	Same as MMP-3, elastin
MMP-14	Fibrillar collagens (I>III), II, gelatin, fibronectin, laminin, nidogen, large tenascin-C, vitronectin, aggrecan, perlecan

Evidence is emerging that secreted or transmembrane MMPs play a crucial role in development, differentiation, and cell motility. We pinpointed above (Processing and Fate) that different MMPs, such as MMPs-1, -2, -3, -7, and -14, were able to activate proMMPs, leading to a MMP-activation cascade and initiating an amplification of ECM breakdown. Almost all MMPs are true serpinases. For instance, α_1-antitrypsin is degraded by MMP-3 [114], MMP-7 [115], and MMP-12 [40]; α_1-antichymotrypsin and antithrombin III are inactivated by MMPs-1, -2 and -3 [116, 117].

MMPs also exert activities against growth factors and/or their receptors. MMPs-1, -2, and -3 cleave insulin-like growth factor-binding proteins (IGFBP)-3 [118] and -5 [119]. MMP-2 releases active soluble ectodomain of FGF receptor 1 which may modulate the mitogenic and angiogenic activities of FGF [120]. It has been suggested that MT1-MMP could process a proTNF-α fusion protein to release mature TNF [121].

Degradation of ECM macromolecules by MMPs allows the release of soluble factors. For instance, digestion of perlecan by MMP-3 liberates basic FGF [122], while degradation of decorin by MMPs-2, -3, and -7 releases TGF-β_1 [123]. Furthermore, peptides derived from ECM component breakdown or cryptic sites unmasked by MMP activity can exhibit properties totally different from those of the intact macromolecule. Cleavage of plasminogen by MMP-7 and MMP-9 [124], but also by MMP-3 [125] or MMP-12 [126] generates angiostatin, a potent inhibitor of angiogenesis. MMP-3 generates a fibronectin fragment that inhibits Schwann cell proliferation [127]. Specific cleavage of laminin-5 by MMP-2 induces migration of breast epithelial cells [128].

In culture, overexpression of MMP-3 in a mammary cell strain that contains both epithelial and mesenchymal cell types produces an involution-like phenotype with induction of apoptosis [129]. Recently, Vu et al [130] using MMP-9-null mice, demonstrated that this MMP was a key regulator of growth plate angiogenesis and apoptosis of hypertrophic chondroblasts.

Together these data indicate that MMPs have a broad spectrum of activities and that, in association with other proteinases, are key modulators of cell behavior.

Role in Vascular Biology

Physiological Function   Matrix metalloproteinases, belonging to the metzincin family, are proteolytic enzymes playing a main contribution to tissue development and tissue repair following injury [8]. Arterial wall homeostasis is conditioned by the correct balance between anabolic and catabolic events. Chemical irritants as in tobacco smoke or shear stress in hypertension can injure endothelium and impair subsequent arterial wall wound healing lead to atherosclerosis. MMPs are involved in all steps of atherosclerosis progression. They participate in endothelium repair and can induce platelet aggregation [131]. MMPs are major determinants in the migration of monocytes-macrophages [132] to the lesion site and actively contribute to the phenotypic modulation of aorta SMCs leading to neointima formation [133]. Finally, they are responsible for plaque rupture and formation of aneurysms [134].

The migration of arterial endothelial cells and SMCs is essential in normal vessel development [133, 135]. The formation of capillaries from preexisting blood vessels, as designated angiogenesis, is required during wound healing; it must be emphasized, however, that in adults, vascular turnover is low and both cell types are in a quiescent state under physiological conditions [136]. The invasive process occurring during angiogenesis is spatially and temporally restricted [137]. The initial fragmentation of basal lamina is attributed mainly to MMPs and the main angiogenic stimuli, i.e. FGF and VEGF, are potent inducers of MMP expression [138, 139]. Although MMP-1 was initially believed to be the major contributor to angiogenesis *in vitro* [140], other MMPs, as well as enzymes of the plasmin system, participate in angiogenesis [134, 141]. For instance, addition of MMP-2 can induce capillary tube formation in matrigel [142]. In recent years, the importance of focalized proteolysis in cell invasion was demonstrated [143]. Concentrating proteolytic events near or at the cell surface can indeed be more effective since enzyme activation processes are amplified and proteinases can escape inhibition by natural inhibitors. Angiogenesis depends on both cell adhesion and proteolytic mechanisms but these two processes can be intimately related. It was recently shown that MMP-2 and integrin $\alpha_V\beta_3$ are functionally associated on the surface of angiogenic blood vessels [103]. The MMP-catalyzed migration of endothelial cells during angiogenesis may be autoregulated by at least four different mechanisms: i) tissue inhibitors of metalloproteinases (TIMPs), also expressed by endothelial cells, may interfere with the activation and activity of MMPs [137], ii) besides degrading extracellular matrix macromolecules, MMPs may proteolyze growth factor receptors, i.e. FGF receptor 1 was shown to serve as a specific target for MMP-2 on the cell surface [120], iii) macrophage-elastase (MMP-12) was found to process efficiently plasminogen, generating angiostatin, a 38-kDa internal fragment which has been shown to have potent antiangiogenic activity [126]; similarly, angiostatin-like fragments may be generated from plasminogen by MMP-3 [125], MMP-7 or MMP-9 [124], iiii) activation followed by autolysis of MMP-2 may lead to enzyme fragment as PEX which was shown to prevent MMP-2 binding to $\alpha_V\beta_3$ and consequently to inhibit cell surface collagenolytic activity [20].

Within the media of fetal arteries, SMCs are in a synthetic state. Following birth, these cells display a phenotypic modulation from a synthetic to a contractile state [133]. Contractile SMCs are characterized by absence of cell division, low metabolic activity and a high actin/myosin content. The main function of these cells in such a state is mechanical: they are essential in maintaining vessel wall rigidity and elasticity and controlling blood pressure.

Pathology   Coronary artery disease and stroke, major causes of death in western societies, are consequences of atherosclerosis. Although atherosclerosis has multiple origins, it can be considered, as a whole, as resulting from impairment of vascular extracellular matrix remodelling over decade(s), a process where MMPs play a pivotal function. Defect in tissue repair may however have different clinical manifestations. At one end are atherosclerotic plaques consisting of a bulk of extracellular matrix, protuding into the lumen and further restricting blood flow: *aortic occlusive disease*. At the other end are lesions with a thin cap but prone to rupture at their shoulder areas: *aortic aneurysmal disease* [144].

Using different technical approaches, several investigators have shown that MMP expression is markedly increased in arteries of patients suffering from atherosclerosis. That holds for MMPs-1, -3, and -9 and enzyme levels appeared more intense in plaque shoulders and regions of foam cell accumulation [145, 146, 147, 148]. Expression of MT1-MMP and activation of MMP-2 occured in rat carotid subjected to balloon catheter injury [149]. With the exception of MMP-2, normal arteries do not express MMPs [145]; within atherosclerotic plaques, two cell types appear to produce those enzymes: the smooth muscle cell and the macrophage.

Reversal of the contractile phenotype of SMC to a motile and synthetic one is currently observed with aging and atherosclerosis. Platelet products as serotonin, PDGF, TGF-β, norepinephrine and histamine are inducers of SMC migration while having no effect on cell prolifera-

tion [133]. Those agents, as well as mechanical forces, could increase MMP expression by SMCs, favoring cell migration through basement membrane and connective tissue barriers [67, 150]. Such an hypothesis was substantiated by the ability of synthetic MMP inhibitors to block SMC migration both *ex vivo* and *in vivo* [151]. It was recently demonstrated, however, that inhibiting SMC migration with MMP inhibitors could not impede lesion growth [152]. Among MMPs, gelatinases, i.e. MMP-2 and MMP-9, probably are key actors in SMC migration and formation of neointima in keeping with their elastolytic activity [153]. Elastic fibers, indeed, can represent the main barrier to cell locomotion.

Margins of fibrous cap are characterized by an important influx of macrophages which accumulate lipid droplets. Such foam cell-rich regions represent vulnerable areas to plaque rupture, formation of hematoma and thrombosis. Macrophages can intervene in tissular proteolysis at two distinct levels. As monocytes differentiate into macrophages, they switch their proteolytic profile from serine proteinases to MMPs. Also, those cells are able to produce a vast panel of cytokines and thus to induce neighbouring cells to produce neutral proteinases. Here again, to our belief, MMPs displaying elastase activity probably play a crucial role in plaque rupture. We initially demonstrated that cholesterol loaded macrophages, using acetylated LDL, secreted high levels of elastase activity [154]. Among metalloelastases produced by macrophages are MMP-9 (gelatinase B), MMP-7 (matrilysin) and MMP-12 (macrophage-elastase). MMP-9 is frequently observed at the site of degenerated internal elastin lamina [155]; recently, it has been shown that MMP-7 is expressed by lipid-laden macrophages at sites of potential rupture in atherosclerotic lesion [156]. Within MMP family, MMP-7 is unique since it lacks hemopexin-like domain; such a property confers to the enzyme a broader specificity but also a reduced propensity to associate with TIMPs [156]. Finally, targeted mutagenesis experiments have clearly demonstrated the requirement of MMP-12 expression for macrophage-mediated extracellular matrix proteolysis and tissue invasion, both *in vitro* and *in vivo* [157]. Thus, MMP-12 is abundantly produced at both the mRNA and protein levels in the aorta of cholesterol-fed rabbits, whereas no expression is observed in the normal aortas [158]. The principal source of MMP-12 is macrophage foam cells that have infiltrated the atherosclerotic intima [158].

Clinical Relevance and Therapeutic Implications   Atherosclerosis can progress either towards occlusion or aneurysmal diseases and the balance between matrix, MMP and TIMP expression will mainly dictate such an evolution. For instance, imbalance favouring matrix deposition contributes to restenosis after angioplasty and endorectomy. A stromelysin (MMP-3) polymorphism in its promoter (insertion/deletion of an adenosin (A) at position approximately 1600 bp upstream from the start of transcription) was recently evidenced. Individuals with the 6A6A genotype showed decreased MMP-3 expression associated with progression of atherosclerosis, as determined by quantitative angiography [159].

As an oversimplification, treatment of occlusive disease characterized by an excess of matrix deposition will require additional proteolysis and MMP expression. On the contrary, control of aneurysms, characterized by extensive tissue degradation, makes it necessary to inhibit proteolysis. Prognostic tools need to be developed in order to evaluate the evolution of the atherosclerotic disease. Measurements of matrix degradation products or/and MMPs and TIMPs in the circulation could be means. We recently analyzed the amount of elastic fibers in the skin and temporal arteries of healthy aged individuals by automated image analysis; a correlation was found between the area fraction occupied by the elastic fibers in the unexposed skin and those in the deep part of the temporal artery, suggesting that skin biopsies might be a valuable diagnostic tool for predicting arterial wall abnormalities of elastic fibers [160].

A series of MMP inhibitors has been synthesized and pseudopeptides such as marimastat are presently used in clinical trial for cancer therapy [161]. Inhibitors against elastase-type MMPs, i.e. MMP-9, MMP-7 and MMP-12, might have therapeutic value in aneurysmal disease. It needs to be emphasized that these enzymes may be controlled at the level of expression, activation or activity. Although a proteolytic cascade is involved in MMP activation, active oxygen species can also process zymogen forms of MMPs. Reactive oxygen species are indeed known to react with the thiol group of cysteines involved in preserving MMP latency. Foam cells *in vivo* were shown to produce superoxide, nitric oxide and hydrogen peroxide [105]. Therefore, combination of antioxidants and MMP inhibitors could be envisaged as a novel perspective for aneurysm treatment.

Acknowledgements: We wish to thank "le Centre de la Recherche Scientifique" (CNRS, UPRESA 6021), "la Fondation pour la Recherche Médicale" (FRM), "l'Association pour la Recherche contre le Cancer" (ARC, n· WH/1236) and Europol.Agro (University of Reims-Champagne Ardenne) for financial support.

Hervé Emonard and William Hornebeck

References

1. Goldberg GI et al (1986) J Biol Chem 261:6600-6605
2. Collier IE et al (1988) J Biol Chem 263:6579-6587
3. Wilhelm SM et al (1989) J Biol Chem 264:17213-17221
4. Wilhelm SM et al (1987) Proc Natl Acad Sci USA 84:6725-6729
5. Muller D et al (1988) Biochem J 253:187-192
6. Shapiro SD et al (1993) J Biol Chem 268:23824-23829
7. Sato H et al (1994) Nature 370:61-65
8. Woessner JF (1991) FASEB J 5:2145-2154
9. Stocker W et al (1995) Protein Sci 4:823-840
10. Birkedal-Hansen H et al (1993) Crit Rev Oral Biol Med 4:197-250
11. Basset P et al (1990) Nature 348:699-704
12. Murphy G et al (1992) J Biol Chem 267:9612-9618
13. Banyai L et al (1996) J Biol Chem 271:12003-12008
14. Emonard HP et al (1992) Cancer Res 52:5845-5848
15. Beranger JY et al (1994) Cell Biol Int 18:715-722
16. Menashi S et al (1998) Int J Cancer 75:259-265
17. Monsky WL et al (1993) Cancer Res 53:3159-3164
18. Strongin AY et al (1995) J Biol Chem 270:5331-5338
19. Zucker S et al (1998) J Biol Chem 273:1216-1222
20. Brooks PC et al (1998) Cell 92:391-400
21. Moll UM et al (1990) Cancer Res 50:6995-7702
22. Omura TH et al (1994) J Biol Chem 269:24994-24998
23. Arkona C, Wiederanders B (1996) Biol Chem 377:695-702
24. Toth M et al (1997) Cancer Res 57:3159-3167
25. Bigg HF et al (1997) J Biol Chem 272:15496-15500
26. Sang QA, Douglas DA (1996) J Prot Chem 15:137-160
27. Takino T et al (1995) Gene 155:293-298
28. Matthews BW et al (1974) J Biol Chem 249:8030-8044
29. Banyai L et al (1994) Biochem J 298:403-407
30. Moy FJ et al (1998) Biochemistry 37:1495-1504
31. Lovejoy B et al (1994) Science 263:375-377
32. Gooley PR et al (1993) Biochemistry 32:13098-13108
33. Browner MF et al (1995) Biochemistry 34:6602-6610
34. Libson et al (1995) Nature Struct Biol 2:938-942
35. Gohlke U et al (1996) FEBS Lett 378:126-130
36. Borkakoti N et al (1994) Nature Struct Biol 1:106-110
37. Gooley PR et al (1994) Nature Struct Biol 1:111-118
38. Spurlino JC et al (1994) Proteins 19:98-109
39. Welch AR et al (1996) Biochemistry 35:10103-10109
40. Gronski TJ et al (1997) J Biol Chem 272:12189-12194
41. Nagase H, Fields GB (1996) Biopolymers (Peptide Sci) 40:399-416
42. Collier IE et al (1988) J Biol Chem 263:10711-10713
43. Sirum KL, Brinckerhoff CE (1989) Biochemistry 28:8691-8698
44. Belaaouaj A et al (1995) J Biol Chem 270:14568-14575
45. Gaire M et al (1994) J Biol Chem 269:2032-2040
46. Collier IE et al (1991) Genomics 9:429-434
47. Huhtala P et al (1991) J Biol Chem 266:16485-16490
48. Knox JD et al (1996) Cytogenet Cell Genet 72:179-182
49. Lin R et al (1996) Cytogenet Cell Genet 72:159-161
50. Sato H et al (1997) Genomics 39:412-413
51. Mauviel A (1993) J Cell Biochem 53:288-295
52. Angel P et al (1987) Mol Cell Biol 7:2256-2266
53. Karin M et al (1997) Curr Opin Cell Biol 9:240-246
54. Gutman A, Wasylyk B (1991) Trends Genet 7:49-54
55. Gutman A, Wasylyk B (1990) EMBO J 9:2241-2246
56. Buttice G et al (1996) Oncogen 13:2297-2306
57. Kerr LD et al (1990) Cell 61:267-278
58. Wilson CL, Matrisian LM (1996) Int J Biochem Cell Biol 28:123-136
59. Sato H et al (1993) J Biol Chem 268:23460-23468
60. Frisch S, Morisaki JH (1990) Mol Cell Biol 10:6524-6552
61. Corcoran ML et al (1996) Enzyme Prot 49:7-19
62. Templeton NS, Stetler-Stevenson WG (1991) Cancer Res 51:6190-6193
63. Bian J, Sun Y (1997) Mol Cell Biol 17:6330-6338
64. Pickering JG et al (1997) Arterioscler Thromb Vasc Biol 17:475-482
65. Sasaguri T et al (1998) Atherosclerosis 138:247-253
66. Jovinge S et al (1997) Arterioscler Thromb Vasc Biol 17:490-497
67. Yanagi H et al (1992) Atherosclerosis 91:207-216
68. Arima N et al (1996) Biochem Biophys Res Commun 225:34-39

69. Wingrove CS et al (1998) Biochim Biophys Acta 1406:169-174
70. Kenagy RD et al (1994) J Clin Invest 93:1987-1993
71. Iwasaka C et al (1996) J Cell Physiol 169:522-531
72. Duhamel-Clérin E et al (1997) Arteriosler Thromb Vasc Biol 17:1931-1938
73. Klein-Soyer C et al (1997) CR Acad Sci Paris Life Sciences 320:857-868
74. Nagase H (1997) Biol Chem 378:151-160
75. Springman EB et al (1990) Proc Natl Acad Sci USA 87:364-368
76. Van Wart HE et al (1990) Proc Natl Acad Sci USA 87:5578-5582
77. Windsor LJ et al (1991) Biochemistry 30:641-647
78. Suzuki K et al (1990) Biochemistry 29:10261-10270
79. Imai K et al (1995) J Biol Chem 270:6691-6697
80. Sang QA et al (1996) J Prot Chem 15:243-253
81. Okada Y et al (1990) Eur J Biochem 194:721-730
82. Crabbe T et al (1994) FEBS Lett 345:14-16
83. Galis ZS et al (1997) Arterioscler Thromb Vasc Biol 17:483-489
84. Zucker S et al (1995) J Biol Chem 270:23730-23738
85. Nagase et al (1990) Biochemistry 29:5783-5789
86. Baramova EN et al (1997) FEBS Lett 405:157-162
87. Desrivières S et al (1993) J Cell Physiol 157:587-593
88. Ogata Y et al (1992) J Biol Chem 267:3581-3584
89. Fridman R et al (1995) Cancer Res 55:2548-2555
90. Pei D, Weiss SJ (1995) Nature 375:244-247
91. Pei D, Weiss SJ (1996) J Biol Chem 271:9135-9140
92. Sato H et al (1996) FEBS Lett 393:101-104
93. Okumura Y et al (1997) FEBS Lett 402:181-184
94. Lee AY et al (1997) Proc Natl Acad Sci USA 94:4424-4429
95. Li L et al (1997) Exp Cell Res 232:322-330
96. Lewalle JM et al (1995) J Cell Physiol 165:475-483
97. Wang H, Keiser JA (1998) J Vasc Res 35:45-54
98. Emmert-Buck MR et al (1995) FEBS Lett 364:28-32
99. Butler G et al (1998) J Biol Chem 273:871-880
100. Kinoshita T et al (1996) Cancer Res 56:2535-2538
101. Sato H et al (1996) FEBS Lett 385:238-240
102. Mazzieri R et al (1997) EMBO J 16:2319-2332
103. Brooks PC et al (1996) Cell 85:683-693
104. Emonard H, Hornebeck W (1997) Biol Chem 378:265-271
105. Rajagopalan S et al (1996) J Clin Invest 98:2572-2579
106. Crabbe T et al (1993) Eur J Biochem 218:431-438
107. Travis J, Salvesen GS (1983) Annu Rev Biochem 52:655-709
108. Pfaff M et al (1995) Exp Cell Res 219:87-92
109. Pan TC et al (1993) J Cell Biol 123:1269-1277
110. Sasaki T et al (1996) Eur J Biochem 240:427-434
111. Preissner KT (1991) Annu Rev Cell Biol 7:275-310
112. Delannet M et al (1994) Development 120:2687-2702
113. Imai K et al (1995) FEBS Lett 369:249-251
114. Winyard PG et al (1991) FEBS Lett 279:91-94
115. Zhang Z et al (1994) Biochim Biophys Acta 1199:224-228
116. Desrochers PE et al (1991) J Clin Invest 87:2258-2265
117. Mast AE et al (1991) J Biol Chem 266:15810-15816
118. Fowlkes JL et al (1994) 269:25742-25746
119. Thrailkill K et al (1995) Endocrinology 136:3527-3533
120. Levi E et al (1996) Proc Natl Acad Sci USA 93:7069-7074
121. D'Ortho MP et al (1997) Eur J Biochem 250:751-757
122. Whitelock JM et al (1996) J Biol Chem 271:10079-10086
123. Imai K et al (1997) Biochem J 322:809-814
124. Patterson BC, Sang QA (1997) J Biol Chem 272:28823-28825
125. Lijnen HR et al (1998) Biochemistry 37:4699-4702
126. Dong Z et al (1997) Cell 88:801-810
127. Muir D, Manthorpe M (1992) J Cell Biol 116:177-185
128. Giannelli G et al (1997) Science 277:225-228
129. Boudreau N et al (1995) Science 267:891-893
130. Vu T et al (1998) Cell 93:411-422
131. Sawicki G et al (1997) Nature 386:616-619
132. Wahl LM et al (1997) Transplant Immunol 5:173-176
133. Van Leeuwen RTJ (1996) Fibrinolysis 10:59-74
134. Carmeliet P et al (1997) Nature Genet 17:439-444
135. D'Amore PA, Thompson RW (1987) Annu Rev Physiol 49:453-464
136. Hobson B, Denckamp J (1984) Br J Cancer 49:405-413
137. Moses MA (1997) Stem Cell 15:180-189
138. Birkedal-Hansen H (1995) Curr Opin Cell Biol 7:728-735
139. Unemori EN et al (1991) J Cell Physiol 153:557-562
140. Fisher C et al (1994) Dev Biol 162:499-510
141. Hanemaaijer R et al (1993) Biochem J 296:803-809
142. Schnaper HW et al (1993) J Cell Physiol 156:235-246
143. Basbaum CB, Werb Z (1996) Curr Opin Cell Biol 8:731-738
144. Thompson RW et al (1995) J Clin Invest 96:318-326
145. Galis ZS et al (1994) J Clin Invest 94:2493-250
146. Henney AM et al (1991) Proc Natl Acad Sci USA 88:8154-8158
147. Knox JB et al (1997) Circulation 95:205-212
148. Newman KM et al (1994) Arterioscler Thromb 14:1315-1320
149. Jenkins GM et al (1998) Circulation 97:82-90
150. James TW et al (1993) J Cell Physiol 157:426-437
151. Southgate KM et al (1992) Biochem J 288:93-99
152. Bendeck MP et al (1996) Circ Res 78:38-43
153. Senior RM et al (1991) J Biol Chem 266:7870-7875
154. Rouis M et al (1990) Arteriosclerosis 10:246-255
155. Nikkari ST et al (1996) Am J Pathol 149:1427-1433
156. Halpert I et al (1996) Proc Natl Acad Sci USA 93:9748-9753
157. Shipley JM et al (1996) Proc Natl Acad Sci USA 93:3942-3946
158. Matsumoto SI et al (1998) Lab Invest 153:109-119
159. Ye S et al (1998) Clin Sci 94:103-110
160. Gogly B et al (1998) Gerontology 44:318-323
161. Rasmussen HS, McCann PP (1997) Pharmacol Ther 75:69-75

MCP

Definition   *Membrane cofactor protein*

See: →Complement system (interaction of vascular cells with)

MCP-1, -2 etc.

Definition   *Monocyte chemoattractant protein-1, -2 etc*

See: →Atherosclerosis; →Thrombosis

MCP-1, -2 etc.

Definition   *Macrophage chemotactic protein-1, -2 etc*

See: →Atherosclerosis; →Thrombosis

Megakaryocytes
Synonym: Platelet precursors

Definition   *Precursors of platelets in the bone marrow*

Introduction   The megakaryocyte was given its name by William Henry Howell in 1890 [1] and recognized as the precursors of platelets by Wright in 1906 [2]. In comparison to other haematopoietic cells in the marrow, megakaryocytes are unique by virtue of the large size and polyploid complement of DNA. Megakaryocytes undergo endomitosis, which results in an increase in DNA content and cell size. These cells are generally heterogeneous in respect to ploid, size, and cytoplasmic and nuclear morphology, and are infrequent in bone marrow, comprising only 0.05 % of all nucleated cells found in the marrow [3, 4].

Megakaryocytes, like other haematopoietic cells, are derived from pluripotent haematopoietic stem cells (PHSC) which possess the genetic capacity to differentiate into different lineage through division and maturation. The whole process of megakaryocyte development and maturation from a pluripotent stem cell is termed megakaryocytopoiesis, which involves the differentiation of PHSC into megakaryocytic lineage, the proliferation of early (BFU-MK) and late (CFU-MK) progenitors of this lineage, and the DNA endoreplication and cytoplasmic maturation of megakaryocytes to give rise to platelets (Figure 1). During this process, synthesis of platelet-specific proteins and packaging of specific organelles as well as production of the demarcation membrane system (DMS) occur and give rise to megakaryocyte endoreplication and cytoplasmic maturation [5-7].

In a normal state, megakaryocytes are primarily located in the bone marrow where they are found lining the vascular sinuses. Megakaryocytes can also be found in the pulmonary circulation. In 1893, Aschoff reported the presence of a significant number of megakaryocytes in histological sections of human lungs. He believed that megakaryocytes were transported from the bone marrow to the lungs and mostly retained there, although a small number passed to other organs [8]. The existence of pulmonary megakaryocytes has since been confirmed by many other investigators. Besides marrow and lung, other tissues such as liver, spleen, peripheral blood and umbilical cord blood contain a small number of megakaryocytes and their progenitors [9-14].

Megakaryocyte progenitors cannot be identified morphologically, and therefore it is important to find some specific markers allowing the recognition and purification of the progenitor cells of this lineage. Like progenitors of other haematopoietic lineages, megakaryocyte progenitors express CD34 and HLA-DR antigens [10, 15, 16]. CD34 is a 115 kDa cell membrane glycoprotein whose expression in the haematopoietic system is limited to pluripotent and lineage-committed progenitors. CD34+ cells purified from bone marrow or peripheral blood are able to develop into megakaryocytes *in vitro* [13, 15, 16]. Studies performed by Fraser et al [17] suggest that platelet glycoprotein IIb/IIIa (GPIIb/IIIa) may be present on cells earlier than the committed megakaryocyte progenitors because the antibody against GP-IIb/IIIa inhibits the proliferation of PHSC and CFU-Mix in a complement-dependent cytotoxic assay. The cells expressing CD34, but not GPIIb/IIIa, have higher proliferative potential than the cells expressing both CD34 and GP-IIb/IIIa [15].

The cells in transitional stages between progenitor and megakaryocytes express GP-IIb/IIIa and GPIb, two specific proteins of megakaryocyte lineage. GPIIb/IIIa can be found not only on surface membrane and cytoplas-

Megakaryocytopoiesis

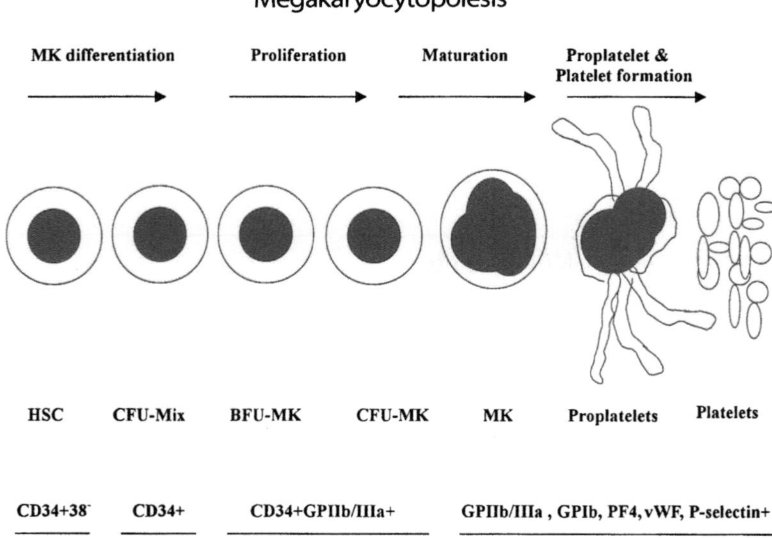

Figure 1. Schematic presentation of megakaryocytopoiesis. HSC, hematopoietic stem cells; CFU-Mix, colony-forming unit-mixture; BFU-MK, burst-forming unit-megakaryocyte; CFU-MK, colony-forming unit-megakaryocyte; MK, megakaryocytes; GP, glycoprotein; PF4, platelet factor 4; TSP, thrombospondin; vWF, von Willebrand Factor.

mic DMS but is also associated with membranes of a-granules. In contrast, labelling for GPIb is present essentially on the DMS. The appearance of these glycoproteins seems to be synchronous with those of cytoplasmic markers including von Willebrand Factor (vWF), platelet factor-4 (PF-4), thrombospondin (TSP). In the mouse, the acetylcholinesterase (AchE) seems a specific marker of megakaryocyte lineage and appears in all stages of megakaryocytes [15, 19-21]. Because GP-IIb/IIIa and AchE appear during the whole process of megakaryocytopoiesis in man and mice, respectively, the GP-IIb/IIIa immunostaining and AchE-staining have been used as conventional methods to detect megakaryocytes and their colonies *in vitro* in both cell smears and cultures (Figure 2). Most of the glycoproteins expressed by platelets have been found in megakaryocytes. Among those more recently demonstrated in megakaryocytes, are VLA-2, VLA-6 [22], GPIV [23] , GPV [23] and FCγ RII [25]. It is therefore reasonable to consider that all platelet membrane glycoproteins are synthesized during megakaryocyte maturation. Recently, Hagiwara et al. observed that megakaryocytes expressed a number of adhesion molecules of the integrin and selectin families as well as those of the immunoglobulin superfamily. The expression of adhesion molecules on megakaryocytes is important for the migration of megakaryocytes and the cytoplasmic processes [26].

Platelets are the end-products of the cytoplasmic fragmentation of mature megakaryocytes. Each megakaryocyte has the capacity to produce from 1000 to 4000 platelets. However, where and how platelets are formed is still subject to controversy. Since the initial observation of Wright [2], evidence has accumulated that production of platelets involves the formation of megakaryocyte projections that cross the vascular endothelium and subsequently pinch off as individual platelets [9, 27]. Other studies have shown that megakaryocytes may shed platelets within the pulmonary microvasculature, which may be the primary site of platelet production [10, 27, 28]. It has recently been shown, using a mouse model, that although thrombopoiesis can occur in the lung or liver, the bone marrow and spleen are the major thrombopoietic organs in the mouse [12]. It is becoming clear that megakaryocytes generate platelets through cytoplasmic processes leading to the release of platelets into the circulation at their terminal stage. Such cytoplasmic processes can be observed *in vitro* in culture. In the studies described by Choi et al [29], human megakaryocytes obtained from CD34+ cells progenitor cells formed proplatelets and then platelet-size particles. The culture-derived platelets were able to aggregate when stimulated by thrombin or ADP. The cytoskeleton such as microtubules play a role in proplatelet formation [26, 30-33].

Structure   In the bone marrow, the earliest cells of this lineage, the promegakaryoblasts, have a diameter of about 7-15 mm, with a non-segmented round nucleus. These cells can be identified by specific membrane or

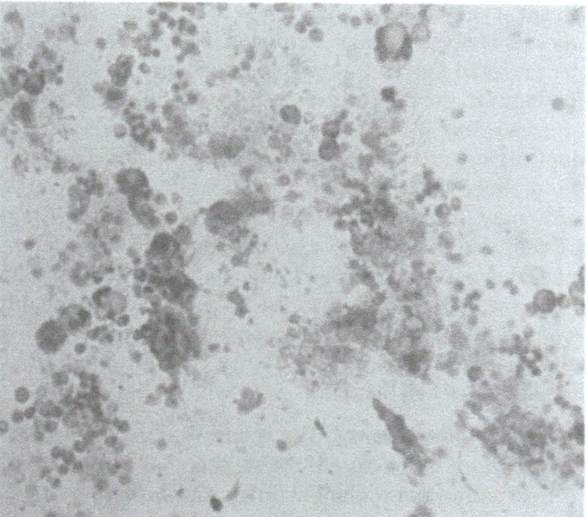

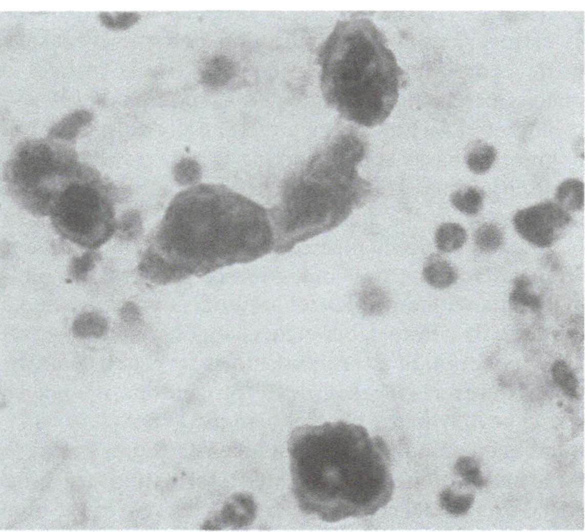

Figure 2. Photomicrographs of human megakaryocytes from immunoperoxidase staining. megakaryocytes which react positively with an antiGPIIb/IIIa monoclonal antibody are represented by brown stain. The upper photo (x 100) shows a colony derived from CFU-Mix containing various megakaryocytes mixed with other lineage cells. The lower photo (x 400) shows several megakaryocytes stained by the antibody.

cytoplasmic markers such as GPIIb/IIIa, PF-4 and vWF, and in rodents, AchE-activity [11, 18, 20]. With conventional methods of light and electron microscopy, the earliest megakaryocyte can be morphologically identified when the cell has a diameter of above 14 mm. The megakaryocyte with such a size corresponds to the 4N and 8N ploidy stages and represents one or two generation cycles from a 2N cell. These cells, called megakaryoblasts or in stage I, contain a bilobulated nucleus with a heavy chromatin network, prominent nucleoli and basophilic cytoplasm without granules.

Megakaryocytes in stage II are basophilic and their nucleus is segmented or lobulated, with increasing condensation of the chromatin and nucleoli of various sizes. At stage III, the post-mitotic stage, cells have a diameter of 20-50 mm with usually multilobulated nucleus, acidophilic cytoplasm and azurophilic granules. The main ploidies are 8N, 16N and 32N but 4N and 64N and rarely 128N may occur during this stage of maturation. Stage III megakaryocytes are also called platelet-producing megakaryocytes. After shedding, the naked nucleus with a narrow cytoplasmic rim can be seen and is probably eventually phagocytized by macrophages [11].

Regulation of Cell Function

Cell to Cell Interactions   The major function of megakaryocytes is to produce platelets that play a central role in maintaining vascular integrity and arresting bleeding. Platelet production depends upon three independent parameters: the number of megakaryocytes, the level of megakaryocyte ploidization and the megakaryocyte cytoplasmic maturation. The whole process of platelet production is regulated at multiple levels with proliferative factors acting preferentially during the early mitotic phases and late acting factors involved in the maturation steps. It can be observed that there is a rapid increase in the number, size, ploidy of megakaryocytes and of newly formed platelets with an augmented mean volume, after induction of an acute thrombocytopenia with anti-platelet antibodies in experimental animals. By contrast, opposite changes can be observed after induction of a thrombocytosis by platelet transfusion. However, megakaryocyte progenitors do not respond acutely to the platelet demand [34, 35].

It has been demonstrated that megakaryocyte proliferation, maturation and proplatelet formation are influenced by stromal cells. Stromal cells present in the bone marrow include fibroblasts, endothelial cells, histiocytes, adipocytes, osteoclasts and monocytes. These cells secrete various cytokines, produce extracellular matrix proteins and mediate cell-cell interactions that regulate haematopoiesis including megakaryocytopoiesis. Although the role of stromal cells in the regulation of megakaryocytopoiesis has been demonstrated for several years, the mechanism of stromal cell action is only recently becoming clear. The stromal control of megakaryocytopoiesis is multifactorial, and these cells synthesize a number of cytokines capable of regulating megakaryocytopoiesis, which include interleukin-1, 6, 11, 13 (IL-1, IL-6, IL-11, IL-13), thrombopoietin (TPO), erythropoietin (EPO), stem cell factor (SCF), granulocyte-macrophage colony stimulating factor (GM-CSF), FLT-3 ligand, leukemia inhibitor factor (LIF) and fibroblast growth factors (FGFs) [36, 37].

Molecular Interactions   Megakaryocytopoiesis is a complex process that is positively and negatively regulated by various cytokines. The identification and cloning of thrombopoietin (TPO), the major humoral regulator of thrombopoiesis, represents a major advance in this area

[38-40]. TPO is known to be able to stimulate megakaryocytic differentiation of CD34+ cells and proliferation and maturation of megakaryocytes. It is not, however, a direct stimulator of proplatelet formation in mature megakaryocytes, although its action at progenitor level is critical for the subsequent proplatelet formation. Several studies have shown that TPO secreted by stromal cells suppress proplatelet formation, and such an effect is probably mediated by inhibiting osteoclastogenesis from bone marrow cells [41-44]. Besides TPO, a number of other cytokines have been demonstrated to be able to act *in vivo* and *in vitro* on megakaryocytopoiesis. These positive cytokines include SCF, Flt3 ligand (FL), IL-3, IL-6, IL-11, IL-13, GM-CSF, platelet-derived growth factor (PDGF), FGF-1 and -2, HST/FGF-4, EPO, LIF and macrophage-stimulating protein (MSP) [45-58]. Furthermore, megakaryocytopoiesis is physiologically controlled by inhibitory proteins as well, including transforming growth factor-β_1 (TGF-β_1) [59], chemokine platelet factor-4 (PF-4) [60], β-thromboglobulin (β-TG) [61] and neutrophil-activating peptide-2 (NAP-2) [62], TSP [63], thrombin and anagrelin [35]. Besides the protein regulators, glycosaminoglycans (GAG) are able to regulate megakaryocytopoiesis by modifying the activities of several cytokines acting on megakaryocytes [64, 65]. Extracellular matrix is also capable of stimulating megakaryocyte proplatelet formation *in vitro* through the interaction between vironectin and its receptor on megakaryocytes [66]. Figure 3 is a schematic presentation of the regulation of megakaryocytopoiesis by a network of positive and negative regulators.

Signalling Mechanisms   Megakaryocytes, their precursors and their progeny have been known to be activated by a number of regulators through several signalling mechanisms. Current studies indicate that protein-tyrosine kinases (PTKs) participates in signal transduction during megakaryocytopoiesis [67]. In cells of megakaryocytic lineage, expression of the mRNA of matk, a megakaryocyte-associated tyrosine kinase has been found to be predominant and can be up-regulated by phorbol esters (PMA) [68]. The PMA-induced megakaryocytic differentiation in K562 cells has been shown to be proceeded by a rapid rise in the activity of MEK(MAP kinase/extracellular regulated kinases) that leads to a sustained activation of ERK2(Extracellular regulated kinase). Blockade of MEK1 activation reverse both the growth arrest and the morphological changes of K562 cells induced by PMA treatment. These changes are associated with a block of the cell-surface expression of the GPIIb/IIIa integrin [69]. Besides PTKs and MEK/MAPK, signal transduction through trimeric G proteins has also been observed in megakaryoblastic cell lines by measurement of the expression of alpha-subunits of trimeric G proteins [70].

The mechanisms by which TPO stimulates megakaryocytopoiesis are beginning to be understood on a molecular level. TPO binds with high affinity to its receptor present on megakaryocytes. Within seconds of

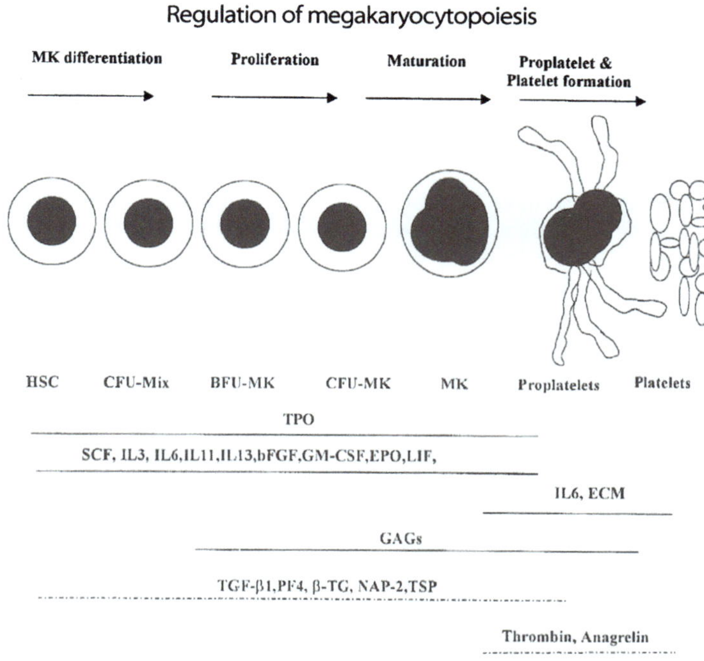

Figure 3. Positive (–) and negative (·······) regulation of megakaryocytopoiesis by various molecules. SCF, stem cell factor; IL, interleukin; TPO, thrombopoietin; EPO, erythropoietin; GM-CSF, granulocyte-macrohage colony-stimulating factor; bFGF, basic fibroblast growth factor; MSP, macrophage stimulating peptide; LIF, leukemia inhibitory factor; ECM, extracellular matrix; GAGs, glycosaminoglycans; TGF-β1, transforming growth factor-β1; PF4, platelet factor 4; β-TG, β-thromboglobin; 1; NAP-2, neutrophil-activating peptide 2; TSP, thrombospondin.

its binding, JAK2 and TYK2 are recruited to the cytoplasmic domain of the Mpl receptor and are activated by tyrosine phosphorylation [71, 72]. These kinases subsequently phosphorylate multiple cellular proteins, including SHC, SHIP, the p85 regulatory subunit of PI3 kinase, STAT3 and STAT5. STAT3 and STAT5 are important because they are part of a growing family of latent transcription factors that, upon phosphorylation dimerize, translocate to the nucleus and activate specific cytokine inducible promoters [73].

Additional Features Megakaryocytes and endothelial cells, two important blood and vascular cells, have been know to share many similar antigens on their cell surface and in the cytoplasm. The two types of cells also share several growth regulators: FGF, GM-CSF, TGF, PF-4, TSP and GAG [74, 75]. These observations suggest that the two types of cells, destined to have close functional interaction as mature cells, may be derived from the same or very close precursor cells during embryonic development and differentiation. They therefore share important similarities and together take the responsibility for the formation and repair of blood vessels. Further studies on the relationship between the cell antigens and related genes of megakaryocytes and endothelial cells will provide new insights to the physiology and pathophysiology of these cells. Recognition of the common factors and studies with them are broadening the understanding of the pathogenesis of megakaryocytic and angiogenic diseases and encouraging attempts to develop new therapeutic strategies for the future.

Role in Vascular Biology

Physiological Function Megakaryocytes are of central importance in a variety of physiological and pathological processes. The physiological roles of megakaryocytes in vascular biology include direct action of megakaryocytes and platelets on vascular cells and the effects of the factors derived from megakaryocytes and their progeny. The major function of megakaryocytes is to produce platelets. Platelets have many physiological and pathological functions. The major physiological function of platelets is to interact with coagulation factors and cellular and extracellular components of the vessels to maintain vascular integrity and to arrest bleeding [76, 77].

Although platelets can take up certain proteins from the circulation [78] the majority of their proteins is synthesized by the parental megakaryocytes. Megakaryocytes synthesize a number of proteins that are involved in physiology and pathophysiology of vascular cells [79-82]. The principal proteins include PDGF, PF-4, TSP, TGF-β_1, bFGF and vascular endothelial growth factor (VEGF). PDGF is a factor that stimulates migration and proliferation of vascular smooth muscle cells [83]. PF-4 is produced only by megakaryocytes and stored in alpha granules of platelets [84]. It is a potent inhibitor of angiogenesis and vascular cell proliferation [85]. TSP is also a potent inhibitor of endothelial cell proliferation [86]. bFGF and VEGF are two most important positive regulators of vascular endothelial cells and angiogenesis [87]. It has been shown that megakaryocytes produce and secrete VEGF in an inducible manner. Within the bone marrow environment, VEGF secreted by megakaryocytes may contribute to the proliferation of endothelial cells. VEGF delivered to sites of vascular injury by activated platelets may initiate angiogenesis [88, 89].

Pathology   Maintenance of vascular integrity requires the normal function of both circulating platelets and the vascular endothelium. Platelets play an essential role in arterial thrombosis and atherosclerosis. They survey the inner lining of the vessel wall without interacting with it under normal circumstances, but respond rapidly to alterations of endothelial cells by attaching firmly to the site of lesion. Circulating platelets are heterogeneous with respect to their size, density and reactivity [90]. Mean platelet volume (MPV) is an important biological variable and there is strong evidence suggesting that large platelets are more active hemostatically [91]. It has been shown that a change in platelet heterogeneity is preceded by changes in megakaryocyte ploidy and cytoplasm volume. Changes in megakaryocyte ploidy distribution may be associated with the production of large platelets. It was found that the mean cytoplasmic volume of megakaryocytes is significantly increased in bone marrow biopsies of patients 2-3 weeks after myocardial infarction compared with controls with non-cardiac chest pain [92]. In patients with coronary artery atherosclerosis, megakaryocyte DNA content is significantly increased. There is also a significant positive correlation between MK ploidy and total serum cholesterol and triglyceride concentrations, suggesting that cholesterol may be a determinant of megakaryocyte ploidy [93]. Apart from DNA content change, megakaryocytes from patients with coronary atherosclerosis express the inducible nitric oxide synthase, suggesting a link between the expression of nitric oxide synthase in megakaryocytes and atherosclerosis [94]. Moreover, it has been observed that patients with stable coronary artery disease have circulating activated platelets, circulating monocyte-platelet aggregates, increased platelet reactivity and an enhanced propensity to form monocyte-platelet aggregates [95].

In diabetes, excess mortality is due predominantly to the vascular complications of the disease. The cause of these complications is unclear but may be related to platelet changes. Patients with diabetes have an altered population of circulating platelets when compared to nondiabetics. Large platelets circulate in an activated state in diabetes mellitus [96, 97]. The reason for these changes are poorly understood but it is likely to be related to changes in megakaryocytes. Indeed, altered megakaryocyte ploidy distribution has been observed in diabetes and particularly those with atherosclerosis. Because these patients have an elevated serum level of IL6, one of the factors capable of stimulating megakaryocytopoiesis, it is reasonable to suggest that the shift in ploidy and platelet count seen in diabetes and atherosclerosis may be due in part to an increase in the circulating level of IL6 [98]. Future work will determine the effect of other factors, particularly TPO on the megakaryocyte-platelet axis in vascular disease.

During acute myocardial infraction, unstable angina and sudden cardiac death, initial rupture of an atherosclerotic plaque, exposure of thrombogenic components in the intercellular matrix of the vessel wall to platelets and subsequent platelet aggregation and adhesion to the vessel wall might be the precipitating events in thrombus formation [89, 90]. In patients suffering from an acute myocardial infarction, the mean platelet volume is increased. The bleeding time is shortened at the time of infarction in these patients probably due to increased synthesis of TxA2, but an increased production of adrenaline may also be of importance. These large, reactive platelets, with more GPIIb/IIIa, may, therefore, be causally related to myocardial infarction [99, 100].

Clinical Relevance Therapeutic Implications   Occlusive thrombus formation in coronary artery diseases probably begins with the deposition of platelets on a damaged atherosclerotic plaque as a result of the interaction of constitutively active platelet surface receptors. Under certain circumstances, the platelet surface receptors on the luminal surface of the adherent platelets are activated and undergo a conformational change that results in their binding to plasma fibrinogen, vWF, or perhaps other glycoproteins with high affinity. This permits the recruitment of an additional layer of platelets, which facilitates thrombin generation, fibrin deposition and leukocyte adhesion and transmigration, ultimately resulting in vaso-occlusion [101].

The rationale for using anti-platelet drugs in humans is mainly to prevent or treat the thromboembolic complications of ischemic cardiovascular diseases. A number of pharmaceutical agents have been developed to inhibit some aspects of platelet function. These agents can be divided into several groups according to their mechanism of action, including drugs increasing cyclic nucleotides, and drugs interfering with arachidonic acid metabolism, and activation or adhesion receptors. Some of these agents have been proven to be clinically effective in humans [101, 102]. As described above, megakaryocytopoiesis and platelet production are stimulated in atherosclerosis and cardiovascular disease. Patients with thrombotic complications of coronary artery disease tend to have large and hyperreactive platelets in the circulation and large megakaryocytes. It would be of interest to investigate whether anti-platelet drugs administered in humans may have an effect on megakaryocytes in the bone marrow or the lung and whether the agents inhibiting megakaryocytopoiesis may be effective in preventing or treating thromboembolic complications of vascular disease.

Another clinical relevance is the possible use of angiogenic or anti-angiogenic factors derived from megakaryocyte-platelets as drugs for the prevention and treatment of ischemic heart – brain vascular diseases or cancer metastasis. Such treatments could be carried out as protein therapies or by genetic biotechnology.

Zhong Chao Han

References

1. Howell WH (1890) J Morphol 4:117-130
2. Wright JH (1906) Boston Medical Surgery Journal 154:643-645
3. Sitar G (1984) Br J Haematol 59:465-472
4. Harker LA (1968) J Clin Invest 47:458-469
5. Radley JM, Haller CJ (1982) Blood 60:213-219
6. 6. Han ZC et al (1991) Int J Hematol 54:3-14
7. Hoffman R (1989) Blood 74:1196-1212
8. Aschoff L(1893)Archiv für Pathologie, Anatomie und Physiologie 134:11-26
9. Eldor A et al (1989) Megakaryocyte function, dysfunction 2:543-568
10. Howell WH, Donahue DP (1939) J Exp Med 65:177-204
11. Hovig T(1989) Bailli⁻⁻⁻res Clinical Haematology 2:503-542
12. Davis RE et al (1997) Exp Hematol 25:638-648
13. Hagiwara T et al (1998) Exp Hematol 26:228-235
14. Nichol JL et al (1994) Stem Cells 12:494-505
15. Debili N et al (1992) Blood 80:3022-3035
16. Xi X et al (1996) Br J Haematol 93:490-496
17. Fraser JK et al (1986) Blood 68:762-769
18. Breton-Gorius, Vainchenker W(1986) Seminars in Hematol 23:43-67
19. Berridge MV, Ralph SJ (1998) Blood 66:76-85
20. Jackson CW (1973) Blood 42:413-421
21. Paulus JM et al (1981) Blood 58:1100-1106
22. Soligo D et al (1990) Br J Haematol 76:323-332
23. Berge G et al (1993) Blood 82:3034-3044
24. Takafuta T et al (1994) Thromb Haemost 72:762-769
25. Wu Z et al (1996) Thromb Haemost 75:661-667
26. Hagiwara T et al (1996) Exp Hematol 24:690-695
27. Stenberg PE, Levin J (1989) Blood Cells 15:23-47
28. Slater DN et al (1983) Thromb Res 31:163-176
29. Choi ES et al (1995) Blood 85:402-413
30. Leven RM, Tablin F (1992) Exp Hematol 20:1316-1322
31. Nagahish H et al (1996) Blood 87:1309-1316
32. Horie K et al (1997) Exp Hematol 25:169-176
33. Tajika K et al (1998) Br J Haematol 100:105-111
34. Vainchenker W et al (1995) In: Critical Reviews in Oncology/Hematology. Elsevier, Dublin, pp 166-192
35. Wendling F, Han ZC (1997) Barlli⁻⁻⁻re's Clinical Haematology 10:47-63
36. Sensebe L et al (1997) Stem Cells 15:133-143
37. Achattner M et al (1998) Stem Cells 16:61-65
38. de Sauvage FJ et al (1994) Nature 369:533-538
39. Wendling F et al (1994) Nature 369:571-574
40. Kaushansky K et al (1995) Proc Natl Acad Sci USA 92:3234-3238
41. Ito T et al (1996) Br J Haematol 94:387-390
42. Horie K et al (1997) Exp Hematol, 25:169-176
43. Nagahisa H et al (1996) Blood 87:1309-1316
44. Wakikawa T et al (1997) 138:4160-4166
45. Avraham H et al (1992) Blood 79:365-371
46. Hunt P et al (1992) Blood 80:904-911
47. Piacibello W et al (1996) Exp Hematol 24:340-346
48. O'Shaughnessy JA et al (1995) Blood 86:2913-2921
49. Yang M et al (1995) Br J Haematol 91:285-289
50. Ishibashi T et al (1989) Proc Natl Acad Sci USA 86:5953-5957
51. Teramura M et al (1992) Blood 79:327-331
52. Xi X et al (1995) Br J Haematol 90:921-927
53. Han ZC et al (1992) Br J Haematol 81:1-5
54. Bikfalvi A et al (1992) Blood 80:1905-1913
55. Konishi H et al (1996) Oncogene 13:9-19
56. McDonald TP et al (1987) Exp Hematol 15:719-721
57. Burstein SA et al (1992) J Cell Physiol 153:305-312
58. Banu N et al (1996) J Immunol 156:2933-294
59. Ishibashi T et al (1987) Blood 69:1737-1741
60. Han ZC et al (1990) Blood 75:1234-1239
61. Han ZC et al (1990) Br J Haematol 74:395-401
62. Gewirtz AM et al (1995) Blood 86:2559-2565
63. Chen Y et al (1997) J Lab Clin Med 129:231-238
64. Han ZC et al (1996) J Cell Physiol 168:97-104
65. Shen ZX et al (1995) Lancet 346:220-221
66. Leven RM, Tablin F (1992) Exp Hematol 20:1316-1322
67. Avraham H et al (1995) Stem Cells 13:380-392
68. Bennett BD et al (1994) J Biol Chem 269:1068-1074
69. Herrera R et al (1998) Exp Cell Res 238:407-414
70. van der Vuurst H et al (1997) Arterioscler Thromb Vasc Biol 17:1830-1836
71. Sattler M et al (1995) Exp Hematol 23:1040-1048
72. Miyakawa Y et al (1995) Blood 86:23-27
73. Ihle JN (1995) Nature 377:591-594
74. Han ZC, Caen JP (1993) J Lab Clin Med 121:821-825
75. Bikfalvi A, Han ZC (1994) Leukemia 8:523-529
76. Caen JP, Rosa JP (1995) Thromb Haemost 74:18-24
77. Perutelli P, Mori PG (1997) Recenti Prog Med 88:526-529
78. Harrison P et al (1989) J Clin Invet 84:1320-1324
79. Witte DP et al (1988) J Cell Physiol 137:86-94
80. Gu XF et L (1995) Eur J Haematol 55:189-194
81. Jiang S et al (1994) Blood 84:4151-4156
82. Sandrock B et al (1996) In Vitro Cell Dev Biol Anim 32:225-233
83. Ross R (1989) Lancet 130:1179-1182
84. Ryo R et al (1980) Thromb Res 17:645-652
85. Gupta SK, Singh JP (1994) J Cell Biol 127:1121-1127
86. Iruela-Arispe ML et al (1990) Proc Natl Acad Sci USA 88:5025-30
87. Klagsbrun M (1991) J Cell Biochem 47:199-200
88. Katoh O et al (1995) Cancer Res 55:5687-5692
89. Mohel R et al (1997) Proc Natl Acad Sci USA 94:663-668
90. van der Loo B, Martin JF(1997) Bailliere's Clinical Haematol 10:109-123
91. Martin JF (1990) In: Martin JF, Trowbridge EA (eds) Platelet Heterogeneity: Biology, Pathology. Springer, London, pp 205-226
92. Trowbridge EA et al Thromb Haemost 52:167-171
93. Bath PM et al (1994) Cardiovasc Res 28:1348-1352
94. de Belder A et al (1995) Arteroscler Thromb Vasc Biol 15:637-641
95. Furman MI et al (1998) J Am Coll Cardiol 31:352-358
96. Tschoepe D (1995) Semin Thromb Hemost 21:152-160
97. Tschoepe D et al (1997) Horm Metab Res 29:631-635
98. Brown AS et al (1997) Arter Throm Vascul Biol 17:802-807
99. Kristensen SD (1992) Dan Med Bull 39:110-127
100. Giles H et al (1994) Eur J Clin Invest 24:69-72
101. Coller BS (1997) J Clin Invest 100:S57-S60
102. Cazanave JP, Gachet C (1997) Barlliere's Clinical Haematology 10:163-180

Mek

Definition   *Mitogen – regulated extracellular kinase*

See: →Signal transduction mechanisms in vascular biology

MHC

Definition *Myosin heavy chain*

See: →Cytokines in vascular biology and disease; →Smooth muscle cells

MIDAS

Definition *Metal ion-dependent adhesion site*

See: →Vascular integrins; →von Willebrand factor

Migration

Definition *Cell movement of adherent cells. Fundamental process in biology involved in ontogeny, physiology and pathology.*

See: →Endothelial cells; Smooth muscle cells; →FGF-1 and -2; →Vascular endothelial growth factor family; →Angiogenesis

Mitogen-Associated Kinases (MAP Kinases)

Definition *Kinases that are activated through the ras/raf pathway. Consist of a family of protein kinases with two main members named ERK1 (p44) and ERK2 (p42). Alternative MAP kinase pathways have also been described (through JNK).*

See: →Signal transduction in vascular biology; →Megakaryocytes; →Platelet stimulus-response coupling; →Smooth muscle cells

MK

Definition *Midkine*

See: →Heparin-affin regulatory protein

MK

Definition *Megakaryocytes*

See: →Megakaryocytes

MLCK

Definition *Myosin light chain kinase*

See: →Vasomotor tone regulation, molecular mechanisms of; →Smooth muscle cells

MLTF

Definition *Major latent transcription factor*

See: →Fibrin/fibrinogen

MMP

Definition *Matrix metalloproteinases*

See: →Matrix metalloproteinases; →Fibrinolytic, hemostatic and matrix metalloproteinases, role of

MPV

Definition *Mean platelet volume*

See: →Megakaryocytes

MRS

Definition *MET related sequence*

See: →Hepatocyte growth factor

MSR

Definition *Macrophage scavenger receptor*

See: →Macrophage scavenger receptor; →Blood cells, interaction with vascular cells; →Cytokines in vascular biology and disease

MT-MMP

Definition *Membrane-type matrix metalloproteinase*

See: →Matrix metalloproteinases

Myocardial Ischemia

Definition *Lack of oxygen supply due to decrease of blood flow in the myocardium often a consequence of atherosclerosis and thrombosis of coronary blood vessels.*

For animal models see: →Fibrinolytic, hemostatic and matrix metalloproteinases, role of

N-CAM

Definition *Neural cell adhesion molecule*

See: →FGF receptors; →Heparin-affin regulatory protein

NAC

Definition *N-acetyl cysteine*

See: →Cytokines in vascular biology and disease

NAD

Definition *Nicotinamide adenine dinucleotide*

See: →FGF-1, FGF-2

NANC

Definition *Nonadrenergic and noncholinergic*

See: →Nitric oxide

NAP-2

Definition *Neutrophil activating peptide*

See: →Megakaryocytes

Neovascularisation

See: →Angiogenesis

Neuropilin

Definition *Found on axons and classically involved in axon guidance. Neuropilin is also a co-receptor for VEGF on endothelial cells.*

See: →Vascular endothelial growth factor family

Neutrophils

Definition *White blood cells involved in phagocytosis and bacterial killing.*

See: →Blood cells, interaction with vascular cells; →Cytokines in vascular biology and disease

NF-κB

Definition *Nuclear factor-κB*

See: →Cytokines in vascular biology and disease

Nitric Oxide (NO)

Synonym: The observation of Furchgott and Zawadzki that acetylcholine-induced vasorelaxation is endothelium dependent led to the discovery of endothelium-derived relaxing factor (EDRF) [1]. EDRF is important in the regulation of vasomotor tone and flow by inhibiting smooth muscle contraction and platelet aggregation. It is a labile, lipophilic, humoral factor that mediates, through activation of soluble guanylate cyclase, the action of a variety of vasodilators such as bradykinin, substance P, histamine, ect. [1]. In 1987 Hibbs et al. demonstrated for the first time the formation of a metabolite of the guanidino group of L-arginine by cells which resulted in the formation of L-cit-

rulline and the end products NO2-/NO3- [2]. The authors speculated that the phenomenon was due to nitrite or oxygenated nitrogen intermediates in the pathway of nitrite and nitrate synthesis [3]. In the same year Ignarro [4] and Moncada [2] in different laboratories experimentally identified EDRF as nitric oxide, as proposed in 1986 by Furchgott and Zawadzki [1].

Definition *Formed through conversion of L-arginine into L-citrulline. The enzymes that catalyze this reaction are the constitutive or inducible NO synthases (cNOS or iNOS).*

See: →Vasomotor tone regulation, molecular mechanisms of; →Cytokines in vascular biology and disease; →Angiogenesis inhibitors; →Vascular endothelial growth factor family

Introduction Until the mid-1980s nitric oxide was regarded as an atmospheric pollutant and bacterial metabolite. Now this free radical gas has been implicated in a wide variety of physiological and pathological processes including vasodilatation, modulation of intestinal mobility, contraction of heart and skeletal muscle, erectile function, neurotransmission and non specific immunity [5,6,7,8,9]. In addition NO is also involved in the pathogenesis of some diseases such as septic shock, inflammation, tissue injury, atherosclerosis, hypertension, diabetes, cerebral ischemia, apoptosis, tumor growth, etc. [10,11]. The precise action of this short-lived mediator is intimately linked to its cellular source as well as its rate of synthesis [5,12]. In blood vessels NO induces endothelium dependent vasodilatation in response to a wide variety of stimuli, nerve dependent vasodilatation and cytokine/ endotoxin induced vasodilatation. In the central nervous system NO is a neurotransmitter implicated in different function such as the formation of memory. In the peripheral nonadrenergic and noncholinergic (NANC) fibers, NO mediates some forms of neurogenic vasodilatation/inflammation and regulates various gastrointestinal, respiratory and genitourinary tract functions. NO also contributes to the control of platelet aggregation and the regulation of cardiac contractility [5]. In mammalian cells NO synthase (NOS) catalyses the sequential five-electron oxidation of the substrate L-arginine to NO and citrulline. These enzymes have been classified as either calcium and calmodulin dependent (constitutive [cNOS]) and calcium calmodulin independent (inducible [iNOS]) [9,13,14]. NOS isoforms have been classified in type I type II and type III [15]. The first type corresponds to neuronal constitutive NOS found in central and peripheral neurons, neuroblastomas, platelets, skeletal muscle, β-pancreatic islets and epithelial cells of bronchioli, alveoli, uterus, and stomach [16]. The third one corresponds to endothelial constitutive NOS expressed in endothelium, neurons, cardiac myocytes and certain subsets of respiratory epithelial cells. cNOS is activated by a rise in intracellular calcium which sustains the binding of calmodulin, leading to NO release over several minutes [15]. The second type is inducible in many cell types after challenge with immunologic or inflammatory stimuli and generates large amounts of NO over a period of days [10]. The tonic activation of this isoform probably reflects the fact that it carries calmodulin as a tightly bound noncovalently attached subunit at the calcium resting-cell level.

NOS are members of the cytocrome P450 enzyme family, however they contain an unusual complement of cofactors – flavin adenine dinucleotide (FAD), flavin mononucleotide (FMN) and tetrahydrobiopterin (BH4) – and all play important roles in shuttling electrons from the substrate NADPH at the carboxy-terminal reductase domain to the heme complex at the amino-terminal oxygenase domain of the enzyme site (Figure 1). Reduced thiols also play an unspecific role in maintaining enzyme activity [15].

L-arginine is the precursor for the synthesis of NO by cells. The conversion of the aminoacid to NO is specific because a number of analogues of L-arginine, including D-arginine are not substrates. Furthermore, L-arginine analogues have been described as inhibitors of NO production in vascular tissue. NG-nitro-L-arginine (L-NA) and its methyl ester, NG-nitro-L-arginine methyl ester (L-NAME), N-iminoethyl-L-ornithine (L-NIO) and NG-monomethyl-L-arginine (L-NMMA) all inhibit the endothelial NO synthase acting as stereospecific inhibitors of NOS (Table 1). The different potency shown by these compounds in vascular tissue *in vitro* and *in vivo* may also be due to differences in uptake, distribution, or metabolism of the compounds. L-NMMA and L-NAME are orally active, they induce an increase in blood pressure when administered this way to rats and rabbits [5,17].

NO acts as a widespread second messenger mediator that stimulates the soluble guanylate cyclase by binding to the heme moiety of this enzyme. The consequent increase of cGMP concentration causes smooth muscle relaxation, inhibition of platelet aggregation or alteration

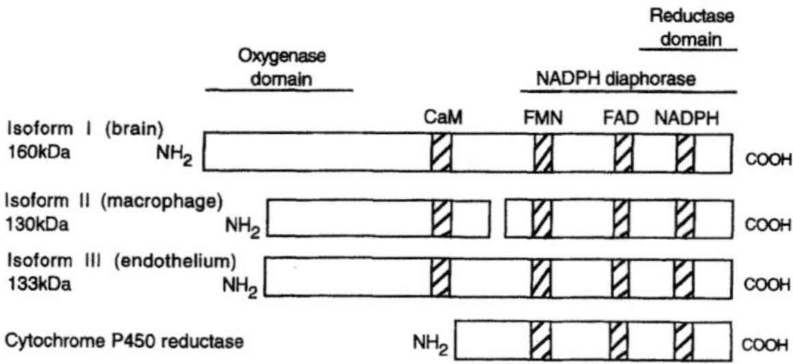

Figure 1. Functional elements in the three cloned isoforms of NOS and cytochrome P_{450} reductase. Alignment of the deduced amino acid sequences of the three isoforms showed 50-60% sequence identity between the enzymes. Consensus sequences for the binding of the cofactors NADPH, FAD, FMN and calmodulin are labelled. The N-terminal region which contains the putative L-arginine binding site, is responsible for the oxygenase activity, whereas the COOH-terminal region is responsible for both the NADPH diaphorase and the reductase activity.

in neurotransmission [2]. There is evidence indicating that the cardiovascular system is in a state of constant active vasodilation dependent on the generation of NO. Because of this, NO can be considered the endogenous nitrovasodilator. Nitrovasodilators such as isoamyl nitrite, nitroglycerin, isosorbide dinitrite, erythridil tetranitrite and pentaerythritol tetranitrate (Table 1) have been used clinically for about 100 years and are still used in conditions such as angina pectoris, congestive heart failure, hypertensive emergency pulmonary hypertension, etc. [17,18]. Some nitrovasodilators such as sodium nitroprusside (NaNP) release NO spontaneously, others such as the organic nitrates require the prior interaction with thiol like cysteine, and sydnonimes release NO subsequent to base-catalysed hydrolysis. In any case, NO is the final common effector molecule of all the nitrovasodilators that activates the soluble guanylate cyclase. Following the increase in cGMP levels, a cGMP-dependent protein kinase is stimulated, with resultant alteration of the phosphorylation of the light chain of myosin, which regulates the contractile state in smooth muscle.

In limiting L-arginine substrate condition the monooxygenase function of NOS is superseded by an oxidase activity with production of superoxide O_2^- and hydrogen peroxide H_2O_2. The same reactive oxygen intermediates generated during normal enzyme activity account for the classical diaphorase activity of NOS in which the dye nitro-blue tetrazolium is converted to blue-diformazan in an NADPH dependent reaction [19].

Characteristics

Molecular Weight   The molecular weight of NO is 30 g/mol.

Binding Sites and Affinity   NO reacts in biological systems with molecular oxygen (O_2), superoxide (O_2^-) and transition metals (M). The products of these reactions, higher nitrogen oxides (NO_x), peroxynitrite ($OONO^-$) and metal-nitrosyl adducts (M-NO) have various biological activities [20]. Indeed the propensity of various NO congeners for toxic or protective actions is determined by the chemistry they undergo in a given biological milieu [21]. Reactions with transition metal centres are central to the biological activities of NO. Differences in the rate constants for association and dissociation of NO lead to important differences in affinities with consequently biological relevance. For example, NO-induced activation of the heme protein soluble guanylyl cyclase is inhibited by addition of haemoglobin. The addition of haemoglobin limits formation of intracellular iron-nitrosyl adducts in cells producing NO [7, 20]. The reaction of NO with oxyhaemoglobin results in the oxidation of NO to nitrate and represents the major metabolic route of nitric oxide metabolism in blood. The binding affinity of haemoglobin for NO exceeds its binding affinity for carbon monoxide by several orders of magnitude [20]. The reaction of NO with haemoglobin to yield methahaemoglobin has been used as a spectrophotometric assay of NO.

NO is known to react with O_2 to form NO_2 (dioxide). NO and NO_2 are the major N-oxides that humans are

Table 1. Nitric oxide

INHIBITORS OF NOS	NITRIC OXIDE DONORS	
	Class	Compound
N^G-monomethyl -L-arginine (L-NMMA)	Organic nitrate	Glyceril trinitrate
		Amyl nitrite
N^G-nitro-L-arginine methyl ester (L-NAME)		Isosorbide
		Erythrityl dinitrate
N^G-amino-L-arginine (L-NAA)		Pentaerythritol
N^G-nitro-L-arginine (L-NA)	Inorganic complex	Sodium nitroprusside
7-Nitroindazole (7-NI)	S-Nitrosothiol	S-Nitroso-L-cysteine
		S-Nitrosoglutathione
N-$^\delta$diminoethyl-L-ornithine (L-NIO)		S-Nitroso-N-acetyl-L-cysteine
		S-Nitroso-N-acetyl-DL-penicellamine
N^G-N^G-dimethyl-L-arginine (L-ADMA)		
	Sydnonimine	3-Morpholino-sydnonimine
Aminoguanidine		
N-(3-(aminomethyl)benzyl)acetamidine (1400W)		

exposed to in ambient air. NO and NO_2 are in equilibrium, the rate of which depends on the presence of dispersed particles, molecular oxygen, ozone, and solar radiation intensity [22]. Kinetic studies indicate that the third-order reaction of NO with oxygen is relatively slow and acts as the rate limiting step in S- and N-nitrosation reactions mediated by NO/O_2 (k=6 x 10^6 /M²/s) [21]. Pure NO, by itself does not react with thiols or amines to yield nitroso derivatives, but the oxides of NO, NO_2 and N_2O_3 do. NO can react with -SH groups on amino acids, amines, organic acids, sugars, peptides and proteins to yield the corresponding S-nitrosothiol [22].

Nitrogen is unique among the elements in forming seven molecular oxides three of which are paramagnetic and all of which are thermodynamically unstable [23]. Because NO contains an unpaired electron and is paramagnetic, it rapidly reacts with O_2^- to form peroxynitrite anion (ONOO⁻) in high yield. Although the constant for the reaction of the free superoxide radical (O_2^-) with NO is high, the concentration of free O_2^- is small, due to the abundant presence of efficient trapping by superoxide dismutase [24]. This reaction represents one of few examples of a radical-radical coupling of O_2^- with another odd-electron species to generate a diamagnetic product.

Signal transduction is elicited through diverse and tissue specific pathways including activation of guanylyl cyclase, the gating of ion channels, modulation of cAMP dependent functions by effect of phosphodiesterases and regulation of calcium homeostasis by activation of protein G kinase [25].

Additional Features   Although its half life *in vivo* is very short, 25ms, the data on NO degradation and metabolism are conflicting. This reflects the complexity of NO metabolism and the different impact of various parameters such as pH, pO_2, concentration of O_2^-, the presence of traps like haemoglobin, the distance of NO generation and its target, etc. (see Processing and Fate for more details). NO diffuses to significant concentration at distances relatively far, on the order of 100-200 μm. This wide diffusibility is consistent also with the high value for its diffusion constant, 3300 μm²/s. These features suggest that the actions of NO are long range and its diffusion is an important determinant of its biological effects [26].

There are several well-defined actions of NO that are independent of cGMP, and metallo-proteins or thiol-containing proteins are likely to regulate these functions. The best examples are the synthesis of N-nitrosamines by lipopolysaccharide (LPS) and interferon gamma (INF-γ) by cytotoxic activated macrophages [26,27]. At the same time the kinetic reality of OONO⁻ formation under normal physiological conditions suggests that certain biological functions of the molecule are not excluded. In fact peroxynitrite has been shown to induce thiol-dependent relaxation of vascular smooth muscle and platelets [26,28].

NO has an extremely small positive electron affinity. Consequently the reduction of NO to NO⁻ is thermodynamically favoured. Nytroxyl converts rapidly to N_2O under physiological conditions through protonation, dimerization and dehydration [29,30]. The addition to thiol groups results in hydroxylamine formation (NH_2OH) [21]. Recently a NO reductase has been identified in eukaryotic systems that reduces NO to dinitrogen monoxide (N_2O). This introduces the possibility that intermediary enzymes may play important biochemical roles in NO metabolism and influence product distributions of NO_x.

Various products of NO reduction may exert biological actions. For example, NH_2OH has been shown to induce vasorelaxation that is cGMP dependent. The identification of NH_2OH in tissues which may support its oxidation by both enzymatic and chemical pathways suggests a potential salvage pathway for NO [31].

Structure

Conformation   The NO molecule containing nitrogen in the +2 oxidation state is second in a sequence of oxides in which the oxidation state of the element ranges from +1 to +5: N_2O NO N_2O_3 (dinitrogen trioxide) NO_2 (nitrogen dioxide) N_2O_4 (dinitrogen tetraoxide) and N_2O_5 (dinitrogen pentoxide). Nitrogen is able to occupy the oxidation state -3 (e.g. in ammonia NH_3), -2 (e.g. in hydrazine N_2H_4), -1 (e.g. in hydroxylamine NH_2OH) and 0 (e.g. in dinitrogen N_2). The capacity of nitrogen to form compounds in all oxidation states from -3 to +5, coupled with the existence of a rich variety of labile redox pathways among these states, is the source of its extraordinary versatility.

The oxides NO, NO_2, NO_3 are radicals, i.e. each contains an odd number of valence electrons in its molecule [6]. Removal of the single unpaired electron from NO results in the oxidation product NO^+. This species, nitrosium ion, is isoelectronic with the molecules N_2, and CO (carbon monoxide). When NO is reduced by addition of an electron the product is NO^-, the nitroxyl ion, isoelectronic with the molecule O_2 [32].

Additional Feature   Like the other nitrogen oxides, NO is thermodynamically instable (Gibbs energy of formation=86.32 KJ/mol). In consequence the synthesis of NO from elements N_2 and O_2 occurs to an appreciable extent only at elevated temperatures [33].

NO exhibits a low level of solubility in water (1.7×10^{-3} mol/l at 25°C and $P_{(NO)}$=1atm) comparable to the solubility of N_2, O_2, and CO [34]. In a reaction of considerable significance for biological systems NO reacts with the superoxide anion O_2^- to form peroxynitrite $^-$OONO and with nitrosothiols to form radical recombinations [35].

Gene

Gene Structure   At least three isoforms of NO synthase are responsible for NO synthesis in the mammalian organism, isoform I (in neuronal and epithelial cells), II (in cytokine-induced cells) and III (in endothelial cells) [36]. Isoform I has been cloned from rat and human brain, isoform II from mouse macrophages, and the third from bovine and human endothelial cells. These isoforms have the same cofactor requirements but show less than 60% sequence omology at the amino acid level and differ in regulation of their activity and in molecular size. The explanation for discordance in amino acid sequence and in different regulation between the three isoforms of NOS is that they represent the products of three distinct genes. Induced enzymes in cells other than macrophages are likely to be highly homologous to the macrophages enzyme. The NOS induced in the rat hepatocytes with a mixture of cytokines and E. coli endotoxin (LPS) is 94% identical at the amino acid level to the mouse macrophage enzyme. The same isoforms in different species show amino acid identities of 90%, indicating high sequence conservation across species [37].

Endothelial cNOS gene (Type III)   The bovine endothelial NOS gene spans approximately 20 kb of DNA and consists of 26 exons and 25 introns. Exon 1 contains the ATG initiation codon and exon 26 contains the entire sequence coding for the 3'-untranslated region of the mRNA. Consensus sequences for FMN, FAD pyrophosphate, flavin, FAD isoalloxazine, NADPH ribose, NADPH adenine and the 3'-untranslated region of the mRNA are localized within single exons (exon 16, 19, 20, 21, 23, 25 and exon 26 respectively). The coding sequence for the Ca^{2+}/calmodulin binding however is split between exon 11 and exon 12. Southern hybridization analysis indicates that bovine endothelial NOS is encoded by a single gene, while primer extension analysis identifies two alternative start sites 170 and 240 bp upstream from the methionine initiation (ATG) codon. Analysis of 2.9 kb of nucleotide sequence upstream from the transcription start sites shows that the bovine endothelial NOS gene lacks a TATA box or CAAT box and appears to be a member of the TATA-less class of RNA polymerase II promoters, known as initiator (Inr) promoters. A potential Sp1 binding site is found in the proximal promoter (-418) and four others are found within 250 bases to the first transcription start site (+133, +155, +219, +228). Fifteen copies of half-palindromic motifs are identified in the 5'-flanking region suggesting an oestrogen responsiveness to the gene. Two potential sterol regulatory elements are found at -625 and -1232 sites and a potential site for binding of the NF-1 transcription factor is found in the promoter at -1014. A total of 6 GATA (or TATC) motifs are present which are known to function in the endothelial cell-type-specific expression of other genes such as those of pre-pro-endothelin-1 and VCAM-1. A consensus sequence for AP-1 binding is found at -441 while the nine copies of the SSRE consensus sequence (shear stress responsive element) are found in the promoter sequence [38,39].

The nucleotide sequence of human endothelial cell NOS is 90% identical to bovine endothelial constitutive NOS. Binding sites for cofactors and nucleotides are highly conserved between human and bovine endothelial NOS. The gene coding for the human endothelial NOS contains 26 exons interrupted by 25 introns and spans approximately 22 kb of DNA. Positions of introns approximately separate the exons bearing individual functional domains of the protein especially in the 3' part of the gene. This separation of the gene reflects the structure of the protein determined with the cloning of the cDNA. Alignment of the bovine endothelial NOS gene promoter with the human endothelial NOS promoter shows a high sequence homology. The 1.6 kb of human 5'-flanking sequence has 75% nucleotide identity with the aligned sequence in the bovine gene with many of the regulatory sequences perfectly matched, including oestrogen half-palindromic motifs, two GATA motifs, an AP-1 site, an NF-1 site, a shear stress responsive element and a sterol regulatory element [38].

Neuronal cNOS gene (Type I)   The 28-exons neuronal NOS gene extends over 100 kb. Characterization of potential regulatory regions in the neuronal NOS gene has not yet been published. Analysis of several neuronal NOS transcripts from adult and fetal human brain, kidney, heart, and skeletal muscle revealed developmental stage and tissue dependent transcriptional heterogeneity generated by the use of several different first exons, followed by splicing within the 5'-untranslated region to a common second exon [40]. Although only one neuronal cNOS protein should be generated by this mechanism, each first exon has a distinct 5'-flanking region and thus is likely to be controlled by its own promoter. In this way the significance of the transcriptional heterogeneity may depend on the accessibility of different promoters to the transcriptor factors activated in different tissues at different times. The notion that neuronal cNOS might be expressed in different sites during specific developmental periods suggests there may be aspects of its physiological function that have not previously been considered [41,42].

Macrophagical iNOS gene (Type II)   iNOS gene consists of 26-exon and 25 introns, and extends over 37 kb. The only NOS gene in which a promoter/enhancer region has been identified is murine iNOS. The 5'-flanking region of the human iNOS gene displays sequences cognate to that of mouse iNOS (66% identical over 400 bp upstream of the transcriptional site), but it has not yet been shown to function as a promoter/enhancer. A 1749 bp fragment from the 5'-flanking region of mouse iNOS contains a TATA box 30 bp upstream of the mRNA transcription site, along with at least 24 oligonucleotide elements homologous to consensus sequences for the binding of transcription factors involved in the inducibility of the other genes by cytokines or bacterial products [43].

Chromosomal Localization

Genomic clones of cNOS   Genomic clones were isolated by plaque hybridization or Southern blot hybridization methods using endothelial cNOS/neuronal cNOS cDNA as probe or via polymerase chain reaction (PCR). The 21kb human endothelial cNOS gene containing 26 exons maps to chromosome 7q35-36. The human neuronal gene containing 29 exons and spanning 160kb of genomic DNA, maps to chromosome 12q24.2 [44].
The genes are structurally related in terms of exon-intron organisation suggesting the common origin by gene duplication from an ancestral gene. Data indicate that the only structurally related gene in the human genome is cytochrome P450 reductase, localized to 7q11.2 [44].

Genomic clones of iNOS   The promoter-enhancer region of the murine iNOS gene was cloned by screening a cosmid library of mouse genomic DNA. The mouse iNOS gene has not been cloned in its entirety, but it has been mapped to chromosome 11 [45,46].
The human iNOS gene was isolated from a foreskin fibroblast genomic DNA library with probes derived from the 5'-end of murine iNOS cDNA. The 37kb human iNOS gene with 26 exons has been mapped to human chromosome 17q11.2-12 [47].

Gene Expression   Interleukin-1 beta (IL-1β) increases iNOS mRNA stability and the addition of cAMP enhances iNOS mRNA half-life over IL-1β exposure alone. Molecular studies of the mouse iNOS gene revealed that the 3'non-coding region of the iNOS mRNA contains the AUUUA motif known as a specific destabilizing sequence [9]. It was supposed that cAMP may indirectly affect this sequence but the exact mechanism has yet to be determined [48].
Oxidized low-density lipoprotein and its metabolic product lysophosphatidylcholine upregulate constitutive NOS mRNA and protein expression in bovine endothelial cells by means of new protein synthesis [48]. Hypoxia significantly suppresses the NO production in endothelial and smooth muscle cells by downregulation of the mRNA. After 48h incubation of endothelial and smooth muscle cells at 0% oxygen environment the mRNA levels of cNOS, its stability and the transcriptional rate of the gene are reduced to one third the level of those in a normoxic environment [49].
IL-1β, tumor necrosis factor-α (TNF-α), INF-γ and LPS upregulate the inducible NOS mRNA expression in human smooth muscle cells up to 88-fold. In contrast, cytokines and LPS dramatically reduce the levels of cNOS mRNA in human endothelial cells up to 10/20-fold [43,50].
TNF-α has been shown to decrease the content of mRNA for the constitutive NOS. A direct effect of TNF-α on transcription/message stability as well as an inhibitory effect of the NO produced by the inducible enzyme on the constitutive enzyme transcription have been hypothesised [43].
The role of transforming growth factor-β, (TGF β) is controversial. It depresses the production of NO in human retinal pigmented epithelial cells while it slightly potentiates NO production in bovine retinal pigmented epithelial cells [51].
Shear stress has been clearly shown to increase NO synthase expression above that observed in the absence of shear [43].

Gene Regulation

Gene regulation of cNOS   The −2835/+240 bp and −1548/+240 endothelial constitutive gene promoter constructs have a similar basal promoter activity in transfected endothelial cells [9]. 5'-deletion of the promoter down to −614 results in a 60-70% reduction in activity and deletion down to -416 results in a 90% loss of full promoter activity. Deletion of 48 bp from the 3'-end of the promoter completely abolishes promoter activity suggesting that the two downstream Sp1 binding sites in this region are required for basal promoter function [9].
Fifteen copies of half-palindromic motifs identified in the 5'-flanking region indicate an oestrogen responsiveness of the gene. Although oestrogen regulation of the NOS gene promoter is to be demonstrated experimentally, oestrogen has been shown to modulate vasomotion of the coronary arteries and it has been speculated

that oestrogen could be involved in the coronary heart diseases in post-menopausal women [38]. Two potential sterol regulatory elements are found at –625 and –1232 sites and are responsible for mediating sterol-dependent regression of the low density lipoprotein receptor gene and of genes for the two rate-limiting enzymes of the cholesterol biosynthetic pathway. A potential site for binding of the NF-1 transcription factor is found in the promoter at –1014 and the sequence has been found to mediate transcriptional activation of different genes by TGF-β.

A consensus sequence for AP-1 binding at –441 could mediate TNF-α down regulation of the gene, while the nine copies of the SSRE consensus sequence (shear stress responsive element) in the promoter could mediate the shear stress up-regulation of the same gene [38,39].

Gene regulation of iNOS   LPS inducibility depends on the unique NF-kB sequence comprised by nucleotides -85 to –76 and on the binding to this region of a cyclo-heximide-sensitive complex containing both p50/c-rel and p50/RelA heterodimers of NF-kB, plus additional nuclear proteins. A second region, position –913 to –1029, mediates the potentiation of the LPS induction by INF-γ although INF-γ alone could not enhance activity [43].

Additional Features

Molecular features of nitric oxide synthase   The mammalian endothelial cNOS protein sequence (1203 amino acids and molecular mass of 133kDa) shows an identity of 57 % with the brain NOS (protein of 1429 amino acids with a relative molecular mass of 155/160 kDa) and 50 % with the macrophage enzyme iNOS (protein of 1144 amino acids with a relative molecular mass of 130.5 kDa) [47]. The N-terminal half of the protein which contains the regulatory elements Ca^{2+}/calmodulin binding and phosphorylation sites appears to be a point of divergence between NOS (Figure 1). Endothelial NOS has a shorter amino terminus than brain NOS or macrophage NOS. There is 71 % and 65 % identity between residues 148 and 469 of endothelial NOS with brain and macrophage NOS respectively. This is likely to represent an L-arginine binding region. The COOH-terminal half of the protein is homologous to the NADPH-Cytochrome P450 reductase and contains the cofactor binding sites [16].

The endothelial constitutive NOS protein has several unique features such as six potential cAMP-dependent protein kinase phosphorylation sites at residues Ser116, –145,–170, –635, –740 and –1.053, some potential phosphate acceptor sites for proline dependent protein kinase, absence of membrane binding regions and one sequence identified to be a substrate for amino terminal myristoylation [39]. It has been suggested that cAMP-dependent protein kinase phosphorylation sites implicated in the regulation of vasomotion [52]. Cellular fractionation has indicated that endothelial NOS is localized to the particulate fraction of endothelial cell preparations and a consensus motif for N-terminal myristoylation is present in both human and bovine endothelial sequences [53].

Endothelial NOS undergoes dual acylation by myristic and palmitic acid, posttranslational modifications that result in targeting of this isoform to cell membranes [54]. After binding of agonists such as bradykinin to endothelial cells, NOS undergoes depalmitoylation, translocation to the cytosol, and phosphorylation with enzyme activation [55]. In endothelial cells, acylated NOS is now known to be localized to caveolae, plasmalemma microdomains that are known to facilitate the transcytosis of macromolecules and the uptake of small molecules by pinocytosis. Caveolae have also been implicated in the compartmentalisation of signal transduction protein, including growth factor receptors, multiple heterotrimeric G proteins, Ca^{2+} channels etc. Knowledge of the cell physiology NOS isoenzymes will be enhanced as the role of caveolae in intracellular protein trafficking in the compartmentalisation and regulation of specific signal transduction pathways becomes better understood.

Chemical and biological features of NOS   NOS isoforms are dimeric-flavin heme enzymes that contain one mole each of FAD, FMN, iron protoporphyrin IX (heme) and BH4 subunit and catalyze the conversion of L-arginine to nitric oxide and citrulline; NADPH and O_2 act as co-substrates in the reaction [16, 56]. BH4 and L-arginine also appear to have a role in forming and stabilizing the dimeric structure of iNOS. In addition the presence of reduced thiols (glutathione, or dithiothreitol) is necessary for full enzyme activity. The inducible and the neuronal cNOS are soluble and found in the cell cytosol while the endothelial constitutive isoform is membrane-associated. Ca^{2+}/calmodulin binding to cNOS activates NO synthesis and the activation mechanism involves a Ca^{2+}/calmodulin-induced conformational change that allows electrons to move from the flavins onto the heme iron, enabling oxygen activation to occur. Sequence differences within the calmodulin-binding sites of the three NOS isoforms result in their displaying different affinities for Ca^{2+} and calmodulin. The moderate affinity of neuronal and endothelial cNOS for calmodulin allows for absence of calmodulin binding at Ca^{2+} in physiological condition. In contrast iNOS has high affinity for calmodulin which remains tightly bound to the enzyme at even low Ca^{2+} concentrations and allows for NO production independently of added Ca^{2+} or calmodulin [57].

Processing and Fate   The metabolic pathway of NO in the intact organism depends on its source of administration or site of formation. The end products of NO metabolism are NO_2^- (nitrite) and NO_3^- (nitrate).

The half-life of NO in blood has been estimated to be as short as 0.46 ms. The NO concentration on the membrane surfaces may vary from submacromolar to micromolar levels. The release of NO is followed by its simultaneous diffusion into the bloodstream. The flux of NO is affected by chemical reactions with such biological constituents as hemoglobin, thiols, and nucleophilic agents. The neutral charge facilitates its diffusibility in aqueous solution and across cell membranes [58].

Most studies have failed to detect NO/NO_2^- in the bloodstream. Furthermore, due to the fact that several metabolic pathways enter the nitrate pool, the determination of nitrate in blood or urine may prove unreliable for assessing the L-arginine/NO pathway. The net synthesis of nitrate ranges from 1200 to 2200 mmol/person per day [59]. The basal rate of NO production by endothelial cells in culture was determined to amount to about 0.8 pmol/min per mg endothelial cells, the total mass of which in the body is 1.5 kg, thus total daily production of NO by the endothelial cNOS would amount to 1728 mmol per person. iNOS may contribute significant amounts of NO in special cases [59].

Two-thirds of nitrite entering the stomach originates from saliva and less than one third stems from nutrients. Ingested nitrate is absorbed from the gastrointestinal tract into the bloodstream which carries it to the salivary glands. Uptake of nitrate from the gastrointestinal lumen, which occurs primarily in the small intestine, appears to be an active process. There is little or no uptake from the stomach. Nitrate and nitrite in the intestine are involved in additional reactions. The conditions in the lower gastrointestinal tract favour bacterial denitrification. About 40%-45% of ingested nitrite is metabolized in the body rather than being excreted directly as nitrite in the urine. Nitrite can be absorbed oxidized to nitrate or reduced to NO in the stomach. Some nitrite is carried to the intestine where it is oxidized to nitrate [60].

Nitrate is the major urinary metabolite of NO. The concentration of nitrate in urine is 250-2000mM whereas nitrite and NO are usually not detectable. Within the urinary tract itself there is no metabolic conversion of nitrite to nitrate [61].

The extremely slow reaction kinetics of NO and oxygen in gaseous phases at low concentration of NO, the high reactivity of NO_x in airway lining fluid, the high diffusion capacity of NO and the extremely rapid reaction of NO with oxyhemoglobin determine the crucial role of NO in the respiratory tract and pulmonary circulation. All of these factors lead to rapid removal or local metabolism of NO in the lungs, limiting systemic side-effects [62].

Recently the presence of NO in the exhaled air of humans has been demonstrated suggesting that both exogenously and endogenously produced NO may modulate bronchial resistance. Exhaled NO derived from L-arginine may be formed either by macrophages present in airways or on the alveolar surface, by the pulmonary epithelium or by bronchial epithelial cells [63].

Biological Activity Nitric oxide regulates a broad range of biologic functions including vascular tone, platelet aggregation, neurotransmission, immunoregulation, microbial killing, and vascular remodelling [5,9] (Figure 2). In the cardiovascular system the endothelium is the main source of NO. Relaxation of vascular smooth muscle cells and inhibition of platelet adhesion and aggregation occur following activation of endothelial cNOS

and represent the most important mechanism for regulating blood pressure and tissue perfusion [12] (Figure 2b).

NO is produced in large quantities during pathophysiological situations such as host defence and immunologic reactions [5]. Nonspecific immunity is associated with the induction of NOS. NO-depending nonspecific immunity involves the reticulo-endothelial system and non reticulo-endothelial cells such as hepatocytes, vascular smooth muscle, and the vascular endothelium. Induced NO synthesis in the vascular wall causes potentially lethal hypotension in severe inflammatory conditions like septic shock or multi-organ failure (Figure 2d). NO also plays a role in tissue damage, interacting with oxygen-derived radicals to generate molecules that could enhance its cytotoxicity [5, 10]. Inhibition of iNOS with some therapeutic compounds such as glucocorticoids may explain the antierythema, antiedema actions and the beneficial effect that they have on endotoxin shock, asthma and rheumatoid arthritis, where NO is responsible for pathological vasodilation and tissue damage [5, 17]. In contrast, it is known that the production of endogenous NO is associated with apoptosis of tumor cells and NO synthesis inhibition may explain how glucocorticoids can facilitate the spread of infections and tumor growth and prevent the consequences of delayed hypersensitivity in conditions such as transplant rejection or vasculitis [64].

NOS activity is normally present in the upper and lower airway epithelium where it plays a major role in host airway defense mechanism and as a potential mediator of the inflammatory response [65].

In addition to its effects on cell viability and proliferation, NO may also play a role in the normal regulation of the response of cells to mitogens, as in vascular remodelling [66,67]. Changes in cGMP have been associated with both the initiation and the control of cell proliferation in many cells. In this context it is important to differentiate between actions resulting from NO released from cNOS and mediated via cGMP and those resulting from NO acting as a cytotoxic/cytostatic agent and mediated via NO released from iNOS [9, 25].

In the central nervous system NO is a neurotransmitter involved in the control of various functions such as long-term potentiation and long-term depression which lie behind the mechanisms of the formation of memory (Figure 2a). NO may also have a physiological role in vision, feeding behaviour, nociception and olfaction. Histochemical studies using antibodies against NOS have shown it to occur widely in the central nervous system, primarily in neurones and also in the vascular endothelium with not glial localisation [5, 68]. Evidence suggests a possible relation between increased production of NO and neuronal damage [12].

In the periphery NO released by NANC nerves mediates some forms of neurogenic vasodilation (Figure 2b) and regulate various gastrointestinal, respiratory and genitourinary tract functions [12] (Figure 2b). Inhibitors of NOS reduce electrically induced relaxation of internal

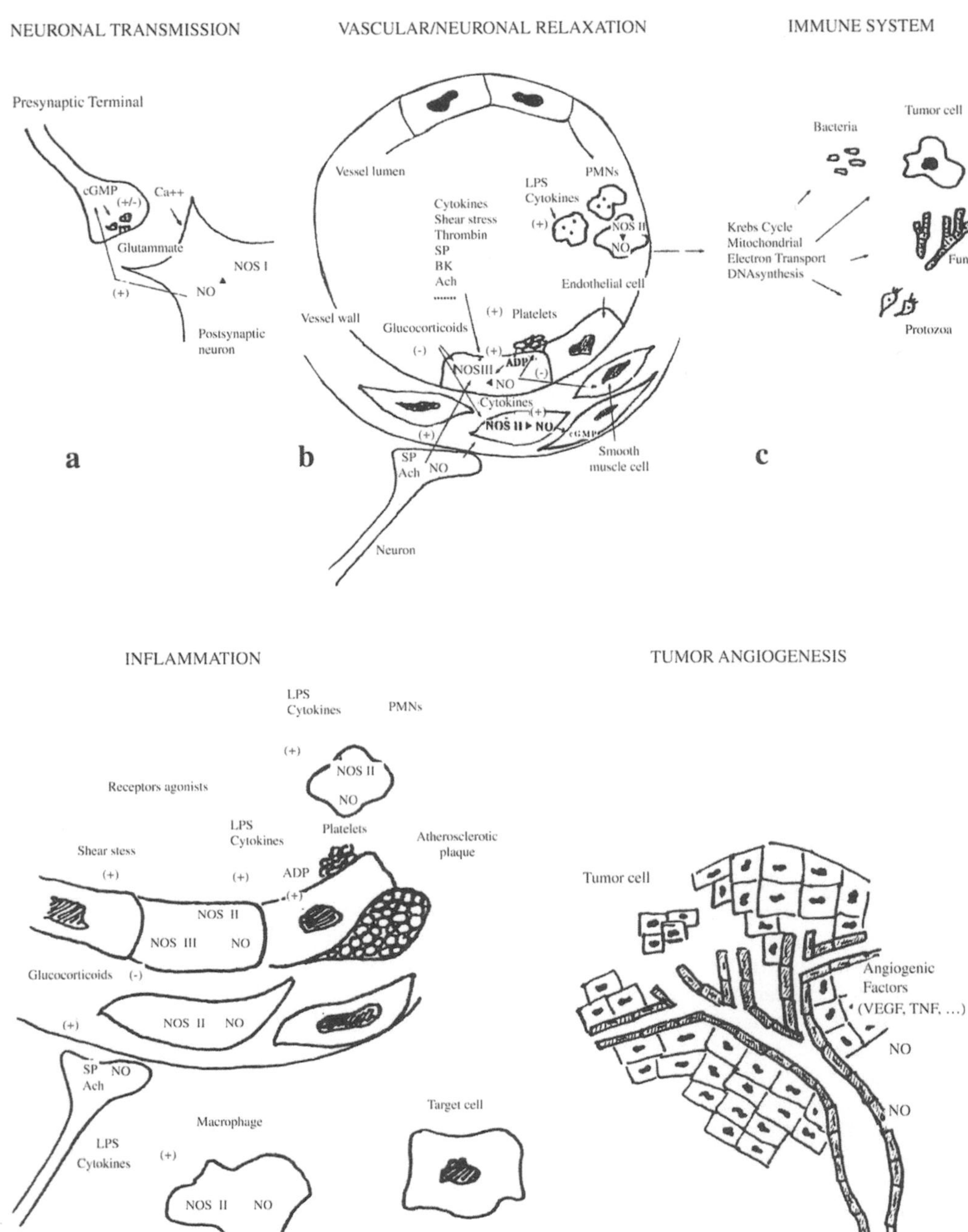

Figure 2. Putative action of NO in the central nervous system (a), in blood vessels (b and d), in macrophages (c) and in tumor angiogenesis (e). a) Neurotransmitters such as glutamate, released by the nerve endings of presynaptic neurones, bind to their receptors, induce Ca²⁺ influx in the postsynaptic neurones and activate NOSI isoform and NO production. NO diffuses back to the presynaptic nerve where it activates the guanylyl cyclase. The cGMP formed can increase (long-term potentiation) or decrease (long-term inhibition) the release of further transmitters. b) In the blood vessels, endothelial cells synthesize NO under basal conditions and the synthesis is enhanced by the shear stress as well as by a variety of receptor agonists (b and d). NOSIII isoform is responsible for basal NO production, which acts in a paracrine fashion on subjacent smooth muscle cells inducing vasodilation. NO is also released toward the lumen of the blood vessel where it prevents platelet and leukocyte adhesion to the vascular endothelium and inhibits platelets aggregation. c and d) Induction of NOSII isoform in vascular smooth muscle cells and macrophages and the subsequent overproduction of NO are implicated in several vascular diseases (d) as well as in cytostatic effects on micro-organisms and tumor cells (c). These effects are mediated by the interaction of NO with protein-bound iron proteins (c). e) NO released by tumor cells, as well as angiogenic growth factors affecting NO production in endothelial cells, are also implicated in vasodilation and hyperaemia of preexisting capillary and in tumor angiogenesis.

anal sphincter muscle. In the colon, NOS is present in the myenteric and submucous plexus in both neuronal cell bodies and fibers. NO also contributes to the NANC vasodilation and relaxation of human tracheal muscle. NO is also responsible for the relaxation of the corpus cavernosum and thus the development of penile erection [69]. Moreover NOS inhibitors in rats induce hyperactivity in the urinary bladder and decrease bladder capacity [12].

The peripheral analgesic effect of acetylcholine has also been proposed to be mediated via NO. Acetylcholine, L-arginine and NaNp induced analgesia in an animal model and L-NMMA, a competitive inhibitor of NOS, prevented the analgesia induced by both acetylcholine and L-arginine [12].

Role in Vascular Biology

Physiological Function   The vasculature is a dynamic organ that can regulate its own tone and structure via different mechanisms. Nitric oxide mainly generated by endothelial cells plays a variety of actions in the cardiovascular system affecting vessel relaxation, platelet adhesion and aggregation, cardiac contractility, vascular remodelling, leukocyte adhesion, renal function and hormone secretion.

Control of cardiovascular tone   Endothelium-dependent relaxation has been demonstrated in isolated arteries, arterioles, veins, microvasculature and lymph vessels. The contribution of NO to vascular tone changes during development. Physiologic vasodilation is regulated by endothelial and neuronal NOS isoforms in healthy blood vessels. Inhibition of NOS with L-arginine analogues induces endothelium-dependent contraction of isolated arteries from many species including humans, indicating that there is continuous release of NO from the endothelium. The basal release of NO counteracts the vasoconstrictor effects of such endothelium- and tissue-derived vasoactive substances as endothelins and angiotensin. This effect is especially evident in middle peripheral vessels where the smooth muscle component is predominant, while in capillaries NO regulates permeability and leakage. Moreover, NO regulates basal vascular tone in resistance vessels but not in veins [5,70]. In isolated human resistance vessels L-NMMA augments nonadrenaline-induced tone and inhibits the vasodilation caused by acetylcholine, whereas the same inhibitor does not cause an increase in basal tone in a variety of venous preparations from animals and humans. Studies *in vivo* support these *in vitro* observations [12].

Vasorelaxant actions of many hormones and autacoids including acetylcholine, bradykinin (BK) and substance P (SP) are mediated by the release of NO from the endothelium (flow-dependent vasodilation) [70] (Figure 2b). NO released from peripheral nitrergic nerves also alters vascular tone in animals and humans. NO synthesized by neurons in the central nervous system has been implicated in the regulation of the sympathetic nervous system in animals, but whether this is also the case in humans remains to be determined. Immunohistochemistry has demonstrated the presence of NOS-containing neurons in nerves supplying cerebral vasculature and the vessels of the adrenal plexus. NO released from these neurones plays a role in the control of the cardiovascular system acting both as a direct vasodilator and a neuromodulator altering the release of other transmitters [12].

In addition NO influences cardiac contractility. Constitutive NOS is present in myocardium and endocardium. Exogenously administered NO shortens contraction time and may be negatively inotropic. In humans NOS activity has been detected in the atrium and ventricels, but a physiologic role for NO remains to be established [71].

Evidence of a physiologic role for NO in the control of the renal vascular bed and renal tubular function suggest an indirect way for cardiovascular control. It has been shown that NO synthase is localized to nonvascular tissues within the kidney including the collecting duct, the macula densa and in cultured renal tubular cells. NOS inhibitors increase basal renal vascular resistence and inhibit endothelium dependent relaxation in the isolated perfused kidney *in vitro* and in intact kidney *in vivo* [72]. In healthy volunteers ingestion of L-arginine increases renal blood flow, glomerular filtration and sodium excretion, suggesting a direct effect of NO on tubular function. In addition NOS inhibitors cause sodium retention and indirectly, hypertension. These studies suggest a direct natriuretic effect of NO within the kidney [72].

NO is implicated in the regulation of the release of several hormones that have direct and indirect effects on the cardiovascular system. For example, NO modulates the secretion of renin [72].

Circulating cell functions   NO inhibits platelet adhesion and aggregation and contributes to the anti-inflammatory properties of the endothelium. NO and prostacyclin act synergistically to inhibit aggregation and to desegregate platelets, suggesting that the release of both molecules by vascular endothelium is involved in its thromboresistant properties [5] (Figure 2b). Aggregating platelets stimulate the release of NO that may be a potent defence against the potentially deleterious effects of activated platelets. The combined action of NO and prostacyclin released from vascular endothelium together with NO released from platelets could result in a synergistic suppression of intracellular calcium elevation and finally inhibit platelet aggregation [70]. Experimental evidence indicate that inhibition of NOS increases platelet accumulation induced by ADP, platelet activating factor (PAF) and thrombin in the lungs of anaesthetized rabbits and enhances PAF-induced accumulation of neutrophils [70]. In fact, in addition to its antiplatelet effects, NO significantly inhibits leukocyte adhesion by affecting the expression of several adhesion molecules such as vascular cell adhesion molecule-1 (VCAM-1), endothelial-leukocyte adhesion molecule-1 (ELAM-1 or E-selectin) and inflammatory cytokine IL-6 and to a lesser extent the leukocyte-specific chemoattractant IL-8 and intercellular adhesion molecule-1 (ICAM-1) [73].

Vascular remodelling   The vascular remodelling is modulated by locally generated growth factors, vasoactive substances, and hemodynamic stimuli and may be an adaptive mechanism that contributes to tissue homeostasis. Endothelial cells play a major role in both vascular remodelling and vascular tone. They have contacts with the bloodstream, sensing changes in flow, pressure, inflammatory signals and levels of circulating hormones. In addition, the endothelium affects cell growth, apoptosis, mobilisation and modulation of extracellular matrix composition by releasing and/or activating molecules involved in these processes [72].

NO maintains vascular integrity by exerting a direct vasculoprotective effect. Its production by endothelial cells inhibits smooth muscle cell proliferation, while it favours in a paracrine way endothelial cell growth and migration to maintain an intact monolayer. Continuous release of NO by endothelial cells thus counteracts the injury of vascular wall and prevents the formation of atheroma by maintaining normal blood vessel responsiveness and flow and by inhibiting smooth muscle cell growth [72]. When tissue homeostasis and nutrient requirement need to be restablished, NO switches the quiescent endothelium to differentiate in angiogenic phenotype [66, 67]. NO has been shown to affect coronary microvasculature in response to long term increases in cardiac work, chronic reduction of oxygen content and partial occlusion of the arterial supply, inducing formation of new microvessels from the pre-existing blood vessels [67, 70]. Furthermore, NO production has been demonstrated to mediate the growth promoting effects of the endothelial-specific growth factor vascular endothelial growth factor (VEGF).

Several reports have implicated a role for NO synthase in angiogenesis. For example, it was shown that an angiogenic activity was released from bacterial endotoxin treated human monocytes in the presence of L-arginine and was blocked by NO synthase inhibitors [74]. More recently the role of NO in mediating the angiogenic activity elicited by different cytokines has been described [75]. Similarly L-arginine has been shown to favour healing and angiogenesis in gastric ulcerations while NO synthase inhibitors delayed it [76]. Conversely, during vascular development of the chorioallantoic membrane an inhibitory effect of NO was reported [77], suggesting a different action of NO on embryo-angiogenesis versus angiogenesis in the adult.

Chemical mediators which activate the constitutive NO synthase, like substance P (SP) as well as NO donors, such as sodium nitroprusside (NaNP), promote endothelial cell proliferation and migration *in vivo* and *in vitro*, while inhibitors of NO synthase suppress these responses [66,78]. The growth promoting effect of NO appears to be linked to cGMP generation in cultured endothelium [78]. *In vivo* NO synthase inhibitors suppress angiogenesis induced by vasodilating effectors like SP and prostaglandin E1 (PGE1) [66]. The growth promoting effects of angiogenic factors, as well as their vasorelaxant action and increase in permeability, seem

to be specifically linked to the NO pathway [79,80]. NO and cGMP mediates the angiogenic effect induced by vascular permeability/vascular endothelial growth factor (VEGF), but not by fibroblast growth factor-2 (FGF-2) [66,81]. Recent data suggest the existence of an autocrine/paracrine loop in NO effects, since NO given exogenously or induced by treatment with SP, increased the expression of mRNA and protein of FGF-2, which in turn mediates NO-induced angiogenesis [67].Thus the "protective" effect of NO on the vasculature appears to be linked to the autocrine/paracrine effect of the molecule on the endothelium.

Pathology

Cardiovascular diseases   The role of NO in cardiovascular diseases including renal and essential hypertension, hypercolesterolemia, atherosclerosis, cardiac failure, septic shock and vascular inflammation has been of considerable interest in recent years [12,72].

Abnormalities of the NO-pathway have been reported in hypertension in animals and humans. In the patients with untreated essential hypertension the response to L-NMMA is reduced in the forearm arterial bed and this reduction appears to be correlated with blood pressure [5]. In addition the NO-system plays a role in the pathogenesis of some forms of secondary hypertension. In humans acute volume expansion causes a rise in forearm blood flow which is abolished by pretreatment of the forearm with L-NMMA. The pathway of L-arginine-NO in hypertension is not yet clear, and possibilities include altered NO synthesis, release, and effect. In the secondary hypertension associated with chronic renal impairment, numerous data suggest an inhibition of NO synthase activity [70,72].

A decrease in endothelium-dependent relaxation has been demonstrated in human atherosclerotic coronary arteries *in vitro* and *in vivo* [5]. Basal and agonist-stimulated release of NO are impaired in arteries obtained from patients with coronary artery disease. In addition, coronary risk factors are associated with loss of endothelium-dependent vasodilation in the coronary arteries of humans with angiographically smooth coronary arteries. Similar results have been shown in hyperlipidaemic animals and humans [70]. The mechanism of NO-dependent relaxation in atheroma and hyperlipidaemic disease is not clear, it is possible that the endothelial dysfunction observed in atherosclerotic blood vessels results from an increase degradation of NO, for example by superoxide anions. In addition, smooth muscle cells in atheromatous blood vessels seem to be less sensitive to NO [70].

NO also contributes to the vasodilatation and plasma extravasation produced by SP released from NANC fibers and to neurogenic inflammation [5].

In animal models, induction of the inducible NOS isoform produces large quantities of NO, which appears to contribute to the pathologic vasodilation observed in septic shock. The effect is to produce vascular relaxation, hyporesponsiveness to vasoconstrictors and even tissue damage. The setting of expression of NOS in vessels may be dependent on the local balance of cytokines. Biochemical evidence shows increased overproduction of NO in patients with sepsis or cytokines induced hypotension [10,12].

At cardiac level, treatment of myocardial cells *in vitro* with LPS and cytokines leads to the induction of iNOS with associated loss of contractility, an effect which is reversed by L-NMMA. In humans iNOS induction has been linked to the pathogenesis of some forms of cardiomyopathy, to heart failure and has been found in patients with cardiac allografts [71].

Evidence suggests that the NO pathway may provide a link between some metabolic and vascular abnormalities in diabetes. NO is implicated in the control of normal insuline release and induction of NOS may play a part in the pathogenesis of early pancreatic damage [70].

Role in trophic disorders   Reduced NO production is linked to atheromatous plaque formation, in which smooth muscle cells actively proliferate and inflammatory infiltrate is a critical component (Figure 2d). Studies using *in vitro* model systems suggest that exogenous NO promotes vascular smooth muscle cell apoptosis via cGMP-mediated mechanism [72]. In contrast, in pathologic situations in which NOS gene is up-regulated, remodelling of vessels with a massive enlargement in lumen diameter is observed.

Pathologic situations with an elevated shear stress are accompanied by an increase in NO synthesis and u-PA production. Consistently, NaNP treatment elicits endogenous FGF-2 expression and u-PA upregulation in coronary endothelium [67], suggesting the triggering of the angiogenic cascade as an additional mechanism for NO donors for overcoming the pathologic distress at the myocardial level.

NO and tumor angiogenesis   Increased NO levels have been found in human tumors which are in a permanent vasodilation state [82]. Most of the cellular components of the tumor mass (the tumor cells themselves and the immune cell infiltrate) have been shown to generate NO *in vitro* (Figure 2e). However the role of NO in tumor biology remains unclear. Transfection of the inducible NO synthase into a colon adenocarcinoma line gives a cell line that induces tumors which grow more rapidly and are more vascularized than wild type cells [83]. The *in vivo* progression of murine haemangiomas, induced by middle T antigen of polyomavirus, is reduced by canavanina, an NO synthase inhibitor [84]. Other observations in agreement with NO being a specific signal for tumor vascularization show that blocking NO synthase activity retards the growth of xenografted tumors [85] and excessive production of NO sustains tumor growth [86]. Consistently, dexamethasone exerts antiangiogenic effect leading to reduced tumor growth [87]. In addition, NO production has been demonstrated to affect vascular permeability and tone modulated by VEGF, that is

highly expressed in tumors and is strongly angiogenic *in vivo* [88]. Transfection of VEGF into human breast carcinoma cell line has been previously shown to enhance tumor growth and vascular density *in vivo* and promotes a strong angiogenic response [81]. This tumor angiogenesis is efficiently blocked by NO synthase inhibition [67]. Although a body of evidence indicates a definite role for NO in tumor growth, the end point of increased NO production within the tumor mass is controversial. In the experimental models reported above NO facilitates tumor growth and vascularization. Conversely, increased production of NO by tumor cells has been shown to cause lysis of bystander cells [64].

Clinical Relevance and Therapeutic Implications   NO has a major role in the physiologic regulation of the cardiovascular system and abnormalities of the pathway exist in cardiovascular diseases. Reduced activity of the pathway could increase vascular tone and promote atherogenesis. In rings of atherosclerotic coronary arteries when compared to rings of normal coronary arteries, the endothelium-dependent relaxation is decreased and the responses to vasoconstrictors are often enhanced [5]. Similar observations have been made in isolated human blood vessels from patients with diabetes or atherosclerosis. *In vivo* the vasodilator response to acetylcholine has been reported to be reduced in the forearm vasculature of patient with primary or secondary hypertension, type II diabetes mellitus, hypercholesterolemia and chronic heart failure [70]. The reponse of forearm blood flow to L-NMMA is reduced in these patients suggesting that impaired generation of NO may contribute to their diseases. In patients with dialysis-dependent renal failure, inhibitors of NOS accumulate in the plasma and may inhibit NO synthesis and white-cell dysfunction, contributing to the development of hypertension and atherosclerosis in this disease. Furthermore, part of the therapeutic effect of angiotensin-converting-enzyme inhibitors may be due to their ability to potentiate the half life of bradykinin, which stimulates the release of NO and increases its concentration in the vascular bed [70]. Increased activity of the NO pathway could contribute to tissue damage and vascular hyporeactivity observed in sepsis and other inflammatory vascular diseases. Expression of the iNOS in smooth muscle cells has been suggested in animal models of atherosclerosis and diabetes mellitus, but the functional consequences remain to be determined. iNOS has also been detected in human myocardium of patients with dilated cardiomyopathy [71]. In humans there is evidence, using NOS inhibitors, that induction of iNOS contributes to the cardiovascular abnormalities associated with clinical sepsis [71]. Consequently, novel methods of NOS gene therapy or NO inhibition might be the basis for new therapies for cardiovascular diseases. Furthermore, selective inhibitors of brain NOS may prove protective in brain ischemia and chronic degenerative diseases of the nervous system and levels of products of the L-arginine-NO pathway in biologic fluids and levels of expression of NOS in tissues may become clinical markers for the monitoring of certain pathologic conditions and the progress of their treatment.

The therapeutic application of NO-related compounds has a double approach:

a. to use the protective effect of NO by the administration of NO donor or NO generating compounds to induce vessel relaxation (in hypertension), to block platelet aggregation (in thrombo-embolic disesase), to inhibit smooth muscle cell proliferation (in atherosclerosis), to induce new vessel formation (during ischemia). Organic nitrates are used in the treatment of ischemic heart disease with angina pectoris, coronary vasospasm, intracoronary thrombosis, atherosclerotic coronary narrowings [18]. NaNP is used as an antihypertensive drug to treat hypertensive emergencies and it is used in many situations when short-term reduction of cardiac preload and/or afterload is desired (aortic dissection, congestive heart failure, acute myocardial infarction). In association with other vasodilators, NaNP and organic nitrates are used in the treatment of acute and chronic congestive heart failure. An additional application of nitrovasodilators is the cholesterol-reducing therapy [18]. NO may be useful in patients with some pulmonary conditions, since inhaled NO gas reverses pulmonary hypertension. Administration of NO for seven days resulted in substained improvement in lung function in affected patients, suggesting that inhaled NO assists lung healing.

b. to block the NO pathway by NOS inhibitors when NOS over-activation leads to tissue damage or threatens host life as during septic shock or tumor growth. In solid tumors monitoring NOS activity could retain prognostic significance in relation to angiogenesis and the selective inhibition of the NOS pathway might prove useful as anticancer-antiangiogenic therapy. Consequently NOS inhibition can affect tumour growth by reducing blood flow via vasoconstriction and also by interfering with the mechanism of endothelial cell replication, ultimately leading to the suppression of the tumour blood supply.

Clinically specific NOS inhibitors without side effects are not currently available. Pharmacological research is now working toward the synthesis of selective NOS inhibitors for the different isoforms and with different tissue-specificity.

Marina Ziche

References

1. Furchgott RF et al (1980) Nature 288:373-376
2. Hibbs JB et al (1987) J Immunol 138:550-565
3. Palmer RMJ et al (1987) Nature 327:524-526
4. Ignarro JJ et al (1990) Annu Rev Pharmacol Toxicol 30; 535-560
5. Moncada S et al (1991) Pharmacol Rev 43:109-142
6. Bredt DS et al (1994) Annu Rev Biochem 63:175-195
7. Nathan C (1992) FASEB J 6:3051-3064
8. Kobzik L et al (1994) Nature 372:546-548
9. Nathan C, Xie QW (1994) Cell 78:915-918
10. Nussler AK, Billiar TR (1993) J Leukocyte Biol 54:171-178
11. Nicotera P et al (1997) Adv Neuroimmunol 5:411-420
12. Moncada S et al (1993) N Engl J Med 329:2002-2012
13. Nakane M et al (1993) FASEB Lett 316:175-180
14. Asano K et al (1994) Proc Natl Acad Sci USA 91:10089-10093
15. Feldman PL et al (1993) Chem Eng News 26-38
16. Marletta MA (1993) J Biol Chem 268:12231-12234
17. Moncada S et al (1997) Pharmacological reviwes 49:149-152
18. Needleman P et al (1984) In: Goodman LS, Gilman AG (eds) The pharmacological basis of therapeutics, 7th edn. McGrawHill, pp 807-826
19. Henry Y et al (1993) FASEB J 7:1124-1134
20. Stamler JS et al (1992) Science 258:1898-1902
21. Stamler JS et al (1994) Nature 367:28
22. Padmaja S, Huie RE (1993) Biochem Biophys Res Commun 195:539-544
23. Butler AR, Williams DLH (1993) Chem Soc Rev 22:233-241
24. Kharitonov VG et al (1994) J Biol Chem 269:5881-5883
25. Schmidt HHHW et al (1993) Biochim Biophys Acta 1178:153-175
26. Knowles RG, Moncada S (1992) TiBS 17:399-402
27. O' Halloran TV (1993) Science 261:715-729
28. Wu M et al (1994) Am J Physiol 266:H2108-H2113
29. Rinden E et al (1989) J Am Chem Soc 3:1203-1210
30. Murphy ME, Sies H (1991) Proc Natl Acad Sci USA 88:10860-10866
31. DeMaster EG et al (1989) Biochem Biophys Res Commun 163:527-533
32. Teillet-Billy D, Fiquet-Fayard F (1997) J Phis 10:L111-L121
33. Melia TP (1965) J Inorg Nucl Chem 27:95-101
34. Wilhelm E et al (1977) Chem Rev 77:219-223
35. Blough NV, Zaffiriou OC (1985) Inorg Chem 24:3504-3510
36. Forstermann U et al (1991) Bioch Pharmacol 42:1849-1857
37. Lowenstein CJ et al (1992) Proc Natl Acad Sci USA 89:6711-6715
38. Venema RC et al (1994) Biochim Biophys Acta 1218:413-420
39. Nadaud S et al (1994) Biochem Biophys Res Commun 198:1027-1033
40. Santiago L et al (1992) Proc Natl Acad Sci USA 89:6348-6352
41. Sessa WC et al (1993) Hypertansion 21:934-938
42. Reiling N et al (1994) Eur J Immunol 24:1941-1944
43. Marsden PA et al (1992) FESEB lett 307:287-293
44. Forstermann U et al (1994) Hypertention 23:1121-1131
45. Nakane M et al (1993) FESEB lett 316:175-180
46. Chartrain NA et al (1994) J Biol Chem 269:6765-6772
47. Marsden PA et al (1993) J Biol Chem 266:17478-17488
48. Hirota K et al (1995) Circ Res 76:958-962
49. MCQuillan LP et al (1994) Am J Physiol 267:H1921-H1927
50. Walter T et al (1994) Biochem Biophys Res Commun 202:450-455
51. Goureau O et al (1993) Proc Natl Acad Sci USA 90:4276-4280
52. Micheal T et al (1993) Biochemistry 62:52-56
53. Pollock JS et al (1993) Am J Physiol 265:C1379-C1387
54. Jianwei L et al (1996) Biochemistry 35:13277-13281
55. Garcia-Cardena G et al (1996) J Biol Chem 271:27237-27240
56. Brets DS et al (1991) Nature 351:714-718
57. Stuehr DJ et al (1991) Proc Natl Acad Sci USA 88:7773-7777
58. Kobzik L et al (1993) Am J Respir Cell Mol Biol 9:371-377
59. Malinski T et al (1993) Biochem Biophys Res Commun 193:1076-1082

60. Cho HJ et al (1992) J Exp Med 176:599-604
61. Mayer B et al (1990) FEBS Lett 277:21-219
62. Kelm M et al (1991) Cardiovasc Res 25:831-836
63. Green LC et al (1982) Anal Biochem 126:131-138
64. Xie K et al (1997) J Natl Cancer Inst 89:421-427
65. Guo FH et al (1995) Proc Natl Acad Sci USA 92:7809-7813
66. Ziche M et al (1994) J Clin Invest 94:2036-2044
67. Ziche M et al (1997) Circ Res 80:845-852
68. Ross C A et al (1990) Trends Neurosci 13:216-222
69. Anderson K-E, Wagner G (1995) Physiological Reviews 75:191-218
70. MacAllister RJ, Vallance P (1996) JIFCC 8:152-158
71. Kelly R et al (1996) Circ Res 79:363-380
72. Gibbons GH (1997) Am J Cardiol 79(5A):3-8
73. De Caterina R et al (1995) J Clin Invest 96:60-68
74. Leibovicz JS et al (1994) Proc Natl Acad Sci USA 91:4190-4194
75. Montrucchio G et al (1997) Am J Pathol 151:557563
76. Brozowski T et al (1995) Digestion 56:463-471
77. Pipili-Synetos E et al (1994) Br J Pharmacol 111:894-902
78. Ziche M et al (1993) Life Sci 53:1105-1112
79. Wu HM et al (1996) Am J Physiol 269(38):C1371-C1378
80. Morbidelli L et al (1996) Am J Physiol 270(39):H411-H415
81. Ziche M et al (1997) J Clin Invest 99:2625-2634
82. Thomsen LL et al (1994) Cancer Res 54:1352-1354
83. Jenkins DC et al (1995) Proc Natl Acad Sci USA 92:4392-4396
84. Ghigo D et al (1995) J Exp Med 181:9-19
85. Orucevich A, Lala PK (1996) Br J Cancer 73:189-196
86. Doi K et al (1996) Cancer 77:1598-1604
87. Wolff JEA et al (1993) Brain Res 604:79-85
88. Ferrara N, Davis-Smyth T (1997) Endocr Rev 18:4-25

NO

Definition *Nitric oxide*

See: →Nitric oxide

OCS

Definition *Open canalicular system*

See: →Vascular integrins

OCT-1

Definition *Octamer binding factor-1*

See: →FGF receptors

Ontogeny of the Vascular System

Definition *Vascular development in the embryo*

See also: →Angiogenesis

Introduction Endothelial cells (EC) constitute the scaffolding of the vascular system, which is completed by differentiation of wall cells and by remodelling of the early vascular tree. In this chapter, we shall consider the cellular and molecular mechanisms underlying the ontogeny of the endothelial tree, actively investigated in recent years in different animal models.

Like forerunners of most lineages, EC precursors migrate extensively during ontogeny and exchange information with homologous and heterologous cell types. These processes are best unravelled by dynamic studies rather than by *in situ* observation of cells or tissues. The avian model has been particularly appropriate for such studies, because experimental manipulations can be performed during the periods of morphogenesis and organogenesis. The contrasting structures of heterochromatin in the cell nuclei of the quail and chicken species serve as cell tracers in these experiments [1]. This so-called "quail/chick marker system" is now more usually exploited by means of species- and lineage-specific monoclonal antibodies (moabs). In the case of the blood system, the MBl/QHl moab [2, 3] recognizes quail endothelial and hemopoietic cells, and has no affinity for any chicken cell type. The QHl moab (available from the Developmental Studies Hybridoma bank, University of Iowa) is currently used to diagnose the origin of EC in quail/chicken combinations by all investigators interested in this problem. It detects single quail cells in a chick host, and has the advantage that EC or their precursors can be identified prior to their integration into endothelial tubes. One drawback is that single QHl+ cells could also be hemopoietic cells.

While cell origins and migrations in the endothelial tree have been analyzed mostly in the avian embryo, the underlying molecular mechanisms have been unravelled in both birds and mammals following the discovery of EC specific growth factors and cognate receptors. In particular, analytic and experimental studies of the angiogenic growth factor VEGF and of its receptors have provided significant insight into the developmental biology of EC. The deletion of corresponding genes addresses the issue of their role during development. In some cases the avian homologues have been cloned. In this respect, a newly obtained moab directed against the avian Vascular Endothelial Growth Factor (VEGF) receptor 2 [4] has brought about particular support for the long held hypothesis of a developmental relationship between endothelial and hemopoietic cells, and more precisely of the existence of a common precursor, the hemangioblast.

Other models such as zebrafish and xenopus offer further opportunities. Several mutations that disturb the development of endothelial and hemopoietic cells have been obtained in mutagenesis screens of zebrafish [5 – 8]. Xenopus studies have contributed information about the establishment of the dorso-ventral axis of the embryo, including regionalization of the mesoderm into muscle dorsally and blood ventrally; data point to the transforming growth factor-β (TGF-β) family member bare morphogenic protein-4 (BMP-4) as a ventralizing agent [see 9; 10]. These aspects, which concern very early events of ontogeny, will not be reviewed in the present chapter.

We shall narrow our focus to avian and mammalian EC ontogeny and will talk about the 'blood system', to refer to both EC and blood cells, which clearly display developmental links, whose nature is currently debated. The term hemangioblast [11] is used to designate the presumptive common precursor for the two lineages, while the endothelial precursor is called the angioblast (from the greek 'angeion': vessel and 'blastos': bud; [12]). Angiogenesis refers to the process by which quiescent blood vessels in the adult resume their growth and invade tumors [13]. The term has been extended to embryonic development, usually with a meaning restricted to a process which involves the migration of EC precursors.

Characteristics   The first blood cells appear in the extraembryonic area, the future yolk sac, in aggregates called blood islands; these soon become interconnected by the first capillaries which then differentiate centripetally. According to His [12], intraembryonic vessels were derived from ingrowth of extraembryonic yolk sac vessels. Early investigators probed this hypothesis by destroying the blood islands or by separating the embryo from the yolk sac [14–16] and found that EC emergence occurred all over the avian blastodisc. This was confirmed later in 'yolk sac chimeras'. These chimeras result from grafting the embryonic body of a quail on a chick yolk sac [17]. The borderline between blood vessels from the two species corresponded to the borderline between chick and quail tissues in these chimeras, even when the surgical association between the two components was performed at stages preceding the assembly of blood vessels [18].

As perceived early in the nineteenth century, the formation of germ layers is central to the development of coelomates. In 1895, Van der Stricht [19] justly ascribed the origin of the blood system to the mesoderm, the median germ layer of the embryo. The pregastrulation avian blastodisc appears to display an extensive capacity to give rise to endothelial and blood cells. EC differentiation can be obtained by transplanting fragments of epiblast (the early embryonic surface cell layer, from which the three primordial germ layers are derived by the gastrulation process) which were prevented from gastrulating by cytochalasin B [20]. These data seemed to indicate that cells in the epiblast have an angioblastic potential, regardless of interaction with a different cell type. However Yablonka-Reuveni [21] and Flamme and Risau [22] found that blood island formation does not occur in cultures of epiblast cells, isolated before gastrulation, i.e., before there is any sign of the primitive streak. Blood island formation is strictly dependent on bFGF added to the culture medium [22].

The endothelial network begins differentiating at the interface of mesoderm with endoderm, as seen in the living embryo [23], in scanning electron micrographs of blastodiscs from which the endoderm has been peeled away [24] or in QHl-stained wholemounts [3, 25] (Figure 1D). Within the embryo proper, single QH1+ cells appear progressively along the ventral and outer side of the somites, in parallel with segmentation and associate to form the paired dorsal aortae; at the same time QH1+

cells appear in the anterior intestinal portal, and soon interconnect to give rise to the V-shaped endocardial anlage. Vessels in the body wall, in particular the two more superficial cardinal veins, begin differentiating a few hours later.

Experimental approaches indicate that the contact with endoderm is important for the emergence of EC at these early stages. In Wilt's culture experiments [26], extraembryonic mesoderm separated from endoderm gave rise to 'blood islands' in culture. However, these collections of erythroid cells were smaller than these derived from endoderm-associated mesoderm and devoid of an endothelial envelope. The role of endoderm in the emergence of vasculogenic and hemopoietic potentials will be discussed later in this review.

Besides QH1, another excellent marker for avian EC is VEGF-R2 [27, 28]. VEGF-R2 is expressed in the mesoderm right after its formation and becomes restricted to angioblasts (defined as vascular EC which have not yet assembled into a tube) at the time when these precursors develop affinity for QH1 in the avian embryo, as determined by double-labelling [27]. In a gastrulation stage blastodisc, VEGF-R2 positive cells are localized in the mesoderm of the caudal and lateral area pellucida (Figure 1A). A few hours later, they become visible in the extraembryonic area (future yolk sac) and begin to form hemangioblastic clusters. VEGF-R2 is still potently expressed in newly recruited mesodermal cells of the caudal region; in addition, a second more cephalic expression domain becomes detectable (Figure 1B, C). The cells in this second domain are angioblasts which will vascularize the embryonic brain, which is devoid of angiogenic potential. At the 7-10-somite stage, the first tubular structures become discernible in the embryonic body (Figure 1D, E). At the 13-somite stage, vessels connect between the yolk sac and the embryonic body and embryonic circulation is established.

VEGF-R2 expression appears to reflect a proliferative state of the endothelium. Overexpression of VEGF achieved by injecting human recombinant factor to explanted quail embryos at the 5-somite stage leads to "hyperfusion" of blood vessels [29]. Overexpression by means of a retroviral vector in quail or chick limb buds [30] leads to VEGF-R2 reexpression and hyperproliferation blood vessels. VEGF-R2 is also upregulated following VEGF application to the quiescent E13 chorioallantoic membrane and extensive capillary sprouting is induced, while lymphatic vessels remain dormant [31, 32].

We have considered above the distribution of EC potential in the early blastodisc, before and during gastrulation. Later, not all compartments of the mesoderm of the avian embryo have equivalent potential regarding the production of EC, as shown by transplantations from quail to chicken. The splanchnic layer of the mesoderm is a highly prolific EC producer [33].

A conspicuous endothelial potential is also an attribute of somites [34–38]. In contrast, somatopleural mesoderm which contributes to the body wall and limbs is

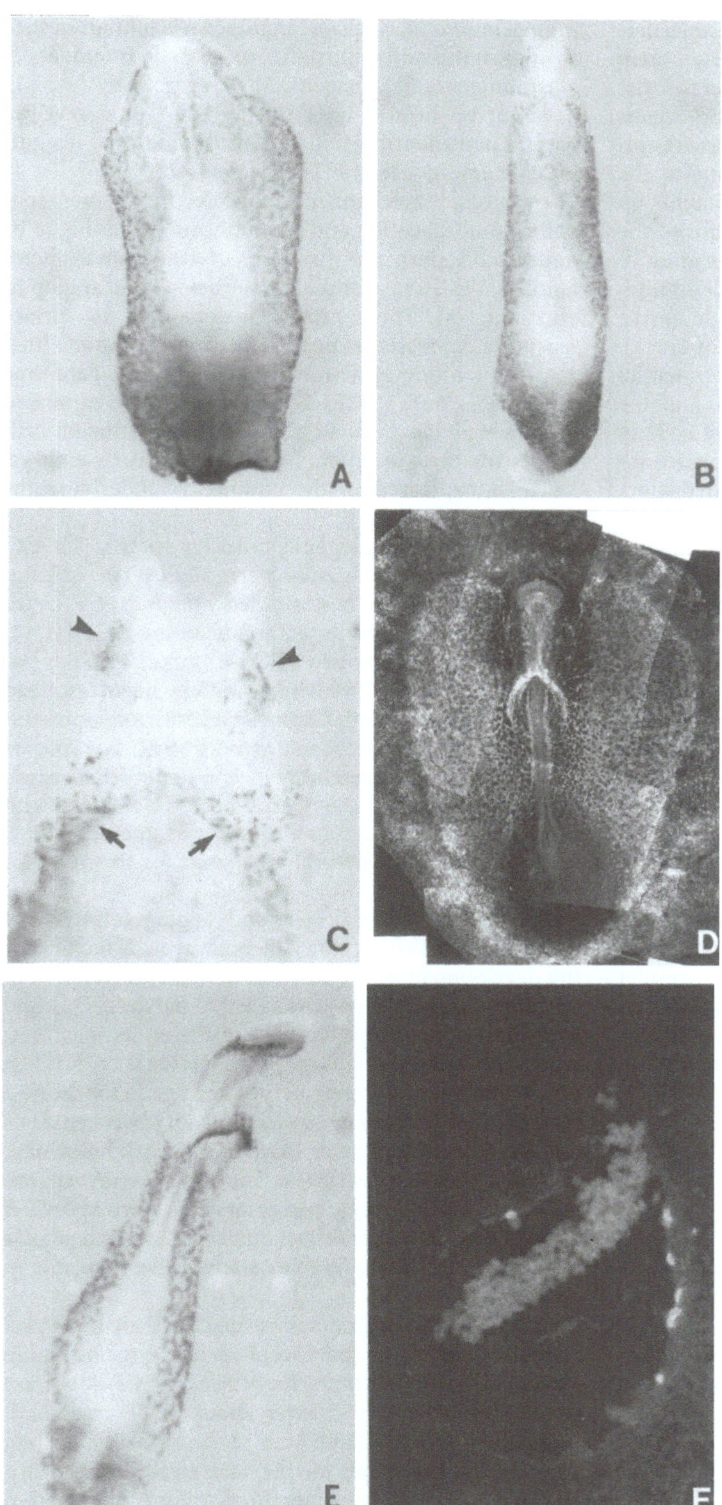

Figure 1. Emergence of the endothelial network as determined by VEGF-R2 and QH1 labelling. A-C, E : whole-mount *in situ* hybridization of quail embryos with a VEGF-R2 antisense riboprobe. Embryonic stages are A: presomitic (stage 4 of Hamburger & Hamilton, 1956), B: 5-somites, C: higher magnification of the anterior region of the embryo shown in B. Note angioblasts located in the region of the anterior intestinal portal (future heart) (arrows) and in the cephalic mesoderm (arrowheads). D: Whole-mount QH1 antibody staining of a 7-somite quail embryo. E: VEGF-R2 labelling of a 10-somite embryo. At 7-10 somite stages, QH1 and VEGF-R2 label all endothelial cells of the developing vascular system (D, E). F: QH1 labeling of quail EC integrated into a chick host aortic endothelium. The chimeric vessel resulted from the orthotopic replacement of a chick somite by a quail somite. This section of the trunk of a 5-day chimera illustrates that single graft-derived QH1+ cells are easily identified with this labelling technique. Figures 1D, F are a courtesy of Luc Pardanaud.

devoid of endothelial potential [39, 40]. As a consequence, bone marrow is colonized by extrinsic EC precursors [39,41]. Somatopleural mesoderm is seeded by endothelial precursors from the somites, which invade it in parallel with myoblasts from the same origin, a derivation demonstrated by orthotopic transplantation of quail somites into chick embryos [36, 40]. A barrier appears however to prevent EC precursors produced by somites from invading splanchnic mesoderm. In contrast, splanchnopleural cells do not display a restricted homing territory. When quail splanchnopleural mesoderm is grafted heterotopically in place of chick somites, quail EC colonize the whole corresponding level of the embryo, that is body wall, kidney, splanchnic organs [40]. As to the aortic endothelium, all its aspects are colonized, roof, sides and floor. In all cases, some of the grafted QH1+ cells bulge in the aortic lumen, making up cell clusters. These clusters, confined to the floor of the aortic endothelium, are a stereotyped manifestation of hemopoiesis in vertebrate embryos. Thus, while somite-derived EC have a restricted potential (angioblastic, homing to body wall, kidney, roof and sides of aorta), splanchnopleural mesoderm has an extended potential (hemangioblastic, homing to all mesoderm), and the aortic endothelium appears to be a mosaic of cells of somitic origin (roof and side) and cells of splanchnopleural origin (floor). In other terms there are two distinct endothelial lineages during ontogeny, one angioblastic and one hemangioblastic (Figure 2). These experiments, which yielded a fate map of endothelial precursors, leave open the question of the mechanism restricting homing of somitic precursors. While splanchnopleural-derived cells are capable of homing to the whole embryo, they display the hemopoietic potential only in the floor of the aorta and, within the 2-3 next days, the mesentery ventral to the aorta. Thus, signalling factors probably present in this particular location must be required for the hemangioblastic potential to be expressed.

Regulation

Molecular Interactions   Tumor angiogenesis in the adult has given a powerful impetus to the search for growth factors involved. Quiescence, which is the normal state of most EC, and resumed growth of EC in tumors results from a balance of angiogenic stimulators and inhibitors [42]. While angiogenic stimulators such as the polypeptide VEGF and related factors have also been described

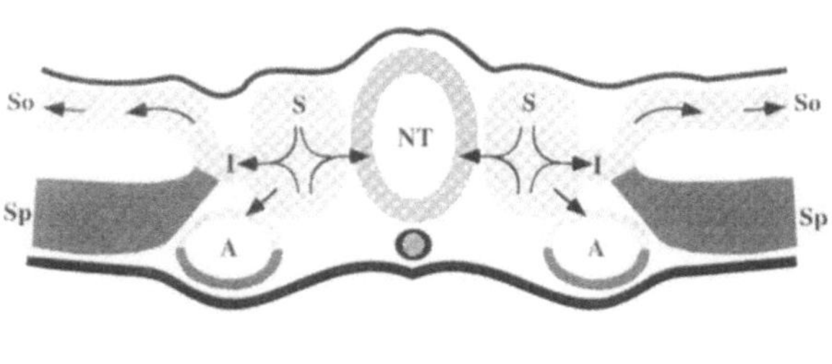

Figure 2.   Scheme depicting the two distinct origins of endothelial cells in the E1.5 avian embryo. Splanchnopleural mesoderm (Sp) gives rise to EC precursors which differentiate *in situ* and give rise to the floor of the aortic endothelium. Somites (S) give rise to another contingent of precursors which vascularize the neural tube (NT) and colonize the somatopleural mesoderm (So) and the intermediate plate (I) which is the forerunner of the kidney. Splanchnopleural mesoderm also produces the hemopoietic cells of the yolk sac and the intraaortic clusters (modified from [40]).

in the embryo (see below), angiogenesis inhibitors such as angiostatin and endostatin, proteolytic fragments of plasminogen and collagen XVIII respectively [43, 44], have not yet been defined during embryonic development. Their identification promises additional insights into the emergence of EC in the embryo.

The molecular mechanisms implicated in EC emergence and/or multiplication and assembly are beginning to be understood since the discovery of two growth factor families, VEGFs [45, for review, see below] and angiopoietins [46–48]. These growth factors bind to surface receptors expressed almost exclusively by the EC lineage. Their mode of action is generally of a paracrine type, that is, both VEGF and angiopoietin are expressed by mesenchymal cells adjacent to the endothelial cells expressing their receptors [45, 47]. In the early embryo, VEGF is expressed in the endoderm underlying the mesoderm [30, 49, 50].

VEGF, a potent angiogenic mitogen specific for EC, which also induces vascular permeability (see chapter about VEGF), is the best studied member of a larger family of related growth factors including PlGF [51, 52] and the recently identified VEGF-B [53, 54], VEGF-C [55, 56] and VEGF-D (57), which has initially been described as c-fos induced growth factor (FIGF) [58]. Angiopoietin-1 does not have any mitogenic activity on EC, but appears to play a role in the assembly of non-endothelial vessel-wall components [46, 47]. Angiopoietin-2 binds with similar affinity to the same receptor as angiopoietin-1, called tie-2, but is a natural antagonist for this receptor [48].

Knockout studies of VEGF and angiopoietin-1 have demonstrated the necessity of both growth factors for the development of the vascular system: inactivation of one allele of the VEGF gene is lethal between E9.5 and E10.5 due to poor development of the endothelial network, indicating that dosage of this factor is critical [59, 60]. Inactivation of angiopoietin-1, the ligand for the tie 2 receptor [46] leads to embryonic lethality between E10.5 and E12.5 due to failure of capillary network branching, the main defect appearing as failure to recruit smooth muscle cells and pericyte precursors [47]. Overexpression of angiopoietin-2 in transgenic mice results in a similar phenotype, indicating its role as a negative regulator *in vivo* [48].

Both VEGFs and angiopoietins mediate their biological actions by binding to endothelial-specific tyrosine kinase receptors. Three receptors for VEGFs, VEGF-R 1-3, are currently known (see chapter about VEGF receptors). Angiopoietin-1 and 2 both bind to the tie 2 receptor [46, 48], whereas the ligand for the structurally related tie 1 receptor has not yet been identified. Among the receptors for VEGFs and angiopoietins, VEGF-R2 appears as the earliest endothelial lineage marker, as determined by *in situ* hybridization studies on mouse and quail embryos [27, 30, 61–64]. Moreover, targeted gene inactivation experiments have shown that the function of VEGF-R2 is crucial for the emergence of both endothelial and hemopoietic cells in the embryo, since both cell types are absent in homozygous null mutant mice [65]. Null mutations of VEGF-R1 [66] or tie 1 and 2 [67–69] also lead to embryonic lethality due to deficiencies in the formation of the vascular system. VEGF-R1 mutants have both endothelial and hemopoietic cells, but fail to elaborate a functional vascular system [66]. Mutations of the angiopoietin receptors tie-1 and 2 appear to affect angiogenesis and vascular remodelling as well as vessel integrity: homozygous tie 2 receptor mutants die at E10.5, they show growth retardation from E9.5 onward and do not elaborate extensive vessel branches, this including a failure to vascularize the embryonic brain [67]. Tie-1 mutants die over a variable period, ranging from E14.5 to birth, as a result of oedema and hemorrhages at late embryonic stages (E13.5 to 18.5) [67]. Thus, each receptor plays a unique role at different stages of the formation of the vascular system; compared to the VEGF-R2 phenotype, all other mutants affect a later stage of vascular development [70].

Cells and Cellular Interactions   The existence of a common precursor for both cell types was postulated at the beginning of the century, based on observations of chick blastoderms cultured *in toto* on coverslips [23] or in explant cultures of the caudal region of blastoderms at the gastrulation stage [11]. In both types of cultures, endothelial and hemopoietic cells developed simultaneously from aggregates of morphologically identical cells. Murray [11] called the whole aggregate 'hemangioblast' and the constituent cells 'hemangioblast cells'. The peripheral hemangioblast cells were found to have a tendency to flatten against the culture surface and to differentiate to EC, whereas the central cells of the cluster differentiated to hemopoietic cells [11]. As mentioned in the introduction, the term 'hemangioblast' is now used with a different meaning, to denominate a cell with the capacity to give rise to both endothelial and hemopoietic progenitors. The existence of the hemangioblast has not formally been proven yet.

A common origin of endothelial and hemopoietic cells is suggested by the existence of shared surface markers including MB1/QH1 [2, 3, 39], CD34 [71, 72], CD31 [73], or tie-2 [74–76]. The transcription factor SCL/tal1 also shows expression in early mesodermal cells and blood islands as well as in angioblasts of mouse and chick embryos [77, 78]. However, when the gene is inactivated, SCL/tal1 mutant mice show normal vascular differentiation, but impaired hemopoietic development [79–82].

Murray [11] mapped the 'hemangioblast' potential of the primitive streak-stage embryo to the caudal two thirds of the embryo. This territory displays strong VEGF-R2 expression [4, 27]. The early expression of VEGF-R2 in the mesoderm of mouse [61, 63, 65] and quail [27, 30, 64] embryos, as well as the absence of both types of cells in homozygous VEGF-R2 mutant mice [65], suggest that VEGF-R2 expressing cells might give rise to both endothelial and hemopoietic cells.

To address this question, VEGF-R2 positive cells, labelled with a monoclonal antibody directed against the extra-cellular domain of the receptor, were sorted out from cell suspensions of caudal territories of chick embryos at the gastrulation stage [4]. 20 % of the total caudal cells expressed VEGF-R2. Positive or negative cells were cultured in semi-solid serum free medium in the absence or presence of the VEGF-R2 ligand VEGF. VEGF-R2 negative cells gave rise to very few colonies in both culture conditions. By comparison with non-sorted cells, a five-fold enrichment of colonies was obtained in the VEGF-R2+ population, indicating that all colony forming precursors segregate with the VEGF-R2+ cells.

In the absence of VEGF, VEGF-R2+ cells gave rise to hemopoietic colonies at a frequency of one hemopoietic colony for 10 precursors seeded. Different hemopoietic lineages could be obtained: erythrocytes as well as thrombocytes formed in the basal conditions (10 % bovine plasma, no growth factors added) [83], whereas macrophages differentiated in medium containing GM-CSF activity. In the presence of VEGF, EC differentiation was induced in a dose-dependent manner, concomitant with a reduction of hemopoietic differentiation. At a VEGF dose inducing maximal endothelial differentiation (one endothelial colony for 10 precursors seeded), hemopoietic differentiation was reduced to about 50 % of control values.

Thus, the population of caudal area VEGF-R2+ cells can give rise to both endothelial and hemopoietic cells and may therefore be considered as a population of hemangioblasts. One VEGF-R2+ cell can, however, give rise to either one or the other cell type. The endothelial differentiation pathway depends on the presence of VEGF, whereas the hemopoietic pathway appears constitutive in these cultures. The VEGF-R2 knockout phenotype in mice had, however, suggested that hemopoietic differentiation also depended on receptor activity [64]. Indeed, when VEGF-R2 positive cells are cultured in the semi-solid medium with the extracellular domain of VEGF-R2, hemopoietic differentiation is reduced in a dose-dependent manner to about 20 % of control values [4]. These data are consistent with the presence of a VEGF-R2 ligand in the medium, which would be responsible for hemopoietic differentiation and titrated out by increasing receptor concentrations. Hemopoietic differentiation thus also requires the presence of functional VEGF-R2.

These results led us to propose that caudal area VEGF-R2 expressing cells are hemangioblasts capable of giving rise to either an endothelial or a hemopoietic daughter cell. It should be noted that the ultimate proof for the existence of the hemangioblast would be a single precursor-derived colony containing both daughter cell types. Such clones are never found in our culture system. Recent approaches in which the endothelial and hemopoietic lineages were traced in the embryo proper yield convergent results. The emergence of the intraaortic hemopoietic clusters of the 3-day old chick or quail embryo was followed by LDL or retroviral vector labelling. Labelled clones were either hemopoietic or endothelial but never both, despite the fact that the label was integrated in the E2 vascular tree, which is purely endothelial at that early stage [84]. Furthermore the VEGF-R2 and CD45 (leukocytic) phenotype appear mutually exclusive. The hemopoietic clusters aggregated onto the floor of the aorta are CD45+ and VEGF-R2, with no underlying VEGF-R2+ EC; in other terms, for a short period the floor of the aorta is lined by hemopoietic cells, not by EC. These experiments suggest that, if the hemangioblast exists, it is faced with a developmental choice, either to the endothelial or the hemopoietic pathway. Both pathways appear to depend on VEGF-R2 activity, as suggested by the VEGF-R2 null phenotype and hemopoietic inhibition by the VEGF-R2 soluble extracellular domain. The pathway could be determined by different ligands binding to the same receptor: with VEGF, the hemangioblast would differentiate to an EC, whereas a still unknown ligand would trigger hemopoietic differentiation.

In the mouse, a common precursor for both cell types has recently been described in ES cell cultures [85, 86]. In one study, ES cells carrying two different lineage markers were mixed together and allowed to develop to blast colonies with hemopoietic potential. It was found that each blast colony was derived from either one of the two types of ES cells, suggesting a clonal origin [87]. Subsequently, it was shown that a single blast cell colony can also give rise to EC, when placed in the appropriate culture conditions [85]. In the other study, ES cells were sorted using a monoclonal antibody directed against mouse VEGFR2 and cultured in medium containing hemopoietic cytokines on a stromal cell layer of OP9 cells. In these conditions, colonies composed of EC and HC developed from a single VEGFR2+ cell [86]. It remains to be determined whether the different results obtained with mouse and chick cells are due to the culture conditions or if these species have evolved differently.

Could VEGF also be responsible for hemopoietic differentiation? Kennedy et al. [87] have reported the emergence of hemopoietic cells from mouse ES cell cultures in response to VEGF and kit ligand. In cultures from chick caudal mesoderm, however, VEGF reduces hemopoietic differentiation [4]. One way to reconcile these results would be to assume the presence of low doses of VEGF in the chick culture system. Indeed, VEGF mRNA expression is already observed in the chick embryo at gastrulation stages [30], suggesting that this growth factor might be bound to VEGF-R2 at the time the cells are isolated from the embryo and/or that VEGF might be produced during the culture period. In heterozygous VEGF mutant mice, the number of endothelial and hemopoietic cells in the yolk sac is dramatically reduced, leading to embryonic lethality at around E10 [59]. These data could be explained by insufficient VEGF-R2 activation. However, when homozygous VEGF null embryos were generated by aggregation of VEGF-/-ES cells with tetraploid wild-type embryos, the entirely mutant cell-derived embryo does contain hemopoietic cells [60]. These observations could be explained by either an effect of maternal VEGF or by the existence of a second ligand for VEGF-R2, which would partially compensate for the absence of

VEGF and allow endothelial and hemopoietic differentiation from VEGF-R2 positive cells in VEGF mutant mice.

The second VEGF-R2 ligand known, VEGF-C [54], was recently cloned in the quail by low-stringency screening with a human cDNA probe and was found to be highly homologous to the human and murine growth factor [88]. Avian VEGF-C is recognized by antibodies directed against the human protein, and both quail and human VEGF-C bind quail VEGF-R2. The effect of human recombinant VEGF-C on avian caudal blastodisc VEGF-R2 + cells was examined. VEGF-C induced endothelial differentiation and reduced hemopoietic differentiation of VEGF-R2 + cells, an effect similar to that of VEGF [88]. The doses of VEGF-C required were 5- to 10-fold higher compared to VEGF. VEGF-C might therefore be the growth factor compensating for the observed endothelial differentiation in VEGF mutant mice, but it is unlikely to be responsible for hemopoietic differentiation.

The VEGF-R2 positive precursors from mesoderm can give rise to at least three different hemopoietic lineages, and are thus pluripotent. They differ from "classical" avian hemopoietic stem cells [83, 89] since they grow in the absence of hemopoietic growth factors. They probably correspond to yolk sac precursors, but some of them may be the forerunners of precursors in subsequent sites of embryonic hemopoiesis, in particular the para-aortic splanchnopleura. In chimeric mouse embryos generated by aggregation of wild-type ES cells with ES cells in which the VEGF-R2 coding region had been replaced by a lacz gene, no β-gal positive cells were ever seen among hemopoietic cells even in the adult [90]. Neither were any present among EC, indicating that gene function is critical for the generation of both cell types. In this study, blue-staining cells were found at gastrulation stages in an ectopic embryonic site, the amnion. According to the authors' interpretation, these cells would be aborted hemopoietic cells incapable of migrating to the correct site of differentiation, the yolk sac, and the defect in hemopoiesis would represent a defect in the migratory potential of the cells.

Do hemangioblastic cells exist at later embryonic stages? Nishikawa et al. [91] recently reported the generation of lymphohemopoietic cells from endothelial cells sorted out from E9.5 mouse embryos or yolk sacs. The endothelial phenotype of the sorted cells was characterized by several markers including VEGFR2 and VE-cadherin, the latter being expressed only by EC which have already assembled into tubes. The generation of B- and T-cells from the sorted EC *in vitro* occurred at a relatively high frequency, but it remains to be shown if EC can give rise to lymphocytes *in vivo*. Asahara et al. [92] have isolated putative endothelial progenitor cells characterized by CD34 and VEGFR-2 expression from adult human peripheral blood. When injected into the tail vein of athymic nude mice, the hindlimb of which had been rendered ischemic by excision of the femoral artery, these cells differentiated into EC which integrated into the neovascularized ischemic hindlimb. A possible hemopoietic differentiation potential of these cells has not yet been reported.

Unexpected findings in the avian embryo converge with Asahara et al's demonstration in the adult. The allantois, one of the embryonic appendages of amniotes, was demonstrated to function like the yolk sac as a primordial hemopoietic organ [93]. In these experiments, the allantoic bud retrieved from a quail embryo prior to vascularisation was grafted into the coelom of a chick host. The graft emitted hemopoietic and also endothelial progenitors which both seeded the host bone marrow. This evidence indicates that angioblasts may be able to migrate through the circulatory pathway. This may constitute a novel developmental mechanism of angiogenesis, suggested earlier by an observation of Christ et al. [94], who grafted quail limb buds on chick embryos and found EC inside the lining of host arteries. Future experiments must settle whether the bone marrow in these experiments was colonized by distinct hemopoietic and endothelial progenitors or by the still elusive hemangioblast, and what is the extent of the allantois contribution to bone marrow formation.

Additional Features

Vascularization of the central nervous system   Quail-chick transplantation experiments have clearly shown that the central nervous system (CNS) and the neural crest are entirely devoid of angiogenic potentialities [35, 95–97]. The CNS, which in vertebrates has evolved in size considerably, thus relies on an external vascular supply. In a systematic study using the quail-chick transplantation method, Couly et al. [97–100] have investigated the respective contributions of the mesoderm and the neural crest to the developing head structures i.e., the skull and masticatory apparatus. At levels anterior to the notocord, the neural crest gives rise to the head dermis, to most of the skull and connective tissue and to the meninges of the prosencephalon. The contribution of the cephalic mesoderm is restricted to few bones of the skull and to the head muscles but, importantly, it gives rise to all EC vascularizing the entire head [97]. These studies support the "New Head" theory [101], according to which the more anterior part of the brain including sense organs, prosencephalon and mesencephalon together with the corresponding part of the skull, is entirely of neurectodermal origin, having evolved *de novo* from protochordate-like ancestors.

Thus, although neural crest-derived mesectoderm and mesoderm share several developmental potentialities, that of producing EC is entirely restricted to the mesoderm. The blood vessels irrigating the anterior part of the embryo (from the prosencephalon to the heart) are composed of cells from two different origins: EC are derived from the mesoderm whereas wall cells (pericytes and smooth muscle) are derived from mesectoderm [95]. In the rest of the body, both EC and wall cells are of mesodermal origin.

The EC that vascularize the mid- and hindbrain are derived from immediately adjacent mesodermal territories [97]. In contrast, the procencephalon (which expands most) recruits its vascularization from a more distant mesodermal territory. In general, however, EC derived from a given mesodermal area vascularize embryonic regions relatively close to their site of origin. EC migration thus appears strictly directed by environmental cues, despite the multidirectional invasive behaviour they display when grafted in less stringent conditions [35, 102].

Paraxial cephalic mesoderm is fated to give rise to muscle, cartilage and EC. A significant proportion of cells in this mesoderm (about 50%) expresses VEGF-R2 from gastrulation stages onward [97], indicating that determination of the angioblastic lineage in the cephalic mesoderm occurs at a very early developmental stage. These VEGF-R2 positive cells can be seen assembling into cephalic vessels and perineural vascular plexus, that vascularise the embryonic brain by sprouting only two days after the first VEGF-R2 positive cells are detectable in the embryo. This angiogenesis process starts at E3.5 in the mesencephalon of the chick embryo and proceeds according to a precise plan, the first sprouts arise from a ventral position, proceed straight towards the ventricular region, and branch sideways when they have reached about half the distance to the ventricle [103].

The molecular mechanisms responsible for this process are now beginning to be understood. In VEGF mutant mice, vascularization of the embryonic brain fails to occur [59]. VEGF expression is first observed in the ventral region of the embryonic brain at stages preceeding its invasion by a few hours [49, 104]. VEGF binding to its receptors thus appears as a key event in the onset of angiogenic sprouting. VEGF expression itself could be regulated by lack of oxygen in the ventral brain region [105]. However, the fact that VEGF is expressed a few hours before the onset of sprouting suggests possible additional mechanisms, either by regulation of the bioavailability of the factor or by other mechanisms, a question addressed by overexpression of VEGF inside the central nervous system. Implantation of growth factor-containing methylcellulose fragments just before the onset of vascularization leads to the formation of abnormally large vessels in the perineural vascular plexus, but not to increased numbers of sprouts in the CNS [104].

Interestingly, the tie2 receptor mutant also shows failure of embryonic brain vascularization [67]. Since tie2 apparently does not act in the formation of endothelial tubes or the proliferation of EC, but in the recruitment of pericytes, a possible role of pericytes in brain vascularization is suggested. However, PDGFB mutant mice, which lack pericytes throughout the microvascular bed, do contain endothelial tubes in their brain, although these capillaries are not growing straight to the ventricle as in normal embryos, but instead deviate in different directions and show frequent microaneurisms (local enlargements) [106]. These data would indicate that pericytes are not required for brain vascularization.

Origin of the lymphatic vascular system   The origin of the lymphatic vascular system is not well understood in the avian embryo (birds in contrast to mammals do not have lymph nodes) [see 64]. Sabin [107] described a venous origin of the lymphatic vessels. But interest has been recently renewed by the description of VEGF-R3 expression in mouse and avian embryos [64, 108, 109]. This receptor, also called flt4 [110], is first expressed in all endothelial cells of the embryo, but becomes restricted to vein's and subsequently to lymphatic vessels, suggesting a role for this receptor and its ligand, VEGF-C, in the development of the lymphatic system. Indeed transgenic mice expressing VEGF-C in the skin develop hyperplasia of lymphatic, but not vascular endothelial cells [111] and application of VEGF-C to the chorioallantoic membrane of chick embryos also induces proliferation of lymphatic but not vascular endothelial cells [112].

Francoise Dieterlen-Lievre, Nicole le Douarin
and Anne Eichmann

References

1. Le Douarin NM (1969) Bull Biol Fr Belg 103:435-452
2. Péault BM et al (1983) Proc Natl Acad Sci USA 80:2976-2980
3. Pardanaud L et al (1987) Development 100:339-49
4. Eichmann A et al (1997) Proc Natl Acad Sci USA 94:5141-5146
5. Stainier DYR et al (1995) Development 121:3141-3150
6. Liao W et al (1997) Development 124:381-389
7. Ranson DG et al (1996) Development 123:311-319
8. Weinstein BM et al (1996) Development 123:303-309
9. Hogan BLM (1995) Nature 376:210-211
10. Kessler DS and Melton DA (1994) Science 266:596-604
11. Murray PDF (1932) Proc Royal Soc B 11:497-501
12. His W (1900) Abhandl Math phys Ges Wiss 26:171-328
13. Folkman J (1974) Adv Cancer Res 19:331-358
14. Hahn H (1909) Roux Arch Dev Biol 27:337-433
15. Miller AM, McWhorter JE (1914) Anat Rec 8:201-227
16. Reagan FP (1915) Anat Rec 9:329-341
17. Martin C (1972) C R Soc Biol Fr 116:283-285
18. Beaupain D et al (1979) Blood 53:212-225
19. Van der Stricht O (1895) CR Soc Biol 12:181-185
20. Christ B et al (1991) Acta Histochem 91:193-199
21. Yablonka-Reuveni Z (1989) Dev Biol 132:230-240
22. Flamme I, Risau W (1992) Development 116:435-439
23. Sabin F (1920) Carnegie Institution Contr Embryol 9:215-262
24. Hirakow R, Hiruma T (1981) Anat Embryol 163:299-306
25. Coffin JD, Poole TJ (1988) Development 102:735-748
26. Wilt FH (1965) Science 147:1588-1590
27. Eichmann A et al (1993) Mech Dev 42:33-48
28. Eichmann A et al (1996) Gene 174:3-8
29. Drake CJ, Little CD (1995) Proc Natl Acad Sci USA 92:7657-7661
30. Flamme I et al (1995) Dev Biol 169:699-712
31. Wilting J et al (1993) Cell Tissue Res 147:207-215
32. Wilting J et al (1996) Dev Biol 176:76-85
33. Meier S (1980) J Embryol Exp Morphol 55:291-306
34. Lance-Jones (1988) Dev Biol 126:394-407
35. Noden DM (1989) Am Rev Resp Dis 140:1097-1103
36. Schramm C, Solursh M (1990) Anat Embryol 182:235-247
37. Wilting J et al (1995) Dev Dyn 202:165-171
38. Pardanaud L, Dieterlen-Lièvre F (1995) Anat Embryol 192:425-435
39. Pardanaud L et al (1989) Development 105:473-485
40. Pardanaud L et al (1996) Development 122:1363-1371
41. Jotereau F, Le Douarin NM (1978) Dev Biol 63:253-265
42. Hanahan D, Folkman J (1996) Cell 86 353-364
43. O'Reilly MS et al (1994) Cell 79:315-328
44. O'Reilly MS et al (1997) Cell 88:277-285
45. Klagsbrun M, D'Amore P (1996) Cytokine Growth Factor Rev 7:259-270
46. Davis SD et al (1996) Cell 87:1161-1169
47. Suri C et al (1996) Cell 87:1171-1180
48. Maisonpierre PC et al (1997) Science 277:55-60
49. Breier G et al (1992) Development 114:521-532
50. Breier G et al (1995) Dev Dyn 204:228-239
51. Maglione D et al (1991) Proc Natl Acad Sci USA 88:9267-9271
52. Park JE et al (1994) J Biol Chem 269:25646-25654
53. Grimmond S et al (1996) Genome Res 6:124-131
54. Olofsson B et al (1996) Proc Natl Acad Sci USA 93:2576-2581
55. Joukov V et al (1996) EMBO J 15:290-298
56. Lee J et al (1996) Proc Natl Acad Sci USA 93:1988-1992
57. Achen MG et al (1998) Proc Natl Acad Sci USA 95:548-553
58. Orlandini M et al (1996) Proc Natl Acad Sci USA 93:11675-11680
59. Ferrara N et al (1996) Nature 380:439-442
60. Carmeliet P et al (1996) Nature 380:435-439
61. Yamaguchi TP et al (1993) Development 118:489-498
62. Millauer B et al (1993) Cell 72:835-846
63. Dumont DJ et al (1995) Dev Dyn 203:80-92
64. Wilting J et al (1997) Cell Tissue Res 288:207-223
65. Shalaby F et al (1995) Nature 376:62-66
66. Fong GH et al (1995) Nature 376:66-70
67. Sato TN et al (1995) Nature 376:70-74
68. Dumont DJ et al (1994) Genes and Dev 8:1897-1909
69. Puri MC et al (1995) EMBO J 14:5884-5891
70. Hanahan D (1997) Science 277:48-50
71. Fina L et al (1990) Blood 75:2417-2426
72. Young PE et al (1995) Blood 85:96-105
73. Baldwin HS et al (1994) Development 120:2539-2553
74. Iwama A et al (1993) Biochem Biophys Res Commun 195:301-309
75. Batard P et al (1996) Blood 87:2212-2220
76. Hashiyama M et al (1996) Blood 87:93-101
77. Kallianpur AR et al (1994) Blood 83:1200-1208
78. Drake CJ et al (1997) Dev Biol 192:17-30
79. Robb L et al (1995) Proc. Natl. Acad. Sci. 92:7075-7079
80. Robb L et al (1996) Cell 86:47-57
81. Shivadsani R et al (1995) Nature 373:432-434
82. Porcher C et al (1996) Cell :47-57
83. Cormier F, Dieterlen-Lièvre F (1988) Development 102:279-285
84. Jaffredo T et al (1998) Development 125:4575-4583
85. Choi K et al (1998) Development 125:725-732
86. Nishikawa SI et al (1998) Development 125:1747-1757
87. Kennedy M et al (1997) Nature 386:488-492
88. Eichmann A et al (1998) Development 125
89. Corbel C et al. (1996) Proc Natl Acad Sci USA 93:2844-2847
90. Shalaby F et al (1997) Cell 89:981-990
91. Nishikawa SI et al (1998) Immunity 8:761-769
92. Asahara T et al (1997) Science 275:964-967
93. Caprioli A et al (1998) Proc Natl Acad Sci USA 95:1641-1646
94. Christ B et al (1990) Anat Embryol 181:333-339
95. Le Lièvre CS, Le Douarin NM (1975) J Embryol exp Morphol 34:125-154

96. Stewart PA, Wiley MJ (1981) Dev Biol 84:183-192
97. Couly GF et al. (1995) Mech Dev 53:97-112
98. Couly GF et al. (1992) Development 114:1-15
99. Couly GF et al (1993) Development 117:409-429
100. Le Douarin et al (1993) Dev Biol 159:24-49
101. Gans C, Northcutt RG (1983) Science 220:268-274
102. Noden DM (1991) Development 111:867-876
103. Feeney JF, Watterson RL (1946) J Morphol 78:231-301
104. Aitkenhead M et al (1998) Dev Dyn in press
105. Kremer C et al (1997) Cancer Res 57:3852-3859
106. Lindahl P et al (1997) Science 277:242-245
107. Sabin F (1909) Am J Anat 9:43-91
108. Kaipainen A et al (1995) Proc Natl Acad Sci USA 92:3566-3570
109. Kukk C et al (1996) Development 122:3829-3837
110. Pajusola V et al (1992) Oncogene 8:2931-2937
111. Jeltsch M et al (1997) Science 276:1423-1425
112. Oh SJ et al (1997) Dev Biol 188:96-109

OSCCS

Definition   *Open surface-connected canalicular system*

See: →Platelet stimulus-response coupling

OSF-1

Definition   *Osteoblast specific factor-1*

See: →Heparin-affin regulatory protein

Ox-LDL

Definition   *Oxidized low density lipoprotein*

See: →Atherosclerosis; →Cytokines in vascular biology and disease; →Thrombosis

PA

Definition   *Plasminogen activator*

See: →Plasminogen activators

PADGEM

Definition   *Platelet-activated derived granule external membrane*

See: →Platelet stimulus-response coupling

PAF

Definition   *Platelet-activating factor*

See: →Platelet activating factor; →Platelet stimulus-response coupling

PAI-1

Definition   *Plasminogen activator inhibitor-1*

See: →Plasminogen activators; →Plasminogen/plasmin

PAR

Definition   *Proteolytically activated receptor*

See: →Fibrinolytic, hemostatic and matrix metalloproteinases, role of; →Platelet stimulus-response coupling; →Thrombin

PCA

Definition   *Procoagulant activity*

See: →Procoagulant activities

PDE

Definition   *Phosphodiesterase*

See: →Platelet stimulus-response coupling; →Vasomotor tone regulation, molecular mechanisms of

PDECGF

Definition   *Platelet-derived endothelial cell growth factor*

See: →Angiogenesis

PDGF

Definition   *Platelet-derived growth factor*

See: →Platelet-derived growth factor

PDTC

Definition   *Pyrrolidone derivative of dithiocarbonate*

See: →Cytokines in vascular biology and disease

PECAM-1

Definition   *Platelet-endothelial cell adhesion molecule. Adhesion molecules of the lectin family present on various cell types including endothelial cells, platelets and leukocytes. Binds in a homotypic fashion to another PECAM-1 molecule.*

See: →Blood cells, interaction with vascular cells

PEX

Definition   *Hemopexin*

See: →Matrix metalloproteinases; →Vitronectin/vitronectin receptors

PF-4

Definition   *Platelet factor-4*

See: →Platelet factor-4; →Angiogenesis inhibitors

PfEMP

Definition   *Plasmodium falciparum erythrocyte membrane protein*

See: →Plasmodium falciparum erythrocyte membrane protein

PGE

Definition   *Prostaglandine E*

See: →Prostacyclin; →Prostaglandins

PGH

Definition   *Prostaglandin H*

See: →Prostacyclin; →Prostaglandins; →Thromboxanes

PGI$_2$

Definition   *Prostaglandin I$_2$*

See: →Prostacyclin

PGIS

Definition   *Prostaglandin I$_2$ synthase*

See: →Prostacyclin

Phosphatidyl inositol-3 kinase (PI-3- Kinase)

Definition   *Phosphorylates position D3 of the inositol ring of phosphatidyl inositol (PI), PI4-phosphate (PI4 P) or PI4,5-bisphosphate (PI4,5 P$_2$) to produce PI 3-P, PI 3,4-P$_2$, or PI 3,4,5-P3.*

See: →Signal transduction in vascular biology; →Platelet stimulus-response coupling

Phospholipase A

Definition   *Enzyme that mobilizes arachidonic acid*

See: →Prostacyclin; →Platelet stimulus-response coupling; →Vasomotor tone regulation, molecular mechanisms of

Phospholipase C

Definition   *Enzymes that cleave sphingophospholipids at sphingosine-phosphate or glycerol-phosphate bonds*

See: →Platelet stimulus-response coupling; →Extracellular matrix; →Vasomotor tone regulation, molecular mechanisms of

Phospholipases

Definition   *Different phospholipases have been described. These include Phospholipase A, C and D. Phospholipase A (A$_2$) generates arachidonic acid from membrane phosphoinositides. Phospholipase C cleaves phospholipids and converts phosphatidyl inositol diphosphate (PIP$_2$) to Inositol triphosphate (IP$_3$) and Diacylglycerol (DAG). Phospholipase D generates phosphatidic acid.*

See: →Platelet stimulus-response coupling; →Vasomotor tone regulation, molecular regulation of; →Prostacyclin; →Prostaglandins

PHSC

Definition   *Pluripotent hemapoietic stem cells*

See: →Megakaryocytes

PI 3,4,5-P$_3$

Definition   *Phosphatidyl inositol 3,4,5 triphosphate*

See: →Phosphatidyl inositol-3 kinase; →Signal transduction mechanisms in vascular biology

PI 3,4-P$_2$

Definition   *Phosphatidyl inositol 3,4 diphosphate*

See: →Phosphatidyl inositol-3 kinase; →Signal transduction mechanisms in vascular biology

PI 3-P

Definition   *Phosphatidyl inositol 3 monophosphate*

See: →Phosphatidyl inositol-3 kinase; →Signal transduction mechanisms in vascular biology

PI 4,5 P$_2$

Definition *Phosphatidyl inositol 4,5-bisphosphate*

See: →Phosphatidyl inositol-3 kinase; →Signal transduction mechanisms in vascular biology

PI 4-P

Definition *Phosphatidyl inositol 4-phosphate*

See: →Phosphatidyl inositol-3 kinase; →Signal transduction mechanisms in vascular biology

PI-3K

Definition *Phosphatidyl inositol 3 kinase*

See: →Phosphatidyl inositol-3 kinase; →Signal transduction mechanisms in vascular biology

PIP$_2$

Definition *Phosphatidyl inositol diphosphate*

See: →Lipid mediators; →Phospholipases

PIP$_3$

Definition *Phosphatidyl inositol triphosphate*

See: →Phospholipases; →Vasomotor tone regulation, molecular mechanisms of

PIVKA

Definition *Protein-induced vitamin K absence*

See: →Bleeding disorders

PKC

Definition *Protein kinase C*

See: →Protein kinase C

PL

Definition *Phospholipase*

See: →Phospholipases

PLA

Definition *Phospholipase A*

See: →Phospholipase A

Plasminogen Activators (PA)

Definition *Proteases that convert plasminogen into plasmin. There are two types namely tissue-type plasminogen activator (tPA) and urokinase-type plasminogen activator (uPA). PAs are inhibited by plasminogen activator inhibitor-1 or -2.*

See: →Fibrinolytic, hemostatic and matrix metalloproteinases, role of; →Thrombosis; →Vitronectin/vitronectin receptors; →Cytokines in vascular biology and disease; →Fibrin/fibrinogen

Plasminogen/Plasmin

Definition *Proteolytic enzymes that degrade fibrin or activate matrix-metalloproteinases*

See: →Fibrinolytic, hemostatic and matrix metalloproteinases, role of; →Thrombosis; →Fibrin/fibrinogen

Plasmodium falciparum

Definition *Causative agent of malaria*

See: →Blood cells, interaction with vascular cells

Plasmodium falciparum Erythrocyte Membrane Proteins (PfEMP)

Definition *Family of high molecular weight proteins located at knob sites on the outer membrane of red blood cells. Two forms have been described: PfEMP1 and PfEMP2.*

See: →Blood cells, interaction with vascular cells

Platelet-Activating Factor (PAF)

Definition *Lipid mediator involved in vascular cell function such as platelet aggregation*

See: →Platelet stimulus-response coupling

Platelet Factor-4

Definition *Platelet-derived CXC chemokine with strong heparin-binding activity*

See: →Angiogenesis inhibitors; →Angiogenesis; →Megakaryocytes

Platelet Glycoproteins

Definition *Membrane proteins important for platelet function. Among these, glycoprotein IIbIIIa and Ib have been the most extensively studied. IIbIIIa binds fibrinogen and Ib von Willebrand Factor.*

See: →Vascular integrins; →Platelets; →Megakaryocytes; →Thrombosis; →von Willebrand Factor

Platelet Precursors

See: →Megakaryocytes

Platelet Stimulus-Response Coupling: Implication in Haemostasis, Thrombosis, and Therapeutics

Synonym: Thrombocytes, stimulus-response coupling.

Introduction Blood platelets are highly dynamic and adhesive cells able to undergo rapid changes of shape. They circulate in a resting discoid state and are prevented from contact with the vessel wall by two vasoprotective factors synthesised by vascular endothelial cells: prostacyclin (PGI_2) and nitric oxide (NO). Platelet activation function contributes to haemostasis in two main forms. At the site of vessel injury, platelets come into contact with subendothelial elements and initiate the haemostatic process by adhering to the newly exposed surface. Platelets spread on the modified surface, secrete granular components and aggregate to each other to form a plug and fill the injured vessel, all platelet events that constitute primary haemostasis. In some conditions platelets participate in coagulation by providing an additional catalytic surface on which two steps of the intrinsic coagulation pathway are greatly accelerated.

During circulation, platelets encounter various molecules. Besides NO and PGI_2 that prevent their activation, they can be in contact with thrombin, ADP, liberated from red blood cells, platelet activating factor (PAF), antibodies, or collagen. Each agonist or antagonist has a specific receptor on the platelet membrane. The receptor that has bound its ligand generates an intracellular signal leading platelet to respond. The present knowledge on the platelet stimulus-response coupling in different situations will be described, followed by the physiological implications.

Structure Platelets are the smallest blood cells, around 3 μm in diameter. Their unique feature is the absence of nucleus, with otherwise all typical cellular characteristics. The plasma membrane is constituted of a bilayer of phospholipids, asymmetrically distributed so that negative charges are maintained inside the cell [1]. The asymmetry is actively controlled at two purposes: to prevent platelet from interaction with its environment and to confine specific phospholipids to the cytoplasmic side as potential substrates for phospholipases to transmit an activation signal. Receptors for circulating molecules are inserted within the phospholipid bilayer.

Platelet structural and dynamic properties rely on its cytoskeleton, which consists of a bundle of coiled microtubules [2] and a loose network of membrane-associated and cytoplasmic actin filaments [3]. The microtubular cytoskeleton underlies the plasma membrane in a so-called marginal band and maintains the discoid shape of resting platelets. The membrane skeleton consists of short actin filaments connected to specific membrane glycoproteins to regulate platelet shape. Cytoplasmic skeleton consists of actin filaments organized in a network of associated structural proteins. The contractile forces of platelets during activation come from actin polymerisation and association with the cytosolic soluble myosin.

Platelet contains a number of intracellular organelles, each of them having a specific and highly specialized function. Besides mitochondria and peroxisomes, three types of storage granules are randomly distributed within the cytoplasm. The smallest are the dense granules, socalled because their high calcium and serotonin (5-hydroxytryptamine) content make them electron dense. They also contain adenine nucleotides. Platelets carry serotonin throughout the body, the amine being avidly incorporated and stored in the dense granules to prevent its catabolism. Alpha-granules are more abundant and heterogeneous, larger and less opaque than dense granules. They contain several proteins, some also present in plasma like fibrinogen and albumin, IgGs, growth factors, and two platelet specific proteins β-thromboglobuline and platelet factor 4. Lysosomes contain several hydrolytic enzymes such as N-acetyl glucosaminidase, β-glucuronidase.

Platelet possesses a functionally important internal membrane system. The open surface- connected canalicular system (OSCCS) forms a channel with the surface membrane through which granule constituents diffuse during secretion. The internal membrane system, the dense tubular system (DTS), is a membranous formation sequestering calcium and enzymes of the prostaglandin metabolism.

Regulation of Cell Function
Cell to Cell Interactions The critical step of platelet activation, adhesion and spreading on damaged vascular wall or interactions with a "primary" circulating inducer, leads to a marked change in shape. From being discoid platelet becomes spherical and emits pseudopods and invaginations. The link between glycoproteins and the membrane skeleton is broken, while other glycoproteins become attached to the cytoplasmic skeleton. Granules are centralized within a reorganized microtubular ring. The following phases tend to strengthen the initial activation. Granules come into contact with the plasma membrane or its invaginations in the OSCCS, membranes fuse and platelet liberates in the extracellular milieu the content of the different granules: the release reaction has started.

The three types of granules are implied in the release reaction. Dense granules secrete proaggregating factors and factors that affect the vessel wall tone. ADP is a major activating agent. Once in the surrounding milieu ADP recruits other circulating platelets. Serotonin induces a local vasoconstriction contributing to the vasospasm. The alpha-granules secrete proteins to consolidate the aggregate (fibrinogen and thrombospondin) or promote vascular cell migration and proliferation (platelet derived growth factor PDGF). Lysosomes liberate the hydrolytic enzymes which will digest the haemostatic plug to clear it. A platelet deficiency in platelet granules leads to a defective release reaction and a bleeding syndrome, the so-called storage-pool disease. A platelet peculiarity is the expression of new surface receptors on the activated platelet membrane. This characteristic results from the release reaction: during fusion of the granule membrane with the plasma membrane, granule glycoproteins are incorporated in the plasma membrane. Hence, such glycoproteins are used as markers of activated platelets. The p53 glycoprotein (CD63) is derived from lysosomes. P-selectine (CD62), also known as GMP-140 (granule membrane protein of 140 kDa) or PADGEM (platelet activated derived granule external membrane), derived from alpha-granules, plays an important role in the interaction of activated platelets with leucocytes [4].

The cytosolic calcium concentration $[Ca^{2+}]i$ is efficiently and precisely controlled by a complex calcium transport ATPases system located on the various membrane systems [5]. During the same time as platelet undergoes the release reaction, the calcium stored in the DTS is "mobilized" during activation, i.e. is liberated in the cytoplasm. The elevation of $[Ca^{2+}]i$ activates a number of enzymes, kinases, phospholipases and calpain. Among the activated enzymes, phospholipase A_2 initiates eicosanoids biosynthesis and thromboxanes (TX) are formed. TX A_2, a major metabolite of this metabolism, is secreted and binds to its specific receptor to further stimulate platelets.

The platelet specific function is to form a platelet aggregate. Platelet aggregation requires the binding of fibrinogen, a symmetrical molecule, to two adjacent platelets on the so-called aggregation receptor, the GPIIb-IIIa glycoprotein complex, α_{IIb}-β_3 in the integrin nomenclature. Circulating resting platelets are unable to bind fibrinogen. An activation-dependent conformational change of GPIIb-IIIa is required for fibrinogen binding. Fibrinogen binds to the activated GPIIb-IIIa in a calcium-dependent way, through two RGD sequences located in the α chain and through a dodecapeptide sequence of the γ chain. The vWF also bears RGD sequences and binds to activated GPIIb-IIIa multiple. Fibrinogen and vWF act in concert to augment the size and the spreading of the platelet aggregate on the damaged vessel.

Finally, platelet ensure efficient coagulation, by exposing a catalytic surface that will adsorb factor X and prothrombin. The assembly of the coagulation complexes (Xase or prothrombinase) is greatly accelerated in the presence of lipid surfaces that contain anionic phospholipids. Anionic phospholipids are normally maintained on the inner leaflet of plasma membranes. During platelet activation, the anionic procoagulant phospholipids are exposed on platelet surface. Thus platelets facilitate the activation of factor Xa and of prothrombin and increase the rate of thrombin generation. The exposure of highly negatively charged phospholipids on the outer surface is accompanied by budding of the membrane and shedding of microparticles. Microparticles offer a great amount of supplementary catalytic surface. The thrombin generated will convert fibrinogen into fibrin all around platelet aggregates, stabilizing the platelet plug.

Molecular Interactions   In the blood flow, platelets can be in contact with various circulating molecules that will either activate or inhibit them, or with damaged vessel wall. Interactions with circulating molecules and adhesion proteins occur through specific receptors present on the platelet membrane. Small molecules bind to receptors that stimulate (thrombin, ADP liberated from red blood cells, PAF, antibodies) or inhibit (NO, PGI_2) platelet aggregation and eventually secretion. Subendothelial structures, adhesines and macromolecular complexes binding to platelets constitute the adhesion and activation processes, ultimately leading to aggregation.

Signalling Mechanisms   The receptor that has bound its ligand generates an intracellular signal leading platelet to respond: the platelet stimulus-response coupling depends on the nature of both the ligand and the receptor. All receptors are glycoproteins, and two general types of signalling mechanisms can be described: one following binding of small molecules to their seven transmembrane domains receptors coupled to a G-protein; and one generated by adhesion or clustered receptors consisting of two or more glycoproteins (Figure 1).

G-protein-coupled receptors   G-protein coupled receptors are integral membrane glycoproteins characterized by seven hydrophobic transmembrane helices, connected by extracellular and intracellular loops, and so-called "heptahelical" or "serpentine" receptors. The N-terminus is extracellular, the C-terminus in the cytoplasm. Ligands for platelet heptahelical receptors are very different both in nature (proteins or lipids) and in origin. Small circulating molecules as different as ADP, adrenaline, PAF, TX or thrombin bind to a specific platelet heptahelical receptor. The signal transduction generated by this category of receptors successively involves 1) a guanine nucleotide-binding regulating protein (G-protein), 2) an effector enzyme generating second messengers, and 3) the activation of kinases and phospholipases. The intensity of events differs from one agonist to the other. Noteworthy, some platelet antagonists also bind to a serpentine receptor: this is the case for PGI_2 and adenosine. The transducing process is again identical: the specificity relies on the receptor, coupled to one or more G-proteins. In all cases, the effector enzyme and the amount of generated second messengers change the level of intraplatelet cytosolic calcium or cAMP: platelet agonists increase the intraplatelet calcium concentration, antagonists increase that of cAMP.

The ligand-binding region of the receptor is specific and determinant. For example, thrombin has a recognition site, called anion binding exosite (ABE_1), which binds to a specific highly acidic region of its receptor, consisting

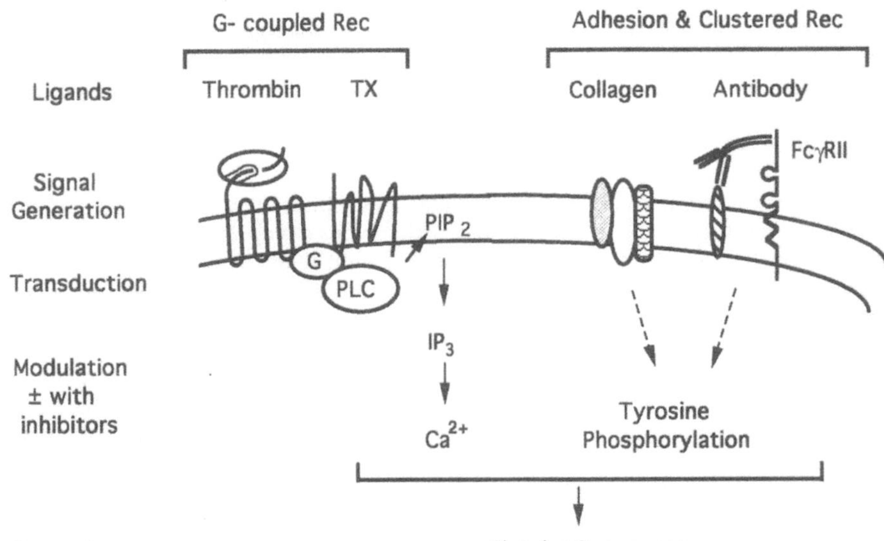

Figure 1. The two general platelet signalling pathways

of a sequence homologous to that of the C-terminus of hirudin. Hirudin thus prevents thrombin binding to this specific region, and hence inhibits thrombin-induced activation of the receptor.

Conformational change of the receptor subsequent to the ligand binding promotes the interaction with one or more G-protein. Among ligands concerned in this section, thrombin displays a unique mechanism of activation of its receptor. By its ABE_1 thrombin binds to the hirudin-like domain of the receptor. By its catalytic site, thrombin cleaves the exodomain of the receptor at a different site than the hirudin-like domain, a process which unmasks a new N-terminus. Thrombin receptor (PAR-1) was the first member of so-called Protease Activated Receptors family. The new N-terminal sequence acts as a classical ligand, except that it is tethered and it has to flip over to interact with a distinct extracellular region of the receptor, on the second extracellular loop [6].

The G-protein contact site is located in the cytoplasmic receptor regions. A domain on the third cytoplasmic loop appears to be determinant for the recognition of the G-protein as recently demonstrated for the PAF receptor [7]. G-proteins are the typical transducing molecules: they couple the receptor to the intracellular effector during an activation/inactivation cycle allowing reversible transmission of the signal. G-proteins, heterotrimers of α, β, and γ subunits, are enzymes bearing a GTPase activity. The interaction with the activated receptor induces the exchange of GDP for GTP on the α-subunit. Hence the heterotrimeric protein dissociates generating a GTP-bound α-subunit and a $\beta\gamma$-dimer. The effector enzyme is modulated until the intrinsic GTPase activity terminates the signal. For a long time it has been thought that only the α-subunit was coupled to the effector enzyme, the $\beta\gamma$-dimer being the anchor of the G-protein at the membrane. Classification of G-proteins

is therefore based on their α-subunits. It is now clear that the two entities, α-subunit and $\beta\gamma$-dimer, contribute to the signal transmission. Presumably, the α-subunit and the $\beta\gamma$ dimer of a G-protein preferentially couple different isoforms of the same platelet effector enzyme [8] or different effector enzymes recruited in platelet cytoplasm [9]. It is fairly probable that $\beta\gamma$-dimer stimulated effector isoforms contribute to secondary irreversible platelet events [10]. About ten G-proteins have been identified in platelets (for review see [11]) belonging to the four classified families: G_s, G_i, G_q, and G_{12}. So far, only members of the first three families have been attributed a role in platelets. Thrombin and TXs receptors are coupled with G-proteins from Gq families. $G_{\alpha q}$-deficient mice show a defective platelet activation [12]. In a patient with a bleeding syndrome, a decreased amount in G_q was related to the abnormal platelet responses to thrombin, PAF and TXs [13].

The effector enzymes modulated by G-protein coupled receptors are phospholipase C (PLC), and adenylyl cyclase (AC). PLC activation leads to calcium mobilization, AC activation or inhibition modulates the level of cAMP. Thrombin and ADP receptors have a dual coupling. Thrombin receptor PAR1 may differentially utilize two G-proteins, one activating PLC and one inhibiting AC [11]. The recent demonstration that a new PAR-4 contributes to platelet activation by thrombin [14] could however modify this concept. In the case of ADP, the presence of two distinct receptors, named P2T on a pharmacological basis [15,16], was recently demonstrated [17]. They both belong to the family of purinoceptors P2, receptors for adenine nucleotides. One of the two P2T is coupled to activation of PLC, the other P2T is coupled to inhibition of AC [18]. The TX receptor (TP) is present in human platelets in two isoforms TPα and TPβ, which differ only in their carboxy terminal tail. The two TP isoforms are coupled to a G-protein of the Gq family [19] to activate PLC, but have distinct effects on AC: TPα activates AC whereas TPβ inhibits it [20].

PLC is a phosphodiesterase that hydrolyses polyphosphoinositides. The breakdown of phosphatidylinositol 4,5-bisphosphate (PIP_2) leads to the generation of two messenger molecules, 1,2-diacylglycerol (DAG) and inositol-1,4,5-triphosphate (IP_3). IP_3 binds to its receptor on the DTS to mobilize calcium and contributes, together with a transient influx through the plasma membrane, to increase the $[Ca^{2+}]_i$. Free cytosolic calcium binds to a cytosolic protein kinase C (PKC) which exposes a binding site for phospholipids and hence translocates to the membrane where it interacts with DAG and becomes fully activated. Activated PKC leads to the phosphorylation of pleckstrin, a cytoskeletal regulatory protein. The increase in $[Ca^{2+}]_i$ also activates several enzymes: myosin light chain kinase to phosphorylate the platelet myosin on its 20 kDa light chain; phospholipase A_2 to liberate arachidonic acid from plasma membrane phospholipids; calpain to proteolyse other enzymes and modulate their activity. Phosphorylation of the myosin light chain allows its interaction with actin so that the actin-myosin assembly can exert a tension and a contractile process, and hence the release reaction. Cleavage of membrane arachidonic acid from membrane phospholipids by phospholipase A_2 has a major consequence: unesterified arachidonic acid is rapidly metabolized by the constitutive platelet cyclooxygenase (COX-1) and converted to endoperoxides (for review see [21]). The cascade of all molecular events following G-protein coupled receptors activation has been extensively documented [22, 23].

Activation of AC yields cAMP out of ATP. Elevation of cAMP activates the dense tubular ATPase and leads to reincorporation of cytosolic calcium in the DTS. AC is coupled to Gs and Gi proteins, for stimulating or inhibiting regulation, respectively. Platelet receptors for PGI_2 and adenosine are coupled to Gs proteins: these two inhibitors stimulate AC, increase intracellular cAMP, and hence prevent platelet activation. As the PLC-calcium pathway activates PKC, the AC-cAMP pathway activates a protein kinase A (PKA). This kinase has various substrates, among which the vasodilator-stimulated phosphoprotein (VASP), and the kinase of myosin light chain to reduce its activity and inhibit contractile phenomenons. The cAMP generated is transformed into inactive AMP by a cyclic nucleotide phosphodiesterase (PDE). Activation of this pathway results in an inhibition of the PLC-calcium pathway. Noteworthy, another effector enzyme contributes to platelet inhibition, the guanylylcyclase (GC). The endothelial-derived relaxation factor, NO, has a platelet serpentine receptor coupled to GC. Activation of GC leads to the accumulation of GMPc as second messenger, and activation of a protein kinase G (PKG) to phosphorylate VASP. In summary, the PLC-calcium-PKC pathway leads to platelet activation, whereas the AC-cAMP-PKA and GC-cGMP-PKG pathways lead to platelet inhibition.

Adhesion and clustered receptors   Adhesion receptors include integrins and leucine-rich proteins. The former consists of one glycoprotein with one transmembrane segment associated to another single chain glycoprotein in a non covalent complex. Engagement of the platelet receptor for the Fc domain of immunoglobulins, FcγRIIa, leads to the cross-link with another FcγRIIa or with an antigenic glycoprotein, thus forming a cluster of receptors. We therefore regrouped in this chapter adhesion and clustered receptors, both characterized by the fact that two or more glycoproteins are coupled by their ligand. Ligands of this group of receptors are glycoproteins (Von Willebrand factor, collagen, immunoglobulins) some of which are multimeric. The common feature of the signal generated by clustered glycoproteins is a cascade of tyrosine phosphorylation. The intracytoplasmic domain of at least one of the engaged glycoproteins being connected to actin filaments through actin binding proteins, the signal transduction is associated with important cytoskeletal rearrangements.

Transmission of the activation signal by adhesion and clustered receptors involves a set of scaffold, anchoring and adaptor proteins, and tyrosine kinases and/or tyrosine phosphatases as the effector proteins, all phosphorylated proteins. The phosphorylation state of the different partners is altered in order to either activate a dormant signalling enzyme or to bring the enzyme into contact with a specific substrate. The signal is transmitted by a cascade of protein-protein interactions which occur through recognition modules (SH2 and SH3, PID/PTB, PH, PDZ, WW), each of which recognizes a different sequence [24]. The recognition modules, located in the non-catalytic regions of enzymes, confer to a given protein the possibility to interact with several partners. As such the modules coordinate the building of a localized signalling complex, and ultimately settle the specificity of the receptor generated signal.

The vWF binds to the platelet membrane leucine-rich glycoprotein complex GPIb-IX-V, on the GPIbα chain. The binding requires a prior change in conformation of vWF. The plasma level of vWF is controlled by endothelial cells, where it is synthesized and stored in the Weibel-Palade bodies. In case of a vessel damage, vWF is secreted and anchors to subendothelial components, like collagen. Thereupon it undergoes a conformational change that allows its binding to platelets: platelet binding to vWF is necessary for platelet adhesion to subendothelial structures. Shear stress also induces vWF binding to GPIb which increases $[Ca^{2+}]_i$ and platelet activation [25]. The situation can be mimicked *in vitro* with the antibiotic ristocetin, and the snake venom botrocetin which induce vWF binding to platelets. Little is known about the signal generated by the binding of vWF to GPIbα. It induces the formation of cytoskeletal signalling complexes involving protein and lipid kinases, such as src and FAK (focal adhesion kinase) tyrosine kinases and phosphatidylinositol-3 kinase (PI-3K) [26]. The process is finally amplified by the activation of the GPIIb-IIIa complex which will recruit other platelets [27]. Collagen is a major subendothelium component that reacts with platelet. The platelet-collagen interaction is complex, mainly because the collagen is a macromolecular protein. *In vivo*, platelets interact with vWF through GPIb, and then with collagen through collagen receptors. In *in vitro* or *ex vivo* experiments, platelets do not interact with collagen monomers but only to immobilized collagen or to collagen polymerized in fibrils, indicating that the interaction occurs through more than one receptor. At present two glycoproteins have been demonstrated to fulfill the criteria of collagen receptors: the GPIaIIa complex or integrin $\alpha_2\beta_1$, and GPVI (for review see [28]). It was established that the collagen-platelet interaction triggers a series of activation steps, including intracellular calcium, activation of PLC, and phosphorylation of numerous proteins, that lead to aggregation. Recently, several data brought evidence that the initial collagen receptor's signalling has common features with immune receptors signalling: clustering of GPVI would lead to the tyrosine phosphorylation of an immunoreceptor tyrosine-based activation motif (ITAM). The ITAM motif containing protein, the GPVI-associated Fc γ chain [29] would interact with the SH2 domains of a tyrosine kinase (syk, a member of ZAP 70 family involved in T-cell signalling) followed by the tyrosine phosphorylation of PLC [30].

The other important ITAM-containing protein present on platelet membrane is the receptor for the Fc domain of IgGs. Among the three classes of receptor for the Fc domain of IgGs, only the low affinity FcγRIIa is expressed on platelet membranes. The majority of activating antibodies activate platelets through FcγRIIa, although some of them induce platelet activation by direct binding on the antigen [31]. The present knowledge concerning immunological platelet activation has been reached with the use of activating monoclonal antibodies (mAbs) or of cross-link of FcγRIIa by aggregated IgGs or by an anti- FcγRIIa mAb cross-linked by an anti- mouse IgG secondary antibody. A polymorphism at amino acid 131 (arginine or histidine) located on the extracellular domain of FcγRIIa appears to be determinant in the affinity of IgGs, at least monoclonal IgG$_1$, for this receptor. For example, low concentrations of an activating mAb are unable to activate platelets from homozygous histidine donors whereas it fully activates those from homozygous arginine donors [32]. Platelet activation mediated by FcγRIIa is characterized by the tyrosine phosphorylation of a number of proteins, among which the receptor itself, phosphorylated on its ITAM sequence. As was later shown for collagen, the phosphorylated ITAM binds to the SH2 domains of tyrosine kinases, among others syk, to increase their activity (for review see [33]). Lyn, a src related tyrosine kinase, is associated with the Fc receptor and could be responsible for the initial phosphorylation of FcγRIIa, although it is not yet established.

The prominent role of syk and PI 3-K in adhesion and cross-linked receptors' mediated signal, at least that mediated by FcγRII, is suggested by the important effects of two specific inhibitors: piceatannol and wortmannin, to inhibit syk and PI 3-K, respectively. Other signal transduction proteins are also involved: the adaptator proteins, Grb2 and Cbl, which recruit the effector enzymes PI 3-K [34] and the PLCγ isoform to mobilize calcium [35]. However, it is not clearly established whether calcium mobilization is the result of a direct effect of the receptor-induced PLC activation or to the amplification by TX. This is still debated following platelet activation by collagen or by FcγRIIa engagement. Noteworthy, the PLC isoforms activated in the adhesion signalling processes involving tyrosine kinases are of the γ forms, whereas those activated by serpentine receptors are usually of the β forms, two isoforms described in platelets. Connections between the two signalling pathways (serpentine receptors and adhesion receptors) remain to be identified in platelets.

Role in Vascular Biology

Physiological Function   Primary haemostasis, i.e. normal platelet activation, occurs in any accidental vascular damage or in normal circulation during a local removal of endothelial cells. On the other hand, if the platelet haemostatic plug formation evades control, platelet physiological function leads to the formation of an arterial thrombus. An excessive platelet activation is implicated in rethrombosis after coronary thrombolysis. In a vascular territory where blood flow is disturbed, such as stenoses, the high shear stress activates platelets. Atherosclerotic thromboembolic complications are initiated by platelet adhesion to a ruptured atherosclerotic plaque. Platelet activation is thus involved in the pathological processes of arterial occlusive diseases: atherosclerosis, thrombosis and vasospasm. Platelets can also be activated by antibodies. A typical example is the heparin-induced thrombocytopenia (HIT). HIT is due to an immunological mechanism involving heparin-dependent antibodies which bind to platelets and cause platelet activation and aggregation. This severe pathology may be complicated by severe thrombotic complications. The latter correlate with the FcγRIIa polymorphism above mentioned [36]. Advances in the platelet activation mechanisms are therefore essential to control platelet involvement in arterial occlusive diseases and develop convenient anti-thrombotic agents.

Pathology   Abnormal primary haemostasis due to deficient platelets, either structurally or functionally, will lead to haemorrhagic manifestations. The common manifestation of all platelet disorders, either qualitative or quantitative, is bleeding, from bruising to more severe hemorrhage. In contrast to provoked bleeding observed in hemophilia or vascular disorders, platelet function defects are characterized by easy and spontaneous bruising, mainly in mucocutaneous tissues. Normal primary haemostasis requires a normal platelet count, intact specific receptors and normal signalling elements, normal granule content, normal metabolism, the presence of specific platelet proteins responsible for interaction with the vessel wall or with each other.

Inherited platelet disorders remain extremely rare (for review see [37]). Besides structural abnormalities concerning membrane glycoproteins (Bernard-Soulier syndrome or Glanzman thrombasthenia) and granule deficiencies (storage pool diseases), inherited bleeding disorders can be linked to any of the above described steps in platelet signalling: defective interaction between ADP and its receptor [38], G-protein dysfunction [13], deficiency in IP_3 formation [39], failure to mobilize intracellular calcium in response to thrombin [40], defective signal transduction mechanisms [41, 42] have been described.

Clinical Relevance and Therapeutic Implications   Normal platelet function is controlled by endothelial cells, which produce NO and PGI_2. Endothelial cells also limit the amplification of platelet activation by ADP released by red blood cells or by platelets themselves. Removal of ADP is performed by an ecto-nucleotidase, on the surface of endothelial cells, identified as CD39 [43], which metabolizes ADP into adenosine. All three compounds, NO, PGI_2 and adenosine have a serpentine receptor on platelets, coupled to AC (PGI_2 and adenosine) or GC (NO). Besides naturally occurring platelet inhibitors, a myriad of factors can affect platelet function: oestrogen increases the plasma level of vWF synthesis which may be determinant of platelet aggregation under shear stress; circulating catecholamine level may be elevated in response to a wide variety of stresses with resultant augmentation of platelet function; different foods and spices, alcohol and cigarettes may alter platelet function. With regard to pharmacological treatment of thrombosis, anti-platelet agents are one of the three types of medicine used, besides anticoagulants that prevent the extension of the thrombus, and fibrinolytics that dissolve the thrombi. Ideal anti-platelet drugs should prevent thrombus formation without interfering with haemostasis, i.e. without causing hemorrhage, and the aim of research is to increase the benefit-risk ratio. The history of the development of drugs that control illegitimate platelet activation has followed the current knowledge on the mechanisms of platelet activation.

In the sixties, a platelet inhibitor had to elevate intraplatelet level of cAMP. Adenosine was an important compound in this category. A weak platelet selectivity was however responsible for important secondary effects. PGI_2 and related prostaglandins also remained disappointing because of a short lasting pharmacological effect following injection and potent vasodilator effects, eventually resulting in hypotension. Other compounds able to increase the platelet cytosolic level of cAMP have been developed, but rather than stimulating AC, inhibiting cAMP degradation, through inhibition of the phosphodiesterase (PDE). Xanthines, dipyridamole, theophylline, papaverine, are the first generation compounds of this type. The lack of platelet selectivity is a high handicap: such inhibitors also inhibit heart and vessel PDEs. Some useful antiplatelet compounds may emerge though, thanks to the identification of distinct PDE isoenzymes,

tissue specific and with different substrate specificities (cAMP or cGMP) and different modulation [44].

In the late sixties and seventies, the targets for the development of anti platelet drugs therapy were the agonists themselves. The interest for aspirin rose when the mode of action of the oldest antiplatelet known was elucidated: aspirin inhibits the biosynthesis of prostaglandins thereby abolishing the formation of the potent platelet agonist TX [45]. The adverse effects of aspirin come from its lack of cell selectivity. Prostaglandin biosynthesis is inhibited both in endothelial cells and in platelets: aspirin thus also prevents PGI_2 production. However that aspirin irreversibly acetylates COX somehow renders the drug specific towards platelets. The anucleated platelet is unable to synthesize a new intact enzyme, and aspirin's inhibitory effect of platelet COX persists for the platelet lifespan albeit a very short half-life time of the drug. Adverse effects of intracranial hemorrhage and gastrointestinal bleeding have been limited by the use of aspirin at the lowest effective doses. Clinically, aspirin is the reference anti-platelet drug, recommended as part of standard therapy for patients with acute myocardial infarction [46]. The efficacy of aspirin led to important research for compounds inhibiting the selective TX synthase rather than the nonselective COX. Most of TX synthase inhibitors are however relatively ineffective, because of an increased production of endoperoxides, TX precursors that cause platelet aggregation. Subsequently, TX receptor antagonists have been developed, in order to antagonize the effects of both TX and endoperoxides at the receptor level. Combined-mode agents that inhibit TX synthesis and block TX receptor have also been developed. So far, none of the compounds represents a significant progress when compared to aspirin.

The other important natural platelet agonists targeted are ADP and thrombin. Thienopyridines fall in this category: ticlopidine and clopidogrel selectively interfere with ADP-mediated platelet activation (for review see [47]). Thienopyridines require several days to attain efficacy and prolong the bleeding time. However, in some types of thrombosis, ticlopidine and its analogues are proving to be more efficient with no more adverse bleeding effects than aspirin. Interestingly, these compounds are totally ineffective *in vitro* and their active principle is unknown, as is the mechanism of the inhibitory effect. Concerning anti-thrombin therapy, the selectivity towards platelets is difficult to reach, considering the numerous biological actions of thrombin. As a serine protease, thrombin cleaves several plasma proteins such as fibrinogen; it is a potent vasoconstrictor and a potent mitogen for vascular cells. Antithrombotic drugs that are direct thrombin inhibitors, like hirudin or hirulog, are not used as anti-platelets. A new category of nucleotide-based molecules, the DNA aptamers [48], that interact with thrombin's ABE, may be useful in the future if they appear to prevent platelet adhesion to the subendothelium.

During the nineties, a new antithrombotic therapy has emerged, thanks to the characterization and advances in the molecular defect concerning Glanzmann's thrombasthenia. This rare autosomal recessive hemorrhagic disease was first characterized by a platelet inability to aggregate. In 1975, Nurden & Caen [49] identified the molecular defect at the GPIIb-IIIa complex level. This finding was at the origin of extensive *in vitro* studies to elucidate the mechanism of aggregation of platelets: fibrinogen binding to GPIIb-IIIa bridging two adjacent platelets. During the following twenty years, *ex vivo* studies have demonstrated the pivotal role of GPIIb-IIIa in the extension of platelet arterial thrombus, and *in vivo* studies in animals provided evidence that blockade of GPIIb-IIIa could protect animals from vascular injury-induced acute thrombosis. Nowadays, blockade of GPIIb-IIIa by rheopro, the Fab fragment of a mouse/human chimeric monoclonal antibody, named abciximab [50], is an important and successful antiplatelet therapy used to prevent ischemic complications of coronary angioplasty or atherectomy. Its use is limited to this clinical situation though, because of the haemorrhagic risk.

Anti-platelet agents of the future will be drugs able to prevent platelet recruitment by preventing platelet adhesion. This blockade may be achieved by agents acting either on adhesive proteins or platelet glycoproteins. This was the case for abciximab directed against GPIIbIIIa. Anti vWF antibodies preventing vWF interaction with GPIb would prove useful.

Francine Rendu and Christilla Bachelot

References

1. Zwall RFA, Schroit AJ (1997) Blood 89:1121-1132
2. Tablin F, Castro MD (1992) Comp Haematol Int 2:125-132
3. Fox JEB (1993) Thromb Haemost 70:884-893
4. Palabrica T et al (1992) Nature 359:848-851
5. Enouf J et al (1997) Platelets 8:5-13
6. Nanewicz T et al (1995) J Biol Chem 267:21619-21625
7. Carlson SA et al (1998) Mol Pharmacol 53:451-458
8. Banno Y et al (1998) Thromb Haemost 79:1008-1013
9. Thomason PA et al J (1994) J Biol Chem 269:16525-16528
10. Banno Y et al (1996) J Biol Chem 271:14989-14994
11. Brass LF et al (1997) Thromb Haemost 78:581-589
12. Offermanns S et al (1997) Nature 361:315-324
13. Gabetta J et al (1997) Proc Natl Acad Sci USA 94:8750-8755
14. Kahn ML et al (1999) J Clin Invest 103:879-887
15. Gachet C (1992) Eur J Biochem 207:259-263
16. Gachet C et al (1997) Thromb Haemost 78:271-275
17. Daniel JL et al (1998) J Biol Chem 273:2024-2029
18. Jin J et al (1998) J Biol Chem 273:2030-2034
19. Kinsella BT et al (1997) J Pharmacol Exp Ther 281:957-964
20. Hirata T et al (1996) J Clin Invest 97:949-956
21. Maclouf J et al (1998) Thromb Haemost 79:691-705
22. Siess W (1989) Physiol Rev 69:58-178
23. McNicol A, Gerrard JM (1993) Blood Coag Fibrinol 4:975-991
24. Pawson T, Scott JD (1997) Science 278:2075-2080
25. Kroll MH et al (1996) Blood 88: 1525-1541
26. Yuan Y et al (1997) J Biol Chem 272, 21847-21854
27. Goto S et al (1998) J Clin Invest 101:479-486
28. Moroi M, Jung SM (1997) Thromb Haemost 78: 439-444
29. Tsuji M et al (1997) J Biol Chem 272:23528-23531
30. Poole A et al (1997) EMBO J 16:2333-2341
31. Rubinstein E et al (1995) Sem Thromb Hemost 21:10-22
32. Bachelot C et al (1995) Thromb Haemost 74:1557-1563
33. Ravetch JV (1994) Cell 78:553-560
34. Saci A et al (1999) J Biol Chem 274:1898-1904
35. Gratacap MP et al (1998) J Biol Chem 273:24314-24321
36. Carlsson LE et al (1998) Blood 92:1526-1531
37. Rendu F, Dupuy E (1991) Blood Cell Biochem, Ed Harris JR, Plenum Corp, vol 2: 77-112
38. Nurden P et al (1995) J Clin Invest 95:1612-1622
39. Mitsui T et al (1997) Thromb Haemost 77:991-995
40. Parker RI et al (1993) J Lab Clin Med 122:441-449
41. Fuse I et al (1993) Blood 81:994-1000
42. Gabetta J et al (1996) Blood 87:1368-1376
43. Marcus AJ et al (1997) J Clin Invest 99:1351-1360
44. Murray KJ, England PJ (1992) Biochem Soc Transac 20:460-464
45. Vane JR (1971) Nature New Biol 231:232-235
46. Sandercock P (1997) Thromb Haemost 78:180-182
47. McTavish D et al (1990) Drugs 40:238-259
48. Bock LC et al (1992) Nature 355:564-566
49. Nurden AT, Caen JP (1975) Br J Haematol 255:720-722
50. Coller BS (1997) J Clin Invest 99:1467-1474

Platelet-Derived Endothelial Growth Factor

Definition *Enzyme with thymidine phosphorylase activity initially isolated from platelets*

See: →Angiogenesis

Platelet-Derived Growth Factor

Definition *Hetero- or homodimeric growth factors that have been initially isolated from blood platelets. Two monomers have been identified, monomer A and monomer B that associate to form PDGF-AA, PDGF-AB or PDGF-BB.*

See also: →Smooth muscle cells; →Atherosclerosis; →Cytokines in vascular biology and disease

Introduction Platelet-derived growth factor (PDGF) was first discovered as a constituent of platelet α-granules, but has more recently been shown to be produced also by a number of other cell types (reviewed by [1, 2]). PDGF is a family of dimeric molecules consisting of disulfide-bonded A- and B-polypeptide chains in homomeric or heteromeric combinations. Early studies showed that PDGF isoforms have potent mitogenic effects on fibroblasts, smooth muscle cells and glial cells and suggested that PDGF is of particular importance for the development of connective tissue; this notion is supported by the finding that topical application of PDGF stimulates wound healing [3]. Another function of PDGF in the connective tissue is to regulate the interstitial tissue pressure [4]. Recent studies have revealed that PDGF also acts on several other cell types, including capillary endothelial cells, mesangial cells, mesothelial cells and neurons. In the central nervous system, PDGF has been shown to have neurotrophic effects for certain types of neurons [5] and to control the differentiation of glial cells [6, 7].

Studies using gene targeting techniques in mice, have revealed important roles for PDGF during the embryonal development. Mice with the PDGF B-chain gene inactivated die at the time of birth with bleedings and show defects in the development of the walls of the heart and large blood vessels as well as of the kidney glomeruli with a total lack of mesangial cells [8]. Mice with the A-chain gene inactivated die shortly after birth with lung emphysema as the most striking feature [9].

PDGF has also been implicated in pathological situations characterized by an abnormal cell growth. The progression of certain tumors involve autocrine or paracrine stimulation by PDGF, and overactivity of PDGF has also been observed in atherosclerotic lesions, rheumatoid arthritis, glomerulonephritis and fibrotic conditions (reviewed by [1, 2]).

PDGF isoforms exert their effects on cells by binding to and dimerizing two structurally related tyrosine kinase receptors. The α-receptor binds both A- and B-chains with high affinity, whereas the β-receptor only binds the B-chain with high affinity. PDGF-AA therefore induces αα receptor dimers, PDGF-AB αβ or αα receptor dimers, and PDGF-BB all three combinations of receptor dimers. Ligand-induced receptor dimerization leads to autophosphorylation of the receptors and interaction with SH_2 domain containing signal transduction molecules, whereby signalling pathways leading to cell growth, chemotaxis and actin reorganization are initiated (reviewed by [10]).

Characteristics
Molecular Weight   The mature PDGF B-chain has 109 amino acid residues. The PDGF A-chain occurs as two forms as a result of differential splicing, the mature forms of which have 124 and 109 amino acid residues, respec-

tively. The A- and B-chains of PDGF show reduced mobilities in SDS-gel electrophoresis, presumably due to their high isoelectric points (about 10); the short form of the A-chain and the B-chain run as components of 17 and 16 kDa, respectively, whereas the dimeric forms run as about 30 kDa.

Domains   The three-dimensional structure of PDGF-BB has been solved at 3.0 Å resolution [11]. It belongs to the cystine knot family of molecules, which is characterized by a tight knot structure of three cystine bonds, one of which passes through the hole formed by the other two and the intervening peptide sequences. Two long loops consisting of twisted β-sheets (loops 1 and 3) extend from the cystine knot and a short loop (loop 2) points in the other direction. The two subunits in the dimer are held together by two cystine bridges and are arranged in an antiparallel manner [11]. Thus, loops 1 and 3 of one subunit will be juxtaposed to loop 2 from the other subunit. Given the sequence similarity between the A- and B-chains, it is likely that also the PDGF A-chain belongs to the cystine knot family.

Binding Sites and Affinity   PDGF-AA binds to α-receptors with a K_d of about 0.2 nM, whereas PDGF-BB binds to α- as well as β-receptors with K_ds of about 0.5 nM [12]. It is likely that these values represent binding to dimeric receptor complexes. The receptor binding epitopes are localized at the tips of loops 1 and 3 with some contribution from the neighbouring loop 2 of the other subunit [13 - 16]. Each dimeric PDGF molecule thus binds two receptor molecules.

Additional Features   The long, but not the short, splice version of the PDGF A-chain, as well as the B-chain have basic sequence motifs which mediate retention of the molecules to the extracellular matrix as well as to structures inside the producer cell [17-19].

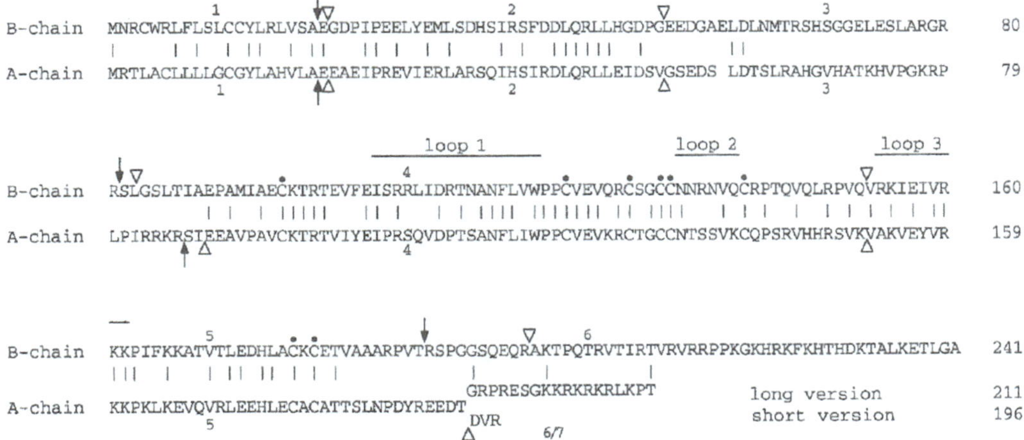

Figure 1. Amino acid sequence of PDGF. The sequences of the B- as well as of the long and short splice versions of the A-chain are shown, as deduced from cDNA sequences [57, 58]. Amino acid residues identical between the A- and B-chains are indicated by vertical lines. Cysteine residues are indicated by dots. Processing sites for removal of the signal sequences, as well as the N- and C-terminal precursor sequences are indicated by arrows. The localization of exon/intron boundaries are indicated by open arrowheads, and the exon numbers are given. Loops 1, 2 and 3, which are involved in receptor binding are indicated by horizontal lines.

Structure

Sequence and Size   The short A-chain, the long A-chain and the B-chain of PDGF are synthesized as precursor molecules of 196, 211 and 241 amino acid residues, respectively (Figure 1). In their mature parts they are about 60 % identical in sequence with a perfect conservation of the eight cysteine residues.

Homologies   PDGF A- and B-chains have sequence homology to members of the vascular endothelial growth factor (VEGF) family (VEGF, VEGF-B, VEGF-C, placenta growth factor and VEGF-related protein) [20]. In particular the eight cysteine residues in the mature part of the molecules are conserved.

Conformation   PDGF is folded in a similar manner to other members of the cystine knot family of proteins, such as transforming growth factor-β and nerve growth factor, but no sequence similarity is apparent with these molecules [21].

Gene

Gene Structure   Both the A- and B-chain genes consist of seven exons [22, 23] (Figure 1). In each case exon 1 encodes the signal sequence, exons 2 and 3 N-terminal sequences which are removed during processing, exons 4 and 5 most of the mature protein, and exon 7 is mainly noncoding. Exon 6 in the B-chain encodes a C-terminal sequence which is removed during processing, whereas exon six of the A-chain gene is subjected to differential splicing giving rise to the short and long forms of PDGF-A. The PDGF B-chain gene is identical to the *sis* protooncogene.

Chromosomal Localization   The A-chain gene is localized on chromosome 7p22 and the B-chain gene on chromosome 22q12.3 – q13.1. (The OMIN database, http://www3.NCBI.NLM.NIH.GOV/OMIM/)

Gene Expression   The A- and/or B-chain genes are expressed in several different cell types, including endothelial cells [24], smooth muscle cells [25, 26], macrophages [27], fibroblasts [28], keratinocytes [29], glial cells [7], neurons [30] and placental trophoblasts [31].

Gene Regulation   Both PDGF ligands and receptors are expressed during the embryonal development. In the mouse embryo the PDGF A-chain and α-receptor are expressed already at the two-cell and blastocyst stages [32, 33]. In the early postimplantation embryos, the A-chain is expressed in the ectoderm whereas the α-receptor is expressed in the mesoderm [32, 34]. The PDGF A-chain is also expressed in the central nervous system. It is striking that the A-chain is often expressed adjacent to cell layers expressing the PDGF α-receptor, suggesting paracrine stimulation.
The PDGF B-chain and β-receptor are not expressed or expressed at low levels in early mouse development [35], but are induced in specific tissues, such as nervous tissues, at later stages of development [30, 36, 37]. There are examples where the expression is modulated in individual cell types, e.g. PDGF expression is increased after thrombin treatment of endothelial cells [38], activation of macrophages [27], and growth factor stimulation of fibroblasts [28]. Since PDGF stimulates fibroblasts, it induces its own synthesis in a feedback manner [28].

Processing and Fate   The precursors of the A- and B-chains are cleaved during secretion from the producer cells first in the N-terminus, and then, in the case of the B-chain, in the C-terminus (Figure 1) [16].
After binding to receptors at the cell surface, PDGF is internalized and degraded in the lysosomes [39]. If PDGF is injected into the blood stream, it is excreted through the kidneys within minutes [40].

Biological Activity   All isoforms of PDGF stimulate cell growth. PDGF-BB and -AB also stimulate chemotaxis, i.e. directed cell movement; however, PDGF-AA inhibits chemotaxis induced by other stimuli in e.g. human fibroblasts and smooth muscle cells. All PDGF isoforms also induce reorganization of the actin filament system, including stimulation of edge ruffling and loss of stress fibers; PDGF-BB and -AB in addition induce the formation of a circular type of ruffles at the dorsal surface of the cells (reviewed by [1, 2]). PDGF also induces anti-apoptotic signals [41].

Role in Vascular Biology

Physiological Function   As far as is known PDGF acts mainly locally at the site where it is produced or released, in a paracrine or autocrine manner. PDGF can be produced by several cell types in the vessels (platelets, monocytes) and in the vessel wall (smooth muscle cells, endothelial cells). It is an important mitogen for smooth muscle cells, and knock-out of the PDGF B-chain gene results in defect formation of large blood vessels [8]. PDGF also acts on capillary endothelial cells, but not on endothelial cells of large blood vessels, and has been shown to have a weak angiogenic activity [42].
PDGF also binds to α-receptors on the platelets and inhibits their aggregation [43, 44]. It is therefore possible that after e.g. thrombin-induced platelet aggregation, PDGF is released from the α-granules of platelets and serves as an important negative feedback.

Pathology   PDGF has been implicated in the atherosclerotic process (reviewed by [45]). After an injury to the endothelial cell layer of an artery, platelets, macrophages and other cell types adhere to the denuded area and release PDGF and other cytokines. Smooth muscle cells are then recruited by chemotaxis from the media layer of the vessel wall into the intimal part, where they proliferate. This results in intimal hyperplasia, an early sign of atherosclerosis.
PDGF is present in the arteries during the atherosclerotic process. The PDGF production by smooth muscle cells and macrophages has been shown to be higher in atherosclerotic lesions than in normal tissue [46, 47]. Moreover, the expression of PDGF-A and -B genes have been found to be increased in circulating mononuclear cells of hypercholesterolemic patients compared with healthy individuals [48]. In an experimental model, bal-

loon catheterization of rabbit femoral arteries was shown to increase the levels of PDGF B-chain and PDGF β-receptor mRNA and protein in the injured vessel [49]. In an injury model of the rat carotid artery and aorta, mRNA for PDGF α-receptor as well as PDGF A- and B-chains were found to be expressed in endothelial cells at the edge of the wound [50]. In support of the notion that PDGF is important for intimal hyperplasia, infusion of PDGF-BB into rats after carotid artery injury was shown to cause intimal thickening [51]. Moreover, antibodies against PDGF inhibit the intimal hyperplasia that follows denudation of the endothelial cell layer of the rat carotis artery [52].

Acute reduction in fluid shear stress have been found to result in proliferation of subendothelial smooth muscle cells. A role for PDGF in this process was suggested by the finding that PDGF-A mRNA and protein was increased in an experimental model in baboons after substantially lowering the blood flow [53].

Clinical Relevance and Therapeutic Implications   PDGF-BB has been shown to stimulate normal healing after topical applications to large wounds of patients [3], and would thus be of potential clinical utility in patients with large wounds or defect wound healing.

Since PDGF is implicated in the atherosclerotic process, PDGF antagonists might be clinically useful in conditions with rapidly progressing atherosclerosis. Important characteristics of such antagonists are that they should be specific and that they should not interfere with important normal functions of PDGF. The known normal functions of PDGF is to regulate embryonal development, to stimulate wound healing and to control platelet aggregability. The role of PDGF during development may not be of any major concern, except maybe during pregnancy. Regarding wound healing, absence of PDGF need not be detrimental for the process since many growth factors and cytokines stimulate wound healing. A potentially more serious complication could be to inhibit the feedback mechanism on platelet aggregation, which could lead to increased risk for thrombosis. Since platelets have only PDGF α-receptors, whereas β-receptors are present in larger amounts on smooth muscle cells and fibroblasts, a possible strategy could be to develop antagonists which would inhibit only β-receptor activation, and not α-receptor activation. That such a strategy could be successful is supported by the observation that antisense oligonucleotides against the PDGF β-receptor suppress intimal thickening [54].

Possible targets for PDGF antagonists could be the interaction between PDGF and its receptors, receptor dimerization, the receptor kinase or components of downstream signalling pathways. Rather specific low molecular weight inhibitors of PDGF receptor kinases have been developed, and are promising candidates for clinically useful PDGF antagonists [55, 56].

Carl-Henrik Heldin

References

1. Heldin C-H, Westermark B (1996) In: Clark RAF (ed) The Molecular and Cellular Biology of Wound Repair. Plenum Press, New York, pp 249-273
2. Raines EW et al (1990) In: Sporn MB, Roberts AB (eds) Peptide growth factors, their receptor. Springer-Verlag, Berlin Heidelberg New York, pp 173-262 (Handbook of Experimental Pharmacology)
3. Robson MC et al (1992) Lancet 339:23-25
4. Rodt S et al (1996) J Physiol (Lond) 495:193-200
5. Othberg A et al (1995) Exp Brain Res 105:111-122
6. Noble M et al (1988) Nature 333:560-562
7. Richardson WD et al (1988) Cell 53:309-319
8. Levéen P et al (1994) Genes Dev 8:1875-1887
9. Boström H et al (1996) Cell 85:863-873
10. Claesson-Welsh L (1994) J Biol Chem 269:32023-32026
11. Oefner C et al (1992) EMBO J 11:3921-3926
12. Östman A et al (1989) Growth Factors 1:271-281
13. Andersson M et al (1995) Growth Factors 12:159-164
14. Clements JM et al (1991) EMBO J 10:4113-4120
15. LaRochelle WJ et al (1992) J Biol Chem 267:17074-17077
16. Östman A et al (199b) J Biol Chem 266:10073-10077
17. LaRochelle WJ et al (1991) Genes and Development 5:1191-1199
18. Raines EW, Ross R (1992) J Cell Biol 116:533-543
19. Östman A et al (1991) Cell Regul 2:503-512
20. Klagsbrun M, D'Amore PA (1996) Cytokine and Growth Factor Reviews 7:259-270
21. Murray-Rust J et al (1993) Structure 1:153-159
22. Bonthron DT et al (1988) Proc Natl Acad Sci USA 85:1492-1496
23. Rorsman F et al (1988) Mol Cell Biol 8:571-577
24. DiCorleto PE, Bowen-Pope DF (1983) Proc Natl Acad Sci USA 80:1919-1923
25. Nilsson K et al (1988) In: Melchers F (ed) Mechanisms of B-cell neoplasia. Springer-Verlag, Berlin Heidelberg New York, pp 172-180 (Current Topics in Microbiology and Immunology)
26. Seifert RA et al (1984) Nature 311:669-671
27. Shimokado K et al (1985) Cell 43:277-286
28. Paulsson Y et al (1987) Nature 328:715-717
29. Ansel JC et al (1993) J Clin Invest 92:671-678
30. Sasahara M et al (1991) Cell 64:217-227
31. Goustin AS et al (1985) Cell 41:301-312
32. Palmieri SL et al (1992) Mech Dev 39:181-191
33. Rappolee DA et al (1988) Science 241:1823-1825
34. Orr-Urtreger A, Lonai P (1992) Development 115:1045-58
35. Mercola M et al (1990) Dev Biol 138:114-122
36. Smits A et al (1991) Proc Natl Acad Sci USA 88:8159-8163
37. Yeh H-J et al (1991) Cell 64:209-216
38. Harlan JM et al (1986) J Cell Biol 103:1129-1133
39. Heldin C-H et al (1982) J Biol Chem 257:4216-4221
40. Bowen-Pope DF et al (1984) Blood 64:458-469
41. Yao R, Cooper GM (1995) Science 267:2003-2006
42. Risau W et al (1992) Growth Factors 7:261-266
43. Bryckaert MC et al (1989) J Biol Chem 264:4336-4341
44. Vassbotn FS et al (1994) J Biol Chem 269:13874-13879
45. Ross R (1993) Nature 362:801-80
46. Ross R et al (1990) Science 248:1009-1012
47. Wilcox JN et al (1988) J Clin Invest 82:1134-1143
48. Billett MA et al (1996) Arterioscler Thromb Vasc Biol 16:399-406

49. Uchida K et al (1996) Atherosclerosis 124:9-23
50. Lindner V, Reidy MA (1995) Am J Pathol 146:1488-1497
51. Jawien A et al (1992) J Clin Invest 89:507-511
52. Ferns GAA et al (1991) Science 253:1129-1132
53. Kraiss LW et al (1996) Circ Res 79:45-53
54. Sirois MG et al (1997) Circulation 95:669-676
55. Buchdunger E et al (1995) Proc Natl Acad Sci USA 92:2558-2562
56. Kovalenko M et al (1994) Cancer Res 54:6106-6114
57. Betsholtz C et al (1986) Nature 320:695-699
58. Josephs SF et al (1984) Science 223:487-491

Platelet-Derived Growth Factor Receptors

Definition   *Tyrosine kinase containing receptors for PDGF. Two receptor types have been identified, the α and the β receptor. The α receptor binds the A and B chains of PDGF whereas the β receptor binds only the B chain.*

See: →Platelet-derived growth factor

Platelets

Definition   *Non-nucleated cells derived from megakaryocytes that are involved in primary hemostasis*

See: →Platelet stimulus-response coupling; →Megakaryocytes; →Atherosclerosis

PLC

Definition   *Phospholipase C*

See: →Phospholipase C

PLD

Definition   *Phospholipase D*

See: →Phospholipases; →Vasomotor tone regulation, molecular mechanisms of

Pleiotropin

Definition   *Same as heparin-affin regulatory protein (HARP)*

See: →Heparin-affin regulatory protein (HARP)

Plg

Definition   *Plasminogen*

See: →Plasminogen/plasmin; →Plasminogen activators

PLGF

Definition   *Placental growth factor*

See: →Ontogeny of the vascular system; →Tyrosine kinase receptors for factors of the VEGF family; →Vascular endothelial growth factor family

PMNL

Definition   *Polymorphonuclear leukocyte*

See: →Angiogenin

PN1

Definition   *Protease nexine-1*

See: →Thrombin

Primary Pulmonary Hypertension (PPH)

Definition   *Hypertension of the pulmonary artery of unknown origin*

See: →Prostacyclin

PRL

Definition   *Prolactin*

See: →Prolactin

Procoagulant Activities

Synonym: Tissue factor, Thromboplastin, CD142

Introduction   The vascular endothelium plays a critical role in the regulation of coagulation through the inducible expression of a procoagulant activity (PCA). While quiescent endothelial cells (EC) are known to provide potent anticoagulant properties, they can rapidly shift the hemostatic balance from an antithrombotic to a prothrombotic state. Exposure to different agents such as lipopolysaccharide (LPS), interleukin 1 (IL1) or tumor necrosis factor (TNF) can rapidly lead to procoagulant behavior of the EC. This inducible endothelial procoagulant activity (PCA) is mainly due to the tissue factor (TF), a transmembrane cell-surface receptor for coagulation factor VII/VIIa.

Characteristics
Molecular Weight   The molecular weight of TF is 47 kDa.

Domains   TF is an integral membrane glycoprotein, member of the cytokine receptor family.
The highly elongated extracellular domain extends from residues S 1 to E 219. It is composed of 2 fibronectin-III domains connected end to end at an angle of about 120 degrees [1]. Surface domain contains two disulfide bonds linking C 49 with C 57 and C 186 with C 209. The first disulfide bridge C 49-C 57 can be abolished by mutation of both cysteine residues to serine with mini-

mal loss of function, whereas the abolition of the C 186-C 209 disulfide bridge results in a significant loss of coagulant activity. Disulfide loops may be of importance in stabilizing the extracellular domain of TF.

The transmembrane region of 23 amino acids extends from I 220 to L 242.

The cytoplasmic portion of the molecule is located on the carboxyl end of the molecule (H 243 to S 263). This portion can be deleted with no apparent loss of function [2].

Binding Sites and Affinity   In the presence of Ca^{2+}, TF reversibly binds Factor VII (F VII) or Factor VIIa (F VIIa). The K_d estimates reported for F VIIa binding to TF vary considerably from 0,1 to 13 nM [3]. The sites of contact on TF are in both fibronectin type-III domains and the interfacial region between them [4]. Crystallographic analysis of the TF/FVIIa complex has identified a ligand-receptor interface area of 1810 $Å^2$, extending onto the concave side of the intermodule angle (5).Thus the domain of contact of factor VIIa and TF extends from the bottom of the carboxyterminal part of extracellular TF up to the top of the aminoterminal domain [6]. Alanine scanning mutagenesis has defined a major cluster of important binding residues, including the surface exposed disulfide bond C 186-C 209, V 207 and W 158 in the carboxyl terminal module which contact the helical aromatic stack region of the FVIIa Gla domain, and residues K 20, F 140, I 22/ E 24 which contact the first FVIIa EGF like domain. In contrast, the interactions between the amino-terminal module of TF and the FVIIa protease EGF2 domain have a lower binding energy which may be related to the activating function of the catalytic domain [7]. Three potential carbohydrate binding sites are located on the extracellular domain (N 11, N 124, N 137).

Additional Features   TF must be inserted into phospholipid vesicles in order to function efficiently [2], but the role of the lipid membrane is still a matter of debate. Complete removal of the membrane anchor dramatically decreases TF procoagulant properties. The replacement of the transmembrane domain with a glycolipid anchor results in a protein that retains full clotting activity [2]. The binding of F VII/VIIa is independent of the lipid composition of the membrane in which TF is embedded. Yet, the ability of the complex F VIIa/TF in initiating coagulation is strongly limited by the availability of phosphatidylserine for association with the substrate.

Structure

Sequence and Size   TF is composed of 263 amino acids. The sequence is schematized on Figure 1 [8].

Homologies   TF belongs to the cytokine receptor family that is characterized by a homologous unit of approximately 200 residues in the extracellular domain. These 200 residues units consist of two modules in which seven strongly conserved blocks and non-conserved intervening sequences were identified. Modules have structural similarity with fibronectin III repeat. Among cytokine receptors, two classes have been identified [9]. Class 2 receptors have a cysteine pair in the carboxy terminal module, whereas class 1 receptors lack a cysteine pair in the carboxyl module. Class 1 includes the growth hormone receptor, prolactin receptor, erythropoietin receptor, interleukin-2 receptor β-chain, interleukin-6 receptor, interleukin-4 receptor, granulocyte-macrophage colony stimulating factor receptor, interleukin-3 receptor, granulocyte colony stimulating factor-receptor, interleukin-7 receptor and interleukin-5 receptor. The class 2 receptors are interferon α receptor, interferon γ receptor and tissue factor.

Conformation   Tissue factor contains two C2-type immunoglobulin-like modules. They are characterized by β-sheets formed by 7 major strands. A hypothetical model derived from the three dimensional structure of the chaperon protein PapD has been proposed [3]. The residues of TF would be predicted to align in two sheets, 3 and 4 β-strands respectively in each module. On the two modules, oriented with an angle of 125°, the solvent exposed surface has a similar number of charged residues with no obvious extended clustering of negative or positive charge.

Additional Features   An extended finger-like region of the extracellular domain contains a short stretch of a helical secondary structure and includes a short antiparallel β-sheet, which protrudes from the side of the molecule at the intermodule surface [1].

A tripeptide WKS is repeated three times in the amino acid sequence. Other proteins containing the tripeptide are factor VIII, von Willebrand factor, high molecular weight kininogen, thrombospondin and antithrombin [10].

Gene

Gene Structure   The gene TF spans about 12,4 kb [11]. It contains 6 exons separated by 5 introns (Figure 2). The 5' region of exon 1 encodes translational initiation while the 3' region encodes the 32 residue leader sequence. The second through fifth exon encode the extracellular domain of the protein. Exons 2, 3 and 4 code for the three WKS sequences. Exon 6 is large and spans 1278 bp. It encodes for the transmembrane and cytoplasmic portion of the protein and all of the extensive 3' untranslated region including alu repeat flanked by 11-nucleic palindromic repeats that contain HindIII sites. The first, third and fifth intron-exon boundaries are type I. The 5' flanking region has been mapped: a typical TATA box is present 26-bp upstream of the transcriptional start site [11].

Chromosomal Localization   The TF gene is located on chromosome 1, at 1p21-22 [12].

Gene Expression   Transcription of the TF gene results in a single 2.2/2.3 kb mRNA [10]. TF mRNA has a short apparent half-life: 48 to 90 min. [13]. This instability might be due to the presence of several AU-rich motifs in the 3' untranslated region.

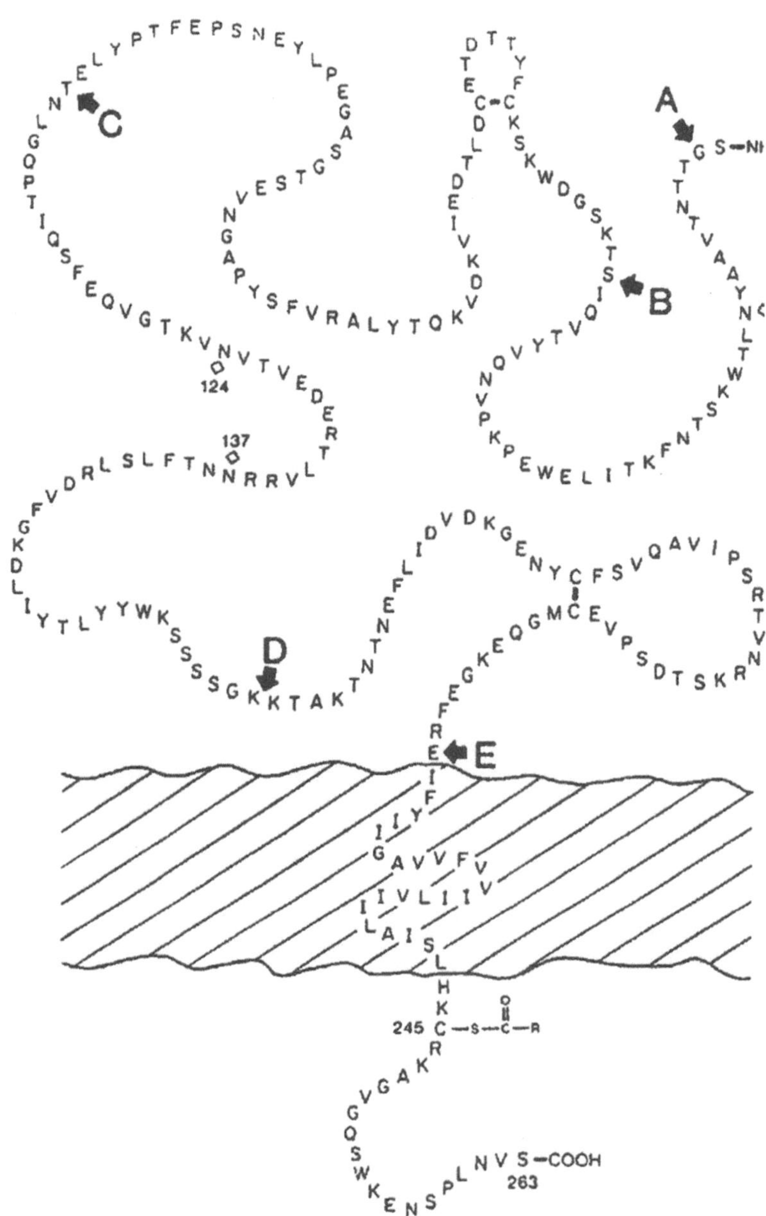

Figure 1. Amino acid sequence for human tissue factor. The arrows in front of A to E indicate the positions of the five introns (from [8]).

Gene Regulation   In vascular cells, the TF gene is inducibly expressed. The basal expression is controlled by 5 Sp1 sites. In cultured cells, a stepwise deletion of the SP-1 sites leads to a progressive reduction of TF promoter activity [14]. *In vivo*, it has been hypothesized that repressor systems operate to limit basal expression.
Induction of the human TF promoter may be mediated by a distal enhancer (−227 to −172) called LPS response element (LRE), containing two AP-1 sites and a NF-kB site. The NF-kB site binds c-Rel/p65 heterodimers. The role of c-Rel/p65 on inducible TF expression is confirmed by the fact that an overexpression of the inhibitor IkBa dramatically reduces TF expression in stimulated porcine endothelial cells [15]. Another region located between -111 and +121 may be involved. This region contains three Egr-1 sites. Egr-1 expression can be induced in human vein umbilical veins (HUVEC) by pharbol ester (PMA).

Additional Features   Two direct repeats of 11 bp and 9 bp have been evidenced on the TF gene. They show some sequence similarities to analogous regions flanking the gene for IL-1, TNFα and PAI-1.
A post transcriptional control may modulate cell surface expression of TF, by the way of mRNA stabilization: in endothelial cells, the half-life of mRNA is 12 times longer 1 h after addition of LPS than 4 hours after.

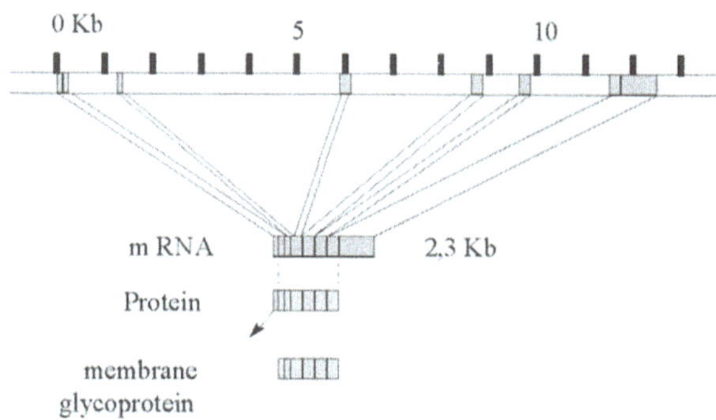

Figure 2. Representation of the human TF gene, mRNA and protein (adapted from [3])

Processing and Fate A primary translation product of 295 amino acid is first synthesized. The 32 residue leader peptide is then removed. It has been found that an additional 2 residues removal occurs frequently, thus leading to a 261 amino acid mature protein starting at residue T3 [10] instead of S1. The molecular masses of these products would be 29,4 to 29,5 kd, considerably smaller than the 47 kd apparent molecular weight estimated from migration rates by SDS-polyacrylamide gel electrophoresis. This discrepancy is partly due to asparagine-linked glycosylation on the extracellular domain. Other post-transcriptional modifications or aberrant migration on SDS gel may also explain this discrepancy.

Some substances are able to induce TF synthesis and surface expression in endothelial cells: interleukin 1 [16], tumor necrosis factor [17], thrombin, bacterial lipopolysaccharide, bacterial infection [18] immune complexes, homocysteine [19], complement fraction [20, 21] or vascular permeability factor [22].

After synthesis, TF is expressed at the cell surface. Unlike the other cofactors of the coagulation protease cascade, TF is fully functional when expressed on cell surface. *In vitro*, expression of TF on endothelial cells is highly polarized on the apical cell surface [23, 24].

Biological Activity TF is the physiologic initiator of blood coagulation. Upon exposure to plasma, TF reversibly binds zymogen FVII or enzymatically active FVIIa. FVII Binding of FVII to TF results in the formation of a binary complex that is rapidly activated by trace amounts of factor Xa (FXa). FVII is a single chain zymogen, composed of an amino-terminal, Gla domain, two EGF-like modules, a short region where the proteolytic activation cleavage occurs, followed by a serine protease domain. FVIIa is a two-chain enzyme. The light chain, bearing Gla and EGF-like modules covalently links to the heavy chain protease domain by a disulfide bridge. FVIIa contacts TF through an extended interface involving the Gla domain with the carboxyterminal module near the TF insertion on membrane, EGF1 with the intermodule interface and a protease EGF2 domain surface contacting the amino-terminal TF module [25]. Both FVII and FVIIa bind the extracellular domain in a 1:1 stoechiometry.

Upon binding to TF, FVIIa undergoes a dramatic enhancement in enzymatic activity. This 20- to 100-fold increased catalytic rate may be a consequence of interactions of specific side chains in TF that stabilize the conformation of corresponding structure located in the protease domain of FVIIa [5]. The TF-FVIIa complex can activate FVII in complex with TF, factor X and factor IX (Figure 3). Once formed, the TF-FVIIa complexes can be inhibited only after they have initiated blood coagulation. The inhibitory reaction requires the presence of factor Xa, lipoprotein and an endothelial derived inhibitor named tissue factor pathway inhibitor (TFPI). TFPI for is a member of the family of Kunitz-type serine proteinase inhibitors. Factor Xa-dependent inhibition of TF-FVIIa involves the formation of a quaternary complex containing factor Xa, TFPI and TF-FVIIa (Figure 4). As a result, factor Xa is rapidly inactivated, whereas factor IXa, released to plasma is more slowly inactivated. Once activated, factor IXa converts factor X to its active form (factor Xa) in the presence of factor VIIIa, phospholipids and calcium. The clotting factors form the "tenase" multimolecular complex on the surface of phos-

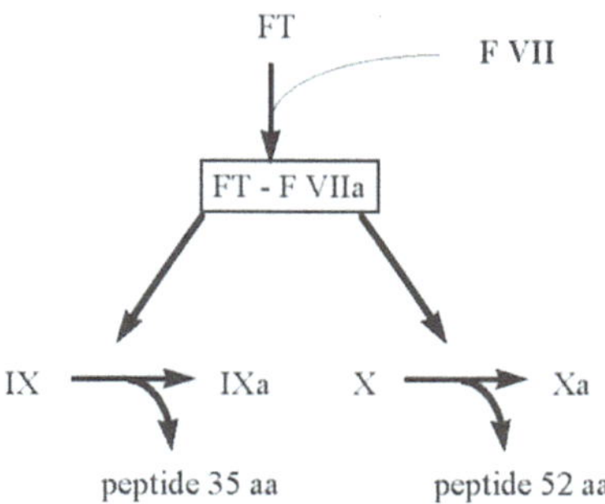

Figure 3. Action of tissue factor on the coagulation cascade. High amount of available TF induces a direct activation of factor Xa by TF/F VIIa complex. Alternatively, factor X activation can result from the formation of a tenase complex in which factor IXa constitutes the active enzyme.

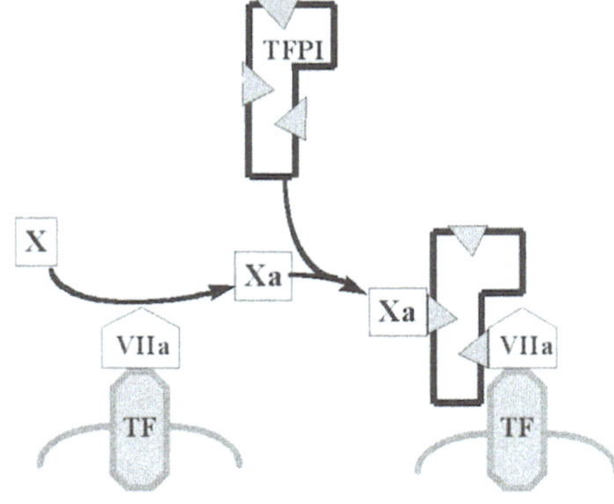

Figure 4. Regulation of factor X activation by tissue factor pathway inhibitor (TFPI).

pholipid. Under physiological conditions, phospholipids for the reaction are provided by platelets. Factor Xa is responsible of the conversion of factor II to thrombin. The rate of prothrombin activation is accelerated about 300 000 fold by the presence of factor Va, phospholipids and calcium ions. Thrombin will then cleave fibrinogen, thus generating fibrin monomers that will polymerize to form the fibrin network. Factor XIIIa will then form covalent bonds on fibrin polymer, leading to an insoluble form of the protein.

The role of cytokine in TF expression has been emphasized. Interleukin 1 [16], tumor necrosis α [17] can induce TF expression on endothelial cells. They can also induce production of various cytokines: colony stimulating factor, IL-6, and IL-1 itself [26]. Another cytokine, the vascular endothelial growth factor/vascular permeability factor (VEGF) can induce tissue factor production on cultured human endothelial cells. This production is supported synergistically by TNF [22]. The notion of synergy is of utmost importance in vascular TF physiology. Cytokines, complement fraction [20, 21], serum, IgG from patients with lupus anticoagulant [27], endotoxin [28], hemoglobin [29] have been found to potentiate TF expression induced by various agonists.

Role in Vascular Biology

Physiological Function   Under normal conditions, blood flow is not in contact with tissue factor, which is the principal initiator of coagulation: TF is present in many tissues, including subendothelium, perivascular fibroblasts, vascular adventicia, cerebral cortex, epidermis, respiratory or bowel mucosa, cardiac myocytes, renal glomeruli and placenta. This distribution represents a kind of hemostatic envelope that is ready to activate coagulation when vascular integrity is disrupted. [30]. Quiescent endothelial cells or blood cells do not express TF. After cytokine or gram negative bacterial endotoxin stimulation, endothelial cells and monocytes change to a procoagulant phenotype by the cell surface expression of TF. Moreover, it has been shown that apoptotic endothelial cells do also express a procoagulant activity. Thus one should hypothesize several different possibilities for a TF dependent activation of coagulation *in vivo*: 1) after vascular wall disruption, blood coagulation may be triggered by TF constitutively expressed on extravascular cells 2) cytokine, endotoxin, or other components inducing cell activation or apoptosis may activate endothelial cells and/or monocytes, leading to TF expression in these cells that are in contact with blood flow. Actually the exact role of endothelial cell in coagulation initiation *in vivo* is unclear: despite clear TF expression in cultured endothelial cells, attempts to document TF expression in endothelial cells *in vivo* have been unsuccessful [3, 31, 32]. In baboons, after a lethal Escherichia coli sepsis, expression of TF by endothelial cells became detectable only in the splenic microvasculature, whereas TF expression could be detected in various epithelial or macrophagic cells [33]. *Ex vivo*, tissue factor mRNA can be expressed in vascular endothelial

cells infected by Rickettsia Rickettsii, but TF antigen remains undetectable on endothelial cell surface [18]. Moreover, apart from TF dependent procoagulant activity, endothelial cells display a variety of prothrombotic activities: factor IX/IXa binding, factor Va or fibrinogen binding, leading to procoagulant macromolecular complexes [34-36] production of coagulation components, such as factor V, synthesis and secretion of adhesive factors, such as von Willebrand factor, thrombospondin, and fibronectin. Finally, it is now accepted that the antithrombotic role of endothelium is an active mechanism, involving EDRF/NO and prostacyclin synthesis, heparin-antithrombin system, protein C-thrombomodulin system, and fibrinolysis.

Pathology   Currently, no specific pathology associated with endothelial tissue factor dysfunction has been described. The physiologic role of tissue factor leads to hypothesize that absence or insufficient expression of tissue factor could be associated with hemorrhagic disease, while overexpression of endothelial TF could induce thrombosis. Tissue factor is widely distributed in the organism and the same mechanisms induce TF expression in monocytes and endothelial cells. Thus it is has not been possible to demonstrate that an abnormal endothelial procoagulant activity could be, by itself, responsible for a pathologic event. However, some data are available concerning TF and pathology. No TF defect has been described in literature. In mice, TF deficient embryos die *in utero* between embryonic days 8.5 and 10.5 because of defective vitello-embryonic circulation, whereas FVII deficient embryos develop normally. FVII deficient mice die perinatally because of fatal haemorrhaging [37]. Thus, in mice, TF plays a crucial role in early embryonic development beyond fibrin formation. It may be indispensable in establishing and/or maintaining vascular integrity in the developing embryo at a time when embryonic and extraembryonic vasculatures are fusing and blood circulation begins [38]. Interestingly, mice with a heterozygous deficiency had no hemorrhagic diathesis and even very low levels (~1%) of tissue factor activity are compatible with embryogenesis development and hemostasis [39].

Clinical Relevance and Therapeutic Implications   Endothelial cell procoagulant activity and TF may be involved in various situations. The more widely studied were disseminated intravascular coagulation (DIC), atherosclerosis, venous thrombosis, angiogenesis and cancer:

TF and disseminated intravascular coagulation: DIC is a situation where the role of TF has been widely studied. In animal models, endotoxin or TF infusion can induce DIC. It has been shown that anti-TF antibodies given before an injection of endotoxin substantially reduce the fall in fibrinogen, factor V and VIII, thus supporting the hypothesis that exposure of blood to TF triggers DIC after endotoxin [40]. Similarly, in baboons, an early treatment by TFPI reduces mortality from Escherichia coli septic shock [41]. Yet, in such models, monocyte TF activity is likely to induce DIC, while it has not been determined whether endotoxin can induce the expression of TF activity on endothelial cells *in vivo* in the rabbit. In humans, TNF infusions also have a procoagulant effect probably due to an increased availability of TF in tissues but, again, it is impossible to determine whether TF expression occurs in endothelial cells [42]. Nevertheless, inhibition of TF or TF pathway by anti TF antibodies [40], TFPI [41] or heparin, is able to reverse the procoagulant properties of stimulated endothelial cells [43] therapeutic inhibition may constitute applications of these features.

TF and arterial disease: The expression of TF antigen and mRNA within the atherosclerotic plaque has been characterized by immunohistochemistry and *in situ* hybridization [44, 45]. It has been recently demonstrated that this TF has a procoagulant activity [46]. TF protein is detected in both cellular and acellular areas. TF mRNA and antigen are detectable in foam cells or macrophages, in mesenchymal intimal cells and in the extracellular matrix. The presence of TF in the extracellular matrix may be due to a release of TF from functional cells or during cell death. TF protein has been detected in human coronary artery lesions. Atherectomy specimens were obtained from patients with stable and unstable coronary syndromes. Tissue factor was readily detected in *de novo* lesions in patients with unstable coronary syndromes, whereas it was rarely detected in patients with restenosis lesions. Recently, it has been shown that the concentration of tissue factor is twice as high in plaques obtained by atherectomy from patients with unstable angina compared with plaques derived from patients with stable angina. [47]. Here again, the cellular origin and localization of TF was not determined, but a monocyte derived procoagulant activity is more likely to be present. However, the role of endothelial cells on monocyte TF expression must be considered. *In vitro*, TF expression by monocytes is significantly increased upon coculture with endothelial cells [28, 48]. The enhancement of TF procoagulant activity may involve cytokines. It has been shown that adhesion of monocytes to cytokine activated endothelial cells increased (approximately 5- to 10- fold) monocyte procoagulant activity [49]. In this view, the question of the cellular origin of TF derived procoagulant activity may be secondary: the association, in atherosclerotic plaques of monocytes and endothelial cells leads to a vascular wall procoagulant activity that could be critical for thrombotic events upon spontaneous or induced plaque rupture. Therapeutic applications of these data are numerous. Inhibition of TF or TF pathway can be achieved by anti TF antibodies [50, 51], retinoic acid [52] or TFPI [53]. Inhibition of TF or of the TF pathway may be useful to prevent arterial thrombosis or reocclusion. TFPI may have a preventive effect on arterial reocclusion [53]. A recombinant soluble tissue factor mutant called hTFAA is a specific inhibitor of the extrinsic pathway of coagulation and displays antithrombotic effects in a rabbit model of arterial thrombosis [51].

TF and venous thrombosis: venous thrombi usually form in regions of slow or disturbed flow. They can be pro-

duced experimentally by a combination of stasis associated to an increase of blood coagulation or to endothelial damage. However the precise initiating event is still questionable. The role of endothelial TF expression has been evoked, but is not yet supported by *in vivo* findings [54].

TF, angiogenesis and cancer: like endothelial cell and monocytes, tumor cells may express tissue factor. Of great interest is the finding that TF has been localized at the surface of vascular endothelial cells within tumors of patients with invasive breast cancer, while TF was not found on endothelial cells within benign tumors [55]. That TF expression in tumors may have a direct and/or indirect effect on neovascularization is supported by other findings. Tumor cells transfected to overexpress TF grow more rapidly and establish more vascularized tumors [56]. Transfection of antisense DNA sequence for TF gene into Meth-A sarcoma cells results in tumors with a markedly reduced angiogenic response when implanted in immunodeficient mice [56]. The mechanism by which TF regulates properties of tumor cells seems distinct from TF activation of the coagulation mechanism. TF may directly alter the production of growth regulatory molecules. Tumor cells over-expressing TF release more mitogenic activity for endothelial cells in parallel with enhanced transcription of vascular permeability factor/vascular endothelial cell growth factor (VEGF/VPF). Thus, endothelial cell TF expression may be a marker for the "angiogenic phenotype" of tumor cells [55]. Whether TF expression can correlate with thromboembolic complications remains to be determined. From these findings, a therapeutic approach using antibody-targeting of tumor vasculature deserves to be studied [57]. Another therapeutic application was recently developed in cancer. A specific antibody against a recombinant form of TF that contained only the cell surface domain was injected intravenously in mice induced to develop neuroblastoma. The truncated TF exhibited limited ability to initiate thrombosis when free in circulation, but became an effective and selective thrombogen when targeted to tumor endothelial cells. Intravenous administration of the antibody-truncated TF complex to mice resulted in complete tumor regressions in 38 percent of the mice [58]. Thus, the antibody-truncated TF conjugate (called "coaguligand" by authors) could selectively thrombose tumor vasculature, illustrating the therapeutic potential of selective initiation of the blood coagulation cascade in tumor vasculature.

Jean-Francois Schved and Muriel Giansily

References

1. Muller YA et al (1994) Biochemistry 33:10864-10870
2. Paborsky LR et al (1991) J Biol Chem 266:2158-2166
3. Edgington TS et al (1991 Thromb Haemost 66:67-79
4. Morrissey JH et al (1997) Thromb Haemost 78:112-116
5. Martin DMA et al (1995) Faseb 9:852-859
6. Banner DW et al (1996) Nature 381:41-46
7. Edgington TS et al (1997) Thromb Haemost 78: 401-405

8. Ichinose A et al (1994) In: Colman RW et al (eds) Hemostasis and thrombosis: Basic principles and clinical practice, 3rd ed. JB Lippincott, Philadelphia, pp 19-54
9. Bazan JF (1990) Immunology today 11:350-354
10. Morrissey JH et al (1987) Cell 50:129-135
11. Mackman N et al (1989) Biochemistry 28:1755-1762
12. Kao FT et al (1988) Somatic Cell Mol Genet 14:407-410
13. Scarpati EM et al (1989) J Biol Chem 264:20705-20713
14. Mackman N (1997) Thromb Haemost 78:747-754
15. Wrighton CJ et al (1996) J Exp Med 183:1013-1022
16. Nawroth P et al (1986) Proc Natl Acad Sci USA 83:4533-4537
17. Bevilacqua MP et al (1986) Proc Natl Acad Sci USA 83:3460-3464
18. Sporn LA et al (1994) Blood 83:1527-1534
19. Fryer RH et al (1993) Atheroscler Thromb 13:1327-1333
20. Carson SD et al (1990) Blood 76:361-367
21. Ikeda K et al (1997) Thromb Haemost 77:394-398
22. Clauss M et al (1996) FEBS Letters 390:334-338
23. Narahara N et al (1994) Atherioscler Thromb 14:1815-1820
24. Camerer E et al (1996) Blood 88:1339-1349
25. Dickinson CD et al (1996) Proc Natl Acad Sci 93:14379-14384
26. Mantovani A et al (1997) Thromb Haemost 78:406-414
27. Hasselar P et al (1989) Thromb Haemost 62:654-660
28. Wharram BL et al (1991) J Immunol 146:1437-1445
29. Roth RI (1994) Blood 83:2860-2865
30. Drake TA et al (1989) Am J Pathol 134:1087-1097
31. Solberg S et al (1990) Blood Coagul Fibrinolysis 1:595-600
32. Courtney MA et al (1996) Blood 87:174-179
33. Drake TA et al (1993) Am J Pathol 142:1458-1470
34. Nawroth P et al (1986) In Gimbrone MA (ed): Vascular endothelium in hemostasis and thrombosis. Churchill Livingstone, Edinburgh, pp 14-39
35. Dejana E et al (1985) J Clin Invest 75:11-18
36. Maruyama I et al (1985) J Clin Invest 74:224-230
37. Rosen ED et al (1997) Nature 390: 290-293
38. Bugge TH et al (1996) Proc Natl Acad Sci USA 93:6258-6263
39. Parry GCN et al (1998) J Clin Invest 101: 560-569
40. Warr TA et al (1990) Blood 75:1481-1489
41. Creasey AA et al (1993) J Clin Invest 91:2850-2856
42. Bauer KA et al (1989) Blood 74:165-172
43. Cadroy Y et al (1996) Thromb Haemost 75:190-195
44. Wilcox JN et al (1989) Proc Natl Acad Sci USA 86:2839-2843
45. Annex BH et al (1995) Circulation 91:619-622
46. Marmur JD et al (1996) Circulation 94:1226-1232
47. Ardissino D et al (1997) Lancet 349:769-771
48. Lewis JC et al (1995) Exp Mol Pathol 62:207-208
49. Lo SK et al (1995) J Immunol 154:4768-4777
50. Ruf W et al (1991) Thromb Haemost 66:529-533
51. Kelley RF et al (1997) Blood 89:3219-3227
52. Ishii H et al (1992) Blood 80:2556-2562
53. Haskel EJ et al (1991) Circulation 84:821-827
54. Semeraro N et al (1997) Thromb Haemost 78:759-764
55. Contrino J et al (1996) Nat Med 2:209-214
56. Zhang Y et al (1994) J Clin Invest 94:1320-1327
57. Folkman J (1996) Nat Med 2:167-168
58. Huang X et al (1997) Science 275:547-550

Prolactin

Definition *Hormone produced in the pituitary gland. The 16 kDa prolactin fragment has anti-angiogenic properties.*

See: →Angiogenesis inhibitors

Proliferation

Definition *Increase of cell number through division*

Prostacyclin

Synonym: Prostaglandin I2, Epoprostenol

Definition *Arachidonic acid metabolite. Has a key regulatory role in vascular tone regulation and thrombosis.*

See also: →Prostaglandins; →Thromboxanes; →Thrombosis; →Platelet stimulus-response coupling; →Thrombin

Introduction Prostacyclin (PGI_2) is a member of the prostaglandin family of lipid mediators that are derived from arachidonic acid (AA). Although discovered by Moncada and Vane in 1976 [1], the function of endogenous PGI_2 *in vivo* has been poorly defined until recently. PGI_2, a potent vasodilator and inhibitor of platelet aggregation *in vitro*, is the most abundant arachidonic acid metabolite formed by macrovascular endothelial cells [2]. Endothelial cells play a key role in the maintenance of normal vascular function, preventing thrombosis by production of mediators such as nitric oxide (NO), PGI_2 and thrombomodulin (TM). Given their effects *in vitro*, it is thought that impaired formation of these compounds, reflective of endothelial dysfunction, might predispose patients to syndromes of vascular occlusion. Indeed, PGI_2, and perhaps NO and TM, produced by endothelial cells may play a homeostatic role as a response to thrombogenic or vasoconstrictor stimuli. Many of these agents evoke PGI_2 release by endothelial cells *in vitro* [3].

Characteristics PGI_2 is synthesized from AA by the sequential actions of a number of enzymes, as shown in Figure 1. Synthesis occurs most notably in vascular endothelial and smooth muscle cells, but PGI_2 generation also occurs in heart, kidney, gastric mucosa, macrophages, lung, brain and small intestine, where its role is less clearly understood ([4] and references therein). AA is first liberated from membrane phospholipids either by a phospholipase (PL) A or by the sequential actions of phospholipase (PL) C and diacylglycerol lipases. The relative contributions of each pathway may vary from cell to cell. In most cell types, including vascular cells, the PLA_2 pathway is considered quantitatively most important. A diverse family of PLA_2 has been described, which includes secretory PLA_2 ($sPLA_2$), calcium independent PLA_2 and the 85kDa calcium-dependent cytosolic PLA_2 ($cPLA_2$)

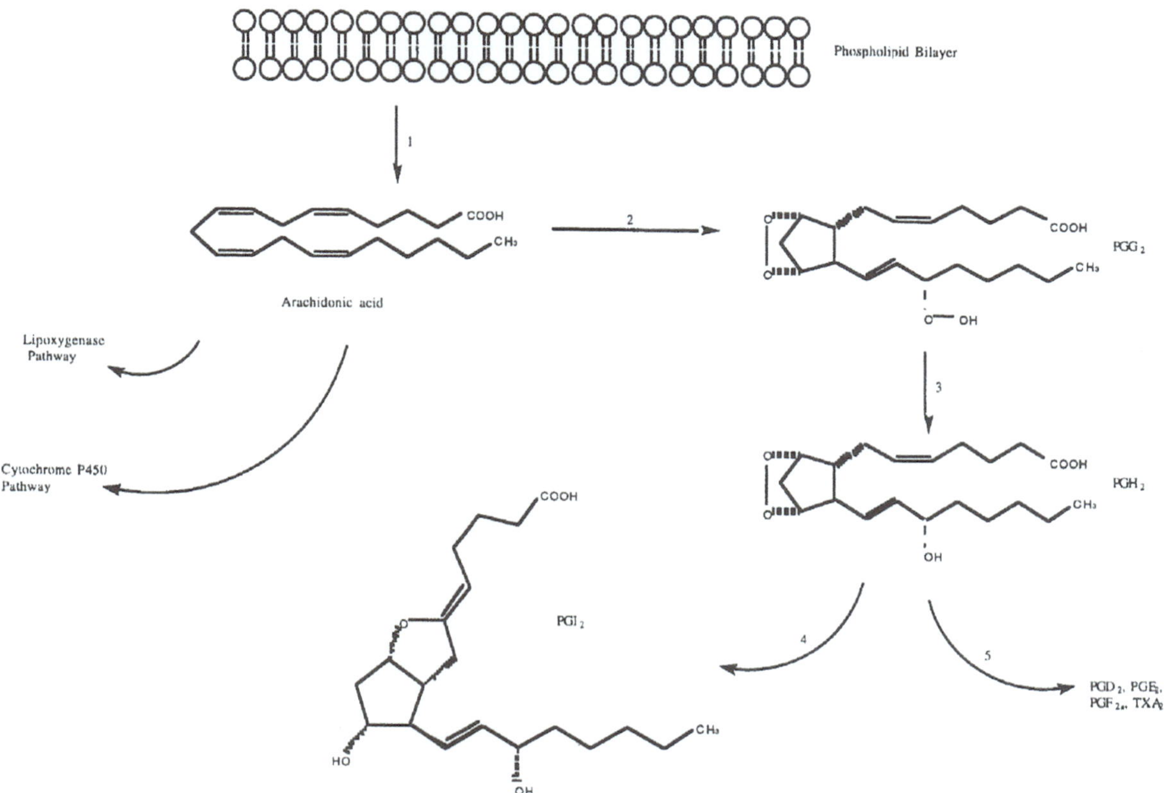

Figure 1.   1. Biosynthesis of PGI₂. The nomenclature of enzymes is as follows: (1) Phospholipases, (2) PG synthase – cyclooxygenase activity, (3) PG synthase – hydroperoxidase activity, (4) PGI synthase, (5) other PG-specific synthases – PGD isomerase, PGE isomerase, PGF synthase, TX synthase.

([5] and references therein). Although cPLA₂ is the only well-characterized PLA₂ which preferentially hydrolyses AA, the other forms of PLA₂ can also mediate AA release, depending on the cell type and agonist involved [6].

Free AA can be converted into the cyclic endoperoxide prostaglandin G₂ (PGG₂), by a cyclooxygenase reaction. PGG₂ then undergoes a peroxidase reaction to form a second endoperoxide, prostaglandin H₂ (PGH₂). Both of these reactions are catalyzed by PGH₂ synthase, an enzyme commonly referred to as COX. This enzyme exists as two isoforms, both of which possess the cyclooxygenase and peroxidase activities [7]. COX-1 is widely expressed and thought to subserve a housekeeping function. COX-2 is generally regulated by cytokines, tumor promoters and growth factors, although it is constitutively expressed in some cells. Although COX-2 is not constitutively expressed in endothelial cells, it is upregulated by laminar shear [8] and this probably contributes substantially to COX-2 expression *in vivo*. Both COX isoforms have been crystallized. They exist as dimers and bear a remarkable structural similarity. AA can access the active site in the body of the enzyme via a tunnel from the ER-bound surface of the enzyme [9, 10]. The tunnel is more accommodating in COX-2 although many of the

residues key to AA catalysis are conserved between the isoforms [11]. It is unclear if the enzymes differ in their subcellular distribution, although this seems unlikely. The structural and other requirements for spatial delivery of the PGH₂ intermediate to downstream enzymes such as PGI synthase are presently unclear.

Recent evidence suggests that two distinct pathways of prostaglandin synthesis, mediated by either COX-1 or COX-2, can co-exist within a single cell. In activated mast cells, the early phase of PGD₂ synthesis is mediated by sPLA₂ and is COX-1 dependent, whereas the late phase of PGD₂ synthesis is mediated by cPLA₂ and is dependent upon COX-2 ([6] and references therein). Once formed, PGH₂ has a short half-life and is rapidly transformed to biologically active prostaglandins or thromboxane A₂, depending on the cell type under study.

The final step in the synthesis of PGI₂ is the isomerisation of PGH₂ by the PGI₂ synthase (PGIS). The gene encoding PGIS, a hemoprotein of the cytochrome P450 type, has been cloned and characterized [12 – 14] and the enzyme itself has been purified to homogeneity from bovine aortic microsomes [14].

The human PGIS gene is approximately 70kb [15] and consists of 10 exons of between 74 and 182 bp each. Intron sizes range from 1 kb (intron 7) to 19 kb (intron 1). The gene encodes a protein of 500 amino acids with a deduced molecular weight of approximately 57 kDa [13]. The enzyme has been classified as a member of a new family of P450 like enzymes, CYP8. Unlike other cytochrome P450s, which generally catalyze oxidation of a variety of hydrophobic substrates, PGIS has a unique isomerase activity.

De Witt and colleagues have shown that conversion of PGH_2 to PGI_2 by PGIS is accompanied by inactivation of the enzyme, concomitant with a bleaching of its heme spectrum [16]. Although the precise mechanism of this catalytic inactivation is unknown, bleaching of the heme spectrum reflects loss or alteration of heme. In contrast, the suicide inactivation of thromboxane synthase (TXS), which appears to involve a tight, non-destructive association of substrate with the prosthetic heme group, is not associated with bleaching of the heme spectrum [17].

Site-directed mutagenesis has demonstrated that PGIS enzyme activity depends on two motifs, a Cys-pocket at Cys^{441} and an EXXR motif at Glu^{347}-Arg^{350} [18]. A three-dimensional molecular model of the synthase enzyme has recently been proposed, based on comparisons with the other P450s [19]. Several amino acids have been identified in the putative substrate channel that regulate substrate access, binding and its interaction with heme. Although most attention has been focused upon PGI_2 produced by vascular endothelial and smooth muscle cells, the tissue distribution of PGIS mRNA is widespread. It is particularly abundant in ovary, heart, skeletal muscle, lung and prostate. Overexpression of PGIS in vascular smooth muscle cells inhibits cellular proliferation [20]. Expression of PGIS mRNA in human aortic endothelial cells is upregulated by several cytokines [13]. Conversely, peroxynitrite has been shown to selectively cause irreversible inhibition of PGIS, by post-translational modification of the enzyme [21]. PGIS activity is also exceedingly sensitive to inhibition by fatty acid hydroperoxides [22]. Accumulation of lipid peroxides in atheromatous plaques could predispose to thrombus formation by inhibiting the generation of PGI_2 by the vessel wall without reducing TXA_2 production by platelets.

Structure   Prostaglandins and related compounds are termed eicosanoids because they are derived from 20-carbon essential fatty acids. PGI_2, in common with the other "classic" prostaglandins (those derived from arachidonic acid by the action of cyclooxygenase) is a 20-carbon unsaturated carboxylic acid (molecular weight 350) consisting of a cyclopentane ring and two side-chains (Figure 2).

PGI_2 is derived enzymatically from prostaglandin endoperoxides of the '2' series, i.e. those containing 2 double bonds in their side-chains. An analogous monoenoic prostacyclin (PGI_1) (Figure 2) has been chemically synthesized and used to make antibodies with which to explore the physiological role of endogenous PGI_2 [23]. However, PGI_1 cannot be produced biosynthetically since monoenoic endoperoxides lack the necessary Δ^5 double bond, which is essential for ring closure [24]. The

characteristic ether bridge between carbons 6 and 9 of the prostanoid chain is indicative of PGI compounds in general (Figure 2).

Eicosapentaenoic acid (EPA) gives rise to prostaglandins of the '3' series. PGI_3 is as potent an antiaggregatory agent as PGI_2 ([4] and references therein). However, as EPA is utilized less efficiently as a COX substrate than AA, the increase in PGI_3 biosynthesis does not replace the decrement in PGI_2 generation during fish oil feeding in humans [25, 26].

Processing and Fate PGI_2 is not stored in the cells, but is transiently produced in response to stimulation and is released into the extracellular milieu to act as an autacoid on adjacent cells. PGI_2 has a half-life of approximately two minutes in blood at physiological pH. It spontaneously hydrolyses to 6-keto-$PGF_{1\alpha}$, which is essentially inactive. Plasma albumin has been shown to stabilize PGI_2 [27]. 6-keto-$PGF_{1\alpha}$ then undergoes further metabolism.

The major enzymatic pathways of prostaglandin metabolism involve dehydrogenation of the carbon-15 alcohol group by the 15-hydroprostaglandin dehydrogenase (15-PG-dehydrogenase), reduction of the Δ^{13} double bond, β-oxidation and ω-oxidation. The enzymes catalyzing these reactions are present largely in liver, lung and kidney. Most PGI_2 entering the pulmonary system escapes inactivation by the lung, but PGI_2 is readily metabolized by both liver and kidney. Both PGI_2 and 6-keto-$PGF_{1\alpha}$ can be acted upon by 15-PG-dehydrogenase and Δ^{13}-reductase, although PGI_2 is apparently a much better substrate for the dehydrogenase. The general scheme of these metabolic transformations of PGI_2 is illustrated in Figure 3. The major route of elimination of PGI_2 is in the urine [28]. The predominant urinary metabolites of both PGI_2 and 6-keto-$PGF_{1\alpha}$ are the 2,3-dinor products. The measurement of urinary metabolites of PGI_2 by gas chromatography/mass spectrometry (GC/MS) provides a non-invasive index of the total body production of PGI_2 [29] and avoids the artifacts caused by artificial activation of platelets during blood sampling. This method possesses both sensitivity and specificity and is widely accepted as the most accurate method for assessing *in vivo* production of PGI_2. However, one cannot specifically attribute a tissue of origin to material measured in urine or plasma.

Biological Activity

PGI_2 production in man   The lack of a specific inhibitor of PGIS and the absence of a specific PGI_2 receptor (IP) antagonist, makes it difficult to determine the role of PGI_2 *in vivo*. Initial studies using bioassay techniques suggested that PGI_2, which differs from other prostaglandins in being subject to pulmonary extraction to only a minor extent, circulates in concentrations sufficient to influence vascular tone and platelet function [30]. However, when estimated by measurement of PGI_2 urinary metabolites, the rate of secretion of PGI into the circulation of normal man appears to be too low for it to act as a circulating vasodilator or platelet inhibitor [29].

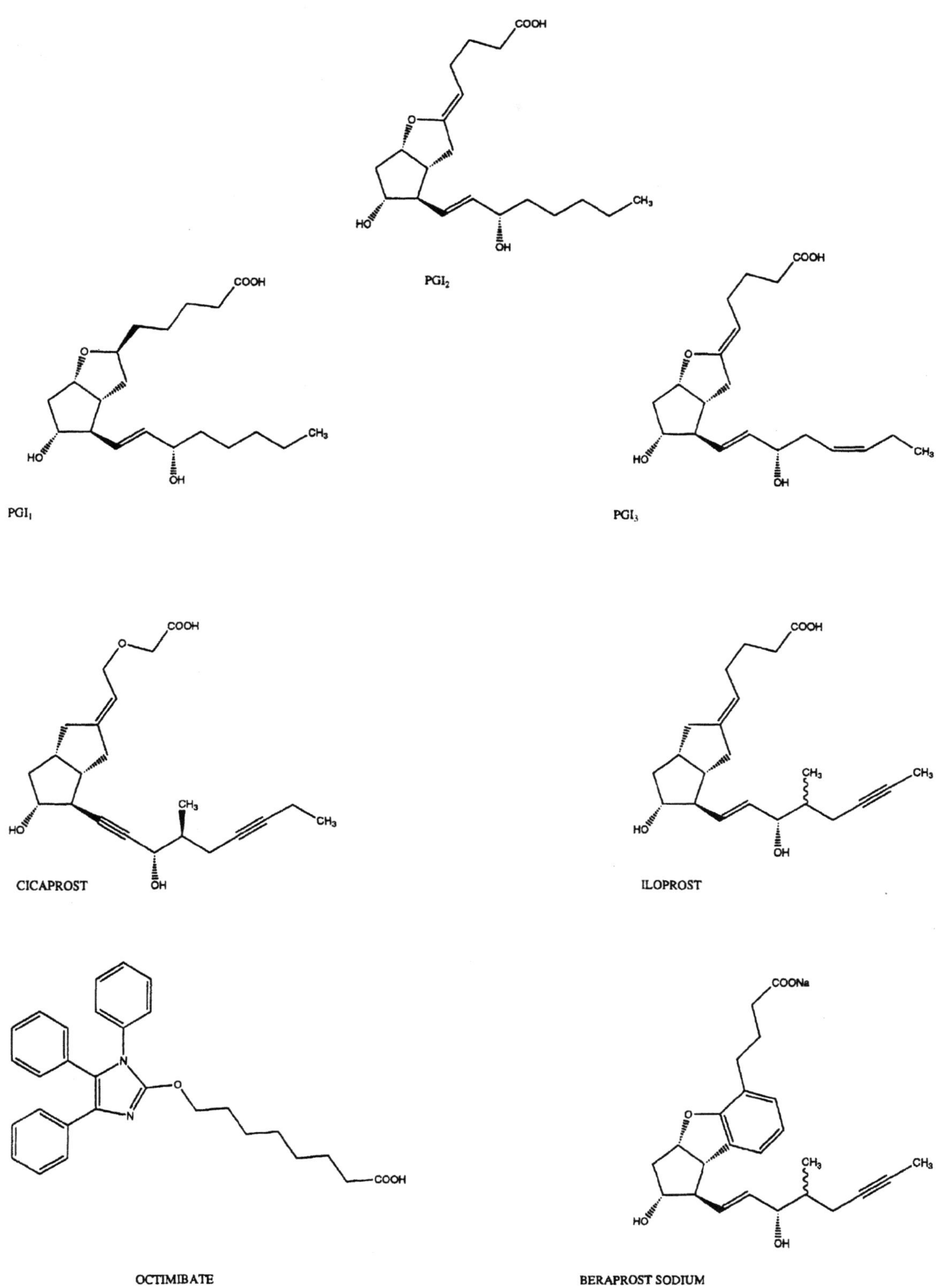

Figure 2.   Chemical structures of PGI$_2$ and related compounds.

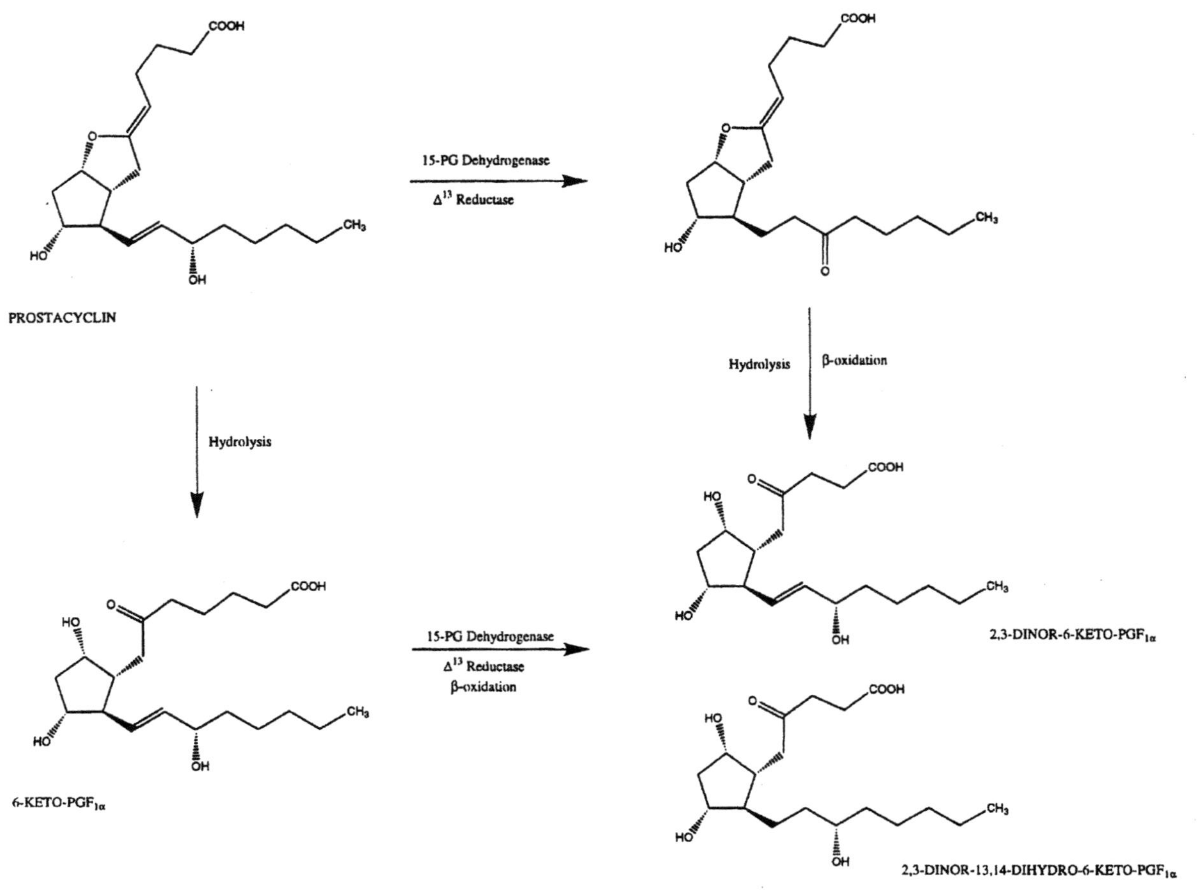

Figure 3. Major pathways of PGI_2 metabolism. At least 16 urinary metabolites of PGI_2 have been identified, but 2,3-dinor-6-keto $PGF_{1\alpha}$ and 13,14-dihydro-2,3-dinor-6-keto $PGF_{1\alpha}$ are the only two that are routinely measured.

For example, whereas the threshold concentration of PGI_2 for an effect on platelets *in vitro* is approximately 30 pg/ml, less than 3 pg/ml circulates under physiological conditions. This does not preclude an important role for PGI_2 in the local regulation of platelet-vessel wall interaction, but it does indicate that systemic circulation of PGI_2 does not represent an important homeostatic mechanism *in vivo*.

Role of PGI_2 receptors   A cell surface PGI_2 receptor (IP) has been cloned from several species [31 – 34]. Sequencing indicates that the IP is a member of the membrane-spanning, G protein-coupled receptor(GPCR) family. Distribution of IP mRNA in human and murine tissue is widespread, with abundant message in kidney, lung, liver, heart and skeletal muscle. In the mouse, levels of IP mRNA are most abundant in thymus, implying a role for PGI_2 in immune modulation [33].

The human IP gene encodes a protein consisting of 386 amino acid residues with seven putative transmembrane domains and a deduced molecular weight of 40,956 (Figure 4). It is located on chromosome 19 and consists of 3 exons and 2 introns spanning a total of 7kb [15]. The binding profile of the cloned receptor (cicaprost=iloprost>PGE_1>>PGE_2, PGF_{2a}, PGD_2, TXA_2) matches the pharmacological description of the human IP. To date, no evidence of a splice variant of the IP has been found and *in situ* hybridization studies using the cloned mouse IP indicates expression of the same receptor on vascular smooth muscle cells and platelets [35]. Expression studies with the cloned human IP have demonstrated the ability of this single isoform to couple to both cAMP and inositol phosphate production [36, 37]. The receptor is rapidly phosphorylated, largely by PKC, in response to agonist stimulation [36].

Recently, Murata and colleagues reported the targeted disruption of the IP gene in mice [38]. The receptor-deficient mice are viable, normotensive and reproduce normally. However, their susceptibility to thrombosis is increased and their inflammatory and pain responses are reduced to the levels of indomethacin-treated wild-type mice. This study confirms the *in vivo* antithrombotic role of PGI_2 and provides evidence for a greater contribution by PGI_2 to the inflammatory process than was previously considered. The role of PGI_2 in inflammation is also supported by studies of PGI_2 expression in normal and

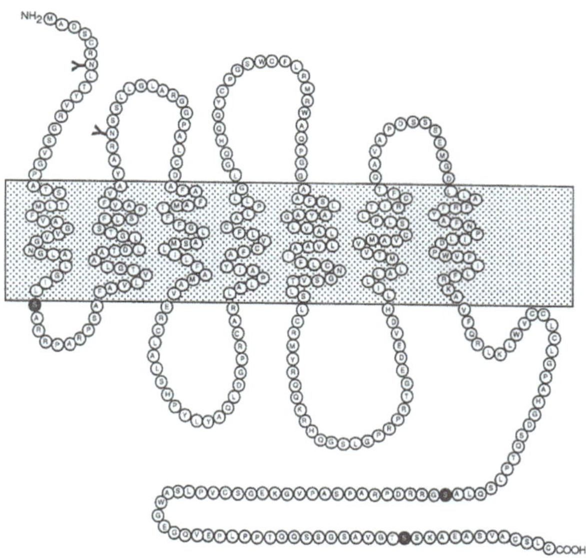

Figure 4. The PGI$_2$ receptor (IP).

leukaemic megakaryocytes [39]. Treatment of mega-karyocyte cell lines with pro-inflammatory stimuli such as IL-1 and TNF$_\alpha$ caused upregulation of IP mRNA expression and potentiated iloprost-induced cAMP production by these cells. Similar upregulation of IP expression was also associated with megkaryocytic maturation.

PGI$_2$ in the gastric mucosa   PGI$_2$ and PGE$_2$ are present in every part of the gastrointestinal tract, where they are thought to have a role in controlling gastric acid secretions, protecting the mucosal barrier and regulating blood flow and motility [4, 40]. Vasodilator prostaglandins probably help to maintain normal perfusion to the stomach, small and large intestine [41].

PGI$_2$ in the kidney   Based on studies using aspirin or indomethacin on healthy volunteers, prostaglandins appear to play a negligible role in maintenance of renal function in the unstressed euvolemic state ([42] and references therein). However, these compounds may serve a major role in the setting of a systemic or intrarenal circulatory disturbance.

The kidney exerts powerful control over arterial blood pressure, blood volume and electrolyte homeostasis through the secretion of renin. The regulation of renin is under the control of both the β-adrenergic system and prostaglandins. PGI$_2$ is the most abundant prostaglandin produced in the cortex and is primarily synthesized in cortical arterioles and glomeruli [43]. Induction of renin release by PGI$_2$ occurs in a dose- and time-dependent manner [44, 45]. All of the primary prostaglandins can be synthesized in the renal cortex, and intrarenal infusion of PGI$_2$, PGE$_2$ or PGD$_2$ increases the rate of sodium and water excretion. Thus, prostaglandins have the potential to regulate sodium hemostasis.

PGI$_2$ in pregnancy   Increased vasodilation and decreased vascular resistance are physiological responses in normal pregnancy, which do not occur in pregnancy-induced hypertension and pre-eclampsia. This results in generalized and local vasoconstriction and platelet aggregation in the utero-fetoplacental tissues [46]. Increased production of TXA$_2$ coupled to decreased PGI$_2$ generation, has been implicated in this disease process. Endothelial cells from placental and umbilical vessels have been shown to produce lower amounts of PGI$_2$ in pre-eclamptic compared to normal pregnancies. Patients who develop pregnancy-induced hypertension exhibit a lesser increment in PGI$_2$ biosynthesis than healthy pregnant subjects [47], although it is unknown whether or not this precedes the development of clinical disease and therefore may be important in disease pathogenesis.

PGI$_2$ in the central nervous system   Expression of IP has been reported in the brain [48], and the possibility of a novel IP subtype in the central nervous system, with different binding properties to those of the known IP, has been examined [49]. The widespread expression of PGI$_2$ receptors in the central nervous system suggests that PGI$_2$ has important roles in neuronal activity. PGI$_2$ and its analogs have been shown to cause dose-dependent enhancement of induced post-synaptic potentials in the CA1 region of the hippocampus at concentrations less than 10^{-6} M [50].

PGI$_2$ analogs   The search for a chemically and metabolically stable compound which would mimic the platelet inhibitory effect of PGI$_2$ but not its vasodilatory effects has led to the synthesis and testing of many PGI$_2$ analogs over the past two decades. The structures of some of these analogs are shown in Figure 2.

A currently studied PGI$_2$ analog is iloprost, in which the enolether oxygen of PGI$_2$ is replaced by a methylene group and the lower side chain is modified to enhance the biological potency of the compound as well as its metabolic stability. PGI$_2$ itself has agonist activity, not only at the IP, but also at PGE$_2$ receptor EP2 and EP3 subtypes and the thromboxane (TP) receptor. Although iloprost is more IP specific, it does possess agonist activity at EP1 receptors [51]. Iloprost has been used extensively in pharmacological and clinical studies and has entered clinical practice as an alternative to PGI$_2$ in the treatment of peripheral occlusive arterial disease [52]. However, bioavailability of iloprost on oral application is only approximately 20 %, due to rapid β-oxidation of the carboxyl side chain.

Cicaprost, another PGI$_2$ mimetic, is a more selective IP agonist than iloprost. In this compound, the β-methylene group of the carboxyl side chain has been substituted by an oxygen atom, thus preventing β-oxidation, and the lower side-chain is further modified to yield a compound that is metabolically stable. Unlike iloprost, it shows no cross-reactivity with EP1 receptors and is a more potent smooth muscle cell relaxant, platelet inhibitor and vasodilator than either iloprost or PGI$_2$ [53].

Another PGI$_2$ analog, Beraprost, has been developed for oral use and is now being studied in the clinic [54, 55]. In this compound (Figure 2), the exoenolether structure has been replaced with a cyclopenta (b) benzofuran structure [56]. Beraprost has recently been shown to possess potent anti-inflammatory properties in mice [57], apparently acting to reduce inflammation by affecting the permeability barrier of the endothelium.

Of recent interest are a group of compounds that activate IP though they are structurally dissimilar to PGI$_2$. Octimibate, which lacks a typical prostanoid ring (Figure 2), exhibits both platelet inhibitory and vasodilatory activity [58]. It is not clear how these non-prostanoid mimetics interact with the IP, but site-directed mutagenesis of the cloned receptor might shed some light on the subject and may eventually lead to the realization of the elusive goal of discovery of an IP receptor antagonist.

Clinical Relevance and Therapeutic Implications PGI$_2$ has increased stability at alkaline pH. It can be stabilized as a pharmaceutical preparation (epoprostenol) by freeze-drying and can be reconstituted for use in man in an alkaline glycine buffer [4].

Thrombosis and peripheral vascular disease PGI$_2$ and its analogs have undergone evaluation as potential antithrombotic agents. Although encouraging results have been reported in observational studies, particularly in patients with peripheral vascular disease [59], the results from controlled studies have been conflicting. The alteration of endothelial function in patients with arterial disease, related to the functional and structural characteristics of the artery wall in atherosclerosis, makes the mechanism by which treatment with PGI$_2$ exerts its effect difficult to interpret [60]. Thus, the correct dosage of PGI$_2$ has been difficult to set. Because of the steep dose-response curve with PGI$_2$ treatment it is difficult to deliver doses that will influence platelet activation *ex vivo* without causing side-effects, such as facial flushing or gastric cramping [4]. This often affects the possibility of blinding trials of PGI$_2$ and its analogs (e.g. [61]). This study did, however, provide convincing evidence of a significant effect of iloprost in the treatment of patients with Buerger's disease. In the largest placebo-controlled study of PGI$_2$ in patients with peripheral vascular disease [62], PGI$_2$ treatment resulted in a significant increase in absolute and relative walking times. It is to be hoped that even more successful treatment will ensue when the mechanism of action of PGI$_2$ is understood more clearly and when the dosage regimens have been defined.

The antithrombotic effect of aspirin is attributed to its inhibition of the enzyme Prostaglandin G/H synthase, which is necessary for the formation of TXA$_2$ in platelets ([63] and references therein). TXA$_2$ is a potent vasoconstrictor and platelet agonist. However, the formation of PGI$_2$ by vascular endothelium also requires prostaglandin G/H synthase, and PGI$_2$ exerts opposite effects on platelet function and vascular tone. It is unknown whether coincidental inhibition of other prostaglandins, particularly PGI$_2$, limits the efficacy of aspirin in cardiovascular disease.

PGI$_2$ has not proven to be effective in the treatment of unstable angina or acute myocardial infarction. Infusion of PGI$_2$ or iloprost has lead in some cases to myocardial ischemia and chest pain in patients with stable coronary disease [64].

Given the aforementioned constraints, PGI$_2$ does appear to cause significant and long-lasting improvements in Raynaud's phenomenon, a disease characterized by vasospasm and platelet aggregation in the digital blood vessels. The results of studies over the last fifteen years have been complicated by the fact that RP can be subdivided into secondary Raynaud's syndrome (RS) where there is an associated disorder (such as systemic sclerosis) and primary Raynaud's disease (RD). It has become clear that primary RD patients do not appear to benefit from treatment with PGI$_2$ or its stable analogs [65, 66]. However patients with RS reportedly show a significant therapeutic response (improved cold and pain tolerance, improved hand warmth) to infusion of PGI$_2$, with an intermittent infusion regimen proving effective and a prolonged beneficial effect lasting weeks after the infusion period [67, 68]. PGI$_2$ may be facilitating endothelial repair in RS patients, whose underlying tissue damage would include impaired endothelial function and adherent White Blood Cells (WBCs) and platelets.

Primary pulmonary hypertension The most successful therapeutic use of PGI$_2$ has been in the treatment of patients with primary pulmonary hypertension (PPH). PPH is a rare, almost uniformly fatal disease, the pathogenesis of which remains poorly understood. Christman and colleagues have determined that alterations in the formation of thromboxane and PGI$_2$ occur in PPH [69]. Increased TXA$_2$ release, suggesting persistent activation of platelets, is coupled to depressed elaboration of PGI$_2$, indicative of endothelial damage. This imbalance between TXA$_2$ and PGI$_2$ release may contribute to the development or maintenance of pulmonary hypertension. The present treatment for this disease combines anticoagulant and vasodilator therapy. Vasodilator therapy is aimed at overcoming pulmonary vasoconstriction, therefore reducing pulmonary vascular resistance, without causing significant systemic hypertension. Continuous infusion of PGI$_2$ has been used successfully in patients with severe forms of the disease [70]. Randomized clinical trials comparing PGI$_2$ infusion with conventional therapy showed a substantial and sustained hemodynamic response to PGI$_2$ in severe PPH, with increased exercise tolerance and prolonged survival of patients [71, 72].

Although PPH is a relatively rare disease at present, its prevalence may grow with the increasing use of appetite-suppressant drugs such as fenfluramine and phentermine, now shown to be associated with an increased risk of PPH [73]) and possibly ischemic stroke [74]. The need for a simple and effective treatment for the disease, such as oral administration of beraprost, would be expected to increase accordingly.

Inhaled, aerosolized PGI_2 has also been used as a pulmonary vasodilator for the treatment of pulmonary hypertension following cardiac surgery or heart transplant [75] and, in a pilot study, has been shown to be as effective as NO as a selective pulmonary vasodilator in patients with acute respiratory distress syndrome [76]. There are a number of actions other than vasodilatation that may be particularly valuable in patients with heart failure associated with pulmonary hypertension. Patients with primary pulmonary hypertension frequently show the presence of small pulmonary arterial thrombi. PGI_2 has also been demonstrated to be of value in decreasing lung injury secondary to platelet microaggregate deposition during cardiopulmonary bypass [77].

Chronic heart failure   Chronic heart failure is considered a major cause of cardiovascular death. Despite advances in the treatment of this syndrome the improvement of the quality of life and of mortality rate have been limited, especially in the advanced stages of heart failure. Attempts to improve the outcome of the more compromised patients by continuous infusion of PGI_2 have been unsuccessful [78].

Metastasis   Cicaprost has been demonstrated, like PGI_2, to inhibit effectively metastatic spread of tumors in several different animal models. Mechanistic studies have revealed that the antimetastatic effects of PGI_2 and its analogs are more related to their interference with tumor cell-host interactions, such as tumor cell adhesion to endothelium and tumor cell induced platelet aggregation, than their direct inhibition of tumor growth [79, 80].

Conclusion   Further studies with the knockout mouse will define more completely the role of PGI_2 *in vivo* in the pathology of disease. A better understanding of the regulation of the PGIS, and of the mechanisms of IP signalling and desensitization, may allow this system to be manipulated to therapeutic advantage. There are clearly many clinical conditions that may benefit from PGI_2 treatment. The place of PGI_2 and its stable analogs in therapeutics is still being defined, over twenty years since PGI_2 was first described.

Sandra C. Austin and Garret A. FitzGerald

References

1. Moncada S et al (1976) Nature 263:663-665
2. Bunting S et al (1976) Prostaglandins 12:897-913
3. Wu KK, Thiagarajan P (1996) Ann Rev Med 7:315-331
4. Dusting GJ et al (1982) In: Oates JA (ed) Prostaglandins and the Cardiovascular System. Raven Press, New York
5. Leslie CC (1997) J Biol Chem 272:16709-16712
6. Reddy ST et al (1997) J Biol Chem 272:13591-13596
7. Smith W L (1992) Am J Physiol 263:F181-F191
8. Topper JN et al (1996) Proc Nat Acad Sci 93:10417-10422
9. Picot D et al (1994) Nature 367:243-9
10. Luong C et al (1996) Nature Struct Biol 3:927-33
11. Smith WL et al (1996) J Biol Chem 271:33157-60
12. Pereira B et al (1994) Biochem Biophys Res Comm 203:59-66
13. Miyata A et al (1994) Biochem Biophys Res Comm 200:1728-1734
14. Hara S et al (1995) Adv Prostag Thromb Leuk Res 23:121-123
15. Nakayama T et al (1996) Biochem Biophys Res Comm 221:803-806
16. DeWitt DL, Smith WL (1983) J Biol Chem 258:3285-3293
17. Jones DA, Fitzpatrick FA (1991) J Biol Chem 266:23510-23514
18. Hatae T et al (1996) FEBS Lett 389:268-272
19. Shyue SK et al (1997) J Biol Chem 272:3657-3662
20. Hara S et al (1995) Biochem Biophys Res Comm 216:862-867
21. Zou MH, Ullrich V (1996) FEBS Letters 382:101-104
22. Ham EA et al (1979) J Biol Chem 254:2191-2194
23. Bunting S et al (1978) Prostaglandin 15:565-573
24. Willis AL (1987) In: Willis A (ed) CRC Handbook of eicosanoids: prostaglandins and related lipids. CRC Press, Boca Raton, FL 1, Part A, 3-46
25. Knapp HR et al (1986) N Engl J Med 314:937-942
26. Knapp HR, FitzGerald GA (1989) N Engl J Med 320:1037-43
27. Wynalda MA, Fitzpatrick FA (1980) Prostaglandins 20:853-861
28. Brash AR et al (1983) J Pharmacol Exp Therap 226:78-87
29. FitzGerald GA et al (1981) J Clin Invest 68:1272-1276
30. Moncada S et al (1978) Nature 273:767-768
31. Boie Y et al (1994) J Biol Chem 269:12173-12178
32. Nakagawa O et al (1994) Circulation 90:1643-1647
33. Namba T et al (1994) J Biol Chem 269:9986-9992
34. Sasaki Y et al (1994) Biochim Biophys Acta 1224:601-605
35. Oida H et al (1995) Brit J Pharmacol 116:2828-2837
36. Smyth EM et al (1996) J Biol Chem 271:33698-33704
37. Katsuyama M et al (1994) FEBS Lett 344:74-78
38. Murata T et al (1997) Nature 388:678-682
39. Sasaki Y et al (1997) Blood 90:1039-1046
40. Whittle BJR, Salmon JA (1983) In: Furnberg L, (ed) Intestinal Secretions. Smith Kline French Publications, Welwyn Garden City
41. Oates J A et al (1988) N Engl J Med 319:689-698
42. Palmer BF (1995) J Investig Med 43:516-533
43. Whorton A R et al (1978) Biochim Biophys Acta 529:176-180
44. FitzGerald GA et al (1979) Life Sci 25:665-72
45. Patrono C et al (1982) J Clin Invest 69:231-239
46. Woodworth SH et al (1994) J Clin Endocrinol Metab 78:1225-1231
47. Fitzgerald DJ et al (1987) Circulation 75:956-963
48. Matsumura K et al (1995) Neuroscience 65:493-503
49. Takechi H et al (1996) J Biol Chem 271:5901-5906
50. Ueno R et al (1982) Proc Natl Acad Sci 79:6093-6097
51. Dong YJ et al (1986) Br J Pharmacol 87:97-107
52. Balzer K et al (1991) Internat Angiol 10:229-32
53. Sturzebecher S et al (1986) Prostaglandins 31:95-109
54. Nony P et al (1996) Can J Physiol Pharmacol 74:887-893
55. Okuda Y et al (1996) Prostaglandins 52:375-384
56. Ohno K et al (1985) Adv Prostaglandin Thromboxane Leukotriene Res 15:279-281
57. Ueno Y et al (1997) Prostaglandins 53:279-89
58. Wise H, Jones RL (1996) Trends Pharmacol Sci 17:17-21
59. Szczeklik A et al (1979) Lancet 1:1111-1114
60. Carter CJ (1996) In: Hull R, Pineo GF (eds) Disorders of Thrombosis. WB Saunders, Philadelphia
61. Fiessinger JN, Schafer M (1990) Lancet 335:555-557
62. Virgolini I et al (1990) Prostaglandins 39:657-664

63. Oates J A et al (1988) N Engl J Med 319:761-767
64. Wennmalm A (1993) In: Vane JR, O'Grady J (eds) Therapeutic applications of Prostaglandins. Edward Arnold, London Boston, pp 277
65. Fitscha P et al (1988) Prostaglandins Leukot Essent Fatty Acids 33:23-27
66. Vayssairat M (1996) J Rheumatol 23:1917-1920
67. Belch JJ et al (1983) Lancet 1:313-315
68. Bellucci S et al (1986) Scand J Rheumatol 15:392-398
69. Christman BW et al (1992) N Engl J Med 327:70-75
70. Cremona G, Higgenbottam T (1995) Am J Cardiol 75:67A-71A
71. Rubin LJ et al (1990) Ann Intern Med 112:485-491
72. Barst R et al (1996) N Engl J Med 334:296-302
73. Abenheim L et al (1996) N Engl J Med 335:609-616
74. Kokkinos J, Levine SR (1993) Stroke 24:310-313
75. Haraldsson A et al (1996) J Cardio Vasc Anesth 10:864-868
76. VanHeerden PV et al (1996) Anaes Inten Care 24:564-568
77. Fessatidis IT et al (1994) Perfusion 9:23-33
78. Califf RM et al (1997) Am Heart J 134:44-54
79. Schneider M R et al (1994) Cancer Metastat Rev 13:349-364
80. Honn KV et al (1992) Seminar Thromb Hemost 18:392-415

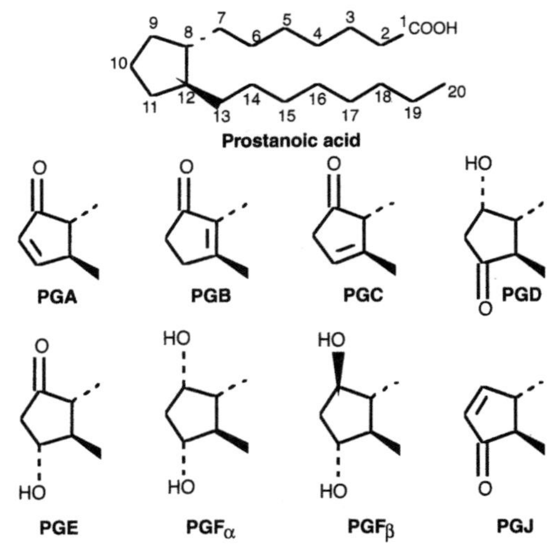

Prostaglandin I₂

See: →Prostacyclin

Prostaglandins

Synonym: Prostanoids

Definition *Derivatives of prostanoid acid isolated from different tissues and organes. The most important types are PGD, E, F, I and thromboxane A₂. Enzymes of prostaglandin synthesis are PGH synthase-1 (Cox-1) and PGH synthase-2 (Cox-2). Prostaglandins play an important role in endothelial cell function, smooth muscle reactivity, permeability and platelet aggregation.*

See: →Prostacyclin; →Thromboxanes

Introduction The discovery of prostaglandins (PG) in the ' 30s is associated with the names of Golblatt and von Euler who described biologically active substances in human semen that could induce strong contraction or relaxation of human uterus and having vasodepressor activity. In 1936 von Euler [1] isolated the active material, lipid soluble acid, which he named *prostaglandin*. It was only twenty years later when von Euler's collaborators Begström and Sjövall isolated two first prostaglandins E₁ (PGE₁) and F₁ (PGF₁) in crystalline form [2] and described the chemical structure [3].They have shown prostaglandins are derivatives of prostanoid acid, a twenty carbon molecule with cyclopentane ring. Since then several other prostaglandins were isolated from different tissues and organs, which were denominated by alphabetical letters from PGA until the last PGJ , isolated recently, see Figure 1.
The most important are PGD,E,F,I and thromboxane A₂ (TxA₂). The common precursor of PG synthesis is arachidonic acid (AA) [4] which is metabolized by the cyclooxy-

9 8 7 6 5 4 3 2 1 COOH
10
11 12 13 14 15 16 17 18 19 20
Prostanoic acid

PGA   PGB   PGC   PGD

PGE   PGFα   PGFβ   PGJ

Figure 1. Prostanoic acid and different types of prostaglandins

genase pathway, see Figure 2. In the 80's it was shown that arachidonic acid could be metabolized by two other enzymatic systems: lipoxygenase pathway generating leukotrienes and monooxygenase pathway generating monooxy derivatives of AA [5] with specific biological activities. All these substances have in common with prostaglandins 20 carbon skeleton of the molecule, therefore a common denomination for all three classes was introduced: *eicosanoids* (eicosa=twenty). It was rapidly recognised that prostaglandins represent a new potent biological system. The prostanoids are of considerable importance for vascular biology because they play an important role in endothelial function, smooth muscle cell reactivity and proliferation, permeability of vascular wall, aggregation of platelets and the adhesion of cells to the endothelium.

Characteristics Prostaglandins are local hormones which are synthetized by virtually all mammalian tissues and act at or near their sites of synthesis. These compounds function in both autocrine and paracrine fashion [6]. They are not circulating hormones as they are rapidly inactivated by the passage in lungs, liver and kidney. The action of prostaglandins on cells is mediated by the binding to receptors coupled to guanine nucleotide-dependent regulatory G proteins [6]. There are distinct receptors for each major prostaglandin [7]. The intracellular response to prostaglandins is mediated by the modification of cAMP concentration and Ca^{++} [6].
Prostaglandins are not stored in cells but are biosynthetized rapidly from cell membrane phospholipids after the hormonal stimulation and the liberation of free substrate which are the following: Dihomo-gamma-linolenic acid (20:3n-6) for PGs of serie 1 (one double bond), arachidonic acid (AA, 20:4n-6) for PGs of serie 2 and eicosapentaenoic acid (EPA, 20:5n-3) for PGs of serie 3. The dominant substrate is arachidonic acid. These essen-

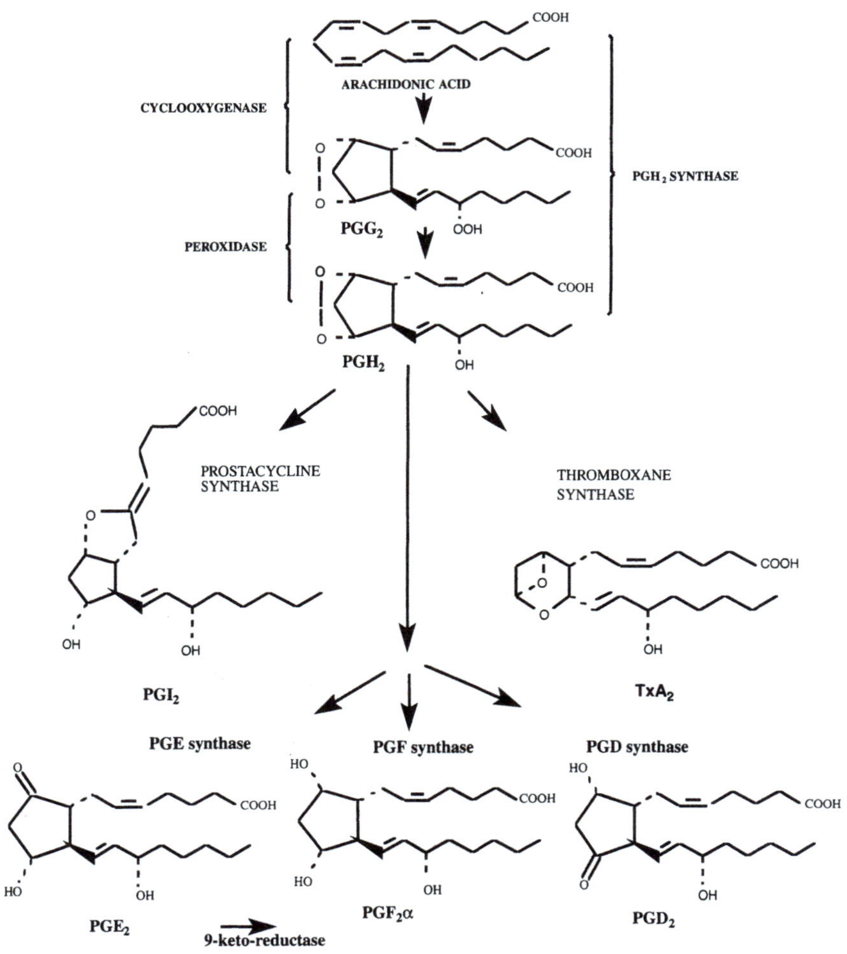

Figure 2 Biosynthesis of prostaglandins by the cyclooxygenase pathway

tial fatty acids are not synthetized *de novo* in cells but derived from dietary linoleic acid (18:2n-6) by desaturation and elongation or ingested as a food constituent. Arachidonic acid is contained in membrane phospholipids esterified at the sn-2 position to glycerol backbone of phosphatidylinositol, phosphatidylcholine and phosphatidylethanolamine. The biosynthesis of PGs is dependent upon the mobilisation of arachidonate by phospholipase A_2, phospholipase C and diacyl- and mono-acyl-glycerol lipase [6]. These phospholipases are stimulated by numerous hormonal, physical or pharmacological agents, which initiate the PG synthesis (see [6, 8]).This latter occurs in 3 phases: 1) mobilisation of AA, 2) conversion of arachidonate in the prostaglandin endoperoxides PGG_2 and PGH_2 by prostaglandine endoperoxide synthetase or PGH synthase [9], enzyme with cyclooxygenase and hydroperoxidase activity and 3) rearrangement or reduction of PGH_2 in biologically active prostanoids – PGD_2, PGE_2, PGF_{2a}, PGI_2 and TxA_2 [6] (see Figure 2). PGH synthase is present in almost all mammalian tissues and organs with the exception of erythrocytes. The synthesis of different PGs depend on specific enzymatic equipment of cells and organs. The PG endoperoxide synthase is an enzyme which auto-destroys after several oxygenation cycles (suicide enzyme) by the covalent binding on hydroperoxylipids produced by cyclooxygenase [6]. The renewal necessitates the new proteosynthesis, which is impossible in anuclear cells (platelets).Recently a new iso-enzyme was identified, the PGH synthase-2 (cyclooxygenase-2, COX-2), which is *inducible* [10], because its synthesis is induced by inflammation, cytokines, LPS and mitogenic stimuli, in contrast to PGH synthase-1 (COX-1), which is *constitutive* and present in numerous tissues. The COX-2 is present in inflammatory cells, fibroblasts, vascular endothelial cells and in several human tissues [11]. PG isomerases are proteins which were isolated from different organs, some of them have multiple form as iso-enzymes [6].

Molecular Weight PGD_2=352.5, PGE_2=352.5, PGF_{2a}=354.5, PGI_2=351,5, TxA_2=356.5. PGH synthase=65.500

Domains   *Prostaglandin endoperoxide synthase* is a membrane glycoprotein containing 590 aminoacids [6, 12] and requiring the heme. This PGH synthase is associated with several intracellular membranes, mainly with endoplasmic reticulum, but also with the outer nuclear membrane, and in some cells with the plasma membrane. *PGI* and *TxA synthase* are heme containing proteins [9]. The critical parts of *prostaglandin* molecules (Figure 2), necessary for biological activity, are: a) Carboxyl (COOH) group at C_1, b) The presence of hydroxyl (OH) group at C_{15} and c) Oxygen, alcool or ketone group at C_9 and/or at C_{11}. *Prostaglandin receptors* (P) have uniformly 7-transmembrane domains as described for DP (receptor for prostaglandin D),subtypes of PGE receptors EP_1, EP_2, EP_3, FP, IP and TP [7, 13].Receptor subtype EP_4 was also described but not yet well identified.

Binding Sites and Affinity   *PHG synthase* exhibits both cyclooxygenase and hydroperoxidase activity (Figure 2), which are associated with the same protein but two activities occur at distinct sites of the PHG synthase molecule [6]. The cyclooxygenase site (Tyr 385) binds preferentially all cis-5,8,11,14-eicosatetraenoate (arachidonate K_m=5μM, V_{max}=1400 molecules arachidonate oxygenated/ min/ subunit) and 8,11,14-eicosatrienoate (γ-homo-linoleate), but also and much less 5,8,11,14,17-eicosapentaenoic acid [6]. The cyclooxygenase site may be competitively inhibited by nonsteroidal anti-inflammatory drugs (AINSD) [14, 15] or irreversibly by aspirin. The best substrate for peroxidase activity is PGG_2, generated by the cyclooxygenase. The presence of heme (Fe^{3+} – Protoporphyrine IX) plays a role of active catalytic site [6]. The binding sites of *prostaglandin molecules* to their receptors were characterized for PGI_2: The presence of C_1 carboxyl group, the C_{5-6} double bond, the 11-OH group and 15-OH group in S configuration are required for high affinity binding [16]. PG molecules bind themselves to specific binding proteins, localised in plasma membranes of many cells. Most of the prostaglandin receptors were isolated and cloned, their affinity for different PGs was established and intracellular signal transduction was determined [7, 13, 16, 17]. They have seven transmembrane domains, an extracellular amino terminus and an intracellular carboxyl terminus. The carboxyl terminus plays an important role in receptor desensitization and down-regulation. Receptors for naturally occurring prostaglandins may also bind other PGs with lower affinity. Potent selective agonists and antagonists for receptors have been identified. For details see Coleman et al. [7].

Structure   *Prostaglandins*: The basic structure -prostanoic acid- consists of a five-membered carbon ring with two side chains, one with carboxyl group at C_1 (Figure 1). Different PGs (A-J) are classified according to the functional groups attached to cyclopentane ring at C_9 and C_{11} [18] (Figure 1). The structure of PGG_2, PGH_2 and PGI_2 is shown in Figure 2.The carboxyl side chain (C_1)is oriented below the plane of cyclopentane ring and by convention is designated α, while the other chain (C_{20}) is out of plane and is designated β. The biologically active PGs always have OH group at C_{15} in a configuration. The structure of *prostaglandin endoperoxide synthase* was described containing 576 aminoacids with heme binding site at His 309 and His 207 or His 388 and with aspirine acetylation site at Ser 530 [6, 18]. Tyr 385 is essential for cyclooxygenase but not peroxidase activity [19]. Human and bovine

prostacyclin synthase is a 500 aminoacid polypeptide showing structural characteristics of cytochrome P450 with 40% identity suggesting a new family of the P450 superfamily [19].The structure of *prostaglandin receptors* was also elucidated, namely by Narumiya et al [17]. They represent proteins containing 341-365 amino acid residues and all are rhodopsin-type receptors with a similar structure but PGE receptor subtypes consist of different molecules [13, 17].

Sequence and Size   The deduced amino acid sequences of human PGH synthase-1(COX-1) were established and start with a methionine, including 576 amino acids, a signal peptide of 24-26 aminoacids, known N-terminus and has a mol. w. 65.500 [6,18]. The PGH synthase-2 (COX-2) contains 602 amino acids and its sequences are similar to COX-1 with 75% identity [18].However it differs significantly from COX-1 at positions prior to amino acids residue 30 and by the NH_2-terminal sequence [20]. Nevertheless both cycloxygenase and peroxidase activities are similar to COX-1. The effect of aspirin is different on both isozymes: COX-1 is inhibited by the acetylation of Ser 530. Cox-2 enzyme is acetylated by aspirine at Ser 516 but is only partially inhibited because arachidonic acid is metabolized in 15-hydroperoxyeicosatetraenoic acid (15-HPETE) [21]. Human and bovine *prostacyclin synthase* contains less amino acids (500) and partial amino acid seqences were determined [22].The sequence and size of *prostaglandin receptors* were described [7, 13, 17].The number of amino acids varies between 343-402, only EP_2 contains more amino acids – 513 [7].According the number of amino acids the molecular mass of different PG receptors lies in the range of 40–45 kDa, only EP_2 receptor has an estimated molecular mass 56 kDa [13].

Homologies   *Enzymes*: COX-1 and COX-2 have similar activity for arachidonic acid (V_{max}) and similar affinity (K_m) but they differ in the following parameters: a) COX-1 is constitutive , b) COX-2 is inducible and glucocorticoids (dexamethasone) inhibit the transcription and the inducibility of the enzyme, c) they are differently inhibited by non steroidal anti-inflammatory drugs, d) the physiological role of both cyclooxygenases seems different: COX-1 is implicated in the regulation of basal physiological functions in response to different hormones, COX-2 seems to play a role in inflammation and/or mitogenesis.

Prostaglandins: There are two epimers of PGF which exhibit similar biological activities, PGF_{2a} and 9a,11b-PGF_2. Recently a new serie of PGF_2-like compounds, termed F_2-isoprostanes, was discovered, which are produced *in vivo* in humans by a *non-cyclooxygenase* mechanism, involving the free radical catalyzed peroxidation of arachidonic acid [24]. 8-iso-PGF_{2a}, which is the most abundant Fa-isoprostane produced *in vivo*, is a potent vasoconstrictor in renal and pulmonary circulation. Recently also, D_2/E_2 isoprostanes were discovered. However the 8-iso-PGE_2 is also a potent vasoconstrictor, in contrast to cyclooxygenase derived vasodilator PGE_2 [24].

Conformation   The general shape of all prostaglandins is characterised as "hairpin" in recognition of the U-shaped arrangement of side chains [25].The flexible structure of α and β chains enables the interaction of PGs with their respective receptors. The interatomic distance between main active sites of PG molecule, which are oxygen atom bonded to C_1, C_9 and C_{15}, plays a role in the physiological action and pharmacological potency. The respective distances vary between 7 – 12 Angströms [26]. The molecular conformation of PGs is essential for their recognition by the receptors.

Additional Features   Prostaglandins are synthesised locally on demand, perform a tissue-specific function, and then are rapidly inactivated by metabolic enzymes. Their half-life is short, seconds or minutes. For this reason different structurally modified analogs have been searched for to obtain longer half-life, peroral use, increased pharmacological activity, dissociate some biological activities or reduce side-effects. The developed analogs concern PGE, used in the treatment of peptic ulcer and cardiovascular diseases. PGF analogs are indicated for gynecological and fertility control. PGI analogs are indicated for cardiovascular diseases. Actually several PGE and PGF analogs have been developed by different pharmaceutical laboratories [27], which are used in clinical practice (see therapeutic implications).

Gene   The gene of human prostaglandin-endoperoxide synthase-1 (PGH synthase-1) was described by Yoko-yama and Tanabe [12] and PGH synthase-2 by Kosaka et al [28]. Cycloxexemide potentiates the expression of PGH synthase mRNA in cultured endothelial cells – phenomenon of superinduction, characteristic of " immediate-early genes" [6]. Prostacyclin synthase gene was identified by Wang and Chen [29] and thromboxane synthase gene by Miyata et al [30]. Prostanoid receptors genes were identified as well [31].

Gene Structure   The structure of human and murine PGH synthase-1 gene was described [12,18] and is 22 kb and consists of 11 exons. The intron/exon structures of the human and murine PGH synthase-1 genes are virtually identical. The human [28] and murine [18] PGH synthase-2 gene is much smaller having 8 kb, but the locations of the intron/exon boundaries are the same except for the first exon [18].The nucleotide sequences of the 5'-end of the genes for PGH synthase-1 and -2 have been determined [18, 28]. Prostacyclin synthase gene is 60kb long and contains 10 exons with an identified transcription start site [29]. Thromboxane synthase gene spans more than 75 kb and consists of 13 exons and 12 introns [30].

Chromosomal Localization   The chromosomal localisation of different enzymes of arachidonic acid cascade was identified. PGH synthase-1 was mapped to chromosome 9 q32-q33.3 while PGH synthase-2 was mapped to chromosome 1q25.2-q25.3, i.e. to distinct chromosomes indicating that these genes are not genetically linked [28].The prostacyclin synthase gene was assigned to chromosome 20 [29] and thromboxane synthese gene to chromosome 7q33-q34 [30] The localisation of human prostanoid receptors was identified by Duncan et al. [31]. The EP_1, IP and TP receptors have been localised by hybridation *in situ* in chromosome 19 at positions 19p13.1, 19p13.3 and 19q13.3 respectively. The FP and EP_3 genes mapped to chromosome 1 at positions 1p31.1 and 1p31.2 respectively, and EP_2 gene mapped to chromosome 5p13.1 [31].

Gene Expression   Factors regulating expression of PGH sythase-1 and -2 genes have been studied in Swiss mouse 3T3 cells, renal mesangial cells, human umbilical cord endothelial cells and other cells [18]. There are positive regulatory agents which promote transcription and PGH synthase mRNA generation as PDGF,Il-1, PMA, EGF, cAMP, LPS , and inhibitory factors as glucocorticoids and possibly also PGH synthase protein, which may contribute by a feedback inhibition of transcription to the "suicide" inactivation of PGH synthase [6]. There are differences in the expression of PGH synthase-1 and -2 genes. In serum stimulated quiescent 3T3 cells, PGH synthase-2 mRNA increases considerably from undetectable levels within 30 mins with a peak at 1-3 h and then rapidly declining, while PGH synthase-1 mRNA increases two- to threefold 3-4 h after stimulation but with elevated levels for 6-8 h [18]. Dexamethasone selectively decreases both basal and induced levels of PGH synthase-2 mRNA.

Gene Regulation   Various transcriptional-regulatory elements of PGH synthase-1 and -2 genes were described [18,28].

Processing and Fate   The principal route of metabolic inactivation of prostaglandins is through enzymatic (or non enzymatic) conversion into biologically inactive products. Several enzymes participate in this process which goes through three principal steps [5], as shown in Figure 3.
1) 15-hydroxy-prostaglandin dehydrogenase converts 15-hydroxy group into a conjugated α,β-unsaturated ketone. 2) 13,14-reductase reduces the double bond and yields 13,14-dihydro-15-keto-prostaglandin metabolites. Both products have minimal or no biological activity and their half-life is about 8 minutes. They could be measured in plasma. These two enzymes are very abundant in the lung, kidney, liver and placenta of mammals and these organs represent main sites of metabolic inactivation. 3) The next step is β-oxidation of the carboxyl chain and ω-oxidation of the methyl end carbon chain yielding dinor or tetranor (loss of 2 or 4 carbons) metabolites on one hand and dicarboxylic acids on the other. The main urinary metabolite of PGE_2 is then 5,11-diketo-7a-hydroxy-tetranorprosta-1,16-dioic acid. The metabolism of PGF_{2a} is similar to PGE_2. Morover, 15-keto-13,14-dihydro-PGE_2 can be catabolized *in vitro* and *in vivo* to bicyclo-PGE_2 metabolite, which is suitable for the measurement in the blood and in urine as an index of PGE_2 production [5]. In conditions of acid or basic pH , PGE_2 is instable, dehy-

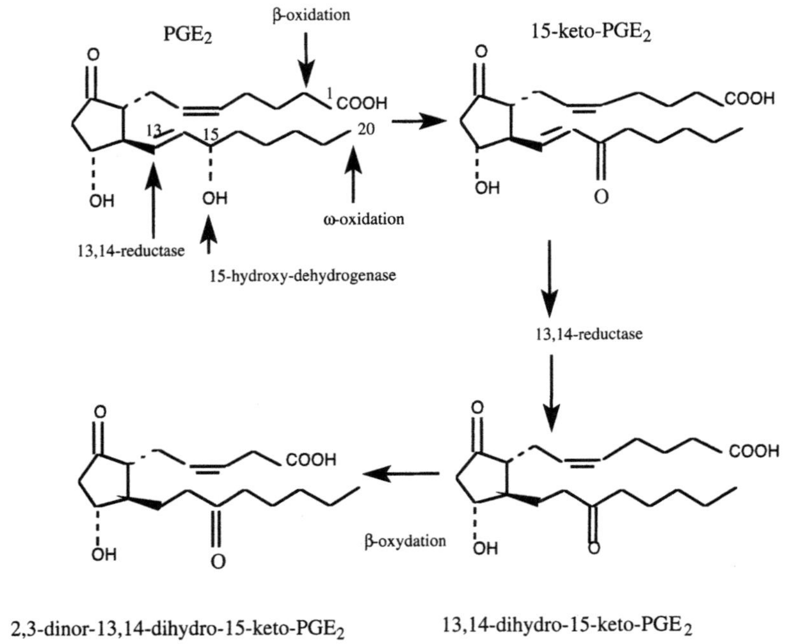

Figure 3 Metabolism of prostaglandin E_2; different enzymes involved in the metabolic inactivation of PGE_2 are indicated.

drates easily and may convert into PGA_2. The basic pH may help the isomerisation of PGA_2 into instable PGC_2 and finally into stable PGB_2. PGs with keto groups at either C_9 or C_{11} can be reduced by specific keto reductases. PGD_2 can be converted by 11-keto-reductase into 9a,11b-PGF_2 as well as into 9a,11b-12-epi-PGF_2. The ω- or 20-hydroxylase activity is mediated by an isozyme of cytochrome P450, present especially in lung and increased during pregnancy. The ω-oxidation completely inactivates the eicosanoids. The pulmonary localisation of this enzyme contributes to the inactivation of PGs entering the pulmonary circulation and protecting the systemic circulation.

Biological Activity   The essential biological activities of PGs are shown in Table 1. (As PGI_2 and TxA_2 are treated in separate chapters, their action is not developed). In summary, PGD_2 is antiaggregant agent, plays a role in the central nervous system, in immuno-allergic reactions, and affects smooth muscle activity [32]. Some effects of PGD_2 are due to the conversion of PGD_2 to 9a,11b-PGF_2 or $\_^{12}$-PGJ_2 [33] (Figure 1). PGE_2 relaxes vascular smooth muscle cells and is implicated in cardiovascular and pulmonary physiology, but contracts or relaxes non vascular smooth muscles and plays an important role in gastro-intestinal and genito-urinary systems. PGE_2 increases vascular permeability and has both pro-inflammatory and anti-inflammatory action. PGE_1 is antiaggregant, vasodilator and inhibits gastric secretion. Since its biochemical synthesis, PGE_1 is widely used in clinical practice with multiple indication.

Role in Vascular Biology

Physiological Function   The physiological role of prostaglandins in vascular biology is ensured by different mechanisms:

1. PGs have a direct action on vascular smooth muscles and myocard contractility, therefore they modulate cardiac output, pulmonary, systemic, capillary circulation and the redistribution of blood flow [34, 36,40,41].
2. PGs contribute to the regulation of blood volume, water and sodium homeostasis through renal diuretic and natriuretic action (PGE1,2, PGI2).
3. PGs modulate vascular tone by inhibiting ($PGE_{1,2}$) or facilitating (PGF_{2a}) sympathetic neurotransmission and by inhibiting vasopressor hormones as noradrenalin, angiotensin II and vasopressin [41].
4. PGs play a role in rheologic eqilibrium protecting the integrity of endothelium (PGI_2), inhibiting platelets adhesiveness (PGE_1,PGI_2) and aggregation.
5. PGs are implicated in immuno-allergic, proliferative and inflammatory reactions. PGD_2 and PGJ_2 inhibit cell proliferation, $PGE_{1,2}$ and PGI_2 inhibit mitogenic effect of angiotensin II, thromboxane A_2 and endothelin, PGE_1 inhibits superoxide anion production. On the other hand, PGE_2 may help the inflammatory reactions by its proinflammatory effects [14] (Table 1).
6. PGs influence vascular biology also by indirect mechanisms through an effect on endocrine secretion and metabolism.

Table 1. Summary of biological actions of D, E and F prostaglandins

System	PG Effect	Ref.
Cardiovascular		
PGD_2	Vasodepressor (vasoconstriction according species)	[34, 35]
$PGE_{1,2}$	Systemic and pulmonary vasodilatation, positive inotrope and chronotrope action, increase of cardiac output, decrease of arterial blood pressure (BP)	[36]
$PGF_{1,2\alpha}$	Vasoconstriction, increases pulmonary and systemic arterial BP and venous pressure	[37, 38]
$9\alpha,11\beta\text{-PGF}_2$	Vasoconstrictor (coronary arteries),8-epi-$PGF_{2\alpha}$ pulmonary vasoconstrictor	[35, 26]
Airway		
PGD_2	PGF1,2, $9\alpha,11\beta\text{-PGF}_2$ Bronchoconstriction	[34, 26]
$PGE_{1,2}$	Bronchodilatation	[36]
Reproduction		
PGD^2	Relaxation of human myometrium	[34]
$PGE_{1,2}$	Inhibition of non-pregnant uterus, increases motility and tonus of pregnant uterus	[36]
$PGF_{1,2\alpha}$	Increases motility and tonus of both non-pregnant and pregnant uterus and motility of Fallopian tube; induces luteolysios; induces abortion and labour; antifertility action [36]	[36]
Gastrointestinal		
PGD_2	Relaxation (or contraction according species)	[34, 35]
$PGE_{1,2}$	Inhibition of gastric secretion, increases gastric and intestinal motility (possible vomiting and diarrhoea	[36, 38]
$PGF_{1,2\alpha}$	Stimulation of both circular and longitudinal smooth muscle of gut, diarrhoea,abdominal cramp	[36, 38]
Nervous		
PGD_2	Induction of sleep, inhibition of hypothalamic-pituitary system (LHRH, prolactin), stimulation of FSH and LH release	[34, 35]
$PGE_{1,2}$	Sedation, catatonia, pulsatile headache, inhibition of noradrenaline release	[36, 37]
$PGF_{1,2\alpha}$	Facilitation of sympathetic neurotrasmission and sympathetic vasoconsctriction	[39]
Renal		
$PGE_{1,2}$	PGE1,2 Increases renal blood flow, diuretic and natriuretic, inhibition of NaCl reabsorption, Inhibition of ADH, angiotensine II and noradrenaline, increases ureteral motility	[36, 37] [40]
$PGF_{1,2\alpha}$	Decreases renal blood flow and renin secretion, increases natriuresis	[36, 40]
$9\alpha,11\beta\text{-FGF}_2$	Diuretic, natriuretic	[35]
Hematological		
PGD_2	Inhibits platelet aggregation	[34]
PGE_1	Inhibits platelet aggregation and adhesiveness, fibrinolytic action	[36]
PGE_2	Induces platelet aggregation	
8-epi-$PGF_{2\alpha}$	Antagonises thromboxane receptors in platelets	[26]
Endocrine-metabolic		
PGE_1	Antagonises hormone stimulated lipolysis in adipose tissue, inhibits cholesterol synthesis and LDL receptor activity	[36]
$PGE_{1,2}$	Increases cAMP and the secretion of thyroid, parathyroid, adrenocortical and ovarian hormones, TSH-like action, stimulation of LHRH,LH, increased glycolysis in response to insulin and stimulation of glucagon=hyperglycemic action; bone resorption	[36, 41]
Immuno-allergic and cellular		
PGD_2	Inhibition of proliferation (by its metabolite PGJ_2=9-deoxy-Δ^9-PGD_2 and by Δ^{12}-PGD2), release of PGD_2 from mastocytes=anaphylactoid mediator	[34, 35]
PGE_1	Stimulation of T lymphocytes, inhibition of B lymphocytes, inhibition of phagocytosis by macrophages, chemotactic to neutrophils, inhibits superoxyde anion production, ionophore action Ca^{++} Mg^{++} monovalent cations	[37]
PGE_2	Pro-inflammatory: increases vascular and capillary permeability, oedema, hyperemia, algesia, pyretic reaction. Anti-inflammatory: Inhibition of liberation of lysozyme enzymes ,inhibition of chemotaxis, inhibition of IL-1 and IL-2,inhibition of lymphocyte B proliferation, "down"-regulation of specific inflammatory cell function	[41]
$9\alpha,11\beta\text{-PGF}_2$	$9\alpha,11\beta\text{-PGF}_2$ Release from mastocytes, similar effect as PGD_2	[35]

7. Prostaglandins have a vital role in the foetal circulation allowing the development and patency of ductus arteriosus (PGE_2).

An important contribution to the physiologic role of prostaglandins was the discovery of inducible cyclooxygenase-2 contrasting with the constitutive cyclooxygenase-1 [10, 11]. COX-1 is implicated in the regulation of principal physiological functions and its inhibition may produce , e.g. peptic ulcer, renal failure etc. COX-2 seem implicated in inflammation or mitogenesis [11]. Its selective inhibition has an anti-inflammatory effect without an effect on gastric circulation or renal function.The development of selective COX-2 inhibitors opens new therapeutic possibilities. The physiological role of prostaglandins depends on stimuli activating phospholipase A_2 and C and/or which may make free substratre for cyclooxygenase and initiate PG synthesis. From all these stimuli, bradykinin is one of the most potent, releasing especially PGE_2 and PGI_2.However, the type of PG released depends on the tissue and its enzymatic equipment.On the other hand, the biosynthesis of prostanoids may be inhibited by endogenous factors as glucocorticoids, eicosapentaenoic acid and other fatty acids (see [14]) or by anti-inflammatory or other drugs [43].

Pathology   The abnormal synthesis of PGs has a pathophysiological role in different diseases or clinical syndromes related to vascular biology. These syndromes may be due to the deficit or excess of particular prostaglandin(s).Again PGI_2 and TxA_2 pathophysiology is treated in respective separate chapters.

Clinical syndromes with hypoprostaglandinism: PGE_1: Defective synthesis of precursors of PG series 1 at the level of Δ-6-desaturase, transforming linoleic acid in dihomo-γ-linolenic acid, seems to be implicated in ageing, diabetes, cardio-vascular disorders, atopic eczema [44]. PGE_2: Significant deficit of renal PGE_2 was demonstrated in essential hypertension [45], Gordon syndrome, prae-eclampsia, and acute glomerulonephritis. *PGS all:* Iatrogenic deficit: Inhibition of PG biosynthesis by non steroidal anti-inflammatory drugs may induce different possible complications: Gastrointestinal: peptic ulcer, bleeding. Renal: water and sodium retention, renal failure. Cardiovascular: increase of arterial pressure, angina pectoris. Central nervous system: headache, vertigo, and other [43].

Clinical syndromes with hyperprostaglandinism: PGD_2: Systemic mastocytosis: proliferation of tissue mast cells with extremely high plasma and urinary levels of PGD_2 and hemodynamic alterations: vasodilatation, facial flushing, headache, hypotension, tachycardia, syncope. Asthma, allergic dermatitis [32]. PGE_2: Renal hypersecretion: Hyperprostaglandin E syndrome, congenital hyperprostaglanduric tubular syndrome with hypercalciuria, associated with a story of premature labor and polyhydramnios [46]. Hydramnios: foetal renal PG hypersecretion. Bartter syndrome: hypokaliemic and hypochloremic alcalosis, hypotension, vascular resist-

ance to vasopressive hormones, renal electrolyte wasting, due to inhibition of tubular NaCl reabsorption by excessive renal PGE_2 (PGF_{2a}) hypersecretion; syndrome reversible by indomethacin (cyclooxygenase) treatment [47]. Gitelman syndrome: variant of Bartter's syndrome with Mg^{++} wasting. Gastro-intestinal hypersecretion was found in ulcerative colitis, irritable bowel syndrome, congenital chloride-loosing diarrhoea of children, secretory diarrhoea associated with rectal villous adenoma. Several of these symptoms are healed or ameliorated by cyclooxygenase inhibitors. Tumors: Neoplastic proliferative diseases are associated often with PGE2 hypersecretion and calcium mobilisation leading to hypercalcemia. Neonatology: Ductus arteriosus persistent after the birth of the child; indomethacin treatment allows the closure [48]. PGF_{2a}: Hypersecretion was described in dysmenorhoe,spontaneous abortus, premature labour and delivery, associated with increased levels in amniotic fluid. Cyclooxygenase inhibitors delay premature labour, reduce symptoms of dysmenorhoe. *PGs all:* Inflammatory reactions are associated with a large release of PGs (E,F,D,I Tx) from macrophages-monocytes, neutrophils, platelets, mastocytes and altered tissue which produce all signs of inflammation as redness, swelling with heat, pain and functional alteration, in general in synergy with bradykinin, histamine and proteolytic enzymes. Inflammatory reactions are treated by non-steroidal anti-inflammatory drugs which inhibit cyclooxygenase [25]. The pathophysiological role of prostaglandins in human pathology is much larger, but limited space does not allow detailed description. For more information see review or books [5, 8, 14, 28, 34, 40, 41].

Clinical Relevance and Therapeutic Implications   The review of physiological and pathological role of prostaglandins in vascular biology shows the importance of these mediators through the whole life of humans, from foetal development until ageing. The disequilibrium in PG balance, induced by a deficit or excess of PGs, was recognized as being the origin of several clinical syndromes and allowed etiopathogenetic treatment. This recognition has been possible by measurement of PG concentration in biological fluids (plasma, serum, urine, sperm, amniotic fluid, cerebrospinal fluid) and is actually used for clinical diagnosis. The most specific and sensitive method is gas chromatography coupled with mass spectrometry. For clinical diagnosis radioimmuno- or immunoenzymo-assay are generally used, which are sensitive and specific enough if coupled with preliminary chromatography. The normal plasma concentrations are very low (5-16 pg/ml) as well as in urine (150-300 pg/min) [47]. Nevertheless, the measurement of PGs allowed to distinguish syndromes with decreased or increased synthesis of PGs and to treat them. In the former case to use synthetic PGs or their analogs or drugs stimulating defective synthesis, in the latter case to use non specific inhibitors of cyclooxygenase I and II, or specific inhibitors of synthases. Finally, there are sit-

Table 2. PGE and PGF analogs used in clinical praxis

Compound	Company	Therapeutic use	Marketed
PGE analogs			
Alprostadil	Upjohn	Ductus arteriosus. patency	Prostine VR
Alprostadil	Upjohn	Induction of erection	Caverject
Misoprostol	Searle	antiulcer	Cytotec
Enprostil	Syntex	antiulcer	Gardrin
Ornoprostol	ONO	antiulcer	Ronak
Rosaprostol	IBI	antiulcer	Rosal
Limaprost	ONO	antihypertensive	Opalmon
Dinoprostone	Upjohn	fertility control	Prostin E_2
Sulprostone	Pfizer Schering	fertility control	Nalador
Gemeprost	ONO	fertility control	Cervagem
PGF analogs			
Dinoprost	Upjohn	fertility control	Prostin $F_{2\alpha}$
Cloprostenol	ICI	veterinary use, luteolytic	Estrumate
Prostalene	Syntex	veterinary use, luteolytic	Synchrocept
Latanoprost	Pharmacia	antiglaucoma	marketed

uations in which prostaglandins are used for their pharmacological properties, vasodilator, anti-aggregant, cytoprotective, anti-atheromatous as e.g. in obliterative arterial diseases or coronary heart disease, where PGE_1 and PGI_2 analogs are widely used. Table 2 presents a short summary of therapeutic use of PGE and PGF analogs. Only marketed drugs are listed from a great number of developed drugs [27, 49].

PGE_1 infusions (alprostadil) are used in new born children with congenital heart diseases to maintain the patency of ductus arteriosus before the surgical intervention. Prolonged and repeated intra-arterial infusions are successfully used for the treatment of peripheral arterial occlusive diseases. Alprostadil is also successfully used in intracavernous injections to help erection in organic impotency. PGE_1 has modest hypotensive action but some analogs as limaprost (Opalmon, ONO) have antihypertensive activity and is used in essential hypertension, acute myocardial infarction, deep vein thrombosis and Raynaud's syndrome. Misoprostol, 15-deoxy-16-methyl-16-hydroxy analogue of PGE_1 (Cytotec, Searle) is a stable analogue allowing peroral treatment and is most widely used in anti-ulcer therapy. It is contraindicated in pregnancy because it induces uterine contractions and may act an as abortive drug. It is used after the application of mifepristone (RU 486) in this indication. Misoprostol is also indicated for the prevention of adverse effects of AINSD therapy , i.e. to prevent gastric ulcer or the decrease of renal function. In spite of vasodilator action of natural PGE_2 , its analogs are mainly used in anti-ulcer therapy or as oxytocic drug in obstetrics for the induction of labour and termination of pregnancy (Table 2). PGF_{2a} analogs are equally used for the induction of labour as oxytocic drugs. In veterinary practice they are used for luteolytic action and the estrus synchronisation. A large part of therapeutic implications of prostaglandins represents the inhibitors of PG biosynthesis because PGs are mediators of inflammation in multiple tissues and organs. The inhibition of PG synthesis may be reached at different levels by inhibiting phospholipases, cyclooxygenases and PG synthases. Since phospholipases are treated in another chapter, only cyclooxygenase inhibitors are reviewed.

Inhibitors of cyclooxygenase: The discovery that aspirin and anti-inflammatory non steroidal drugs (AINSD) inhibit biosynthesis of prostaglandins [50] has opened a large field for the development of new generation of AINSD with high inhibitory activity. Aspirin and all AINSD are largely used in all medical practice. Aspirine (acetylsalicylic acid) inactivates enzymes by irreversible acetylation of Ser 530 in COX-1 and only partially inhibits COX-2 by the acetylation of Ser 516. The persistent activity of COX-2 transforms arachidonic acid in 15-HETE) [23], a substance inhibiting the synthesis of proinflammatory leukotrienes B_4,C_4, and D_4 and therefore aspirin has a supplementary anti-inflammatory action. Aspirine is widely used for its analgesic, antipyretic and anti-iflammatory action, but recently in cardiovascular prevention as antiaggregant and antithrombotic agent because it inhibits irreversibly the platelet cyclooxygenase for the whole life of platelets and therefore inhibits the generation of TxA_2. AINSD were developed initially as derivatives of salicylic acid, but several new drugs were introduced, most of them acids. They could be divided into different classes according to their structure as shown in Table 3 [45]. Only essential AINSD are included.

Table 3. Classificatiuon of anti-inflammatory non steroidal drugs

1) Arylcarboxylic acids:	Salicylic acids: aspirin, diflunisal; Anthranilic acids (fenemates): flufenamic acid, mefenamic acid, meclofenamic acid
2) Aryl alkanoic acids:	Aryl acetic acids: diclofenac, fenclofenac, alclofenac; Aryl propionic acids (profens): ibuprofen, ketoprofen, naproxen, tiaprofenic acid; Heteroaryl acetic acids: tolmetin; Indole and indene acetic acids: indomethacin, sulindac
3) Enolic acids:	Pyrazolidinediones: phenylbutazone, oxyphenbutazone; Oxicams: piroxicam
4) Non-acidic agents:	Proquazone, fluproquazone, diftalone, nabumetone

Most of them are competitive inhibitors of cyclooxygenase receptor site for the substrate. Inhibitory potency differs according to the class of the drug and according the sensibility of the tissue [14]. The kidney is very sensitive to AINSD and in different conditions as dehydration, hypovolemia, stimulated renin-angiotensin system, may be nephrotoxic. Another vulnerable tissue is gastro-intestinal system. In both systems the predominant cyclooxygenase is constitutive and its inhibition by non specific AINSD is often associated with side effects. AINSD differ in inhibitory potency of COX-1 and COX-2. The comparative inhibitory studies *in vitro* of both iso-enzymes have shown equipotent inhibition for several drugs but nabumetone (Relafen®) inhibited preferentially COX-2, and COX-1 only by 14 % [23]. These new drugs, e.g. nimesulide (Nexen®) or meloxicam (Mobic®) open the possibility to treat inflammatory diseases with more selective AINSD sparing the gastrointestinal and renal system. These perspectives are important especially for rheumatology and geriatric care. In general, the introduction of AINSD in clinical practice improved considerably the treatment of different diseases , sometimes with spectacular beneficial results. Actually the therapeutic implications of eicosanoids seem important and will increase even more in the future for the benefit of ill people.

Antonin Hornych

References

1. Euler US von (1936) J Physiol (Lond) 88:213-234
2. Bergström S, Sjövall J (1957) Acta Chem Scand 11:1086-1087
3. Bergström S et al (1963) J Biol Chem 238:3555-3564
4. Bergström S et al (1978) Pharmacol Rev 29:1-48
5. Murphy RC (1989) In: Watkins WD et al (eds) Prostaglandins in Clinical Practice. Raven Press, New York, pp 1-20
6. Smith WL et al (1991) Pharmac Ther 49:153-179
7. Coleman RA et al (1994) Pharmacol Rev 46:205-229
8. Hyslop S, DeNucci G (1993) Prostaglandins Leukotr Essent Fatty Acids 49:723-760
9. Eicosanoid Nomenclature Committee (1989) Prostaglandins Leukotr Essent Fatty Acids 37:1-6
10. Fu JY et al (1990) J Biol Chem 265:16737-16740
11. O'Neil GP, Ford-Hutchinson AV (1993) FEBS Letters 330:156-160
12. Yokoyama C, Tanabe T (1989) Biochem Biophys Res Commun 165:888-894
13. Negishi M et al (1993) Prog Lipid Res 32:417-434
14. Flower RJ (1974) Pharmac Rev 26:33-67
15. Vane JR, Botting R (1987) FASEB 1:89-96
16. Halushka PV et al (1989) Annu Rev Pharm Tox,10:213-239
17. Narumiya S et al (1993) J Lipid Mediators 6:155-161
18. Smith WL (1992) Am J Physiol 263:F181-F191
19. Miyata A et al (1994) Biochem Biophys Res Commun 200:1728-1734
20. Kujubu DA et al (1991) J Biol Chem 266:12866-12872
21. Holtzman MJ et al (1991) Clin Res 39:182
22. Hara S et al (1994) J Biol Chem 269:19897-18903
23. Meade EA et al (1993) J Biol Chem 268:6610-6614
24. Roberts LJ, Morrow JD (1995) Advances in Prostaglandin Thromboxane and Leukotriene Res 23:219-224
25. DeTitta GT et al (1980) Advances in Prostaglandin and Thromboxane Res 6:381-384
26. Murakami A, Akahori Y (1977) Chem Pharm Bull 25:3155-3162
27. Collins PE, Djuric SW (1993) Chem Rev 93:1533-1564
28. Kosaka T et al (1994) Eur J Biochem 221:889-897
29. Wang LH, Chen L (1996) Biochem Biophys Res Commun 226:631-637
30. Miyata A et al (1994) Eur J Biochem 224:273-279
31. Duncan AMV et al (1995) Genomics 25:740-742
32. Giles H, Leff P (1988) Prostaglandins 35:277-300
33. Ito S et al (1989) Prostaglandins Leukotr Essent Fatty Acids 37:219-234
34. Nakano J (1972) In: Cuthbert MF (ed) The Prostaglandins. W Heinemann Medical Books Ltd, London, pp 23-124
35. Kirtland SJ (1988) Prostaglandins Leukotr Essent Fatty Acids 32:165-174
36. Eklund B, Carlson LA (1980) Prostaglandins 20:333-347
37. Hedqvist P (1977) Ann Rev Pharmacol Toxicol 17:259-279
38. Levenson DJ et al (1982) Am J Med 72:354-374
39. Willey SC, Chernow B (1989) In: Watkins WD et al (eds) Prostaglandins in Clinical Practice. Raven Press, New York, pp 227-246
40. Hornych A (1980) In: Ramwell P (ed) Prostaglandin synthetase inhibitors: New clinical applications. AR Liss, New York, pp 231-256
41. Lipton HL et al (1988) In: Curtis-Prior PR (ed) Prostaglandins. Churchill Livingstone, London, pp 217-237
42. Hornych A (1991) Cor Vasa 33:492-505
43. Peplow P (1988) Prostaglandins Leukotr Essent Fatty Acids 33:239-252
44. Horrobin DF (1990) Rev Contemp Pharmacother 1:1-45
45. Hornych A et al (1989) Arch Mal Coeur 82 (IV):85-90
46. Seyberth HW et al (1985) J Pediatr 107:694-701
47. Hornych A et al (1987) Nephron 46:137-143
48. Coecani F et al (1975) Prostaglandins 9:299-308
49. Caton MPL, Crowshaw K (1988) In: PB Curtis-Prior (ed) Prostaglandins. Churchill Livingstone, London, pp 671-683
50. Vane JR (1971) Nature New Biology 231:232:235

Prostanoids

See: →Prostaglandins

Proteases

Definition   *Enzymes that cleave proteins such as extra-cellular matrix molecules or coagulation factors*

See: →Matrix metalloproteinases; →Coagulation factors; →Thrombin; →Fibrinolytic, hemostatic and matrix metalloproteinases, role of

Protein Kinase C (PKC)

Definition   *Monomeric serine/threonine kinases that have a large number of protein substrates. It is activated by diacylglycerol (DAG). A number of isoforms have been described (PKC α, β, γ, δ, ε, ζ).*

See: →Platelet stimulus-response coupling; →Extracellular matrix; →Vasomotor tone regulation, molecular mechanisms of

Proteoglycans

Definition   *Extracellular matrix proteins that are glycosylated*

See: →FGF-1 and -2; →FGF receptors; →Extracellular matrix

Proteoheparan Sulfate

Definition   *Proteoglycan with heparan sulfate side chains*

See: →FGF-1 and -2; →FGF receptors; →Extracellular matrix

Prothrombin

See: →Thrombin

PT

Definition   *Prothrombin time*

See: →Coagulation factors; →Thrombin

PTB

Definition   *Phosphotyrosine binding*

See: →Signal transduction mechanisms in vascular biology

PTH

Definition   *Paratyroid hormone*

See: →Thrombospondins

PTK

Definition   *Protein tyrosine phosphatase*

See: →Megakaryocytes

PTN

Definition   *Pleiotropin*

See: →Pleiotropin; →Heparin-affin regulatory protein

RAG-1

Definition   *Recombinase-activating gene-1*

See: →Cytokines in vascular biology and disease

RAGE

Definition   *Receptors for advanced glycated end products (AGE). Two RAGE have been identified.*

See: →Blood cells, interaction with vascular cells

RAGE

Definition   *Receptor for advanced glycosylated end products*

See: →RAGE; →Blood cells, interaction with vascular cells

RBC

Definition   *Red blood cells*

See: →Erythrocytes/red blood cells; →Blood cells, interaction with vascular cells

Receptor Tyrosine Kinases

Definition   *Receptors with intrinsic tyrosine kinase activity*

See: →FGF-1 and -2; →FGF receptors; →Signal transduction mechanisms in vascular biology; →Platelet stimulus-response coupling; →Cytokines in vascular biology and disease; →Extracellular matrix

Renin-Angiotensin System

Definition   *System composed of active renin secreted by the juxtaglomerular apparatus in the kidney and angiotensinogen mainly secreted in the liver. Angiotensin I is converted into angiotensin II by the converting enzyme (ACE). Angiotensin II binds an AT1 receptor on smooth muscle cells and induces phospholipase activity and calcium mobilisation.*

See: →Vasomotor tone regulation, molecular mechanisms of

RI-HB

Definition   *Retinoic acid-induced heparin-binding*

See: →Heparin-affin regulatory protein

Ribonuclease Inhibitors

Definition   *Inhibitors of ribonuclease isolated from different organs. Placental ribonuclease inhibitor inhibits angiogenin activity.*

See: →Angiogenin

RNA

Definition   *Ribonucleic acid*

RNase 5

Definition   *Pancreatic-type ribonuclease 5*

See: →Angiogenin

RON

Definition   *Receptor of nantes*

See: →Hepatocyte growth factor

RTK

Definition   *Receptor tyrosine kinase*

See: →Receptor tyrosine kinases

SAPK

Definition   *Stress-activated protein kinase*

See: →Matrix metalloproteinases; →Signal transduction mechanisms in vascular biology

Scatter Factor

Definition   *Same as hepatocyte growth factor (HGF)*

See: →Hepatocyte growth factor (HGF)

SCP-2

Definition   *Sterol carrier protein-2*

See: →Lipoproteins

Selectins

Definition   *Molecules involved in cell to cell adhesion that interact with sugar residues. E-selectin (ELAM-1) is expressed in endothelial cells. P-selectin (CD62P) is expressed in platelets and endothelial cells. L-selectin (CD62L) is expressed in leukocytes.*

See: →Blood cells, interaction with vascular cells

SERP-1

Definition   *Serpin-1*

See: →Fibrinolytic, hemostatic and matrix metalloproteinases, role of

Serum-Spreading Factor

See: →Vitronectin/vitronectin receptors

SF

Definition   *Scatter factor*

See: →Scatter factor; →Hepatocyte growth factor

SH-2

Definition   *Src-homology-2*

See: →Src-homology domain-2 (SH-2) and -3 (SH-3)

SH-3

Definition   *Src-homology-3*

See: →Src-homology domain-2 (SH-2) and -3 (SH-3)

Shc

Definition   *Src homology collagen-like*

See: →Signal transduction mechanisms in vascular biology

SHIP

Definition   *SH2 domain-containing inositol 5' phosphatase*

See: →Megakaryocytes

SHR

Definition   *Spontaneously hypertensive rats*

See: →Smooth muscle cells; →Vasomotor tone regulation, molecular mechanisms of

Sickle Cell Anemia

Definition   *Hemoglobinopathy caused by hemoglobin-S (HBS). HBS polymerizes causing the deformation of red blood cells (sickle cell shape). Frequent in people from African descent.*

See: →Blood cells, interaction with vascular cells

Signal Transduction Mechanisms in Vascular Biology

Synonym: Mitogenic signalling in vascular endothelial cells

Definition   *Signalling network triggered by cell surface receptors. The biochemical reactions involve association between molecules and phosphorylation or dephosphorylation, reactions catalyzed by protein kinases or phosphatases.*

See also: →Platelet stimulus-response coupling; →FGF-1 and -2; →FGF receptors; →Cytokines in vascular biology and disease; →Smooth muscle cells; →Extracellular matrix

Introduction   In a normal adult vessel, only 1 of every 10,000 endothelial cells is in division at any given time. Thus, the turnover time for the endothelium is measured in years [1]. Vasculature is therefore one of the most quiescent tissues of adult mammals. However, in response to the appropriate stimulus, the quiescent endothelium is competent to produce new vessels. This happens in normal situations: during embryonic development, wound healing, and in the highly orderly processes in the female reproductive cycles (ovulation, menstruation, implantation, pregnancy). Under these conditions, angiogenesis (the formation of new blood vessels from pre-existing vasculature) is highly regulated; being turned on for brief periods and then completely inhibited. However, many diseases are characterized by persistently activated vessel growth, occurring during tumor expansion, diabetic retinopathy, or arthritis [2]. In the last few years, several studies have led to the discovery of inducers and inhibitors of the angiogenic process. aFGF (acidic fibroblast growth factor), FGF-2 (or bFGF, basic fibroblast growth factor), VEGF (vascular endothelial growth factor), are among the most potent inducers; thrombospondin-1, angiostatin, endostatin, etc, as potent inhibitors. Folkman's group has suggested that the balance between inhibitors and inducers governs the angiogenic switch [1]. However the molecular mechanisms regulated by these factors to control the reversible growth arrest state of endothelial cells are not well understood. This review will focus on the factors that stimulate the proliferation and the differentiation of endothelial cells, and the signalling pathways implicated in these processes. In particular, we will concentrate our attention on the importance of the Mitogen Activated Protein (MAP) kinase cascade in the induction of endothelial cell proliferation.

Characteristics   Cell proliferation and differentiation are often induced by extracellular signals, the final response depending on both the cell lineage and the mitogens or differentiating agents. Cells respond to these extracellular signals by changes in gene expression, cytoskeletal architecture, protein trafficking, adhesion, migration and metabolism, amongst others. The response of the cell is initiated by the interaction of the extracellular signal with a cellular receptor. Plasma membrane receptors can be classified into three major groups as a function of the systems they utilize to transduce the signal inside the cell: receptors with associated kinase activity in their cytoplasmic domain, G-protein-coupled receptors or cytokine receptors. Each class of receptors can stimulate cellular proliferation or differentiation, depending on both the cell type and the receptor expression level, and each class of receptor shares the capacity to stimulate the Ras-MAP kinase cascade, suggesting a central role for this pathway in the regulation of these processes. In addition, in the last few years other signalling cascades have been described, which control the activity of Akt (PKB) and p70S6 kinase. These two protein kinases, downstream of the PI3 kinase cascade, have been demonstrated to play a key role in controlling proliferation, survival and differentiation.

Regulation

Molecular Interactions

Map Kinase Cascades   MAP kinase cascades are constituted by a series of protein kinases activated sequentially by phosphorylation and playing a major role in transmitting extracellular signals to the cellular machinery [3, 4]. They are remarkably well conserved throughout evolution, indicating their central role in the signalling network of eukaryotic cells. In the last few years, different MAP kinase cascades have been described: the ERKs (extracellular signal-regulated kinases) also known as p42/p44 Mitogen Activated Protein Kinase cascade, was the first cascade to be discovered, JNKs (c-jun N-terminal kinases) and p38 kinases form a sub-group of SAPKs (Stress Activated Protein Kinase). All the cascades share a common module of three sequential protein kinases MAPKKK > MAPKK > MAPK as schematized in Figure 1. The activation of the MAPK cascade starts at the level of membrane receptors and involves activation of members of the low molecular weight GTP-binding proteins (Ras, etc). Activated p21Ras will then initiate the translocation of the first member of the MAP kinase cascade, a MAP kinase kinase kinase or MAPKKK (Raf-1) to the plasma membrane where it becomes phosphorylated and in turn activates the MAP kinase kinase or MAPKK (MEK-1/2, MKK3 to 7). These are dual specificity kinases, capable of phosphorylating the MAP kinases on threonine and tyrosine residues and leading to their activation. MAP kinases are then able to phosphorylate cytoplasmic enzymes, cytoskeletal proteins or after rapid translocation to the nucleus, transcription factors. A set of inducible MAP kinase phosphatases (see later) is then able to terminate the signal.

The first and perhaps most widely studied MAP kinase module is that mediating the activation of ERKs (Erk1 and Erk2 also referred to as p42 and p44 MAPK) in response to growth factors [5, 6]. This cascade plays a central role in the regulation of cell growth, differentiation and protection against apoptosis. Following receptor activation at the plasma membrane, by either receptor tyrosine kinase, G protein-coupled receptor or integrin engagement, activation of p42/p44 MAPK is initiated by the binding to the receptor of an adaptor protein that contains an SH2 domain (as Grb2 or Shc) (Figure 1). This complex binds to a guanine nucleotide exchange factor (Sos), protein that catalyses nucleotide exchange for the binding of GTP to Ras. GTP-loaded Ras recruits members of the Raf family (B-Raf, c-Raf, A-Raf) to the plasma membrane where they will be activated by a mechanism still not fully resolved that subsequently will initiate the cascade of phosphorylations.

The phosphorylation on tyrosine 183 and threonine 185 residues of the members of the MAP kinase family (p42 MAPK sequence) leads to their activation and phosphorylation of their substrates. In quiescent cells, p42/p44 MAPK are cytoplasmic. Their sustained activation in

fibroblasts is accompanied by their translocation to the nucleus [7], whereby phosphorylation of different transcription factors, such as members of the Ets family, Elk-1 or Sap1α, ATF-2, etc, they exert a critical role in gene expression [3]. In addition to nuclear factors, p42/p44 MAPKs have cytoplasmic substrates, for example, cytoskeletal proteins (MAP, caldesmon, etc), phospholipase A2, tyrosine hydroxylase, and upstream members of the cascade, including the EGF receptor, Sos, Raf-1 and MEK. Interestingly, the exclusive activation of the p42/p44 MAPK cascade is sufficient to promote the expression of MKP-1 and MKP-2, two MAP kinase phosphatases involved in the down-regulation of MAP kinase activity when cells progress through the G1 phase of the cell cycle [8].

A strong correlation has been established between the ability of extracellular agents to stimulate components of the p42/p44 MAPK cascade and the ability to induce a proliferative response. Upstream elements of this cascade (Ras and Raf) have been shown to be crucial for cell proliferation since expression of permanently active forms of these proteins switches on the cascade [9] and leads to cellular transformation. On the other hand, expression of antisense or dominant-negative mutants of Ras and Raf suppress the growth factor response. More recently, similar results could be extended to the downstream members of the cascade: MAPKK (MEK1) and p42/p44MAPKs. Indeed, expression of a constitutively active mutant of MEK1 (S218D/S222D) raised the basal MAPK activity and induced oncogenicity when expressed in fibroblasts [10, 11]. Alternatively, interruption of the p42/p44 MAPK cascade by expression of p44MAPK antisense, a dominant-negative p44MAPK mutant (T192A), or overexpression of MKP-1 prevent quiescent fibroblast entering S phase in response to growth factors [12, 13]. These results lead to the notion that the entire cascade has, if any, very minor bifurcation, and that p42/p44 MAPK are major key players in controlling cell growth. A second notion that has emerged from these studies is that a sustained activation of p42/p44 MAPK is required for fibroblasts to pass the G1 restriction point and enter S-phase [12, 13]. This sustained activation of p42/p44 MAP kinases is always accompanied by the translocation of both isoforms to the nucleus [7].

At what level of the cell cycle machinery is MAP kinase activation important for the control of cell growth? Progression through the G1 phase of the cell cycle is controlled by complexes composed of cyclins and cyclin-dependent kinases (Cdks) [14]. Activity of these complexes can be inhibited by the presence of cyclin kinase inhibitors (CKIs), which can be grouped into two families, p16, p15, p18 and related proteins of 15-20 kDa in one, and p21Cip1, p27KIP1 and p57KIP2 in another. The decision to proceed beyond late G1 appears to be executed by hyperphosphorylating and thereby inactivating the retinoblastoma protein (pRb), thus liberating key transcription factors required for turning on S-phase specific genes [15, 16]. The critical phosphorylation of pRb in mid to late G1 is

MITOGENIC PATHWAYS IN MAMMALIAN CELLS

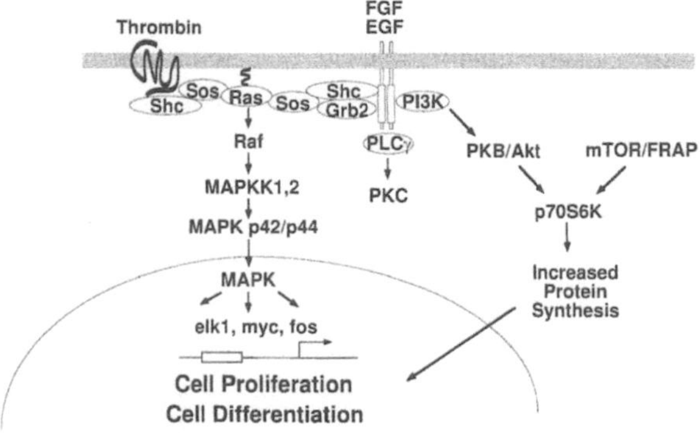

Figure 1. Mitogenic pathways in mammalian cells. The figure is explained in the text.

most likely carried out by the D-type cyclins (cyclin D_1, D_2 and D_3) and their partner cyclin-dependent kinases cdk4 and cdk6. Cyclin D_1 expression is positively controlled by p42/p44 MAPK in fibroblast cells [17], and it has been shown that different mitogenic stimuli from distinct signalling cascades (activated by growth factors with tyrosine kinase receptors, by estrogen receptors, and by cAMP-elevating agents via G-protein-coupled receptors) all converge and strictly require cyclin D-cdk activity to promote S-phase onset [18].

An additional response that appears to be regulated through the Ras-MEK-MAPK signalling pathway is cellular differentiation. Different members of the MAPK cascade have been implicated in the differentiation of monocytes, neurite outgrowth of PC12 cells, T cell maturation and mast cell development. One of the differentiation models most widely explored is the neurite extension in PC12 cells. In these cells both EGF and NGF activate the MAPK cascade, yet EGF treatment enhances proliferation while NGF treatment induces differentiation and cessation of cell division [19]. EGF-stimulated MAPK activity is transient, whereas NGF-stimulated MAPK activity is sustained for several hours and promotes MAPK nuclear translocation. If the number of EGF receptors is increased by stable transfection of PC12 cells, then EGF becomes capable of promoting a sustained MAP kinase activation and differentiation. As proliferation and differentiation correspond to alternative cellular fates, the requirement of sustained MAPK activation for both cellular states strongly indicates that the p42/p44 MAPK signalling cascade does not determine the cell fate, but "throws the switch" which is predetermined. The set of transcription factors expressed by a given cell, in an appropriate cellular environment is crucial in specifying the cellular engagement. In both situations however, MAPK activation is required to maintain the cellular fate: proliferation or differentiation, presumably *via* the phosphorylation of nuclear factors.

PI3K Phosphatidylinositol 3' kinase (PI3K) phosphorylates position D3 of the inositol ring of either phosphatidylinositol (PI), PI 4-phosphate (PI 4-P) or PI 4,5-bisphosphate (PI 4,5 P2) to produce PI 3-P, PI 3,4-P2, or PI 3,4,5-P_3. These lipids act as second messengers, and their levels are acutely modulated by hormones and growth factors. These products of PI3K can bind to multiple downstream effectors and modulate their activity. Such effectors include serine/threonine protein kinases such as specific isoforms of PKC, and PKB (also known as Akt). All these protein kinases bind the PI3K products by their SH2 (Src-homology-2) or the PH (Pleckstrin homology) domains. Other proteins regulated by these lipids are various cytoskeletal proteins (actin-binding proteins) and vesicle-associated proteins (synaptotagmin, centaurin-α) [20].

Multiple forms of PI3K exist in mammalian cells with distinct mechanisms of regulation and different substrate specificities. They have been divided into three families depending on their structure and mechanism of regulation [21]. The first class of PI3K are all heterodimers, with a regulatory subunit (of 55, 85 or 101 kDa) which serves as an adaptor for the catalytic p110 subunit. The second class is formed by larger catalytic subunits (170-220 kDa) which are not known to have regulatory subunits. The last class is formed by the catalytic subunit Vps34pa and its binding partner Vps15/p. This isoform only phosphorylates PI producing PI 3-P, and has been implicated in vesicular trafficking. Class I PI3K activation occurs by the binding of the adaptor subunit directly to tyrosine phosphorylated receptors (by the SH_2 domain of the p85 adaptors, for example). The binding of the phosphopeptide to the adaptor subunit results in the direct activation of the lipid kinase activity, and moreover, brings the catalytic subunit close to its lipid substrate. Heterotrimeric G-protein coupled receptors can activate specific members of the PI3K family, by the binding direct of the $\beta\gamma$ subunits of the G-protein to the catalytic subunit of PI3Kγ.

PI3K activation has been implicated in distinct cellular functions: mitogenic signalling, cell differentiation, inhibition of apoptosis, intracellular vesicle trafficking/secretion, regulation of actin and integrin functions, etc [22]. The role of PI3K in the regulation of cell growth is cell type-dependent and stimulus-dependent. This conclusion is based on the fact that antibodies to PI3K block DNA synthesis in response to PDGF and EGF, but not in response to colony-stimulating factor (CSF)-1, bombesin or lysophosphatidic acid [22]. The relation between the MAP kinase cascade and PI3K activation is not clear [20]. Several studies have shown that inhibitors of PI3K can block activation of Raf and MAPK, and constitutively active mutants of p110 can activate MAPK, placing PI3K upstream of the MAPK cascade. In addition, Ras binds directly to the p110 subunit of PI3Kα and β, but it is not clear if Ras is downstream or upstream to PI3K. Thus, dominant-negative Ras blunted the increase in levels of PI 3,4-P2 and PI 3,4,5-P3 in response to EGF or NGF in PC12 cells, suggesting that PI3K acts downstream of Ras. But other reports have described that the expression of a constitutively active PI3K results in Fos induction, which is blocked by dominant-negative forms of Ras and Raf, placing PI3K upstream of Ras. All these observations are complicated by the fact that PI3K activates directly a number of PKC isoforms, and PKCs can activate Raf by phosphorylation [20]. Given the complexity of these pathways it may be that the products of PI3K affect multiple steps in this cascade. Moreover, PI3K activates PKB, a kinase that is upstream of the p70S6 kinase and as a consequence, the regulation of protein synthesis, as we will see later. Lastly, with regard to the role of PI3K in cell differentiation, it has been observed that PI3K might be involved in neurite outgrowth in PC12 cells [23], and in differentiation of skeletal muscle cells [24].

S6 KINASE p70S6 kinase, first discovered by its capacity to phosphorylate the ribosomal protein S6 in response to mitogens, plays an important role in progression through the G1 phase of the cell cycle [25]. This enzyme is rapidly activated by all mitogenic stimuli, including growth factors, cytokines, and phorbol esters. Agents that block p70S6K activation (the immunosuppressant agent rapamycin or neutralizing antibodies) prevent proliferation, most notably in lymphocytes. Mitogenic stimulation induces serine/threonine phosphorylation on multiple sites within p70S6K, resulting in its activation, but the kinases implicated in this activation are not at present identified. MAP kinase phosphorylates p70S6K *in vitro*, but experiments *in vivo* have demonstrated that MAPK is neither necessary nor sufficient to elicit p70S6K activation. Indeed, the p70S6K activation is a complex process requiring multiple signalling inputs [26]. PI3K is a crucial mediator of p70S6K activation in response to growth factors. The mediators of the activation *via* PI3K could be PKC isoforms, PKB, or p65PAK (p21 activated kinase), kinase that binds to the low-molecular weight GTP-binding proteins Cdc42 and Rac-1 in a GTP-dependent manner. Another activator of p70S6K is the target of the immunsuppressant rapamycin, mTOR/FRAP, member of the PIK-related family of protein kinases, but the exact role of this protein in the activation of p70S6K has not yet been totally elucidated [26]. p70S6K regulates protein synthesis, presumably through ribosomal S6 protein phosphorylation. S6-phosphorylated ribosomes will recruit a class of mRNAs containing polypyrimidine-rich tracts in their 5' untranslated regions (UTRs). This class of mRNAs encodes many of the components of the protein synthesis apparatus and can represent up to 20 % of the total mRNA in the cell. Failure to recruit these mRNAs into polysomes reduces the biogenesis of translational machinery required for cell cycle progression [26]. p70S6K also participates in the phosphorylation of PHAS-1, a protein that associates with the eukaryotic initiation factor 4E (eIF4E). Phosphorylation of PHAS-1 induced by mitogens, releases PHAS-1 from this complex, allowing eIF4E to initiate the translation process [25, 27]. p70S6K can also regulate gene expression phosphorylating CREM, a member of the family of CAMP-responsive transcription factors.

Cells and Cellular Interactions   In common with many cells *in vivo*, endothelial cells are exposed to opposing signals that stimulate or repress the cells capacity to divide and to form new capillaries, the angiogenic phenotype. The final balance between these positive and negative regulators will determine whether the endothelial cell will divide or not. Different angiogenic factors have been described, including the growth factors FGF-2 and VEGF, which bind to plasma membrane receptors with intrinsic tyrosine kinase activity. In contrast, much less is known about inhibitors of endothelial cell proliferation, such as thrombospondins, angiostatin or endostatin. We will describe below the signal transduction pathways that have been characterized in response to the most potent angiogenic factors.

Growth Factors Activating Receptor Tyrosine Kinases
FGFs   FGF-2 or basic fibroblast growth factor is a potent chemotactic and mitogenic factor for endothelial cells. Human umbilical vein endothelial cells (HUVEC) are absolutely dependent upon the presence of FGF for *in vitro* proliferation [28], and its activity has been implicated during wound healing, differentiation and tumor angiogenesis. The cellular effects of FGF-2 are transduced by its interaction with a family of high-affinity tyrosine kinase receptors, the FGF receptors (FGFR). Four members of this family have been described (FGFR1-4) [29]. All of them are single transmembrane proteins, with an extracellular segment composed of three immunoglobulin-like domains, the transmembrane domain, and a cytoplasmic domain with tyrosine kinase activity. The activation of FGFR causes the autophosphorylation of the receptor on specific tyrosine residues, as well as the phosphorylation of downstream molecules. The requirement for an increase in tyrosine phosphorylation activity for the angiogenic properties of FGF-2 has been suggested by experiments using inhibitors of tyrosine phosphatases. The treatment of endothelial cells with the cell permeable tyrosine phosphatase inhibitor vanadate, elicited morphologic and biochemical changes similar to those induced by FGF-2, such as the formation of tubules or the increase in the levels of plasminogen activator expressed by the cells [29].

Proteins which show an increase in their levels of tyrosine phosphorylation in response to FGF are phospholipase Cγ1, Src, cortactin, and Shc, but only phospholipase Cγ1 is known to be a direct substrate of the FGFR. Phospholipase Cγ1 is phosphorylated in tyrosine during the entire G1 phase of the cell cycle, but it has been shown that its activity is not essential for FGF-induced mitogenesis in L6 transfected cells [29]. Src is a cytosolic tyrosine kinase that after activation can be redistributed from cellular membranes to focal adhesion sites. Src has been implicated in the control of cell adhesion and motility, and senescent endothelial cells that do not respond to FGF have impaired activation of Src in response to FGF-2 [30]. Cortactin is a substrate of Src that binds F-actin, and can be phosphorylated on serine/threonine and tyrosine residues. In fibroblasts, an association between cortactin and Src during the entire G1 phase of the cell cycle after stimulation with FGF-1 has been demonstrated [31], but the importance of this association to mitogenesis or the differentiation of endothelial cells is not known.

It has been shown that activation of the FGFR stimulates the MAP kinase cascade in endothelial cells [32, 33, 34]. This activation is due in part to the interaction of the FGFR with the adaptor protein Shc [34, 35]. Shc is a SH_2-phosphotyrosine-binding (PTB) domain adaptor that links receptor activation to Ras signalling by recruiting the Grb2-Sos complex to the plasma membrane in a tyrosine phosphorylation-dependent manner [36]. For mechanisms dependent or independent of Shc, the stimulation of the FGFR causes the activation of Ras, which in turn will stimulate the MAP kinase cascade in HUVEC cells. This cascade activates DNA synthesis, and inhibitors of MEK1 inhibit the FGF-2-stimulated incorporation of thymidine into DNA [37]. In addition, the microinjection of dominant-negative forms of Ras in bovine corneal endothelial cells blocks the ability of FGF-2 to induce both Fos expression, cell migration and DNA synthesis [38]. In BAEC cells, the microinjection of a neutralizing Ras-specific antibody completely blocks FGF-2-stimulated cellular movement, implicating this cascade in the control of cellular migration [39].

In vascular endothelial cells, as for fibroblasts, a sustained activation of the p42/p44 MAPK cascade correlates with mitogenic potency, and this is accompanied by the translocation of p42/p44 MAPK to the nucleus. Thus, in a lung mouse endothelial cell line, 1G11 cells [40], FGF-2 is only poorly mitogenic, stimulating thymidine incorporation to less than 10 % of the stimulation elicited by 20 % fetal calf serum (Figure 2, part A). This weak effect of FGF-2 on DNA synthesis is accompanied by a small and transient activation of p44 MAPK activity (Figure 2, part B). In contrast, fetal calf serum strongly stimulates MAPK activity, and this activation is maintained for at least 4 hours following serum addition. Serum stimulation of MAPK activity is accompanied by

the translocation of p42/p44 MAPK to the nucleus (Figure 2, part C right) however in the case of FGF-2 that stimulates only slightly MAPK, no MAPK nuclear translocation is detectable (Figure 2, part C left).

FGF-2 stimulates p70S6K activity in murine brain endothelial cells [41]. This activation is inhibited by the immunosuppressant rapamycin. Moreover, rapamycin blocks endothelial cell proliferation, but not differentiation in collagen gels. Hence, the activation of p70S6K in response to FGF-2 drives cells to proliferate, but not to tube formation [41].

A role for calcium in FGF-2 mediated angiogenesis has recently been proposed. Thus, Presta et al. have shown that calcium-dependent pathways are important for the increase in plasminogen activator activity FGF-2-mediated [42]. In addition, experiments employing inhibitors of either calcium influx or calcium-mediated signalling pathways, have underlined a role for calcium in the stimulation by FGF-2 of tyrosine phosphorylation of several intracellular proteins, and in the proliferation, migration, and adhesion of human endothelial cells in culture, as well as capillary outgrowth *in vivo* [43]. Moreover, it has been reported that the activation of protein kinase

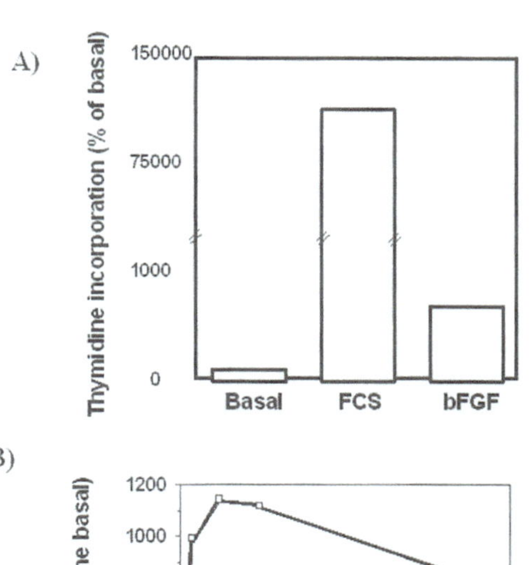

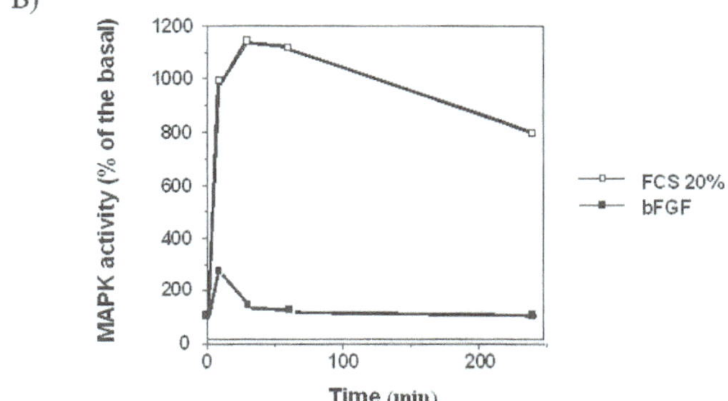

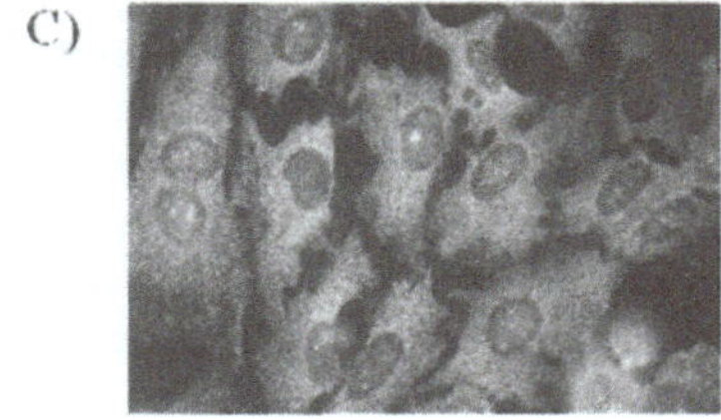

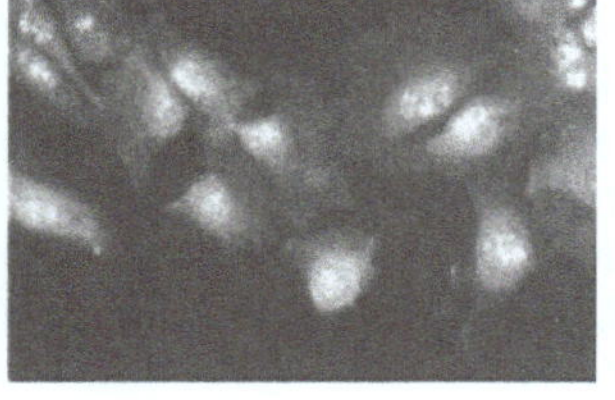

Figure 3 Stimulation of MAPK activity, MAPK translocation and DNA synthesis in endothelial cells: the effects of fetal calf serum and bFGF. The 1G11 lung mouse endothelial cell line was cultured in DMEM medium containing 20 % fetal bovine serum, 150 µg/ml endothelial growth supplement, 100 µg/ml heparin, 1 % non-essential amino acids and 2 mM sodium pyruvate. After 48 hours of serum depletion in a 1:1 mixture of DMEM and Ham's F12 medium, cells were stimulated with 20 % fetal calf serum or 25 ng/ml of bFGF for the times indicated. *A)* Cells were stimulated for 24 hours in presence of 0.25 µCi/ml (methyl-3H)thymidine. Cells were then fixed and washed three times with ice-cold trichloroacetic acid (5 %), and harvested with 0.1 N NaOH, and the radioactivity counted. *B)* Cells were lysed in Triton X-100 lysis buffer and MAPK activity was measured as described [8]. *C)* Cells were stimulated with 20 % serum (right) or 25 ng/ml bFGF (left) for 3 hours. Cells were fixed and processed as described [7].

C by FGF-2 is necessary for the stimulation by FGF-2 of endothelial cell movement and proliferation [44]. The overexpression of different forms of PKC can provoke the activation of the MAP kinase cascade, but this activation is not in itself sufficient for the induction of cell proliferation, indicating the requirement of additional signals [45]. Several investigators have shown a dual role for PKC in the proliferation of endothelial cells: thus, PKC activation potentiates growth-factor-induced DNA synthesis when activated during the early G1 phase, whereas PKC activity completely inhibits the initiation of DNA synthesis in the late G1 phase [46]. This apparent "dual-role" of PKC is closely associated with either potentiation or inhibition of pRb hyperphosphorylation and cdk4 and 2 activities.

VEGF   Vascular endothelial growth factor (VEGF) also known as vascular permeability factor, is a dimeric glycoprotein member of the PDGF family [47]. Four isoforms have been described: VEGF121, VEGF165, VEGF189 and VEGF206 which result from alternative splicing of a single gene. VEGF has several properties expected of a candidate regulator of angiogenesis: it is secreted by normal and tumor cell lines; it is a mitogen specific for endothelial cells; and it is angiogenic in *in vivo* test systems such as the chorioallantoic membrane assay (CAM) and the rabbit cornea [47, 48].

Three direct transmembrane receptors for the VEGF have been described (VEGFR), which all possess intrinsic tyrosine kinase activity, and are highly expressed in endothelial cells [48]: a) Flk-1 (for fetal liver kinase), and its human homolog KDR (kinase insert domain-containing receptor), with a K_d (apparent dissociation constant) for VEGF between 400 and 1000 pM; b) Flt-1 (fms-like tyrosine kinase) with a K_d of 16-114 pM; and c) Flt-4 which is the receptor for a novel VEGF-related endothelial growth factor, VEGF-C. All the receptors are structurally related, with seven immunoglobulin-like repeats in their extracellular domain. VEGFR are principally expressed in endothelial cells although certain hematopoietic cells also express VEGFR. Moreover, only endothelial cells have been reported to proliferate in response to VEGF. Recently, neuropilin-1 has been described as a new receptor for $VEGF_{165}$ [49]. Its role seems to be to serve as a coreceptor that enchances binding of VEGF to KDR and subsequent bioactivity. The role of different receptors for the VEGF has been more extensively developed in another chapter.

The stimulation of bovine aortic or bovine brain capillary endothelial cells by VEGF provokes the autophosphorylation of Flt-1 and Flk-1 receptors, and leads to the phosphorylation of several SH_2-containing proteins, including phospholipase Cγ, phosphatidylinositol 3-kinase, GAP (Ras GTPase-activating protein), and the adaptor protein Nck [33, 50]. VEGF-stimulated porcine aortic endothelial cells overexpressing the Flt-1 receptor have an increase in the level of tyrosine phosphorylation of the Src family members Yes and Fyn [51]. In HUVEC cells, VEGF stimulates the phosphorylation on tyrosine of p125 focal adhesion kinase (p125 FAK) and paxillin, a 68 kDa focal adhesion-associated protein [52]. This phosphorylation implicates the activity of PKC and the integrity of the actin filament network. Moreover, in GAP immunoprecipitates of VEGF-stimulated endothelial cells tyrosine-phosphorylated Nck, phospholipase Cγ, and two GAP-associated proteins, p190 and p62 are present, whereas tyrosine-phosphorylated Nck was detected in phosphatidylinositol 3-kinase immunoprecipitates [50]. This suggests that VEGF, as for other agonists which function through the activation of tyrosine kinase receptors, promotes the formation of multimeric aggregates of VEGF receptors with proteins that contain SH_2 domains and in this way activates various signalling pathways.

MAP kinases are activated in bovine brain endothelial cells [33], and in rat liver sinusoidal endothelial cells [53] by VEGF. The exact mechanism that leads to the stimulation of the MAP kinase cascade is unknown, although most of the adaptors and exchange factors signalling *via* Ras have been identified. VEGF stimulates only weakly the phosphorylation of Shc, and the stimulation of Grb2-Sos complex formation could be due to the formation of a complex with the p62 and p190-GAP associated proteins, as in the case of the FGF receptor [53]. The importance of the activation of the MAP kinase cascade by VEGF for the stimulation of cell division may be anticipated although this has not been formally demonstrated. It has been shown that Ras activity has a role in the stimulation of the production of metalloproteinases and VEGF in transformed endothelial cells, and PI3K participates in this pathway [54]. VEGF has been shown to modulate cell migration and can induce the disorganization of actin stress fibers. As we have described, the activation of Ras has been implicated in the control of motility of endothelial cells in culture [38, 39], and it is possible that the stimulation by VEGF of its receptors may modify the GTP-bound levels of Ras causing endothelial cells to change their migration and adherence [55]. In favor of this model, endothelial cells from mice lacking GAP fail to organize and to form a highly developed vascular network, implicating Ras-GAP in the processes of motility and remodelling of endothelial cells [56]. Finally, it is important to note that mice lacking the B-Raf gene, die *in utero* by embryonic day 12.5 [57]. The B-Raf-/- embryos display defects in vascular endothelial cell differentiation and survival, indicating that B-Raf plays a critical role in the mechanisms that regulate differentiation of angioblasts, the development of large vessels, and the survival of endothelial cells.

Adhesion Molecules   The extracellular matrix exerts complex effects on cell function in addition to promoting cell adhesion. Normal cells require anchorage to the extracellular matrix in order to proliferate. However, for some cell types this interaction promotes exit from the cell cycle and cellular differentiation [58]. These diverse effects are often mediated via the integrins and other adhesion molecules. Vascular cells proliferate in an anchorage-dependent manner, which suggests that adhesion-mediated signals may be important in the growth of new blood vessels, both for the proliferative and for the differentiation signals. Integrins are the cellular receptors that mediate cellular adhesion to the extracellular matrix. The integrin family is composed of 15 α and 8 β subunits that are able to form more than twenty different $\alpha\beta$ heterodimeric combinations. Integrins bind to extracellular matrix proteins and transduce signals across the membrane. Among the integrin-generated signals are increases in intracellular pH, intracellular calcium, inositol lipid synthesis, tyrosine phosphorylation of a tyrosine kinase associated with focal contacts (pp125FAK or focal adhesion kinase), activation of protein kinase C, p21Ras, p42/p44 MAP kinase, PI3K, and activation of p34/cdc2 and cyclin A [59]. Endothelial cells express several members of the integrin family which attach to a wide variety of adhesive proteins, such as collagen, vitronectin, or fibronectin.

Integrins activate the p42/p44 MAP kinase cascade in HUVEC endothelial cells [34, 60, 61]. This stimulation is dependent on actin filaments, protein kinase C activity and a herbimycin-sensitive tyrosine kinase activity, which may be a member of the Src family [60]. The work of Wary and colleagues has shown that integrins activate the MAP kinase cascade by the formation of a complex between the α subunit of the integrin and the adaptor protein Shc and thereby with Grb2. The authors showed that complex between the integrin and Shc is necessary and sufficient for the activation of the MAP kinase pathway in response to integrin ligation [34]. Adhesion to fibronectin potentiates FGF-2, VEGF and EGF activation of MAPK cascade [61], and blocking MAPK activation by integrins with a dominant negative Shc molecule causes a partial inhibition of S phase entry [34]. Thus, soluble mitogens and integrins cooperate in efficient propagation of signals and to promote the transit through the G1 phase of the cell cycle. Moreover, activation of the MAP kinase pathways by integrin engagement protects cells from apoptosis. In contrast, adhesion mediated by integrins not linked to Shc and that do not stimulate the MAPK cascade results in apoptosis and cell cycle arrest [34].

Cheresh's group demonstrated that two different integrins participate in a positive manner in the angiogenic pathways stimulated by FGF-2 or VEGF. Thus, the integrin $\alpha_V\beta_3$ participates in the *in vivo* angiogenesis induced by FGF-2 or TNF-α, whereas angiogenesis initiated by VEGF, TGF-α or phorbol esters depends on the integrin $\alpha_V\beta_5$ and also on protein kinase C activity [62]. Integrin $\alpha_V\beta_3$ is poorly expressed in resting blood vessels, but is significantly upregulated in vascular cells within human tumors and in response to FGF-2 addition [59]. MAPK activation by FGF-2 in the angiogenic blood vessels of the chick chrioallantoic membrane is more persistent than in cells in culture, and the late phase is dependent on $\alpha_V\beta_3$ ligation [63]. This sustained activation of MAPK is important for endothelial cell migration during the angiogenic process. The same group have shown that blocking the function of the $\alpha_V\beta_3$ integrin induces apoptosis of proliferating endothelial

cells and thereby blunts angiogenesis *in vivo* [64]. Inhibition of the activity of integrin $\alpha_V\beta_3$ increases the activity of p53 and the production of the cell cycle inhibitor p21WAF1/CIP1. In contrast, the activation of this integrin and not for example that of $\alpha_V\beta_5$, elicits the inhibition of p53 activity, blocks p21WAF1/CIP1 expression and increases the expression of the survival protein Bcl-2 [65]. All these findings indicate that $\alpha_V\beta_3$ and the sustained activation of MAPK provide a survival signal to proliferating vascular cells during the growth of new blood vessels.

In addition to integrins, cadherins have been shown to regulate the cell growth and differentiation of endothelial cells. Cadherins are single chain transmembrane glycoproteins that mediate cell-to-cell adherent contacts, and the physical attachment between the cell membrane and an intracellular network of cytoplasmic proteins and actin microfilaments [66]. The intracellular domain of cadherins interacts with catenins, which transmit the adhesion signal and contribute to the anchorage of the protein to the actin cytoskeleton. Endothelial cells have been found to express both, specific (VE-cadherin or cadherin 5) and non tissue-specific cadherins (N- and E-cadherins). It has been shown that overexpression of VE-cadherin has a negative effect on the proliferation of endothelial cells [67]. Moreover, decreased cadherin and catenin expression is correlated with increased tumor cell invasion and metastasis. The mechanisms implicated in this inhibition of cellular proliferation by cadherins is unknown. In other cellular models, catenins participate in several signalling pathways [68]. Beta catenin interacts with the tumor-suppressor APC (adenomatous polyposis coli). Mutations in this protein contribute to colorectal tumorigenesis, and it has been shown that overexpression of this protein blocks cell cycle progression [69]. Moreover, β-catenin binds to the Tcf-Lef (T cell-factor-lymphoid enhancer factor) family of transcription factors, and activates gene expression [70]. Thus, cadherins could influence gene expression by modifying the cellular levels of free catenins. In favor of this model it has been shown that a dynamic exchange of catenins exists at endothelial cell-to-cell junctions associated to cadherins as a function of the state of confluence and proliferation of the cells [66].

G-Protein Coupled Receptors   Heterotrimeric GTP-binding protein (G protein) coupled receptors participate in cellular signalling and regulate a variety of physiological processes. All such receptors have seven transmembrane regions, and can activate the MAP kinase cascade by mechanisms still not clear [71]. It appears that the G protein β/γ-subunits are able to stimulate MAP kinase activation in a Ras-dependent manner. Furthermore, the receptors and β/γ-subunits are able to induce tyrosine phosphorylation of Shc by an unidentified tyrosine kinase. In endothelial cells, classical agonists that stimulate G protein-coupled receptors (thrombin or endothelin-1) are poor mitogens. Endothelin-1 has been described to be a potent mitogen for several endothelial cell models, such as rat brain capillary endothelial cells, but not for others, such as bovine aortic endothelial cells [72]. Other authors have described a mitogenic effect of adenosine in HUVEC cells [37]. In these cells, adenosine stimulates Ras, p42/p44 MAP kinases and induces their translocation to the nucleus, an action that correlates with adenosine mitogenic effect.

A role for the G protein $G\alpha_{13}$ in angiogenesis has recently been suggested, following the disruption of the gene encoding this G protein [73]. The $G\alpha_{13}$ homozygous mutant embryos die *in utero* at day 10.5, due to an impaired ability of endothelial cells to form the organized and functional vascular network. The lack of this G protein did not affect the differentiation of progenitor cells into endothelial cells (vasculogenesis), which were apparent throughout the embryo, rather, the defects occurred during the subsequent process of angiogenesis. The authors showed that the migration of fibroblasts derived from mutant embryos in response to thrombin or lysophosphatidic acid was reduced, while the mitogenic and metabolic responses to these agonists were normal. This suggests that a failure of cell movement and orientation in response to specific extracellular stimuli is the mechanism responsible for the observed vascular system defects in $G\alpha_{13}$ homozygous mutant embryos [73]. A G protein-coupled receptor, EDG-1, was cloned as an immediate-early gene induced during differentiation of human endothelial cells into capillary-like tubules *in vitro* [74]. Recently, sphingosine-1-phosphate has been identified as the ligand for this receptor, and its specific binding induces morphogenetic differentiation via a Rho-dependent signalling pathway [75]. The link between $G\alpha_{13}$ and EDG-1 is still unknown.

Angiogenesis Inhibitors   The angiogenic phenotype appears to be the result of a balance between positive and negative factors. Although a considerable body of evidence exists concerning the signalling mechanisms mediating the positive signals, much less is known about the mechanisms that instigate the negative regulation. In the last few years, several inhibitors of proliferation or angiogenesis have been identified. These include α and β interferons, interleukin-12, TGF-β, TNF-α, platelet factor-4, fragments of the hormone prolactin, thrombospondin, angiostatin and endostatin [1]. For the majority of these molecules, the mechanism by which they mediate the inhibition of angiogenesis is not known. For thrombospondin-1 the receptor has been identified: is CD36, a transmembrane glycoprotein found on microvascular endothelial cells [76]. In the case of angiostatin or endostatin, the existence of a cellular receptor is unknown [77], and it is possible that these proteins may interact with membrane proteins, such as integrins, blocking angiogenesis. Moreover, some of these molecules, notably TGF-β or TNF-α, can function as inhibitors of the proliferation of endothelial cells but also as inducers of capillary formation. This difference may reflect the fact that these factors favor the differentiation process,

thus inhibiting proliferation. In fact, it has been shown that the different effects of TGF-β1 on proliferating endothelial cells or on collagen induced differentiation could be explained by differential expression of TGF-β-receptor subtypes in these cells [78]. Thus, the type II TGF-β receptor is expressed in proliferating endothelial cells, and is repressed in cells growing on collagen gels. The authors have shown that the type II receptor mediates the antiproliferative effects of TGF-β, while the type I receptor mediates the collagen differentiating effect. Other authors have shown that TGF-β1 down-regulates the VEGF receptor Flk-1, decreasing the cell surface binding of VEGF [79]. This down-regulation may be responsible for the inhibitory effect of TGF-β1 on VEGF-induced proliferation. A 16-kDa N-terminal fragment of prolactin has been shown to inhibit both VEGF and FGF-2-induced endothelial cell proliferation by inhibiting the activity of p42/p44 MAP kinases, without affecting the VEGF-receptor autophosphorylation, or the stimulation of phospholipase Cγ , but the mechanism implicated in the inhibition of the MAP kinase activity has not been resolved at the present time [33]. Clearly, with the recent progress made on the elucidation of the signalling pathways regulating cell cycle entry, differentiation and apoptosis in vascular endothelial cells, we may expect an explosion of results on the mechanisms that will antagonize the neo-vascularisation.

Acknowledgments: We thank Dr. Fergus McKenzie for his editorial support, and to all the members of the laboratory for their support. This work was supported by research grants from CNRS (Centre National de la Recherche Scientifique), the Ministere de la Recherche et de la Technologie (ACC-SV09), INSERM (Institut National de la Santé et de la Recherche Medicale), ARC (Association pour la Recherche contre le Cancer) and from the Ligue Nationale de la Recherche contre le Cancer. FV was the recipient of postdoctoral fellowships from ARC and from Ministerio de Educacion y Cultura (Spain).

Francesc Vinals and Jacques Pouysségur

References

1. Hanahan D, Folkman J (1996) Cell 86:353-364
2. Folkman J, Shing Y (1992) J Biol Chem 267:10931-10934
3. Hill CS, Treisman R (1995) Cell 80:199-211
4. Fanger et al (1997) Curr Op Genet Develop 7:67-74
5. Cobb MH, Goldsmith EJ (1995) J Biol Chem 270:14843-14846
6. Seger R, Krebs EG (1995) FASEB J 9:726-735
7. Lenormand P et al (1993) J Cell Biol 122:1079-1088
8. Brondello JM et al (1997) J Biol Chem 272:1368-1376
9. Wood et al (1992) Cell 68:1041-1050
10. Brunet et al (1994) Oncogene 9:3379-3387
11. Mansour et al (1994) Science 265:966-970
12. Pages G et al (1993) Proc Natl Acad Sci USA 90:8319-8323
13. Brondello JM et al (1995) Oncogene 10:1895-1904
14. Sherr CJ (1996) Science 274:1672-1677
15. Weinberg RA (1995) Cell 81:323-330
16. Bartek J et al (1996) Curr Op Cell Biol 8:805-814
17. Lavoie JN et al (1996) J Biol Chem 271:20608-20616

18. Lukas J et al (1996) Mol Cell Biol 16:6917-6925
19. Marshall CJ (1995) Cell 80:179-185
20. Toker A, Cantley LC (1997) Nature 387:673-676
21. Domin J, Waterfield MD (1997) FEBS Lett 410:91-95
22. Carpenter CL, Cantley LC (1996) Curr Op Cell Biol 8:153-158
23. Kimura et al (1994) J Biol Chem 269:18961-18967
24. Kaliman et al (1996) J Biol Chem 271:19146-19151
25. Chou MM, Blenis J (1995) Curr Op Cell Biol 7:806-814
26. Pullen N, Thomas G (1997) FEBS Lett 410:78-82
27. Brown EJ, Schreiber SL (1996) Cell 86:517-520
28. Maciag T (1981) J Cell Biol 91:420-426
29. Friesel RE, Maciag T (1995) FASEB J 919-925
30. Garfinkel S et al (1996) J Cell Biol 134:783-791
31. Zhan X et al (1994) J Biol Chem 269:20221-20224
32. Sa G (1995) J Biol Chem 270:2360-2366
33. D'Angelo G et al (1995) Proc Natl Acad Sci USA 92:6374-6378
34. Wary KK et al (1996) Cell 87:733-743
35. Mohammadi M et al (1996) Mol Cell Biol 16:977-989
36. Pawson T (1995) Nature 373:573-580
37. Sexl V et al (1997) J Biol Chem 272:5792-5799
38. Sosnowski RG et al (1993) J Cell Biol 121:113-119
39. Fox PL et al (1994) Oncogene 9:3519-3526
40. Dong QG et al (1997) Arterioscler Thromb Vas Biol 17:1599-1604
41. Kanda et al (1997) J Biol Chem 272:23347-23353
42. Presta M et al (1989) J Cell Biol 109:1877-1884
43. Kohn EC et al (1995) Proc Natl Acad Sci USA 92:1307-1311
44. Daviet I et al (1990) FEBS Lett 259:315-317
45. Harrington EO et al (1997) J Biol Chem 272:7390-7397
46. Zhou WZ et al (1993) J Biol Chem 268:23041-23048
47. Breier G, Risau W (1996) Trends Cell Biol 6:454-456
48. Mustonen T, Alitalo K (1995) J Cell Biol 129:895-898
49. Soker S et al (1998) Cell 92: 735-745
50. Guo D et al (1995) J Biol Chem 270:6729-6733
51. Waltenberger J et al (1994) J Biol Chem 269:26988-26995
52. Abedi H, Zachary I (1997) J Biol Chem 272:15442-15451
53. Seetharam L et al (1995) Oncogene 10:135-147
54. Arbiser JL et al (1997) Proc Natl Acad Sci USA 94 861-866
55. Merenmies J (1997) Cell Growth, Differ 8:3-10
56. Henkemeyer M et al (1995) Nature 377:695-701
57. Wojnowski L (1997) Nature Genetics 16:293-297
58. Lin CQ, Bissell MJ (1993) FASEB J 7:737-743
59. Varner JA, Cheresh DA (1996) Curr Op Cell Biol 8:724-730
60. Takahashi M, Berk BC (1996) J Clin Invest 98:2623-2631
61. Short SM et al (1998) Mol Biol Cell 9: 1969-1980
62. Friedlander MF et al (1995) Science 270:1500-1502
63. Eliceiri BP et al (1998) J Cell Biol 140: 1255-1263
64. Brooks PC et al (1994) Cell 79:1157-1164
65. Stromblad S et al (1996) J Clin Invest 98:426-433
66. Dejana E et al (1995) FASEB J 9:910-918
67. Caveda L et al (1996) J Clin Invest 98:886-893
68. Miller JR, Moon RT (1996) Genes Dev 10:2527-2539
69. Baeg GH et al (1995) EMBO J 14:5618-5625
70. Peifer M (1997) Science 275:1752-1753
71. Inglese J et al (1995) TIBS 20:151-156
72. Vigne P et al (1990) Biochem J 266:415-420
73. Offermanns S et al (1997) Science 275:533-536
74. Hla T, Maciag T (1990) J Biol Chem 265: 9308-9313
75. Lee MJ et al (1998) Science 279: 1552-1555
76. Dawson DW et al (1997) J Cell Biol 138: 707-717
77. Cao Y et al (1996) J Biol Chem 271 29461-29467
78. Sankar S (1996) J Clin Invest 97:1436-1446
79. Mandriota SJ et al (1996) J Biol Chem 271:11500-11505

Smooth Muscle Cells

Definition *Mesenchyme derived cells that are positive for smooth muscle α actin (α-SM-actin).*

See also: →Vasomotor tone regulation, molecular mechanisms of; →Atherosclerosis

Introduction Smooth muscle cells (SMCs), which constitute the wall of the vessels, trachea, bronchioles, intestine, uterus etc, play a major role in physiological processes such as the maintenance of arterial pressure, peristalsis and parturition. The normal SMC is a highly specialized cell that exhibits unique contractile properties, which characteristically depend on a highly developed system of contractile and cytoskeletal elements, as well as electrical activity and agonist responsiveness that distinguish it from any other cell type. SMCs also synthetize and secrete components of the connective tissue matrix, including several types of collagen, elastic fiber proteins and proteoglycans. Although SMCs are considered to be contractile and connective tissue-forming cells, they also have the capacity to synthetize a number of growth-regulating substances, such as prostaglandins, growth factors and cytokines. SMCs have diverse mesenchymatous origins [1] and these differences in lineage could be important in pathogenesis [2]. Recent studies using transgenic mice have revealed the existence of different SMC differentiation programmes in veins and arteries, which raises the possibility of further subdivision of programmes in visceral muscles [3]. During the periods of growth and development, arterial SMCs contain an extensive rough endoplasmic reticulum and a well developed Golgi complex that enable the synthesis and secretion of connective tissue molecules. These organelles are diminished or lacking in the adult. Conversely, proteins such as calponin and caldesmon (87Kda), which are specific markers of SMC differentiation, are synthesized later than other contractile proteins such as desmin, tropomyosin and myosin [4]. Indeed, SMCs exist in a multitude of forms that may bear little resemblance to the highly differentiated cell type found in normal vascular media [5]. On the other hand, differentiation and maturation of SMCs are highly dependent on environmental conditions, even in mature organisms, rather than being relatively fixed and inherent to the cell itself, and the ultimate phenotype depend on local environmental cues. Differentiated SMCs are characterized by their capacity to synthesize a variety of contractile proteins, i.e. muscle isoforms of actin, myosin heavy chains, myosin light chains and α-tropomyosin, that are important for their ultimate function (review in [6]). Smooth muscle cells are responsible for the formation of the elastic lamellae within the developing arterial wall as well as of the extracellular matrix proteins that surround each cell (review in [7]). However, the embryological origins of vascular smooth muscle cells (VSMCs) and the timing of the events which lead to a SMC lineage are undetermined and there are no ways of recognizing the potential but undifferentiated SMCs. Arterial SMCs demonstrate characteristic changes in phenotype after injury and when established *in vitro*. This process includes a major structural reorganization with partial loss of filaments and formation of a large endoplasmic reticulum and Golgi complex. The cell loses contractility, as a result, but becomes capable of division and secretion of extracellular matrix components [8].

Structure   SMCs are mononucleated and fusiform with a parallel orientation. When isolated from neonatal rats, they become polyploid in culture while tetraploid cells from adult rat aorta maintain their phenotype. Formation of polyploid cells is associated with cell growth and does not appear to be caused by cell fusion (review in [9]), but by aberrant mitosis and/or endoreplication. SMCs are connected by a scaffold of connective tissue containing collagen and elastin and attach to their neighbors by attachment plaques and junctional complexes that permit cell-cell communications. The adjective "smooth" arises from the difficulty in distinguishing contractile structure using normal light microscopy. Immunocytochemical approaches and high resolution fluorescence microscopy have contributed greatly to our present knowledge of the structure of these cells. Their mechanical properties derive from two mutually interactive filament systems; the contractile apparatus and the cytoskeleton, the latter playing a central role in subcellular organization (review in [10]). The contractile apparatus is composed of thick (14nm), and thin (6-8 nm) filaments, the thin/thick ratio being about 15:1 in rabbit portal vein. These contractile elements are organized obliquely with respect to the cell axis. They are anchored at cytoplasmic level to ovoid and electron-dense structures (dense bodies) which anchor actin filaments to the contractile machinery and at the adhesion plaque, or adherens junctions, of the membrane skeleton. Dense bodies are scattered within the cytoplasm and are rich in α-actinin, an actin binding protein. These structures, which can be compared to the Z lines of striated muscle, are responsible for cell motility. Adhesion plaques are associated with invaginated structures or caveolaes, at the sarcolemmal level, which are particularly rich in Ca^{2+} and dystrophin. Two types of thin filaments exist. One is contractile and mainly composed of α-SM actin and caldesmon, the other is cytoskeletal and mainly composed of non-muscle actin and filamin. Indeed, SMCs express four actin isoforms: two smooth muscle-specific and two cytoplasmic, typically associated with the cytoskeleton of non-muscle cells. The relative amounts of each isoform and the total actin content vary according to smooth muscle type, the stage of development, culture conditions, and the presence or absence of pathology. Proteins, such as gelsolin and profilin influence the polymerization state of actin. Calponin, another thin filament protein, is associated with the contractile apparatus, dense bodies and adhesion plaques, while caldesmon and tropomyosin maintain actin filament stability. Thick filaments are mostly composed of myosin. The myosin-linked regulation of actin-myosin interaction is accompanied by phosphorylation of the regulatory light chain and dynamic assembly and diassembly of the thick filament. SMCs express at least four types of myosin heavy chain (MHC), two are smooth muscle-specific while two are of non-muscle type. Smooth muscle MHC SM-1 and SM-2 characterize the SMC phenotype and analysis of their expression has made a significant contribution to our understanding of SMC phenotypes in development and disease. Although the physiological significance of each isoform is incompletely understood, their differential expression suggests that they may have distinct roles in smooth muscle development and physiology (review in [11]). The cytoskeleton is a structural lattice, made up of cytoplasmic (non-muscle) actin filaments, of unknown length that are tethered into longitudinal fibrils by "dense bodies". The fibrils contain mostly colinear arrays of "intermediate filaments", whose main constituent is desmin (but also vimentin) and are predominantly organized into loosely packed bundles oriented parallel to the long axis of the cell (review [10]). Filamin is associated with actin and could contribute to the maintenance of tone. Smoothelin, a novel marker of the contractile phenotype, is colocalized with actin filaments and is generally coexpressed with desmin, but its distribution appears to be limited to blood vessels that are capable of pulsatile contraction. The structural framework of the cytoskeleton pervades not only the cell interior but also coats the sarcolemma, where it anchors the cell surface to the surrounding matrix, via transmembrane molecules. Among these molecules, the integrins together with ECM proteins, proteases and growth factors participate in the differentiation of SMC [12].

Regulation of Cell Function
Cell to Cell Interactions   There is mutual SMC cross talk between connective tissue and overlying differentiated epithelium or endothelium (review in [1]). Interestingly, cultured epithelium on different SMC-derived substrates dedifferentiates into different tissue types (cited in [5]). The endothelium lines all vessels of the body and is the most important communication structure between blood cells and the vessel wall. Interactions between vascular endothelial cells and SMCs contribute to normal vascular wall function and to the pathogenesis of atherosclerotic lesions and fibrointimal hyperplasia (review in [5]). The endothelium is also a mechanically sensitive signal transduction interface which maintains constant wall shear stress. Indeed, shear stress resulting from blood flow and transmural plasma flux alters the function of vascular cells (primarily endothelial cells), leading to adaptive tissue responses. Transmission of the shear stress signal throughout the vascular endothelial cells and smooth muscle cells involves a complex interplay between cytoskeletal and biochemical elements resulting in changes in structure, metabolism and gene expression. The endothelium is thought to play an important role in the remodelling process through the release of products such

as nitric oxide (NO) and growth factors, which affect the differentiation and growth of SMCs. For example, endothelin-1 can convert phasic developing SMCs to the tonic type *in vitro*, suggesting a role in determining the contractile phenotype [13]. Endothelial and SMC interactions are probably involved in transduction of mechanical signals; high flow rates modulate arterial growth through stimulation of endothelial cells which secrete NO and thereby induce muscle relaxation and an increase in vascular diameter [14]. A complex network of interacting cells underlies the process of atherogenesis; three circulating cellular components – monocytes, T lymphocytes and platelets – and two cell types within the arterial wall – endothelium and smooth muscle. Specific genes are expressed in each of these cell types resulting in the formation and release of numerous factors, including small molecules such as NO and eicosanoids, growth-regulatory molecules (both stimulatory and inhibitory) and chemotactic agents. The net result of these interactions determines whether the lesions progress, regress, or remain stable (review in [15]). Cocultures of SMCs with activated, intact, viable T lymphocytes induce metalloproteinases, MMPs, (MMP-1, MMP-3, MMP-9), although activated T lymphocytes release an inhibitor of MMP expression, interferon-γ (IFN-γ).

Molecular Interactions   Interactions between SMCs and other cell types depend essentially on cell-specific cytokines, and VSMC growth and proliferation require the synergistic action of multiple factors. The extracellular matrix and its components interact with many other molecules, including cytokines and/or growth factors, metalloproteinases and membrane receptors, to regulate SMC function. Our current concepts derive from cultured cell models. Among examples, heparin inhibits, and thrombospondin (a glycoprotein secreted into matrix by endothelial and smooth muscle cells) potentiates, the proliferative effects of epidermal growth factor (EGF) on SMCs. Tissue type plasminogen activator (t-PA) appears to be an important activator of platelet-derived growth factor-BB (PDGF-BB) with regard to the proliferation and migration of SMCs, whereas urokinase (u-PA) activates the effect of fibroblast growth factor-2 (FGF-2) on SMCs [16]. Matrix metalloproteinases (MMPs) upregulate SMC synthesis of tenascin-C, by generating β_3 integrin ligands in type I collagen. In turn, interactions $\alpha_v\beta_3$ integrin with tenascin-C alter SMC shape and increase EGF receptor clustering and EGF-dependent growth. Conversely, suppression of MMPs downregulates tenascin-C and induces apoptosis [12]. Cytokines and growth factors determine cell-cell interactions; their effect on SMC phenotype has been mainly analysed *in vitro*. Isolated adult rat media SMCs cultured in the presence of fetal calf serum, proliferate and show a decreased expression of α-smooth muscle actin, desmin and SM-MHC. These cells acquire cytoskeletal features of fetal or atheromatous SMCs. On this basis it is generally assumed that cultured arterial SMCs are a reliable model to study modifications of SMC phenotype (review in [17]). Inter-

actions between cytokines, including PDGF, IFNγ, interleukin-1 (IL-1), tumor necrosis factor-α (TNF-α), transforming growth factor-β (TFG-β), prostaglandins, leukotrienes derivatives and connective tissue macromolecules such as proteoglycans, collagen and elastic fiber proteins, have been particularly analysed during atherosclerosis. The interactions of these molecules with the different cells involved in atherosclerotic process (endothelial and smooth muscle cells, macrophages, lymphocytes and leukocytes) can lead to further lesion progression, lesion stability, or regression depending on the circumstances. Activated T lymphocytes products modulate MMP expression by atheroma-associated cells. Soluble inflammatory mediators, such as TNF-α and IL-1 activate MMP expression in SMCs as well as in macrophages. A T-lymphocyte surface molecule CD40 ligand, recently localized in atherosclerotic plaques, regulates the expression of a series of matrix metalloproteinases MMPs – interstitial collagenase (MMP-1), stromelysin (MMP-3), and gelatinase B (MMP-9) – in human arterial VSMCs [18].

Signalling Mechanisms   Regulation of SMC activity has been implicated in diverse signalling pathways *in vitro*, including those mediated by extracellular matrix/integrin interactions, G-protein coupled receptors and oncogene products. SMCs can be stimulated or inhibited by many hormones and transmitters that activate an intricate complex of signal transduction mechanisms. A given neurohormone may relax one but contract another SMC through different mechanisms (review in [19]). The SMC contractile response to α-adrenergic, muscarinic and angiotensin agonists, depends on activation of seven-transmembrane-helix receptors associated with G protein(s) that activate phospholipase C, to release InsP3. Similarly, smooth muscle relaxation results from G protein receptor activation coupled to adenyl cyclase, the generation of cAMP and activation of c-AMP and c-GMP-dependent protein kinases. Growth factors may also influence cell attachment, migration, survival, production of extracellular matrix, thrombosis, vasoconstriction and regulation of cytokine synthesis. Integrins play an essential role in the intracellular transmission of mechanical signals [20]. Indeed, stretch increases DNA synthesis in VSMCs cultured on collagen, fibronectin, or vibronectin but not on elastin or laminin. The cell response thus depends on specific integrin-ECM interactions. For example, increased mitotic response to strain is abrogated by anti β_3-integrin but not by anti-β_1 integrin antibodies, whereas the expression of SMMH-2 is stretch-responsive. The intracellular integrin domain is associated with cytoskeleton proteins and a unique non-receptor protein-tyrosine kinase (FAK) localized in focal adhesion plaques. FAK expression is particularly abundant in the arterial media of early embryo arteries. Phosphorylation of these components could be involved in transduction of the mechanical signal. Mitogen-activated protein (MAP) kinases are rapidly activated in cells stimulated with various extracellular signals by dual phosphorylation of tyrosine and threonine residues.

They are thought to play a pivotal role in transmitting transmembrane signals required for cell growth and differentiation. Indeed, two MAP kinases, JNK (c-jun amino-terminal protein kinase) and ERK (extracellular signal-regulated kinase) are transiently activated in rat arteries in response to acute elevation of blood pressure induced by restraint or administration of phenylephrine or Ang II. Kinase activation is followed by an increase in c-fos and c-jun expression and enhanced activating protein 1 (AP-1) DNA binding activity. Activation of ERK and JNK could contribute to SMC hypertrophy/hyperplasia during arterial remodelling due to frequent and/or persistent elevation of blood pressure [21]. Integrin-regulated SMC growth depends on pathway(s) which are parallel to, but distinct from, the growth-mediated extracellular signal-regulated (ERK) pathway [22]. Ang II has multiple transduction pathways which are activated in a tissue-dependent manner [23]. Growth-promoting effects are mediated by the AT1 receptor. Signal transduction events are similar to those stimulated by growth factors and cytokines, and include stimulation of phospholipase C to generate the second messengers InsP3 and diacylglycerol (DG), and inhibition of adenyl cyclase. InsP3 mobilies calcium from endoplasmic reticulum, whereas DG activates protein kinase C and ultimately Na^+/H^+ exchange, leading to intracellular alkalinization. One of the immediate consequences of AT1 receptor binding is activation of MAP kinases (ERK1/ERK2) and p70S6K. AT1 receptor activation also leads to increased tyrosine phosphorylation of multiple proteins, including those of the focal adhesion plaque, which is inhibited by cAMP [24]. *In vivo*, however, VSMC phenotype appears to be triggered via the AT1 receptor, while trophic effects of Ang II on VSMCs are mediated via the AT2 receptor [25]. It now appears that growth factor signal transduction pathways are not identical. Heparin inhibits serum-induced SMC MAPK activation but not activation induced by PDGF. It also stimulates MAPK in the presence of FGF, but only activates MEK-1, although MEK-2 is present in comparable amounts within the SMC [22]. Similarly, the signal transduction pathways which mediate the chemotactic effects of PDGF-BB and other VSMC migration factors are poorly understood, though several cellular components have been implicated, including focal adhesion components, small GTP-binding proteins of the rho family, and certain substrates of the PDGF-β receptor. Activation of the cytosolic protooncogenes tyrosine kinase pp60c-src and p21 ras is crucial for G protein-dependent factors in SMC proliferation and DNA synthesis, as Ang II, endothelin-I or thrombin, but not for more classical growth factors such as PDGF-BB [26].

Additional Features   The mechanisms underlying the multiplicity of SMC functions and phenotypes are unclear. The SMC has unique morphology and expresses a repertoire of ion channels, receptors, signal transducing molecules and contractile proteins and proteins of the extracellular matrix that are necessary for its highly specialized cellular functions. An important characteristic of SMCs is their spectrum of phenotypes and diversity between organ systems and within a given tissue. The mechanisms regulating smooth muscle development and differentiation remain poorly understood. It is possible that their heterogeneity extends to a subset of "stem" cells, preferentially stimulated to migrate and proliferate following injury. Cloning and identification of characteristic smooth muscle genes could enable dissection of the molecular mechanisms regulating smooth muscle myogenesis and gene expression.

Role in Vascular Biology

Physiological Function   The vascular wall is an active, integrated organ composed of numerous cell types. Under normal circumstances, VSMCs constitute the single medial cell type. This compartment, which is responsible for vessel vasomotion and remodelling, receives signals from both endothelium and adventitia. Under physiological conditions, endothelial cells secrete nitric oxide (NO) which relaxes SMCs and ensures vessel patency. When damaged or activated, they secrete vasoconstrictor factors such as endothelin-1 (ET-1), as well as factors that affect SMC differentiation and growth. Smooth muscle contractile force, coupled with the elasticity of the elastic lamina, facilitates blood flow and SMC tone dampens the pulsatile flow induced with each systole. *In vivo*, normal SMC growth during ontogenesis is associated with a change in phenotype. A well-differentiated phenotype predominates in vessels when cells contract in response to chemical and mechanical stimuli and are involved in the control of blood pressure and flow. At this stage, αSM-actin and SM2 are the major contractile proteins expressed. In contrast, phenotypic shift in expression of contractile and extracellular matrix proteins towards the form expressed in the fetus is observed in the VSMCs of hypertensive rats and during the early phases of atherogenesis [reviews in 6, 11, 15, 17]. The functional significance of the variable expression of SM-MHC is unclear.

Vascular SMCs perform several functions in the blood vessel wall, according to their specific location and the presence or absence of disease. In normal vessels, SMCs have a contractile phenotype and the ability of smooth muscle to develop maximal force is equal to or greater than that of striated muscle [review in 19]. SMCs respond to numerous vasoactive agents, including adrenergic components, cholinergic compounds and substances such as Ang II, each of which can modulate cell contractility and thus alter vascular tone [7]. Ovarian steroids play an essential role in the vascular system and are involved in multiple phenomena, controlling vasculogenesis during the menstrual cycle and pregnancy, and inhibiting SMC proliferation in the arterial wall, thus explaining their role in atherosclerosis, venous insufficiency, and the modification of hemostasis [27]. SMCs also play a role in the maintenance of medial architecture and in the remodelling of the arterial wall matrix. As a consequence of their ability to synthesize and secrete all the principal components of the connective tissue matrix, they are responsible for the formation of the elastic lamellae within the developing arterial wall as well as of collagen and proteoglycans that surround each lamella cell. Although SMCs are generally considered to be contractile and connective-tissue forming cells, they also express a number of growth-regulatory substances and may thus participate in inducing their own vasoactive responses.

SMC contraction is driven by a cyclical interaction between the globular head domain of myosin and actin filaments, coupled to the breakdown of ATP (see for reviews [28, 29]). In vertebrate smooth muscle it is generally accepted that actomyosin ATPase and subsequent, tension generation is regulated primarily by reversible Ca^{2+} calmodulin dependent phosphorylation of the 20Kda myosin light chain, mediated by myosin light chain kinase (MLCK) [28]. In vascular smooth muscle, calmodulin kinase (CaM kinase) may regulate Ca^{2+} dependent force generation via phosphorylation of the actin-associated regulatory protein, caldesmon. Caldesmon binds actin and myosin through its C- and N- terminals, respectively, and has been proposed to structurally cross-link actin and myosin filaments, thereby decreasing myosin ATPase and preventing cross bridge cycling. Phosphorylation of caldesmon via CaM kinase decreases its binding to myosin, which may relieve its inhibitory effect on myosin ATPase activity and allow cross-bridge cycling following elevation of intracellular Ca^{2+}. Importantly, the binding of actin and myosin by caldesmon may further serve to maintain the contractile filaments in a three-dimensional meshwork capable of force generation. (CaM kinase, when activated via Ca^{2+}/calmodulin, phosphorylates MLCK). Phosphorylation of MLCK at a site within the calmodulin binding region (Ser 512) decreases its affinity for activator calmodulin and would thus decrease its sensitivity to activation following a stimulus, thus representing a regulatory mechanism for desensitizing smooth muscle to subsequent contractile stimuli. In vessels, the main function of SMCs is the regulation of blood pressure through changes in vascular tone. Schematically, constrictors increase cytosolic free Ca^{2+} concentration and the Ca^{2+} sensitivity of the contractile elements, while relaxant agonists have opposite effects [for a review see 30]. The entry of Ca^{2+} into VSMCs via voltage-dependent Ca^{2+} channels is modulated by the membrane potential, itself modulated by Ca^{2+}-activated channels such as Cl^- (depolarization) or K^+(hyperpolarization). Capacitative Ca^{2+} entry and ligand-gated channels ($P_{2\alpha}$-purinoceptors) also allow extracellular Ca^{2+} to flow into the cytosol. The intracellular source of Ca^{2+} depends on the storage capacity of the sarcoplasmic reticulum (SR), which involves intraluminal Ca^{2+} binding proteins, such as calsequestrin and calreticulin. Ca^{2+} is released from the SR into the cytosol through inositol 1,4,5-triphosphate (InsP3) and ryanodine receptors (InsP3-induced Ca^{2+} release and Ca^{2+}- induced Ca^{2+} release respectively). During relaxation, cytosolic Ca^{2+} concentration is reduced by pumping into the SR by Ca^{2+}ATPases (SER-

CA). Modulation of the force at constant Ca^{2+} concentration results from altered activity of kinases and phosphatases which act on myosin light chain (MLC20) phosphorylation. MLC20 phophatase activity depends on intracellular messengers such as arachidonic acid, protein kinases and G proteins (Rho p21, ras p21) [19]. However, the simple on/off system based on myosin phosphorylation does not explain all aspects of vascular smooth muscle contractile activity, particularly the tonic slow cycling "latch-bridge" responses, characteristic of sustained isometric force production (see for review [31]). Caldesmon and calponin are major elements in the phosphorylation-independent mechanism of Ca^{2+}-dependent maintenance of tension. The two proteins bind to the major proteins of the actomyosin system and inhibit the actin-activated Mg^{2+}ATPase activity of myosin, which may be reversed upon interaction with Ca^{2+} dependent proteins such as calmodulin, caltropin and S-100, or through changes in phosphorylation. These proteins are likely to be involved in an actin-linked Ca^{2+}-sensitive secondary regulatory process. The interplay between the thick and thin filament-based regulatory systems may explain some of the unique properties of SMCs, in particular the phenomenon of the "latch state". Variation in the relative expression of each actin isoform or total actin content does not contribute to contractile diversity. Increased understanding of SMC contractility will depend not only on better knowledge of the contractile apparatus but also of the underlying signalling mechanisms. Indeed, vasorelaxation could depend on a series of peptide signals and the interactions of the guanylate cyclase-stimulatory monoxide gases, NO and CO. For example, exogenous and endogenous NO stimulate heme oxygenase 1 (HO-1) gene expression, which catalyzes CO production in vascular smooth muscle cells. Furthermore, CO could modulate high conductance KCa channels in vascular SMCs [32]. On the other hand, vasoconstriction is likely to depend on Ang II regulation via the AT1 receptor and/or endothelin-1. Furthermore vasodilatation could result from AT2 receptor activation [33].

Migration of vascular SMCs is an important means by which arterial intima thickens during the course of vascular disease. Potential regulators include chemotactic cytokines and growth factors, as well as dynamic integrin interactions, organisation of the actin cytoskeleton and focal adhesion turn over. PDGF-BB released from activated platelets adherent to subendothelial connective tissue is a principal chemotactic factor. FGF-2 potentiates PDGF-dependent SMC migration by promoting the upregulation of $\alpha_2\beta_1$ integrin and dissassembly of actin filaments [34].

A growing body of evidence suggests that vascular remodelling occurring in the context of post-natal development and post-angioplasty restenosis results from parallel amplification of the antagonistic processes of SMC growth (hypertrophy and/or hyperplasia) and apoptosis. For example, during development, arterial remodelling, in response to altered arterial blood flow depends on coordinated regulation of both cell death and cell proliferation [35]. Mechanical and non-mechanical factors, including growth factors, cytokines, hormones (Ang II), passive stretch and tension, mediate SMCs growth during other remodelling phenomena [36], but factors inducing apoptosis are poorly understood. Vessels are permanently exposed to two forms of stress: tensile stress originating from arterial pressure and shear stress resulting from blood flow. Blood flow regulates arterial diameter through reorganization of vascular wall cellular and extracellular components. Multiple receptors and signal transduction pathways of many of these ligands have been identified. As mentioned above, contractile agonists and mechanical stretching stimulate vascular growth [36]. Mechanical stimulation of organotypic culture of endothelial cell-denuded aorta increases protein synthesis, particularly fibronectin and collagen. Cyclic stretch of cultured rabbit aortic SMC grown on elastic membranes increases the synthesis of total protein, collagen and elastin, prevents myofilament loss and phenotypic alterations that normally occur in culture. It also induces SMC growth via autocrine effects of PDGF, activates phophoinositol metabolism, increases the concentration of AngII, which has a serum and PGDF independent growth stimulatory effect on VSMCs *in vitro*. AngII produces both trophic and phenotypic changes in SMCs. In culture, AngII produces a 20% to 50% increase in cellular protein content of quiescent rat aortic SMCs and a 50% increase in cells with 4C DNA content, but no increase in cell number (review in [9]). Trophic effects are largely mediated via the AT1 receptor. Contradictory data concern the AT2 receptor which has an anti-growth (anti-hypertrophy/ hyperplasy or pro-apoptotic) action, both *in vitro* and *in vivo*, in situations with high mitotic index, such as restenosis post-angioplasty [23]. On the other hand, it has a trophic effect in arterial hypertension (situations with a quasi loss of mitotic activity) [25]. In the latter study, changes in SMC phenotype were mainly triggered via the AT1 receptor. Ang II binding to the AT1 receptor elicits rapid activation of the proto-oncogenes c-myc and c-jun, and many G protein-dependent cellular events, which are also stimulated by mitogens such as PDGF and EGF. However, the growth-promoting effects of Ang II, serum and PDGF are not additive. Ang II may also stimulate VSMC growth via an autocrine/paracrine loop involving PDGF release [23].

SMC proliferation is dependent on both anchorage to the extracellular matrix by integrins and the presence of growth factors. Proliferation and migration of SMCs can be positively or negatively regulated as a result of various interactions between growth factors, extracellular matrix components and expression of proteases (review in 37). *In vitro*, laminin and other membrane components promote the expression of a differentiated smooth muscle phenotype, whereas fibronectin stimulates cells to adopt a proliferative and secretory phenotype [8]. This process is characterized by structural and functional changes of cells, including cytoskeletal reorganization, formation of a large secretory apparatus and the

acquisition of proliferative capacity. These phenotypic changes depend on the assembly of focal adhesion plaques with associated tyrosine-kinase activity. Other proteins, such as calponin and tenascin-C, an extracellular matrix glycoprotein prominent during tissue remodelling, appear to be important in growth factor-dependent SMC proliferation [12]. Cytokines, such as interleukin-1β (IL-1), IL-4 and IL-8, induce increased VSMC DNA synthesis and proliferation, as well as stimulation of 12 lipoxygenase production, which modulates arachidonate metabolism [38].

Apoptosis (programmed cell death) is a physiological counterpart of cell replication with shared as well as specific pathways. According to Geng et al. TNF-α or IL-1β induced apoptosis in cultured human SMCs only when combined with IFN-γ (a cytostatic lymphokine) [39]. The same authors reported that this cytokine-induced apoptosis was partly dependent upon a rise in NO production. In contrast, according to Pollman et al. the apoptotic effects of NO in rat, rabbit and human SMCs are mediated via an inhibition of cGMP-specific phosphodiesterase, which is antagonized by AT1 receptor activation [40]. Thus, substances such as NO and AngII via AT1 receptor, have opposite actions on vasomotion, cellular proliferation and apoptosis. Apoptosis depends on an equilibrium between the activity of different genes and the accumulation of their products such as the tumor-suppressor p53, a potential mediator of cell cycle arrest and programmed death, the Bax protein and the protein product of proto-oncogenes bcl-2 and c-myc, the adeno-virus product EA1 (review in 37). Indeed, increased apoptosis in SMCs overexpressing c-myc in a constitutive and unregulated manner, was associated with a higher mitotic index and hyperplasia which were both independent of IFN-γ, heparin or cyclic nucleotides (cAMP and cGMP analogs). In normal SMCs the low rate of apoptosis after serum deprivation is independent of p53. In contrast, in SMCs overexpressing c-myc or E1A (a c-myc adenoviral analog), apoptosis is dependent on and mediated by p53; in both cases, Bcl-2 can induce apoptosis. Thus, two regulatory pathways for apoptosis may exist; one depending on p53, the other probably functioning as a causal agent. Furthermore, RB protein (a product of the retinoblastoma gene) regulates human SMC proliferation via an interaction with p53 [41]. On the other hand, PKC antagonizes the effects of verapamil (a Ca^{2+} channel antagonist) and dibutyryl cAMP, which both induce apoptosis [42].

Pathology Pathological smooth muscle modifications involve cell-cell and cell- extracellular matrix interactions which are mediated by growth factors, cytokines, and extracellular matrix components. Vascular remodelling occurs when pressure or blood flow are altered and may thus contribute to the pathophysiology of arterial diseases, affecting both large and small vessels, i.e. hypertension, heart failure, atherosclerosis, angioplasty restenosis and pulmonary hypertension. SMC apoptosis is likely to make an important contribution to the evolution of these diseases. The basic feature of these pathologies is the proliferation/migration of medial SMCs to the intima, accompanied by marked morphological, biochemical and functional changes [reviews in 2, 15, 17]. Indeed, SMCs display remarkable plasticity in terms of differentiation, proliferation and motility – characteristics that are particularly evident when adult arteries are subjected to wall injury. Smooth muscle cells recapitulate some aspects of ontogenesis during hyperplastic or hypertrophic growth occurring as part of an adaptative response to pathophysiological stimuli. However, recent observations demonstrate that the vascular wall comprises phenotypically heterogeneous subpopulations of endothelial cells, SMCs and fibroblasts. Subsets of cells respond to injury and thus contribute to vascular remodelling in hypertension [2]. Folkow was the first to appreciate the significance of VSMC proliferation in the pathogenesis of hypertension (reviews in [6, 9]), demonstrating that isolated vessels from spontaneously hypertensive rats (SHR) had increased resistance compared to normotensive control vessels, even at maximum dilatation. Large arteries undergo adaptation or remodelling in response to elevated blood pressure and develop medial hypertrophy and/or intimal hyperplasia of the arterial wall. It is unknown whether all arterial media SMCs have the same potential to migrate, grow and change their phenotype [17]. By definition, remodelling during hypertension is associated with structural changes in resistance vessels but not with net growth. Increasing arteriolar resistance in hypertensive patients is due to structural rearrangement of otherwise normal material. SMC mass can increase in two ways in response to chronic hypertension; by either hypertrophy of SMC in large capacitance vessels, with the development of increased ploïdy, or hyperplasia of cells in medium-sized resistance vessels. According to Gibbons and Dzau, vascular remodelling is "an active process of structural alteration that involves changes in at least four cellular processes – cell growth, cell death, cell migration and production of extracellular matrix- and is dependent on a dynamic interaction between locally generated growth factors, vasoactive substances and hemodynamic stimuli" [43]. Increased shear stress induces expansive remodelling in relation to flow-dependent vasodilatation, whereas increased tensile stress is responsible for medial hypertrophy and fibrosis [14]. In spontaneously hypertensive rats, SHR, SMC apoptosis might be inhibited in small arteries as a consequence of an excess of the protein bcl-2. Chronic angiotensin-converting enzyme inhibition might restore susceptibility of SMCs to apoptosis through stimulation of the protein bax (see [41]).

SMC in atherosclerotic lesions are acted on by cytokines and growth factors (PDGF and TGF-β) secreted by macrophages, lymphocytes, platelets and endothelial cells (reviews in [15, 37]). PDGF, a major mitogen and chemoattractant secreted by macrophages, stimulates both smooth muscle migration and proliferation of pre-existing intimal SMCs. Furthermore, SMCs can be induced to synthesize factors, such as granulocyte-macrophage-colony stimulating factor (GMCSF) or IL-1 that initiate replication and activation of monocyte-derived macrophages. SMC could thus play a role in inducing macrophages to replicate in their immediate vicinity. Studies performed on monkeys, swine and hypercholesterolemic rabbits have provided insight into the main features of arterosclerosis. Initial changes consist of increased entry of lipid and lipoprotein particles into the subendothelial intimal space, rapidly followed by leukocyte penetration and migration between the endothelium and adherence throughout the arterial tree. Leucocytes may scavenge altered lipoproteins (oxidised LDL) and remove the offending substance(s) from the site. The progression of fibrofatty lesions is associated with the continuing entry and migration of macrophages, T lymphocytes and SMC. SMCs may be derived from intimal " cushions " that form during normal growth or following chemotactic attraction and migration from the underlying media. The final fibrous plaque results from a series of events within the microenvironment of the fibrofatty lesion including migration and proliferation of numerous SMCs to form a connective tissue matrix of collagen and proteoglycans that covers the surface of the lesion. Intimal SMC and macrophage proliferative events are principally responsible for lesion expansion. Growth-regulatory peptides and cytokines, such as PDGF, TGFα, M-CSF, TNFα, IFNγ and IL-2, can have a stimulatory or inhibitory effect on neighbouring cells within the arterial wall. SMCs migrate into the vascular intima, proliferate and lay down extracellular matrix, thereby forming an important component of atherosclerotic plaque in both native arteries and saphenous grafts. SMC viability and the integrity of the surrounding matrix also determine the liability of plaques to rupture. Since SMC migration and proliferation depend on interactions between growth factors, the extracellular matrix and apoptosis, an extensive and complex network of interacting cells is included in the process of atherogenesis. Based on the local microenvironment, specific genes can be expressed in each of these cell types, resulting in the formation and release of numerous factors, including small molecules such as NO and eicosanoids, growth-regulatory molecules (both stimulatory and inhibitory) and chemotactic agents. Various compounds such as reactive oxygen species, lipids, growth factors, and the extracellular matrix are involved in apoptosis. Activation of the membrane protein fas/apo-1/CD95 contributes to SMC apoptosis during atherogenesis and provides a mechanism whereby immune cells and their cytokines promote the process of cell death in relation to vascular remodelling and plaque rupture [44]. Once SMCs are attracted into the lesion, interactions between macrophages and smooth muscle, endothelium and smooth muscle, and between SMCs themselves, could play critical roles in generating intermediate, and ultimately advanced, lesions of atherosclerosis. Changes in the SMC expression of ECM components, such as fibronectin and collagen type VIII after arterial injury may contribute to vascular remodelling through the promotion of VSMC migration. Degeneration of SMCs in the fibrous cap of the atherosclerotic lesion is an important factor in plaque rupture and oxidized lipids play a role in plaque cell death.

Restenosis is a major complication that limits the long term efficacy of coronary angioplasty. Neointimal thickening involves SMC activation, proliferation, migration and proliferation, and the expression of cytoskeletal and extracellular matrix proteins (NM-MHC B, fibronectin, and osteopontin) which are controlled by factors, such as serotonin and thromboxane A_2, or hormonal and mechanical factors (review in [45]). The expression of smooth muscle MHC isoforms provides insight into the biology of healing after angioplasty or fibrosis and improves understanding of the pathological role of SMC differentiation and phenotypic modulation. For example, in the early period of restenosis after angioplasty, neointimal SMCs show features of an undifferentiated state (indicated by altered expression of SM-MHC) and undergo redifferentiation in a time-dependent manner. Many of the SMCs in advanced proliferative atherosclerotic lesions have a highly developed protein secretory apparatus which is associated with the formation of extensive amounts of connective tissue. The fibrous cap of the plaque is formed by such active and proliferating SMCs.

According to recent data, airway SMCs express cyclooxygenase-2 cytokines in addition to their contractile function and contribute to the inflammatory response seen in diseases such as asthma [46].

Insulin dependent diabetic patients have a high risk of vascular dysfunction. Arterial SMCs in diabetic rats and rabbits have unique properties including altered expression of growth factors, such as PDGF and TGF and extracellular matrix proteins, such as fibronectin, typifying the shift from a contractile to a synthetic phenotype. Among other effects, heparan sulfate is important in regulating intimal SMC growth and loss of proteoglycan heparan sulfate in the vasculature may explain the widespread nature of the disease [47].

Pulmonary hypertension is associated with functional and structural changes in the pulmonary vascular wall which include progressive hypertrophy of the muscular coat within the arterial media and the abnormal extension of smooth muscle into the normally thin-walled pulmonary arterioles, with consequent luminal narrowing [48]. Tenascin-C is induced in pulmonary vascular disease and colocalizes with EGF in proliferating SMC. The protein could modulate EGF-dependent neointimal SMC proliferation and expression of fibronectin would provide a gradient for SMC migration from media to neointima [49]. Impaired NO production contributes to pulmonary vasoconstriction and vascular remodelling in several forms of pulmonary hypertension, notably during the neonatal period and following hypoxia.

Clinical Relevance Therapeutic Implications   VSMC contractility is controlled in a complex manner by both extracellular and intracellular messages. The vascular endothelium integrates intravascular signals and controls contractility of underlying SMCs by releasing paracrine factors with contractile or relaxant properties. Vasoconstrictors trigger a cascade of interacting cellular signals that initiate and maintain contraction. Each step of these signalling pathways is a possible site for potential therapeutic interventions. Gene therapy now appears to be a practical and effective form of molecular intervention for proliferative arterial diseases [50]. Furthermore, enhanced SMC growth and proliferation may be the primary abnormality in genetic hypertension and therefore a cause, as well as a consequence of elevated blood pressure [9]. To elucidate the mechanisms of enhanced vascular SMC growth in hypertension it is necessary to understand the interactions of the various growth factors and contractile agonists, their signalling pathways and their effects on SMCs. Studies on the regulation of vascular growth will provide important insights into the pathogenesis of both spontaneous and acquired hypertension. The design of pharmaceutical agents that can block SMC growth could also provide an important approach to antihypertensive therapy. Moreover, since intimal SMC proliferation is the key event that characterizes the advanced atherosclerotic lesions, understanding the basis of the SMC proliferative process is critical to our knowledge of atherogenesis. Genetic approaches may provide new insights. For instance, genetic hypertension is controlled by a fairly small number of genes some of which are likely to be located in SMC. Finally, SMC apopotosis may represent a novel therapeutic target for the control of hypertensive vessel remodelling.

Lydie Rappaport, Catherine Chassagne
and Jane-Lyse Samuel

References

1. Gittenberg-de Groot AC et al (1995) In: Schwartz SM, Mecham RP (eds) The vascular Smooth Muscle Cell. Molecular and Biological Responses to the Extracellular Matrix. Academic Press, pp 17-33
2. Schwartz SM (1997) J Clin Invest 99: 2814-2817
3. Moessler H et al(1996) Development 122(8): 2415-2425
4. Frid MG et al (1992) Dev Biol 153: 185-193
5. Schwartz SM et al (1997) In: Schwartz SM, Mecham RP (eds) The Vascular Smooth Muscle Cell. Molecular and Biological Responses to the Extracellular Matrix. Academic Press, pp 81-140
6. Owens GK (1995) Physiol Rev 75: 487-517
7. Ross R (1992) In: Fozzard HA et al (eds.) The Heart and Cardiovascular System, 2nd edition. Raven Press, New York, pp 163-185
8. Thyberg J et al (1997) J Histochem Cytochem 45: 837-846
9. Oparil S et al (1992) In: Fozzard HA et al (eds) The Heart and Cardiovascular System, 2nd edition. Raven Press, New York,pp 295-331
10. Small JV et al (1995) In: Schwartz SM, Mecham RP (eds) The Vascular Smooth Muscle Cell. Molecular and Biological Responses to the Extracellular Matrix. Academic Press, pp 169-188
11. Sartore S et al (1997) Arterioscler Thromb Vasc Biol 17: 1210-1215

12. Jones PL et al (1997) J Cell Biol 139: 279-293
13. Fisher SA et al (1997) Circ Res 80; 885-893
14. Ben Driss A et al (1997) Am J Plysiol 272: H851-H858
15. Ross R (1995) Ann Rev Physiol 57: 791-804
16. Herbert JM et al (1997) J Biol Chem 272: 23585-23591
17. Desmoulière A et al (1995) In: Schwartz SM, Mecham RP (eds) The Vascular Smooth Muscle C.ell Molecular and Biological Responses to the Extracellular Matrix. Academic Press, pp 329-360
18. Schonbeck U et al (1997) Circ Res 81: 448-454
19. Somlyo AP (1997) Nature 389 (6654): 908-909
20. Wilson E et al (1995) J Clin Invest 96: 2364-2372
21. Xu QB et al (1996) J Clin Invest 97: 508-514
22. Hedin Ul et al (1997) J Cell Physiol 172: 109-116
23. Berk BC et al (1997) Circ Res 80: 607-616
24. Giasson E et al (1997) J Biol Chem 272: 26879-26886
25. Sabri A et al (1997) Arterioscler Thromb Vasc Biol 17: 257-264
26. Schieffer B et al (1997) Am J Physiol 272: C2019-C2030
27. Perrot-Applanat M (1996) Steroids 61: 212-215
28. Allen BG et al (1994) Trends Biochem Sci 19: 362-368
29. Periasamy M et al (1995) In: Schwartz SM, Mecham RP (eds) The Vascular Smooth Muscle Cell. Molecular and Biological Responses to the Extracellular Matrix. Academic Press, pp 189-213
30. Loirand G et al (1997) Medecine/Sciences 13: 766-76
31. Fattoum A (1997) Médecine/Sciences 13: 777-789
32. Wang R et al (1997) Pflugers Arch 434: 285-291
33. Tsutsumi Y et al (1999) J Clin Invest 104:760-765
34. Pickering JG et al (1997) Circ Res 80: 627-637
35. Cho A et al (1997) Circ Res 81 328-337
36. Tedgui A et al (1997) Médecine/Sciences 13: 790-798
37. Newby AC (1997) Coronary Artery Dis 8: 213-224
38. Natarajan R et al (1997) Hypertension 30: 873-879
39. Geng YJ et al (1995) Am J Pathol 147: 229-273
40. Pollman MJ et al (1996) Circ Res79: 748-756
41. Bennet MR et al (1997) Circ Res 77: 266-273
42. Leszczynski D et al (1995) Am J Pathol 147: 229-234
43. Gibbons GH et al (1994) N Eng J Med 330:1431-1438
44. Libby P et al (1997) Ann N Y Acad Sci 811:134-142
45. Bauters C et al (1997) Prog Cardiovasc Dis 40: 107-116
46. Belvisi MG et al (1997) Br J Pharmacol 120: 910-916
47. Jensen T (1997) Diabetes suppl 2: S98-S100
48. Riley DJ (1991) In: Crystal RG, West JB (eds) The Lung. Raven Press, New York, pp 1189-1198
49. Jones PL et al (1997) Am J Pathol 150:1349-1360
50. Kaneda Y et al (1997) Ann NY Acad Sci 127 (2) 221-228

Somatostatin

Definition *Cyclic tetradecapeptide that inhibits the release of peptides such as gastrin, insulin, or substance P. It inhibits endothelial cell proliferation and angiogenesis.*

See: →Angiogenesis inhibitors

SOS

Definition *Son of sevenless*

See: →Signal transduction mechanisms in vascular biology

SPARC

Definition *Secreted protein, acidic and rich in cysteine*

See: →Matrix metalloproteinases

SR

Definition *Sarcoplasmic reticulum*

See: →Vasomotor tone regulation, molecular mechanisms of

Src-Homology Domain-2 (SH-2) and -3 (SH-3)

Definition *SH domains are involved in protein/protein interaction. SH-2 is a 100 amino acid domain that binds certain phosphotyrosine residues. SH-3 is a 50 amino acid domain that binds proline-rich sequences.*

See: →Signal transduction mechanisms in vascular biology; →Platelet stimulus-response coupling

SSRE

Definition *Shear stress responsive element*

See: →Nitric oxide

STAT

Definition *Signal transducers and activators of transcription*

See: →Cytokines in vascular biology and disease

TAF

Definition *Tumor angiogenesis factor*

See: →Vascular endothelial growth factor family

TAM

Definition *Tumor-associated macrophage*

See: →Angiogenesis

Tcf-lef

Definition *T cell factor-lymphoid enhancer factor*

See: →Signal transduction mechanisms in vascular biology

TEK

Definition   *Transmembrane endothelial cell-specific kinase, same as Tie-2*

See: →Endothelial cells; →Tyrosine kinase receptors for factors of the VEGF family; →Ontogeny of the vascular system

TF

Definition   *Tissue factor*

See: →Tissue factor

TFPI

Definition   *Tissue factor protein inhibitor*

See: →Tissue factor

TGF-β

Definition   *Transforming growth factor-β*

See: →Transforming growth factor-β

TGRL

Definition   *Triglyceride-rich lipoprotein*

See: →Lipoproteins

Thrombin
Synonym: Activated prothrombin

Definition   *Proteolytic enzyme that converts fibrinogen to fibrin, and activates cells by cleaving protease-activated receptors PAR-1, PAR-3 and PAR-4.*

See also: →Platelet stimulus-response coupling; →Coagulation factors; →Fibrinolytic, hemostatic and matrix metalloproteinases, role of; →Thrombosis; →Atherosclerosis; →Fibrin/fibrinogen

Introduction   Thrombin is the final enzyme produced by the blood coagulation cascade [1] that consists of successive proteolytic conversions of plasma zymogens into enzymes (Figure 1). Thrombin is generated from its zymogen, prothrombin. When thrombin was discovered at the end of the last century, it was assumed to be an enzyme and referred to as a fibrin ferment. For a long time, thrombin has been studied as the proteolytic enzyme involved in the final step of blood coagulation, conversion of fibrinogen into fibrin. However, for the last two decades thrombin has also been recognized as an important cell agonist. The cellular aspects of thrombin activity have found particular developments follow-

ing the discovery of a specific cell receptor for thrombin, widely expressed in most tissues. The multiplicity of potential thrombin activities, even in the extravascular domain, needs to be considered in view of the capacity of thrombin to be provided to these tissues or to be produced locally by a still unidentified mechanism. However, it must be pointed out that several of the cellular actions of thrombin have been defined in cell culture systems and that their *in vivo* significance remains to be determined.

During the last twenty years, important progress has been made in the comprehension of thrombin: its structure has been elucidated, mutagenesis studies have assigned a function to many residues, new physiological effects have been described and thrombin receptors have been cloned. This research has been conducted for the development of new direct antithrombins such as hirudin and small synthetic inhibitors.

Characteristics
Molecular Weight   The molecular mass of thrombin is 36,500 daltons, corresponding approximately to half the size of its precursor, prothrombin.

Domains   Human α-thrombin consists of two polypeptide A and B chains connected through a disulfide bond [2]. The A-chain forms an integral structural part of the enzyme, but has not been demonstrated to be involved in catalysis. The B-chain is homologous to the reactive domains of other trypsin-like serine proteases. Three functional domains are identified: the catalytic domain and two ligands binding sites also called exosites (Figure 2) [3, 4]. The active site cleft, is deep and narrow, with the active site triad His^{43}, Asp^{99} and Ser^{205} at the base. The bottom of the primary specificity pocket in which binds the Arg side chain of the scissile peptidic bond is occupied by Asp^{199}. An extended hydrophobic pocket lies near the entrance to the specificity pocket, lined by Ile^{179}, Trp^{227}, segment 93-96, His^{43}, with the mainly hydrophobic Tyr^{47}-Trp^{50} loop on one side and Ile^{179} on the other. One large surface of negatively charged residues extends around the active site of thrombin generating a strong negative field. A dipolar charge distribution results from the presence of a surface region of high positive charge density located in the same groove

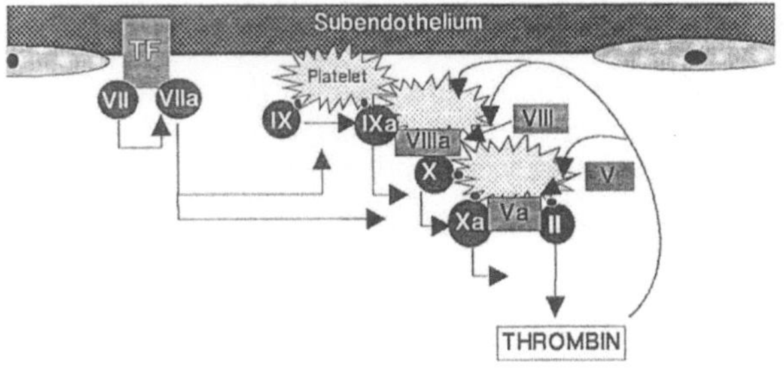

Figure 1. Thrombin formation. The initiation of the coagulation cascade occurs following vascular injury and exposure of tissue factor to the blood.

but remote from the catalytic site and known as anion binding exosite 1 or fibrinogen binding site [5]. Thrombin exosite 1 is constituted by the loop comprising segment Lys^{65}-Glu^{76} and its surrounding groups. The surface of this loop has positively charged amino acid side chains, in particular, Arg^{68}, Arg^{70}, and Arg^{73}, surrounded by other positively charged residues. A second patch of positively charged residues (exosite 2) is located at the top of the thrombin molecule, near the C-terminal B-chain helix. Four residues (Arg^{123}, Lys^{248}, Lys^{252} and Arg^{89}) surrounded by other basic residues give rise to a very strong positive electrostatic field [3, 4].

Binding Sites and Affinity Thrombin is characterized by a relatively strict specificity of substrate proteolysis. Structures within the catalytic domain and the exosites are involved in this specificity. Substrate recognition involves the presence of a basic P1 residue as in trypsin but there is an obvious requirement for an Arg in the primary specificity pocket since in most cases of mammalian proteins, thrombin cleaves a peptide bond C-terminal to an Arg residue. A number of residues can occupy the P'1 position (C-terminal to Arg) but they are by far generally small and hydrophylic (Gly or Ser). The P2 position, (N-terminal to Arg) is usually preferentially occupied by a Pro but not exclusively. With respect to the P'2 specificity, thrombin prefers bulky hydrophobic side chains [6]. The P3 preferences of thrombin are not fully characterized but Phe appears to be one of the most favorable residues whereas acidic side chains are clearly detrimental as they are in P'3 position [6].

Thrombin's anion binding exosite 1 is a positively charged surface region located at distance from the catalytic triad and implicated in the binding of thrombin to fibrin-(ogen), thrombomodulin, glycoprotein Ib and the thrombin receptor [3, 5, 7]. Thrombin exosite 1 is constituted by the loop comprising segment Lys^{65}-Glu^{76} and its surrounding groups. The surface of this loop has positively charged amino acid side chains: in particular, Arg^{68}, Arg^{70} and Arg^{73}, surrounded by other positively charged

residues. The dipolar charge distribution due to the presence of the negatively charged surface around the active site, is thought to influence the orientation of molecules during their binding to thrombin. Crystal structure analysis of the thrombin-hirudin complex revealed that thrombin exosite 1 interacts with the C-terminal segment of hirudin, the thrombin-specific anticoagulant from leech saliva [4]. Exosite 1 plays a critical role in thrombin interaction with many of its substrates or acceptor proteins. Hydrolysis of fibrinopeptides requires a recognition step between exosite 1 and fibrinogen [5]. Exosite 1 is also involved in activation of factor V [8], in thrombin interaction with the endothelial cell membrane protein thrombomodulin and with the platelet membrane glycoprotein Ib, and in activation of the thrombin receptor [7]. The exposed 70-80 loop is particularly susceptible to tryptic and to autocatalytic attack [9]. Upon the cleavage of the Arg^{73}-Asn^{74} bond, human β-thrombin is formed and the whole loop structure unfolds. The disruption of exosite 1 has as a consequence a tremendous reduction in clotting and cell-stimulating activity as well as loss of affinity for fibrin, thrombomodulin, GPIb and hirudin. The binding sites for these different proteins appear to share common structures as indicated by the fact that point mutations within the exosite 1 simultaneously impair the binding of several of these proteins [10]. In addition, these proteins reciprocally exert competitive inhibition on the binding of each other to thrombin. However, it has also been reported that some point mutations unequally impair thrombin interaction with the exosite-binding proteins [10], and the thrombin receptor and platelet membrane GPIb that both use exosite 1 for their binding to thrombin do not compete for interaction with thrombin [11]. This indicates that the different binding sites are specified by distinct residues with unique function. Thus, thrombin utilizes the same general surface for substrate recognition regardless of substrate function, although the critical contact residues may vary.

Exosite 2, the second patch of positively-charged residues [3, 12], interacts with polyanions such as heparin,

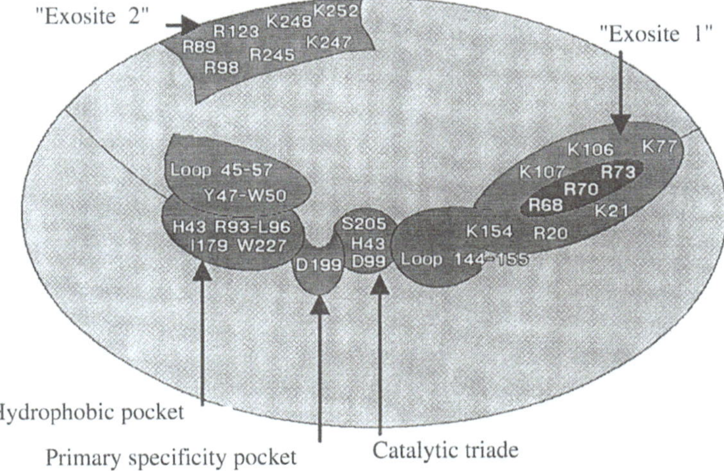

Figure 2. Schematic representation of human α-thrombin (Taken from reference [7] with the permission of the editor).

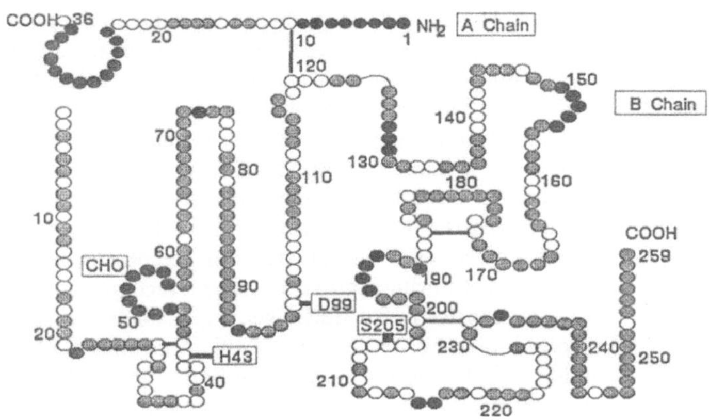

Figure 3. Primary structure of thrombin. By comparison to chymotrypsin, open circles correspond to identical residues, gray circles to substitutions and black circles to insertions. Residues from the catalytic triad (His43, Asp99, Ser205) and carbohydrates (CHO) are in boxes.

Table 1. Thrombin A-chain amino acid sequence

1	2	3	4	5	6	7	8	9	10	11	12	13
T	F	G	S	G	E	A	D	C	G	L	R	P
1H	*1G*	*1F*	*1D*	*1C*	*1B*	*1A*	*1*	*2*	*3*	*4*	*5*	*6*
14	**15**	**16**	**17**	**18**	**19**	**20**	**21**	**22**	**23**	**24**	**25**	**26**
L	F	E	K	K	S	L	E	D	K	T	E	R
6	*7*	*8*	*9*	*10*	*11*	*12*	*13*	*14*	*14A*	*14B*	*14C*	*14D*
27	**28**	**29**	**30**	**31**	**32**	**33**	**34**	**35**	**36**			
E	L	L	E	S	Y	I	D	G	R			
14E	*14F*	*14G*	*14H*	*14I*	*14J*	*14K*	*14L*	*14M*	*15*			

Table 2. Thrombin B-chain amino acid sequence

1	2	3	4	5	6	7	8	9	10	11	12	13
I	V	E	G	S	D	A	E	I	G	M	S	P
16	*17*	*18*	*19*	*20*	*21*	*22*	*23*	*24*	*25*	*26*	*27*	*28*
14	**15**	**16**	**17**	**18**	**19**	**20**	**21**	**22**	**23**	**24**	**25**	**26**
W	Q	V	M	L	F	R	K	S	P	Q	E	L
29	*30*	*31*	*32*	*33*	*34*	*35*	*36*	*37*	*37A*	*38*	*39*	*40*
27	**28**	**29**	**30**	**31**	**32**	**33**	**34**	**35**	**36**	**37**	**38**	**39**
L	C	G	A	S	L	I	S	D	R	W	V	L
41	*42*	*43*	*44*	*45*	*46*	*47*	*48*	*49*	*50*	*51*	*52*	*53*
40	**41**	**42**	**43**	**44**	**45**	**46**	**47**	**48**	**49**	**50**	**51**	**52**
T	A	A	H	C	L	L	Y	P	P	W	D	K
54	*55*	*56*	*57*	*58*	*59*	*60*	*60A*	*60B*	*60C*	*60D*	*60E*	*60F*
53	**54**	**55**	**56**	**57**	**58**	**59**	**60**	**61**	**62**	**63**	**64**	**65**
N	F	T	E	N	D	L	L	V	R	I	G	K
60G	*60H*	*60I*	*61*	*62*	*63*	*64*	*65*	*66*	*67*	*68*	*69*	*70*
66	**67**	**68**	**69**	**70**	**71**	**72**	**73**	**74**	**75**	**76**	**77**	**78**
H	S	R	T	Y	R	Y	E	R	N	I	E	K
71	*72*	*73*	*74*	*75*	*76*	*77*	*77A*	*78*	*79*	*80*	*81*	*82*
79	**80**	**81**	**82**	**83**	**84**	**85**	**86**	**87**	**88**	**89**	**90**	**91**
S	L	M	E	K	I	Y	I	H	P	R	Y	N
83	*84*	*85*	*86*	*87*	*88*	*89*	*90*	*91*	*92*	*93*	*94*	*95*
92	**93**	**94**	**95**	**96**	**97**	**98**	**99**	**100**	**101**	**102**	**103**	**104**
W	R	E	N	L	D	R	D	I	A	L	M	K

96	*97*	*97A*	*98*	*99*	*100*	*101*	*102*	*103*	*104*	*105*	*106*	*107*
105	**106**	**107**	**108**	**109**	**110**	**111**	**112**	**113**	**114**	**115**	**116**	**117**
L	L	K	K	P	V	A	F	S	D	Y	I	H
108	*109*	*110*	*111*	*112*	*113*	*114*	*115*	*116*	*117*	*118*	*119*	*120*
118	**119**	**120**	**121**	**122**	**123**	**124**	**125**	**126**	**127**	**128**	**129**	**130**
V	C	L	P	D	R	E	T	A	A	S	L	L
121	*122*	*123*	*124*	*125*	*126*	*127*	*128*	*129*	*129A*	*129B*	*129C*	*130*
131	**132**	**133**	**134**	**135**	**136**	**137**	**138**	**139**	**140**	**141**	**142**	**143**
Q	A	G	Y	K	G	R	V	T	G	W	G	N
131	*132*	*133*	*134*	*135*	*136*	*137*	*138*	*139*	*140*	*141*	*142*	*143*
144	**145**	**146**	**147**	**148**	**149**	**150**	**151**	**152**	**153**	**154**	**155**	**156**
L	K	E	T	W	T	A	N	V	G	K	G	Q
144	*145*	*146*	*147*	*148*	*149*	*149A*	*149B*	*149C*	*149D*	*149E*	*150*	*151*
157	**158**	**159**	**160**	**161**	**162**	**163**	**164**	**165**	**166**	**167**	**168**	**169**
P	S	V	L	Q	V	V	N	L	P	I	V	E
152	*153*	*154*	*155*	*156*	*157*	*158*	*159*	*160*	*161*	*162*	*163*	*164*
170	**171**	**172**	**173**	**174**	**175**	**176**	**177**	**178**	**179**	**180**	**181**	**182**
R	P	V	C	K	D	S	T	R	I	R	I	T
165	*166*	*167*	*168*	*169*	*170*	*171*	*172*	*173*	*174*	*175*	*176*	*177*
183	**184**	**185**	**186**	**187**	**188**	**189**	**190**	**191**	**192**	**193**	**194**	**195**
D	N	M	F	C	A	G	Y	K	P	D	G	P
178	*179*	*180*	*181*	*182*	*183*	*184*	*184A*	*185*	*186*	*186A*	*186B*	*186C*
196	**197**	**198**	**199**	**200**	**201**	**202**	**203**	**204**	**205**	**206**	**207**	**208**
K	R	G	D	A	C	E	G	D	S	G	G	P
186D	*187*	*188*	*189*	*190*	*191*	*192*	*193*	*194*	*195*	*196*	*197*	*198*
209	**210**	**211**	**212**	**213**	**214**	**215**	**216**	**217**	**218**	**219**	**220**	**221**
F	V	M	K	S	P	F	N	N	R	W	Y	Q
199	*200*	*201*	*202*	*203*	*204*	*204A*	*204B*	*205*	*206*	*207*	*208*	*209*
222	**223**	**224**	**225**	**226**	**227**	**228**	**229**	**230**	**231**	**232**	**233**	**234**
M	G	I	V	S	W	G	E	G	C	D	R	D
210	*211*	*212*	*213*	*214*	*215*	*216*	*217*	*219*	*220*	*221*	*221A*	*222*
235	**236**	**237**	**238**	**239**	**240**	**241**	**242**	**243**	**244**	**245**	**246**	**247**
G	K	Y	G	F	Y	T	H	V	F	R	L	K
223	*224*	*225*	*226*	*227*	*228*	*229*	*230*	*231*	*232*	*233*	*234*	*235*
248	**249**	**250**	**251**	**252**	**253**	**254**	**255**	**256**	**257**	**258**	**259**	
K	W	I	D	Q	K	V	I	D	F	G	E	
236	*237*	*238*	*239*	*240*	*241*	*242*	*243*	*244*	*245*	*246*	*247*	

a sulfated glycosaminoglycan which accelerates thrombin interaction with antithrombin [13]. The prothrombin fragment 2 also binds to this site [14]. In addition, the chondroitin sulfate chain on thrombomodulin binds to thombin exosite 2 and contributes to the high affinity interaction between the two proteins [15].

The ability to allosterically modulate thrombin-substrate interactions by exosites ligands has been described particularly concerning exosite 1 [16]. Very recently it was also shown that sodium can induce an allosteric switch, the sodium bound fast form of thrombin being procoagulant while the sodium-free slow form more specifically activates protein C [17].

Structure

Sequence and Size   The sequence of human a-thrombin is numbered in bold characters with the topological chymotrypsinogen equivalence indicated in italics. The light A-chain contains 36 residues and the heavy B-chain is composed of 259 residues and is glycosylated at Asn53. The primary structure of human thrombin is schematized in Figure 3.

Homologies   Thrombin sequence presents similarities with bovine chymotrypsin and trypsin. However, thrombin B-chain is 27 residues longer than chymotrypsin (Figure 3), most insertions forming loops involved in thrombin specificity.

Conformation   An important structural feature for thrombin specificity involves the multiplicity of interaction sites for its substrates. At the end of the eighties, crystals of thrombin in complexes with the small inhibitor D-Phe-Pro-Arg chloromethylketone [3] or with hirudin [4] were obtained, allowing the elucidation of the thrombin structure (Figure 2). The thrombin molecule is a prolate ellipsoid (45 x 45 x 50 Å) characterized by the presence on one of its faces, of a deep, narrow, canyon-like incision containing the active site cleft at one extremity. The active-site residues and the entrance of the specificity pocket are partially obstructed by two surrounding loops. In particular, the "60" (residues 47 to 57) and "149" (residues 144-155) insertion loops to the north and south rims of the active site cleft (Figure 2) limit the access of many macromolecular substrates or inhibitors to the catalytic residues. For example, thrombin cannot be inhibited by typical serine proteinase inhibitors such as bovine pancreatic trypsin inhibitor (BPTI), unless one or both of these loops has been removed by mutagenesis [6]. The Leu^{144}- Gly^{150} loop is very flexible and exposed to the surface. This is the reason why it is easily proteolysed by several enzymes including cathepsin G and chymotrypsin (cleavage at Trp148), elastase (cleavage at Ala^{150}), trypsin or thrombin (cleavage at Lys^{154}). Also within the groove, but at the opposite extremity to the catalytic cleft, are located the structural elements of the exosite 1. Looking this face of thrombin, exosite 2 is located at the upper part of the molecule.

Gene

Gene Structure   The thrombin precursor, prothrombin is encoded by a 21-kb-long gene [18]. The prothrombin gene is organized in 14 exons separated by 13 introns with the 5' upstream and 3' untranslated regions which may play regulatory roles in gene expression. The prothrombin cDNA encodes a multidomain protein that includes a signal peptide, a propeptide, a domain rich in γ-carboxyglutamic acid (Gla), a short aromatic amino acid stack domain, two kringle domains and the serine protease (thrombin) domain.

Chromosomal Localization   The prothrombin gene is located on chromosome 11, position 11p11-q12 [19].

Gene Expression   Prothrombin is synthesized by liver cells and circulates in plasma at the relatively high concentration of 1.4 µM. Prothrombin mRNA has also been detected in other tissues such as in the normal brain [20].

Gene Regulation   A genetic variant of prothrombin was recently described as being associated with an increased risk of venous thrombosis [21]. The variant is located in the 3'-untranslated region of the gene, at position 20210 where one nucleotide is changed (a G to A transition). Carriership of this variant is associated with elevated levels of prothrombin in plasma, which are related to an increased risk of thrombosis.
Hereditary prothrombin deficiency is a rare disorder expressed either as hypoprothrombinemia or as dysprothrombinemia. Several families with hereditary hypoprothrombinemia (~ 20) or dysfunctional prothrombin variants (~ 25) have been described but the responsible mutations have been reported for only few of them [22].

Processing and Fate   Thrombin is generated from its zymogen prothrombin, which (as factors VII, IX, and X as well as proteins C and S) is a vitamin K-dependent protein. Vitamin K is required for the post-transcriptional carboxylation of glutamic acid residues (Gla) located in the N-terminal region of the zymogen. Prothrombin is a single chain protein of 71.6 kDa composed of 579 residues arranged in four domains consist-

ing of the N-terminal Gla domain followed by two tandem kringle domains and the catalytic domain. Prothrombin activation is catalyzed by prothrombinase [23], an enzymatic complex consisting of equimolar factor Xa (a serine protease) and its protein cofactor, factor Va, bound to a procoagulant surface in a Ca^{2+}-dependent manner. In comparison to FXa alone, formation of the complete complex enhances the rate of prothrombin activation by 10^5 fold. Therefore, the rate of thrombin generation depends on the simultaneous availability of these components. Dramatic increase in catalytic efficiency of prothrombinase arise from a 100 fold decrease in apparent Km resulting from interaction of negatively charged phospholipids with proteins and a 10^3 fold increase in kcat due to the cofactor effect of factor Va. Prothrombin activation requires two proteolytic cleavages, one at Arg ^{271} to release the activation fragment, fragment 1+2, which correspond to roughly half the mass of prothrombin, and the other at Arg^{320} to generate the two chain enzyme thrombin [23]. Depending on the order of the cleavages, two intermediates can be formed, prethrombin 2 or meizothrombin (Figure 4). In the absence of factor Va, the first cleavage occurs at Arg^{271} and the major intermediate is prethrombin 2. In contrast, in the presence of factor Va and of phospholipids, the first cleavage occurs at Arg^{320} resulting in the formation of meizothrombin as the major intermediate. Meizothrombin hydrolyses well small peptide chromogenic substrate but does not clot fibrinogen effectively . Subsequent cleavage at Arg^{271} releases the mature two-chains procoagulant α-thrombin in the blood vessel. Due to an autocatalytic cleavage at Arg^{285}-Thr^{286}, the human α-thrombin A chain is further truncated to 36 residues. The conversion of prothrombin to thrombin is physiologically catalyzed by factor Xa in the prothrombinase complex, but prothrombin activation can also be efficiently performed by proteases derived from snake venoms such as *Echis carinatus, Dipholidus typus* and *Oxyuranus scutellatus*.

Autoregulation of thrombin formation   One striking characteristic of thrombin is that it controls the rate of its formation. The thrombin paradox lies in the fact that thrombin can both promote and prevent blood clotting [24]. Actually, thrombin contributes to make coagulant surfaces available by activating platelets. Full platelet activation results in the relocation of anionic aminophospholipids, in particular phosphatidylserine, from the inner leaflet of the membrane to the outer surface allowing the Ca^{2+}-dependent binding of vitamin K-dependent clotting factors via their Gla residues [25]. Platelet activation also results in factor V release. Factor V is activated by two proteolytic cleavages that are efficiently performed by thrombin at low concentration [26]. Finally, thrombin contributes to the generation of factor Xa by activating factor VIII, the cofactor for "tenase", the intrinsic factor X activating enzyme complex [26]. Amplification of thrombin formation by thrombin itself is evidenced *in vitro* by the sigmoidal

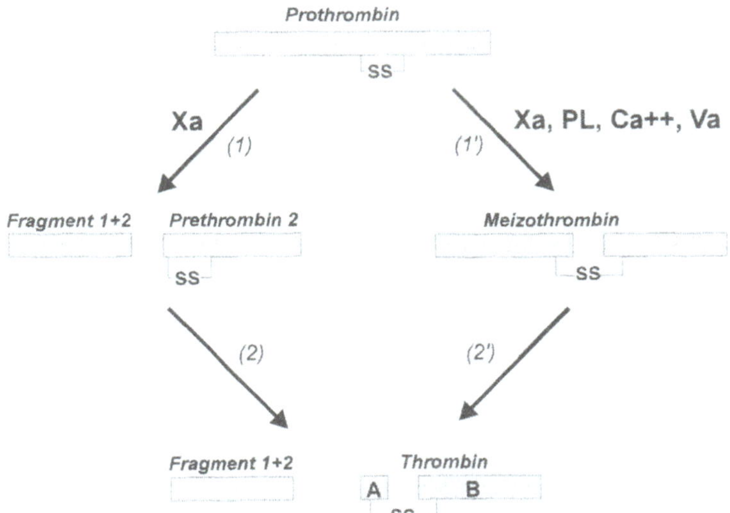

Figure 4. Prothrombin activation by factor Xa.

aspect of the thrombin generation curve and by the dramatic slow-down effect of thrombin inhibitors on prothrombin activation.

In contrast, thrombin decreases prothrombin activation by activating a major anticoagulant pathway. In the presence of the endothelial cell surface cofactor thrombomodulin, thrombin activates protein C in a Ca^{2+}-dependent reaction [15]. Activated protein C in turn, inactivates factors Va and VIIIa by limited proteolysis, resulting in the decrease of cofactor concentration and thus of the rate of factor Xa and thrombin production. The physiological importance of this regulatory mechanism is demonstrated by the occurrence of life-threatening thrombosis in neonates with homozygous protein C deficiency [27]. In addition, a single mutation of factor Va at a peptide bond cleaved by activated protein C is associated with an increased risk of thromboembolic disease [28].

Biological Activity

Thrombin interaction with fibrin(ogen)   Fibrinogen is a soluble plasma protein that consists of three pairs of polypeptide chains linked by disulfide bridges to form the $(A\alpha,B\beta,\gamma)_2$ hexamer. Thrombin catalyzes the conversion of fibrinogen to insoluble fibrin. Thrombin cleaves 4 short fibrinopeptides (fibrinopeptides A and B), thus producing fibrin monomers that spontaneously polymerize to form the nascent insoluble but unstable clot. Thrombin cleaves only 4 of a potential 376 Arg/Lys-X fibrinogen bonds, underlining the high specificity that results from the multiple interactions between fibrinogen and thrombin, involving both the catalytic domain and exosite1 [29].

After cleavage of fibrinopeptide A, thrombin remains associated to fibrin via interactions at exosite 1. Limited diffusion of the thrombin molecule could then enhance cleavage of the remaining fibrinopeptides. In addition, colocalisation of thrombin and factor XIII promotes the activation of factor XIII by thrombin and in turn the crosslinking of fibrin [30].

Thrombin sequestered in the clots is the subject of much interest, for the binding of thrombin to fibrin is thought to have critical consequences. Clinically significant thrombotic disease has been found in patients with congenital dysfibrinogens that demonstrate impaired thrombin-fibrin interactions *in vitro* [31]. Thus the thrombin-fibrin interaction probably restricts the active enzyme to the site of injury *in vivo* and limits further pathologic thrombin formation. On the other hand, fibrin-bound thrombin remains catalytically active and is protected from inactivation by its main inhibitor antithrombin in the presence of heparin [32]. This means that clots have a thrombin-dependent procoagulant activity that results in more fibrin formation and also in platelet activation. In addition, sequestration means that active thrombin may be released upon fibrin degradation by plasmin.

Protein C-thrombomodulin system   Anticoagulatory function of thrombin resides in the activation of protein C, which in turns inactivates factors Va and VIIIa thereby inhibiting further production of thrombin [15]. Protein C is a poor substrate for thrombin on its own, due to the presence of two acidic unfavorable residues (Asp) at P3 and P'3 that might collide with thrombin putative S3 and S'3 residues Glu^{25} and Glu^{202} [13]. Binding of thrombin to the endothelial cell surface protein thrombomodulin causes a dramatic enhancement of the ability of the enzyme to cleave protein C (Figure 5). Thrombomodulin binding to thrombin exosite 1 induces a conformational change in thrombin resulting in the transition of an acidic to a basic residue (Arg 20) at S'3 thus permitting the accomodation of protein C in the active site cleft. The thrombomodulin-induced modification of thrombin specificity from procoagulant to anticoagulant results not only from the capacity of the complex to activate protein C but also from the fact that the high affinity binding (K_d=5 nM) of thrombomodulin to thrombin

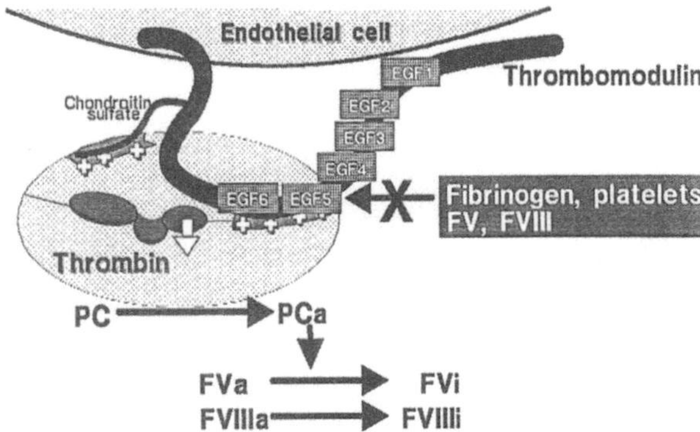

Figure 5. Modification of thrombin specificity by thrombomodulin. Thrombomodulin is represented with its EGF5-6 domains docked into the thrombin exosite 1 and the chondroitin sulfate chain bound to the exosite 2.

inhibits thrombin interaction with fibrinogen and with factor V and also thrombin-induced platelet activation. Since at the site of lesions, the inflammatory reaction can reduce thrombomodulin expression via the action of cytokines such as TNFα, one can assume that the anticoagulant effect of the thrombin-thrombomodulin complex should be maximal at distance of vascular lesions and control the dissemination of the coagulation. Because protein C activation by the thrombin-thrombomodulin complex is a relatively low affinity reaction, it has been believed to occur only in microcirculation. However, a novel endothelial receptor for protein C/activated protein C (EPCR) has recently been identified. EPCR greatly accelerates protein C activation by the thrombin-thrombomodulin complex. EPCR is found to be abundantly expressed by the endothelium of large vessels. This suggests that the protein C anticoagulant pathway is also important for the regulation of blood coagulation in large vessels [33].

A novel biological activity of the thrombin-thrombomodulin complex has recently been recognized in fibrinolysis. TAFI (thrombin activatable fibrinolysis inhibitor), the precursor of a plasma carboxypeptidase B, can be converted by thrombin to an active enzyme capable of eliminating the C-terminal Lys and Arg residues from fibrin thus reducing plasmin production and fibrinolysis. Thrombomodulin increases the rate of TAFI activation by thrombin by about three orders of magnitude [34].

Mechanism of thrombin-induced cell activation

Thrombin receptors   Most effects of thrombin on cells appear to rely on its proteinase activity. This important characteristic has been explained by the identification of a novel family of receptors activated by proteolytic cleavage rather than by ligand binding. The thrombin receptor is the prototype of proteolytically activated receptors (PARs). The thrombin receptor PAR1 cDNA was isolated by expression cloning in *Xenopus* oocytes [35, 36]. The resulting amino acid sequence revealed that PAR1 belongs to the family of the seven transmembrane domain, G protein-coupled receptors, being largely similar to receptors for neuropeptides and glycoprotein hormones. However, its mechanism of activation is unique. Thrombin cleaves the receptor amino terminal exodomain, unmasking a new N-terminus that then acts as a tethered ligand by binding to the body of the receptor and causing receptor activation (Figure 6). The unusually long N-terminal extracellular domain of the receptor thus contains structural determinants which are essential for the correct functioning of the receptor. The thrombin cleavage site is defined by the sequence LDPR/SFLL, thrombin cleaving the Arg^{41}-Ser^{42} bond. As it is the case for protein C, the presence of an unfavorable acidic residue in position P3 (Asp^{39}) should be associated to a low catalytic efficiency. However, a second site of interaction with thrombin resides between residues 53 and 64. This sequence shows similarity to the C-terminal region of hirudin and is involved in interaction with thrombin exosite 1. The K51YEPF sequence appears to be the primary area of interaction with exosite 1 as indicated by the observations that the substitutions for Tyr^{52}, Glu^{53} and Phe^{55} in the receptor result in loss of receptor response to thrombin. The initial contact between thrombin exosite 1 and the receptor sequence KYEPF would orientate the binding of the cleavage site into the active site cleft thus allowing the cleavage to occur. Synthetic peptides identical with the sequence N-terminal to the receptor cleavage site (SFLLRN...) are full agonist for activation of the thrombin receptor and are thus called thrombin-receptor-activating peptides (TRAPs). TRAPs can cause platelet activation, endothelial cell secretion and synthesis of prostacyclin and a number of effects associated with stimulation by thrombin [37]. PAR1 from different species have also been cloned. The thrombin cleavage and exosite-binding sequences are highly conserved between the mammalian thrombin receptors. The features of TRAPs required to mimic the effects of thrombin lie within the first 5-6 N-terminal residues of the "tethered ligand" SFLLRN and the side chains of Phe, Leu and Arg in position 2, 4 and 5 respectively are essential for full activity.

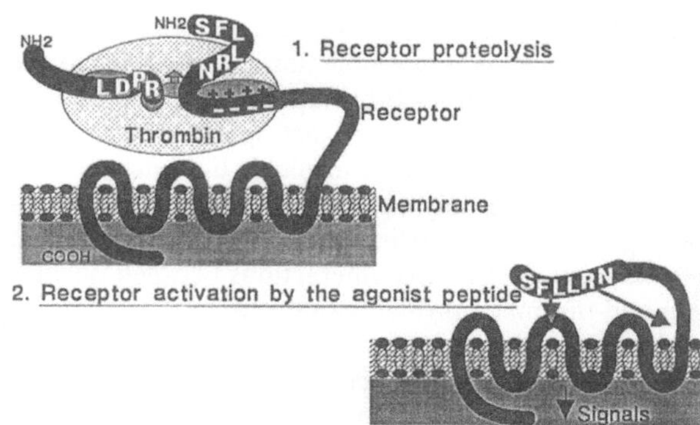

Figure 6. Activation of the thrombin receptor PAR1 by thrombin.

Two extracellular structural domains appear to be critical for the docking of the agonist peptide: region 82-90 in the amino-terminal exodomain and 259-262 in the second extracellular loop [38, 39]. The importance of the second extracellular loop in the transmission of the information across the cell membrane to cause signalling has been underlined by the observation that mutation within this domain results in the constitutive activation of the receptor [39].

Although TRAPs can substitute for thrombin in the promotion of many cellular events, such as second messenger generation, secretion, and mitogenesis, there are several lines of evidence that the agonist peptide may not evoke all the responses seen with thrombin, or even necessarily the same pattern of responses as those which are seen with thrombin [40]. These apparent discrepancies suggested the existence of a second platelet receptor. This has recently been supported by the observation that deletion of the mouse PAR1 had no obvious effect on the response of mouse platelets to thrombin even though it did eliminate thrombin responses in other cell types [41,42]. Very recently, a third receptor PAR has been cloned and characterized as a new thrombin receptor PAR3 [43]. Although PAR3 can mediate thrombin-triggered phosphoinositides hydrolysis and is expressed in a variety of human tissues, its biological role is presently unknown. However, very recent studies have indicated that responses to thrombin were largely diminished in platelets from PAR3-deficient mice but did persist. These responses are mediated by a new thrombin receptor PAR4 which has been cloned and appears to be functional in human platelets. The mechanism of PAR4 activation by thrombin appears to be very similar to that of PAR1 but to require higher thrombin concentrations [44, 45]

The second known member of the PAR family, PAR2, closely resembles PAR1 in both structure and mechanism of activation but is activated by trypsin and tryptase and not by thrombin [46]. Peptides corresponding to tethered ligand domain of PAR2 can activate PAR2, as can peptides corresponding to the PAR1 sequence, but the reverse is not true. The capacity of PAR1-activating peptides to activate platelets is thus not a sufficient criteria to conclude to the presence of PAR1 on these cells. PAR2 does not appear to be present in human platelets [47] but is expressed in at least some endothelial cells, in keratinocytes, intestinal epithelium and some smooth muscle cells. The human PAR2 and PAR3 genes as the PAR1 gene are present as a single-copy locus and mapped to chromosomal region 5q13, whereas PAR4 gene is mapped to 19p12.

Like other G protein coupled receptors, PARs are active briefly and become refractory to subsequent activation. However, PARs are characterized by the fact that they can be used only once since they are irreversibly cleaved by their activating proteases. Following activation, PAR1 is rapidly internalized in platelets and in endothelial cells. In platelets, there is no recovery due to the absence of an intracellular reserve of PAR1 and to the little capacity for protein synthesis [48]. In contrast, recovery of endothelial cells responsiveness to thrombin occurs in parallel with the re-emergence on cell surface of intact receptors from the intracellular pool and to the replenishment of the stock by protein synthesis [47]. Several peptide bonds within the extracellular N-terminal domain of PAR1 are potential cleavage sites for proteases but only one cleavage leads to receptor activation. Different proteases (cathepsin G, elastase, chymotrypsin, plasmin) have been shown to inactivate the receptor by cleaving PAR1 N-terminus, at a disabling site downstream from the activating site [47], suggesting that PARs activation may be modulated *in vivo* by such mechanisms.

Activation of the receptor by thrombin or agonist peptides results in dual coupling to phospholipase C and adenylyl cyclase (reviewed in [37]). In most cells, thrombin stimulates phospholipases C, A_2 and D, activates protein kinase C, PI3 kinases, MAP kinases and tyrosine kinases, transiently raises the cytosolic free Ca^{2+} concentration, gates the movements of ions across the membrane. In endothelial cells, thrombin can increase cAMP levels whereas in platelets, thrombin inhibits adenylyl cyclase activity. The b PLC form is activated by two potential routes in response to thrombin. In platelets, the predominant route appears to be pertussis toxin-sensitive Gi-derived Gbγ [49] whereas in endothelial cells, fibroblasts and most other cells, phospholipase C activation is largely unaffected by pertussis toxin and appears mediated predominantly by a member of the Gqα family [50].

Cell surface proteins modulating thrombin-triggered responses
The sensitivity of different cells to thrombin is related to the amount of receptor at their surface, to the efficiency of the signal transduction pathway and also to the amount of thrombin locally available to cleave the receptor. The presence of thrombin-binding proteins at the cell surface can thus modulate the activity of thrombin on these cells. On endothelial cells thrombomodulin is susceptible to decrease cell responses triggered by thrombin. The addition of recombinant thrombomodulin to HUVEC decreases the responses to thrombin whereas in contrast, thrombomodulin blockade by antibodies increases the responses [51]. In addition, coexpression of thrombomodulin with PAR1 in vascular smooth muscle cells results in the decrease of the responses to thrombin [52].

On the other hand, one protein has long been known to play a potentiating role on cell activation. This is platelet membrane glycoprotein Ib (GPIb) [53]. Numerous studies have contributed to demonstrate that GPIb is a high affinity binding-site for thrombin on platelets and that in the absence of GPIb such as platelets from the patients suffering from the Bernard-Soulier syndrome, platelets sensitivity to thrombin is decreased and all the responses are delayed. The mechanism of thrombin interaction with GPIb involves, on the one hand the thrombin exosite and on the other a negatively charged region within the N-terminal globular domain of the GPIbα chain. The potentiating effect of GPIb on platelet activation by thrombin requires its anchorage to the membrane [11]. A GPIb-

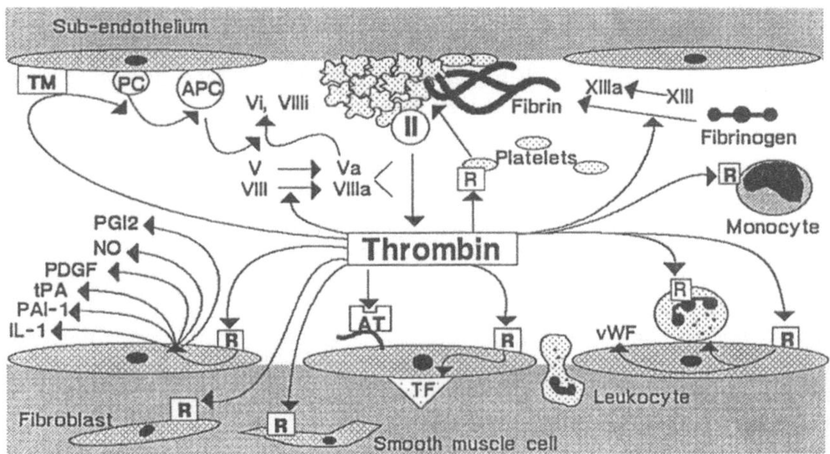

Figure 7. Thrombin effects on circulating cells and cells from the vascular wall. Proteins are symbolized by: TR, thrombin receptor; TF, tissue factor; TM, thrombomodulin; AT, antithrombin. On this schema, antithrombin is represented bound to the glycosaminoglycans expressed at the endothelial cell surface.

dependent thrombin-induced increase in intracellular calcium concentration has been reported [54] but the precise mechanism by which the GPIb-thrombin interaction potentiates the PAR1-coupled responses remains to be determined.

Thrombin inhibition   Thrombin activity is regulated in plasma by several antiproteinases. In the circulation, thrombin is essentially controlled by antithrombin, which is a member of the protein superfamily of serpins, and also inhibits other coagulation serine proteases mainly factor Xa [13]. Actually, antithrombin deficiencies are associated to an increased risk of thromboembolic disease. Antithrombin acts by forming an equimolecular complex with thrombin in which the enzyme is inactive. The reaction is initiated by the proteinase attacking a specific reactive bond in antithrombin. However, at some intermediate step of the catalytic process the cleavage stops, and the enzyme is trapped. The resulting complex only very slowly dissociates to free enzyme and inactive, cleaved inhibitor. The rate of thrombin inhibition by antithrombin is low but is greatly accelerated (up to 10^4) in the presence of glycosaminoglycans such as heparan sulfate. Thrombin inhibition is accelerated by thrombin and antithrombin being brought together by binding to the same chain of heparin. These properties make heparin a powerful anticoagulant largely used in the prophylaxis and the treatment of venous thrombosis. Another serpin, heparin cofactor 2 is specific for thrombin. Thrombin inhibition by heparin cofactor 2 is also accelerated by glycosaminoglycans but the mechanism is different from that involved with antithrombin [55]. Binding of heparin or dermatan sulfate to heparin cofactor 2 demasks a site on the inhibitor that binds to thrombin exosite 1.

If thrombin is efficiently inhibited in solution by these inhibitors and particularly antithrombin, this is not the case when thrombin is immobilized within a fibrin clot or on the extracellular matrix where bound thrombin resists inactivation by antithrombin-heparin complex [32, 56]. Cells also contain a serpin, protease-nexin 1 (PN1) that is a potent inhibitor of thrombin. This is illustrated by the inhibitory effect of PN1 on the thrombin-triggered retrac-

tion of the processes on neurons and astroglial cells [57]. Several cell types are expressing PN1 such as vascular endothelial cells [58]. However, the possible involvement of PN1 in the regulation of thrombin interaction with its receptor on theses cells has not yet been indicated.

Role in Vascular Biology   Thrombin has widespread *in vitro* cellular effects concerning cells located within the vascular bed (Figure 7) and in extravascular tissues.

Blood and vascular cells   Thrombin is the most potent physiologic platelet agonist [59]. *In vitro*, thrombin concentrations below 0.5 nM (0.05 U/ml) induce full platelet activation evidenced by shape change, secretion of the content of dense, α- and lysosomial granules and platelet aggregation. Thrombin has growth promoting and chemotactic effects on macrophages [60] and is chemotactic for monocytes and neutrophils [61].
Thrombin induces production of a wide variety of compounds by endothelial cells (reviewed in [62]). It enhances the synthesis of NO, prostacyclin and platelet-activating factor, induces tissue factor expression, increases transcription and release of platelet-derived growth factor, tissue plasminogen activator and its inhibitor, and promotes endothelial cell release of von Willebrand factor. Furthermore, thrombin evokes the expression of specific leukocyte adhesion molecules on the endothelial surface, inducing adherence of neutrophils and monocytes to the cells, and increases cell permeability and gap formation. In addition, thrombin is mitogenic for fibroblasts and smooth muscle cells [39,63]. These observations suggest an important role for thrombin not only in mediating haemostatic events, but also in controlling the inflammatory and proliferative responses associated with wound healing. In addition, thrombin bound to the extracellular matrix remains functionally active and protected from inactivation by antithrombin, allowing prolonged effects on the vessel wall [56]. Actually, thrombin activity is thought to be involved in post-clotting vascular events observed *in vivo*, as restenosis after angioplasty.

Non vascular cells   Thrombin is also active on non vascular cells such as mesangial cells [64],osteoblasts and keratinocytes. Interestingly, thrombin has important effects on neural cells; it has a potent inhibitory effect on neurite outgrowth [65] and on astrocytes stellation [66]. These non-vascular effects of thrombin raise the question of the origin of the enzyme in tissues. Indeed tissues can enter in contact with thrombin following a disruption of the blood barrier and the deleterious effect of hemorrhages on neural tissues is well known. However, the observations that prothrombin mRNA [20], thrombin receptor [65] and thrombin regulatory proteins such as thrombomodulin [67] and protease-nexin [57] are expressed in developing and mature cells in the nervous system, suggest that thrombin may be locally produced and involved in development. This, in turn, suggests that tissues have, in place of prothrombinase, a still unknown prothrombin activator. The involvement of thrombin in processes other than haemostasis and in particular in development is supported by the recent observation that the disruption of the prothrombin gene in mice is lethal *in utero* [68].

Pathology   Except for the rare cases of dysprothrombinemia for which the molecular abnormality affects the thrombin moiety of the molecule, pathological situations result mainly from a desequilibrium between thrombin production and thrombin inhibition, a deficit in thrombin being associated with a risk of hemorrhages while an excess of thrombin being associated with a risk of thrombosis.

Clinical Relevance and Therapeutic Implications   Due to its pivotal role in the coagulation cascade, thrombin is an important target for antithrombotic therapies. Although heparin is the current antithrombin widely used in clinical practice, it presents some limitations. It catalyzes the inactivation of fluid-phase thrombin by antithrombin (thus depending on the plasma antithrombin level) but is unable to inhibit clot-bound thrombin. Heparin has several other disadvantages including a wide variation in its anticoagulant effect, its binding to several plasma proteins and endothelial cells thus reducing its availability and potency, its inhibition by platelet factor 4, and the development of immune thrombocytopenia. Direct thrombin inhibitors have several theoretical advantages over heparin: they are independent of the antithrombin level. They do not bind to plasma or platelet proteins and they effectively inhibit fibrin-bound thrombin. That is why the direct inhibition of thrombin is hypothesized to produce a more complete and consistent inhibition of thrombin activity with reduced rates of adverse events [69]. The prototypic direct antithrombin is hirudin, a 65-amino acids compound derived from the medicinal leech, which binds selectively via its carboxy terminus to thrombin exosite 1 and via its amino terminus to the catalytic center of the enzyme. Large clinical trials comparing recombinant hirudin or bivalirudin to heparin in patients with coronary angioplasty or thrombolysis for acute myocardial infarction, have failed to demonstrate a long term benefit over heparin when given at safe levels. In contrast, hirudin has been observed to be more effective than unfractionated and low molecular weight heparin in prevention of postoperative deep venous thrombosis after total hip replacement [70]. Other direct antithrombins are found in small synthetic molecules that block the catalytic site. Several are in development and clinical trials are in progress for some of them. Another interesting possibility to antagonize the prothrombotic effects of thrombin is to modify the balance between its pro and anticoagulant effects in favor of its anticoagulant activity. Molecules able to inhibit the coagulant activity of thrombin but to respect or even to increase its capacity to activate protein C would be of theoretical therapeutic interest.

The identification of the thrombin receptor in relevant cell types, together with the realization that many of the pathological actions of thrombin on these cells appear at least in part mediated by the thrombin receptor, suggest a role for this receptor in the pathological processes of thrombosis, inflammation, atherosclerosis, and fibroproliferative disorders. Thrombin receptor antagonists may thus have value as therapeutic agents by specific inhibition of the cellular actions of thrombin. Such molecules are in development.

Martine Jandrot-Perrus, Marie-Christine Bouton
and Marie-Claude Guillin

References

1. Davie EW (1995) Thromb Haemostasis 74:1-6
2. Butkowski RJ et al (1977) J Biol Chem 252:4942-4957
3. Bode W et al (1992) In: Berliner LJ (ed) Thrombin: Structure and Function. Plenum Press, London, pp 3-61
4. Rydel TJ et al (1990) Science 249:277-280
5. Fenton JW et al (1988) Biochemistry 27:7106-7112
6. Le Bonniec B et al (1996) Biochemistry 35:7114-7122
7. Guillin MC et al (1995) Thromb Haemost 74:129-133
8. Esmon CT, Lollar P (1996) J Biol Chem 271:13882-13887
9. Guillin MC, Bezeaud A (1992) Semin Thromb Hemostasis 18:224-229
10. Tsiang M et al (1995) J Biol Chem 270:16854-16863
11. Bouton MC et al (1995) Biochem J 305:635-641
12. Gan ZR et al (1994) J Biol Chem 269:1301-1305
13. Olson ST, Björk I (1994) Semin Thromb Hemost 20:373-409
14. Arni RK et al (1993) Biochemistry 32:4727-4737
15. Esmon CT (1995) FASEB J 9:946-955
16. Ye J et al (1991) J Biol Chem 266:23016-23021
17. Dang QD (1996) Proc Natl Acad Sci USA 93:10653-10656
18. Degen SJF, Davie EW (1987) Biochemistry 26:6165-6170
19. Royle NJ et al (1987) Somat Cell Mol Genet 13:285-289
20. Dihanich M et al (1991) Neuron 6:575-581
21. Poort SR et al (1996) Blood 88:3698-3703
22. Degen SJF (1995) In: High KA, Roberts HR (eds) Molecular basis of thrombosis and haemostasis. Marcel Dekker, New York, p 75
23. Mann KG (1994) In: Coleman RW, Hirsh J, Marder VJ, Salzman EW (eds) Hemostasis and Thrombosis, 3rd edition. Lippincott, Philadelphia, pp 184-199
24. Griffin JH (1995) Nature 378:337-338
25. Rosing J et al (1985) Blood 65:319-322
26. Kane WH, Davie EW (1988) Blood 71:539-555
27. Seligsohn U et al (1984) N Engl J Med 310:559-562
28. Dahlbäck B et al (1993) Proc Natl Acad Sci 90:1004-1008

29. Fenton JW et al (1991) Blood Coagulation and Fibrinolysis 2:69-75
30. Greenberg CS et al (1987) Blood 69:867-871
31. Liu LH et al (1979) J Biol Chem 257:10421-10425
32. Weitz JI et al (1990) J Clin Invest 86:385-391
33. Laszik Z et al (1997) Circulation 96:3633-3640
34. Bajzar L et al (1996) J Biol Chem 271:16603-16608
35. Vu TKH et al (1991) Cell 6775-6779
36. Rasmussen UB et al (1991) FEBS Let 288:123-128
37. Grand RJA et al (1996) Biochem J 313:353-368
38. Bahou WF et al (1994) Blood 84:4195-4202
39. Nanevicz T et al (1996) J Biol Chem 271:702-706
40. Vouret-Craviari et al (1992) Mol Biol Cell 3:95-102
41. Conolly AJ et al (1996) Nature 381:516-519
42. Darrow AL et al (1996) Thromb Haemost 96:860-866
43. Ishihara H et al (1997) Nature 386:302-306
44. Xu WF et al (1998) Proc Natl Acad Sci USA 95:6642-6646
45. Kahn ML et al (1998) Nature 394:690-694
46. Nystedt S et al (1994) Proc Natl Acad Sci USA 91 9208-9212
47. Brass LF, Molino M (1997) Thromb Haemost 78:234-241
48. Molino M et al (1997) J Biol Chem 272:6011 6017
49. Brass LW et al (1988) J Biol Chem 263:5348-5355
50. Garcia J et al (1991) Cell Physiol 142:186-193
51. Parkinson JF et al (1993) Atheriosclerosis and thrombosis 13:1119-1123
52. Grinnel BW et al (1996) Am J.Physiol 270:H603-H609
53. Jandrot-Perrus M et al (1996) Thromb Haemost 22:151-156
54. Greco NJ et al (1996) Biochemistry 35:906-914
55. Sheehan JP et al (1994) J Biol Chem 269:32747-32751
56. Bar Shavit R et al (1989) Cell Regul 1:453-463
57. Cunningham DD et al (1993) Thromb Haemost 70 168-171
58. Leroy-Viard K et al (1989) Exp Cell Res 181:1-9
59. Shuman MA (1986) Ann NY Acad Sci 485:228-239
60. Bar Shavit R et al (1986) Cell Biochem 32:261-272
61. Joseph LV, Mc Dermot J (1993) Biochem J 290:571-577
62. Garcia JGN (1995) Blood Coagulation and Fibrinolysis 6:609-626
63. McNamara CA et al (1992) J Clin Invest 91:94-98
64. He CJ et al (1992) J Am Physiol 150:475-483
65. Suidan et al (1992) Neuron 8:363-375
66. Cavanaugh et al (1990) J Neurochem 54:1735-1743
67. Pindon A et al (1997) Glia 19 259-268
68. Xue J et al (1998) Proc Natl Acad Sci USA 95:7603-7607
69. Verstraete M (1997) Thomb Haemost 78:257-363
70. Eriksson BI et al (1997) N Engl J Med 19:1329-1335

Thrombocytes

Definition *same as platelets*

See: →Platelet stimulus-response coupling; →Megakaryocytes

Thrombomodulin

Definition *Protein of the blood clotting system that induces a conformational change in thrombin resulting in the activation of protein C*

See: →Thrombin; →Fibrinolytic, hemostatic and matrix metalloproteinases, role of

Thromboplastin

See: →Procoagulant activities

Thrombopoietin

Definition *Peptide factor that stimulates megakaryocytopoiesis*

See: →Megakaryocytes

Thrombosis, Arterial
Synonym: Atherothrombosis

Definition *Clot formation within arteries; clincial complication of atherosclerosis*

See also: →Fibrinolytic, hemostatic and matrix metalloproteinases, role of; →Coagulation factors; →Prostacyclin; →Prostaglandins; →Vascular integrins

Introduction The factors implicated in arterial thrombosis (endothelium, hemostasis, coagulation, fibrinolysis) are also implicated at all the stages of atherosclerosis. This has been demonstrated for the initial stages of atherosclerosis. It has been even more extensively studied for the subsequent steps of intra-arterial evolution of atherosclerotic lesion, of their complications and of those of angioplasty. The early theory of response to endothelial physical injury [1] as an initiating factor for atherosclerotic lesions attributed a key role to platelet reaction. More recent theories consider an early functional lesion of the endothelium with a cellular response led by monocytes. Monocytes infiltrate the vascular wall becoming macrophages. Macrophages and smooth muscle cells (SMC) which have migrated from the media and proliferated in the intima, endocytose lipids mainly from oxidized LDL and become foam cells. Most of these cells die during this process and the remainder form the lipid core of the lesion lined on the surface by a fibrocellular cap, mostly formed by SMC but infiltrated, especially in the shoulders of the cap by macrophages, which again are implicated in the rupture of the plaque. Since the earliest stages, all the components, and cofactors of the tissue factor pathway of coagulation are assembled in the atherosclerotic lesions. This intratissular "coagulation" leads to thrombin generation and fibrin formation within the lesion in the arterial wall. These steps precede and cooperate with SMC-proliferation and extra-cellular lipid deposition. The factors involved in fibrinolysis (uPA pathway) participate to cellular migration and proliferation. In the later stages of evolution, this evolving lesion will activate intravascular hemostasis, coagulation and fibrinolysis. These intravascular systems develop differently to intratissular systems, and lead to evolution and clinical complications of atherosclerostic lesions. These intravascular and intratissular systems are implicated both in thrombosis and in the initiation, evolution and complications of atherosclerosis. Consequently, this disease is often known as athero-thrombosis. The understanding of the mechanisms involved in athero-thrombosis is necessary for the identification of biological markers and the design of therapeutic strategies for each of the various stages of the disease. This is made even more complex since in the same patient numerous atherosclerotic lesions at various stages of evolution are present at the same time.

Characteristics In physiological conditions, blood is not exposed to activating (thrombogenic) surfaces nor submitted to extreme flow conditions. It circulates inactivated because endothelial cells at the interface between vascular lumen and parietal tissues constitute a surface passively and actively non-reactive and non-activating for the cellular and molecular mechanisms of hemostasis. These mechanisms of hemostasis become activated in a thrombogenic reaction when surfaces are altered and/or the flow conditions are drastically changed. These two components (cellular and molecular) of the hemostatic/thrombogenic reaction are differently activated (Figure 1).

- The molecular components of the coagulation cascade are activated by exposition to tissue factor (TF) which triggers a succession of amplification mechanisms which lead to thrombin generation and ultimately fibrin polymerization.
- The cellular components mainly formed by platelets recognize a foreign surface, adhere, activate and aggregate. Monocytes/macrophages but also cellular components of the arterial wall can participate in the thrombogenic reaction.
- These two molecular and cellular components closely cooperate in all the thrombogenic mechanisms. Thrombin is the most potent agent of platelet recruitment and activation; platelets liberate coagulation factors, expose an ideal phospholipidic surface for

assembly of coagulation factors complexes and bind to fibrinogen and the fibrin network to form an aggregate; monocytes (and endothelial cells) express TF and also expose an anionic phospholipidic surface when activated.

The differential participation of circulating cellular or molecular factors depends mostly on flow conditions. In slow-flow conditions, platelets do not have sufficient kinetic energy to reach the surface of activation in significant numbers, while the undisturbed coagulation cascade is activated and overcome its potent but quantitatively limited inhibitor systems. Conversely, in high-flow conditions, sufficient numbers of platelets can reach the surface and form cellular aggregates which resist the shear stress, while the activated coagulation factors are dispersed and the inhibitor systems are replenished by the flow. Higher shear conditions can even by themselves activate platelets. This explains why an arterial thrombus at an initial stage is mostly formed by platelets. Flow disturbances are induced by the atherosclerotic lesions and by the thrombus itself growing in the lumen. In this case the coagulation cascade is activated only by the local flow conditions.

These explain the type of thrombotic reaction developing on an advanced atherosclerotic lesion. However, this does not fully explain the participation of thrombosis the atherosclerotic disease. The cellular and molecular mechanisms of thrombosis are implicated at each stage of the atherosclerotic disease from the earliest anatomical pre-clinical lesions until their clinical ischemic complications and the complications of invasive therapies (thrombosis, restenosis) (Figure 2). The involvement of coagulation, hemostasis and fibrinolysis in the initiation, evolution and complications of the atherosclerotic lesion is known as the atherothrombotic disease. Even more interestingly, the same mechanisms as those that take place in the arterial lumen can occur within the arterial wall (where they are present without previous

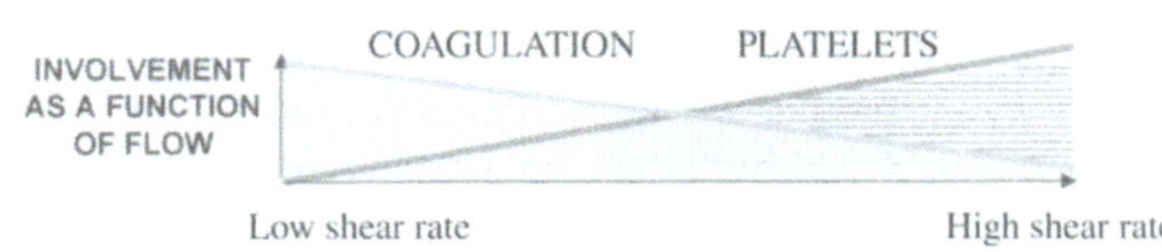

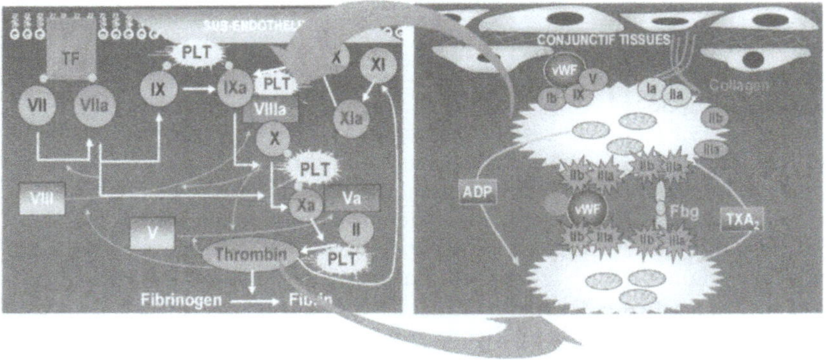

Figure 1. Relative involvement of platelets and coagulation in the thrombotic reaction are highly dependent on the flow conditions: low-flow conditions favour coagulation involvement while high shear rates allow mostly the development of cellular (platelet) reactivity. The situation is often complex because platelet and plasmatic coagulation cooperate and the flow conditions vary in the same patient from one vascular lesion to an other, in the different parts of a lesion and at a given place during the evolution of thrombosis.

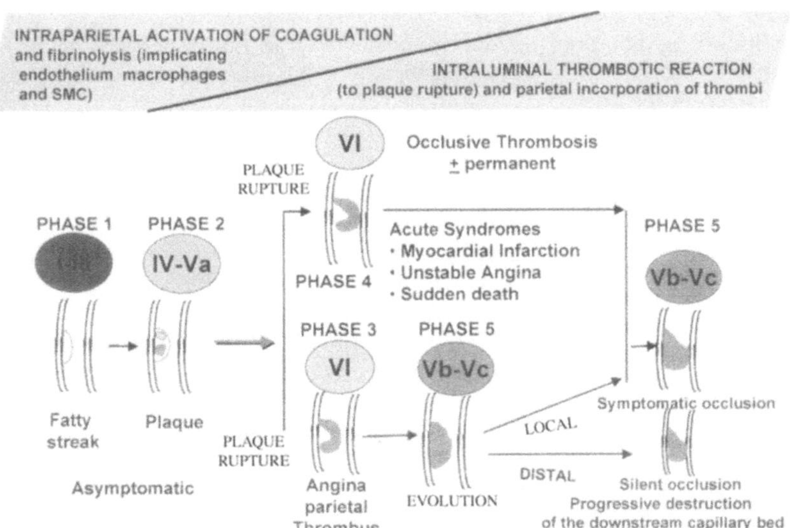

Figure 2. Implication of cellular and molecular factors of thrombosis in the initiation and progression of atherosclerosis (from the phases defined by the AHA (American Heart Association), modified and adapted from V. Fuster, ca. 1994; 90:2126). Distinction between parietal and intraluminal implications of the thrombosis mechanisms.

lesion of the endothelium) and play a role in the initiation and the evolution of the atherosclerotic lesions.

The early stages of the lesions used to be explained by the theory of response to endothelial injury [1]. Platelets are deposited and the platelet growth factors (PDGF, TGF-β) carried in platelet α granules are released at the site of injury in smooth muscle cells (SMC) proliferation in reaction to endothelial lesion. This theory was supported by a number of evidences such as the protection against atherosclerosis conferred by severe von Willebrand's disease [2], this deficiency preventing platelet deposition and activation onto exposed reactive surfaces. Recent observations no longer support this hypothesis, at least as initially formulated. The main role in the initiating mechanisms is now conferred to other cell types (monocytes), to enzymatic systems (coagulation and fibrinolysis), and to the different types of endothelial lesions, and/or dysfunction due to metabolic injury (diabetes, homocystinemia...), or toxic injury (smoking). Atherosclerosis is still often presented as an abnormal lipid deposition. In fact part of the lipid metabolism abnormalities, hemostasis or thrombosis factors are principally involved.

Regulation

Molecular Interactions

Von Willebrand Factor (vWF)   The proposed role of vWF in atherogenesis [2,3] has not been confirmed by other studies [4,5]. VWF may be implicated as a cofactor of platelet adhesion on arterial wall (endothelial) lesions with subsequent platelet activation and release of factors, stored in or synthesized by the platelets, and active on the vessel wall. vWF has several potential implications since it is of endothelial origin. Increased plasma vWF levels (and more specifically higher molecular weight multimers) are a consequence of endothelial lesion (or stimulation). The increase of highly reactive forms of vWF favors or even triggers platelet adhesion

and aggregation [6]. Conversely, fragments of vWF inhibit platelet reactivity to connective tissues from human atherosclerotic lesions [7]. In patients with peripheral arterial disease (PAD) the intima-media thickness of the carotid (a pre-clinical marker of atherosclerosis) is correlated with the plasma concentration of vWF [8]. This suggests an implication of endothelial lesion, and increased plasma vWF in the progression of the atherosclerotic disease. Chronic renal insufficiency is characterized by accelerated atherosclerosis, implicating an endothelial cell lesion as evidenced by increased vWF plasma level [8] and generation of atherogenic oxidized LDL [9]. Diabetes, a condition with accelerated atherosclerosis is characterized by an increased plasma level of vWF (and higher molecular forms of vWF).

Implication of coagulation factors   In hemodynamic conditions of low flow or disturbed flow (as in and around ruptured atherosclerotic plaques), plasma coagulation (no longer the platelets) is the main factor in the thrombogenic reaction. Activation of the plasma coagulation system is favored by:

- the slower flow conditions (stasis),
- the exposition of anionic phospholipid surfaces (such as membranes of activated cells) necessary for the stoichiometric organization of the molecular complexes of the coagulation cascade, and
- the availability of an adequate phospholipid environment and of the initiator of the coagulation system, the tissue factor (TF).

Tissue factor (TF) (Figure 3)   TF plays a key role in atherogenesis, hemostasis and thrombosis [10]. As a physiological and pathological activator of the coagulation cascade, it leads to thrombin generation. TF is found (as antigen and activity) at every stage of the atherosclerotic lesions [11], but in greater amounts in the lipid core (rich in oxidized LDL) in advanced lesions [12]. At pre-atherosclerotic stages, an endothelial lesion (as induced

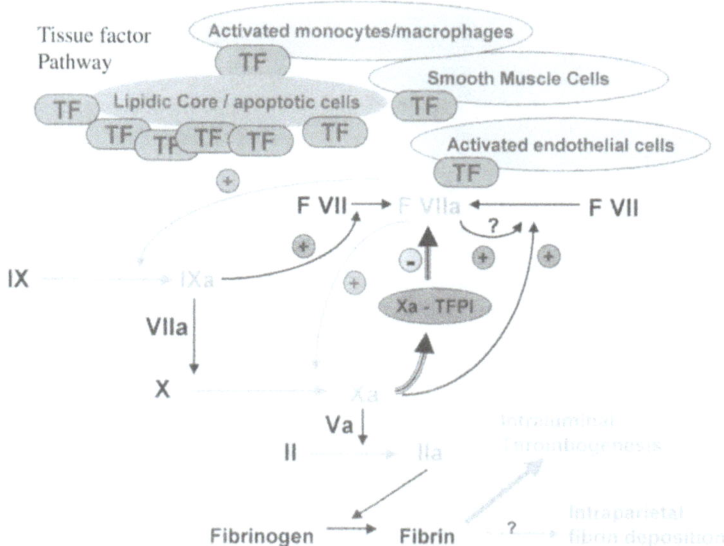

Figure 3. When exposed to circulating blood, the various parietal sources of Tissue Factor (TF) in an adequate phospholipid environment activate the coagulation cascade. The same system probably participates to intraparietal fibrinogen/fibrin deposition and subsequently lipid deposition and lesion initiation.

by hyperhomocystinemia) induces pro-coagulant activity mainly by expression of TF. Expression of TF is triggered in the arterial wall by mechanical lesion (such as angioplasty). TF accumulates in the neointima secondary to the lesion. TF induces smooth SMC proliferation similarly to the effect of PDGF or basic FGF. In atherosclerotic lesions, macrophages are the major source of TF. TF accumulation is an early event associated with monocytes migration in the arterial wall (in response to chemoattractants such as MCP-1) and their transformation as macrophages. SMC proliferating in the plaque are the second source of TF. Macrophages and SMC accumulate lipids, become foam cells and degenerate to form the necrotic core riche in TF. *In vitro* macrophages express TF activity in response to oxidized LDL or endotoxins. Plaque rupture (either spontaneous or induced by angioplasty) exposes the TF from the lipid core to circulating blood, triggering thrombosis. In experimental animal models, balloon-induced expression of TF by medial SMC from normal vessels does not lead to a sufficient generation of thrombin to allow significant fibrin deposition, whereas the endothelial lesion secondary to ballooning, exposes TF on the intima surface to circulating blood thus generating a sufficient amount of thrombin to allow fibrin deposition. The different phospholipid environments explain the differences in the activity of TF which requires anionic phospholipids in optimal amounts and configurations to allow binding of the vitamin K-dependent coagulation factors. In resting conditions, monocytes do not expose these phospholipids on their external membrane, but only when they have been activated to become macrophages, whereas SMC external membranes express such an activity in resting conditions. There are membrane changes with overexpression of this activity when the cells become apoptotic at the end of their cycle in the plaques. In advanced human atherosclerosis, the arterial walls – even in regions which do not include plaques (stable or

unstable) – are not fully normal and appear more like the experimental vessels which have been previously injured with an active intima including apoptotic cells expressing TF in a phospholipidic configuration allowing activation of coagulation. Endovascular injury such as angioplasty, atherotomy or stenting exposes parietal TF in a phospholipid configuration allowing the activation of coagulation and leading to fibrin deposition. It has recently been demonstrated that TF of blood-born origin may play a main role in arterial thrombosis. This TF may be leukocyte origin and concentrated at the platelet surface. TF is a potential target for pharmacological agents designed for the prevention of thrombotic complications of atherosclerotic lesions and even for the prevention of the atherosclerotic lesions themselves. Clinical trials currently evaluate the efficiency of Tissue Factor Pathway Inhibitor (TFPI) [13] as an anticoagulant, whereas studies in animal models evaluate its potential activity in prevention of intimal hyperplasia [14]. Direct factor Xa inhibitors such as the anticoagulant peptides from the tic (TAP) or from the leech (APS) are also in evaluation [15,16]. The crystal structure of TF [17] and of the complex: TF- activated factor VII [18] should allow the designing of direct inhibitors of TF which would be valuable tools to precisely determine the role of TF in the various steps of athero-thrombosis.

Plasma TFPI activity is decreased in stroke patients. This decrease in TFPI is independent of other hemostatic or fibrinolytic modifications such as consumption coagulopathy and may be induced by atherosclerotic lesions [19]. Cellular- and copper-induced oxidation inhibits LDL-associated TFPI activity [20].

Besides the well established intraluminal activation of coagulation, an activation of coagulation in the arterial wall at the onset of the atherosclerotic lesion (well before plaque rupture) leading to tissular thrombin generation independently of any direct contact with circulating blood has never been demonstrated. This is sus-

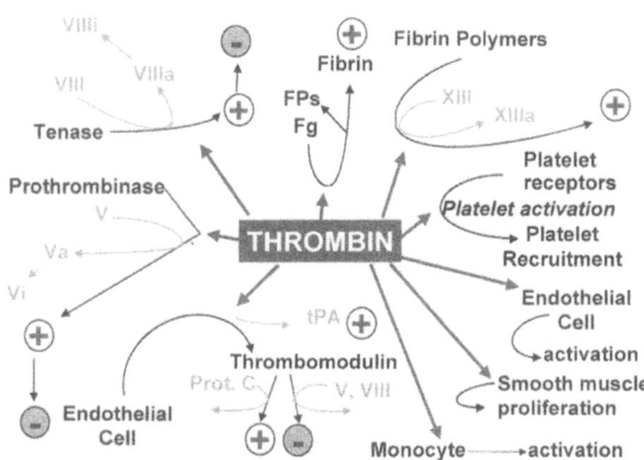

Figure 4. Thrombin, the last enzyme of the coagulation cascade, is the key factor in the coagulation system due to its multiple direct and indirect effects on the regulation of coagulation activation and inhibition. Additionally, it has numerous cellular effects participating in both thrombotic and atherogenic reactions.

tained by the observation of fibrinogen/fibrin and thrombine-ATIII complexes since the earliest stages of lesions (fatty streaks).

Thrombin (Figure 4)   Thrombin, the enzyme generated during the final step of coagulation, has multiple effects and activates each of the cell types present in the arterial wall. Thrombin is mitogenic for SMC. Factor Xa stimulates DNA synthesis and SMC proliferation through release of PDGF. PDGF acts on specific tyrosine kinase receptors [21] and increase expression of LDL receptors on SMC [22]. Thrombin, through its specific proteolytic activation of its receptor on SMC, antagonizes expression of inducible NO synthase by SMC [23], induces MCP-1 expression by these cells [24], and increases the chemotactic activity for monocytes produced by rat aortic rings. Thrombin, by induction of the expression of a 95 KD gelatinase (metaloproteinase 9 or gelatinase B) by macrophages without increase of tissue inhibitors of metaloproteases (TIMP), stimulates basal membrane degradation and SMC migration [25]. Oxidized LDL increases endothelial pro-coagulant activity by the stimulation of TF activity but also decreases the protein C system (a potent inhibitor system of coagulation) through TNF-induced thrombomodulin endocytosis [26].

Insulin-like growth factor (IGF) exerts an autocrine activity on vascular SMC. Thrombin activates its specific receptor linked to G-proteins but has also a mitogenic effect on SMC through activation a tyrosine kinase type receptor. A functional pathway through the IGF receptor is essential for thrombin-induced intracellular signaling. These observations sustain the hypothesis of a cooperation between the G-proteins pathway and the tyrosine kinase receptors pathway [27]. It has been recently shown [28] that heparan sulfate-rich proteoglycans present on the cell surface and in extracellular matrices contribute to thrombin inhibition through heparin cofactor II which,

in the presence of heparan or dermatan sulfate, is a potent inhibitor of thrombin [28]. Hirudin prevents restenosis triggered by angioplasty, mostly by inhibition of thrombin-induced SMC proliferation [29,30].

Homocystein, by inhibition of thrombomodulin expression at the endothelial cell surface, inhibits the protein C system. Homocystein also inhibits the binding of anti-thrombin III to proteoglycans and heparan sulfates of endothelial cells. Through such effects, homocystein reduces the anti-coagulant properties of the endothelium [31]. Homocystein potentiates cholesterol-induced atherosclerosis in normal as well as in hypertensive rats [32].

Fibrinogen (Figure 5)   Fibrinogen (and/or fibrin) is present in atherosclerotic plaques from the earliest stage of the lesion [33]. The mechanisms of this early deposition is probably different to that of latter stages, when most of the deposition is due to internalisation of parietal thrombi. Fibrinogen (and/or fibrin)

– participate in numerous cellular functions (monocytes adhesion [34], platelets aggregation [35] but also adhesion [36], erythrocytes aggregation..),
– form a matrix for SMC proliferation,
– is the final substrate of the plasma coagulation cascade,
– is the preferential site for the stoichiometric organization of fibrinolysis complexes,
– is the main determinant of plasma viscosity [37],
– gives rise to fibrin-split products are mitogenic for parietal SMC.

Due to these multiples reasons, fibrinogen is implicated in atherogenesis, thrombogenesis and in every stage of athero-thrombosis.

Increased plasma fibrinogen is a potent and independent risk factor:

– for the severity of atherosclerosis (such as the intima-media thickness) [38],
– for the evolution of atherosclerosis [39, 40],
– for the complications of atherosclerosis (restenosis, acute and subacute thrombosis…) [41],
– for invasive intra-arterial procedures (angioplasty, stenting…) [42, 43],
– for every types of population (general population [41], or patients (primary or secondary events) [39],
– for every localization of the atherothrombotic disease [41] (coronaries [39, 44], stroke [44], PAD [45], poly-vascular [40],
– for every condition of the atherothrombotic disease (acute [46], unstable [47] or chronic [48].

Fibrinogen plasma levels depend on genetic factors (such as race), on non-alterable factors (such as age, gender), on environmental factors (smoking, diet (alcohol or fish…), social and economical conditions), on metabolical conditions (weight, dyslipidemia, diabetes, hyperhomocystinemia), on pathological conditions (inflammation, condition of teeth, chronic and acute infection …), and on drugs and treatment regiments (oral contraception, fibrates, pentoxyfilline, Ticlopidine…) [49-52]. In plasma, fib-

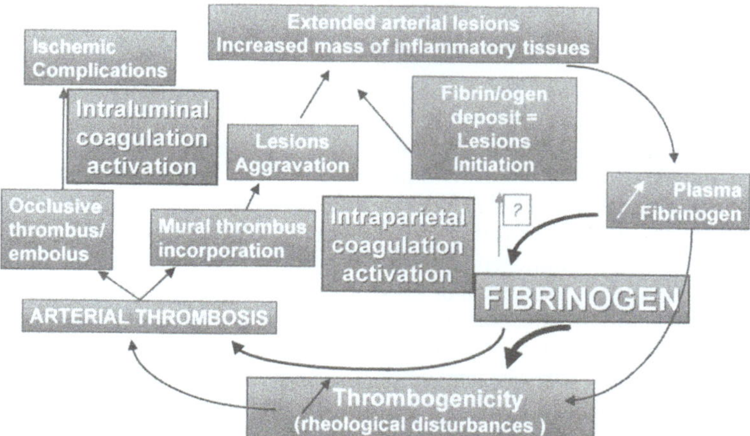

Figure 5. Implications of fibrinogen at the successive stages of the atherothrombotic disease. Fibrinogen is both a marker for the extent of the (inflammatory) lesions and a factor implicated in the molecular and cellular mechanisms of atherogenesis and of thrombogenesis, not only intraluminal but also intraparietal.

rinogen concentrations less than 50 % are of genetic determinism [53]. This part of heritability has been calculated by genetic studies on large populations. The synthesis of each of the three chains of fibrinogen (A-α, B-β, and γ) depends on a different gene. The amount of B-β chain synthesized determines the total number of fibrinogen molecules synthesized. Several genetic polymorphisms have been reported on each of the genes of fibrinogen. These polymorphisms may directly or indirectly (by linkage to a mutation) be associated to quantitative or qualitative changes in the fibrinogen synthesized. Associations have been reported between some polymorphisms, especially in the region of the promoter (which include binding sites for transcription factors) of the β-chain and the plasma fibrinogen level [54]. The epidemiological studies on the cardiovascular risk associated with increased plasma fibrinogen levels have unveiled a genetic predeterminism [55]. Case-control studies and transversal populations studies have shown a specific distribution of the polymorphisms for the fibrinogen genes in cardiovascular patients. Large prospective studies have to be performed including subjects of various geographic origins and taking into account the existence of other known risk factors and the genetic background of these populations. Studies on fibrinogen genetic polymorphisms give arguments for a causal implication of fibrinogen in the atherosclerotic process showing, for example, that the statin-induced regression of atherosclerotic plaques is modulated by the fibrinogen genotype [56].

Fibrinogen is an acute phase-reactant protein. Its plasma level is stimulated by the main cytokines involved in inflammation such as IL-6 [48]. Pathological conditions associating an inflammatory state are characterized by an increased plasma fibrinogen level. The smoking status is correlated with the plasma fibrinogen level. The effect of smoking as of other factors of hyperfibrinogenemia is modulated by the polymorphisms of fibrinogen genes [57]. This is a clear illustration by the intricate influences of genetic and acquired determinisms on the fibrinogen level. The close dependency between fibrinogen levels and inflammation raises the question of the

significance of plasma fibrinogen levels as a cardio-vascular risk factor. Atherothrombosis is an inflammatory disease (all the cellular and molecular factors of inflammations are found and produced in atherosclerotic plaques). Is the increase of plasma fibrinogen associated with the cardio-vascular disease a cause and/or a consequence? In the atherothrombotic disease, ischemic events are the consequence of a plaque rupture. Among the mechanisms of plaque rupture, one of the best established is the activation of macrophages in the shoulders of the plaque [58]. The enzymatic activities of the macrophages degrade the extracellular matrix and, more specifically, of the fibrous cape, destabilizing the plaque and leading to its rupture. The more enzymatic activity is found in the plaques, the more at risk is the patient. The association of all these sites of inflammation forms a mass of inflammatory tissues, enough to significantly increase plasma markers of inflammation, such as fibrinogen, which may constitute a marker of the extent and of the "activity" of the disease. This hypothesis is reinforced by the observation that other markers of inflammation, such as CRP, bear a prediction of cardiovascular events similar to fibrinogen [59]. When the risks factors were equilibrated for CRP, fibrinogen maintained its predictability, while, after equilibration for fibrinogen, CRP lost its predictability. Fibrinogen, but not CRP, is an independent risk marker. As a multifunctional molecule, fibrinogen could be both a marker and a factor of risk. The inflammatory component of atherothrombosis increases the level of the fibrinogen and fibrinogen accelerates the (thrombotic) evolution of the atherothrombotic disease in an auto-amplified cycle. The recent observation that the mouse knock-out for the gene of fibrinogen exhibits atherosclerotic lesions induced by deletion of the apoE gene, is an argument against a causative role of fibrinogen in the genesis of atherosclerosis [60].

Two different pools of fibrinogen are circulating. Usually considered to be the most important is the plasma pool, which is synthesized by the liver, but also the platelet pool (stored in platelet α-granules), which is not synthesized by the megacaryocytes but coming from plasma fibrinogen (of hepatic origin), which is actively incorporated in the α-granules during the platelet life. Fibrinogen is not a unique molecule but a family of molecules. By alternative splicing of the γ-chain's gene, the C-terminal end of this chain can differ in amino acid sequence and number. The molecular type most frequently found is the γ-chain whose last 12 amino acids form the specific binding sites for platelet membrane glycoprotein IIb/IIIa (the other binding site is the more ubiquitous RGDS sequence of the α-chain). The splicing of the gene of the γ-chain can alternatively lead to a γ'-chain shorter and of different sequence which no longer supports platelet aggregation. In plasma fibrinogen, the γ'/γ ratio is lower than 0.1. How is this determined; does this vary among different types of patients? Genetic polymorphisms affecting the coding sequence of the gene can modify the amino acid sequence and/or site of post-transcriptional modifications such as glycosylation sites. In circulation, fibrinogen is affected by various enzymatic activities, mostly plasmatic and cellular proteases. C-terminal ends of α-chain are sensitive to these proteases with fundamental consequences since these parts of the molecule are highly implicated in the initial steps of fibrin polymerization, especially lateral polymerization, which has important structural and functional consequences on fibrin, such as sensitivity to the fibrinolytic system.

Much less is known about genetic and environmental factors responsible for qualitative changes on fibrinogen than about those responsible for quantitative changes, but they are, perhaps, at least as important and possibly more important.

The molecular diversity of fibrinogen has a clinical relevance due to the functional reactivity of the molecules. It is obvious that the techniques for determination of fibrinogen level should not only quantitatively assess fibrinogen but also be sensitive to structural and/or functional characteristics of the molecule.

Factor VII   Factor VII is also both influenced by several environmental [49, 50] and genetic conditions [61]. Factor VII behaves as a cardiovascular risk factor, mostly when it is evaluated by functional techniques. It is implicated in thrombotic phenomenona immediately downstream of TF.

Fibrinolysis   Activators and inhibitors of fibrinolysis are implicated in the proteolytic systems of the arterial wall and intervene in SMC proliferation and atherogenesis and in the fibrinolytic reaction to thrombogenesis [62]. It has to be noted that a hyperactivity of the parietal proteolytic system lead to an accelerated atherosclerosis, whereas a hyperactivity of luminal fibrinolytic systems lead to a reduced atherothrombotic reactivity.

The parietal fibrinolytic system depends on the relative concentrations and availability of the active forms of tissue plasminogen activator (tPA), urokinase (UK) and PAI-1 [63]. Earliest observations noted an increase in PAI-1 during the evolution from a normal vessel wall to fatty streaks and to the atherosclerotic plaque. In the atherosclerotic lesions, PAI-1 co-localizes with vitronectin. Its binding protein tPA is decreased in complex advanced

lesions. These observations suggested that inhibition of the fibrinolytic system would favor progression of atherosclerotic lesions [64-66]. A more recent observation [67] sustains the opposite hypothesis: intraparietal overexpression of PAI-1 could have an anti-atherogenic effect, inhibiting SMC migration by blockage of integrin $\alpha_V\beta_3$ to vitronectin. Implication of PAI-1 in complexes with tPA and UK reduces the affinity of PAI-1 to vitronectin, restoring its effect on cell migration. The local synthesis of plasminogen activators appears to play a critical role in the exposition of sites for the cellular adhesion necessary for SMC migration [67]. PAI-1 in the presence of vitronectin efficiently inhibits thrombin, and complexes are metabolized by interaction with proteins linked to LDL receptors. This catabolic pathway for thrombin could be important in decreasing the mitotic activity of thrombin in arterial lesions [68]. Cytokines induce the exposition of UK receptors by monocytes, a step in monocyte adhesion. PAI-1 inhibits monocyte adhesion by inactivating UK [69].

Plasmin generation at the endothelial surface induces PAF synthesis, P-selectin expression and increased leukocyte adhesion to the endothelium [70]. Localized overexpression of UK and tPA is only partially inhibited by PAI-1 and could contribute to local proteolytic degradation and remodeling of extracellular matrices. This reduces fibrous cape resistance and favors plaque rupture. Lipoprotein a (Lp-a) is an independent cardiovascular risk factor. Based on the structural homology between Lp-a and plasminogen, it has been hypothesized that Lp-a is an inhibitor of fibrinolysis. Lp-a, but also LDL compete with plasminogen for binding to extracellular matrices. Inhibition depends on lipoproteins levels and on the amount of exposed extracellular matrices [71]. Lp-a induces SMC proliferation and this effect is correlated with inhibition of transforming growth-factor β (TGFβ) activation [72].

Atherogenesis is associated with a decreased vessel wall fibrinolytic activity, which could favor thrombotiques reactions. The decreased proteolytic activity in the arterial wall should favor extracellular matrice accumulation during the initial steps of atherosclerotic lesions. The increased proteolytic activity in the arterial wall should favor SMC migration and proliferation. The follow-up of large cohorts of general population (such as ARIC [73]), of coronary-diseased patients (such as ECAT [74]) and of diabetics [75] has demonstrated a positive correlation between increased plasma PAI-1 levels, and the risk of cardiovascular ischemic events, of myocardial infarction or sudden death. In similar pathological conditions, tPA appeared increased. This increase is a technical artefact. The technique measuring tPA is also sensitive to the tPA-PAI-1 complex. As the half-life of PAI-1 is much longer than that of tPA, the increase of PAI-1 induces an increase of tPA antigenic material (complexes TPA-PAI-1). PAI-1 is increased during insulin resistance, dysmetabolism (with obesity) and hypertriglyceridemia. The increased levels of PAI-1 in patients with type IIA, IIB and IV hyperlipidemia are correlated with the level of triglycerids, apolipoprotein B and cholesterol. These correlations depend on apoliprotein B lev-

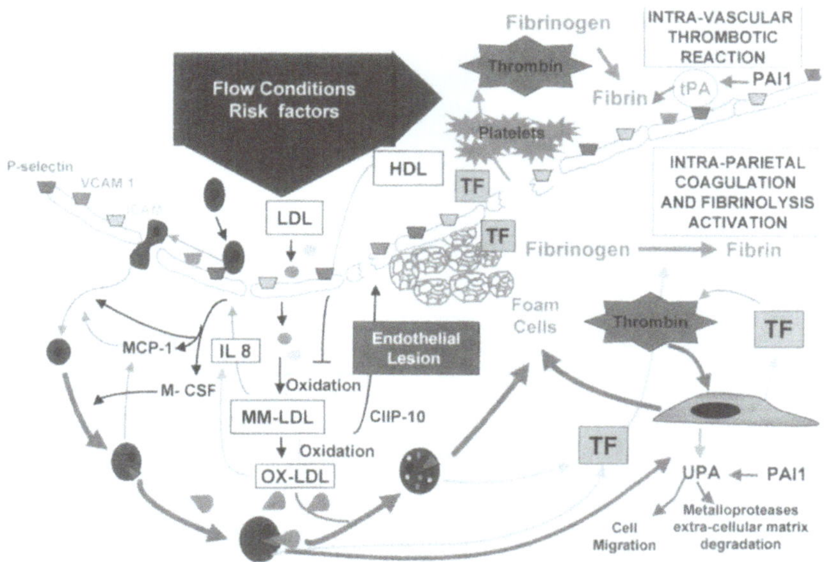

Figure 6. Implication of cellular and molecular factors of thrombosis in the initial stages of atherosclerosis. The endothelium and macrophages play the major role. Parietal and intraluminal implications of factor of thrombosis, coagulation and fibrinolyse. Modified and adapted from D. Steinberg, ca.1991; 84: 1420.

el and could reflect the correlation between PAI-1 and LDL [76]. *In vitro,* VLDL induces expression of PAI-1 mRNA and protein [77]. This is inhibited by niacine, suggesting a link between pharmacologically-induced lipid reduction and fibrinolysis potentiation which could slow down the evolution of atherosclerosis by the decrease of the mitogens associated with microthrombi [78,79]. Plasma TFPI is negatively correlated to triglycerid levels in hypertriglyceridemic patients. Hypercoagulability and hypofibrinolysis could contribute to the prothrombotic tendency in these patients [80]. VLDL also decreases plasminogen binding to endothelial cells, resulting in an inhibition of fibrinolysis at the cellular surface, contributing to initial fibrin deposits and to the increased thrombotic risk associated with triglycerids.

Cells and Cellular Interactions

Endothelial dysfunctions since the earliest stages of atherosclerosis (Figure 6) Endothelium physiologically exposes a non-reactive luminal surface to circulating blood. This surface not only does not activate the molecular (coagulation) and cellular (platelets, monocytes) components of the hemostatic and thrombotic systems, but additionally expresses or releases inhibitors or cytokines which specifically inhibit the activation process. These endothelial properties, known as thromboresistancy, are altered by numerous aggressions that can injure the endothelium. These aggressions are hemodynamic (hypertension), metabolic (diabetes, hyperhomocystinemia, dyslipidemia...) immunological, inflammatory or infectious. In these various pathologies, a dysfunction of the endothelium can be evidenced long before the first clinical manifestations of atherosclerosis. The dysfunctions of vasoactive reaction are easily explored in clinical settings [81] but the lesions affect also other functions of the endothelium. The endothelium loses some of its characteristics of non-reactivity towards circulating elements and even can exhibit an acti-

vated phenotype including activation of coagulation, activation of inflammation and immunology, adhesion and aggregation of platelets, adhesion and inter-endothelial transfer of leukocytes, or anti-fibrinolytic activities. One of the earliest events of athero-thrombosis is adhesion and penetration of monocytes in the arterial wall at the site of endothelial lesion. The first step is rolling [82] of monocytes which is loose, reversible, cellular adhesion mediated by selectins. P- and E-selectin are expressed by endothelial cells triggered by inflammation and L-selectin expressed by leukocytes. β_2-integrins are implicated in the secondary step of adhesion which requires activation of the leukocytes and subsequent conformational modification of the β_2-integrins [82]. Activation of monocytes is triggered by cytokines, mainly interleukin 8 (IL-8) and monocyte chemoattractant protein-1(MCP-1) which are secreted by activated cells from the arterial wall (endothelial cells but also SMC) [83]. Monocytes, interacting with endothelial cells, trigger synthesis of granulocyte macrophage colony stimulating factor (GM-CSF) which in turn stimulates proliferation of monocytic stemcells and stimulates synthesis of inflammatory cytokines increasing the overall phenomenon [84].

At the luminal surface of atherosclerotic plaques, endothelial cells express P-selectin and intercellular adhesion molecule-1 (ICAM-1) [85]. Monocytes adhesion to this surface is inhibited by antibody against P-selectin [86]. Monocytes, after active adhesion to the endothelial cells, enter the arterial wall through gaps between the endothelial cells. In the arterial wall they actively incorporate lipid vesicles. The initial lipid deposition in the arterial wall is intracellular. Platelets could be implicated in these phenomena since they can adhere to monocytes when P-selectin expression is stimulated by potent inducers such as thrombin (which is locally generated, as detailed in the following section). Expression of P-selectin by the endothelium is an active phenomenon inhibited by protein kinase C (PKC) antagonists and by NO donors [87].

Oxidized LDL (but not native LDL) induces expression of P-selectin in the endothelium (released from Webel-Palade bodies) and leukocytes adhesion at very low concentrations [88] independently from other NO-dependent endothelial functions. Interleukin-1β (IL-1β) induces vascular cell adhesion molecule-1 (VCAM-1) expression by human umbilical vein-derived endothelial cells by a mechanism modulated by antioxidants. Antioxidants decrease cellular adhesion dependent on TNF-induced VCAM 1 expression [89]. Investigations of genetic predisposition in these reactive systems leading to inappropriate endothelial response to environmental stimulation and possibly conferring a higher risk of atherosclerosis have started. The first candidate genes were those encoding adhesive proteins E-, P-, L-selectin, ICAM-1, VCAM-1. Genetic polymorphisms on E-selectin have been found with higher frequency in patients younger than 40 years affected by severe atherosclerosis [90]. Serum-soluble ICAM-1 is increased in patients affected by ischemic heart disease, while serum and leukocyte L-selectin are decreased. These correlations between the stages of the cardio-vascular disease and serum levels of ICAM-1 and L-selectin are additional arguments for the implication of a chronic inflammatory process and of leukocyte activation in the pathogenesis of atherothrombotic disease [91]. Cholesterol-loaded macrophages from human atheromatous lesions express IL-8 which induces the irreversible adhesion of polymorpho nuclear leukocytes (PMN) [92]. The cytokine interferon-inducible protein-10 (CIIP-10) which induces mononuclear cell and T-lymphocyte infiltration in peripheral tissues [93] also induces SMC proliferation *in vitro* and neo-intima formation *in vivo* after balloon angioplasty. These suggest that CIIP-10 is actively involved in vascular remodeling [94]. During the inflammatory reaction, several cell types synthesize and release phospholipase A2 (PLA2). In human atherosclerotic lesions and in macrophage-derived foam cells, PLA2 is overexpressed. As, *in vitro*, PLA2-modified lipoproteins accumulate in macrophages, PLA2 could be involved in foam cell formation [36]. Inflammatory cells recruited in atherosclerotic lesions synthesize metalloproteinases such as gelatinase A (metalloproteinase 2) and gelatinase B (metalloproteinase 9), which contribute to migration and proliferation of SMC at early stages of atherosclerosis and to plaque rupture at advanced stages of these plaques [95]. Cells from atherosclerotic lesions are affected by apoptosis. Apoptosis of inflammatory cells (macrophages and T-lymphocytes) may reduce the inflammatory reaction. Apoptosis of SMC may contribute to plaques instability [96]. Cellular apoptosis increases the thrombogenic potential of the tissues inside the plaques as detailed in the section on coagulation.

Endothelial cells are implicated in these vasoactive functions because:

– they express specific receptors (mecano-receptors sensitive to shear stress and pressure, chemo-receptors sensitive to circulating factors),

– they metabolize vasoactive substances (adenylic nucleotides, biogenic amines, bradykinine...) and

– they synthesize and release of NO, endothelium derived hyperpolarizing factor (EDHF) and prostacycline (PGI$_2$).

Some of these vasoactive functions are accessible through clinical tests and allow the evaluation of endothelial cell functions at the early stages of atherosclerosis [97]. The wall shear stress modulates synthesis of NO which opposes the vasoconstricting factors from neuronal and SMC origin. This mechanism allows the adaptation of vessel diameter to blood flow. Endothelial cells-derived NO is implicated in other functions besides vasomotricity. NO down-regulates the expression of various genes such as MCP-1, P-selectin, and VCAM-1, leading to an inhibition of platelet and leukocyte adhesion. NO decreases the synthesis of EDHF which induces relaxation of underlying SMC. EDHF may contribute to the regulation of vascular tone, especially in pathological situations in which NO production is decreased, such as in endothelial cell dysfunction [98].

Hypercholesterolemia decreases NO production in the endothelium from the microvasculature, increases rolling and adhesiveness of leukocytes, and increases P-selectin and VCAM-1 expression. NO donors reduce these changes in hypercholesterolemic patients. These data suggest that NO is involved in early endothelial cell dysfunction affecting hypercholesterolemic patients [99]. Shear stress modifies leukocyte adhesion to the endothelium, mostly through a flow-dependent secretion of NO and to a lesser degree of PGI$_2$. This is probably more due to an alteration of signal transduction than to an expression of adhesion molecules by the endothelium. Oxidized LDL inhibit endothelium-dependent relaxation by inhibition of NO and EDHF production [100]. Oxidized LDL decreases anti-platelet properties of endothelial cells more likely by direct interaction with NO than by inhibition of its synthesis [101].

PGI$_2$ antagonizes platelet interaction with monocytes, inhibits platelet activation and aggregation, induces SMC relaxation and reduces lipid accumulation in SMC [102]. Hypoxia [103], reperfusion injury [104] and advanced products of glycation [105] inhibit PGI$_2$ synthesis, whereas estrogens [106] stimulate PGI$_2$ synthesis. HDL stabilize PGI$_2$ but this function is decreased for HDL from diabetic rats [107]. The expression of PGI$_2$-stimulating factor is decreased in endothelial cells and SMC from diabetic rats and in human atherosclerotic lesions [108].

Ischemic lesions, due to hyperproduction of free radicals during oxidative stress, activate both the cyclooxygenase-dependent pathway of prostaglandin synthesis in endothelial cells [109] and cyclooxygenase-independent pathway, which synthesize PGF$_2$-type compounds. F$_2$ isoprostane is a potent vasoconstrictors and platelet pro-aggregating agents [110-112]. NO and antioxidants inhibit F$_2$ isoprostane production [113,114]. Activated platelets produce aldehydes which can modify LDL infiltrating the intima and atherosclerotic lesions [115].

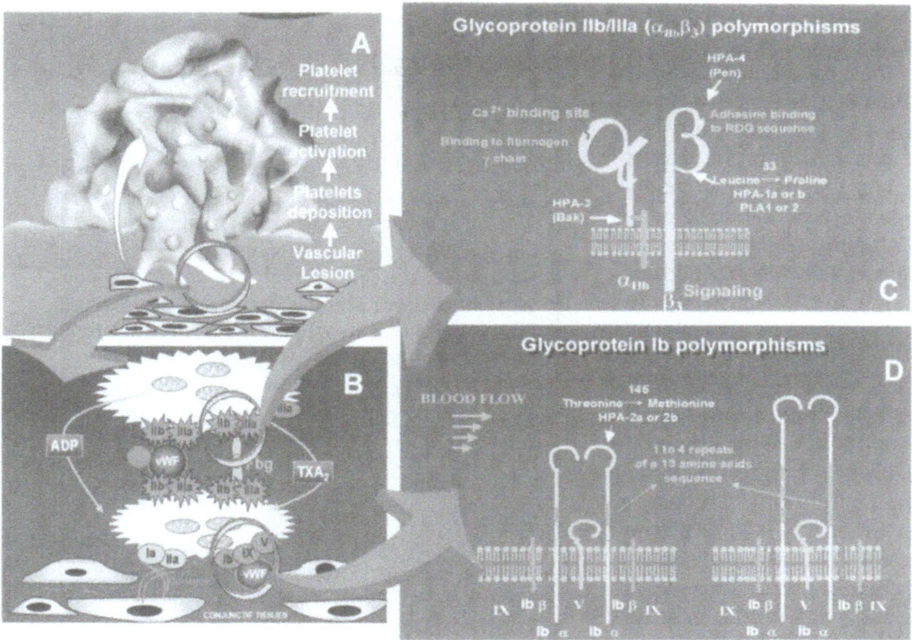

Figure 7. Role of the platelet in intraluminal thrombogenesis. Platelets implicated in the thrombotic reaction (A) establish molecular bridges (B) (fibrinogen, von Willebrand Factor...) anchored on their membrane glycoproteins. Polymorphisms of the genes of glycoproteins IIb/IIIa (C) or Ib (D) can affect platelet reactivity and their implication in thrombotic reactivity.

Implication of hemostasis at the various stages of atherosclerotic lesions

Platelets (Figure 7)   Whereas the implication of platelets in the evolution and thrombotic complications of advanced atherosclerotic lesions is fully established, their implication in the initial steps of atherosclerosis is still a matter for debate. Platelets participate in a succession of phases which are integrated in space and in time. These successive phases are adhesion, activation, recruitment (of circulating platelets), cooperation with coagulation, aggregation (formation of molecular bridges between platelets) and clot retraction (formation of bridges between the platelets and the fibrin network). In conditions of rapid blood flow (mainly arterial conditions), platelets are the major contributors to the thrombotic reaction. Antiplatelet agents efficiently prevent thrombotic events under these flow conditions [116] but their effect on the evolution of atherosclerotic lesions remains to be established. Platelets are implicated in the thrombotic reaction to a primary lesion. Indeed, constitutive modification of platelet functions are related to arterial thrombotic events more significantly in the youngest patients. Among these constitutive modifications of platelet function, the easiest to study are those related to genetic polymorphisms. Among genetic polymorphisms for platelet membrane glycoproteins, glycoprotein β_3 [117, 118] and GPIbα [119] polymorphism have been statistically-associated with an increased risk of ischemic events but only for patients of the youngest age classes. Among younger atherosclerotic patients, the lesions tend to be less advanced and the overall disease less widespread. Ischemic complications tend to be more frequently due to an explosive thrombotic reaction and to the rupture of a moderately stenotic (but highly unstable) plaque. These deleterious polymorphisms has more as thrombotic than atherosclerotic consequences. If the thrombotic reaction remains limited to a non-occlusive mural thrombus, this thrombus, stabilized by a fibrin network, is rapidly invaded by SMC and results in a step-wise evolution of the atherosclerotic lesion. If the thrombotic reaction occludes the arterial flow then ischemic consequence make this event clinically apparent.

Macrophages and macrophages-derived foam cells are potential sources of PAF at the sites of inflammatory lesion in the arterial intima where this phospholipid mediator acts both as a atherogenic and thrombotic agent [120]. Activated platelets induce the expression of thrombin receptors on SMC. They potentiate the vasoconstrictive and mitogenic response of these cells to the stimulatory action of thrombin [121]. Thrombin activated platelets induce monocytes to release IL-8 and MCP-1 [122] and therefore monocyte adhesion and arterial wall penetration. Activated platelets express P-selectin (stored in their α-granules) at the outer surface. They can bind both leukocytes and endothelial cells and in doing so, are involved in leukocyte adhesion to endothelial cells at high wall shear rates, which usually do not allow direct adhesion of leukocytes to endothelial cells [123]. Platelets are sensitive to the anti-activating and anti-aggregating activities of NO produced by the arterial wall (mainly endothelial cells). Platelets also express and activate an endogenous NO synthase which acts as a negative feed-back loop during platelet activation [134].

Clinical Relevance   Macrophages not only incorporate oxidized LDL leading to formation of foam cells, but also express TF, expose sites for fibrin polymerization, are part of the inflammatory reaction in the plaques and – mainly through expression of collagenases – are one of the main contributors to plaques instability. Activated platelets contribute to atherosclerosis by cooperative activity in leukocyte adhesion, by contribution to LDL oxidation, by induction of SMC proliferation, and by cooperation in the thrombotic reaction. If a decrease in the fibrinolytic system contributes to atherosclerosis by a defective degradation of microthrombi, local over-expression of PAI-1 could protect against atherosclerosis by inhibition of plasminogen activators involved in leukocyte adhesion and by inhibiting SMC proliferation triggered by thrombin. Intraparietal fibrinolysis increases plaque instability directly or by activation of metallopro-teases. Most of the plaque ruptures have no clinical man-ifestation but lead to thrombotic reactions, both inside the plaque (usually referred as hematoma beneath the plaque) and at the luminal surface of the plaque. The rel-ative composition in platelet and fibrin of this thrombus depends on local flow conditions. Thrombi inside the arterial wall are mostly composed of fibrin. Thrombi at the intimal surface in the lumen are initially mostly con-stituted by platelets. Coagulation and fibrin deposition depend on its evolution. The organization of these throm-bi and their colonization by SMC explain the step-wise evolution of the atherosclerotic lesion. The intraluminal thrombotic reaction is an active process associating thrombogenesis and thrombi degradation during which thrombotic fragments are embolized. The thrombus is then remodeled by action of the fibrinolytic system. Most often these events have no clinical manifestations. Even if they are not clinically manifested, these events are not without significant consequences since they progressive-ly destroy the downstream capillary system leading to a decrease of tissue capillarization. At advanced stages, the disease is bifocal by the evolution of the parietal lesions, but also in the downstream-irrigated tissues which become more and more hypoxic and sensitive to any re-duction of arterial flow. The thrombotic processes inter-vene evidently at a late stage (sometimes terminal) of the atherosclerotic lesions. The thrombotic reaction into a plaque (most often rupture) can have clinically manifes-tations, either because it grows to an occlusive size or because it embolizes thrombotic fragments of significant size. These early and late thrombotic phenomena are extremely frequent but their clinical manifestations (occurrence of ischemic event) are exceptional and delayed.

Due to the implication of the mechanisms of hemosta-sis, coagulation and fibrinolysis in the atherothrombot-ic disease, it becomes evident that

- from a diagnostic point of view, we must look for con-stitutive or acquired modifications of these factors to predict and diagnose the state and the evolutivity of the atherothrombotic disease.
- from a therapeutic point of view, we must use today's best-adapted anti-thrombotic agents to prevent throm-botic complications (late and secondary even if they are the first clinical manifestations) [97] and to limit the evolution of the atherosclerotic disease itself.

Jacques Caen and Ludovic Drouet

References

1. Ross R et al (1977) Am J Pathol 86:675-684
2. Fuster V et al (1991) Mayo Clin Proc 66:818-831
3. Badimon L et al (1993) Thrombos Haemost 70:111-118
4. Nichols TC et al (1992) Am J Pathol 140:403-415
5. Nichols TC 1998 Arterioscler Thromb Vasc Biol 18:323-330
6. Lip GY, Blann AD (1995) Br Heart J 74:580-583
7. Sixma JJ et al (1996) Arterioscler Thromb Vasc Biol 16:64-71
8. Cortellaro M et al (1996) Stroke 27:450-454
9. Holvoet P et al (1996) Thromb Haemost 76:663-669
10. Sato Y et al (1996) Thromb Haemost 75:389-392
11. Taubman MB et al (1997) Thromb Haemost 78:200-204
12. Lewis JC et al (1995) Am J Pathol 47:1029-1040
13. Broze Jr GJ (1995) Thromb Haemost 74:90-93
14. Jang Y et al (1995) Circul 92:3041-3050
15. Waxman L et al (1990) Science 248:593-596
16. Dunwiddie CT et al (1989) J Biol Chem 264:16694-16699
17. Harlos K et al (1994) Nature 370:662-666
18. Banner DW et al (1996) Nature 380:41-46
19. Abumiya T et al (1995) Thromb Haemost 74:1050-1054
20. Lesnik P et al (1995) Arterioscler Thromb Vasc Biol 15:1121-1130
21. Ko FN et al (1996) J Clin Invest 98:1493-1501
22. Weaver AM et al (1996) J Biol Chem 271:24894-24900
23. Schini-Kerth VB et al (1995) Thromb Haemost 74:980-986
24. Wenzel UO et al (1995) Circ Res 77:503-509
25. Fabunni RY et al (1996) Biochem J 315:335-342
26. Weis JR et al (1991) FASEB J 5:2459-2465
27. Delafontaine P et al (1996) J Clin Invest 97:139-145
28. Shirk RA et al (1996) Arterioscler Thromb Vasc Biol 16:1138-1146
29. Sarembock IJ et al (1996) J Vasc Res 33:308-314
30. Ragosta M et al (1996) Circulation 93:1194-1200
31. Harpel PC et al (1996) J Nutr 126:1285S-1289S
32. Matthias D et al (1996) Atherosclerosis 122:201-216
33. Smith EB et al (1990) Arteriosclerosis 10 263-275
34. Languino LR et al (1995) Proc Natl Acad Sc USA 92:1505-1509
35. Farrel DH et al (1992) Proc Natl Acad Sc USA 89:10729-32
36. Menschikowski M et al (1995) Atherosclerosis 118:173-181
37. Lowe GDO et al (1993) Circulation 87:1915-1920
38. Sosef MN et al (1994) Thromb Haemost 72:250-254
39. Thompson SG et al (1995) N Engl J Med 332:635-641
40. Mazoyer E et al. (1999) Thromb Res 95:163-176
41. Ernst E, Resch KLR (1993) Ann Intern Med 118:956-963
42. Montalescot G et al (1995) Circulation 92:31-38
43. Karrillon GJ et al (1996) Circulation 94:1519-1527
44. Kannel WB et al (1987) JAMA 258:1185-1186
45. Reinecke H et al (1999) Arterioscler Thromb Vasc Biol 19:2355-2363
46. Danesh J et al (1999) BMJ 319:1157-1162
47. Toss H et al (1997) Circul 96:4204-4210
48. Amrani DL (1990) Blood Coag and fibrinol 1:443-446
49. Balleisen L et al (1985) Thromb Haemost 54:475-479
50. Balleisen L et al (1985) Thromb Haemost 54:721-723
51. Lee AJ et al (1993) Brit J Haematol 83:616-621
52. Iso H et al (1993) Arterioscler Thromb 13:783-790
53. Hamsten A et al (1987) Lancet ii:988-991
54. Scarabin PY et al (1993) Arterioscler Thrombosis 13:886-891
55. Di Minno G et al (1997) Thromb Haemost 78:462-466
56. de Maat MPM et al (1998) Arteriocler Thromb Vasc Biol 18:265-271
57. Thomas AE et al (1991) Thromb Haemos 65:487-490

58. Falk E, Shah PK, Fuster VCoronary plaque disruption Circulation, 1995, 92, 657-671
59. Ridker PM et al (1998) Circulation 97:425-428
60. Qing Xiao et al (1998) J Clin Invest 101:1184-1194
61. Meade TW et al (1986) Lancet 2:533-537
62. Schneider DJ et al (1997) Arterioscler Thromb Vas Biol 17:3294-3302
63. Loskutoff DJ et al (1995) Ann NY Acad Sc 748:177-184
64. Robbie LA et al (1996) Arterioscler Thromb Vasc Biol 16:539-545
65. Shireman PK (1996)J Vasc Surg 23:810-817
66. Padro T 81995) Arterioscler Thromb Vasc Biol 15:893-902
67. Stefansson S, Lawrence DA (1996) Nature 383:441-443
68. Stefansson S et al (1996) J Biol Chem 271:8215-8220
69. Waltz DA et al (1993) J Clin Invest 91:1541-1552
70. Montrucchio G et al (1996) Circulation 93:2152-2160
71. Hajjar KA, Nachmann RL (1996) Annu Rev Med 47:423-442
72. Grainger DJ, Metcalfe JC (1995) Curr Opin Lipidol 6:81-85
73. Salomaa V et al (1995) Circulation 91:284-290
74. ECAT Angina pectoris study group (1993) Eur Heart J 14:8-17
75. Thompson SG et al (1995) N Eng J Med 332:635-641
76. Zitoun D et al (1996) Arterioscler Thromb Vasc Biol 16:77-81
77. Sironi L (1996)Arterioscler Thromb Vasc Biol 16:89-96
78. Brown SL 81995) Circulation 92:767-772
79. Holvoet P, Collen D (1995) Circulation 92:698-699
80. Li XN et al (1996) Biochemistry 35:6080-6088
81. Celermajer DS et al (1993) J Am CollCardiol 22:854-858
82. Zimmerman GA et al (1996) J Clin Invest 98:1699-1702
83. Ben-Baruh A et al (1995) J BiolCell 270:11703-11706
84. Takahashi M et al (1996) Circul 93:1185-1193
85. Johnson Tidey RR et al (1994) Am J Pathol 144:952-961
86. Poston RN et al (1996) Am J Pathol 149:73-80
87. Murohara T et al (1996) Circ Res 78:780-789
88. Mehta A et al (1995) Arterioscler Thromb Vasc Biol 15:2076-2083
89. Marui N et al (1993) J Clin Invest 92:1866-1874
90. Wenzel K et al (1996) Hum Genet 97:15-20
91. Haught WH et al (1996) Am Heart J 132:1-8
92. Wang N et al (1996) J BiolChem 271:8837-8842
93. Taub DD et al (1996) Blood 87:1423-1431
94. Wang X et al (1996) J BiolChem 271:24286-24293
95. Freestone T et al (1995) Arterioscler Thromb Vasc Biol 15:1145-1151
96. Bjorkerud S, Bjorkerud B (1996) Am J Pathol 149:367-380
97. Simons LA et al (1998) Atherosclerosis 137:197-203
98. Fleming I et al (1996) Mol Cell Biochem 157:137-145
99. Gauthier TW et al (1995) Arterioscler Thromb Vasc Biol 15:1652-1659
100. Heizawa H et al (1995) Circul 92:3520-3526
101. Minuz P et al (1995) Thromb Haemost 74:1175-1179
102. Wu KK, Thiagarajan P (1996) Annu Rev Med 47:315-331
103. Watkins MT et al (1996) J Vasc Surg 23:95-103
104. Myers SI et al (1996) Ann Surg 224:213-218
105. Yamagishi S et al (1996) FEBS Lett 384:103-106
106. Farhat MY et al (1996) FASEB J 10:615-624
107. Ginon I et al (1995) Acta Diabetol 32:170-175
108. Umeda F et al (1996) Diabetes 45:S111-Sl13
109. Farber HW, Barnett HF (1991) Circ Res 68:1446-1457
110. Lynch SM et al (1994) J Clin Invest 93: 998-1004
111. Morrow JD et al (1992) Proc Natl Acad Sc USA 89:10721-10725
112. Morrow JD, Roberts LJ (1996) Biochem Pharmacol 51:1-9
113. Laskey RE, Mathews WR (1996) Arch Biochem Biophys 330:193-198
114. Pratico D, FitzGerald GA (1996) J Biol Chem 271:8919-8924
115. Holvoet P et al (1995) J Clin Invest 95:2611-2619
116. Anti Platelet Trialists' Collaboration (1994) Brit Med J 308:81-106
117. Ridker PM et al (1997) Lancet 349:385-388
118. Newman PJ (1997) Lancet 349:370-371
119. Murata M et al (1997) Circulation 96:3281-3286
120. Dentan C et al (1996) Eur J Biochem 236:48-55
121. Schini-Kerth VB et al (1996) Circul 93:2170-2177
122. Weyrich AS et al (1996) J Clin Invest 97:1525-1534
123. Kuijper PH et al (1996) Blood 87:3271-3281
124. Chen LY, Mehta JL (1996) J Cardiovasc Pharmacol 27:154-158

Thrombospondins

Synonym:

Thrombospondin-1 = thrombin-stimulated protein = TSP1
Thrombospondin-2 = corticotropin-induced secreted protein = TSP2

Definition *Large, multimodular glycoproteins that inhibit angiogenesis. The main prototypes involved in vascular biology are thrombospondin-1 (TSP1) and thrombospondin-2 (TSP2).*

See also: →Angiogenesis inhibitors

Introduction Thrombospondin (TSP) is a large, trimeric, multimodular glycoprotein initially purified from thrombin-activated platelet releasates and later detected as a secretion product from a wide variety of epithelial and mesenchymal cells. In the early 90s it became clear that platelet thrombospondin is just one member of a family of structurally related proteins that presently comprises five distinct members [1]. Our current appreciation of the involvement of thrombospondins in diverse biological processes extends far beyond its initially proposed role in platelet aggregation and coagulation [2]. In particular, the role of thrombospondins in modulating endothelial cell function has been widely studied *in vitro* since the 70s. The recent finding by Bouck and colleagues that thrombospondins possess anti-angiogenic activity has generated exceptional interest [3, 4]. More precisely, the discovery that altered expression of certain tumor suppressor genes is correlated with the loss of TSP1 expression and the acquisition of an angiogenic phenotype by tumor cells, provides an excitingly new paradigm for tumor progression.

Characteristics
Molecular Weight Human TSP1 was initially purified as a 420 kDa disulfide-bonded homotrimeric glycoprotein released from the α-granules of thrombin-stimulated platelets. Although the size of the individual monomers was initially reported to be 140 kDa, apparent molecular weights ranging from 145,000 to 180,000 have been further reported from SDS-PAGE analyses, depending on the particular gel system employed. The predicted amino-

acid composition of TSP1, deduced from translation of its cDNA, provides for a molecular mass of 126,500 Da. The remaining mass can be accounted for largely by the presence of N-linked and O-linked carbohydrate chains. Whereas native platelet TSP1 behaves as a 420 kDa protein when analyzed by sedimentation equilibrium analysis [5], it migrates as a 540 kDa protein during electrophoresis in non reducing SDS- polyacrylamide gels [6]. Bovine TSP2 has been purified as a 600 kDa disulfide-bonded homotrimeric glycoprotein from the conditioned medium of adrenocortical cell primary cultures [7]. Each individual monomer migrated as a 195 kDa band on SDS-polyacrylamide gels. In the adrenal cortex tissue, however, the size of TSP2 monomers was found to be 180 kDa (personal observations) just like the reported size of recombinant mouse TSP2 [8]. The discrepancy between the calculated native mass of TSP2 (126,500 Da) and its apparent mass observed by PAGE-SDS also appears to be due to glycosylation (contributing to 5-15 kDa). In addition, the low isoelectric points of both thrombospondins cause reduced binding of SDS and anomalously slowered migration on SDS-polyacrylamide gels.

Domains Thrombospondins are multimodular proteins [1]. The structures of TSP1 and TSP2 are shown diagrammatically in Figure 1. TSP1 and TSP2 each contain an amino-terminal heparin-binding domain, a linker region containing the two cysteine residues involved in interchain disulfide bonding, a procollagen-homology domain, three type I (TSP) repeats, three type II (EGF-like) repeats, seven type III (Ca^{++}-binding) repeats and a globular C-terminal domain. As pointed out in Figure 1, there exists an increasing gradient of sequence homology between TSP1 and TSP2 extending from the N- to C-termini of the molecules.

Binding Sites and Affinity The characterization of thrombospondin receptors has been determined essentially for TSP1, the prototypic member of the family. The known receptors are summarized in Figure 2. Heparin-binding sites have been identified in the N-terminal domain and in the type I repeats of TSP1. The sequence encompassed by residues 60-94 in the globular N-terminal domain appears to constitute the glycosaminoglycan binding site [9]. Distinct heparin-binding sequences have been identified in the type I repeats. The consensus sequence WSXW (that is present in each type I repeat) has been shown to bind heparin and sulfated glycolipids and to prevent the binding of melanoma cells to components of the extracellular matrix. Additional studies with synthetic peptides have shown that basic amino acids flanking this consensus sequence enhance the adhesion and chemotaxis of melanoma cells [10]. The binding of TSP1 to heparan sulfate proteoglycans appears to be a prerequisite for its binding and internalization by the low density lipoprotein receptor – related protein (LRP) [11]. Although the sequence homology between TSP1 and TSP2 is weakest in their N-terminal domains, TSP2 also appears to bind heparin and heparan sulfates, although with a slightly weaker affinity and this property has been widely used for its purification [7, 8]. Both human TSP1 and human TSP2 possess duplicate CSVTCG sequences within their type I repeats. These motifs represent binding sites for the receptor CD 36 [12] and for a 60 kDa receptor termed CBP [13].

An RGD integrin recognition motif (RGDA in TSP1 and RGDI in TSP2) is present in the last type III repeat. This motif is probably responsible for the binding of thrombospondins to the platelet integrin aIIBb3 and to the endothelial cell integrin $\alpha_v\beta_3$ (often referred to as the fibronectin receptor) [14].

The C-terminal domain of TSP also contains repeated binding motifs containing the tripeptide VVM. The TSP1 peptides RFYVVM and IRVVM were shown to support attachment of melanoma, erythroleukemia, fibrosarcoma and endothelial cells through the integrin-associated protein [15].

Additional Features In addition to cell surface receptors, several extracellular matrix proteins present a significant affinity for TSP. These include fibronectin, laminin, osteonectin, SPARC and the fibrillin-related latent TGF-ß binding protein (LTBP) [16]. The fibronectin binding site has been identified in TSP1 as the sequence GGWSHW (436-441) whereas those for other matrix proteins are still unknown. TSP1 also binds to several components of the coagulation and fibrinolytic system including fib-

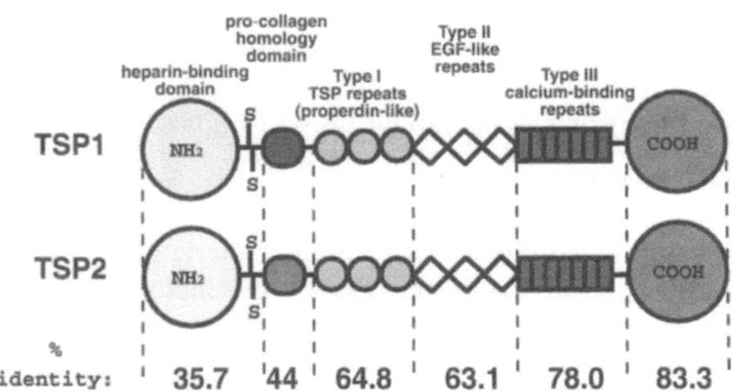

Figure 1. Structural organization and sequence homology of TSP1 and TSP2 monomers

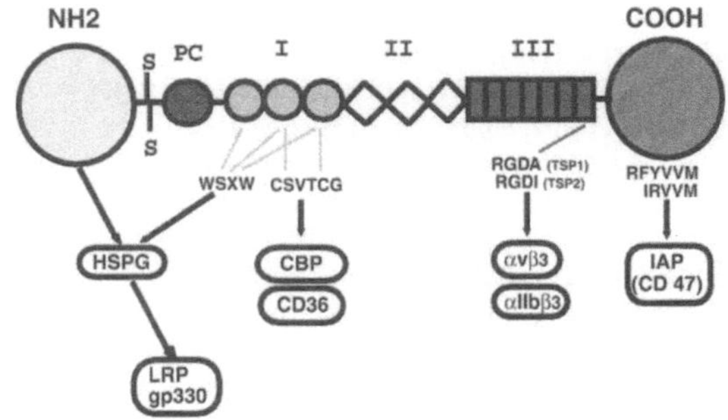

Figure 2. The thrombospondin receptors

rinogen, kininogen, plasminogen and plasmin, urokinase and tissue-plasminogen activator, and thrombin. A resulting inhibition of their proteolytic activities has been reported in a number of cases [16,17].

Structure

Sequence and Size   TSP1 is a 1170 amino acid-long protein in the human and murine species whereas human and mouse TSP2 comprise 1172 amino-acids. The multimodular organization of both molecules is extremely similar (Figure 1) and both TSP1 and TSP2 exist as disulfide-bonded trimers. The two cysteine residues forming interchain disulfide bonds are located between the N-terminal heparin-binding domain and the procollagen-homology domain and are separated by three amino-acids in both thrombospondins. Although disulfide bridges in TSP1 bond residues 270 and 274, the recombinant fragment TSP1 1-277 poorly trimerizes whereas the fragment 1-333 does [18]. This indicates that the 60 amino-acids located in the C-terminal vicinity of the disulfide bonded cysteines contain an essential sequence for trimer formation. In fact, the amino acid sequence surrounding these cysteines can form an amphipathic a helix that could result in the formation of coiled coils [18]. Although the corresponding sequence in TSP2 differs considerably, the spacing of the cysteines is conserved and the surrounding sequence also lends itself to the formation of a coiled coil [19]. It is important to notice that, *in vivo*, TSP trimerization takes place during translation before the synthesis and folding of each chain is completed.

Homologies   The TSP family now comprises 5 distinct members. Three structurally related members of the TSP family (TSP3, TSP4 and TSP5/COMP) resemble one another but differ from TSP1 and TSP2 in that they lack the procollagen domain and type I repeats and contain four instead of three type II repeats [19]. In addition, TSP3, TSP4 and TSP5/COMP also differ from TSP1 and TSP2 in their secondary structure, being organized as pentamers instead of trimers.

Conformation   The overall shape of the thrombospondin molecules has been visualized by rotary shadowing electron microscopy. TSP1 and TSP2 appear to be composed of three equally electron-dense nodules

and of a fourth nodule formed by the close association of three smaller fragments [20, 7]. When calcium ions are chelated by EDTA or EGTA, linker arms become visible between these nodules and the overall shape of the molecules becomes less compact. These observations led to the structural model in which each subunit has the form of a dissimetric halter and Ca^{++} favors the compaction of the molecule (Figure 3). The calcium-induced conformational change of TSP1 also results in a change in circular dichroism spectra, proteolytic susceptibility, monoclonal antibody-binding properties, dynamic light-scattering and sedimentation properties of the molecule.

Gene

Gene Structure   Both mouse TSP1 and mouse TSP2 genes contain 22 exons whose size class and intron boundaries are highly conserved [1, 21]. Strikingly, exon sizes, from exon 9 to exon 21, are identical in the two genes. As observed in other gene families, the sequences of orthologous genes, i.e. the same gene in two different species, bear a higher degree of nucleotide identity than the sequences of paralogous genes, i.e. the sequences of two different homologous genes in a single species. As mentioned previously for the protein sequence, there exists a progressive increase of identity in the 3'-direction in the nucleotide sequence of TSP1 and TSP2, ranging from 40-50 % in the 5'-region to 75-80 % in the 3'-region. Construction of the sequence-based evolutionary tree of members of the thrombospondin family has revealed that the gene duplication producing TSP1 and TSP2 would have occurred about 600 million years ago whereas the gene duplication leading to the two subfamilies (TSP1/ TSP2 vs TSP3/TSP4/TSP5) would have occurred approximately 900 million years ago [19].

Chromosomal Localization   The chromosomal positions of TSP1 and TSP2 genes in the human and murine genomes are reported in Table 1. The two family members are encoded by distinct genes located on distinct chromosomes.

Gene Expression   The expression of TSP1 and TSP2 during murine embryonic development has been studied in great detail by Iruela-Arispe and collaborators [22]. The TSP1 and TSP2 mRNAs exhibit spatial and temporal dif-

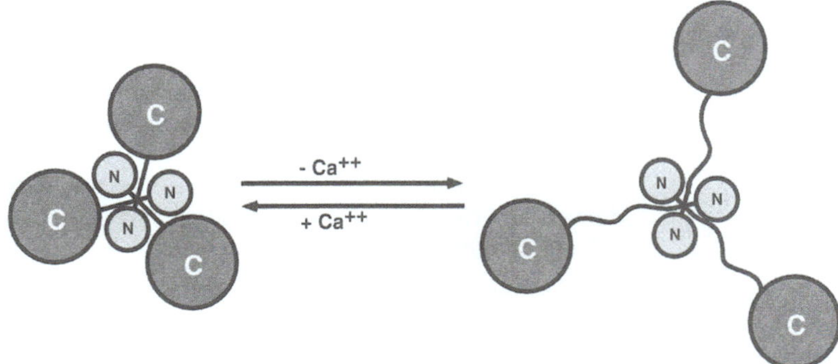

Figure 3. Ca^{++}-dependent modifications of TSP conformation

Table 1. Chromosomal locations of TSP1 and TSP2 genes

		Chromosomal location (gene)	Gene size	Number of amino-acids (protein)
TSP1	human	15 q 15		1170
	mouse	2 F	19 kb	1170
TSP2	human	6 q 27		1172
	mouse	17 A3	29 kb	1172

ferences in their pattern of expression [19]. In 14-day embryos, the expression of TSP1 is predominant in the mesenchyme, brain, heart and liver whereas TSP2 is prominent in the vasculature. Both transcripts are expressed in the brain. At later stages of mouse embryonic development (18 days), constitutive expression of TSP1 and TSP2 occurs in most tissues. The distribution of TSP1 and, to a lesser extent, of TSP2 proteins has been studied by various techniques including immunohistochemistry and Western blotting. In several adult tissues including human placenta, porcine skin and artery, TSP1 is localized to the junction between the basement membrane and the adjacent connective tissue [23]. Few studies have addressed the distribution pattern of TSP2. In the bovine species, using a polyclonal antibody that does not cross-react with TSP1, my colleagues and I observed that TSP2 is present in several endocrine tissues including adrenal cortex, ovary, testis, placenta, as well as in heart, spleen, brain and kidney. In the adult mouse, TSP2 was immunolocalized in dermal fibroblasts, articular chondrocytes, Purkinje cells in the cerebellum, Leydig cells in the testis and glandular adrenocortical cells [24].

Gene Regulation   The strong and very rapid induction of TSP1 gene expression by serum and PDGF and its independance of protein synthesis have qualified this gene as a member of the immediate early gene family. Framson and Bornstein [25] have determined that this serum response required the coordinate function of a serum-response element (SRE) and an NF-Y binding site located respectively at positions -1280 and -65 in the human TSP1 promoter. In the mouse TSP1 promoter, an Egr-1 binding site appears to have substituted for the NF-Y binding site of the human promoter and to be dispensable for mediating the serum response [26]. The binding of Egr-1 and SP1 to a GC-rich region from the proximal promoter mediates the constitutive expression of the mouse gene [26]. In contrast to TSP1, whose serum-induced mRNA level increase approaches 50-fold, TSP2 is poorly stimulated by serum [1]. In accordance with this observation, the mouse TSP2 promoter appears to lack an upstream serum-response element, and an NF-Y binding site or an Egr-1 binding site in the proximal promoter [21]. Under conditions of more stringent serum deprivation, mouse Swiss 3T3 mRNA levels were shown to be induced by stimulation with either FGF-2, PDGF or IL-1 [27]. However, TSP2 mRNA

levels were not superinduced by cycloheximide, confirming that TSP2 is not an immediate early gene. Other potential regulatory elements have been identified in the promoter of human TSP1 including response elements for heat shock and cAMP, and an AP2 binding site. Similarly, three common factor 1 (CF1) binding sites, two potential AP-1 binding sites, a ras-responsive factor 1 (RRF-1) binding site and two GATA (NF-E) binding sites have been identified in the mouse TSP2 gene promoter [21].

In addition to the information brought by the genetic analyses of the promoters, several studies have reported either parallel or differential regulations of the TSP1 and TSP2 mRNA levels. In bovine adrenocortical cells, TGF-β1, TGF-β2 and activin stimulate both TSP1 and TSP2 expression [28]. In the same cells, the pituitary hormone ACTH and its second messenger cAMP strongly induce TSP2 expression and simultaneously repress TSP1 expression [29]. Within 24 h, the pattern of TSP expression is thus shifted from TSP1 to TSP2. Negative regulation of TSP1 expression by cAMP-inducing pituitary hormones appears to be a general feature of endocrine cells since TSP1 mRNA levels are decreased in response to TSH in thyroid epithelial cells [30], in response to FSH in ovarian granulosa cells [31] and in response to PTH in rat osteosarcoma cells (Lafeuillade B and Feige J. J., unpublished observations). So far, cAMP stimulation of TSP2 expression has not been observed in other endocrine tissues than the adrenal cortex. Steroid hormones are another class of potential regulators of TSP expression. Whereas corticosteroids do not affect TSP1 and TSP2 expression in adrenocortical cells, progesterone stimulates TSP1 expression in human endometrium during the secretory phase of the hormonal cycle [32]. It should also be mentioned that TSP1 and TSP2 exhibit distinct patterns of tissue distribution both in embryo and adult animals [22, 27]. This suggests that distinct cis-acting elements act on each gene to specify tissue-specific expression.

Additional Features   Preliminary evidence for alternative splicing of TSP1 has been presented by Frazier [33]. A first isoform, called TSP50 was identified in melanoma cells. It encodes a 50 kDa protein comprising the N-terminal heparin-binding domain, the procollagen-homology domain and the first type I repeat of standard TSP1. A second isoform of TSP1, named TSP140, which may be generated by alternative splicing, encodes a 140 kDa monomeric protein that begins at TSP1 residue 293 and lacks the cysteine residues involved in trimerization. It has been reported to be expressed in human tissues in a developmentally regulated manner [33] and is similar to the hamster glycoprotein gp140 that was first identified by Good et al [3] as a tumor suppressor-controlled inhibitor of angiogenesis.

Processing and Fate   Binding, internalization and degradation of both TSP1 and TSP2 are energy-dependent saturable processes. Using mutant Chinese hamster ovary cells defective in glycosaminoglycan synthesis, Mosher and collaborators have clearly demonstrated that heparan sulfate proteoglycans are required for these processes [34]. Recent studies have established that TSP1 and TSP2 internalizations are mediated by the low density lipoprotein receptor-related protein (LRP), a multiligand receptor that mediates the cellular uptake and subsequent degradation of tissue-type plasminogen activator (tPA), urokinase-type plasminogen activator (uPA), α_2-macroglobulin, apolipoprotein E, lipoprotein lipase as well as apoE- and lipoprotein lipase-enriched VLDL [11, 34]. Internalization results in degradation into trichloracetic acid-soluble peptides. No recycling and no limited proteolysis of internalized TSP have been described.

Biological Activity The term "matricellular" has been coined by Paul Bornstein to refer to a group of multimodular extracellular proteins, of which TSP1 and TSP2 represent prototypes [35]. Matricellular proteins possess a diversity of cell surface receptors specific for their diverse modules and thereby mediate a diversity of biological functions. Clearly, a given cell type expresses only a subset of thrombospondin receptors and the level of expression of a given receptor varies between diverse cell types. Biological responses to TSPs are thus combinatorial in nature. This may explain some of the conflicting results present in the literature. Biological functions of TSP1 have been reviewed in detail by several authors [2, 19, 33]. In this section, we will overview the main domains of action of thrombospondins, except for their effect on endothelial cell function and angiogenesis that will be detailed in the next section.

Thrombospondin 1 has been originally named after its property to be released from platelet α-granules upon activation by thrombin. It has been clearly established that released TSP1 incorporates into the fibrin clot and plays an important role in the secondary phase of aggregation. Antibodies to TSP1 or a recombinant heparin-binding domain of TSP1 effectively block aggregation and secretion induced by low concentrations of thrombin in washed platelet suspensions [36]. These studies suggest that these effects are mediated by the interaction of the N-terminal heparin-binding domain of TSP1 with membrane-bound fibrinogen. Other sequences such as CSVTCG in the type I repeats and RGD in the last type III repeat, that bind to CD36 and integrin $\alpha_{IIB}\beta_3$ respectively, have also been proposed to play a role in these processes. TSP2 does not appear to be involved in these processes since it is not expressed by megakaryocytes and is not present in thrombin-induced platelet supernatants.

TSP1 and TSP2 present adhesive or anti-adhesive properties for a variety of cell types. Since the combinatorial expression of the various receptors has not been extensively examined in most of these studies, it is still quite difficult to decipher which portion of the molecule mediates the positive or negative effect observed. As an example, Tuszynski et al. found that TSP1 promotes the cell-substratum adhesion of a variety of cell types including the endothelial cell line BFAE-39/6-3 [37] whereas Lahav

later reported that adsorption of TSP1 to a substratum prevents adhesion of bovine aortic endothelial cells [38]. Besides cell attachment, cell spreading is another function that is stimulated by TSP1 in diverse cell types. It is possible however that these two functions (attachment and spreading) are triggered by distinct modules of the TSP1 molecule. When comparing the effects of TSP1 and TSP2 on adhesion of bovine adrenocortical cells, my laboratory observed that TSP1 stimulates both attachment and spreading whereas TSP2 stimulates attachment but prevents cell spreading in a concentration-dependent manner [39]. The N-terminal end of the TSP molecules, which is the least conserved between the two family members, thus probably bears the sequence controlling cell spreading. This functional difference between TSP1 and TSP2 represents a first indication that the biological functions of these two isoforms, are not completely redundant. Murphy-Ullrich and colleagues have shown that the anti-adhesive effect of TSPs on bovine aortic endothelial cells is correlated with a loss of focal adhesions. Interestingly, a 19 amino acid-long peptide (residues 17-35) from the heparin-binding domain of TSP1 and a peptide from the analogous sequence in TSP2 both contain the focal adhesion- labilizing activity [40].

TSP1 has also been characterized as an inhibitor of several distinct proteolytic enzymes. These activities have been reviewed by Hogg [17]. Briefly, TSP1 is a fast-acting tight-binding competitive inhibitor of neutrophil elastase and cathepsin G and the active binding site has been localized to the calcium-binding type III repeats. Recombinant TSP2 appears as a potent competitive inhibitor of cathepsin G and a weak inhibitor of neutrophil elastase [17]. TSP1 is also a slow-acting tight-binding inhibitor of plasmin and uPA, and particularly of uPA-mediated fibrinolysis.

Another important function recently ascribed to TSPs is their ability to activate latent TGF-β. TGF-β_1 is the prototype of a family of peptide regulators which are synthesized under a latent form that requires activation to become fully biologically active. Since TGF-β receptors are quite ubiquitous, the key step in controlling TGF-β activity is its activation. The most common latent form of TGF-β_1 is a non-covalent association between the dimerized N-terminal part of pro- TGF-β_1 (called latency-associated protein or LAP) and the dimerized C-terminal end representing mature TGF-β_1. In the conditioned medium of various cell types, this small latent TGF-β is covalently associated with a large glycoprotein called LTBP (latent TGF-β binding protein). The best characterized physiological mechanism of latent TGF-β activation is the proteolytic attack of LAP by plasmin that triggers the release of mature TGF-β. Murphy Ullrich and collaborators have reported that TSP1 binds mature TGF-β without inhibiting its activity and that TGF-β-depleted TSP is able to activate both the small and the large latent TGF-β complexes. They further reported that the activation was independent of proteolysis and was mimicked by the tripeptide RFK or the tetrapeptide KRFK present between the first and the second type I repeats of human

TSP1 [41]. A peptide with the corresponding sequence in TSP2, RIR, was found to be inactive, as was recombinant mouse TSP2 [41]. Using a slightly different experimental design with partially purified natural molecules, we reached a different conclusion. We observed that coincubation of large latent TGF-β, partially purified from the conditioned medium of bovine adrenocortical cell primary cultures, and either purified TSP1 or purified TSP2 generated active TGF-β [42]. Interestingly, both TSPs were unable to activate the latent TGF-β complex formed by the association of mature TGF-β with α_2-macroglobulin. It should be noted that the possibility exists that the sequence KKFK present in human, mouse and bovine TSP2 (residues 344–347) can substitute for the KRFK sequence of TSP1 (residues 430–433) to activate the latent TGF-β complex. Using the surface plasmon resonance technique (BiaCore), we could show that purified TSP1 and TSP2 do not bind directly to reconstituted or preformed small latent TGF-β whereas plasmin does bind and degrade this complex [43]. Moreover, using this technology, no binding of TSP1 or TSP2 to mature TGF-β or recombinant LAP was detectable whereas binding to heparin was clearly detectable. This result clearly shows that activation of latent TGF-β by thrombospondin does not occur through a direct molecular interaction but rather requires a cofactor (LTBP or another cell surface molecule ?).

Role in Vascular Biology

Physiological Function Angiogenesis, the process by which new microvessels sprout from pre-existing capillaries and post-capillary venules is a multistep process that implies disruption of the endothelium basement membrane, endothelial cell proliferation and migration, morphogenic organization into a tubular structure and stabilization of the newly formed microvessels. These diverse functions are stimulated by a number of angiogenic factors and inhibited by several angiostatic factors that have been characterized over the last decade. Noel Bouck and collaborators provided a first demonstration of the anti-angiogenic property of TSP1 *in vivo*, when they characterized that a 140 kDa monomeric protein (gp 140) secreted by a hamster cell line in a tumor suppressor-dependent manner, was indistiguishable from a portion of TSP1 [3]. In this model, secretion of a neovascularization inhibiting activity and secretion of TSP1 were tightly linked to the presence of an active cancer suppressor gene in both transformants and revertants. gp 140 and homotrimeric TSP1 inhibited neovascularization *in vivo* and endothelial cell migration toward FGF2 *in vitro* [3]. Recombinant mouse TSP2 and natural bovine TSP2 were almost as active as TSP1 in these biological assays whereas TSP5/COMP, a pentameric member of the TSP family, was inactive [4]. Using proteolytic fragments and synthetic peptides, the angioinhibitory activity of TSP1 has been mapped to the procollagen homology domain and to the type I repeats [44]. Two separate non-overlapping peptides bearing the antiangiogenic activity map to the type I repeats; one of them

comprises the latent TGFβ-activating sequence. Since TGFβ is an inhibitor of endothelial cell proliferation, the possibility existed that the angiostatic activity of TSP1 could be mediated by its ability to activate latent TGFβ. However, neutralizing anti-TGFβ antibodies have no effect on the thrombospondin inhibition of bFGF-induced endothelial cells migration whereas anti-thrombospondin monoclonal antibodies that do not recognize TGFβ block the angiostatic effects of TSP1 and gp 140 *in vitro* and *in vivo* [3]. In addition, although TGFβ inhibits endothelial cell proliferation *in vitro*, it is angiogenic *in vivo*. The anti-angiogenic activities of TSP1 and TSP2 thus appear independent from their abilities to activate latent TGFβ.

Human platelet TSP1 specifically inhibits the proliferation of endothelial cells from various tissue origins. Using an *in vitro* model in which strains of endothelial cells spontaneously organize into cords and tubes with patent lumen, a prominent role of TSP1 in this morphogenetic process could clearly be established [45]. In the presence of anti-TSP1 antibodies, the production of endothelial cords was increased by 33-55 %. The maximal expression of TSP1 in confluent monolayers and in a limited number of proliferating cells adjacent to the developing cords and tubes confirmed its suggested angiostatic function. It should be noted however that, using a distinct assay consisting in serum-free cultures of aortic explants in a collagen gel, Nicosia and Tuszynski reached opposite conclusions. They observed that the presence of TSP1 in the gel promoted the number and length of the microvessels outgrowing the aortic explants [46]. *In vivo*, during mouse embryonic development, TSP2 appears to be expressed in capillaries and large vessels, from day 10 post- conception continuously through parturition on day 18 whereas TSP1 mRNA was only detected in capillaries of day 16-18 embryos [22]. The invalidation of the TSP1 gene does not alter however the normal embryonic development of the vasculature, since knock-out mice are viable and appear to develop normally [47]. After 4 weeks of life, these mice develop however extensive acute pneumonia. The phenotype of TSP2 knock-out mice is more dramatic [48]. Although viable, these mice develop alterations in the tensile strength of skin and tendons due to an abnormal pattern of collagen fiber organization. In addition, mutant mice display an abnormal bleeding time and an increased density of blood vessels in many tissues. This latter feature is indicative that TSP2 is more dramatically implicated than TSP1 in the regulation of vascularization.

Pathology Acquisition of an angiogenic phenotype is one of the key changes that transformed cells must go through in order to develop into solid tumors. This angiogenic switch can result either from the overexpression of an angiogenic peptide (e.g. VEGF in gliomas) or from the loss of expression of angiostatic proteins such as thrombospondins. The correlation between the loss of expression of a tumor suppressor gene, the decreased expression of thrombospondin, and the development of an angiogenic phenotype has now been observed in a number of pathological situations [49]. In cultural fibroblasts from Li-Fraumeni patients, the loss of the wild type allele of the p53 tumor suppressor gene correlates with the reduced expression of TSP1 and the switch to an angiogenic phenotype [50]. Wild type p53 was shown to positively regulate the TSP1 promoter. A similar correlation between p53 and TSP1 expressions was also established in bladder cancer [51] and in the breast carcinoma cell line BT 549 [52]. Transfer of wild-type chromosome 10, which carries a number of tumor-suppressor genes frequently lost in glioblastoma multiform, into three different glioblastoma cell lines (lines U251, U87 and LG11) results in the loss of their ability to form tumors in nude mice and the acquisition of an angiostatic phenotype [53]. This switch in the angiogenic phenotype could be attributed to the increased secretion of TSP1 and was TGF-β-independent [53].

From a recent study, it appears that the expression of TSP1 and TSP2 mRNAs is significantly increased in invasive breast carcinoma as compared to the levels observed in normal and benign tissues [54]. These levels appear to correlate with the quantity of stroma present in the tumors, which is in agreement with the immunohistochemical localization of TSP1 in the fibroblasts from the desmoplastic stroma of the breast tumors.

Clinical Relevance and Therapeutic Implications The evaluation of TSP1 or TSP2 levels as markers of pathological evaluation or as prognostic markers has been carried out in a number of cancers and other diseases. Given the abundance of TSP1 in platelets, measurement of TSP1 levels in serum is not informative and measurement in plasma requires extreme care and reproducibility in the collection of platelet- poor biological samples. TSP2 being absent from platelets, its serum levels may appear to be more informative but no data have been published so far. The results from the determination of TSP1 concentrations in cancer tissues show some discrepancies. In human adrenal cortex carcinomas, the concentrations of TSP1 are decreased by a factor larger than 3 as compared to the concentrations in normal tissues and adenomas (our unpublished observations). Bladder cancer patients with low levels of TSP1 (measured in tumor cytosols) exhibit increased tumor vascularization, increased disease recurrence rates and decreased overall survival [51]. As elegantly proposed by Hanahan and Folkman in the balance hypothesis, the angiogenic switch that is necessary for cancer growth may result either from an induced expression of angiogenic factors or from the loss of expression of angiostatic factors such as thrombospondins [55]. This second mechanism is more likely to occur in tissues where the vasculature is in direct contact with the differentiated cells (such as endocrine organs) than in tissues such as epidermis which are separated from the blood vessels by a basement membrane and a stromal support [55]. The observation that highly vascularized breast tumors display elevated tissular levels of TSP1, and a TSP1 localization in the stromal tissue [54] may indicate that, in this type of cancer, TSP1 is a marker of the stromal response and does not play an important role in the angiogenic switch. In those tissues where a correlation between the loss of TSP expression and the angiogenic switch toward tumor progression has been established, it is conceivable that TSP or, even better, active TSP fragments, could be used as therapeutic agents. Several angiostatic factors including interferon α_{2A}, platelet factor 4, metalloproteinase inhibitors or interleukin 12 are under current study in clinical trials as treatment for a wide spectrum of solid tumors. In *in vitro* experiments, overexpression of TSP1 in a human breast carcinoma cell line results in the reduction of the primary tumor growth, the metastatic potential and angiogenic response of the transfected cells [56]. Similarly, expression of TSP1 in the virally transformed endothelial cell line b.END.3 suppresses their tumorigenic potential and restores saturation density growth inhibition [57]. This opens the possibility to use this protein or shorter active fragments in the treatment of hemangiomas.

Jean-Jacques Feige

References

1. Bornstein P (1992) FASEB J 6:3290-3298
2. Lahav J (1993) Biochim Biophys Acta 1182:1-14
3. Good DJ et al (1990) Proc Natl Acad Sci USA 87:6624-6628
4. Volpert OV et al (1995) Biochem Biophys Res Commun 217:326-332
5. Margossian SS et al (1981) J Biol Chem 256:7495-7500
6. Pellerin S et al (1993) J Biol Chem 268:18810-18817
7. Pellerin S et al (1993) J Biol Chem 268:4304-4310
8. Chen H et al (1994) J Biol Chem 269:32226-32232
9. Clezardin P et al (1997) Biochem J 321:819-827
10. Guo NH et al (1992) J Biol Chem 267:19349-19355
11. Mikhailenko I et al (1995) J Biol Chem 270:9543-9549
12. Li WX et al (1993) J Biol Chem 268:16179-16184
13. Tuszynski GP et al (1993) J Cell Biol 120:513-521
14. Lawler J, Hynes RO (1989) Blood 74:2022-2027
15. Gao AG et al (1996) J Biol Chem 271:21-24
16. Lahav J (1993) In: Lahav J (ed) Thrombospondin, CRC Press, pp 63-71
17. Hogg PJ (1994) Thromb Haemostasis 72:787-92
18. Sottile J et al (1991) Biochemistry 30:6556-6562
19. Bornstein P, Sage EH (1994) Meth Enzymol 245:62-85
20. Lawler J et al (1985) J Biol Chem 260:3762-3772
21. Shingu T, Bornstein P (1993) Genomics 16:74-84
22. Iruela-Arispe ML et al (1993) Dev Dynam 197:40-56
23. Arbeille BB et al (1991) J Histochem Cytochem 39:1367-1375
24. Kyriakides TR et al (1998) J Histochem Cytochem 46:1007-1015
25. Framson P, Bornstein P (1993) J Biol Chem 268:4989-4996
26. Shingu T, Bornstein P (1994) J Biol Chem 269:32551-32557
27. Laherty CD et al (1992) J Biol Chem 267:3274-3281
28. Negoescu A et al (1995) Exp Cell Res 217:404-409
29. Lafeuillade B et al (1996) J Cell Physiol 167:164-172
30. Bellon G et al (1994) J Cell Physiol 160:75-88
31. Dreyfus M et al (1992) Endocrinol 130:2565-2570
32. Iruela-Arispe ML et al (1996) J Clin Invest 97:403-412
33. Frazier WA (1991) Curr Op Cell Biol 3:792-799
34. Chen H et al (1996) J Biol Chem 271:15993-15999
35. Bornstein P (1995) J Cell Biol 130:503-506

36. Legrand C et al (1994) Arterioscler Thromb 14:1784-1791
37. Tuszynski GP et al (1987) Science 236:1570-1573
38. Lahav J (1988) Exp Cell Res 177:199-204
39. Pellerin S et al (1994) Mol Cell Endocrinol 106:181-186
40. Murphy-Ullrich JE et al (1993) J Biol Chem 269:26784-26789
41. Schultz-Cherry et al (1995) J Biol Chem 270:7304-7310
42. Souchelnistskiy et al (1995) Endocrinol 136:5118-5126
43. Bailly S et al (1997) J Biol Chem 272:16329-16334
44. Tolsma SS et al (1993) J Cell Biol 122:497-511
45. Iruela-Arispe ML et al (1991) Proc Natl Acad Sci USA 88:5026-5030
46. Nicosia RF and Tuszynski GP (1994) J Cell Biol 124:183-193
47. Lawler J et al (1998) J Clin Invest 101:982-992
48. Kyriakides TR et al (1998) J Cell Biol 140, 419-430
49. Stellmach V et al (1996) Eur J Cancer 32:2394-2400
50. Dameron KM et al (1994) Science 265:1582-1584
51. Grossfeld GD et al (1997) J Natl Cancer Inst 89:219-27
52. Volpert OV et al (1995) Breast Cancer Res Treat 36:119-126
53. Hsu SC et al (1996) Cancer Res 56:5684-5691
54. Bertin N et al (1997) Cancer Res 57:396-399
55. Hanahan D, Folkman J (1996) Cell 86:353-364
56. Weinstat-Saslow DL et al (1994) Cancer Res 54:6504-6511
57. Sheibani N, Frazier WA (1995) Proc Natl Acad Sci USA 92:6788-6792

Thromboxanes

Synonym: Thromboxane A2; TxA2; 9a,11a- Epoxy-15(S)-hydroxy-thromba-5Z-13E-dienoic acid

Definition *Members of the prostanoid family generated through the cyclooxygenase pathway. The main thromboxane involved in vascular biology is thromboxane A_2 (TxA$_2$).*

See also: →Prostacyclin; →Prostaglandins; →Platelet stimulus-response coupling

Introduction Thromboxane A_2 is a member of prostanoid family, an important mediator of the action of polyunsaturated fatty acids, generated by cyclooxygenase pathway, discovered in 1975 by Samuelsson's group [1]. The name was given according to its origin and structure: TxA$_2$ was first isolated from thrombocytes and contains oxane-oxetane ring [1,2]. Since it contains 20 atoms of carbon as prostaglandins, it is also an eicosanoid (eicosa=twenty). TxA$_2$ induces irreversible platelet aggregation and strong smooth muscle contraction. Hamberg et al [2] have identified RCS – rabbit aorta contracting substance, released from guinea pig lung after anaphylactic shock, as being TxA$_2$. Thromboxane A$_2$ is formed from prostaglandin endoperoxide PGH$_2$ by thromboxane synthase, an enzyme which was isolated and described. TxA$_2$ is an unstable substance with a short half-life of 30 sec, but with considerable biological activity and a very important role in vascular biology. It is a strong vasoconstrictor and aggregating agent, increasing intracellular Ca and inhibiting intracellular cAMP. Its action is mediated by specific membrane thromboxane receptors (TP) [3]. TxA$_2$ has quite opposite biological actions to prostacyclin (PGI$_2$), thus the balance between both prostanoids has an important homeostatic role. TxA$_2$ is implicated in the pathophysiology of several cardiovascular diseases and for this reason considerable research efforts have been made to prevent the harmful action of this eicosanoid. A number of drugs were developed, either inhibitors of thromboxane synthase or antagonists of TxA$_2$ receptors or drugs with both activities. The clinical impact seems promising. However the most widely used drug at this time is aspirin because it inactivates irreversibly platelet cyclooxygenase and therefore the synthesis of TxA$_2$. The discovery of TxA$_2$ and its known implications in blood coagulation, coronary and peripheral arterial diseases, pulmonary and renal diseases, opens new important fields in vascular biology and pathology.

Characteristics *Thromboxane biosynthesis* is a result of the enzymatic cascade of arachidonic acid (AA), which generates the PG endoperoxide PGH$_2$, a substrate for thromboxane synthase, as shows Figure 1. TxA$_2$ is formed from PGH$_2$ by cleavage of the cyclopentane ring between C-11 and C-12 and of the endoperoxide bridge between the two oxygen atoms. The new bonds are also formed: the oxygen atom at C-9 becomes attached to C-11and the oxygen atom at C-11 becomes attached to C-12. The resulting bicyclic oxane,oxetane compound is TxA$_2$, which is rapidly hydrolyzed in aqueous solutions to a hemiacetal derivative, thromboxane B$_2$. At the same time, the thromboxane synthase generates another metabolite, which is 12-hydroxy-5,8,10-heptadecatrienoic acid (HHTrE) and three carbon fragment, malondialdehyd (MDA) (Figure 1) by a fragmentation of PGH$_2$ in a ratio about 1 [4]. The *thromboxane synthase* is a microsomal enzyme, which has been purified to homogeneity from human platelets [5], its composition was identified and characterized as a cytochrome P450, with one heme/59kDa polypeptide. The enzyme is present especially in platelets, monocyte/ macrophages, lung, liver, brain, renal cortex, vascular endothelial cells [6,7]. The substrate of thromboxane synthase may be PGH$_1$, PGH$_2$ and PGH$_3$, generated by PG endoperoxide synthase from dihomo-γ-linolenic, arachidonic and eicosapentaenoic acid (EPA) respectively. However endoperoxides, lacking a cis double bond at C5-C6, which is the case of PGH$_1$, are very poor substrates [4]. The predominant substrate is PGH$_2$, to a lesser extent PGH$_3$, but the biological activity of TxA$_3$ is 5-10 times less important in comparison with TxA$_2$ [8]. Since PGI$_3$, generated from the same endoperoxide PGH$_3$, is fully biologically active, then the diet rich in EPA favours the predominant effect of prostacyclins over thromboxanes [8,9]. Thromboxane contracts most of tested arteries, therefore it is regarded as one of the potent endogenous substances responsible for coronary, cerebral and peripheral vasospasms.

Molecular Weight TxA$_2$=353; TxB$_2$=370; Thromboxane synthase=60.684, Thromboxane receptor=37.429

Domains *Thromboxane synthase*: The human thromboxane synthase is a member of the cytochrome P450 superfamily [9,10] containing one heme per molecule.

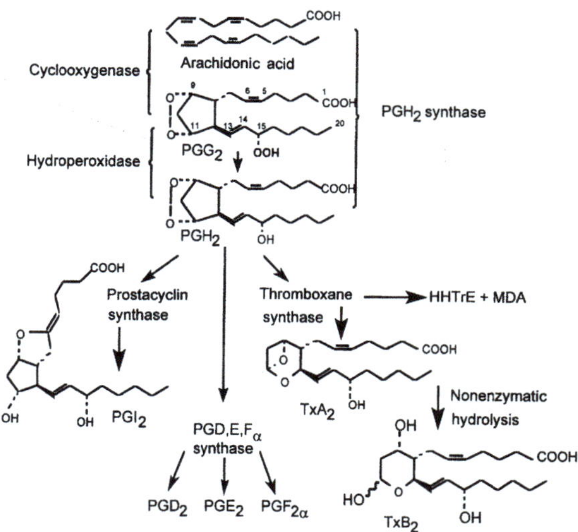

Figure 1. Metabolism of arachidonic acid by the cyclooxygenase pathway

The membrane topology of microsomal cytochromes P450 correspond to the large cytoplasmic domain anchored to the membrane by NH_2 transmembrane segment. The bulk of the protein is exposed on the cytoplasmic surface of the endoplasmic reticulum [11]. *Thromboxane A₂* : Critical parts of TxA_2 molecule are again, as in other prostanoids, C1 carboxyl group, C15 hydroxyl group, the a orientation of carboxyl chain (under the plane of oxane ring) and 5,6 Z and 13,14 E double bonds. Oxane ring is essential for the biological activity because the replacement by a pinane or bicycloheptane ring or the hydratation of the ring to TxB_2 results in loss of biological activity [12]. *Thromboxane receptor* : This receptor is designated as TxA_2 / PGH_2 receptor because PGH_2 has the same pharmacological properties as TxA_2 in many tissues. The human and mouse TxA_2 receptors are rhodopsin-type receptors with seven transmembrane domains. These receptors have an N-terminal extracellular portion of less than 30 amino acids and C-terminal is present free in the cytoplasm. The cytoplasmic loops are short, consist mainly of hydrophobic amino acids [12].

Binding Sites and Affinity *Thromboxane synthase-1*, isolated from human platelets, designated as a longer protein, contains 534 amino acids and cysteine, that serves as the axial heme ligand. *Thromboxane synthase-2*, isolated from the human lung as a shorter protein, contains 460 amino acids and lacks the conserved cysteine [11]. Thromboxane synthase contains one heme per polypeptide chain. The critical active binding site is the sixth ligand position of the iron center of the heme [9]. The K_m for PGH_2 is 10-24 µM. *Thromboxane receptor:* The seven transmembrane segments present different binding sites. Ligand binding site occurs in the outside half of the seven transmembrane segments. An arginine residue at the seventh segment serves as the binding site for the carboxyl group of TxA_2. The prostanoid ring is

likely to interact with the third transmembrane segment. The affinity of the Tx receptor (TP) was tested with selective TxA_2 receptor ligand S-145, K_d being with (^{3}H)-S 145 1.2 – 3.3 nM [12] and with (^{125}I)-S 145 0.23 nM [3].

Additional Features Several studies have shown that TxA_2 analogs stimulate phosphoinositide metabolism and increase the intracellular Ca^{2+} concentration. TxA_2 agonist-stimulated GTPase activity was blocked by an antibody against the carboxyl(C)-terminal peptide of $Gq\alpha$ in human platelet membranes. Therefore TxA_2 receptor is coupled to Gq, leading to activation of phospholipase C. It was also shown that TxA_2 receptor stimulation inhibits adenylate cyclase [13] and reduces cAMP-mediated inhibition of ADP-evoked responses in human platelets. Thus TxA_2 receptor seems to be coupled with two signal transduction systems, one linked to phospholipase C activation, responsible for platelet aggregation, and the other one mediating an increase in cytosolic Ca^{2+} and responsible for platelet shape change [14].

Structure The primary structure of *thromboxane synthase*, isolated from human platelets, has recently been determined by Yokoyama et al [10]. It contains 534 amino acids, has a m. w. 60.684 and shares primary and secondary structure with other cytochromes P450 [9,10]. The structure of *TxA₂* shows Figure 1 Thromboxane A₂ has a third double bond C17=C18, as well as TxB_3 [8]. The structure of *TxA₂/PGH₂ receptor* was identified by Narumiya et al [12]. It is a protein with seven transmembrane domains resembling to other prostanoid receptors [12,14]. There are two conserved potential C-kinase phosphorylation sites in the second cytoplasmic loop, which together with several serine and threonine residues in the terminal may be involved in phosphorylation mediated receptor desensitisation [12].

Sequence and Size *Thromboxane synthase* (Tx synthase) cDNA coding for TxA_2 synthase has been isolated from human platelets [10] and human lung [15] libraries and the amino acid sequence was deduced. Two cDNA clones of Tx synthase were found: 1) The longer predicted protein, designated as Tx synthase-I, contains 534 amino acids and has m. w. of 60.684. 2) The shorter protein, designated as Tx synthase-II, contains 460 amino acids and has m. w. of 52.408. The latter synthase lacks the conserved cysteine that serves as the proximal heme ligand in other cytochromes P450 [11,15]. The primary structure of human platelet Tx synthase has also been deduced [10]. The predicted human lung Tx synthase-I and human platelet synthase amino acid seqences are identical. In general, the amino acid sequence of Tx synthase has considerable similarity with other cytochromes P450,especially with the family 3 [15]. *Thromboxane receptor* (TP) cDNA coding for human and mouse TxA_2/PGH_2 receptor was cloned by Narumiya et al [12]. Sequencing, homology and hydrophobicity analyses have shown that these receptors are proteins containing 343 and 341 amino acids respectively. The estimated m.w. is 37.429. Tx receptor has seven transmembrane domains indicating the TP receptor belongs to the family of G-protein-coupled rhodopsin-type receptors. The comparison of the amino acid sequence of Tx receptor and three subtypes of EP receptor was also published [14].

Homologies Two thromboxane synthases have been identified but it is not clear whether Tx synthase-II is a functional protein. Thromboxane A_3 is generated by PGH synthase from eicosapentaenoic acid and PGH_3 substrate, but they did not aggregate human platelets, in contrast to PGH_2 and TxA_2. On the contrary, PGH_3 and TxA_3 increase platelet cAMP and thereby a) inhibit aggregation by other agonists, b) block the ADP-induced release reaction and c) suppress platelet phospholipase A_2 activity [8].

Conformation PGH_2 is the same substrate for similar P450 enzymes, which are TxA_2 synthase and PGI_2 synthase. The different isomerisation products can be explained on the same catalytic principle by a different ligation of the heme centers with the two epidioxy oxygen atoms. This requires different conformation of substrate binding at the active site. This difference resides in a) Fe^{III} ion coordinated to the 9-position in TxA_2 synthase and 11-position in PGI_2 synthase, b) parallel orientation of the carboxylate and alkyl chains of PGH_2 to active binding site of TxA_2 synthase and antiparallel orientation of both chains of PGH_2 to the active binding site of PGI_2 synthase. This binding site is the sixth ligand position of iron center in the heme [9]. TxA_2 and similarly PGH_2 have intramolecular H-bond between C15 and COOH groups at C1 position, because carboxyl group exists as COO- in biological pH. There is also an intramolecular H-bond between double bond at C5 position and COO- group at C1 position [16]. The number of charge distribution in the triangle in the structures of platelet aggregatory agents is two for TxA_2 and three for PGH_2 [16]. Thromboxane A_2 platelet receptor (stimulatory receptor, responsible for platelet aggregation) has recess of 11 x 12 Å in diameter, in which there are three charges, one negative and two positive for TxA_2. Thromboxane A_2, as an agonist of stimulatory receptor, may have 3-10 stimulating conformations [16].

Additional Features The conformation studies of thromboxane synthase, TxA_2 molecule and corresponding Tx receptors allowed the development of different inhibitors of TxA_2 synthesis or anatagonists. Several of these drugs have been introduced in clinical trials and will be treated in the chapter of Clinical Relevance.

Gene The gene encoding human thromboxane synthase was isolated from a human genomic library using human platelet thromboxane synthase cDNA as a probe [17]. The gene of thromboxane A_2 receptor was also characterized [18].

Gene Structure The nucleotide sequencing revealed that human TxA_2 synthase gene spans more than 75 kb and consists of 13 exons and 12 introns. The exon-intron boundaries of the TxA_2 synthase gene is similar to those of human cytochrome P450 nifedipine oxydase gene, except for introns 9 and 10, although the prime sequences of these enzymes exhibited 35% identity to each other. Primer-extension analysis indicated multiple transcription sites, and the major start site was identified as an adenine residue located 142 bases upstream of the translation-initiation site. Southern-blot analysis revealed the presence of one copy of TxA_2 synthase gene per haploid genome [17]. The structure of TxA_2 receptor gene was also described [18].

Chromosomal Localization The human gene for TxA_2 synthase has been localised to the long arm of chromosome 7, band q33-q34 [17] and the gene for TxA_2 receptor to chromosome 19 p 13.3 [18,19].

Gene Expression Thromboxane synthase mRNA is widely expressed in human tissues, especially in peripheral blood leukocytes, spleen, lung, liver; lower levels are in kidney, placenta and thymus [17]. Increased expression of TxA_2 synthase gene has been found in placental and decidual tissues from preeclamptic pregnancies [20]. The mouse TxA_2 receptor mRNA is most abundantly expressed in thymus, followed by spleen, lung, kidney, heart and uterus; less in brain. In human tissues the same mRNA is most rich in thymus, placenta and platelets [12].

Gene Regulation Lipopolysacharides (LPS) stimulate the levels of PG endoperoxide synthase-1 (Cyclooxygenase-1, COX-1) mRNA 1.6 times during 36 hours, and COX-2 mRNA 20 times with a peak at 12 hours in human monoblastoid cell line U 937. LPS also increases, but to a lesser extent, Tx synthase mRNA and protein levels. Therefore LPS may modulate Tx synthase gene expression.

Processing and Fate Thromboxane A_2 is very labile substance with a half-life only 30 sec in aqueous medium at pH 7.4 and is rapidly inactivated by a non enzy-

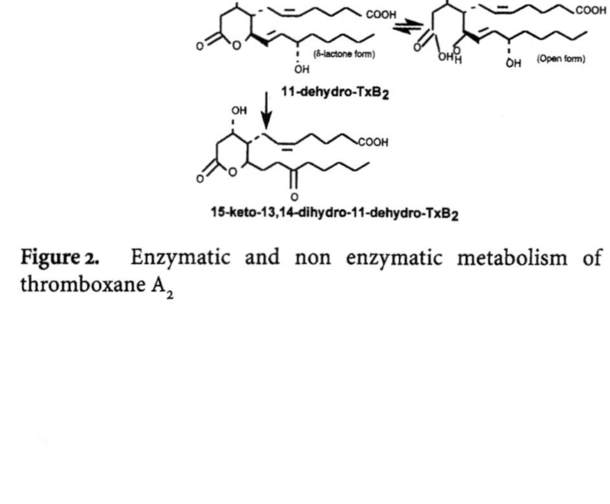

Figure 2. Enzymatic and non enzymatic metabolism of thromboxane A_2

matic degradation (hydrolyse) to *TxB$_2$*, which is the major mechanism of TxA$_2$ inactivation, see Figure 2. TxB$_2$ is devoid of biological activity [21]. TxB$_2$ may exist in buffer solution in two forms, hemiacetal and acyclic, which are in equilibrium (Figure 2). TxA$_2$ formed in plasma may be covalently or non-covalently bound to serum albumin. The former binding inactivates it, while the latter may prolong its biological half-life [21]. TxB$_2$ metabolism in man was studied after intravenous infusion of tritiated TxB2 in healthy volunteers and twenty different metabolites were identified in urine [21]. Two principal metabolic pathways were predominant: 1) The major urinary metabolite was 2,3-dinor-TxB$_2$, generated by a single step β-oxydation. This metabolite has a half-life 15 min. and is eliminated in urine during one hour. 2) The second most abundant metabolite was the initial product formed by dehydrogenation of the C-11 alcohol group of TxB$_2$ and was identified as 11-dehydro-TxB$_2$ (Figure 2) [21]. This enzymatic transformation takes place in liver and the half-life is 45 min. 3) The third possible way is the classical metabolic pathway of prostaglandins (see chapter Prostaglandins) generating by 15-hydroxy-dehydrogenase and by 13,14- reductase 15-keto-, and 15-keto-13,14-dihydro-PG metabolites. These metabolites were not found in man and non-human primates. TxB$_2$ is not, under normal circumstances, a good substrate for 15-hydroxy-PG-dehydrogenase [21]. However the 11-dehydro-TxB$_2$ may be converted in liver by this pathway also to 15-keto-13,14-dihydro-11-dehydro-TxB$_2$ (Figure 2) [21]. All four principal metabolites are measured in biological fluids reflecting TxA$_2$ synthesis. TxB$_2$ is measured in urine reflecting predominantly renal TxA$_2$ synthesis, but not exclusively. The normal values are in order of 8 – 200 pg/min [22, 23]. The measurement in plasma is more difficult because minimal disruption of platelets during blood sampling may increase plasma concentration of TxB$_2$. The measurement of TxB$_2$ in human serum (whole blood is allowed to clot at 37°C for 1 hour) is interesting because it allows to test different inhibitors of TxA$_2$ synthesis. The normal values are very high, 250-400 ng/ml

[22, 23]. The measurement of urinary 2,3-dinor-TxB$_2$ gives good information about a time-integrated whole body TxA$_2$ synthesis. Since large quantities of TxA$_2$ are generated by platelets, urinary levels of this metabolite reflect mainly platelet origin. The normal values in healthy male volunteers are 10.3 ng/hr (4.5–24) or 171 pg/min (75-400) [23]. The measurement of another metabolite, which is 11-dehydro-TxB$_2$, is very important because this metabolite has a longer half-life, it is present in the circulation (plasma) as well as in urine and no artifactual formation occurs in blood samples [24]. It reflects mainly TxB$_2$ generated by platelets. The measurement of urinary 11-dehydro-2,3-dinor-TxB$_2$ in man was also published. In summary, the appreciation of TxA$_2$ synthesis is of great importance because this prostanoid plays an important role in human pathology.

Biological Activity   The biological activity of TxA$_2$ and PG endoperoxides is often difficult to dissociate because they act on the same PGH$_2$/ TxA$_2$ receptors. However TxA$_2$ is in general much more potent [25], but when TxA$_2$ synthase is inhibited, then PGH$_2$ is fully active and may be responsible for platelet aggregation. Biological activity of TxA$_2$ is summarised in Table 1.

Factors stimulating TxA2 synthesis   Humoral mediators: angiotensin II, arginine vasopressin, endothelin, adrenaline, neuropeptide Y. Cell derived mediators: PAF, Il-1, Complement membrane attack complex, immune complexes, endotoxins, LPS, cholesterol. Drugs: cyclosporine. Hypoxia [29]. Age [30].

Interaction between platelets and blood-vessel walls   The vessel wall endothelial cells synthetize prostacyclin (PGI$_2$) from its own precursors, but they are also able to produce PGI$_2$ from endoperoxides released by blood platelets. Prostaglandin endoperoxides are precursors of both pro-aggregatory vasoconstrictor TxA$_2$, generated by platelets, and the anti-aggregatory vasodilator PGI$_2$, produced by vascular endothelial cells. Vascular damage leads to platelet adhesion and generation of PG endo-

Table 1. Biological effects of TxA$_2$

Tissue, Organ	Effect	Ref.
Vascular smooth muscle	– Strong vasoconstrictor, especially in coronary and pulmonary circulation	[25]
Non vascular smooth muscle	– Powerful constrictor	[25]
Airway smooth muscle	– Strong constrictor	[25]
Platelets	– Rapid irreversible aggregation and release of humoral mediators, inhibition of cAMP and mobilisation of Ca^{++}	[25] [13]
Kidney	– Contraction of glomerular mesangial cells and afferent arteriole, decrease of glomerular filtration rate and renal plasma flow, lowering of ultrafiltration coefficient, antinatriuretic action, mesangial cell proliferation	[26]
Urinary bladder	– Stimulation of water flow in toad urinary bladder	[27]
Endothelial cells	– Stimulation of PGI$_2$ synthesis	[26]
PMN leukocytes	– Chemotactic action; leukotriene B$_4$ synthesis	[25]
Brain	– Vasoconstriction, pain, ischemia	[28]

peroxides in very close proximity of the endothelial surface, thereby enhancing PGI_2 formation and preventing further clumping of platelets to form a thrombus [31].

Role in Vascular Biology Thromboxane A_2 plays an important role in vascular biology through different mechanisms of action:

1. TxA_2 is a powerful platelet aggregating agent and its release is amplified by positive feedback regulation.
2. TxA_2 constricts vascular smooth muscles and interferes with general and regional hemodynamics.
3. TxA_2 releases vasoactive humoral mediators.
4. TxA_2 stimulates the proliferation of several cells having a potential mitogenic effect.

Physiological Function TxA_2 is a very labile substance with a short half-life. Its physiological function was analysed by the use of synthetic agonists (for ex. U-46619 of Upjohn Co) or TxA_2 synthase inhibitors or inhibitors of TxA_2/PGH_2 receptors. The physiological function of TxA_2 in vascular biology is dependent on the main sources of TxA_2 and the localisation of TxA_2/PGH_2 receptors. The principal sources of TxA_2 are platelets, lung, spleen, liver, heart, kidney, thymus, brain [25], umbilical circulation, endothelial cells [7], macrophages, monocytes, mastocytes [25]. TxA_2/PGH_2 receptors are abundant in vascular and respiratory smooth muscles, platelets, myofibroblasts of umbilical tissue, renal glomerular mesangial cells, epithelial gastro-intestinal cells [3,14]. According to the localisation of these receptors, the main action of TxA_2 is involved in the regulation of vascular tone and arterial blood pressure in equilibrium with vasodilator humoral mediators as PGI_2, NO, bradykinin. TxA_2 also plays a role in pulmonary circulation because hypoxia stimulates its release [29]. After parturition TxA_2 helps the closure of umbilical vessels. One very important function of TxA_2 is the platelet aggregation, which may be beneficial in hemorrhage with simultaneous vasoconstriction, but harmful in occlusive, thrombotic or embolic situations. Its vasoconstrictor and antinatriuretic renal effect may be useful again in situations with reduced blood volume, but excessive vasoconstriction may lead to renal failure. The humoral-metabolic action of TxA_2 is also important: it inhibits renin secretion, augments adrenergic neurotransmission, stimulates hydrolysis of phosphoinositides and subsequent formation of IP_3 leading to an increase of the intracellular Ca^{2+} concentration. In platelets TxA_2 decreases cAMP. Another function of TxA_2 includes the stimulation of the smooth muscle cell proliferation by increasing the production of matrix metalloproteinase and DNA synthesis [32].

Pathology Thromboxane A_2 has an important physiopathological role in several diseases, especially in cardiovascular pathology, morbidity and mortality. *Cardiovascular diseases: Unstable angina.* Phasic activation of platelets in patients with coronary heart disease is associated with a large rise in TxA_2 release and appearance of coronary spasm-unstable angina [33], which may lead, when not treated, to coronary obstruc-

tion by platelet thrombi and to myocardial infarction. *Acute myocardial infarction* is associated with increased levels of TxA_2 synthesis [34], especially in patients with severe cardiac arrythmias as ventricular fibrillation or ventricular tachycardia [35]. TxA_2 is one of the mediators of myocardial ischemia induced *arrythmias*. Interestingly, the clinically used antiarrythmic agents from class I,II and IV have a significant inhibitory effect on TxA_2 synthesis [36]. *Reocclusion* after coronary thrombolysis is often the consequence of platelet activation with a large increase of TxA_2 secretion as reflected by high urinary excretion of 11-dehydro-TxB_2. *Stable angina.* Patients with stable angina have chronically increased TxA_2 synthesis as assessed by the excretion of urinary metabolites [37]. *Pulmonary diseases: Primary pulmonary hypertension* is associated with only moderate increased levels of TxB_2 in arterial blood. *Secondary pulmonary hypertension* due to chronic pulmonary diseases is associated with a high concentration of TxB_2 in arterial blood, decreasing after the respiration of 100% oxygen [38]. Acute high altitude hypoxia stimulates TxA_2 release participating in the appearance of symptoms of acute mountain sickness [29]. TxA_2 is involved in asthmatic allergen bronchoconstriction [39] and its synthesis in lung is also stimulated by leukotriene C_4, which is another eicosanoid implicated in allergic reactions (Leukotrienes C_4, D_4 and E_4 were identified as Slow Reacting Substance of anaphylaxis). Finally TxA_2 is implicated in *pulmonary embolism.* Increased levels of TxB_2 were observed in chronic smokers. *Arterial hypertension*: TxB_2 levels were found significantly increased in vena caval blood and in renal venous blood of patients with *primary and secondary hypertension* [40]. *Renal diseases*: TxA_2 is implicated in the *nephrotoxicity* of pharmacologic, immune and toxic insults, in the progression of renal disease and in renal allograft rejection [41]. TxA_2 synthesis is increased in *hemolytic-uremic syndrome*, in *hepato-renal syndrome*, in patients with *nephrotic syndrome* and in patients with *systemic lupus erythematosus*. It plays a role in the decrease of glomerular filtration rate and in the progression of the nephropathy. Acute ureteral obstruction and *hydronephrosis* stimulates considerably TxA_2 synthesis in altered kidney with a decrease of renal function, but its origin is not from resident cells but from an influx of immune cells, mainly infiltrating leukocytes and mononuclear cells. TxA_2 is also responsible for the *cytotoxicity of ciclosporine* in transplanted patients. *Pregnancy*: TxA_2 seem implicated in *pregnancy induced hypertension, preeclampsia and eclampsia.* Increased urinary levels of 2,3-dinor-TxB_2 and 11-dehydro-TxB_2 were found in preeclamptic women, as well as increased levels of TxB_2 in maternal blood, in umbilical cord plasma and in placental tissues, often with decreased levels of PGI_2 synthesis [42]. *Cerebrovascular diseases*: The vasoconstrictor activity of TxA_2 was described in migraine, in the etiology of stroke and cerebral vasospasm. Increased biosynthesis of TxA_2 was found in patients with acute cerebral ischemia [43]. *Platelets*: Increased release of TxA_2 from platelets was observed in atherosclerosis, in patients with ischemic heart disease, under the effect of low density lipoproteins (but not of HDL), in smokers, in diabetes of type II.

Clinical Relevance and Therapeutic Implications

Thromboxane A_2 represents a powerful humoral mediator which is critically implicated in major cardiovascular pathology, i.e. atherosclerosis, coronary heart disease, myocardial infarction, platelet aggregation and thrombosis, cerebrovascular accidents, renal diseases, pregnancy complications and others. Thus its excessive release should be prevented or reduced so as to avoid possible fatal and non fatal events. The measurement of TxA_2 metabolites in plasma, serum or urine is very helpful in different clinical situations and in preventive treatment. It should be pointed out that pathophysiological action of TxA_2 may be limited by another eicosanoid which is prostacyclin. PGI_2 is secreted mainly by vascular endothelial cells and has quite opposite biologic actions: it is a powerful antiaggregant, vasodilator, antihypertensive, cytoprotective and anti-atheromatous agent. Therefore the balance between PGI_2 – TxA_2 is of great clinical importance and all therapeutic interventions inhibiting TxA_2 synthesis should be careful to avoid the simultaneous suppression of PGI_2 synthesis. The therapeutic implications are numerous: The principal objective is to limit TxA_2 synthesis or to prevent its action at the level of its receptors. Since the greatest quantity of TxA_2 is produced by platelets and platelet aggregation and thrombi represent an important danger, the target of therapeutic interventions is especially the biosynthetic capacity of platelets. Several drugs were developed in this field, but only drugs used in clinical practice or in clinical trials will be treated. They can be divided in four groups according to their mechanism of action: 1) Inhibitors of cycloxygenase, 2) Inhibitors of thromboxane synthase, 3) Antagonists of TxA_2 receptors, 4) Mixed drugs.

Inhibitors of cyclooxygenase   There are three different groups of inhibitors: a) Drugs with irreversible inactivation of cycloxygenase (COX) whose type is aspirin; b) drugs with competitive reversible inhibition of COX, represented by non steroidal anti-inflammatory drugs; c) Drugs with other mechanisms of action. In clinical practice the most widely used is aspirin (acetylsalicylic acid). It inactivates constitutive cycloxygenase-I by irreversible acetylation of serine 530. This effect persists 24h after a single dose but is not identical in all tissues. The most important is the inhibition of platelet COX, where the TxA_2 synthesis is inhibited by 95% after a single dose of 75 mg/day and by 97-99% with a prolonged treatment [44]. The irreversible inactivation of platelet COX persists for the entire life of platelets (5-7 days), because platelets cannot renew the synthesis of enzyme lacking the nucleus and corresponding DNA. The result is clinically important: Aspirin suppresses TxA_2 synthesis and acts as an anti-aggregant and secondary as an anti-thrombotic agent. However, aspirin is a non specific COX inhibitor, therefore decreases also PGI_2 synthesis, but to a lesser extent and for a shorter time, because endothelial cells renew the synthesis of a new enzyme within 6-12 hours. Aspirin is now largely used in preventive clinical practice with very satisfactory results as shown in Table 2 (Data from recent publications, see ref. [45-47]).

The initial high doses of aspirin (1000-1300 mg/day) decreased to 75 mg/day and to 325 mg/day in invasive cardiology. Recently a new benefit of aspirin therapy was discovered when its effect on inducible cyclooxygenase-II was analyzed. The acetylation concerns Ser 516, but the inhibition of enzyme is only partial and the persistent activity transforms arachidonic acid in 15-HETE, a substance which inhibits the synthesis of pro-inflammatory and vasoconstrictive leukotrienes B_4, C_4 and D_4.

Table 2. Aspirin and cardiovascular diseases

Pathology	Aspirin mg/day	Month of treatment	Result
– Unstable angina	324	3	– 50-70% reduction of risk of death
	1300	18	– 50% reduction of myocardial infarction(M.I.) and death
– Stable angina	75	50	– 34% reduction of M.I. or subite death
– Acute M.I.	160	24	– ISIS-2: Study in 17.187 patients: 23% reduction of mortality,49% reduction of non fatal M.I., 46% reduction of risk of cerebrovascular accident
– Late reocclusion			– Meta-analysis of 32 studies: Efficacy of aspirin (ASP) versus placebo (PLAC) 11% incidence vs 25%
– Aorto-coronary	325–975		– Significant better permeability with ASP vs PLAC, bypass 90-93% vs 85%
– Coronary angioplasty	330		– Aspirin with Dipyridamole 75 mg; reduction of risk of M.I. to 1.6 % vs 6.9%
– Restenoses	100	6	– 25% reduction of number
– Secondary prevention	300–1500	12-36	– Meta-analysis showing significant reduction (13%) of mortality and 29% reduction of non fatal M.I.
– Primary prevention	325	60	– USA study in 22.071 physicians: 44% reduction of risk of M.I., but non significant increase of cerebrovascular accidents

Table 3. Thromboxane synthase inhibitors

Drug	Name
UK 37248	Dazoxiben
UK 38485	Dazmegrel
CGS 13080	Pirmagrel
U 63557 A	Furegrelate
OKY 1581, OKY 046, CV 4151	

Table 4. Thromboxane receptor antagonists in clinical trials

Drug	Name
GR 32191	Vapiprost
BM 13177	Solutroban
BM 13505	Daltroban
AH 23848, SQ 28668, ICI 192605	

Thromboxane synthase inhibitors   Three principle groups of inhibitors have been developed: 1) Imidazole derivatives, 2) Pyridine derivatives and 3) Analogs of PGH_2/TxA_2. The main objective was to prevent platelet aggregation, thrombotic complications, vasoconstriction and to avoid the inhibition of PGI_2 synthesis. Several drugs have been synthetized, but only those introduced in clinical trials are summarized in Table 3 (according to Fiddler and Lumley, ref. [48]).

These peroral inhibitors, introduced for the first time in 1981, inhibited TxA_2 synthesis by 90-99%, but their effect decreased after 4-6 hours to 50% inhibition, the bleeding time moderately increased, but there was no marked effect on platelet aggregation. In general, these compounds tested in different conditions as stable or unstable angina, were without significant clinical effect. It was due to incomplete blockade of thromboxane synthase and to persistent effect of endoperoxides PGG_2 and PGH_2, non-consumed for the TxA_2 synthesis. These endoperoxides are also agonists of TxA_2 receptors. On the other hand, the treatment with TxA_2 synthase inhibitors was associated with an increased synthesis of PGI_2 and other PGs, produced by the redirection of endoperoxides. The modest results obtained with Tx synthase inhibitors reoriented intensive research to the development of thromboxane receptor blockers which may antagonize deleterious effect of both TxA_2 and PGH_2 equally.

Competitive antagonists of TxA2/PGH2 receptors   Again several drugs have been developed and those succesfully tested in clinical trials are shown in Table 4 (according ref. [48]). They block the action of TxA_2/PGH_2 on platelet TxA_2 receptors, but also in vascular and airway smooth muscles. They do not modify the prostanoid synthesis. These agents are efficient inhibitors of platelet aggregation and increase bleeding time [49].

These antagonists have been tested in healthy volunteers and in patients. Positive results were observed with AH 23848, which prevented platelet deposition onto the vascular grafts. Solutroban prevented significantly the occlusion of coronary bypasses versus placebo. Solutroban significantly improved renal functions in patients with lupus nephritis. However there was no clinical benefit in anginous coronaropathy. In general, these drugs are efficient but clinical and experimental studies have shown some limits: 1) Most of these substances are competitive antagonists, therefore they can be displaced from their binding to the receptor by TxA_2, if the latter is generated in excess by massive platelet activation. 2) Their half-life is relatively short (with the exception of vapiprost), therefore the receptor antagonism is not permanent. 3) These drugs do not block platelet aggregation induced by other agonists TxA_2 independent as thrombin, collagen or ADP.

Compound mixtures   Thromboxane synthase inhibitors and TxA2 receptor antagonists; dual blockers. Recently drugs with both inhibitory actions have been developed and two of them have been tested in humans: R 68070 (Ridogrel) and G 137 (Picotamide) [49]. The experimental results have shown significant inhibition of platelet aggregation, an increase of PGI2 and PGE2 synthesis, suppression of intraplatelet Ca increase, induced by arachidonic acid, and a rise in intraplatelet cAMP- a mechanism reducing platelet aggregation induced by all agonists. Ridogrel (pyridine derivative) is a more potent inhibitor of TxA2 synthase ($IC_{50}=1.5 \times 10^{-8}$ M) than as TxA2 receptor antagonist ($IC_{50}=2 \times 10^{-6}$ M). Picotamide is a non-prostanoid substance and exerts Tx synthase inhibition and receptor antagonism at equivalent concentrations. However its effect is relatively weak ($IC_{50}=1 \times 10^{-4}$ M). Ridogrel has been used in a large clinical study as adjunct to thrombolysis with streptokinase in 907 patients with acute myocardial infarction in comparison with aspirin [50]. The results have shown a successful reperfusion in both groups but in the Ridogrel treated group the incidence of new ischemic events (reinfarction, recurrent angina, ischemic stroke) was significantly lower (13%) in comparison with the aspirin treated group (19%), i.e. 32% reduction. There were no major bleeding complications including hemorrhegic stroke. Ridogrel is not superior to aspirin in enhancing the fibrinolytic efficacy of streptokinase but might be more effective in preventing new ischemic events. In conclusion, thromboxane A2 has a very important clinical impact, mainly in cardiovascular pathology and fatal and non fatal cardiovascular events. The comprehension of its pathophysiological action allowed the introduction of efficient widely used preventive treatment by aspirin and the development of new drugs. The progress in preventive medicine also comes from dietic interventions with w-3 polyunsaturated fatty acid rich diets, supplying eicosapentaenoic acid. This latter acid is also substrate for the cyclooxygenase, generating biologically inactive TxA3 at the expense of TxA2, and allowing at the same time the generation of fully active PGI3. Both pharmacological and non-pharmacological treatment opens promising therapeutic possibilities for the future.

Antonin Hornych

References

1. Hamberg M et al (1975) Proc Nat Acad Sci USA 72:2994-2998
2. Samuelsson B, Hamberg M (1978) Prostaglandins 16:857-860
3. Coleman RA et al (1994) Pharmacol Rev 46:205-229
4. Hammarström S (1982) Arch Biochem Biophys 214:431-445
5. Haurand M, Ullrich V (1985) J Biol Chem 260:15059-15067
6. Ingerman-Wojenski C et al (1981) J Clin Invest 67:1292-1296
7. Oravec S et al (1995) Nephrol Dial Transpl 10:796-800
8. Needleman P et al (1979) Proc Natl Acad Sci USA 76:944-948
9. Ullrich V, Brugger R (1994) Angew Chem Int Ed Engl 33:1911-1919
10. Yokoyama C et al (1991) Biochem Biophys Res Commun 178:1479-1484
11. Ruan K-H et al (1993) J Biol Chem,268:19483-19490
12. Narumiya S et al (1993) J Lipid Mediators 6:155-161
13. Gorman RR (1979) Federation Proc 38:83-88
14. Negishi M et al (1993) Prog Lipid Res 32:417-434
15. Ohashi K et al (1992) J Biol Chem 267:789-793
16. Ojima M, Tokuhiro T (1992) Prostaglandins Leukotr Essent Fatty Acids 47:69-76
17. Miyata A et al (1994) Eur J Biochem 224:273-279
18. Schwengel et al (1993) Genomics 18:212-215
19. Duncan AM et al (1995) Genomics 25:740-742
20. Woodworth SH et al (1994) J Clin Endocrinol Metab 78:1225-1231
21. Roberts LJ et al (1982) In: Oates JA (ed) Prostaglandins and the Cardiovascular system. Raven Press, New York, pp 211-225
22. Hornych A et al (1987) Nephron 46:137-143
23. Lawson JA et al (1985) Analyt Biochem 150:463-470
24. Kumlin M, Granström E (1986) Prostaglandins 32:741-767
25. Moncada S, Vane JR (1979) Pharmacol Rev 30:293-331
26. Remuzzi G et al (1992) Kidney Int 41:1493-1493
27. Burch RM, Halushka PV (1980) J Clin Invest 66:1251-1257
28. Wolfe LS (1988) In: Curtis-Prior PB (ed) Prostaglandins. Churchill Livingstone, London, pp 411-421
29. Richalet J-P et al (1991) Respiration Physiol 85:205-215
30. Hornych A et al (1991) Prostaglandins Leukot Essent Fatty Acids 43:191-195
31. Moncada S, Vane JR (1979) New Engl J Med 300:1142-1147
32. Takagishi T et al (1995) Biochem Molec Biol Int 35:265-273
33. FitzGerald DJ et al (1986) N Engl J Med 315:983-989
34. Walinsky P et al (1984) Am Heart J 108:868-872
35. Friedrich T et al (1985) Am Heart J,109:218-222
36. Lucas J et al (1987) Prostaglandins Leukot and Med 26:105-113
37. Montalescot G et al (1991) Circulation 84:1054-2062
38. Daum S (1989) Medicina Thoracalis 42:483-497
39. Sladek K (1990) Am Rev Resp Dis 141:1141-1145
40. Hornych A et al (1981) In: Blaufox MD, Bianchi C (eds) Secondary forms of hypertension. Grune & Stratton, New York, pp 73-88
41. Remuzzi G et al (1992) Kidney Int 41:1483-1493
42. Zahradnik HP et al (1991) Eicosanoids 4:123-136
43. Koudstaal PJ et al (1993) Stroke 24:219-223
44. Clarke RJ et al (1991) N Engl J Med 325:1137-1141
45. Antiplatelet Trialists' Collaboration (1994) Brit Med J 308:81-106
46. Antiplatelet Trialists' Collaboration (1994) Brit Med J 308:159-168
47. Hornych A (1997) In: Mamas S (ed) Prostaglandins and Thromboxanes Applications cliniques et pharmacologiques. Masson Paris, pp 289-315
48. Fiddler GI, Lumley P (1990) Circulation 81 (Suppl I) I-69 - I-78
49. Gresele P et al (1991) Trends in Pharmacol Sci 12:530-548
50. The RAPT investigators (1993) Circulation 89:588-595

Thymidine Phosphorylase

Definition *Enzyme that converts thymidine into 2-deoxy-D-ribose and that is chemotactic for endothelial cells.*

See: →Angiogenesis

Tie-1

Definition *Tyrosine kinase with immunoglobulin and epidermal growth factor homology domains. Role in vascular development in the embryo.*

See: →Tie-1 or Tie-2 tyrosine kinases

Tie-1 or Tie-2 Tyrosine Kinases

Definition *Endothelial cell specific tyrosine kinase containing receptor. Tie-2 binds angiopoietin-1 or 2. The ligand for the Tie-1 kinase has not yet been identified.*

See: →Angiogenesis; →Ontogeny of the vascular system

Tie-2

Definition *Tyrosine kinase with immunoglobulin and epidermal growth factor homology domains. Role in prenatal and postnatal vascular development.*

See: →Tie-1 or Tie-2 tyrosine kinases

TIMP

Definition *Tissue inhibitor of metalloproteinase*

See: →Tissue inhibitors of metalloproteinases

Tissue Factor (TF)

Definition *Protein that activates the extrinsic pathway of blood coagulation. Functions as a cofactor for factor VII or VIIa. Tissue factor is inhibited by the tissue factor protein inhibitor (TFPI).*

See: →Procoagulant activities; →Coagulation factors; →Fibrinolytic, hemostatic and matrix metalloproteinases, role of; →Thrombosis

Tissue Inhibitors of Metalloproteinases (TIMPs)

Synonym: Tissue Inhibitors of Metalloproteinases (TIMPs) were initially identified as factors influencing the growth of cells of the erythroid lineage: Erythroid Potentiating Activity (EPA).

Definition *Matrix metalloproteinase (MMP) activity is inhibited by TIMPs. These include TIMP-1, TIMP-2, TIMP-3 and TIMP-4.*

See also: →*Matrix metalloproteinases*

Introduction Matrix metalloproteinases (MMPs) are involved in the physiological turnover of extracellular matrix (ECM) during growth and development. These neutral endopeptidases belong to the metzincin family and are synthesized by cells as inactive precursors [1]. Following activation, MMP activity is regulated by specific tissular inhibitors designated as Tissue Inhibitors of MetalloProteinases, TIMPs [1]. This balance between MMPs and TIMPs is precisely controlled in space and time and it is now considered that any disturbance in such an equilibrium leads to disease states: tumor invasion and metastasis, osteoarthritis, athero-arteriosclerosis.

TIMP designates a family of genes consisting so far of four members: TIMP-1 to -4, presenting an overall 40 % homology. They can be defined according to several common characteristics:

- they present a conserved gene structure,
- they inhibit specifically matrix metalloproteinases and are secreted extracellularly,
- twelve cysteine residues, forming six intrachain disulfide bonds, define their secondary structure,
- the N-terminal domain of all TIMPs appears highly conserved.

Besides acting as MMP inhibitors, TIMPs have also been shown to exhibit other biological functions distinct from their enzyme inhibitory capacity. TIMP-1 and TIMP-2 influence cell growth rate while TIMP-3 is involved in cell cycle regulation, differentiation and senescence.

Characteristics

Molecular Weight The molecular weight of the four TIMPs ranges between 21,000 and 28,500 [1].
TIMP-1 has a Mr of 28,500 [2]. TIMP-2 is a protein with a calculated Mr of 21,755 [3]. On protease substrate gel, e.g. reverse zymography, it exhibits a Mr of 21,000. TIMP-3 has a predicted size of 21,665 [4] and its position on reverse zymography is intermediate between those of TIMP-1 and TIMP-2. The deduced amino acid sequence of TIMP-4 indicates a mature protein having a molecular weight of 22,609 [5].

Domains Most structural information to date has been obtained for TIMP-1. The TIMP molecule has twelve cysteine residues that are totally conserved within the family and across species. The precise spatial preservation of these residues suggests that the disulfide bonding and the gross structural features of each TIMP are comparable.

The cysteine residues form six intrachain disulfide bonds that fold the protein into a two-domain, six-loop structure [6]. The use of TIMP-1 and TIMP-2 mutants indicates that the N-terminal domain (loops 1-3) alone is an efficient MMP inhibitor, while the C-terminal domain (loops 4-6) has at least two separate enzyme binding sites [7].

Binding Sites and Affinity TIMPs bind tightly to each MMP with a 1:1 stoichiometry and with K_d values of less than 1 nM [7]. These complexes can be dissociated following polyacrylamide gel electrophoresis in presence of sodium dodecyl sulfate, both protease and inhibitor being recovered from the complex in an unaltered fully active form [8]. Since low-Mr synthetic inhibitors can compete with TIMP-1 for interstitial collagenase binding, inhibition of enzyme activity probably occurs via binding of the natural inhibitor to the enzyme active site [9]. Formation of an initial reversible complex which is slowly transformed to an inactive tight complex with K_d=0.1 nM at 3 nM concentration of enzyme and inhibitor has been described [10]. Mutation of the essential Glu residue in the active site of MMP, although totally suppressing enzyme activity, was found not to influence TIMP-1 binding [11].

TIMP-1 and TIMP-2 can also form tight complexes [12] with precursor forms of gelatinase B [13] and gelatinase A [14] respectively. Similarly, TIMP-4 can bind to progelatinase A [15].

Additional Features TIMP-1 contains approximately 13 hexose residues per mole [16]. Glycosylation of TIMP-1 was found not to be a prerequisite for secretion of the protein from cells; also, it does not influence either its MMP inhibitory capacity or its susceptibility to proteolytic cleavage [17]. TIMP-1 is remarkably stable: its biological activity is unaltered following incubation at pH 2.0 or heating at 100 °C for 30 min., but reduction of disulfide bonds inactivates TIMP-1 [16]. Neutrophil elastase, but not cathepsin G or plasmin, can also destroy the MMP inhibitory potential of TIMP-1 [18].

The main characteristics of TIMPs at the protein level are summarized in Table 1.

Structure

Sequence and Size The secreted TIMP-1 consists of 184 amino acid residues; its cDNA sequence predicts a 23-amino acid leader peptide [2]. TIMP-1 is glycosylated heterogeneously at two sites: Asn30 and Asn78 [17]. The cDNA for TIMP-2 codes for a mature protein of 194 amino acids, without glycosylation site, and a leader sequence of 26 amino acids [3]. TIMP-3 cDNA codes for a protein of 211 amino acids; mature TIMP-3 is composed of 188 residues [4]. It is secreted as a nonglycosylated form although a potential glycosylation site can be evidenced at its C-terminus [4]. The cDNA for TIMP-4 encodes a 224-amino acid precursor including a 29-residue secretion signal; no N-glycosylation motif was reported in its sequence [5].

Table 1. The main characteristics of TIMPs at the protein level

	TIMP-1	TIMP-2	TIMP-3	TIMP-4
Number of amino acids	184	194	188	195
Molecular weight (SDS-PAGE)	28.5 kDa	21 kDa	24 kDa	22 kDa
Disulfide bonds	6	6	6	6
Isoelectric point	8.0	6.45	9.04	7.34
Glycosylation sites	2	0	1	0
Other features			High Tyr content	Large number of His (12) and Tyr

Homologies   The degree of identity in the sequences of TIMP-2 and TIMP-1 mature proteins is 38 % overall, but is higher within certain regions (71 % for amino acids 1-24 and 59 % for amino acids 83-111) [3]. A comparison of its amino acid sequence with that corresponding to the other human TIMPs indicates that TIMP-3 shares 40 % identity with TIMP-1 and 45 % identity with TIMP-2 [4]. A similar study shows that TIMP-4 shares 37 % sequence identity with TIMP-1 and 51 % identity with TIMP-2 and -3, including the twelve cysteine residues which are conserved among all members of the family [5].

The N-terminal moiety of TIMPs (126 amino acids) appears critical for inhibition of MMPs and is highly conserved (70 per cent identity) among inhibitors. Notably, it contains a consensus sequence CXCXPXHPQXAFC-NXDX-*VIRAK* (X, any amino acid) with VIRAK peptide present in all TIMPs [5, 19]. Single mutation within this sequence greatly affects the interactions between TIMP-1 and MMPs. That holds for His7 and Gln9 but probably, in contrast to serine proteinase inhibitors, several amino acid residues would define the primary specificity site of interactions between TIMPs and MMPs [20]. However, His7 and His95 are totally conserved among all TIMP-1 and-2 whose sequences have been determined and mutation of those residues nearly totally abolished the capacity of murine TIMP-1 to inhibit cell migration, matrix invasion and tumor formation of individual clones [21]. Recent data underline the critical importance of cysteine-1, in its MMP inhibitory capacity [22].

Conformation   Analysis of the active N-terminal domain of TIMP-2 by nuclear magnetic resonance spectroscopy allowed determination of both the secondary structure of the domain and also a low-resolution tertiary structure defining the protein backbone topology. The protein contains a five-stranded antiparallel β-sheet that is rolled over on itself to form a closed β-barrel, and two short helices which pack close to one another on the same barrel face [23].

Gene
Gene Structure   The murine TIMP-1 gene, homologous to human TIMP-1 gene, comprises five exons and four introns extending over 4.3 kb of DNA [24]. The human TIMP-2 gene is 83 kb long; it is composed of five exons separated by four introns of 54.8, 2.7, 9.1, and 1.7 kb [25].

The initiator ATG codon and the TAA stop codon are in exons 1 and 5, respectively [25]. The human TIMP-3 gene consists of five exons spanning a distance of approximately 30 kb [26]. Exon 1 contains both the transcription and translation start sites, whereas the translation termination codon is located in exon 5 [26]. Intron 1 is large, being 17-18 kb in size [26]. Other authors indicate that the human TIMP-3 gene is encoded by 5 exons spanning approximately 55 kb of genomic DNA [27]. The human TIMP-4 has been recently identified; its full-length cDNA sequence contains 1189 bp with a 672-bp open reading frame: 59 bp in the 5'-untranslated region and 458 bp of a 3'-untranslated sequence [5]. Like other members of the TIMP family, the TIMP-4 protein is encoded by five exons; these span 6 kb of genomic DNA [28], so that TIMP-4 gene is considerably smaller than TIMP-2 and TIMP-3 genes. The exon-intron boundaries of TIMP-4 gene are at locations very similar to those of the other TIMP genes [28].

Chromosomal Localization   The TIMP-1 gene was found to map to the proximal short arm of the human X chromosome at Xp11.23 → Xp11.4 [29]. The involvement of TIMP-1 in X-linked connective tissue disorders, such as Ehlers-Danlos disease type V and cutis laxa has been suggested. The TIMP-2 gene was assigned to the terminal region of chromosome 17, 17q25 [30]. The TIMP-3 gene maps to chromosome 22 at 22q13.1 [26]. The TIMP-4 gene is localized to chromosome 3p25 [28].

Gene Expression   TIMP-1 is encoded within a single 0.9 kb mRNA [31]. TIMP-2 gene gives rise to two distinct mRNAs of respectively 1.2 and 3.8 kb; the difference in size could reflect the use of different polyadenylation signals [25]. Northern blot analysis shows that the most prominent transcripts of TIMP-3 gene are 2.4, 2.8 and 4.5-5 kb in size, probably by alternative polyadenylation signal usage [26]. TIMP-4 transcripts have a size of 1.4 kb [5].

Gene Regulation   TIMP-1 transcription is highly stimulus-responsive and transformation-sensitive. A 38-bp region, located between -1093 and -1010 bp of the human TIMP-1 promoter, is conserved in both murine and human genes [32]. This region contains an enhancer element responsive to the phorbol ester, 12-O-tetradecanoylphorbol-13-actetate (TPA), TPA-responsive element (TRE) [32, 33]. AP-1 proteins, such as *c-jun*, *c-fos*, *junB* and *junD* bind to this region [33]. In addition, an inverted repeat containing a potential Ets binding site is also present in the 38-bp region [33]. This region is also sufficient to direct a response to TGF-β_1 in the presence of serum [32]. Adult T-cell leukemia cell lines infected with human T-cell leukemia virus type 1 (HTLV-1) express high levels of a TIMP-1 transcript [33]. Tax1, a transcriptional activator of the virus, controls this expression [33], and the AP-1 binding site is required for activation by Tax1 [33].

TIMP-2 expression is largely constitutive. The TIMP-2 promoter contains several regulatory elements: 5 SP1, 2 AP-2, 1 AP-1 and 3 PEA3 binding sites. However, in most cases, the expression of TIMP-2 is differently affected by agents such as TPA or TGFβ, as compared to TIMP-1. It is often produced by cells in a constitutive manner and it provides therefore a basal level of inhibitory activity in tissues. The AP-1 site located at -288/-281 in TIMP-2 promoter occupies a distal position to TATA box, a particularity which might explain the lack of a transcriptional response following AP-1 site occupancy [25]. TIMP-2 promoter also exhibits high G/C content but methylation did not appear to play a prominent function in TIMP-2 expression. In several instances, gelatinase A (MMP-2) and TIMP-2 expressions are found coregulated. Variations in the levels of cAMP intracellular levels were incriminated as a determinant in enzyme and inhibitor expressions. In this respect, TIMP-2 promoter contains two AP-2 binding sites, known to mediate cAMP response in many genes [25].

Presence of a TATA box in the TIMP-3 promoter sequence remains controverted [26, 27]. The TIMP-3 gene contains three [27] or four [26] GC-boxes (Sp1 sites). Other putative transcription factor binding sites (NF-1 and C/EBP) have been also described [26]. TIMP-3 is highly induced during the recruitment of quiescent cells into the cell cycle [26].

Additional Features

Tissue distribution of TIMP-1 and TIMP-2 mRNAs is ubiquitous [34]. By Northern analysis, transcripts for TIMP-3 gene are identified in a broad cross-section of tissues examined from both embryonic and adult origin [35]. However, point mutations have been observed in the TIMP-3 gene in Sorsby's fundus dystrophy [36] and TIMP-3 mRNA expression is enhanced in retinas affected by simplex retinitis pigmentosa [37], suggesting that TIMP-3 might have an essential, particular function in the eye. By Northern analysis, only the adult heart shows abundant TIMP-4 transcripts; very low levels of the transcripts were detected in the kidney, placenta, colon and testes [5]. This unique expression pattern suggests that TIMP-4 may function in a tissue-specific fashion during extracellular matrix homeostasis.

Processing and Fate

TIMP-1 and respectively TIMP-2 are generally secreted as complexes with gelatinase B [13] and gelatinase A [14]. These interactions involve the C-terminal domains of proenzymes and inhibitors [38, 39]. Binding is not associated with conformational changes of any entity but it enhances the interaction between the N-terminal domains of both molecules. In that sense, as far as gelatinases A and B are concerned, TIMP-1 and TIMP-2 appear to control not only the activity but also activation of these enzymes.

Unlike other TIMPs, TIMP-3 has an affinity for the extracellular matrix (ECM). TIMP-3 is quite basic, with an isoelectric point of 9.04 [35]. Its amino acid sequence exhibits three motifs rich in basic amino acids (lysine or arginine) that have been shown to be involved in the association between several proteins and hyaluronic acid [40] and between cytokines and heparan sulfate [41]. One of these motifs is localized to the N-terminal domain and two are juxtaposed close to the C-terminus of TIMP-3. They are not found in TIMP-1 and TIMP-2, suggesting that TIMP-3 binding to ECM may involve interaction between TIMP-3 and either hyaluronic acid or other glycosaminoglycans such as heparan sulfate.

Biological Activity

TIMP-1 was initially identified as an homologue of erythroid potentiating activity, EPA [31], a protein isolated from the conditioned medium of a T-lymphoblast cell line [42]. Such a property is also shared by TIMP-2 [43] and maximum EPA was observed for concentrations of TIMP-1 and TIMP-2 respectively equal to 80 pM and 50 pM [43]. TIMP-1 is mitogenic for a number of cell types including keratinocytes [44], fibroblasts and various hematopoietic cell types [45]. TIMP-2 also has broad spectrum growth promoting activity [46, 47]. Several experimental features support the contention that the antiproteolytic activity and growth effect of TIMPs are independent phenomena:

- Reductive alkylation of TIMP-2, which completely destroyed its anti-MMP function, has no influence on its growth stimulatory activity [46].
- Its complexation with gelatinase A does not impede its effect on cell proliferation [46].
- TIMP-1 mutants, with N-terminal point mutations at His7, Gln9 and a double His7/Gln9, loose their MMP inhibitory capacity, while maintaining their ability to bind to active MMPs and EPA activity [48].
- Finally, synthetic MMP inhibitors are found not to display any EPA over a wide range of concentrations used [48].

Therefore, TIMPs can be considered as bifunctional molecules. It needs to be emphasized that TIMP influence on cell growth occurs at concentration one order of magnitude lower than those required for inhibiting MMPs. Human pancreatic secretory trypsin inhibitor is another example of a mitogenic inhibitor that enhances DNA synthesis at a concentration far below those necessary for inhibiting serine proteases [49].

The interaction of TIMPs with several cell types is mediated via specific receptors which have not yet been characterized [44, 46, 50]. Recent investigations strongly suggested that the putative TIMP-2 receptor might be MT1-MMP [51]; also a G-protein seems to be involved in the signalling pathway. Either an increase in cAMP depend-ent kinase activity [52] or enhancement of tyrosine phosphorylation [53] were reported as signalling events following binding of TIMP-1 and -2 to various cell types. TIMP-3, in turn, has been involved in the transformation of chick embryo fibroblasts [54] and in progression of cell cycle.

Role in Vascular Biology

Physiological Function   The maintainance of the struc-tural integrity of aorta necessitates that their main fibril-lar components, collagen(s) and elastin, be protected from degradation. Therefore, the activities of interstitial collagenase, stromelysin-1 and gelatinases A and B, two true elastases [55], must be tightly controlled. Endothelial cells [56], and aorta smooth muscle cells [57], secrete, under basal conditions, TIMP-1, TIMP-2 and TIMP-3. In normal arteries, most TIMP-1 appears matrix-associated [58] and it could thus assume a protective function against further matrix proteolysis. For instance, overex-pression of TIMP-1 increased elastin content in rat carot-ic artery intima by inhibiting a 28-kDa metalloelastase [59]. For any tissue however, the physiological signifi-cance of any TIMP production implies knowledge of the biological activity required in the particular context: i) MMP control; ii) binding to ECM; iii) growth promoting activity.

It was recently shown that adenovirus-mediated gene transfer of tissue inhibitor of metalloproteinases (TIMP-1, -2, -3) resulted in inhibition of smooth muscle cell migration [60, 61]. In addition TIMP-2 could reduce cell proliferation and overexpression of TIMP-3 was found to promote smooth muscle cell apoptosis [60]. These data suggest that individual TIMPs, besides their inhibitory potential towards MMPs, exhibit other diverse physiological functions in arterial wall.

Pathology   Under normal physiological conditions, arte-rial smooth muscle cells are in a quiescent growth state; following vessel wall injury, a modulation of their pheno-type is observed: they proliferate and migrate from media to intima. This phenomenon, as an attempt to tissue repair, requires focal proteolysis catalyzed mostly by matrix metalloproteinases [62]. MMP expression (inter-stitial collagenase, stromelysin-1, gelatinases A and B) by aorta smooth muscle cells can be induced by thrombin, interleukin-1, PDGF in combination with IL-1α, ... [63, 64]. Recent investigations have shown that none of these agents could influence the basal expression of TIMP-1 and TIMP-2 from rabbit aorta smooth muscle cells [63]. TIMP-3 mRNA was, on the contrary, increased by PDGF and TGFb especially in combination [63]. These data indi-cated that MMPs and their natural inhibitors were differ-entially regulated in aorta smooth muscle cells.

Human atherosclerotic plaques are characterized by an increased expression of interstitial collagenase, stromel-ysin-1 and gelatinase B [65]. Both foam cells [66] and migrated smooth muscle cells [62] could contribute to the overexpression of these neutral proteinases. However, an increased level of TIMP-1 is recovered in extracts from restenotic lesions, as compared to TIMP-1 amounts in normal tissue [56]. In the normal coronary vessel, in situ hybridization studies indicate that TIMP-4 mRNA is only observed in the fibroblasts outside of the vessel wall while increased TIMP-4 mRNA is demonstrated in the media layer of the vessel wall in the atherosclerotic lesion [67]. Tissue from aortic abdominal aneurysms, a genetic pre-disposition with male predominance, contains low levels of TIMP-1 [68].

Clinical Relevance and Therapeutic Implications   Forough et al [69] have shown that the local seeding of smooth muscle cells overexpressing TIMP-1 into balloon-injured rat carotid artery could inhibit intimal hyperplasia. It suggested that TIMPs could be of significant therapeu-tic value in atherosclerosis and in preventing restenosis following angioplasty.

It has been demonstrated that migration of smooth mus-cle cells, and intimal thickening *in vivo* could be partly inhibited using synthetic MMP inhibitors [70, 71]. Such an approach is promising, and the usual criteria for uti-lization of these substances are their bioavailability, absence of toxicity, potency and specificity. Informations related to the important determinants in TIMPs respon-sible for MMP inhibition will be of great importance for designing potent inhibitors. However, it must be under-lined that less potent and specific MMP inhibitors as doxycycline may have a potential therapeutic application in abdominal aortic aneurysms [72].

Acknowledgements: We wish to thank "le Centre de la Re-cherche Scientifique" (CNRS, UPRESA 6021), "la Fonda-tion pour la Recherche Médicale" (FRM), "l'Association pour la Recherche contre le Cancer" (ARC, n· WH/1236) and Europol.Agro (University of Reims-Champagne Ar-denne) for financial support.

William Hornebeck and Hervé Emonard

References

1. Cawston TE et al (1996) Pharmacol Ther 70:163-182
2. Carmichael DF et al (1986) Proc Natl Acad Sci USA 83:2407-2411
3. Boone TC et al (1990) Proc Natl Acad Sci USA 87:2800-2804
4. Uria JA et al (1994) Cancer Res 54:2091-2094
5. Greene J et al (1996) J Biol Chem 271:30375-30380
6. Williamson RA et al (1990) Biochem J 268:267-274
7. Willenbrock F, Murphy G (1994) Am J Respir Crit Care Med 150:S165-S170
8. Murphy G et al (1989) Biochem J 261:1031-1034
9. Lelièvre Y et al (1990) Matrix 10:292-299
10. Taylor KB et al (1996) J Biol Chem 271:23938-23945
11. Crabbe T et al (1994) Biochemistry 33:6684-6690
12. Olson MW et al (1997) J Biol Chem 272:29975-29983
13. Wilhelm SM et al (1989) J Biol Chem 264:17213-17221
14. Stetler-Stevenson WG et al (1989) J Biol Chem 264:17374-17378
15. Bigg HF et al (1997) J Biol Chem 272:15496-15500
16. Stricklin GP, Welgus HG (1983) J Biol Chem 258:12252-12258
17. Stricklin GP (1986) Collagen Rel Res 6:219-228

18. Itoh Y, Nagase H (1995) J Biol Chem 270:16518-16521
19. Apte SS et al (1995) J Biol Chem 270:14313-14318
20. O'Shea M et al (1992) Biochemistry 31:10146-10152
21. Walther SE, Denhardt DT (1996) Cell Growth Differ 7:1579-1588
22. Gomis-Rüth FX et al (1997) Nature 389:77-81
23. Williamson RA et al (1994) Biochemistry 33:11745-11759
24. Gewert DR et al (1987) EMBO J 6:651-657
25. Hammani K et al (1996) J Biol Chem 271:25498-25505
26. Wick M et al (1995) Biochem J 311:549-554
27. Stohr H et al (1995) Genome Res 5:483-487
28. Olson TM et al (1998) Genomics 51:148-151
29. Willard HF et al (1989) Hum Genet 81:234-238
30. De Clerck Y et al (1992) Genomics 14:782-784
31. Docherty AJP et al (1985) Nature 318:66-69
32. Campbell CE et al (1991) J Biol Chem 266:7199-7206
33. Uchijima M et al (1994) J Biol Chem 269:14946-14950
34. Edwards DR et al (1996) Int J Obesity 20:S9-S15
35. Wilde CG et al (1994) DNA Cell Biol 13:711-718
36. Felbor U et al (1996) J Med Genet 33:233-236
37. Jones SE et al (1994) FEBS Lett 352:171-174
38. O'Connell JP et al (1994) J Biol Chem 269:14967-14973
39. Willenbrock F et al (1993) Biochemistry 32:4330-4337
40. Yang B et al (1994) EMBO J 13:286-296
41. Lortat-Jacob H, Grimaud JA (1991) FEBS Lett 280:152-154
42. Gasson JC et al (1985) Nature 315:768-771
43. Stetler-Stevenson WG et al (1992) FEBS Lett 296:231-234
44. Bertaux B et al (1991) J Invest Dermatol 97:679-685
45. Hayakawa T et al (1992) FEBS Lett 298:29-32
46. Hayakawa T et al (1994) J Cell Sci 107:2373-2379
47. Nemeth JA, Goolsby CL (1993) Exp Cell Res 207:376-382
48. Chesler L et al (1995) Blood 86:4506-4515
49. Niinobu T et al (1990) J Exp Med 172:1133-1142
50. Emmert-Buck MR et al (1995) FEBS Lett 364:28-32
51. Zucker S et al (1998) J Biol Chem 273:1216-1222
52. Corcoran ML, Stetler-Stevenson WG (1995) J Biol Chem 270:13453-13459
53. Yamashita K et al (1996) FEBS Lett 396:103-107
54. Yang TT, Hawkes SP (1992) Proc Natl Acad Sci USA 89:10676-10680
55. Senior RM et al (1991) J Biol Chem 266:7870-7875
56. Tyagi SC et al (1996) Can J Cardiol 12:353-362
57. Pickering JG et al (1997) Arterioscler Thromb Vasc Biol 17:475-482
58. Brophy CM et al (1990) Biochem Biophys Res Commun 167:898-903
59. Forough RL et al (1998) Arterioscler Thromb Vasc Biol 18:803-807
60. Baker AM et al (1998) J Clin Invest 101:1478-1487
61. George SJ et al (1998) Hum Gene Ther 9:867-877
62. Kenagy RD et al (1996) Arterioscler Thromb Vasc Biol 16:1373-1382
63. Fabumni RP et al (1996) Biochem J 315:335-342
64. Yanagi H et al (1992) Atherosclerosis 91:207-216
65. Galis ZS et al (1994) J Clin Invest 94:2493-2503
66. Galis ZS et al (1995) Proc Natl Acad Sci USA 92:402-406
67. Douglas DA et al (1997) J Prot Chem 16:237-255
68. Brophy CM et al (1991) J Surg Res 50:653-657
69. Forough R et al (1996) Circ Res 79:812-820
70. Bendeck MP et al (1996) Circ Res 78:38-43
71. Zempo N et al (1996) Arterioscler Thromb Vasc Biol 16:28-33
72. Boyle JR et al (1998) J Vasc Surg 27:354-361

TM

Definition   *Thrombomodulin*

See: →Thrombomodulin

TNF-α

Definition   *Tumor necrosis factor-α*

See: →Tumor necrosis factor

TP

Definition   *Thymidine phosphorylase*

See: →Thymidine phosphorylase

tPA

Definition   *Tissue-type plasminogen activator*

See: →Plasminogen activators; →Plasminogen/plasmin

TPA

Definition   *12-O-tetradecanoylphorbol-13-actetate*

See: →Tissue inhibitors of metalloproteinases

TPO

Definition   *Thrombopoietin*

See: →Thrombopoietin

Transcription Factors

Definition   *Proteins that regulate gene transcription. The most extensively studied in vascular biology are nuclear factor-κB (NF-κB), AP-1, STATs, fos or jun.*

See: →Cytokines in vascular biology and disease; →Signal transduction mechanisms in vascular biology

Transforming Growth Factor β
Synonym: TGF-β

Definition   *Family of growth factors with modulatory properties according to the cell type. The major TGF-β prototype involved in vascular biology is TGF-β1. The precursor is known as latency associated peptide (LAP). The latter is complexed to latent TGF-β binding protein (LTBP). Transforming growth factor β receptors include the serine/threonine kinase TGF-β receptor 1 and 2, and also betaglycan.*

See also: →Signal transduction mechanisms in vascular biology; →Thrombospondins; →Atherosclerosis; →Cytokines in vascular biology and disease

Introduction   TGF-β was first identified as a distinct molecular entity in 1981 independently by two laboratories [1, 2]. Since then three isoforms, TGF-β1, β2 and β3 have been identified in mammals, of which TGF-β1 is now viewed as the prototype of a much larger family of related molecules involved in the regulation of growth and differentiation of many cell types [3, 4]. These other family members, such as Bone Morphogenesis Proteins (BMP), Activins, Inhibins, Müllerian Inhibitory Substance (MIS) and TGF-β-like molecules in Drosophila, will not be considered here in any detail (see section on homologies). As is often the case in scientific endeavour the relatively small field of biological research which led to TGF-β's discovery has burgeoned into incredibly wider areas of investigation, involving not only basic research but also the therapeutic implications of these molecules in a host of different pathologies. Historically, it was in pursuit of characterization of proteins secreted by retrovirally-transformed Rat fibroblasts which rapidly led to the isolation of TGF-β [1, 2]. It soon became clear that cells transformed by DNA viruses, or by chemicals and, more importantly, most normal cells and tissues from both embryonic and adult sources, all secreted various amounts of TGF-β. Actually, two main areas of research are on-going, with inevitable overlap between them: one aimed at determining the action mechanism(s) by which these molecules inhibit or (more rarely) stimulate cell growth and one developing therapeutic applications.

Characteristics
Molecular Weight   All three mammalian TGF-β's are 25kD homodimers in their mature, biologically active form. Very small amounts of 25kD heterodimers of TGF-β1/β2 and of TGF-β2/β3 are produced by some cells [3, 4].

Domains   By appropriate molecular techniques several mutant or hybrid TGF-β molecules have been constructed to investigate the functional importance of particular sequences (domains) of the 25kD form (cited in [5]). Thus, the replacement of glycine in position 71 by tryptophan resulted in a mutant TGF-β1 having quasi-identical properties to authentic TGF-β1, whereas the deletion of five amino acids (positions 69-73) produced a mutant molecule with growth-inhibitory activity similar to TGF-β2 but different from that of TGF-β1. A chimeric dimer molecule whose two monomeric units consist of amino acids 1-39 of TGF-β2, 40-82 of TGF-β1 and 83-112 of TGF-β2 possesses an activity very similar to TGF-β1; the use of selective bioassays showed that the region 40-82 conferred a TGF-β1-like activity on this otherwise essentially TGF-β2 molecule. By the replacement of amino acids 45-47 of TGF-β2 with the corresponding amino acids of TGF-ß1, it was shown that this site is important for the 10-fold more effective binding

of the serum protease inhibitor, α_2-macroglobulin, by TGF-β2 than by TGF-β1. These results bring out the importance of particular domains in explaining some of the differences in the biological effects which exist between the TGF-β isoforms.

Binding Sites and Affinity   Of the many hundreds of cell types and cell lines examined for the binding of TGF-β, only a small proportion (see section Pathology) do not bind one or other of the TGF-β isoforms [3, 6]. From early studies on the number of high affinity binding sites for TGF-β it was found that these extended from as little as 1000 to 50000 per cell, with epithelial cells having higher affinity sites but lower numbers of sites than fibroblastic cells [6]. Crosslinking studies revealed three major types of transmembrane receptors (Figure 1). Type I (53kD) and type II (70-80kD) receptors both possess a cytoplasmic domain with serine/threonine kinase activity [7]. The type III receptor (approx. 300kD), is now usually called Betaglycan and has no known enzymatic activity [3, 8]. Betaglycan is the most abundant of the three receptors on most cells (up to 200000 per cell) and this is reflected on autoradiographs from crosslinking experiments where it appears as a long smear, preceded by a usually distinct spot (band) corresponding to the type II receptor (TβRII), which is itself preceded by a more faint spot for the type I receptor (TβRI) [8]. Affinity binding constants are in the range of 5-50pM for TβRI and TβRII and 30-300pM for Betaglycan. Most cells show higher affinity binding of TGF-β1 and β3 than of TGF-β-2 to TβRI and TβRII, whereas TGF-β2 binds with marginally higher affinity than the other two isoforms to Betaglycan [8].

From expression cloning studies considerable data have been assembled on the protein structure of each of the three receptor types [7]. Although a sub-family of about ten type I receptors have been cloned, only one (called ALK-5 or R-2) has been shown to be a functional TGF-β receptor [9]. The ALK-5 type I receptor (TβRI) is a 503 amino acid transmembrane protein: after a putative signal sequence of 25 residues there follows an extracellular domain of 100 amino acids containing ten cysteines and one N-glycosylation site; a hydrophobic region corresponding to the transmembrane domain (positions 126-147) is followed by the cytoplasmic domain which possesses a ser/thr kinase activity. This kinase domain is interrupted by two so-called kinase inserts (i.e., peptide sequences not having homology to those known to possess kinase activity). Proximal to the first kinase insert in the cytoplasmic domain is a SGSGSG (serine/glycine) sequence, termed GS domain (see later). The type II receptor (TβRII) is 567 amino acids long: after a presumed signal sequence of 23 amino acids, the sequence ensues with an extracellular region of 136 residues, containing a cysteine-rich domain and three potential N-linked glycosylation sites, followed by a hydrophobic stretch of 30 amino acids indicative of a transmembrane domain and completed by a 378 amino acid sequence containing a ser/thr kinase domain with two kinase

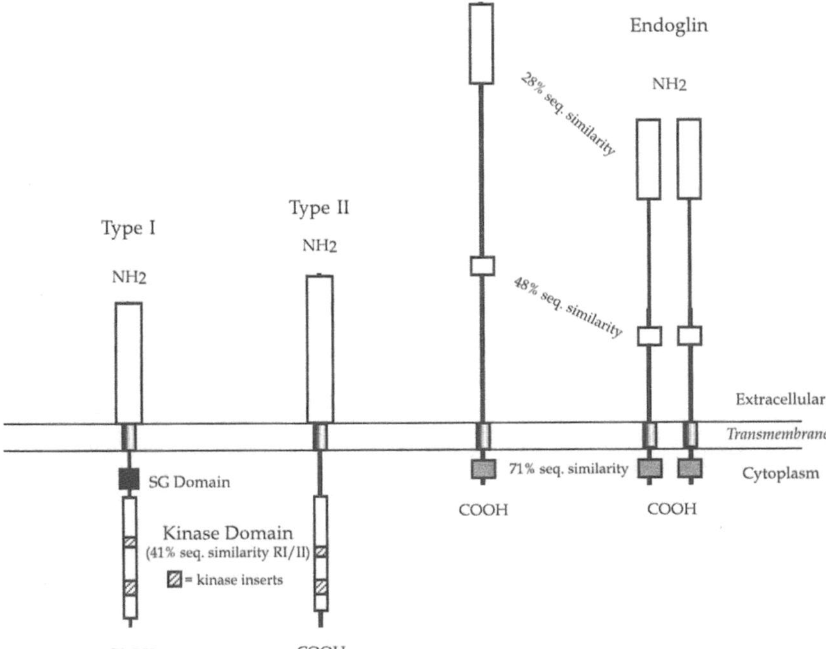

Figure 1. Schematic representation of TGF-β receptors. The two signaling receptors type I (TβRI) and type II (TβRII), Betaglycan and Endoglin are shown approximately according to the length of their respective amino acid sequences. It is to note the disproportion between TβRI and TβRII on the one hand and on Betaglycan and Endoglin on the other hand concerning the relative lengths of the extra- and intracellular portions of these molecules. The percent sequence similarity between the kinase domains of TβRI and TβRII and between various corresponding parts of the Betaglycan and Endoglin molecules is indicated. For references see [7-16].

inserts analagous to those in TβRI [10]. Using the yeast two-hybrid system which detects protein-protein interactions, a second putative type II receptor, T-ALK, has been identified [11]. T-ALK differs from the previously characterized type II receptor in having a very long C-terminal tail region extending beyond the kinase domain. The mRNA for TβRI in human tissues (5,5kb) is more highly expressed in placenta than in the brain and heart [9]. TβRII mRNA (5,5kb) expression has been shown to be well correlated with responsiveness to ligand [9]. Three mRNA forms, 11, 8, and 5kb were found for the putative type II receptor, T-ALK, of which the 11kb species was particularly abundant in the heart and liver [11]. Type III receptor, (Betaglycan), is composed of a 100kD core protein associated with heparan and chondroitin sulfate glycosaminoglycan chains [8]. Only the 100kD core protein is necessary for ligand binding. The Betaglycan core protein is 853 amino acids long: following a typical signal sequence of 24 amino acids, almost all of the rest of the core protein molecule represents the extracellular domain, which contains 16 cysteines, two major and four other possible sites for glycosaminoglycan attachment, seven potential N-linked glycosylation sites and a proline-rich domain close to the transmembrane region. The last 68 residues correspond to a membrane-spanning domain of 25 amino acids, followed by a cytoplasmic domain of 43 amino acids rich in serine/threonine residues. Data from protein sequence Banks reveal that the N-terminal third of the extracellular domain and the transmembrane and cytoplasmic domains of Betaglycan show homology to Endoglin (Figure 1), a dimeric protein highly expressed in human vascular endothelial cells [8]. Endoglin binds TGF-β1

and β3 but not TGF-β2, unlike Betaglycan which binds almost equally well to all three isoforms of TGF-β. Two TGF-β binding sites exist in the extracellular domain of Betaglycan, one in the N-terminal, Endoglin-related part of the molecule and one near the transmembrane domain [12]. It is noteworthy that Endoglin in its reduced form migrates electrophoretically like TβRII, from which it cannot therefore be distinguished by cross-linking methods alone, but requires the additional use of antibodies specific to one or other of these molecules.

The cloning of the TGF-β receptors has greatly accelerated elucidation of their signalling properties. It has been established for most cell types that a biological response to TGF-β requires co-expression of the type I and type II receptors [7, 9, 10, 13]. The functional signalling complex requires a heteromeric, possibly tetrameric, complex of TβRI and TβRII. A direct interaction between the cytoplasmic domains of these two receptors, initiated by ligand binding to their extracellular domains is a requisite for effective signalling. TβRII is a constitutively active ser/thr kinase but undergoes autophosphorylation only on serine residues. Upon binding of ligand, TβRII associates with and transphosphorylates TβRI on serine and threonine residues particularly in the GS domain. This ligand-dependent phosphorylation in the GS domain of TβRI allows downstream progression of TGF-β signals. Three different patterns of phosphorylation of the serine/threonine residues were observed in the GS domain, perhaps corresponding to different levels of kinase activation. Another serine residue, ser165, located about mid-way between the transmembrane and GS domains, was also found to be a ligand-dependent phosphorylation site. By site-directed mutagenesis the replacement of ser165 to alanine or glutamic acid caused a reduced effect of TGF-β1 on apoptosis, yet enhanced its effect on growth inhibition and extracellular matrix formation [14].

Whilst TGF-β-signalling depends predominantly on the type I and type II receptors, a role for the type III receptor, Betaglycan, has been found [8]. First, it must be recalled that the TGF-β2 isoform shows much lower binding affinity than TGF-β1 or β3 to TβRI and TβRII. Betaglycan binds almost equally well to all three TGF-β isoforms and is able to form a heteromeric complex with TβRII, allowing it to present ligand to this latter receptor. Subsequent association of this complex with TβRI would displace Betaglycan, producing the effective signalling TβRI/TβRII complex. This mechanism would be particularly important in enhancing the biological response to TGF-β in those cells expressing the appropriate receptors. It is to note that cells lacking Betaglycan, such as some vascular endothelial cell lines and hematopoietic precursor cell lines, are relatively insensitive to the growth-inhibitory effect of TGF-β2 [8].

Additional Features   Unlike, for example, the Epidermal Growth Factor receptor (EGF-R), TβRI and TβRII do not show any noticeable ligand-induced downregulation (i.e., endocytosis and intracellular degradation of receptor) [13]. However, TβRI and TβRII are differentially processed and have different half-lives: TβRI has a half-life of about 12 hours compared to only some 60 minutes for TβRII, at least in CCL64 Mink lung epithelial cells [15]. Other members of the type I and type II receptor families exist, which bind for example to TGF-β-related molecules like the Activins. The diversity of these receptor families and their binding specificity to their respective ligands create a complex system of binding proteins, allowing a cell to respond to the most subtle combinations of growth factors. On the other side of the coin however, the inherent complexity of such a system provides many points where it could jump tracks as a result of particular gene mutations.

Betaglycan, in addition to the membrane-anchored form described above, also exists in a soluble form composed of only the extracellular domain [16]. This soluble form, while capable of binding ligand, does not enhance binding to TβRII. On the contrary, soluble Betaglycan strongly inhibits ligand binding to TβRII, thereby acting as an antagonist of the TGF-β response, particularly that of TGF-β2. The relative proportions of membrane-anchored and soluble forms of Betaglycan present in the immediate vicinity of cells could therefore regulate TGF-β action. These forms of Betaglycan may not only play a role in mediating TGF-β action, but also that of basic Fibroblast Growth Factor (bFGF), because bFGF binds to the heparan sulfate glycosaminoglycan side chains associated with the Betaglycan core protein.

A number of other membrane bound proteins have been evidenced which bind the TGF-'βs in particular cell types [17]; only the mannose-6-phosphate / Insulin-like-Growth Factor II (M6P/IGF-II) will be mentioned (see section on processing and fate). Apart from the soluble form of Betaglycan, several other soluble proteins capable of binding TGF-β have been characterized [17, 18]. They can be placed in two categories, those found in body fluids such as plasma and those associated with the extracellular matrix. α_2-Macroglobulin (α_2-M) is an important plasma protein in humans and binds TGF-β2 with 10 fold greater affinity than TGF-β1. This binding blocks the biological activity of these TGF-β isoforms. Acting as a molecular trap for proteases, α_2-M is a very large tetrameric protein (725kD) and exists in two forms, a slow form capable of protease binding which is converted to a fast form after binding. This fast form is the clearance complex, its removal from the circulation being assured essentially by liver hepatocytes. However, α_2-M is not only secreted by liver cells and several cell types express an α_2-M receptor, so that this molecule might participate in the differential regulation of TGF-β isoforms around the periphery of cells as well as in the circulation. Heparin has been shown to dissociate the α_2-M/TGF-β1 complex and may explain the anti-proliferative effect of Heparin, this effect being distinct from its anti-coagulant activity. TGF-β2 has also been shown to bind to α-fetoprotein (AFP) in murine amniotic fluid. Here, such binding does not suppress TGF-β2's immunosuppressive activity. AFP plasma levels increase during pregnancy and as TGF-β2 expression is maximum around day 15 in the Mouse uterus, a complex of AFP/TGF-β2 might be involved in maternal tolerance of the fetal allograft. As AFP plasma levels go up in several disease states, ex., cirrhosis, hepatitis, gastrointestinal, liver and testicular cancers, it might play a role as a transporter of TGF-β2. It is to note that KCl treatment of the cytosol from human breast cancer revealed previously undetected amounts of AFP, so the existence of a complex AFP/TGF-β in this model should be looked into. β-amyloid precursor protein (APP) is yet one more protein which binds to TGF-β2. Such binding was initially found on analysing Glioblastoma cell supernatant: this brain cancer secretes two forms of TGF-β2, a free, biologically active form and one in which the growth factor, though retaining biological activity, is stably complexed to APP [19]. The proteolytic processing of APP to amyloid or non-amyloid peptides is determinant in the development of Alzheimer's disease and Down's syndrome. β-amyloid peptide (βA4) appears to be the causative agent in the cognitive decline associated with these last two diseases. TGF-β1 treatment of Rat hippocampal neurons showed a protective effect against βA4-induced neuron degeneration [20]. Several extracellular matrix proteins or proteins associated therewith are known to bind to the TGF-β's. Decorin is a proteoglycan containing chondroitin/dermatan sulfate chains and binds to all three TGF-β isoforms, although the consequence of this binding on the bioactivity of the TGF-ß's appears to depend on the biological system. Experimentally-induced glomerulonephritis in the Rat could be attenuated by systemic administration of Decorin [21]. Two other extracellular matrix proteins, Fibronectin and Thrombospondin are known to bind the TGF-β's (see section on processing and fate).

Structure

Sequence and Size   All three mammalian TGF-β's are normally secreted by most cells in the form of much larger molecular weight precursors [3, 4], which are biologically inactive [18 – see section on processing and fate and Figure 2]. These precursors exist as dimers and, per monomer, have 390 amino acids for TGF-β1 and 412 amino acids for both TGF-β2 and β3, of which the last 112 C-terminal amino acids are present in each monomer of a biologically active 25kD dimer, for all three isoforms [3, 4]. Each TGF-β precursor sequence starts with a stretch of 20-30 hydrophobic amino acids, constituting the N-terminal peptide signal typical of secreted molecules. Whilst the C-terminal portion of the TGF-β monomers

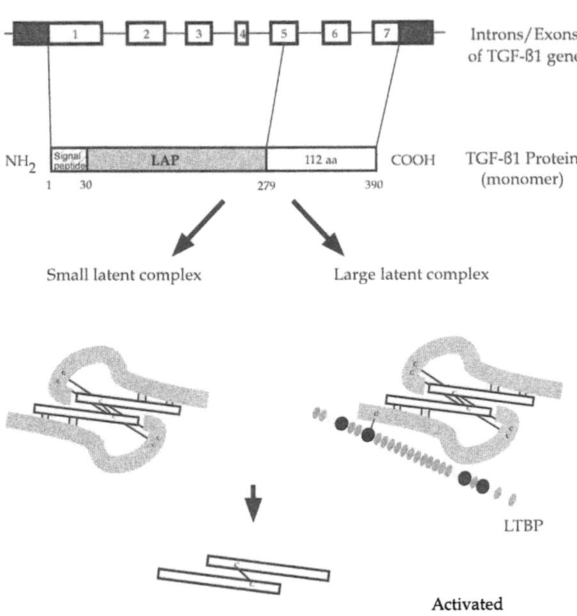

Figure 2. Schematic representation of the intron/exon structure of the TGF-β1 gene, the corresponding protein monomer and of the latent complexes. The dark boxes (top of figure) are exterior to the TGF-β1 coding sequence. The last half (approx.) of exon 1 through to the first part of exon 5 corresponds to the N-terminal part of the protein (amino acids 1-278), which is present as a dimer, β1LAP (minus the peptide signal sequence). The last part of exon 5 through to the first part of exon 7 corresponds to the C-terminal segment (amino acids 279-390), present as a dimer (25kd) in active, mature TGF-β1. In the small and large latent complexes, β1LAP is shown as a thick curvy line and the TGF-β1 dimer, still associated with β1LAP, is shown as two empty parallel rectangles. The two thick parallel lines joining C-C indicate disulfide bonds. The speckled boxes represent hydrogen bonds holding β1LAP to mature TGF-β1. The black circles on LTBP indicate the 8-cysteine repeat motif and the shaded ovals indicate EGF repeats. Activation of the latent forms lead to dissociation of the 25kd mature TGF-β1 from β1LAP (see text).

shows 72% sequence identity between TGF-β1/β2, 77% for β2/β3 and 79% for β1/β3, the N-terminal precursor sequence shows little similarity (33%-45%) between the three isoforms, except for the sequence just after the peptide signal where a stretch of 30 amino acids show 19 residues in identical positions. The number of cysteine residues in the N-terminal part varies, there being 3 in TGF-β1, 6 in TGF-β2 and 5 in TGF-β3; of these, three are in identical positions, 33, 223 and 225 using the TGF-β1 numbering, and are important in conferring latency [17]. Both the TGF-β1 and β2 precursors have three glycosylation sites and that of TGF-β3 has four sites: the most N-terminal site is at an identical position in all three precursors. In TGF-β1, at least, all three of these glycosylation sites are used. Evidence has been reported for phosphorylation of the TGF-β1 precursor, the phosphate residues mostly being attached to the carbohydrate moiety [17]. A common feature for the TGF-β1 and β3 precursors, but not that of TGF-β2, is the presence of a RGDL site, which is a recognition signal in many extracellular matrix proteins for interaction with cells [3, 4]. As mentioned above, the C-terminal part of the three TGF-β isoforms is closely similar in sequence: in particular 9 cysteine residues are conserved in identical positions (see section on Conformation) and the last five amino acids of the C-terminal sequence, SCKCS, are also conserved. The dimerised N-terminal precursor, now normally called β1-LAP, is cleaved from the 25kD dimer representing the C-terminal portion just after a dibasic RR or KR site, but the two parts of the molecule remain non-covalently associated (see section on processing and fate) [3, 4, 17].

Homologies   At present the TGF-β super-family contains over thirty distinct molecules spread over a large phylogenetic span, from Man to Drosophila [3, 4, 18]. They are all implicated in pivotal roles in normal growth and development. A large selection of these molecules is listed in Table 1 and are grouped into sub-families according to their degree of homology. Characteristically, molecules of the TGF-β super-family have an N-terminal signal sequence for secretion, a proteolytic cleavage processing site and a C-terminal sequence with at least 7 cysteine residues in conserved positions. Amino acid sequence identities between the mature C-terminal regions of the family members vary from as little as 16% (TGF-β1/GNDF) to about 80% (TGF-β2/TGF-β3), with a large majority showing around 40% homology. It is to note that the mature TGF-β1 sequence is identical in man, monkey, pig, cow and chicken and differs by only one amino acid in the mouse, serine being substituted for alanine at residue 75. From phylogenetic considerations five sub-families can be discerned; the TGF-β, Activin, dpp, 60A and GNDF/Neurturin sub-families. The dpp branch includes dpp itself and BMP-2 and BMP-4, that of 60A includes BMP-5 to -8 and 60A; these last two branches fall into a wider category called DVR including nodal and the group BMP-3 to dorsalin. No doubt some of these family members, e.g., TGF-β's and BMP's, could interplay with each other modulating their overall effect. As mentioned briefly above, there exists in parallel a TGF-β receptor family, all of whose members are type I or type II transmembrane ser/thr kinases [7].

Conformation   Two independent determinations have been made by X-ray crystallography of the 3-dimensional structure of the TGF-β2 isoform (cited in [5]). Both of these analyses concurred in representing the TGF-β monomer as a flat, slightly curved four-fingered hand, consisting of two pairs of anti-parallel β strands, projecting out from and perpendicular to the heel of the hand which consists of a long α-helix with nearly four

Table 1. A selection of molecules from the TGF-β super-family

Protein	present in
TGF-β homodimers	
TGF-β1 (prototypical member)	mammals
TGF-β2	mammals
TGF-β3	mammals
TGF-β5	Xenopus
Inhibin/Activin	
Inhibin-βA	mammals
Inhibin-βB	mammals
DVR group	
nodal	mammals
dpp	Drosophila
BMP-2	mammals
BMP-4	mammals
BMP-5	mammals
BMP-6 (Vgr-1)	mammals
GDF-8	mammals
BMP-7 (OP-1)	mammals
BMP-8 (OP-2)	mammals
60A	Drosophila
BMP-3	mammals
Vg1	Xenopus
GDF-1	mammals
GDF-3 (Vgr-2)	mammals
dorsalin	Chicken
Neurotrophic factors	
GNDF	mammals
Neurturin	mammals
miscellaneous members	
MIS	mammals

DVR, decapentaplegic, vgr-1-related; dpp, decapentaplegic; BMP, bone morphogenesis protein; GDF, growth and differentiation factor; Vg, vegetal pole; GNDF, glial-derived neurotrophic factor; MIS, Müllerian inhibitory substance.

turns. The β strands of the third and fourth fingers are crossed-over. The overall dimensions of the monomer are 60Å x 20Å x 15Å and it displays an unusual fold at one end of the long α-helix. Another unusual feature is the presence of a so-called cystine knot comprising two disulfide bridges in a rigid central core of the monomer. Association of two monomers to form the dimer occurs in anti-parallel fashion with the α-helix heel of one monomer lying against the β strands of the third and fourth fingers as well as parts of the first and second fingers. In a central cavity of the dimer there are four molecules of water. The 3-dimensional structure in solution of TGF-β1 has been analysed by nuclear magnetic resonance spectroscopy: whilst the basic structure of TGF-β1 closely resembles that found for TGF-β2, there are some important differences in structure and in flexibility which could explain the functional specificity of these two isoforms. X-ray analysis of crystals of another, distantly related member of the TGF-β super family, Osteogenic Protein 1 (OP-1, also called BMP-7) reveals that it is structurally similar to TGF-β2 (and thus to TGF-β1), despite having only 36 % sequence similarity to TGF-β2 (considering only the mature C-terminal regions of the molecules) (cited in [5]). Overall, these different analyses have established two structural features which are likely to be common to all member ligands of the TGF-β super-family, the unusual fold and the cystine knot.

Gene

Gene Structure   Each of the mammalian TGF-β isoforms is coded by a separate gene [3, 4]; the structure of all three comprises seven exons and six introns (Figure 2). For TGF-β1 and β3 all the intron-exon junctions are conserved, except for the first intron, where the intron-second exon junction differs by three nucleotides. The 3' and 5' untranslated regions are quite different for each of the three TGF-β isoforms; those of TGF-β1 are GC-rich, those of TGF-β2 are AT-rich and those of TGF-β3 are intermediate.

Chromosomal Localization   By Southern blot analysis of somatic cell hybrids and by *in situ* chromosomal hybridization, the gene for each TGF-β isoform has been localised as follows; TGF-β1 localises on chromosome 19 at band q13, TGF-β2 on chomosome 1 at band q41 and TGF-β3 on chromosome 14 at band q24 [3, 4]. The genes coding several other members of the TGF-β super-family have been localised and it is clear that they have become widely dispersed during evolution. These data support the notion that an ancestral TGF-β gene gave rise by successive duplications and mutations to the various family members.

Gene Expression   The mRNA transcript of TGF-β1 was initially shown to be 2,5kb, but a more detailed analysis revealed that the major transcript uses a polyadenylation signal 381 nucleotides further upstream than previously recognized resulting in a 2,1kb mRNA [22]. Several reports have indicated that some minor TGF-β1 transcripts occur in certain human cells [4]. The major TGF-β1 mRNA is fairly ubiquitously expressed. As many as five distinct mRNA transcripts of TGF-β2 have been evidenced in some cells (e.g., A549 human lung carcinoma cell line), of 2,8, 3,8, 4,0, 5,1 and 6,0kb, but in others two major bands of 4,0 and 6,0kb are seen. TGF-β2 mRNA expression was observed in both epithelial and mesenchymal cells. For TGF-β3 a single 3,2kb mRNA transcript was found, especially in mesenchymal cells.

Gene Regulation   At the 5' end of the human TGF-β1 gene two promoter regions have been identified; one 5' to the upstream transcriptional start site and one situated between the two major start sites, which are located 271 nucleotides apart [4]. The more 5' promoter sequence does not contain typical TATA or CAAT boxes, is G+C rich and has several binding sites for the Sp1 transcription factor. TGF-β1 autoinduces its own gene and this effect is mediated by AP-1 binding sites located in both the TGF-β1 promoters. In addition to the two promoter regions, two strong negative regulatory sequences are also present at the 5' end of the gene. At the 3' end of the human TGF-β1 gene a 16-base pair fragment downstream of the last TGF-β1 exon contains three phorbol ester-responsive elements (TREs), which might participate in regulation of TGF-β1 expression via activation of protein kinase C [22].

The human TGF-β2 and β3 promoters both possess a TATA box and a cyclic-AMP-responsive element (CRE) [23]. The CRE site in the TGF-β3 promoter is in close proximity to a AP-2 site; it is thought that this AP-2 site could influence the binding of the relevant transcription factors to the CRE site. Although the TGF-β2 promoter contains AP-1, AP-2 and NF-1 sites, they are not functional. From the analysis of the promoters of the three TGF-β genes it is clear that they are quite distinct, indicating that each isoform is differentially regulated. Apart from regulation at the level of transcription, there are probably other levels of control affecting mRNA stability and translation.

Additional Features   In the wake of the extensive characterization of the TGF-β receptors (see section on binding proteins), considerable advances have been made in understanding the signal transduction pathways utilized by the TGF-betas [24, 25]. It has long been known that TGF-β1, in its inhibitory mode, causes a blockage of the cell cycle in late G1. The G1 → S transition requires that the Retinoblastoma protein (Rb) undergoes phosphorylation. Upon phosphorylation, Rb releases the E2F transcription factors, essential for S phase progression. TGF-β1 keeps Rb in an under-phosphorylated state, thereby halting cell-cycle progression into S phase. Among the molecular participants controlling phosphorylation of Rb and other proteins are the cyclin-dependent kinases (CDKs, containing a catalytic subunit), cdk-2, cdk-4 and cdk-6, associated in a complex with an essential activator subunit called cyclin (ex. Cyclins A, D and E). TGF-β1 downregulates various members of these elements, depending on cell type (ex. cyclin E, cdk-2 and cdk-4 are

reduced in human keratinocytes). To be functional these CDKs must be phosphorylated on a conserved threonine (residue 160 or 161 depending on the cdk) by the cdk-activating kinase (CAK). Negative regulation of the kinase activities of these CDKs can occur by decreased synthesis/increased degradation of cyclins, by binding various CDK inhibitors (CKIs) or by tyrosine phosphorylation at the tyr15 site. Two families of CKIs are known based on sequence similarity; the CIP-KIP family so far contains three proteins, p21, p27 and p57 and the INK-4 family contains at present four proteins, p15, p16, p18 and p19. These two families of CKIs differ in their range of specificity, in that the CIP-KIP group inhibit cdk-2, cdk-3, cdk-4 and cdk-6, whereas the INK-4 group target cdk-4 and cdk-6. TGF-β1-mediated growth inhibition has been correlated with induction of these CKIs and a reduction in kinase activity of various CDKs. It is to note that the CKI p21, which was first found as a p53-transactivated product, can be induced by TGF-β1 in cells growth-inhibited by this factor without the presence of a functional p53 protein [26]. In fibroblasts growth-stimulated by TGFβ1 p21 is repressed. The CKI p15 is an important mediator of growth-inhibitory effects of TGF-β, but is not essential, since several p15-deficient human tumor cell lines are still inhibited by TGF-β [27]. In these p15-deficient cells TGF-β treatment led to growth arrest by inhibition of cdk-4 and cdk-6 activity due to an increase in their level of tyrosine phosphorylation. This increased level of tyrosine phosphorylation stemmed from the repression by TGFβ of the cdk tyrosine phosphatase cdc25A. Thus TGF-β-mediated-inhibition of cdk-4 and cdk-6 (which prevents Rb phosphorylation and provokes growth arrest) can be achieved independently, either by repression of the tyrosine phosphatase cdc25A or by induction of the CKI p15. It should be noted that some reports indicate the existence of a pRB-independant pathway for TGF-β1-mediated inhibition of human keratinocytes and breast cancer cells (cited in [28]).

A more recently identified family of proteins called the SMAD family has been shown to constitute a TGF-β signalling pathway from the TGF-β receptors to the target genes [25]. The SMAD proteins were discovered following genetic studies of the TGF-β homologue, dpp, in Drosophila. The SMADs are activated by serine phosphorylation (except Smad-4); Smad-1 by the BMP receptor I, Smad-2 and Smad-3 by TβRI and by the Activin receptor IB. Further propagation of the signal depends on the formation of hetero-oligomeric complexes between (in the case of TGF-β) Smad-2 or Smad-3 and Smad-4. Subsequent activation of gene transcription may involve the formation of a complex of Smad-2 and Smad-4 with a constitutively nuclear DNA-binding protein called Fast-1. Figure 3 summarizes the proposed sequence of events. Two other SMAD family members, Smad-6 and Smad-7, act as decoys by binding to TβRI, thereby preventing phosphorylation of Smad-2 and Smad-3. However, it remains to be shown that the SMAD pathway controls transcription of all or only some TGF-β-sensitive genes.

Since TGF-β1 exerts growth inhibition by causing cell cycle arrest in late G1, genes coding for cell cycle components would be among these TGF-β-sensitive genes. It is important to note that among the above signalling components, p16INK4 and Smad-4 (also called DPC4 for "deleted in pancreatic cancer"), are defective in particular human tumors, which allows the attribution to their corresponding genes of the term, tumor suppressor gene. Concerning the growth-stimulatory effects of the TGF betas on certain cells, no clear evidence has been provided for known mitogen pathways being activated directly by the TGF-β receptors. Several other signalling molecules have been implicated in TGF-β-mediated effects, including certain G proteins, c-*myc*, Protein kinase C and Map kinases [3, 4, 18]. It may be that the TGF-betas can direct regulation via several signal transduction pathways, depending on cell type and on the state of cell differentiation. Points of overlap between such pathways may provide cellular machinery for producing flexibility in the response.

Processing and Fate   As mentioned above the three TGF-β isoforms are secreted in the form of high molecular weight complexes, lacking the biological activities of the mature 25kD peptides. Such complexes are referred to as latent TGF-β [29, 30]. At the time that the latency phenomenon was discovered, only one isoform of TGF-β was known. It was shown that latent TGF-β could be converted to the lower molecular weight, 25kD, biologically active form by exposure to acid pH (pH 3,0), to alkaline pH (pH 11,0), to heating 3 minutes at 100°C and to urea. From these results it was proposed that activation of latent TGF-β *in vivo* would be a crucial regulatory step in controlling the biological activity of this peptide [29, 30]. Subsequently it was shown that the TGF-β2 and β3 isoforms are also released as latent forms [31]. The acid and alkaline pH profiles of activation and those obtained by heating (65°C-100°C) are fairly simi-

lar for all three isoforms. Activation is most frequently obtained *in vitro* by acidification of samples to below pH 3,0. Before considering other, and more obviously physiological, activation treatments of latent TGF-β complexes, their structural biochemistry will be outlined. Whilst the essential structure of the latent complexes of all three TGF-β isoforms is no doubt very similar, virtually all the work on this matter concerns latent TGF-β1. The latter can exist in two major forms, called the small and the large latent complexes [17]. In both cases latency is conferred by the non-covalent association of the N-terminal precursor remnant, cleaved of its peptide signal, to the C-terminal part of the TGF-β molecule, each part being present as a dimer (Figure 2). Thus the N-terminal precursor dimer is known as the latency-associated peptide (LAP). The β 1-LAP dimer is approx. 80kD, so that in association with the mature 25kD peptide, the size of the small latent complex is around 100kD. There are three asparagine residues (positions 82, 136 and 176) in β1-LAP; they are all N-glycosylated and residues 82 and 136 bear mannose-6-phosphate groups, which may allow the latent complex to bind to Mannose-6-phosphate/Insulin-like-Growth Factor II receptors. β1-LAP contains three cysteine residues, positions 33, 223 225, of which the last two are essential for dimerization of two LAP chains. Mutation of either cysteine 223 or 225 prevents disulfide bonding of two LAP monomers, which in turn leads to improper assembly of the complex and secretion of active, mature TGFβ-1. Most cell types secrete small latent TGF-β1.
Many others also secrete large latent TGF-β1, which is the form found in human blood platelets [17]. This larger latent form contains a third component, latent TGF-β binding protein (LTBP-1), whose molecular size varies from 125 to 160kD, depending on the cellular source (Figure 2). The overall size of the large latent complex of TGF-β1 is 235-265kD. LTBP-1 is disulfide-linked to β1-LAP, cysteine 33 of the latter being involved in this cova-

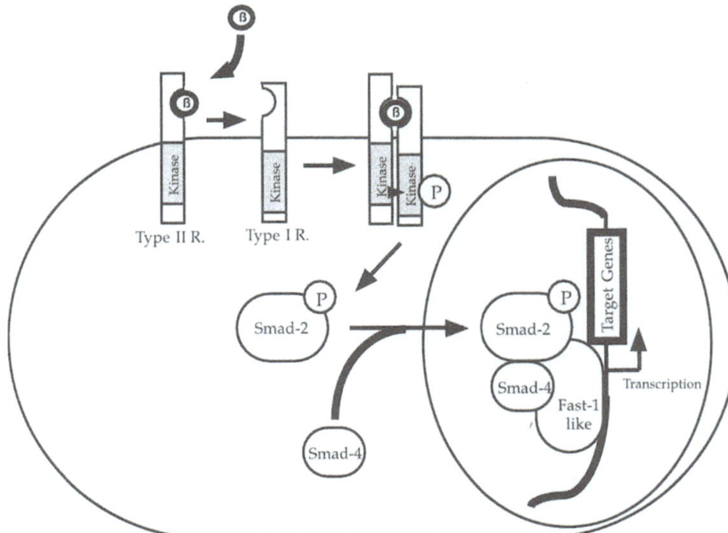

Figure 3. The SMAD pathway of TGF-β signal transduction. This pathway can be decomposed into seven steps, the last two taking place in the nucleus: binding of ligand to TβRII, recognition by ligand-bound TβRII of TβRI, heterodimerisation of the two signaling receptors (probably a heterotetramer), TβRII-dependent phosphorylation (indicated by P within a circle) of TβRI, phosphorylation by TβRI of a downstream effector (SMAD-2 or -3), complex formation between SMAD-2 or -3 and SMAD-4, together with Fast-1, leading to transcription of target genes.

lent linkage. To date, two other LTBP-like molecules, LTBP-2 and LTBP-3, have been cloned from human, bovine or murine sources: they all show homology to the Fibrillin family of microfibrillar proteins [17, 18]. LTBP-1, -2 and -3 contain two types of cysteine-rich repeats, namely, the Epidermal Growth Factor (EGF)-like repeats and repeat stretches of a 8-cysteine motif. This last motif is found in human Fibrillin, an extracellular matrix protein which is defective in Marfan's syndrome. The components of the large latent complex are assembled inside the cell [17]. Its secretion to the extracellular space is facilitated by LTBP-1, since the latent complex lacking LTBP-1 is secreted more slowly. A free form of LTBP-1 is also secreted. LTBP-1 and TGF-β1 co-locate in the extracellular matrix, where LTBP-1 might play a role in storage and release of active TGF-β1. A similar scenario can be anticipated for the other TGF-β isoforms, with isoform specificity depending on cell type and tissue localization. An additional role for LTBP-1 as a structural matrix protein in bone formation has been suggested. Recombinant (small) latent TGF-β1 has a 90 minute half-life in Rats, whereas active TGF-β1 is cleared from the circulation via the liver in 2-3 minutes. Thus, one may suppose that the function of latency is to provide, simultaneously, stability, a locally activatable source of TGF-β and a control point of its powerful biological effects.

Ever since the latency phenomenon was discovered, much attention has been given to determining *in vivo* activation mechanisms. It is important to emphasize again that certain cells secrete both small and large latent complexes, so that their biological effects will depend on each cell's capacity to activate these different latent forms. As will become evident below from the many, diverse means of obtaining active TGF-β1 (and TGF-β2 and β3) it is more than likely that several activation mechanisms exist *in vivo*. Among the initially discovered activating treatments, that of acidification may still be retained as a possible *in vivo* activating mechanism. Thus all three latent TGF-β isoforms are completely activated at pH 3,0 and examples of acidic microenvironments close to this pH are known *in vivo*, (osteoclasts, activated macrophages, stroma of solid tumors). In support of this notion, transient acidification of fibroblasts in culture was shown to activate latent TGF-β *in situ* and to render the cells more responsive to EGF [32]. Several, more obviously physiological mechanisms have been found and will be outlined hereafter.

One of the most heuristic approaches in this area has involved the use of co-culture experiments: when pericytes or smooth muscle cells were grown in the presence of aortic endothelial cells activation of latent TGF-β occurred *in situ*, yet each cell type alone secreted latent TGF-β but could not activate it [17, 18, 33]. Upon closer examination using appropriate antibodies and inhibitors, this co-culture activation was shown to involve Plasmin. Subsequently it was shown that Retinoids could activate latent TGF-β, perhaps by increasing basal levels of Plasmin. Activation of latent TGF-β by co-culture and by Retinoids takes place on the cell surface and involves the

participation of type II Transglutaminase (TGase-II). TGase-II has a wide distribution among both normal and tumor cells and one of its functions is to form cross-links in erythocyte membrane proteins. High cellular levels of TGase-II correlate with growth inhibition and/or induction of cell differentiation, which would support its participation in activating latent TGF-β [33]. The treatment of normal human osteoblast-like cells with the anti-inflammatory glucocorticoid, Dexamethasone, led to activation of latent forms of TGF-β [34]. Conditioned medium from such cultures retained the capacity to activate latent TGF-β in the absence of cells, suggesting that Dexamethasone-induced activation did not occur on the cell surface. Dexamethasone treated cells increased the secretion of lysosomial proteases (Cathepsin B), which activate latent TGF-β secreted by the same cells. Among other enzymatic activating treatments one may cite, Cathepsin D and certain glycosidases like Endoglycosidase F and N-acetylneuraminidase [17].

A potentially most important way of activating latent TGF-β is that obtained using the extracellular matrix protein, Thrombospondin-1 (TSP-1), since activation occurs by a protease- and cell-independent mechanism [35]. TSP-1 is a trimeric, disulfide-linked, multiple domain glycoprotein containing characteristic repeat motifs and is found in the α-granules of blood platelets (like TGF-β1) and in wound fluids. A tripeptide, RFK (arg, phe, lys), located in the so-called type I repeats of TSP-1, is sufficient to activate both the small and large latent TGF-β1 complexes. TSP-2, an isotype of TSP-1, was reported to activate only the large latent TGF-β complex in conditioned medium from adrenocortical cells [36]. In these cells TSP-2 was induced by ACTH, so that the synthesis of steroids also induced by ACTH might be negatively regulated by TSP-2-activated latent TGF-β. Of great interest is the finding of active TGF-β1 complexed to IgG inside B cells and plasma cells in a murine model of Systemic Lupus Erythematosus [37]. TGF-β1 complexed with IgG was some 500 times more potent than free TGF-β1 in suppressing neutrophil function *in vitro* and host defence against *Staphylococcus aureus* infection *in vivo*. Before closing this section on latency it is of considerable importance to the radiotherapy field to mention that within one hour after 60Co-γ radiation, active TGF-β1 was found in the mammary gland of mice [38]. Such locally active TGF-β might participate in the remodelling of stromal collagens which occurs in irradiated mammary gland.

Biological Activity   A plethora of excellent reviews have provided ample coverage of the biological activities of the TGF-betas, so to avoid unnecessary repetition the present section will briefly summarize their three principal effects, before giving an account of the more clinically orientated data in the later sections of this chapter.

Many of the *in vitro* effects of the TGF-βs are manifest by all three isoforms [3, 4], but their tissular and temporal expression and the results of TGF-β gene knockouts or overexpression studies clearly indicate some isoform specificity [39]. The TGF-betas

1) inhibit the proliferation of most cell types, but can stimulate growth of some mesenchymal cells, notably chondrocytes, osteoblasts, Schwann cells and the WI-38 human fetal lung fibroblast [18]. A strong and normally reversible growth inhibition can be obtained with 1ng/ml of TGF-β in cell culture. *In vivo*, the implantation of slow-release pellets of TGF-β1 reversibly inhibited the proliferation of terminal epithelial buds in the Mouse mammary gland [4]. However, in another study TGF-β1 was shown to induce androgen-dependent apoptosis of Rat prostate cells both *in vivo* and *in vitro* [40]. Related to their bifunctional nature on growth, the TGF-βs can affect, positively or negatively, cell differentiation [3, 4]. This bifunctionality is also manifest in the synthesis of steroid hormones. Intriguingly, the proliferative effect of TGF-β1 can be inversed in certain instances: thus adherent B lymphocytes are normally growth-inhibited by TGF-β1, but are stimulated when these cells carry Epstein-Barr virus [18] and certain human and murine fibroblasts which are growth-stimulated by TGF-β1, are slightly growth-inhibited when stably transfected with Large T antigen of SV40 virus [26, 28, 41]. Furthermore, NIH3T3 murine fibroblasts are growth stimulated by TGF-β1 when the P53 protein is wild type, but this response switches to one of growth inhibition when P53 is inactivated by mutation or by complexing with viral oncoproteins which bind P53, whereas pRB function did not affect such responses [28].

2) exert powerful immunosuppressive effects [3, 4, 39], in part as a consequence of their anti-proliferative properties. TGF-β1 and β2 (and no doubt TGF-β3 as well) block the proliferation of thymocytes and T and B lymphocytes: in consequence, activated B lymphocytes secrete lower quantities of antibodies. An important exception however is that exogenous TGF-β1 increases the secretion of IgA, the major isotype in mucosal secretions. Among other immune cell activities affected by the TGF-betas, one may cite the blockage of the natural killer activity of large granu-lar lymphocytes and the reduced production, but not activity, of cytotoxic T lymphocytes. Certain effects of the TGF-betas on immune cell functions derive from their opposition or antagonism to other cytokines, like the Interleukins, Interferons and Tumor Necrosis Factor. TGF-β1 participates in the inflammatory response by the chemotactic attraction of fibroblasts, monocytes and neutrophils to the wound site, and later, after macrophage maturation, TGF-β1 contributes to resorption of inflammation by deactivating macrophages. Both TGF-β1 and β2 downregulate synthesis of the histocompatibility complex class II antigen, HLA-DR, in human glioma and melanoma cell lines. HLA-DR is of crucial importance in the immune response as it participates in antigen recognition by helper T lymphocytes.

3) enhance the formation of extracellular matrix (ECM).

Cellular interactions, like cell migration and tissue morphogenesis, are critical determinants in embryogenesis, wound repair and metastasis [3, 4, 21]. These cellular events depend on such ECM components as the glycosaminoglycan Hyaluronic acid (Hyaluranon), the glycoproteins collagen, fibronectin, tenascin and on the proteoglycans, which contain glycosaminoglycans like chondroitin and dermatan sulfates in complex with a "core" protein. The TGF-βs exert four different actions which cumulatively enhance the synthesis and maintenance of ECM components (Figure 4). Thus, they stimulate the synthesis of most ECM constituents, increase the synthesis of inhibitors of proteases which would normally degrade these constituents, while they reduce the synthesis of these same proteases and they increase the synthesis of integrins (a family of heterodimeric membrane receptors through which cells are enabled to recognize specific ECM molecules in basement membranes and in other cells).

Role in Vascular Biology

Physiological Function Most studies in this area concern TGF-β1 and their results have revealed important effects of this growth factor on endothelial cells, vascu-

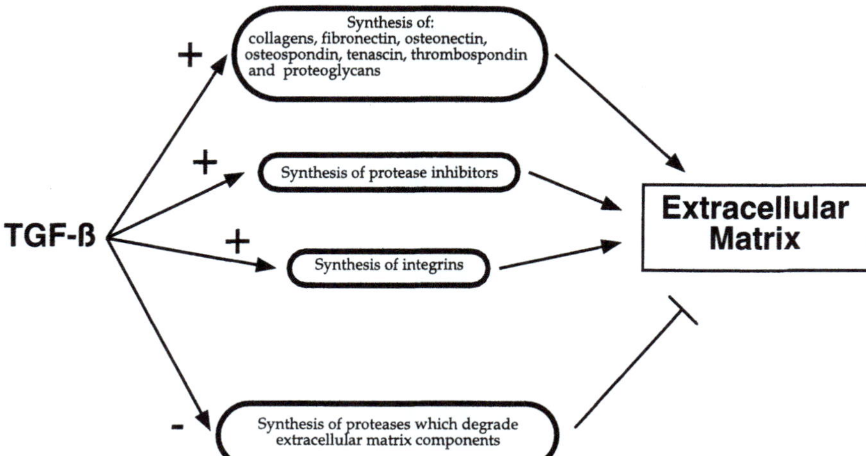

Figure 4. TGF-β effects on the synthesis and maintenance of the extracellular matrix.

lar smooth muscle cells and on haematopoietic progenitors [42, 43]. The differentiation of mesodermal precursors into endothelial cells (vasculogenesis) and the migration and maturation of endothelial cells (angiogenesis) are crucial events in early embryogenesis and in subsequent development of the foetus as well as tissue repair in the adult, respectively. TGF-β1 strongly inhibits the proliferation and migration of endothelial cells *in vitro* in two-dimensional assays. This inhibitory effect has been attributed in one instance to the TGF-β1-mediated downregulation of EGF receptors, thereby preventing EGF-induced *c-myc* expression (c-*myc* protein being a positive regulator of cell growth – [3, 4]). A similar inhibition of endothelial cell proliferation is obtained by the addition of exogenous Fibronectin. TGF-β1 increases the expression of Fibronectin (and other ECM components) in such cells, so this may explain the TGF-β1-growth inhibition. Although many ECM molecules are up-regulated in endothelial cells by TGF-β1, the bifunctionality of this factor is again evident here in that the expression of E-selectin, a cell adhesion molecule and VEGFR-2 (Vascular Endothelial Growth Factor Receptor-2), for example, is decreased by TGF-β1 [42]. The Selectins are involved in adhesion of leukocytes to activated endothelium and E-Selectin seems to facilitate extravasation through the vascular wall into lymphoid tissue and sites of inflammation. VEGF has been clearly recognized as a positive regulator of angiogenesis. Thus the reduced expression of E-Selectin and of the VEGF receptor can be seen as manifestations of the anti-inflammatory repertoire of TGF-β1. Although TGF-β1 is inhibitory in two-dimensional assays, this is not the case when endothelial cells are grown in three-dimensional assays (e.g., collagen gels). In the latter condition TGF-β1 induces a differentiated phenotype in endothelial cells, which form capillary tube-like aggregates. This change of endothelial cell phenotype depending on culture conditions may derive from an alteration in TGF-β-receptor expression; thus in two-dimensional culture there is a higher ratio of type I to type II TGF-β receptors [42]. Overall, the results from *in vitro* studies suggest that TGF-β1 exerts a biphasic action on endothelial cells depending on whether they are activated or in a quiescent state. Thus, it actually potentiates the action of VEGF on activated endothelial cells grown in collagen gels, but once these cells enter the resolution phase it participates in maintaining them in a quiescent state. However, the significance of these *in vitro* results remains in question, since to date, hardly any of the TGF-β-mediated effects on endothelial cells *in vitro* have been reproduced *in vivo*. Indeed, endothelial cells *in vivo* do not seem to express any of the three TGF-β receptors, TβRI, TβRII and Betaglycan, but do express Endoglin which binds TGF-β1 and β3 but not TGF-β2.

TGF-β1 has been reported to inhibit or stimulate vascular smooth muscle (VSM) cell proliferation *in vitro* depending on cell density, culture conditions [43] and on the method of obtaining the cells [trypsin dispersal or explant culture – [44]). Different authors have found that TGF-β1 potentiates growth-stimulation of VSM cells mediated by PDGF-AA, PDGF-BB, bFGF and EGF, but with delayed kinetics compared to the last four growth factors alone. VSM cells from injured arteries secreted more fibronectin and proteoglycans into culture medium and contained more TGF-β1 mRNA than did normal VSM cells, suggesting that the accumulation of ECM components in neointimal lesions might be caused by autocrine TGF-β1 activity released by VSM cells [45]. The latent TGF-β binding protein, LTBP, has been reported to stimulate migration of Rat VSM cells in culture, suggesting a role in vascular remodeling [46]. Here again, the confirmation *in vivo* of TGF-β-mediated effects observed *in vitro* largely remains to be shown.

Steel Factor (SF), produced by bone marrow stromal cells, is an important factor which stimulates haematopoiesis *in vivo*, in association with other haematopoietic factors [47]. SF mediates this action via a membrane-bound receptor, c-kit, present on haematopoietic progenitor cells. TGF-β1 inhibits the SF/c-kit effect on haematopoiesis *in vitro* by inhibiting transcription of the SF gene and by accelerating the degradation of c-kit mRNA. However, whilst SF / c-kit may also be critical *in vivo* for haematopoiesis in later developmental stages of the mouse embryo and in the adult, this is not the case in the earliest phase of haematopoiesis which is independent of SF / c-kit.

Cogent information on the role of the TGF-βs *in vivo* has come from gene knockout studies in mice [48-51]. Previously, it had been shown that the first notable expression of TGF-β1 in the normal mouse embryo occurred at 7,5 days post coitum, especially in the extraembryonic blood islands of the yolk sac. Later on in embryonic development TGF-β1 was expressed in endothelial precursors and in fetal blood cells of the liver, suggesting that the factor was involved in vasculogenesis, angiogenesis and in haematopoiesis [52]. These data on TGF-β1 expression fit well with the prenatal lethality observed in TGF-β1-/- mice from knockout studies. Indeed, 50% of such TGF-β1-/- animals die at about 10 days pc: they show reduced formation and increased fragility of the yolk sac vasculature and/or defective haematopoiesis. No endothelial or haematopoietic cell hyperplasia was seen in these TGF-β1-/- mice, which might have been anticipated from the absence of a negative growth regulator. The other 50% of TGF-β1-/- mice are born apparently normal, but die rapidly at about 3 weeks of age from tissue necrosis and massive inflammatory infiltration of essential organs, especially the heart. Those TGF-β1-/- animals which reach parturition and live for a few weeks appear to be "rescued" by maternal transfer across the placenta and via lactation prior to weaning, respectively. A phenotype strikingly similar to that of the TGF-β1-/- mice has been observed in knockout mice lacking the type II TGF-β receptor [42]. So it appears that the major role of TGF-β1 in fetal development is to ensure correct differentiation, rather than proliferation, of the extraembryonic vasculature and haematopoiesis. As the TGF-β1-/- embryos themselves showed no particular developmental defects, other than those that may be attributed to the yolk sac abnormalities, these studies suggest that embryonic and extraembryonic vasculogenesis and haematopoiesis are differentially regulated. TGF-β2 null and TGF-β3 null mice also show perinatal mortality due to multiple developmental defects [49, 50, 51]. These knockout studies indicate that the TGF-β isoforms cannot substitute for each other *in vivo*. Finally, the increased angiogenesis which follows subcutaneous injection of TGF-β1 and that occurring in settings such as wound healing, where TGF-β1 as well as TGF-β2 and β3 are expressed, would be indirect and a consequence of the local recruitment of inflammatory cells.

Pathology   Despite some incertitude as to which of TGF-β-mediated effects on vascular cells observed *in vitro* are applicable to the *in vivo* situation, one can confidently predict that the TGF-betas play crucially important roles in this domain. The following examples will amply bear out this statement. The TGF-β1 knockout studies in mice have clearly shown the importance of this growth factor in proper differentiation of the extraembryonic vasculature and in haematopoiesis [48, 52]. Although further studies will be necessary to confirm that TGF-β1 also plays a crucial role in the adult vasculature, the presently available evidence indicates that a lack of this growth factor leads to vessel wall fragility.

Atherosclerosis associated with diabetes mellitus frequently affects the arteries below the knee (infragenicular arteries) and insulin is often considered as the causative agent in lesion development by stimulating the proliferation of VSM cells. The insulin-mediated increase in growth of VSM cells from human infragenicular arteries can be blocked by exogenous TGF-β1 *in vitro* [53]. However, although TGF-β1 may limit VSM cell proliferation, it also causes increased ECM synthesis leading to thickening of the intima, which occurs naturally in response to arterial injury. Indeed, as mentioned previously, cultured VSM cells from injured arteries of Rats showed higher levels of TGF-β1 mRNA and synthesized more fibronectin and proteoglycans than control cells [45].

A high level of lipoprotein(α) in the blood circulation is a known risk factor for atherosclerosis [54]. Lipoprotein(α) is composed of an LDL particle covalently bound to a specific protein, apolipoprotein(α), which shows sequence-relatedness to plasminogen. The atherogenic activity of lipoprotein(α) is considered to derive from the competitive inhibition of plasminogen by apolipoprotein(α). Both lipoprotein(α) and apolipoprotein(α) stimulate the proliferation of VSM cells by inhibiting the activation of plasminogen to plasmin, thereby preventing the proteolytic activation of latent TGF-β to the active form, which acts as an *in situ* inhibitor of VSM cell proliferation. In serum from patients with severe atherosclerosis fivefold less active TGF-β was found compared to controls, which adds further evidence to support the notion that TGF-β plays an important role in this disease [55]. It is of interest that the chemotherapeutic drug frequently employed in breast cancer treatment, tamoxifen, as well as aspirin, increase the levels of active TGF-β in mice and humans, respectively [54]. Other studies have suggested roles for TGF-β-1 in contributing to hypertension, in attenuating the consequences of myocardial and cerebral ischemia [56] and in regulating stem cell proliferation [47].

Clinical Relevance and Therapeutic Implications   Given the wide range of the currently known cellular activities of the TGF-βs, (inhibition, or more rarely, stimulation of cell proliferation, chemoattraction of inflammatory cells, immunosuppression and enhancement of ECM synthesis), there is no doubt that they participate in crucially important regulatory events in both normal and pathological situations. Thus, current evidence indicates that lack of a particular TGF-β isoform, overexpression, defective activation of latent forms or mutation of their signalling receptors, or mutation of particular SMADS can lead to a pathological situation [25, 39, 48-52, 54, 57, 58]. Clearly, each of the TGF-β isoforms, despite their structural similarities and often their biological equivalence *in vitro*, play distinct roles *in vivo* and require tight regulation to ensure correct function.

One of the keys to understanding the *in vivo* effects of the TGF-βs is the difference observed between systemic and local injection of exogenous growth factor [39]: systemic injection tends to be anti-inflammatory, whereas local injection has the opposite effect. In the former case, TGF-β would encounter vascular endothelial cells, downregulate synthesis of E-selectin and thereby reduce leukocyte adhesion at sites of inflammation. On the contrary, subcutaneous or intra-articular injection provokes or exacerbates localised inflammation, by chemotactic attraction of neutrophils, macrophages and fibroblasts. Such local administration leads to an angiogenic response in an inflammatory setting, whereas systemic delivery would allow expression of the immunosuppressive action of the TGF-βs.

The first clinically relevant use of the TGF-βs concerns their effects in wound healing [21, 56]. A voluminous literature documents their use in accelerating repair of various types of surgical wounds. Fetal wounds repair without scarring, which is not the case in adults where scarring can cause a functional handicap. By the careful timing of their application, it has been shown that a combination of neutralizing antibodies against TGF-β1 and TGF-β2, coupled with application of TGF-β3 improves repair of skin wounds with reduced scarring [59]. TGF-β2 has been found to enhance repair of venous ulcers and macular holes [60]. The results of another study with rats suggest that, if TGF-β1 production could be satisfactorily controlled locally, it might be possible to obtain nerve outgrowth after penetrating brain injuries [61]. Excessive production of TGF-β1 occurs following experimental brain injury and is responsible for the formation of fibrous (glial) scars, which prevent nerve regeneration. All three TGF-βs isoforms are present in bone matrix and TGF-β2 in particular (see above), appears to be an important positive regulator in bone morphogenesis and repair [58]. In regard to bone repair it is known that the active form of vitamin D3 is important, so it should be noted that this vitamin increases the secretion of TGF-β [56]. This last point underscores the complex interactions between different families of molecules, which require clarification as to the finality of their effects, before these substances can be administered with confidence for therapeutic benefit.

The TGF-βs have been implicated in several inflammatory auto-immune disorders [18, 56, 62], e.g., multiple sclerosis, uveitis and rheumatoid arthritis. In some instances, at least, the inflammatory events are mediated by nitric oxide (NO) and as TGF-β1 inhibits NO synthesis, this growth factor would appear as an important regulator of inflammation [63]. In many chronic fibrotic diseases [21, 56], e.g., glomerulonephritis, scleroderma, hepatic cirrhosis, pulmonary fibrosis and proliferative vitreoretinopathy, there occurs an accumulation of extracellular matrix, in part due to overexpression of the TGF-betas. Many reports have implicated the TGF-betas in the field of cancer: e.g., overexpression of TGF-β2 is associated with immune failure in glioblastoma [19, 64] and a functional type II TGF-β receptor is absent in some human colon cancers [57]. A gene therapy approach using antisense TGF-β2 has been validated in a Rat model of glioblastoma [64]. An *in vitro* study [65] showed that TGF-β1 enhanced cell death of a human cancer cell line treated with DNA-damaging agents (γ rays, UV-C, methotrexate, 5-flurouracil, cisplatin), raising the possibility that, *in vivo*, TGF-β1 might affect the curative and/or undesirable secondary side effects of cancer therapy. Other instances where the TGF-betas appear to be of clinical relevance include septic shock, myocardial and cerebral ischemia, AIDS and in oral tolerance [18, 39 56].

Owing to the multifunctional nature of the TGF-βs and their complex interactions with other substances, e.g., various cytokines, vitamins, steroids, etc., striking the right balance between them will be no easy task. However, we now have a clearer understanding of the *in vivo* roles played by this family of growth factors, as well as their structure. Several clinical trials are in progress with natural and modified forms of the TGF-βs, their binding proteins and anti-TGF-β antibodies and the results are eagerly awaited.

David A. Lawrence

References

1. Moses HL et al (1981) Cancer Res 41:2842-2848
2. Roberts AB et al (1981) Proc Natl Acad Sci USA 78:5339-5343
3. Massagué J (1990) Ann Rev Cell Biol 6:597-641
4. Roberts AB, Sporn MB (1990) In: Roberts AB, Sporn MB (eds) Peptide Growth Factors and their Receptors. Springer, Berlin Heidelberg New York, pp 419
5. Hinck AP et al (1996) Biochemistry 35:8517-8534
6. Wakefield LM (1987) Methods in Enzymol 146:167-173
7. Derynck R (1994) Trends Biochem Sci 19:548-553
8. Lopez-Casillas F et al (1993) Cell 73:1435-1444
9. Franzén P et al (1993) Cell 75:681-692
10. Lin HY et al (1993) Cell 68:775-785
11. Kawabata M et al (1995) J Biol Chem 270:5625-5630
12. Pepin M-C et al (1995) FEBS Letters 377:368-372
13. Ventura F et al (1994) EMBO J 13:5581-5589
14. Souchelnytskyi S et al (1996) EMBO J 15:6231-6240
15. Koli KM, Arteaga CL (1997) J Biol Chem 272:6423-6427
16. Lopez-Casillas F et al (1994) J Biol Chem 124:557-568
17. Miyazono K et al (1993) Growth Factors 8:11-22
18. Lawrence DA (1996) Eur Cytokine Netw 7:363-374
19. Bodmer S et al (1990) Biochem Biophys Res Comm 171:890-897
20. Prehn JH et al (1996) Mol Pharmacol 49:319-328
21. Border WA, Ruoslahti E (1992) J Clin Invest 90:1-7
22. Scotto L et al (1990) J Biol Chem 265:2203-2208
23. Noma T et al (1991) Growth Factors 4:247-255
24. Massagué J, Polyak K (1995) Current Opinion in Genetics & Development 5:91-96
25. Heldin C-H et al (1997) Nature 390:465-471
26. Raynal S, Lawrence DA (1995) Int J Oncol 7:337-341
27. Ivarone A, Massagué J (1997) Nature 387:417-422
28. Dkhissi F et al (1998) Oncogene 17:in press
29. Lawrence DA et al (1984) J Cell Physiol 121:184-188
30. Pircher R et al (1986) Biochem Biophys Res Comm 136:30-37
31. Brown PD et al (1990) Growth Factors 3:35-43
32. Jullien P et al (1989) Int J Cancer 43:886-891
33. Kojima S et al (1993) J Cell Biol 121:439-448
34. Oursler MJ et al (1993) Endocrinology 133:2187-2196
35. Crawford SE et al (1998) Cell 93:1159-1170
36. Souchelnitskyi S et al (1995) Endocrinology 136:5118-5126
37. Caver TE et al (1996) J Clin Invest 98:2496-2506
38. Barcellos-Hoff MH et al (1994) J Clin Invest 93:892-899
39. McCartney-Francis NL, Wahl SM (1994) J Leukocyte Biol 55:401-409
40. Hsing AY et al (1996) Cancer Res 56:5146-5149
41. Raynal S et al (1994) Growth Factors 11:197-203
42. Pepper MS (1997) Cytokine & Growth Factor Reviews 8:21-43
43. Newby AC, George SJ (1996) Current Opinion in Cardiology 11:574-582
44. Kirschenlohr HL et al (1995) Cardiovasc Res 29:848-855
45. Rasmussen LM et al (1995) Am J Pathol 147:1041-1048
46. Kanzaki T et al (1998) Biochem Biophys Res Comm 246:26-30
47. Heinrich MC et al (1995) Blood 85:1769-1780
48. Shull MM et al (1992) Nature 359:693-699
49. Sanford LP et al (1997) Development 124:2659-2670
50. Proetzel G et al (1995) Nature Genetics 11:409-414
51. Kaartinen V et al (1995) Nature Genetics 11:415-421
52. Dickson MC et al (1995) Development 121:1845-1854
53. Forsyth EA et al (1997) J Vasc Surg 25:432-436
54. Grainger DJ et al (1996) Nature Med 2:381-385
55. Grainger DJ et al (1995) Nature Med 1:74-79
56. Roberts AB, Sporn MB (1993) Growth Factors 8:1-9
57. Markowitz S et al (1995) Science 268:1336-1338
58. Erlebacher A, Derynck R (1996) J Cell Biol 132:195-210
59. Shah M et al (1995) J Cell Sci 108:985-1002
60. Rowe PM (1994) Lancet 344:72-73
61. Logan A et al (1994) Eur J Neurosci 6:355-363
62. Beck J et al (1991) Acta Neurol Scand 84:452-455
63. Vodovotz Y, Bogdan C (1994) Progr Growth Factor Res 5:341-351
64. Fakhrai H et al (1996) Proc Natl Acad Sci USA 93:2909-2914
65. Raynal S et al (1997) Int J Cancer 72:356-361

Transglutaminase

Definition *Transglutaminase stabilizes the fibrin meshwork by introduction of covalent bonds between adjacent fibrin molecules. It is also involved in TGF-β activation.*

See: →Coagulation factors; →Fibrin/fibrinogen; →Transforming Growth Factor-β (TGF-β)

TSP

Definition *Thrombospondin*

See: →Thrombospondins

Tumor Angiogenesis

Definition *Blood vessel development in tumors*

See: →Angiogenesis; →Angiogenesis inhibitors; →Vascular endothelial growth factor family; →FGF-1 and -2; →Fibrinolytic, hemostatic and matrix metalloproteinases, role of

Tumor Necrosis Factor

Definition *Cytokines with multiple functions in vascular cells. Two forms are described, TNF-α and TNF-β. TNF-α is produced in macrophages, natural killer cells and eosinophils. TNF-β is synthetized in T cells. Both forms bind to identical receptors.*

See: →Cytokines in vascular biology and disease

Tx

Definition *Thromboxane*

See: →Thromboxanes

Tyrosine Kinase Receptors for Factors of the VEGF Family

Synonym: The human VEGFR family comprises three distinct but related members : FLT1, KDR and FLT4. They tend more and more to be called VEGFR1, VEGFR2 and VEGFR3, respectively.

Genes orthologous to human genes have been identified in rat, mouse and quail. TKrC, Flk1 and Quek1 are rat, mouse and quail VEGFR2, respectively. Quek2 is the quail FLT4 gene [1-2].

Introduction Vasculogenesis is the birth of blood vessels from the differentiation and proliferation of endothelial cell precursors (angioblasts) whereas angiogenesis defines the outgrowth of new blood vessels from a preexisting vascular tree. Both events describe critical multistep phenomena involved in physiological processes such as embryogenesis, wound healing, corpus luteum formation or pathological processes such as tumorigenesis and inflammation. A number of adhesion molecules,

proteolytic enzymes, and soluble factors including cytokines (IL-8, TNFα), peptides (angiogenin) and growth factors, induce endothelial cell growth and vessel formation [3]. Some angiogenic growth factors are ligands for endothelial cell surface receptors endowed with a tyrosine kinase activity (RTK). They are principally represented by the different PDGF, FGF and VEGF molecules, and by angiopoietins [4-5]. In contrast to FGF and PDGF, which represent pleiotropic growth factors involved in a wide range of biological events, VEGF molecules target almost exclusively endothelial cells and participate mainly in angiogenesis and vasculogenesis. Over the past ten years, a considerable amount of experiments have demonstrated the crucial role of VEGF and their receptors (VEGFR) during development and diverse physiological and pathological processes.

Originally described as a soluble factor promoting vascular permeability, VEGF – or rather VEGFa (sometimes also called VPF for Vascular Permeability Factor) since it is the founding member of the family – is the major endothelial growth factor [6-7]. VEGFa is a 20 kD glycoprotein, found as a 40 kD disulfide-linked dimer, which shares 18 % identity with PDGFa, including conserved cysteine residues required for proper folding and binding to receptors. The human *VEGFa* gene is located on chromosome 6 (band p12), and composed of eight exons coding for various protein isoforms of 121, 145, 165, 189 and 206 amino acids derived from alternative splicing events [8-10]. The shortest isoforms are secreted in the extracellular compartment, whereas the two longest remain associated with the cell membrane through an arginine/lysine-rich C-terminal region [11]. Like many other growth factors, VEGFa has a tendency to associate with proteoglycans from the extracellular matrix [11-12]. Loss of a single *VEGFa* allele leads to impaired angiogenesis and blood-island formation, and is lethal by day 11 post conceptus (pc) in the mouse, indicating that normal vascular development is concentration-dependent [13].

Three novel VEGF, called VEGFb, VEGFc and VEGFd (also known as FiGF), have been characterized recently. They are about 35 % identical with VEGFa. *VEGFb* and *VEGFc* genes map on human chromosomal regions 11q13 and 4q34, respectively. In addition to an overall structural similarity with the other VEGF, VEGFc and VEGFd have a cystein-rich carboxy-terminal tail of unknown functional significance [14-18]. PlGF (Placenta Growth Factor) is the fifth ligand for VEGFR. The human *PlGF* gene is localized on chromosome 14 [10] and encodes three glycoproteins of 131 (PlGF-1),152 (PlGF-2) and 203 (PlGF-3) amino acids, respectively, that share more than 50 % identity with VEGFa. Only PlGF-2 is able to bind heparin [19-21]. In addition to the ability of VEGF and PlGF to homodimerize, functional VEGFa/PlGF heterodimers have also been found in cell culture supernatants and in the serum [22-23].

Characteristics
Molecular Weight Several apparent molecular masses have been obtained depending on the recipient cells

used in the different studies. For example, the FLT1 cDNA for the long isoform expressed in COS, NIH3T3 or PAE cells, encodes proteins of 160, 180 or 205 kDa respectively. This probably reflects different levels of receptor glycosylation between the various cell types [24-26]. Likewise, FLK1 migrates as a 235 kDa polypeptide in CMT3 cell extracts or as a 205 kDa protein in PAE cell extracts [26-27]. The long and short FLT4 isoforms are 170 and 175 kDa proteins in COS cells but 190 and 195 kDa in NIH3T3 cells, respectively [28]. The tunicamycin-treated FLT4 protein migrates with an apparent molecular mass of 140/150 kDa, consistent with the 143 kDa predicted molecular mass [29].

Domains   The three characterized VEGF receptors belong to the large family of tyrosine kinase receptors (RTK). These type I integral proteins have an extracellular region that bind ligands, linked to a short peptide spanning the cellular membrane (transmembrane domain) and a cytoplasmic region endowed with a tyrosine kinase activity. RTK are classified according to their structural characteristics. The VEGFR family represents the class V. In this particular class, the extracellular region is composed of seven Immunoglobulin-like (Ig-like) domains [30-32] and the kinase domain is split by a long hydrophilic sequence named kinase insert (ki).

Binding Sites and Affinity   The specificities of interaction between VEGF ligands and their receptors are similar in complexity to those found in the FGF and PDGF families (Figure 1). The FLT1 and KDR/FLK1 receptors are high affinity receptors for VEGFa (K_d 1-20 and 75-770 pM, respectively) [22, 24, 27, 33-34]. Among the five VEGFa isoforms, VEGF165 has the highest affinity for binding to VEGFR. VEGF165 but not VEGF121, also interacts with a cell-surface protein that constitutes a high affinity coreceptor [35]. It has been identified as neuropilin-1 [36]. PlGF binds FLT1 (Kd 200 pM) apparently at the same site as VEGFa, but has no affinity for KDR/FLK1 or FLT4. VEGFc and VEGFd interact with FLT4/VEGFR3 and to a lesser extent, with KDR/FLK1 [14, 17]. VEGFb binds to FLT1/VEGFR1 [37].
The second Ig-like domain of FLT1 and KDR/FLK1 is almost exclusively responsible for the specificity of the interaction with VEGFa, but requires adjacent Ig domains to be fully functional. Transposition of this domain to the FLT4 receptor confers on this receptor the ability to bind VEGFa [38]. VEGFa specificity was resolved by mutagenesis analysis: a positively charged stretch of amino acids present in the growth factor is crucial for KDR/FLK1 binding, whereas the interaction with FLT1 is driven by a negatively charged region [39]. Even though the VEGF/VEGFR system shares numerous similarities with the PDGF/PDGFR system, heterodimerization between FLT1 and KDR/FLK1 has not yet been described in cells. VEGFa/PlGF have been characterized in cell culture supernatants.

Additional Features   Contrasting with the two other VEGFR, FLT4 is processed in two polypeptides of 70 and 120 kDa [29]. The functional relevance of FLT4 processing remains to be established.

Structure

Sequence and Size   In humans, the FLT1 transmembrane isoforms (1272 aa and 1338 aa) are encoded by a transcript of 7.7kb while the FLT4 short and long isoforms (1298 aa and 1363 aa) are encoded by two transcripts of 4.5 and 5.8kb, respectively.
The human *KDR* gene gives a transcript of 7kb corresponding to a protein of 1356 aa.

Homologies   VEGF receptors are a family of closely related RTK. The overall amino acid sequence identity is as follows: 38% for FLT4 versus (vs) FLT1, 40% for FLT4 vs KDR/FLK1 and 41% for FLT1 vs KDR/FLK1, respectively. VEGFR have a higher amino acid similarity in their intracellular region (around 50%), especially in the conserved tyrosine kinase domain, than in the extracellular ligand-binding region (around 30%) [40].

Additional Features   The complex organization of RTK is increased by the presence of isoforms for some members, generated mostly by alternative exon splicing of genes. Exon splicing can affect both the extra- and intracellular regions of receptors, and lead to the synthesis of isoforms devoid of some regions, as previously described for example, for FGF receptors. In the case of VEGFR, FLT1 may be produced as a soluble secreted receptor lacking its transmembrane and cytoplasmic regions but retaining its ability to bind VEGFa. This naturally occurring soluble receptor could sequester a VEGF or act as a dominant negative kinase-dead receptor, after ligand dimerization between the soluble and the membrane spanning receptor [41-

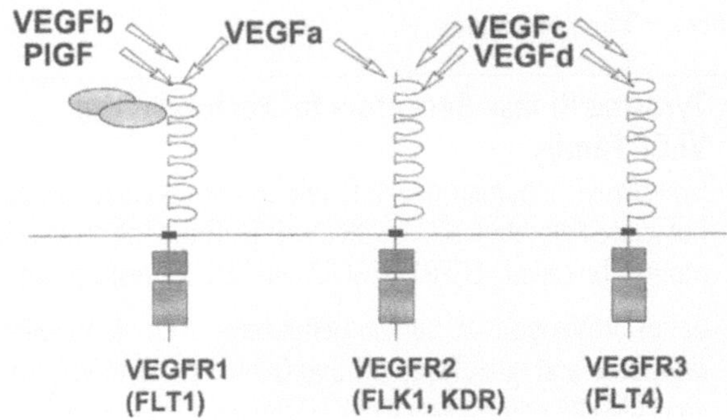

Figure 1. Schematic representation of ligand/receptor interactions in the VEGF system. The five ligands known to date (VEGFa, VEGFb, VEGFc, VEGFd and PlGF) bind the three VEGF receptors with distinct or overlapping specificities. VEGFa and VEGFb on the one hand, and VEGFc and VEGFd on the other hand, are the most closely related proteins. VEGFa, VEGFc and PlGF binding affinities have been determined precisely. This system, together with the TIE receptors and angiopoietins, is involved in the regulation of vasculogenesis and angiogenesis.

42]. A soluble FLT1 protein can efficiently abrogate VEGF-dependent mitogenesis in HUVE cells [41]. Both human FLT1 and FLT4 receptors are found as two isoforms that differ in their cytoplasmic tail by the presence or absence of 65 amino acids [24, 28, 43]. These isoforms may regulate the receptor activity by varying the binding sites for cellular substrates [43-44]. Therefore, isoforms represent a system by which an organism can efficiently modulate the range of interactions of a given receptor, hence its biological activities.

Gene

Gene Structure   *VEGFR* genes remain poorly characterized.

Chromosomal Localization   The three *VEGFR* genes are located on three different chromosomes, close to the location of *PDGF* receptors and hematopoietic RTK (*FLT3, KIT* and *FMS*). *KDR/FLK1* is clustered with *PDGFRA* and *KIT* on human chromosome 4 [45] and mouse chromosome 5 [46], *FLT1* lies in the vicinity of *FLT3* in chromosomal region 13q12 [47]. The human *FLT4* gene is slightly more telomeric than the *PDGFRB/FMS* tandem on the long arm of chromosome 5 [48]. In the mouse, *FlT4* maps on chromosome 11 (region 11a5-11b1) while the *Pdgfrb/Fms* tandem is located on chromosome 18 [49]. Given the worldwide efforts to clone genes encoding tyrosine kinases and the enormous amounts of randomly sequenced cDNA (Expressed Sequence Tags), the existence of a fourth *VEGFR* gene is unlikely.

Gene Expression   The *VEGFR* genes are predominantly expressed in cells of the endothelial compartment and their embryonic progenitors. A study on 17 to 20 weeks old human embryonic tissues has shown spatially distinct expressing patterns for the three genes. For example, *FLT1* and *KDR/FLK1* but not *FLT4* are found in endocardium, whereas strong *FLT4* and *KDR* signals are present in most vessels of the mesenterium [43, 50]. Later, *FLT4* is most predominantly expressed in lymphatic endothelium [51]. This is corroborated by the pattern of expression of its ligand, *VEGFc*, and suggests a paracrine mode of action [52]. In the mouse, *Flk1* is ubiquitously expressed in the endothelial compartment; its mRNA is found as early as E7 in the primitive blood islands [53]. The early expression of *Flk1* leads to the idea that it may constitute a cell-surface marker of angioblasts, the precursors of endothelial cells and possibly, a marker of hemangioblasts, the hematopoietic and endothelial cells precursors [2, 53]. In addition to its endothelial expression, FLT1 is also detected in monocytes while KDR/FLK1 is not [54].

Gene Regulation   One of the striking features of VEGFR is represented by their restricted pattern of expression in the endothelium. The molecular mechanism underlying this expression has been resolved by several groups showing an endothelial specific activity of FLT1 and KDR/FLK1 gene promoters [55-56]. Although the promoter regions of the *VEGFR* genes contain several potential transcription factor binding sites, ETS motifs represent the most interesting sites since ETS proteins have a role in blood vessel formation and angiogenesis associated with tumor growth [57]. Promoters of genes whose expression is restricted to endothelial cells contain ETS binding sites and ETS is implicated in matrix-degrading protease production. This accounts for a possible coordinate expression by ETS factors of proteins involved in angiogenic processes. In the case of *FLT1*, the first intron contains a sequence very similar to the transcription arrest site of the adenosine deaminase gene, and this sequence could play a role in the negative control of *FLT1* transcription. These endothelial-restricted gene promoters will represent useful tools to target transgenes expression in specific vascular-associated compartments.

Biological Activity   Receptor dimerization by an appropriate ligand is responsible for RTK activation. This event turns on the catalytic activity, leading to phosphorylation of tyrosine residues present on the receptor itself (auto/trans-phosphorylation) and on its cytoplasmic substrates. A complex network of interactions takes place after the enzymatic event, mostly regulated by phosphorylation/dephosphorylation of cytoplasmic proteins and receptors. The integrity of the kinase domain conditions the biological effects of the receptor and a kinase-dead receptor can no longer transduce a normal effective downstream signal. Although FLT1 and KDR/FLK1 share the same ligand, they promote different cellular responses, as shown by the mitogenic and chemotactic activities of KDR/FLK1, but not of FLT1, in PAE endothelial cells [26]. However, VEGFa and PlGF induce tissue factor production, chemotaxis and increase in intracellular calcium level in FLT1 expressing monocytes [55]. Stimulation with PlGF of HUVE cells, which possess both FLT1 and KDR receptors, elicit a dose dependent tissue factor production. One possible explanation for this phenomenon is that PAE cells lack an important intracellular pathway required for FLT1 signalling. The results of the *Flt1* and *Flk1* gene inactivation also demonstrate the lack of functional redundancy among VEGF receptors (see physiological function) and their separate functions during vascular development. FLT4 is endowed with a mitogenic and chemotactic activity when expressed in endothelial or fibroblastic cells [14, 44, 29].

The endothelial-specific functions of FLT1 and KDR/FLK1 are not well defined at the cellular level. The signalling cascade of VEGFR has begun to be investigated by several groups in both endothelial and non-endothelial cellular contexts. Until now, the described cellular targets of VEGFR are common to many tyrosine kinases, and include rasGAP, SRC family PTK (Fyn, Yes), NCK and PLCgamma1 molecules [26, 30, 58-60]. Among these VEGFR substrates, rasGAP seems to play an important role since embryos lacking rasGAP are unable to organize endothelial cells into a highly vascularized network [60]. This result is consistent with *in vitro* studies indicating that RAS is required for the motility of endothelial cells and oncogenic RAS mutants induce a random walk trajectory. KDR/FLK1 and FLT4 are able to activate the RAS pathway, in accordance with their proliferative effect. In the case of FLT4, such activity depends tightly on the SHC adaptor protein [44, 29]. FLT1 provokes only weak phosphorylation of the SHC protein [30]. Its interaction with PI3 kinase is a matter of discussion [26, 59]. In summary, it is likely that the three VEGFR interact both with common and distinct substrates and induce different responses in endothelial cells.

Role in Vascular Biology

Physiological Function   The central role of VEGFR in the earliest steps of blood vessel formation and development was clarified by studying homozygous deficient mice for this receptors.

Flk1 -/- embryos die *in utero* at day 9.5 pc. They are characterized by a profound defect in vasculogenesis and blood island formation [61]. This confirms that KDR/FLK1 is necessary for the very early steps of maturation of endothelial cells and is in agreement with its early expression in the embryo. The absence of hematopoietic precursors in the *Flk1 -/-* mice could either be due to FLK1 deficiency or be secondary to the endothelium defect.

Flt1 inactivation is also responsible for embryonic lethality at day 9.5 pc; the blood islands are present but disorganized, with angioblasts and hematopoietic cells intermixed [62].

The *VEGFa* gene knock out shows defects at the heterozygote stage in endothelial cells differentiation and blood vessel formation, but there is a delay in lethality (occurring at day 10.5 pc) compared to VEGFR deficiency, suggesting a possible rescue by other ligands [13].

Pathology   When secreted in a regulated fashion from tissues adjacent to blood vessels, VEGFa is a potent mitogenic and chemotactic factor for endothelial cells and promotes angiogenesis *in vivo*. It also induces the release of proteolytic factors that facilitate infiltration of tissues by newly formed blood vessels. During pathological events, VEGFa is produced by tumor cells and acts in a paracrine manner on the vascular endothelium [63]. Hypoxia, a feature of tumoral environment, is a major stimulus that upregulates *VEGFa* expression [64]. It has also been suggested that tumor-derived VEGF can act on circulating dendritic cells and compromise the mounting of antitumoral immunity by affecting their antigen presentation potential. In adults, *VEGFR* transcripts are found at low levels in endothelium but become abundant in the ingrowing vessels of neovascularized tumor tissues [65]. This is most likely mediated by cytokines, growth factors and inflammatory second messengers present in the neighborhood of tumors. Stimulation of endothelial VEGFR promotes tumor neovascularization which improves the transport of nutrients and oxygen to cancer cells. Furthermore, the newly formed network of blood vessels allows the dissemination of metastasic embols from the tumor throughout the body, and thus worsens the vital prognosis [66]. VEGF/VEGFR are also involved in inflammatory diseases that have an angiogenic phenotype, such as psoriasis and rheumatoid arthritis [66]. In such cases, monocytes chemotaxis promoted by FLT1 may participate to inflammation [54, 67]. In the same manner, during wound healing, VEGF/VEGFR participate to neoangiogenesis after upregulation by a network of stimuli and factors.

The central role of VEGF and their receptors in tumorigenesis has been shown using several approaches. For example, injection of an anti-VEGF antibody is able to reduce the growth of tumors grafted in *nude* mice, and *VEGFa* null embryonic stem cells have a reduced ability to form tumors [13]. Inhibition of glioblastoma growth can be obtained in animals by ectopic expression of a tyrosine kinase-deleted FLK1/KDR able to dimerize with the endogenous full length receptors. A reduction of growth, vascularization and oedema is observed in these tumors [68]/Lately, Albini et al. rationalized the angiogenic activity of the HIV-1 Tat protein by showing that it binds and activates FLK1/KDR on endothelial cells. As the Tat protein induces growth of the highly angiogenic Kaposi's sarcoma in HIV patients, it is speculated that the virus mediates this specific effect through VEGFR activation [69]. In spite of a lack of clues concerning the involvement of FLT4 and its ligand in pathological situations, their restricted expression in the lymphatic endothelium allows to speculate on a potential role in lymphatic permeability and exchanges between endothelium and blood [51-52] and possibly in metastasis.

These effects of the VEGF receptors are mediated through upregulation. Although they are endowed with a potential transforming activity, there have been no reports of activating structural alteration of VEGFR in cancers.

Clinical Relevance and Therapeutic Implications   In addition to the classical anticancer drugs, anti-angiogenic therapy is now viewed as an interesting adjuvant to decrease dissemination and growth of tumors [66]. Several molecules are studied for their angiogenic inhibiting activity in clinical trials. VEGF and VEGFR are ideal targets for such anti-tumor therapies. A better knowledge of signal transduction of VEGFR should allow the development of tyrosine kinase inhibitors targeting specifically their enzymatic activity or the cytoplasmic factors of their transducing cascade [68]. On the other hand, an appreciation of ligand-receptor specificities between VEGFR and their ligands gives the possibility to selectively inhibit individual receptors. For example, a VEGFa/toxin fusion protein can specifically kill FLK1/KDR expressing cells and suppress angiogenesis *in vivo* [71]. The same considerations apply to ligands (angiopoietins) and tyrosine kinase receptors of the TIE family (TIE1, TIE2/TEK). They function in complement to VEGF/VEGFR activities to ensure the proper development of blood vessels. Their specific role is being established [5, 72].

In pathology, angiogenesis must not be considered as a deleterious event in all circumstances. For example, revascularization of ischemic limbs considerably improves the prognosis and recovery of patients. Peripheral artery diseases also represent a good indication for therapeutic angiogenesis [73]. In such cases, VEGF implants could help considerably the reendothelisation of necrotic tissues.

Emmanuel Fournier, Daniel Birnbaum
and Jean-Paul Borg

References

1. Sarzani R et al (1992) Biochem Biophys Res Comm 186:706-714
2. Eichmann A et al (1993) Mech Dev 42:33-48
3. Folkman J et al (1987) Science 235:442-447
4. Mustonen T (1995) J Cell Biol 4:895-898
5. Davis S et al (1996) Cell 87:1161-1169
6. Senger DR et al (1983) Science 219:983-985
7. Gospodarowicz D et al (1989) Proc Natl Acad Sci USA 86:7311-7315
8. Keck PJ et al (1989) Science 246:1309-1312
9. Tischer E et al (1991) J Biol Chem 266:11947-11954
10. Mattei MG et al (1996) Genomics 32:168-169
11. Thomas KA et al (1996) J Biol Chem 271:603-606
12. Gitay-Goren H et al (1992) J Biol Chem 267:6093-6098
13. Ferrara N et al (1996) Nature 380:439-442
14. Joukov V et al (1996) EMBO J 15:290-298
15. Olofsson B et al (1996) Proc Natl Acad Sci USA 93:2576-2581
16. Paavonen K et al (1996) Circulation 93:1079-1082
17. Achen et al (1998) Proc Natl Acad Sci USA 95:548-553
18. Orlandini et al (1996) Proc Natl Acad Sci USA 93: 11675-11680
19. Maglione D et al (1991) Proc Natl Acad Sci USA 88:9267-9271
20. Hauser S et al (1993) Growth Factors 9:259-268
21. Cao et al. (1997) Biochem biophys Res Comm 235:493-498
22. Park JE et al (1994) J Biol Chem 269:25646-25654
23. DiSalvo J et al (1995) J Biol Chem 270:7717-7723
24. De Vries C et al (1992) Science 255:989-991
25. Seetharam L et al (1994) Oncogene 10:135-147
26. Waltenberger J et al (1994) J Biol Chem 269:26988-26995
27. Terman BI et al (1992) Biochem Biophys Res Comm 187:1579-1586
28. Pajusola K et al (1993) Oncogene 8:2931-2937
29. Pajusola K et al (1994) Oncogene 9:3545-3555
30. Shibuya M et al (1990) Oncogene 5:519-524
31. Terman BI et al (1991) Oncogene 6:1677-1683
32. Galland F et al (1993) Oncogene 8:1233-1240
33. Millauer B et al (1993) Cell 72:835-846
34. Quinn TP et al (1993) Proc Natl Acad Sci USA 90:7533-7537
35. Soker S et al (1996) J Biol Chem 271:5761-5767
36. Soker et al (1998) Cell 92: 735-745
37. Olofsson et al (1998) Proc Natl Acad Sci USA 95: 11709-11714
38. Davis-Smyth T et al (1996) EMBO J 15:4919-4927
39. Keyt BA et al (1996) J Biol Chem 271:5638-5646
40. Borg J-P et al (1996) Mammalian Development. Harwood Academic Publishers, pp 233-248
41. Kendall RL et al (1993) Proc Natl Acad Sci USA 90:10705-10709
42. Kendall RL et al (1996) Biochem Biophys Res Comm 226:324-328
43. Borg JP et al (1995) Oncogene 10:973-984
44. Fournier E et al (1996) J Biol Chem 271:12956-12963
45. Spritz R et al (1994) Genomics 22:431-436
46. Nagle D et al (1994) Proc Natl Acad Sci USA 91:7237-7241
47. Imbert A et al (1994) Cytogenet Cell Genet 67:175-177
48. Pebusque MJ et al (1993) Genes Chromosom Cancer 8:119-126
49. Galland F et al (1992) Genomics 13:475-478
50. Kaipainen A et al (1993) J Exp Med 178:2077-2088
51. Kaipainen A et al (1994) Proc Natl Acad Sci USA 92:3566-3570
52. Kukk E et al (1996) Development 122:3829-3837
53. Yamaguchi TP et al (1993) Development 118:489-498
54. Clauss M et al (1996) J Biol Chem 271:17629-17634
55. Morishita K et al (1995) J Biol Chem 270:27948-27953
56. Patterson C et al (1995) J Biol Chem 270:23111-23118
57. Wernert N et al (1992) Am J Pathol 140:119-127
58. Guo D et al (1995) J Biol Chem 270:6729-6733
59. Cunningham SA et al (1995) J Biol Chem 270:20254-20257
60. Henkemeyer M et al (1995) Nature 377:695-701
61. Shalaby F et al (1995) Nature 376:62-66
62. Senger D et al (1993) Metastasis Rev 12:303-324
63. Fong G-H et al (1995) Nature 376:66-70
64. Shweiki D et al (1992) Nature 359:843-845
65. Plate KH et al (1993) Cancer Res 53:5822-5827
66. Folkman J (1995) Nature Med 1:27-31
67. Barleon B et al (1996) Blood 87:3336-3343
68. Millauer B et al (1994) Nature 367:576-579
69. Albini (1996) Nature Med 2:1371-1375
70. Levitski A et al (1995) Science 267:1782-1788
71. Ramakrishnan S et al (1996) Cancer Res 56:1324-1330
72. Folkman J et al (1996) Cell 87:1153-1155
73. Isner JM et al (1996) Human Gene Therapy 7:959-988

u-PAR

Definition   *Urokinase-type plasminogen activator receptor*

See: →Plasminogen/plasmin; →Plasminogen activators

uPA

Definition   *Urokinase-type plasminogen activator*

See: →Plasminogen/plasmin; →Plasminogen activators

USF

Definition   *Upstream stimulatory factor*

See: →Fibrin/fibrinogen

Vascular Cell Adhesion Molecule-1 (VCAM-1)

Definition   *Adhesion molecule of the immunoglobulin superfamily expressed on endothelial. Cell surface expression on endothelial cells is only detectable after cytokine induction.*

See: →Blood cells, interaction with vascular cells

Vascular Cells

Definition   *Cells that compose the vessel wall*

See: →Endothelial cells; →Smooth muscle cells

Vascular Endothelial Growth Factor Family (VEGF)

Synonym: Vascular permeability factor, vasculotropin

Definition   *Angiogenic factors that have homologies to Platelet-derived Growth Factors (PDGF). There are four major VEGF prototypes (VEGF-A, VEGF-B, VEGF-C, VEGF-D). Placental growth factors (PlGF) share more than 50 % homology with VEGF-A and also binds VEGF receptors.*

See also: →Vascular endothelial growth factor receptors; →Signal transduction mechanisms in vascular cells; →Angiogenesis; →Ontogeny of the vascular system; →Hormonal regulation of vascular cell function

Introduction   In adult mammals the vasculature is quiescent except during the physiological cycles of reproduction. Endothelial cells are among those exhibiting the lowest replication in the body with only 0.01 % cells engaged in cell division at any time. Additional requirements in oxygen or nutrients, as in the course of tumor progression, diabetic retinopathy or rheumatoid arthritis, result in the sprouting of new capillaries from pre-existing vessels, or angiogenesis. This local hypervascularization is thought to result from the release by the tissues involved of soluble mediators which induce the switch of the quiescent endothelial cell phenotype to the activated one, so that the endothelial cells are then able to respond to mitogenic signals. The release of mitogenic growth factors switches the activated phenotype to an angiogenic phenotype. The interactions of angiogenic growth factors with their receptors provide the signal for cell migration, proliferation and differentiation to form new capillaries.

It was postulated half a century ago that hypoxia preceded the retinal neovascularizations observed in diabetic retinopathy or retinopathy of the premature and led to the release of soluble mediators of retinal angiogenesis (Factor X) [1,2]. The hypothesis that tumoral progression was dependent on angiogenesis also led to the concept of "tumor angiogenic growth factors", TAF [3]. The search for "Factor X" and "TAF" proved misleading until purification and cloning of the genes encoding angiogenic growth factors was achieved. During the past decade it has become clearer that the angiogenic growth factors which are upregulated during pathological angiogenesis are similar to those promoting physiological angiogenesis or vasculogenesis during embryogenesis. Several candidates have been identified such as acidic fibroblast growth factor, basic fibroblast growth factor, transforming growth factor a, hepatocyte growth factor and interleukin 8, all of which are able to

induce neovascularization in avascular organs such as cornea and to mimic *in vitro* all the steps of the angiogenic cascade (protease induction, cleavage of the extracellular matrix, migration, proliferation and differentiation into capillary-like tubes). It is now evident that the biological factor purified in 1989, and designated vascular endothelial growth factor [4], vascular permeability factor [5] or vasculotropin [6], represents a key regulator of angiogenesis.

Characteristics   Vascular endothelial growth factor (VEGF) has been purified from the conditioned medium of a variety of cell types including bovine folliculostellate cells [4], tumoral cell line AtT20 derived from mouse anterior pituitary [6], guinea-pig tumor [5] Its bioactivity has been purified 10,000 fold by cation exchange, heparin affinity and reverse-phase high performance liquid chromatography.

Molecular Weight   VEGF is a glycosylated homodimer of 45 kDa. This growth factor, also known as vasculotropin, was primarily described as a specific mitogen for vascular endothelial cells derived from large or small vessels, regardless of their species origin. It is also angiogenic *in vivo* in the chick chorioallantoïc membrane assay and the corneal pocket assay [6]. VEGF is identical to the previously described vascular permeability factor [7].

Binding Sites and Affinity   Two classes of high-affinity binding sites were initially described on microvascular derived endothelial cells, with dissociation constants (K_d) values of 2-5 pM and 50-100 pM respectively [8]. Cross-linking experiments showed that the molecular mass of these cellular binding sites corresponds to 180-200 kDa [8]. These two classes of binding sites have also been found in cells which are not of endothelial origin, namely retinal pigment endothelial cells [9], stromal cells cultured from neonatal hemangiomas [10] or hair dermal follicle cells [11]. The content or the structure of membrane heparansulfate proteoglycans likely modulate the affinity for VEGF. Although heparin increases the binding of iodinated VEGF to endothelial cells [12], it does not modulate the biological activity of VEGF [13]. However other cells, such as lens epithelial or corneal endothelial cells, bind VEGF on a single high-affinity binding site with a K_d in the 10 pM range and a molecular mass of 120 kDa. These cells migrate or differentiate, but do not proliferate when VEGF is added. Recent reports have stated that other binding sites specific for exon 7-containing VEGF isoforms exist such as the 120 kDa molecular species. This receptor has been recently identified as identical to neuropilin-1, the neuronal cell guidance receptor of the semaphorin. Neuropilin-1 modulates the interaction of VEGF and VEGFR2, and therefore the mitogenic activity of VEGF and it might also been involved in endothelial cell guidance [14].

Ligand autoradiography studies of fetal or adult tissues depicted a specific binding of VEGF on endothelial cells regardless of their proliferation status [15]. The distribution of VEGF and its binding sites during cyclical growth of corpus luteum suggests that they are hormonally regulated.

Receptors and Signal Transduction   Two VEGF tyrosine kinase receptors have been identified. The fms-like-tyrosine kinase [16], Flt-1 also called VEGFR1, and the kinase domain region, KDR also called VEGFR2, or its murine homologue fetal liver kinase-1, Flk-1 [17]. Both of these tyrosine kinase receptors bind VEGF with high affinity. Both receptors have seven immunoglobulin-like domains in the extracellular domain, a single transmembrane region and a consensus tyrosine kinase sequence that is interrupted by a kinase insert domain (Figure 1). The two promoters of VEGFR1 and VEGFR2 contain a 5' flanking sequence essential for endothelial specific expression [18].

VEGF binds to VEGFR1 with a K_d of 20 pM [16], whereas the K_d observed for PlGF1 or PlGF2 is significantly higher, in the 100-200 pM range (Figure 1). It is expressed in cells which do not proliferate but migrate in response to VEGF, such as corneal endothelial cells [19] or monocytes. Constitutive expression of VEGFR1 allows cells to migrate in response to VEGF or PlGF [20].

Lower affinity (K_d of 100-700 pM) has been reported for the binding of VEGF to VEGFR2 in transfected cells [21]. Downregulation of VEGFR2 achieved either by selective internalization by anti-idiotypic antibodies or by VEGFR2 antisense oligonucleotide transfection reduced the overall binding of iodinated VEGF to human umbilical vein endothelial cells or stromal cells derived from neonatal hemangiomas by only 10 %, but this reduction appeared to lie on a high-affinity (K_d=1 pM) binding site [10]. VEGF-C, providing it is cleaved by proteases, binds and activates VEGFR2. Recently the HIV-1 trans-activator tat-1 has also been demonstrated as a VEGFR2 ligand inducing its phosphorylation. VEGFR2 constitutive expression allows cells to proliferate upon addition of VEGF [21].

A third member of this family of type III tyrosine kinases is represented by Flt-4, also called VEGFR3 [22]. Its expression is restricted in adults to lymphatic endothelial cells which suggests a role in lymphangiogenesis. It does not bind VEGF or PlGF but binds VEGF-D and VEGF-C.

Although pioneer experiments using endothelial cells expressing each VEGFR constitutively led to the conclusion that VEGFR2 mediates cell proliferation, migration and actin reorganization whereas VEGFR1 does not signal for VEGF [21], it is now emphasized that VEGFR1 expression constitutive mediates cell migration [23,24].

In response to VEGF, both VEGF receptors are tyrosine phosphorylated [21]. Several proteins are phosphorylated in endothelial cells (Figure 2) in response to VEGF [25]. Both receptors have the potential to signal through PLCγ (Figure 2) via phosphotyrosine residues located in the juxtamembrane and carboxyl tail regions [26]. Several adaptations of the Src family, such as Fyn and Yes are phosphorylated in response to VEGF/VEGFR1 but not to VEGF/VEGFR2 interactions. Phosphorylated VEGFR2 associates with Shc, Grb2 and Nck, and also with the phosphatases SHP-1 and SHP-2 [27], activates the phosphorylation of the MAPK cascade via the stimulation of Raf [28] and the tyrosine phosphorylation and recruitment of focal adhesion kinase and paxillin [29].

Domains   The production of specific VEGFR2 agonists by anti-idiotypic strategy led to the hypothesis that critical VEGF determinants for each of its receptors should

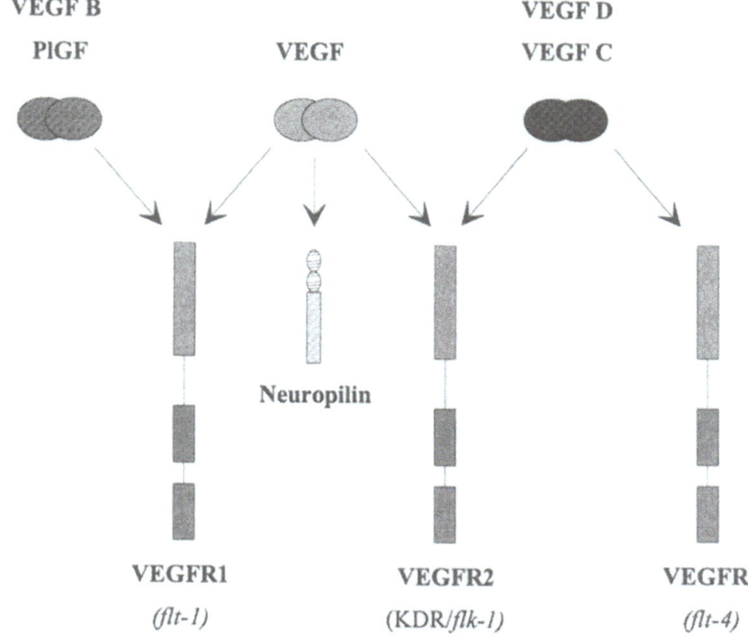

Figure 1. *Interactions of the five VEGF ligands with the four VEGF receptors.* Schematic representation of the distinct binding of the members of the VEGF family to the 3 tyrosine kinase receptors containing 7 immunoglobulin-like loops. Spliced variants which contain basic domains (VEGF 165 aa, VEGF 189 aa and PlGF 152 aa) bind also to Neuropilin-1.

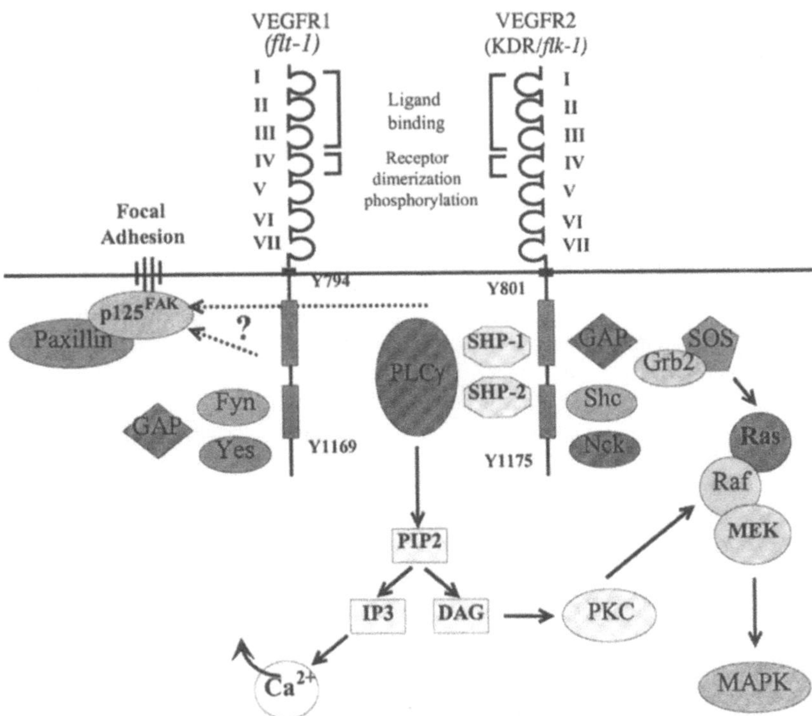

Figure 2. *Signal transduction of the VEGF receptors.* Schematic representation of the interactions of cellular adapters to VEGF tyrosine kinases receptors.

be distinct [30]. Site-directed mutagenesis was used to localize the determinants that mediate VEGF binding to its two receptors [31]. A model based on the crystal structure of PDGF-BB was used to perform alanine-scanning analysis. Although single mutations of Lys^{82}, Arg^{84} and His^{86} located in a hairpin loop displayed a moderate decrease in KDR binding, the triple mutation yielded to a strong inhibition of VEGF binding. Another study revealed that a second region (Ile^{46}, Ile^{43}) is critical for binding to the receptor and to a neutralizing monoclonal anti-VEGF antibody [32]. Mapping of these determinants to the structure of the VEGF dimer shows that these two hot-spots are located at each pole of the molecule. Similarly, negatively-charged Asp^{63}, Glu^{64} and Glu^{67} are responsible for VEGFR1 binding. These mutants are still mitogenic whereas those which do not bind to VEGFR2 are not.

The construction of various mutants of the Ig-like domains of each VEGF receptor has provided useful tools for analysis of their ligand-binding domains. VEGFR1 mutants deleted of the second Ig-like domain fail to bind VEGF or PlGF, whereas the exchange of this domain with the second Ig-like domain of VEGFR2 restores VEGF binding similar to that observed in native VEGFR2, which means that this mutant no longer binds PlGF [33]. However maximal VEGF binding to VEGFR1 only requires the presence of the 1-3 Ig-like domains [33] or only the domains 2 and 3 [34] whereas maximal transphosphorylation also requires the presence of the 4-7 Ig-like domains [35].

The crystal structure of VEGF showed that its dimerization is similar to that of PDGF and symmetrical binding sites for VEGFR2 are located at each pole of the homod-imer [32]. Each site contains two functional hot-spots composed of binding determinants presented across the subunit interface. The crystal structure of VEGF complexed with the second domain of VEGFR1 shows predominant hydrophobic interactions with the poles of the VEGF dimer. Five of the seven VEGF residues critical for tight binding to VEGFR2 are buried in the interface with the domain 2 of VEGFR1 in the complex [34].

VEGF Homologues   Over the past few years, four VEGF-related genes have been identified (Figure 3): VEGF-B, VEGF-C or VEGF-related protein (VRP), VEGF-D and placenta growth factor (PlGF). These four VEGF-related genes generate glycosylated dimers which are secreted after cleavage of their signal peptide. They encode the eight cystein residues that are highly conserved within the VEGF and the PDGF family.

PlGF was primarily identified as a vascular endothelial growth factor [36]. This molecule shares a 37% homology with VEGF. The encoded protein is expected to have 131 amino acids after the cleavage of a signal peptide. An additional isoform of 152 amino acids was later discovered. These two isoforms are generated by alternative splicing of the pre-mRNA, the shorter being deleted of a 21 basic sequence encoded by exon 6 and now referred to as PlGF1 and PlGF2 respectively. A recent report describes the presence in placenta of a third isoform of 201 amino acids generated by alternative insertion in a donor site of exon 4. This isoform has been called PlGF3. Further studies of PlGF biological functions gave conflicting results. Although there is a good agreement concerning the lack of mitogenic activity of PlGF1, it seems that PlGF2 might

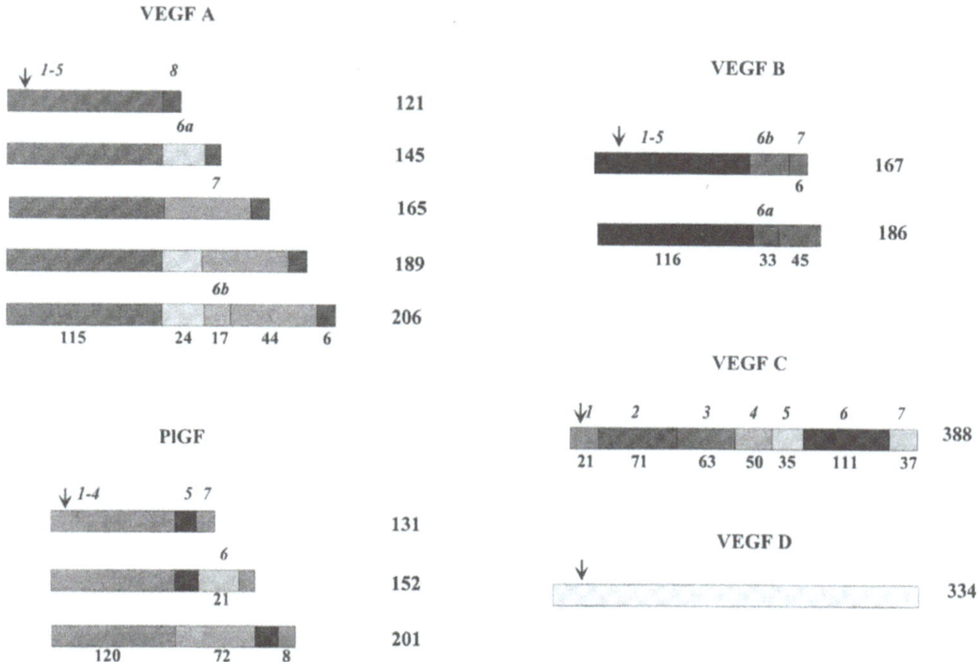

Figure 3. *Schematic representation of the organization of VEGF-related genes.* The optional utilization of exons 6, 7 and 8 of VEGF pre-messenger RNA can generate at least 5 messenger RNAs encoding polypeptides of 206, 189, 165, 145 and 121 amino acids. Arrow indicates the site of signal peptide processing. The number of amino acids encoded by each exon is written beneath the scheme of the VEGF isoform and the total length of the mature isoform is written on the right.

be mitogenic [37]. Surprisingly, Park et al [38] showed that PlGF1 and PlGF2 were able to potentiate the mitogenic effect of low doses of VEGF. However good agreement exists regarding the chemotactic effects of both isoforms on monocytes [23] or Flt-1 transfected cells [20]. Conflicting results have been reported about the effects of PlGF *in vivo*: PlGF1 being angiogenic for some authors [39] but not for others. In our laboratory we found that PlGF1, despite its absence of mitogenic effect on vascular endothelial cells *in vitro*, could induce a strong angiogenic effect in the corneal pocket assay, whereas PlGF2 was mitogenic *in vitro* (through FGF-2 signaling pathways) and angiogenic *in vivo* (Ortéga, in preparation). In contrast to Park et al, we found that both PlGF isoforms induced a dose-dependent effect on permeability assayed by the Miles assay, although their specific activity was 10-fold lower than that of VEGF. Synthetic or natural heterodimers of VEGF and PlGF possess biological activities similar to that of VEGF, although their relative potency is reduced by an order of magnitude.

VEGF-B is a 186 amino acids protein after cleavage of the signal peptide [40] sharing a 23% homology with VEGF. It has been reported as a cell membrane bound protein which can be released by heparin and exert a mitogenic effect on vascular endothelial cells. An alternative splice variant with a different splice acceptor site in exon 6 introduces a frameshift and a partial use of different reading frames encodes a 167 amino acids with an entirely different COOH-terminal sequence. Its colocalization with VEGF in skeletal muscle and myocardium led to the hypo-

thesis that it might modulate angiogenesis in these structures.

VEGF-C [41], also known as VEGF-related protein or VRP, is a 388 amino acids protein after cleavage of the signal peptide with an overall homology of 16% with VEGF, located in the sequence encoded by exons 3 and 4. Exons 5 and 6 encode cystein-rich domains similar to a protein present in silk produced by the larval salivary gland of the midge *C. tentanus*. Despite a potency reduced 100-fold as compared to that of VEGF, VEGF-C might induce a mitogenic effect on endothelial cells which would require proteolytic maturation, similar to that observed for V189.

VEGF-D [42], also identified as a c-fos-induced gene [43], encodes a protein of 334 amino acids after cleavage of the signal peptide. It contains a cystein-rich domain in its C-terminus similar to that observed in VEGF-C. Its pro-eminent expression in fetal and adult lung suggests a specific control in lung angiogenesis.

Intriguingly, two sequences exhibiting significant homology with the VEGF gene [44] have been identified in different strains of orf virus, a parapoxvirus which infects goats and occasionally humans and induces extensive microvascular proliferation in the skin.

Gene

Gene Structure Molecular cloning of the cDNA showed that VEGF shares an overall homology of 18% with the B chain of platelet-derived growth factor [45-47]. The human VEGF gene is organized in eight exons separated by seven introns. Alternative splicing of a single gene transcription product can generate multiple species. The full transcript encodes a 189 amino acids isoform (V189). A longer molecular species, V206, which contains an additional 17 codons following the 24-codon encoded by exon 6, appears to be expressed only in embryonal tissue [48]. The transcript encoding the 165 amino acids form, deleted of exon 6, is expected to generate the 46 kDa peptide following signal peptide cleavage (V165). A fourth transcript, deleted of exon 7, encodes a 145 amino acids isoform [49]. A shorter transcript, deleted of exons 6 and 7 encodes a 121 amino acids isoform which, compared with V165, presents a 44 amino acids deletion between positions 116 and 159. The promoter region contains a single major transcription start site near a cluster of potential Sp1 factor binding sites (Figure 4). The promoter also contains one Hypoxia Inducible Factor and several AP1 or AP2 binding sites. The sequence within the 3' untranslated region reveals several consensus structures known to be involved in the regulation of mRNA stability.

Gene Expression   During embryonic development, VEGF is first expressed in the trophoblast suggesting a role in the vascularization of the placenta, and later widely distributed in the embryo. Lung, kidney and spleen are the human organs expressing the most VEGF mRNA. The protein is mainly detected in epithelial cells and myocytes. VEGFR1 and VEGFR2 [17] are selectively expressed in embryonic endothelial cells, first in the yolk sac and intraembryonic mesoderm, later in angioblasts, endo-

cardium and in small and large vessels. The observation that VEGFR2 expression decreases at the end of gestation, whereas VEGFR1 persists in adult quiescent endothelial cells led to the hypothesis that they might be involved in different functions: VEGFR2 mediating the vasculogenic and angiogenic activities and VEGFR1 mediating vessel survival. This hypothesis was confirmed by gene knockout experiments demonstrating that both receptors were essential for development of the vasculature. Null embryos for VEGFR1 fail to organize normal vessel channels, although endothelial cells proliferate, and die between day 8.5 and 9.5 [50]. Both endothelial and hematopoietic cells fail to develop in VEGFR2 null-mutant embryos which die at the same stage from distinct malformations [51]. Complementation experiments have shown that VEGF is required for endothelial differentiation [52], and the analysis of null VEGFR2 embryo stem cells has shown that endothelial cells fail to proliferate and migrate from the primitive streak to the yolk sac and to intraembryonic sites of early angiogenesis.

VEGF knock-out experiments have confirmed that VEGF is essential for vascular development and that the VEGF related genes, which might activate VEGFR1 (PlGF, VEGF-B) or VEGFR2 (VEGF-C, VEGF-D) play a minor role, if any. In deficient embryos the forebrain region, aorta and the thickness of the ventricular wall appeared underdeveloped. The vitelline veins failed to fuse with the vascular plexus of the yolk sac. Surprisingly the phenotype did not result only from gene invalidation but also from gene dosage since heterozygotus embryos died in utero two

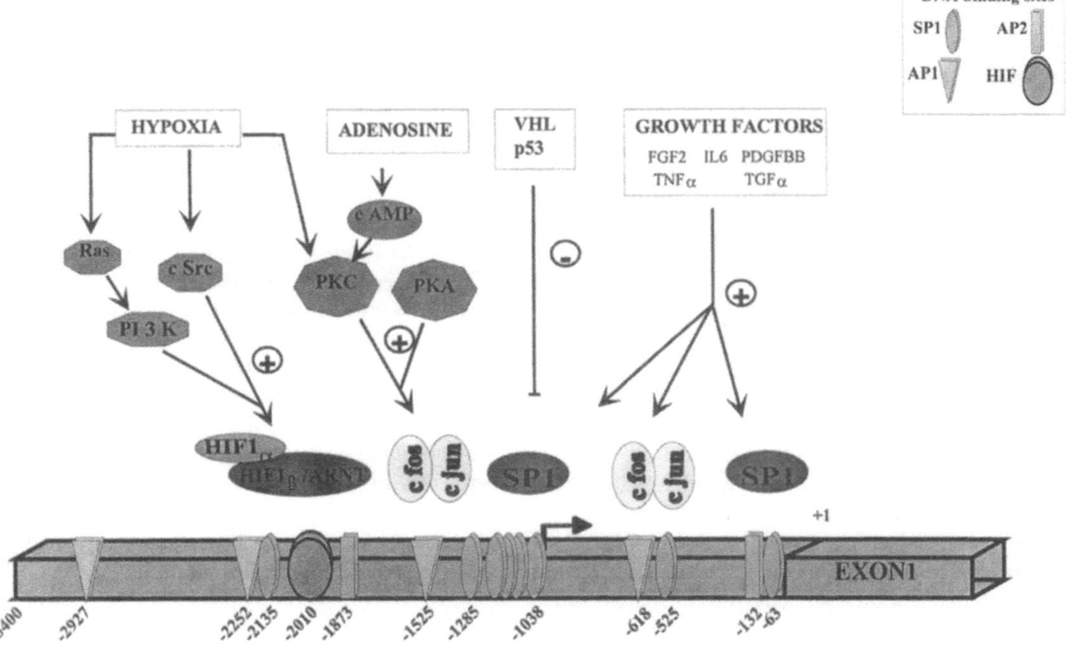

Figure 4. *Schematic representation of the organization of the VEGF promoter.* The number on the bottom line indicates the localization of the base by comparison with the +1 translation start.

days later than the VEGFRs null mutants [53,54]. Interestingly the phenotypes of null mutants for VEGF and for the arylhydrocarbon-receptor nuclear translocator, which is one of the major regulators of VEGF transcription, are very similar [55]. These findings suggest that an increase in mass tissue in a developing organ will induce hypoxia and hypoglycemia in surrounding regions, which would in turn activate this transcription factor, leading to up-regulation of VEGF and subsequent vascularization to supply oxygen and glucose.

Conversely, the inappropriate supply of VEGF during embryogenesis induces vessel and cardiovascular malformations and hyperperfusion leading to embryonic lethality. Accordingly, transgenic mice overexpressing VEGF in photoreceptors under an opsin promoter develop intra-retinal and subretinal angiogenesis [56]. These findings underline the necessity for a tight control of VEGF bio-availability during embryogenesis.

Gene Regulation   It was postulated half a century ago that retinal neovascularization was a consequence of hypoxia, and that hypoxic retina contained angiogenic factors [1]. Similarly several tumors with extensive vascularization, such as multiform glioblastoma, exhibit hypoxic areas in the periphery of necrotic foci. These physiopathological observations led to a search for hypoxia-regulated angiogenic factors. Low oxygen tension dramatically up-regulates VEGF expression in tumoral cell lines cultured on plastic or forming spheroids as well as *in vivo* [57,58]. Erythropoietin [59] and the glucose transporter Glut-1 [60] which also hypoxia-regulated genes with contain similar Hypoxia Inducible Factor (HIF) binding sequences of their promoters. This sequence binds the heterodimeric transcription factor composed of basic helix-loop-helix hypoxia inducible factor 1 and aryl hydrocarbon-receptor nuclear translocator [61]. c-Src [62] and Ras pathways can activate HIF-1/ARNT association and thus VEGF transcription (Figure 4). This mechanism of hypoxia regulation is not unique. The protein kinase A-mediated pathway can also be activated by adenosine accumulation and protein kinase C-mediated pathway activation results in fos-jun activation and AP1 binding. VEGF mRNA is also stabilized via sequences identified in the 3'-untranslated region [60]. Hypoxia does not seem to increase the transcription of VEGF-B or VEGF-C mRNAs but decreases the transcription of angiopoietin-1 and increases that of angiopoietin-2.

It is not clear whether hypoxia regulates VEGF receptors. It seems that VEGFR1 is upregulated whereas the expression of VEGFR2 is not [63]. Hypoxia induces the expression of VEGF in human endothelial cells which in turn activates VEGFR2 phosphorylation and cell proliferation [64]. The upregulation of VEGFR2 observed in endothelial cells surrounding hypoxic tissues seems to require the release of a yet unidentified paracrine factor [65].

Several cytokines or growth factors upregulate VEGF and VEGFRs. Exposure of several normal growth-arrested cultured cells to various growth factors induce an upregulation of VEGF at the mRNA and protein levels. For instance, VEGF expression is increased in keratinocytes stimulated by EGF or TGF-β, or smooth muscle cells by FGF-2 or IL-1α. Although VEGF is up-regulated in tumoral cell lines, its expression might be stimulated by PDGF through the protein kinase C pathway, TGF-α through the AP2 pathway, TGF-β, EGF or IL-6.

The upregulation of VEGF that occurs in virtually any tumoral cell line has led to many studies to decipher the mechanism of VEGF overexpression during the switch of a normal cell to a transformed one. Spontaneously immortalized human retinal pigment epithelial cells secrete more VEGF than their normal counterparts. Phorbol ester, forskolin and luteotrophic hormone, which activate adenylate cyclase, induce the expression of VEGF mRNA, as would be expected from the presence of AP_1 and AP_2 sites in the VEGF promoter. Several oncogenic mutations such as Ras, v-Raf, v-Src [62] or mutations inactivating a tumor suppresser gene such as the product of the von Hippel Lindau gene [66] or p53 upregulate VEGF mRNA by blocking protein kinase C pathways. These oncogenic mutations might synergize the hypoxia-induced upregulation of VEGF transcription. Interestingly the expression of the other members of the VEGF family, namely VEGF-B and VEGF-C, are not modulated by hypoxia and Ras or p53 mutations [67].

Although the VEGF promoter does not contain consensus hormone response elements, its mRNA accumulation can be increased by oestradiol, luteotropic hormone or testosterone [68]. Although both VEGF receptors are expressed in normal adult endothelial cells, their proliferation rate is very low. This has yet to be explained. However, VEGF overexpression might in turn activate the expression of VEGFR1.

Processing and Fate   In contrast to V121, which is secreted and freely soluble in the tissue culture medium, the bioavailability of V165 and more specifically of the V189 form appears to be regulated by binding to the extracellular matrix (Figure 5). This would provide a storage site of growth factor which could be released by heparin, heparan sulfate and heparinase or cleaved by uPA within the sequence encoded by exon 6. The relationships between intact V189 and cell membranes are apparently complex and involve more than electrostatic interactions between the anionic charges of the heparan sulfates and the basic sequences encoded by exons 6 and 7, as suggested by Houck et al [69]. Another type of binding site would be sequence-specific. Surprisingly, native V189 binds to VEGFR1 but not to VEGFR2 unless it has been processed into 38 or 34 kDa homodimers by urokinase or plasmin respectively [24]. Similar findings have been reported for VEGF-C which binds VEGFR3 but requires maturation to activate VEGFR2.

Recombinant VEGF expressed in mammalian, insect or prokaryotic cells binds with similar affinity to heparan sulfate-deficient cells, although prokaryotic cells only compete with 30 % of the binding sites occupied by glycosylated VEGF, and corresponding to heparan sulfate. Although glycosylation does not affect the mitogenic

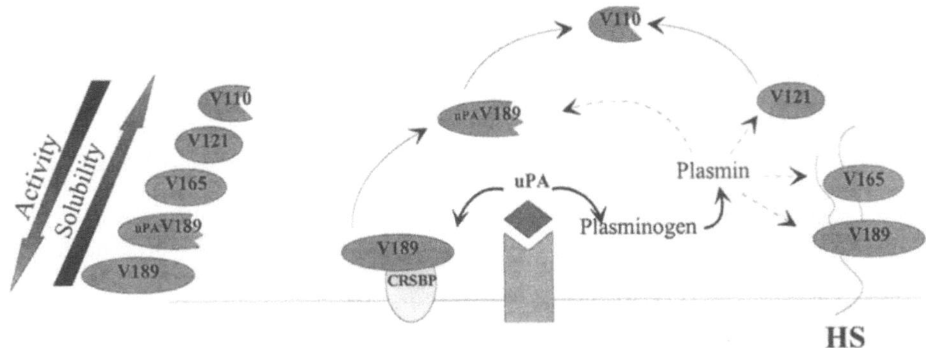

Figure 5. *VEGF storage in the extracellular matrix and protease processing.* The binding of V189 to both proteoheparansulfates and retention sites (HS and CRS BP) induces its sequestration into the extracellular matrix. The disruption of its interaction with either storage site allows urokinase to cleave VEGF 189 into 2 soluble fragments of 140 aa and 49 aa. When urokinase converts plasminogen into plasmin, all the VEGF isoforms are processed in a very soluble 110 aa isoform.

effect of VEGF *in vitro*, it enhances its angiogenic activity in the corneal pocket assay by allowing it to bind to proteoheparan sulfate and thus protecting it from plasmin inactivation. Similarly, VEGF glycation, which might occur in diabetes mellitus, enhances its angiogenic activity in contrast to FGF2 which is inactivated by such treatment.

Although plasmin and urokinase-matured V189 bind to VEGFR2 with the same affinity, the shorter fragment exerts a mitogenic effect that is reduced 50-fold compared to that of urokinase-matured V189, thereby suggesting that exon 6 behaves as a mitogenic enhancer. Surprisingly, V189 can induce thymidine incorporation on corneal endothelial cells which do not express VEG-FR2. In fact V189, but not V165, induces a mitogenic effect only on target cells which express FGF2 and store it in their extracellular matrix, and releases FGF2 *in vitro* and *ex vivo*. The ability of exon 6 containing VEGF isoforms to do this was confirmed by the observation that their angiogenic activity was inhibited by anti-FGF2 neutralizing antibodies whereas that of V165 was not [20]. When endothelial cells are quiescent, V189 does not induce their proliferation but can sustain other functions such as survival [70]. When the cells are stimulated by various stimuli such as hypoxia or growth factors, they respond by increasing VEGF transcription, which in turn may upregulate uPA in the most ischemic areas of tumors as well as in tumor-associated macrophages. Thus tumor upregulated uPA cleaves V189 within the sequence encoded by exon 6 into a 40 kDa-diffusible isoform which is able to activate VEGFR2 and to induce cell proliferation [24]. When plasmin is generated by uPA, the resulting 34 kDa V189 isoform cleaved within the sequence encoded by exon 5 is less active and diffuses more easily [71]. The heterogeneity of the molecular species generated from V189 is in good agreement with the biological actions expected from an angiogenic/survival factor such as VEGF: inactive mitogen but efficient for survival, strong mitogen with a weak ability to circulate and finally a weak but diffusible mitogen.

Biological Activity   The identification of VEGF transcripts in cultured vascular smooth muscle cells and the exquisite response of endothelial cells suggested that VEGF was a typical paracrine factor involved in proliferation and/or survival of the endothelium. Although most endothelial cells do not express VEGF, endothelial cells derived from brain and retina capillaries secrete and respond to exogenously-added VEGF [72]. This functional autocrine loop suggests that in neural tissues, VEGF might contribute to maintenance of the fenestration in the blood brain barrier [73]. VEGF upregulates the expression of plasminogen activator and plasminogen activator type I inhibitor, tissue factor [13], interstitial collagenase [74] and promotes tube formation of endothelial cells cultured in three dimensional collagen gels [13,75] or fibrin gels. Although VEGF and FGF2 do not exert any synergistic effect on the proliferation of cells cultured on plastic, this kind of synergy has been observed on cells cultured on collagen gel [75]. However several lines of evidence suggest that both growth factors signal angiogenesis *in vivo* through distinct mechanisms: (a) specific neutralizing antibodies against VEGF or FGF2 selectively inhibit the angiogenic effect of their own immunogen [20], (b) inhibition of NO production inhibits only VEGF-driven angiogenesis, and not FGF2-driven angiogenesis [39]. VEGF promotes the expression of VCAM-1 and ICAM-1 [76], a phenomenon which might explain the preferential adhesion of natural killer cells to the tumor vasculature.

The pioneer observation by Dvorak's group that several tumoral cell lines increased microvessel permeability, leakage of plasma proteins and formation of a fibrin gel facilitating endothelial cell sprouting led to the discovery of the permeability action of VEGF [7]. This effect is thought to constitute a crucial step in tumoral angiogenesis and wound healing. It is not yet clear which VEGF receptor mediates the permeability effect of VEGF.

Since adult endothelial cells express VEGFR, it is tempting to speculate that VEGF may be involved in endothelial survival. Indeed, recent reports demonstrate that VEGF delays the senescence of human dermal endothelial cells. A unique effect of VEGF has been observed on the deposition of a scaffolding allowing these cells to attach, proliferate and escape apoptosis [77]. When endothelial cells are coated on gelatin gels, VEGF prevents cell apoptosis induced by TNFα [78].

Further studies have demonstrated that the vascular endothelial cell is not the sole target of VEGF which is a true interleukine since it stimulates the proliferation of IL2-dependent lymphocytes [79], the colony formation of granulocyte-macrophage progenitor cells [80] and inhibits the differentiation of dendritic cells [81]. Our group has demonstrated that VEGF is also an autocrine growth factor for retinal pigment epithelial cells, hair dermal papilla cells and stromal cells cultured from neonatal hemangiomas [10]. In addition, VEGF is chemotactic for monocytes, lens epithelial or corneal endothelial cells [82], as well as a differentiation factor for osteoblasts and 3T3 cells.

A recent report focused attention on the ability of VEGF to induce cell transformation. Constitutive expression of VEGF in rat retinal pigment epithelial cells leads to cell transformation, which is dramatically increased by FGF2 [83]. This transformation is linked to a major overexpression of FGFR1. However neutralizing antibodies directed against VEGF or FGF2 have no effect on FGFR1 expression or cell transformation, suggesting that the target site of VEGF is intracellular.

Role in Vascular Biology

Physiological Function   As early as 1989, VEGF was described as a potent angiogenic factor in the chick chorioallantoic membrane [6] and the rat corneal pocket assay [5]. Angiogenesis can be mediated by the homodimerization of each VEGF receptor or by heterodimerization of both. PlGF1 or PlGF2 alone may elicit angiogenesis as well as specific VEGFR2 agonists obtained by the anti-idiotypic strategy [30]. In contrast to VEGF, VEGF-C fails to induce angiogenesis but stimulates lymphangiogenesis in the chick chorioallantoic membrane assay.

Systemic injections of VEGF do not induce angiogenesis, unless the endothelial cells of large vessels have been activated by a traumatism such as artery ligation or angioplasty. The ability of VEGF to bind to the proteoheparan-sulfate of the vascular wall prevents it from reaching tumors and modulating tumoral angiogenesis.

One crucial step in initiating angiogenesis is vasodilatation. *In vivo* VEGF induces vasodilatation through an NO-dependent pathway leading to acute and severe hypotension. This effect is distinct from its angiogenic effect since VEGFR2 agonists mediate angiogenesis but not hypotension [84].

Topical administration of VEGF in the dermis induces plasma leakage following final fenestration of the endothelial cells [73]. Intravitreal injection of VEGF in the monkey vitreous is sufficient to induce iris neovascularization [85]. Accordingly, the potential role of VEGF in iris neovascularization after veinous occlusion has been demonstrated *in vivo*: two days after the induction

of veinous occlusion, the previously undetectable intra-camerular and intravitreal VEGF levels rose and reached a peak during the second week at the time of maximal iris neovascularization. VEGF was undetectable in the serum while its intravitreal level was constantly higher than that measured in the anterior chamber. These data support the hypothesis that VEGF is synthesized in the posterior segment with secondary diffusion into the anterior chamber. The intravitreal level of VEGF mRNA also appeared to be significantly elevated in hypoxic retinas with venous occlusion compared to normal contralateral retinas. The sites of VEGF synthesis in this model could be the Muller cells or retinal pericytes since *in situ* hybridization revealed the presence of VEGF mRNA at the internal plexiform layer level in hypoxic retinas. Another retinal neovascularization model consists of exposing 12-day old rats to hyperoxia for 24 hours. Hyperoxia reduces the expression of VEGF mRNA, leading to vasoconstriction followed by apoptosis of the developing retinal endothelial cells, thus creating hypoxia in the developing tissues. Hypoxia increases VEGF mRNA in the Muller cells and neovascularization is observed within 5 days [86]. Intravitreal injections of VEGF prior to the onset of hyperoxia induce a vasodilatation which prevents apoptosis of the retinal endothelial cells [70].

The persistence of VEGF and VEGF receptors in organs where angiogenesis does not occur, such as brain choroid plexus and ventricular epithelium or kidneys has led to numerous hypotheses. The most widespread hypothesis speculates functions of permeability or survival. However long-term treatment of mice with neutralizing antibodies reduces tumor growth but has no effect on kidney or brain ultrastructure [30]. *In situ* hybridization demonstrates that VEGF expression is spatially and temporally expressed during the menstrual cycle in the ovaries and uterus [87]. The concomitant expression of VEGF receptors suggested that the VEGF system plays a role in hormonally regulated angiogenesis (see review by Applanat). This hypothesis has been recently confirmed by the same group by the demonstration that the selective decrease of VEGF bioavailability impairs corpus luteum angiogenesis [88]. Hair follicle growth also requires cyclical angiogenesis to initiate the proliferation and survival of the dermal papilla. Immunohistochemical studies have provided insight into the cyclical expression of VEGF in the hair follicle where this latter may act additionally as an autocrine factor for hair dermal papilla cells. This hypothesis is in good agreement with the report of a concomitant decrease in VEGF expression and the onset of alopecia.

Pathology   Due to its angiogenic and permeability activity, VEGF has been believed to be involved in ocular neovascularization. Analysis of pathological samples has confirmed this hypothesis. VEGF immunoreactivity has also been detected in the endothelium of choriocapillaries as well as in intravascular leukocytes and migrating retinal pigment epithelium whereas VEGF is not detected in normal retinas [89]. VEGF mRNA has been identified as the sole angiogenic factor expressed in fibrovascular membranes collected from diabetic patients and its levels are significantly elevated in the vitreous [90].

Counting endothelial cells has been proposed as an independent factor reflecting the metastatic potential of prostate and breast cancer [91]. The demonstration that VEGF mRNA is expressed in cancer cells, whereas the VEGF protein is also accumulated in endothelial cells located in the vicinity of cancer cells [92] has paved the way for numerous studies confirming the essential role of VEGF in tumoral angiogenesis [93,95]. Indeed a good correlation between vascularity and VEGF expression has been found, and VEGF seems to represent a useful prognosis marker. VEGF immunoreactivity is increased in the plasma of cancer patients [95] and its decrease might indicate an efficacy of the chemotherapy employed [96]. VEGF expression has been found in several pathologies in which angiogenesis is a pro-eminent feature such as rheumatoid arthritis, bullous pemphigoid and psoriasis. In neonatal hemangiomas VEGF and VEGFR2 are co-expressed in the stromal cells. This is the first example to date in which VEGF confers a growth advantage *in vitro* and *in vivo* to a benign tumor [10]. VEGF is not always upregulated in pathologies associated with angiogenesis, but rather with stress. For instance, tears contain high concentrations of VEGF following refractive keratotomy.

Clinical Relevance and Therapeutic Implication Several attempts to inhibit VEGF bioavailability have proved to be successful, including monoclonal neutralizing antibodies [97], antisense oligonucleotides [98], soluble VEGFR1 fusion proteins [99], anti-VEGFR2 antibodies [100], blocker of VEGFR2 tyrosine phosphorylation (SU 5416) or immunotoxins targeted on VEGFR2 (our laboratory) have been shown to inhibit tumoral progression or retinal neovascularization, as well as metastasis [101]. It seems that the immunoneutralization of VEGF decreases the permeability of tumoral vessels without affecting the maintenance of the normal vasculature [30]. Phase II and III clinical trials of humanized anti-VEGF antibodies and SU 5416 are currently in progress.

The angiogenic activity of VEGF delivered either through protein infusion or the intramuscular injection of naked DNA seems to improve the recovery of limb arteritis or myocardial ischemia.

Perspectives Experimental angiogenic assays have already ascribed an angiogenic activity to so many factors that it would be illusive to attempt to control angiogenesis by interfering with only one angiogenic pathway. However, the growth factors tested in these assays act in an inflammatory context created by a surgical traumatism which might overstate the actual role of a compound in pathological angiogenesis. It has been proved that several so-called angiogenic factors act by releasing endogenous angiogenic factors from macrophages or from extracellular matrix. These assays therefore constitute a better reflection of controlled angiogenesis, as

observed in wound healing or ovarian cycle, than the uncontrolled pathological angiogenesis observed in tumors or diabetic retinopathy. The discovery of VEGF led to major insights in the understanding of the development of the vascular system as well as its role in physiological and pathological angiogenesis. Further studies using conditional gene knock-outs of one or all the VEGF isoforms, and then tempting to recover the phenotype by activating VEGFRs signal transduction by injection of circulating receptors agonists such as anti-idiotypic antibodies should help to determine where and when the VEGF system is critical. The elucidation of the signal transduction cascade of the tyrosine kinases VEGFR1, VEGFR2, VEGFR3 and their modulation by the recently identified neuropilin should also help to understand the mechanism of VEGFR2 silencing in quiescent endothelial cells or VEGFR2 transduction in angiogenic endothelial cells. Clinical trials of VEGF inhibition in cancers and VEGF infusion in myocardial ischaemia should also delineate the beneficial and adverse effects of each treatment.

Acknowledgements. Supported by the Association de la Recherche pour le Cancer, the Fondation de France and the Association Française "Retinitis Pigmentosa". Sylvie Sordello* and Bernard Malavaud· are supported by Fellowships from the Ligue Nationale Contre le Cancer * and the Fédération de la Recherche Médicale. We would like to thank Mrs Isabelle Pingree and Diana Warwick for editorial assistance.

Sylvie Sordello, Pierre Fons, Bernard Malavaud
and Jean Plouet

References

1. Michaélson IC et al (1948) Trans Ophtalmol Soc UK 68:137-180
2. Ashton N (1957) Am J Ophtalmol 44:7-24
3. Folkman J et al (1971) N Engl J Med 285:1182-1186
4. Ferrara N et al (1989) Biochem Biophys Res Commun 161:851-858
5. Connolly DT et al (1989) J Clin Invest 84:1470-1478
6. Plouèt J et al (1989) EMBO J 8:3801-3806
7. Senger DR et al (1986) Cancer Res 46:5629-5632
8. Plouèt J et al (1990) J Biol Chem 265:22071-22074
9. Guerrin M et al (1995) J Cell Physiol 164:385-394
10. Bérard M et al (1997) Am J Pathol 150:1315-1326
11. Lachgar S et al (1996) J Invest Dermatol 106:17-23
12. Gitay-Goren H et al (1992) J Biol Chem 267:6093-6098
13. Bikfalvi A et al (1991) J Cell Physiol 149:50-59
14. Soker S et al (1998) Cell 92:735-745
15. Jakeman LB et al (1992) J Clin Invest 89:244-253
16. De Vries C et al (1992) Science 255:989-91
17. Millauer B et al (1993) Cell 72:835-846
18. Morishita K et al (1995) J Biol Chem 270:27948-27953
19. Bednarz J et al (1995) Cornea 14:372-381
20. Jonca F et al (1997) J Biol Chem 272:24203-24209
21. Waltenberger J et al (1994) J Biol Chem 269:26988-26995
22. Pajusola K et al (1994) Oncogene 9:3545-3555
23. Barleon B et al (1996) Blood 87:3336-3343
24. Plouèt J et al (1997) J Biol Chem 272:13390-13396

25. Guo D et al (1995) J Biol Chem 270:6729-6733
26. Cunningham SA et al (1997) Biochem Biophys Res Commun 240:635-639
27. Kroll J et al (1997) J Biol Chem 272:32521-32527
28. D'Angelo G et al (1997) J Cell Biochem 67(3):353-66
29. Abedi H et al (1997) J Biol Chem 272:15442-15451
30. Ortéga N et al (1997) Am J Pathol 151:1215-1224
31. Keyt BA et al (1996) J Biol Chem 271:5638-5646
32. Muller Y et al (1997) Proc Natl Acad Sci USA 94:7192-7197
33. Davis-Smyth T et al (1996) EMBO J 15:4919-4927
34. Wiesmann C et al (1997) Cell 91:695-704
35. Tanaka et al (1997) Jpn J Cancer Res 88:867-876
36. Maglione D et al (1991) Proc Natl Acad Sci USA 88:9267-9271
37. Sawano A et al (1996) Cell Growth Differ 7:213-221
38. Park JE et al (1994) J Biol Chem 269:25646-25654
39. Ziche M et al (1997) Lab Invest 76:517-531
40. Olofsson B et al (1996) Proc Natl Acad Sci USA 93:2576-2581
41. Joukov V et al (1996) EMBO J 15:290-298
42. Yamada Y et al (1997) Genomics 42:483-488
43. Orlandini M et al (1996) Proc Natl Acad Sci USA 93:11675-11680
44. Lyttle DJ et al (1994) J Virol 68:84-92
45. Keck PJ et al (1989) Science 246:1309-1312
46. Leung DW et al (1989) Science 246:1306-1309
47. Tischer E et al (1989) Biochem Biophys Res Commun 165:1198-1206
48. Houck KA et al (1991) Mol Endocrinol 5:1806-1814
49. Poltorak Z et al (1997) J Biol Chem 272:7151-7158
50. Fong GH et al (1995) Nature 376:66-70
51. Shalaby F et al (1995) Nature 376:62-66
52. Eichmann A et al (1997) Proc Natl Acad Sci U S A 94:5141-5146
53. Carmeliet P et al (1996) Nature 380:435-439
54. Ferrara N et al (1996) Nature 380:439-442
55. Maltepe E et al (1997) Nature 386:403-407
56. Okamoto N et al (1997) Am J Pathol 151:281-291
57. Shweiki D et al (1992) Nature 359:843-845
58. Plate KH et al (1992) Nature 359:845-848
59. Goldberg MA et al (1994) J Biol Chem 269:4355-4359
60. Stein I et al (1995) Mol Cell Biol 15:5363-5368
61. Forsythe JA et al (1996) Mol Cell Biol 16:4604-4613
62. Mukhopadhyay D et al (1995) Nature 375:577-581
63. Gerber HP et al (1997) J Biol Chem 272:23659-23667
64. Waltenberger J et al (1996) Circulation 94:1647-1654
65. Brogi E et al (1996) J Clin Invest 97:469-476
66. Mukhopadhyay D et al (1997) Mol Cell Biol 17:5629-5639
67. Enholm et al (1997) Oncogene 14:2475-2483
68. Sordello S et al (1998) Biochem Biophys Res Commun 251:287-290
69. Houck KA et al (1992) J Biol Chem 267:26031-26037
70. Alon T et al (1995) Nat Med 1:1024-1028
71. Keyt BA et al (1996) J Biol Chem 271:7788-7795
72. Simorre-Pinatel V et al (1994) Invest Ophthalmol Vis Sci 35:3393-3400
73. Roberts WG et al (1995) J Cell Sci 108:2369-2379
74. Unemori EN et al (1992) J Cell Physiol 153:557-562
75. Pepper MS et al (1992) Biochem Biophys Res Commun 189:824-831
76. Melder RJ et al (1996) Nat Med 2:992-997
77. Watanabe Y et al (1997) Exp Cell Res 233:340-349
78. Spyridopoulos I et al (1997) J Mol Cell Cardiol 29:1321-1330
79. Praloran V et al (1991) CR Acad Sci 313:21-26
80. Broxmeyer HE et al (1995) Int J Hematol 62:203-215
81. Gabrilovich DI et al (1996) Nat Med 2:1096-1103
82. Moukadiri A et al (1992) In Biotechnology of growth factors Basels Karger123-128
83. Guerrin M et al (1997) Oncogene 14:463-471
84. Malavaud B et al (1997) Cardiovasc Res 36:276-281
85. Tolentino JM et al (1996) Arch Ophthalmology 114:964-970
86. Pierce EA et al (1995) Proc Natl Acad Sci U S A 92:905-909
87. Phillips HS et al (1990) Endocrinology 127:965-967
88. Ferrara N et al (1998) Nat Med 4:336-340
89. Lutty GA et al (1996) Arch Ophthalmol 114:971-977
90. Malecaze F et al (1994) Arch Ophthalmol 112:1476-1482
91. Weidner N et al (1993) Am J Pathol 143:401-409
92. Dvorak HF et al (1991) J Exp Med 174:1275-1278
93. Dvorak HF et al (1995) Am J Pathol 146:1029-1039
94. Ferrara N et al (1997) Endocr Rev 18:4-25
95. Gasparini G et al (1997) J Natl Cancer Inst 89:139-147
96. Dirix LY et al (1997) Br J Cancer 76:238-243
97. Kim KJ et al (1993) Nature 362:841-844
98. Robinson GS et al (1996) Proc Natl Acad Sci USA 93:4851-4856
99. Aiello LP et al (1995) Proc Natl Acad Sci U S A 92:10457-10461
100. Skobe M et al (1997) Nature Med 3:1222-1227
101. Warren RS et al (1995) J Clin Invest 95:1789-1797

Vascular Endothelial Growth Factor Receptors

Definition *Tyrosine-kinase domain receptors with seven IGG-like domains. Three main tyrosine-kinase receptors are described, VEGFR1 (FLT1), VEGFR2 (FLK1, KDR) and VEGFR3 (FLT4). A coreceptor specific for VEGF 165 has also been described (neuropilin).*

See: →Tyrosine Kinase receptors for factors of the VEGF family; →Vascular endothelial growth factor family; →Signal transduction mechanisms in vascular cells; →Angiogenesis; →Ontogeny of the vascular system; →Hormonal regulation of vascular cell function

Vascular Integrins

Synonym: A large number of vascular cell surface glycoproteins that function as cell adhesion receptors belong to the integrin gene family. The most frequently encountered synonyms for these receptors are listed in Table 1.

Introduction Cellular interactions in the vasculature are constantly exposed to hemodynamic forces generated by the flow of blood, and rely on a complex interplay of different classes of membrane-bound adhesion receptors necessary to counteract these dispersive forces. In addition, following physiologic stimuli, circulating cells in the blood flow switch from non-adherent to adherent cells, a process that relies essentially on dynamic adhesion events that can either restrict or promote cell movement. Several families of cell adhesion molecules have

Table 1. Integrin Synonyms

Integrin	CD Cluster	Synonyms
$\alpha 1$	CD49a	VLA-1
$\alpha 2$	CD49b	VLA-2, platelet GPIa
$\alpha 3$	CD49c	VLA-3
$\alpha 4$	CD49d	VLA-4
$\alpha 5$	CD49e	VLA-5, platelet GPIc
$\alpha 6$	CD49f	VLA-6, laminin receptor α, platelet GPIc'
$\alpha 7$		
$\alpha 8$		
$\alpha 9$		
$\alpha 10$		
α IIb	CD41	Platelet GPIIb
α v	CD51	Vitronectin receptor α
α L	CD11a	LFA-1alpha; LeuCAMa
α M	CD11b	CR3A; Mac1; Mo1; LeuCAMb
α X	CD11c	Surf. Ag. p150,95 α; euM5; Myel. membr.Ag.; LeuCAMc; CR4
α D		
α E	CD103	Lymphocyte HLM-1
$\beta 1$	CD29	VLA-beta, fibronectin receptor β subunit, platelet GPIIa
$\beta 2$	CD18	
$\beta 3$	CD61	Platelet GPIIIa
$\beta 4$	CD104	Tumor Specific Protein 180
$\beta 5$		
$\beta 6$		
$\beta 7$		
$\beta 8$		
$\beta 9$		

evolved to serve these specialized functions. Members of one of these families, the integrins, are key vascular cell adhesion receptors, as they control platelet aggregation during hemostasis, leukocyte interaction with the vessel wall and extravasation to surrounding tissues during inflammation, lymphocyte binding to antigen-presenting cells or endothelium during the immune response, as well as endothelial cell migration during angiogenesis [1].

Integrins are heterodimeric cell surface glycoproteins composed of an α and a β subunit, that mediate both cell-cell and cell-matrix interactions. To date, 16 α-subunits and 8 β subunits have been identified in man, that can combine to form at least 22 different integrin receptors, varying in their ligand-binding specificity, intracellular signalling properties and tissue distribution. An initial classification of integrins into subclasses was based on β subunits that associate with one or more α subunits. The most widely distributed integrins belong to the $\beta 1$ class, also known as the very late antigens (VLA), because of their late appearance on the surface of lymphocytes following antigen stimulation [2,3]. The $\beta 1$-containing integrins are essentially receptors for extra-cellular matrix proteins and play an important role in cell-matrix interactions. The second class of integrins are leukocyte specific [4,5]. They share a common $\beta 2$ subunit that associates with one of four α subunits (αL, αM, αX, αD). β_2 integrins mediate cell-cell interactions and mainly bind cell surface anchored counter-receptors of vascular cells. Cytoadhesins represent a class of integrins that share the common β_3 subunit [6]. Two members belong to this class, the platelet specific $\alpha_{IIb}\beta_3$ integrin, also known as the platelet fibrinogen receptor GPIIb-IIIa, and the widely distributed vitronectin receptor $\alpha_v\beta_3$. The β_4 subunit, which only associates with the α_6 subunit, is essentially expressed in epithelial cells [7], while the β_7 subunit associates with two α subunits, αE and $\alpha4$, to form the lymphocyte activation and homing receptors HML-1 (αE$\beta7$) and LPAM-1 ($\alpha4\beta7$). The β subunit classification of integrins has become obsolete with the discovery that α subunits can also associate with several distinct β subunits. The most promiscuous α subunit in this respect is αv, able to form heterodimers with five different β subunits ($\beta1$, $\beta3$, $\beta5$, $\beta6$, $\beta8$), and so far, $\beta5$ and $\beta8$ subunits have only been found associated with αv (reviewed in [8]) (see Table 2).

Table 2. Structural Characteristics of Integrin α and β Subunits

Integrin	Sequence of mature protein	Extracell.	Transmemb.	Cytoplasmic	Mr. KDa Reduced	Non Reduced	Reference (*)
α 1	1151	1112	24	15	185	200	Briezewitz, 1993
α 2	1152	1100	25	27	165	160	Takada, 1989
α 3	1019			37	110/30	150	Takada, 1990
α 4	999	942	25	32	(150) 80/70	145	Takada, 1989
α 5	1008	955	25	28	130	150	Argraves, 1987
α 6	1050	991	23	36	125	145	Tamura, 1990; Hogervorst, 1991
α 7	1106	1006	23	77	100/35	120	Song, 1992
α 8	1021				155/25	160	Schnapp, 1995
α 9	1006	947	26	33	150	140	Palmer, 1993
α 10	1167	1098	25	22	160	160	Camper, 1998
α IIb	1009	963	26	20	125/23	136	Poncz, 1987
α v	1017	956	29	32	130/27	150	Suzuki, 1987
α L	1145	1063	24	58		180	Larson, 1989
α M	1133	1088	21	24		170	Arnaout, 1988; Corbi, 1988
α X	1144	1089	26	29	150	150	Corbi, 1987
α E	1160	1105	23	32	150/25	175	Shaw, 1994
β 1	778	708	23	47	130	120	Argraves, 1987
β 2	747	678	23	46	95	95	Kishimoto, 1986; Law, 1987
β 3	762	692	23	47	108	97	Fitzgerald, 1988; Rosa, 1988, Zimrin, 1988
β 4	1778	683	23	1072	220	210	Hogevorst, 1990; Suzuki, 1990
β 5	771	711	23	57	100	95	McLean, 1990
β 6	788			58		100–110	Sheppard, 1990
β 7	780	707	21	52		105	Erle, 1991
β 8	727	639	30	58	97	95	Moyle, 1991

(*) Not all references indexed in the bibliography section

Characteristics

Domains

Domain structure of the a subunits Purified integrin receptors viewed by rotary shadowing electron microscopy consist of two domains, a globular head of 8 x 10 nm, composed by the N-terminal part of the α and β subunits of the integrin heterodimer, and two rod-like tails, extending from one side of the globular domain and comprising the transmembrane and cytoplasmic tail of each subunit [9,10]. The structural organization of integrin α and β subunits is shown in Figure 1. The N-terminal portion of mature integrin α subunits comprises seven homologous, tandemly repeated domains each about fifty amino acids long. The three or four last repeat domains, depending on the subunit, contain cation-binding sequences similar to the EF hand motif found in Ca^{2+}-binding proteins such as calmodulin [11]. The classical EF hand motif consists of a thirteen amino acid long loop between two α-helices, in which six acidic or hydrophilic amino acid side chains in position 1,3,5,7,9

and 12 form an octahedric coordination sphere surrounding a Ca^{2+} ion in the center. In contrast to this classical EF hand structure, the integrin metal-binding sites have only five of the six predicted metal coordination sites, with the general structure DXDXDGXXD, and are flanked by two β strands [12]. Still this incomplete integrin EF hand motif has been shown to bind divalent cations, and it is now generally accepted that the aspartate or glutamate residue, commonly found in all binding motifs of integrin-ligands, provides the sixth missing coordinating position to complete the EF hand structure [13,14]. These Ca^{2+} and ligand binding motifs are now referred to as "metal ion dependent adhesion sites" (MIDAS).

Eight integrin α subunits (αl, α2, α10, αE, αL, αM, αX and αD) contain a domain of about 200 amino acids inserted between the second and third N-terminal repeats. This domain, which appears to have been inserted by exon shuffling and is therefore referred to as the I domain (I for insertion), is homologous in sequence to the "A" domain of von Willebrand factor [15]. Interestingly, this I domain

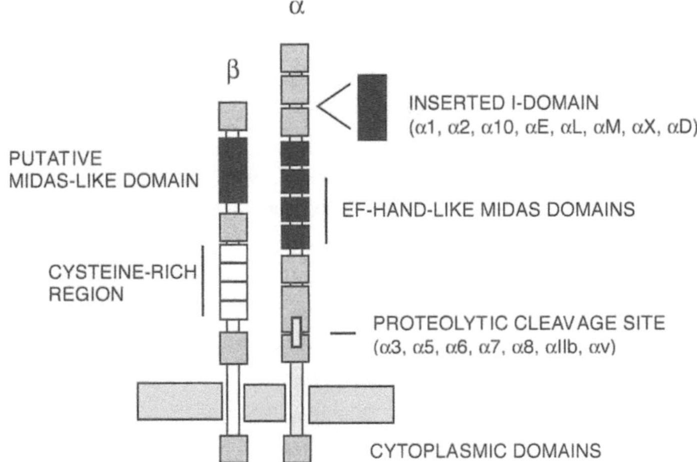

Figure 1. Schematic diagram of the structural organization of the integrin αβ heterodimer (adapted from [159]).

also has the unusual divalent cation-binding site, which is not present in the von Willebrand factor A3 domain [16-19]. There is now growing evidence that I domain is directly involved in integrin-ligand binding. A stalk region links the globular head with the transmembrane domain. The α subunit is anchored in the plasma membrane by an α helix, about 25 or more amino acids in length, that spans the membrane once.

The cytoplasmic tails of all integrin α subunits are relatively short ranging from 15 to 77 amino acids in length. They share little homology among each other, suggesting unique functional roles in intracellular signalling processes. This variability is further increased by splice variants, that have been identified for the α3 (α3A and α3B), α6 and α7 subunits (reviewed in [20]). All α subunits, however, share a heptapeptide consensus sequence KXGFFKR adjacent to the transmembrane domain, except for the α9 subunit which has a KXGFFRR sequence [21]. This GFFK/RR motif, which is also found in members of the steroid hormone receptor family, has been shown to function as an integrin-binding site for the intracellular protein calreticulin [22]. In addition, this KXGFFKR consensus sequence might also be important for inter-subunit interactions, necessary for heterodimerization and for maintaining integrin receptors in a low affinity state [23].

Domain structure of the β subunits The integrin β subunits show higher similarities among each other than do α subunits. A conserved domain in the N-terminal part of all β subunits is a 250 residue long sequence about 100 amino acids from the N-terminus. The presence within this domain of a consensus DXSXS sequence identical to that found in the α subunit I domain, as well as a striking similarity between the hydropathy plot and secondary structure of this region in the β3 integrin subunit and the αM subunit I domain suggested that the β subunit possibly shares many of the structural features of the a subunit I domain [24] and also bind divalent cations [17,24]. This domain appears to be involved both in lig-and binding and in the association with the α subunit

[24]. Contiguous to the conserved ligand-binding region is a cystein-rich region connecting the globular head to the transmembrane domain. This domain is composed of four cystein-rich repeat units, each 45 amino acids in length, with a cystein content of about 20 % [4]. Calvete et al have characterized the position of the disulfide bonds for the 56 cysteines in the β3 subunit and shown that all cysteines are engaged in disulfide bonds, thus conferring an extreme rigidity to the β subunit [25].

A single α helical transmembrane domain is followed by a short and highly conserved cytoplasmic tail, ranging in length from 47 to 58 amino acids. Three conserved regions within the cytoplasmic tail, termed cyto1, cyto2 and cyto3, have been implicated in the recruitment of β_1 integrins to focal adhesion plaques and regulation of adhesive functions of these integrins [26]. The membrane proximal cyto-1 motif contains the interface to the α subunit. The cyto-2 motif bears a consensus NPXY tyrosine kinase recognition sequence [27,28] and may be involved in internalization of integrin receptors [29], as well as integrin signalling and affinity regulation [30]. The cyto-3 motif at the C-terminus of the β-subunits has an NXXY sequence, less conserved among the different β subunits, whose functional role is still unclear. A number of integrin β subunit variants have been identified for β1 (β1B, β1C, β1D), β3 (β3B, β3C) and β4 (β4B, β4C, β4D), resulting from alternative splicing events in the cytoplasmic tail (reviewed in [20]).

Integrin Ligands Integrins bind to a large variety of ligands, including components of the extracellular matrix, cell surface IgG superfamily receptors, plasma proteins, as well as bacterial and viral proteins. The first target amino acid sequence capable of inhibiting integrin-dependent cell adhesion to fibronectin was identified more than 12 years ago and corresponds to the tripeptide motif Arg-Gly-Asp (RGD) present in the tenth type III repeat of fibronectin [31]. This RGD motif, which has been found subsequently in a large number of extracellular matrix proteins, blood proteins as well as cell surface proteins, has evolved as a basic unit of an evolutionary conserved cellular recognition system (reviewed in [32]). RGD is however not the only motif recognized by integrins, and over the last few years, other cell-adhesion motifs have been identified, many of which represent slight variations of the canonical RGD sequence. As shown in Table 3, at least half of the currently known integrins are RGD-binding integrins. It is noteworthy that the α subunits of these RGD-binding integrins do not contain the I domain, and show more sequence similarity to one another than to the other α subunits [33]. For some RGD-binding integrins, a second recognition site in the adhesive protein is necessary and acts in synergy with the RGD site. In fibronectin, the RGD loop in the tenth FN-III repeat and the PHSRN synergy domain in the ninth FN-III repeat are only 30 to 40 Å apart on the same surface of the fibronectin molecule, suggesting that the binding site defined by these two regions could easily be spanned by a single integrin molecule [34]. Similarly, fibrinogen contains multiple integrin recog-

nition sequences, two of which have been identified as crucial for fibrinogen-$\alpha_{IIb}\beta_3$ interaction, the fibrinogen γ chain C-terminal KQAGDV [35], and the α chain RGD motif [36].

Structural analysis by NMR and crystallography of recombinant polypeptide domains containing an RGD sequence, such as the tenth type III repeat of fibronectin [37], the third type III repeat of human tenascin [38], or the small snake venom disintegrin decorsin [39] provide evidence that the RGD sequence is located at the apex of a solvent-exposed loop formed by two anti-parallel β strands. Structural analysis of the RGD motif in various disintegrins has further shown that the RGD motif is more or less flexible depending on the surrounding amino acid sequence. Based on these data, a common scheme proposes that the RGD sequence protrudes out of the molecular scaffold of the adhesive protein in a fingertip-like structure, able to fit into the ligand-binding pocket of the integrin receptor, allowingF ligand – inte-

Table 3. Major Integrin Ligands and Ligand Binding Sites

Receptor	Ligand	Binding Sequence
$\alpha 1\beta 1$	Native collagens, Laminins	
$\alpha 2\beta 1$	Native collagens, Laminins	
$\alpha 3\beta 1$	Laminins	KQNCLSSRASFRGCVRNLRLSR
	Collagen I	?
	Fibronectin	RGD
$\alpha 4\beta 1$	Fibronectin,	IDAPS/LDV/REDV/X6Asp-Y
	VCAM-1	?
$\alpha 5\beta 1$	Fibronectin	RGD
$\alpha 6\beta 1$	Laminins	E8
$\alpha 6\beta 4$		
$\alpha v\beta 1$	Fibronectin, Vitronectin	RGD
$\alpha L\beta 2$	ICAM-1, ICAM-2	KELLLPGNNRKV
	ICAM-3	?
$\alpha M\beta 2$	Fibrinogen,	KGWTVFQKRLDGSV
	C3bi, ICAM-1	?
$\alpha x\beta 2$	Fibrinogen	GPRP
	C3bi	?
$\alpha IIb\beta 3$	Fibrinogen,	KQAGDV/RGD
	Fibronectin, Vitronectin, vWF	RGD
	Thrombospondin	
$\alpha v\beta 1$	Fibronectin, Vitronectin	RGD
$\alpha v\beta 3$	Vitronectin, Fibronectin, Fibrinogen, Denatured collagen, Thrombospondin, Thrombin, Osteopontin, Tenascin.	RGD
$\alpha v\beta 5$	Vitronectin	RGD
$\alpha v\beta 6$	Fibronectin	RGD
$\alpha v\beta 8$	?	RGD
$\alpha 4\beta 7$	MAdCAM-1, VCAM-1, Fibronectin	
$\alpha E\beta 7$	E-cadherin	

grin interaction (reviewed in [34]). However, in some adhesion molecules, RGD sites appear to be cryptic and are accessible only following conformational changes of the ligand. Soluble plasma fibronectin and fibrinogen have only a low affinity for integrins; immobilization of these ligands results in their "contact activation" allowing cell adhesion and spreading. Conformational changes in fibrinogen have been detected after its immobilization or binding to $\alpha_{IIb}\beta_3$, resulting in the surface exposure of the α chain RGDF site [40]. As mentioned earlier, a feature common to all the integrin-binding motifs identified so far is the presence of an aspartic acid or the closely related glutamic acid residue. This aspartate residue appears to be important because of its potential to contribute to integrin divalent cation binding by providing the missing coordination site [13].

For non-RGD-binding integrins, ligand binding appears to involve essentially the α chain I domain (reviewed in [41]). It has been shown that $\alpha M\beta 2$ – fibrinogen binding is mediated by the fibrinogen γ chain derived peptide KGWTVFQKRLDGSV, which also inhibits I domain fibrinogen interactions [42]. ICAM-1 binding to $\alpha L\beta 2$ requires an essential glutamate residue within an IETP sequence [43], and ICAM-3 binding to $\alpha L\beta 2$ relies on a glutamate residue in the homologous sequence LETS [44-46]. As the I domains of αL and αM are involved in binding to at least some members of the ICAM family, the LETS/IETPS sequence appears as a candidate binding site. I domain containing integrins of the $\beta 1$ subclass, such as $\alpha 1\beta 1$ or $\alpha 2\beta 1$ interact with collagens or laminins, and the motifs YGYYGDARL within laminin and FYFDLR within collagen type IV have been identified as $\alpha 2\beta 1$ recognition sites [47], while the $\alpha 2\beta 1$-collagen type I interaction is mediated through a tetrapeptide DGEA [48].

Integrin Ligand-Binding Sites   Identification of the site(s) within integrins involved in ligand recognition has been relatively slow, due to the complexity of these large heterodimeric molecules and the requirement of the $\alpha\beta$ complex for ligand binding. Potential ligand-binding sites have initially been identified through a combination of immunological, biochemical and mutational approaches (reviewed in [49]). These studies have identified the N-terminal portion of both the α and the β subunits as the integrin ligand-binding domain, and demonstrated that high affinity ligand recognition requires both subunits. Most of the initial progress made relied on RGD peptide cross-linking studies, mutagenesis, as well as the study of patients with naturally occurring mutations affecting the ligand-binding properties of $\beta 2$ and $\beta 3$ integrins. So far, three major domains have been identified, two in the N-terminal part of the α subunits, corresponding to the 7 N-terminal repeats and the I domain, and one in the N-terminal part of the β subunits corresponding to an I domain-like structure. The difficulties in determining the crystal structure of integrins has hampered the progress in characterizing their ligand-binding pocket. However, atomic resolution crystal structures have been obtained for the I domain, and

hypothetical computer models have been generated for the two other domains.

The α-chain sevenfold repeats   Ligand-binding sites in α chains devoid of the I domain have been identified by chemically cross-linking ligand-derived peptides. A γ chain peptide from fibrinogen was cross-linked to the second metal-binding site of α_{IIb} (residues 294-314) [50], and the RGD peptide to an overlapping domain in αv [51]. In addition, a recombinant α_{IIb} peptide encompassing the four Ca^{2+}-binding repeats bound to fibrinogen [52], and peptides from the second metal-binding repeat of α_{IIb} as well as antibodies to this peptide blocked fibrinogen binding to $\alpha_{IIb}\beta_3$ [53]. Further attempts to map α chain ligand-binding sites relied on the characterization of epitopes recognized by monoclonal antibodies that inhibit ligand binding [54]. Recently, a hypothetical atomic model of the seven N-terminal repeats of the α subunits has been proposed [55], based on the crystal structure of the heterotrimeric G proteins. This model predicts that the α subunit folds into a β propeller structure, a cyclic structure made of seven modules or blades. The ligand contact points are located on the upper surface of the propeller, and may include loops from more than one blade. Interestingly, in this model, the EF-hand-like Ca^{2+}-binding motifs of the α subunit lie on the lower surface of the propeller, far from the ligand contact sites. Mould et al. have provided evidence that the RGD sequence of the tenth type III repeat binds to the I domain-like structure of $\beta 1$, while the synergistic motif in the ninth type III repeat of fibronectin PHSRN interacts with the propeller portion of $\alpha 5$ [56].

The α subunit I domain   The α subunit I domain appears to be critically involved in ligand binding. A number of blocking or activating antibodies map to the I domain of $\alpha L\beta 2$ [57], $\alpha M\beta 2$ [58], $\alpha 2\beta 1$ [59] and $\alpha 1\beta 1$ [60], and recombinant I domain fusion proteins have further been shown to bind to their respective counterreceptors. The crystal structure of the I domain of αM and αL has been determined [17,18], as well as that of the A3 domain of von Willebrand factor [19]. These domains adopt a classic "Rossmann" fold and consist of hydrophobic β sheets in the middle and hydrophilic α helices on both sides of the β-sheets. Within the I domain, several Asp residues that are critical for ligand binding have been shown to be also involved in the coordination of divalent cations [16]. This domain is now called "metal ion-dependent adhesion site" (MIDAS) [38].

The β subunit MIDAS-like domain   The β subunit ligand-binding site has been localized to a 250 amino acid stretch located near the N-terminus of the β subunit. Strong support for this region of $\beta 3$ in ligand-binding function came from the observation that RGD peptides cross-link to this region [61-63], and that naturally occurring mutations within this region abolish ligand binding [49]. A 23 amino acid peptide from the $\beta 3$ subunit, encompassing the DXSXS motif (aa 113-153) is capable of binding both divalent cations and RGD peptides [64], and a further recombinant β3 peptide, containing part of the I domain-like sequence, binds directly to fibrinogen [65]. Antibodies that inhibit ligand binding to the β1 integrins also map to this conserved region [66], and mutations in this corresponding region of other β subunits similarly block ligand binding [67]. Interestingly, this region also contains epitopes that stimulate integrin ligand binding function. Mutagenesis studies of candidate oxygenated residues in the invariant DXSXS sequence of β1 [68], β2 [69] and β3 subunits [63] further support the hypothesis that this region of the β subunit adopts a similar structure as the α subunit I domain and might interact with ligands via a MIDAS-like motif.

Structural aspects of integrin ligand interaction

Activation-dependent integrin receptor function   Integrin receptor functions are tightly regulated, and a large body of evidence suggests that at least some of the integrin ligand-binding sites are cryptic and become exposed only following integrin activation, increasing the integrin affinity for extracellular ligands or counter-receptors. This activation process, called inside-out signalling, is thought to involve the propagation of conformational changes from the cytoplasmic domains to the extracellular ligand-binding site in response to intracellular signalling events [30]. Changes in the affinity of integrins in response to cytoplasmic signals have been observed in integrins of the β1 [70], β2 [71], β3 [72], and β7 subclass [73], and monitored with monoclonal antibodies that distinguish between inactive and active forms of integrins [74,75]. Furthermore, some monoclonal antibodies have also been shown to stimulate the high affinity, ligand-binding state of integrins, suggesting that antibody binding by itself induces the conformational change necessary for ligand binding [76-78].

For example, on circulating platelets, $\alpha_{IIb}\beta_3$ is in an inactive state as it binds little soluble fibrinogen, but activation of platelets by agonists such as thrombin results in a dramatic increase in the binding affinity of $\alpha_{IIb}\beta_3$ for fibrinogen, allowing platelet aggregation (reviewed in [79]. Platelet activation also leads to increased surface exposure of $\alpha_{IIb}\beta_3$ receptors through fusion of secretory α granules with the platelet membrane as well as surface exposure of intracellular membranes of the open canalicular system (OCS). Another example of affinity regulation are leucocyte integrins. Activation of leucocytes with stimulating agents such as phorbol esters, chemo-attractants or by the aggregation of functionally relevant surface receptors such as the antigen receptor/CD3 complex [80] or CD2 [81] causes αLβ2 to bind to purified ICAM-1 within minutes. Dependent on the cell type and the stimulus used, the increase in adhesion is transient or permanent. An activation requirement has also been observed for the αMβ2 receptor found on the surface of cells of the myeloid lineage. However, similar to platelets, an auxiliary regulatory mechanism is also present in these cells, as an upregulation of αMβ2 at the cell surface occurs after stimulation, due to translocation to the cell membrane of secretory granules containing an intracellular pool of αMβ2 integrins [82]. Finally the β1 integrins on hematopoietic cells can be activated and bind to activated endothelial cells and to ECM proteins and play a major role in leukocyte diapedesis and cell migration. Phorbol esters, chemotactic proteins that bind to seven transmembrane receptors have been shown to induce β1-integrin-binding to fibronectin, laminin or VCAM-1 [82,83]. As shown for platelet $\alpha_{IIb}\beta_3$ and leukocyte β2 integrins, vascular cells use both integrin affinity modulation and integrin avidity modulation (integrin upregulation or clustering) to regulate ligand-binding. Changes in affinity and avidity are not mutually exclusive, and the relative contributions of each to integrin activation is still unclear. There is some evidence that avidity modulation could represent a major regulatory process in the enhanced αLβ2-mediated adhesiveness of leukocytes (reviewed in [84]).

Role of divalent cations in integrin receptor function Integrin-ligand interaction is completely inhibited by divalent ion chelators such as EDTA, demonstrating the importance of metal ions in integrin-ligand binding. The three major integrin domains involved in ligand binding are also close to cation-binding sites. Although the precise mechanism of integrin-ligand interaction is still unclear, a possible model linking ligand binding with cation binding emphasizes the role of the divalent cation in acting as a bridge between the integrin and an acidic residue in the ligand that provides the missing coordination site; cation, ligand and receptor appear to form initially a quaternary complex, followed by subsequent displacement of the cation from the ligand-binding site [64]. However, it is not clear so far why various cations have different effects on integrin-ligand interactions, as Mn^{2+} is a potent activator of many integrins, while Ca^{2+} can inhibit integrin receptor function [85].

Integrin dependent intracellular signalling Ligation of integrin receptors through ligand binding initiates a variety of cellular responses, including cytoskeleton assembly, migration, differential gene expression, differentiation, and proliferation. This integrin dependent outside-in signalling process appears to be initiated through conformational changes, as ligand binding to the activated integrin induces the exposure of neoepitopes called "ligand-induced binding sites" (LIBS). These epitopes can be monitored with monoclonal antibodies that bind with high affinity only to the occupied integrin receptor [74,86]. Adhesion-dependent clustering of integrins further leads to the activation of non-receptor tyrosine kinase pathways such as the focal adhesion kinase pathway, or the Ras/mitogen activated pathway [87]. These and other integrin-dependent biochemical pathways, such as inositol lipid metabolism, Ca^{2+} fluxes, increase in intracellular pH, or activation of PKC, are beyond the scope of this review, and are discussed in excellent reviews [88,89].

Role of integrin cytoplasmic domains in integrin receptor-function Integrin cytoplasmic domains present a major link between integrin engagement and the cellular signal transduction machinery. Experimental approaches using recombinant mutant integrins have been instrumental in demonstrating that both the α and β subunit cytoplasmic tails are required for receptor affinity modulation. The membrane proximal sequences of the α (GFFKR) and β (WKLLXXXHDR) subunits are highly conserved among all integrin subunits and deletions of these motifs or point mutations of charged residues within these motifs activate integrins. The high affinity state of these deletion mutants is independent of cell type and physiological signalling pathways, suggesting that these motifs interact to form a structural constraint, referred to as membrane proximal hinge, that locks the integrin receptor in a low affinity state [73,90,91]. In some integrins, such as α6Aβ1, the membrane proximal motif of the α subunit also appears to be required for αβ heterodimerization [92]. Sequences C-terminal to the membrane proximal GFFKR domain of the integrin α subunits are highly divergent and are essentially involved in cell type or receptor-specific integrin activation [93]. In contrast, the integrin β subunit C-terminal parts are more conserved and contain two motifs, NPXY and NXXY, that have been shown to play a predominant role in almost all integrin functions, including receptor affinity modulation [30], cell adhesion to the extracellular matrix [29], cell spreading [94], as well as signal transduction [95].

The use of chimeric receptors containing solely the β subunit cytoplasmic domain fused to the extracellular domain of a reporter protein, demonstrates that the β subunit by itself can translocate into focal adhesion plaques and activate integrin-dependent signalling events [96]. Also, deletion mutations or point mutations in these motifs that disrupt the structural integrity of the β3 subunit C-terminal part prevent cell spreading [29,94] focal contact localisation as well as p125Fak and paxillin phosphorylation [95]. Recent studies have suggested that the highly conserved tyrosine residues in integrin cytoplasmic domains are phosphorylated and could represent targets for PTB domains of adapter proteins such as Shc involved in tyrosine kinase signalling events. Indeed, tyrosine phosphorylation of the β3 integrin subunit occurs during $\alpha_{IIb}\beta_3$-mediated platelet aggregation [97,98] or Mn^{2+} induced $\alpha_v\beta_3$-dependent CHO cell agglutination [99]. Association of Shc and Grb2 with tyrosine-phosphorylated β3 integrins has been documented *in vitro* and *in vivo*, suggesting a link of β3 integrins to the intracellular Ras signalling pathway. Although *in vitro* studies have demonstrated that phosphorylation of the 2 tyrosine residues is necessary for Grb2 binding, mutation of the membrane distal Y residue into F in the β3 integrin subunit does not affect $\alpha_v\beta_3$-mediated cell spreading or intracellular Fak and paxillin phosphorylation [95] (Figure 2).

Integrin intracellular ligands   Given the importance of integrin cytoplasmic domains in integrin receptor function, the search for proteins that interact directly with integrin cytoplasmic domains and are involved in both inside-out and outside-in signalling has been intense, although progress has been slow due to the low affinity of these ligands for integrins and hence the difficulty for their co-purification. However, with the use of the two-hybrid screen technology, the list of new proteins that can bind to integrin cytoplamic domains, at least *in vitro*, has been growing (Table 4). Some of these proteins are structural cytoskeletal proteins such as α-actinin [100], F-actin [101], filamin [102], talin [103], while others possess kinase activity such as p125Fak [104] and p59ILK [105], guanine nucleotide exchange activity (cytohesin-1) [106], or function as adapter proteins (paxillin) [107] or even as a receptor for activated protein kinase C (Rack-I) [108]. Some of these proteins are integrin specific, namely CIB, a Ca^{2+} and integrin-binding protein interacting specifically with αIIb [109], or ICAP-1 [110], cytohesin-1 [106] and β_3-endonexin [111] that specifically bind to the β1, β2 and β3 cytoplasmic tails, respectively. Although the mechanisms by which these intracellular proteins are implicated in integrin receptor function remain essentially unknown, some at least appear to be involved in integrin affinity modulation. Indeed, cytohesin-1 and β_3-endonexin, which both require a structurally intact C-terminal NXXY motif for integrin binding, have been shown to modulate β1 [106] and $\alpha_{IIb}\beta_3$ ligand-binding affinity [112].

Structure

Sequence and size   Integrin α subunits are 999 to 1160 amino acids long, while the integrin β subunits range in length from 727 to 788 residues, with the exception of β4, which is about 1770 amino acids long. The amino acid sequence homologies among integrin α subunits range from 25 to 60%, with the highest homologies found in the N-terminal repeat structures, the transmembrane domain and the short 6 amino acid long cytoplasmic sequence next to the membrane. Based on their sequence, structural and functional relationship, integrin α subunits have been divided into three distinct subfamilies [1, 113] (Figure 3). The first subfamily consists of the α subunits that undergo cleavage near the transmembrane domain. The second subfamily includes the α subunits that contain an I domain; within this subfamily, alpha-E is exceptional in that it contains an I domain, followed by a unique extra domain of 55 amino acids, with no counterpart in other integrin α subunits. This domain contains a stretch of 18 charged residues and includes a proteolytic cleavage site. Thus alpha-E is the only I-domain containing integrin α subunit that is also cleaved [113]. The third subfamily consists of 2 subunits, α4, which is cleaved into 2 fragments of similar size, while α9 contains neither the I domain, nor undergoes cleavage yielding disulfide-linked fragments.

Integrin β subunits show an even higher degree of similarity in their structural and functional organization, although β4 and β8 are clearly more divergent than the other β subunits. The highest homologies are found for β1, β2, β3, β5 and β6 in the N-terminal part corresponding to the MIDAS-like domain, as well as the cytoplas-

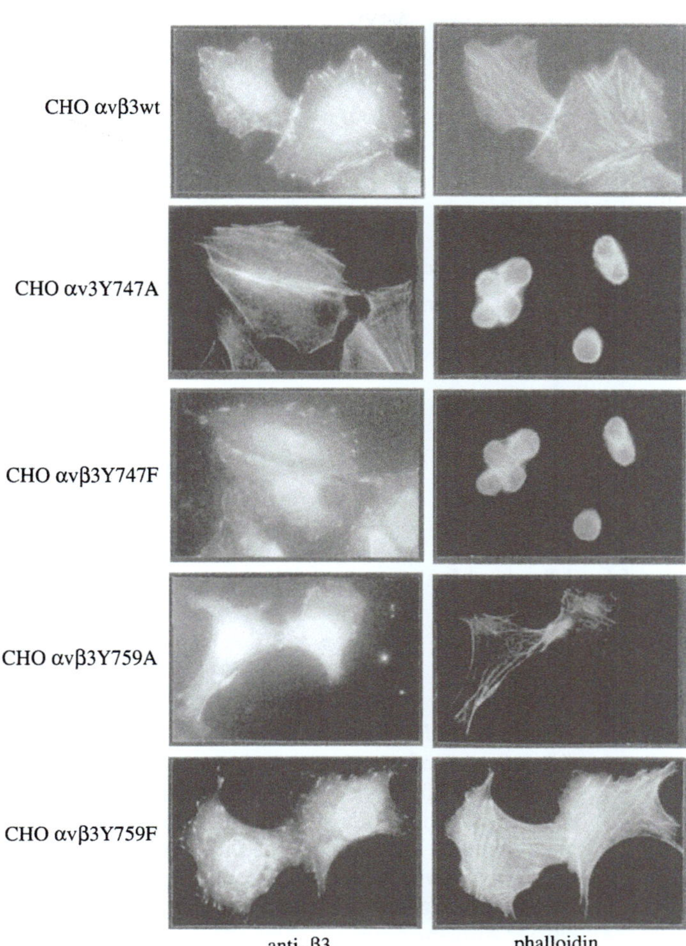

Figure 2. Immunofluorescence staining of recombinant wild-type and mutant αvβ3 integrins expressed in CHO cells. Transfected CHO cells grown on immobilized fibrinogen were co-labeled with an anti-β3 monoclonal antibody and phalloidin. The β3 mutations correspond to an Ala (A) or Phe (F) substitution of the Tyr (Y) residue in the respective NPXY[747] and NXXY[759] motifs of the β3 subunit cytoplasmic tail.

Table 4. Integrin Cytoplasmic Domain Ligands

Intracellular protein	Integrin subunit	Reference (*)
Cytoskeletal proteins		
F-actin	α2 only	Kieffer, 1995
Taline	αIIbβ	Horwitz, 1986; Knezevic, 1996
α-actinine	β	Otey, 1993
Filamin	β	Sharma, 1995
Kinases		
pp125 FAK	β	Schaller, 1995
p59 ILK	β	Hannigan, 1996
Adapter proteins		
Paxillin	β1	Tanaka, 1996
Regulatory proteins		
Cytohesin-1	β2 only	Kolanus, 1996; Meacci, 1997
β3-endonexin	β3 only	Shattil, 1995; Kashiwagi, 1997
ICAP-1	β1 only	Chang, 1997
CIB ?	αIIb only	Naik, 1997
Calreticulin	α	Rojiani, 1991

(*) Not all references indexed in the bibliography section

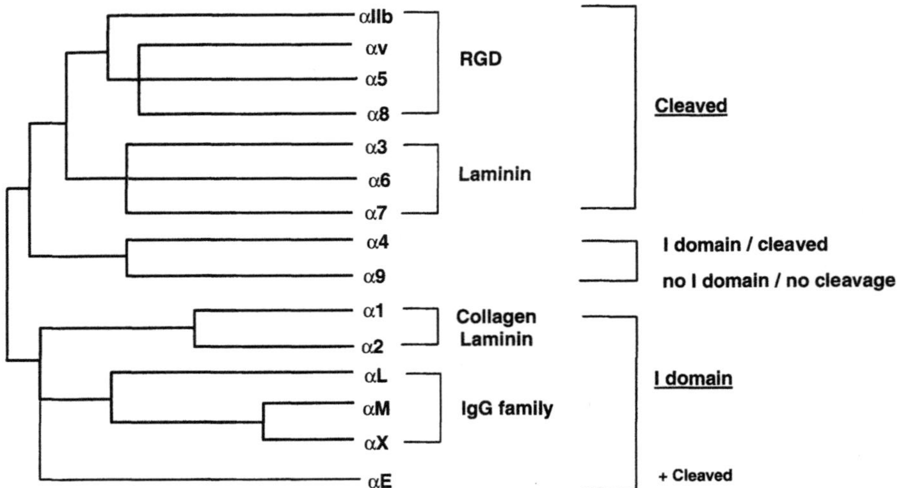

Figure 3. Family tree of the integrin α subunits. The evolutionary relationship between the different α subunits is based on sequence and structure similarities as determined by Hynes [1] and Shaw et al. [113].

mic domain. These subunits have 56 conserved cysteine residues, in contrast to β4, β7 and β8, which have 48, 54 and 50, respectively.

Gene

Gene structure   Integrin α and β subunits are genetically unrelated and are encoded by distinct genes found as single copy genes in the human genome. Analysis of the cDNA sequence and genomic structure of representative members of the two integrin α subunit subgroups, i.e. the I domain containing αM subunit and the cleavage site containing αIIb subunit, has revealed a high degree of conservation in the location of their intron/exon junctions [114]. As shown in Figure 4, both subunits are encoded by 30 exons, spanning over 50 kb of genomic sequence. The first four exons of αIIb correspond precisely to the five first exons of αM. The three EF-hand cation-binding domains in αM are encoded by three exons, that correspond to four exons in αIIb. The highest conservation of exon/intron structure between αM and αIIb is found 3' to the cation-binding exons, with precise alignment of amino acids between exons 16 to 30 of the leukocyte integrin and exons 15 to 30 of platelet αIIb [115-118]. The I domain of αM is encoded by 4 exons (6-9) not found in αIIb, and exon 25 of αIIb, which contains the proteolytic cleavage site, is not found in leukocyte integrin α subunits.

To date, the complete genomic organization of human integrin β subunits has been determined for β2 [119], β3 [120, 121], β4 [122,123] and β7 [124]. The genes vary in size from 10 to 50 kb and comprise 14 to 41 exons. Structural and genetic evidence suggests at least two major branches in the β subunit evolutionary tree, with the β1, β2, β7, and β8 subunits in one branch and β3, β5 and β6 in the other [124].

Chromosomal localization   The chromosomal localization of most of the cloned integrin subunits has been determined (Table 5). With the progress in sequencing the human genome, most of the integrin genes have also been mapped. Interestingly, the three leukocyte-specific integrin α subunits map to chromosome 16p 11 -p 13. 1 [13], suggesting that this area represents a possible gene cluster. Also, the genes encoding the platelet specific integrin $α_{IIb}β_3$ map to chromosome 17q21-22 and are physically linked within a 260 kilobase fragment [125]. These data suggest that coordinate expression of some integrins may depend on physical proximity.

Gene expression   In many cell types, the expression of integrins is a regulated process, modulated in part by peptide growth factors. Regulatory signals include PDGF, TGF-β, EGF, VEGF, IL-1, and IL-6 (reviewed in [126]). Also, PMA or hormones such as retinoic acid, 1,25-dihydroxyvitamin D3 and corticosterone have been reported to up-regulate integrin expression. For a number of integrin subunits, the promoter region has been cloned and characterized (see table 5), and the identification of regulatory regions in the promoters has allowed the unravelling of the complex control mechanisms that regulate integrin gene transcription. As reviewed by Kim and Yamada [126], a common feature of all these promoters is the absence of both the TATA and CAAT boxes, except for αIIb, which only contains a CAAT box [116], and α4, which contains both TATA and CAAT boxes [118]. In most eukaryotic promoters deficient of a TATA box, transcription is thought to be initiated through the binding of TFHD to an Sp-1 binding site instead of the missing TATA box [127]. Indeed, all characterized integrin gene promoters except α4 and mouse β7 contain Sp-1 binding sites. A number of transcription factor binding

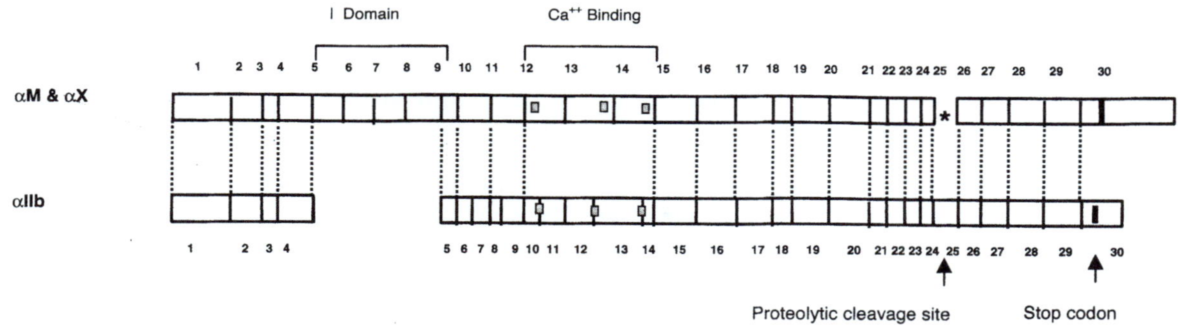

Figure 4. Comparison of the genomic structure of the human integrin αM and αIIb subunits (adapted from [114]). The numbers indicate exons, the dotted vertical lines the exon/intron junctions that align with those in αIIb. The putative Ca²⁺ binding sites are indicated by hatched bars, and the exon containing the proteolytic cleavage site by an arrow.

Table 5. Integrin Genes

Integrin		Chromosomal localisation	Exons	Gene size		Promoter characterized
Protein	Gene			kb	Reference (*)	Reference (*)
α1	ITGA1	5				
α2	ITGA2	5q23-q31				Zutter, 1994
α3	ITGA3		26	36.3	Jones, 1998	
α4	ITGA4	2q31-q32				Rosen, 1994
α5	ITGA5	12q11-q13				Birkenmeier, 1991
α6	ITGA6	2				
α7	ITGA7	2q13				
α8	ITGA8					
α9	ITGA9					
α10	ITGA10					
αIIb	ITGA2B	17q21.32	30	22	Prandini, 1988	Prandini, 1988; Uzan, 1991
αv	ITGAV	2q31-q32				Donahue, 1994
αL	ITGAL	16p11.2				Cornwell, 1993
αM	ITGAM	16p11.2	30	55	Fleming, 1993	Hickstein, 1992; Pahl, 1992
αX	ITGAX	16p11.2				Lopez-Cabrera, 1993
αD	ITGAD					
αE	ITGAE	17				
β1	ITGB1	10p11.2				Cervella, 1993
β2	ITGB2	21q22.3	16	40	Weitzman, 1991	Agura, 1992
β3	ITGB3	17q12-q21	15	>50	Lanza, 1990; Zimrin, 1990	Villa-Garcia, 1994
β4	ITGB4	17q25.1-q25.5			Iacovacci, 1997; Pulkkinen, 1997	
β5	ITGB5	3				
β6	ITGB6					
β7	ITGB7	12q13	14	10	Jiang, 1992	
β8	ITGB8	7				
β9	ITGB9					

(*) Not all references indexed in the bibliography section

sites have been identified that are involved in tissue-specific integrin gene expression. Megakaryocyte-specific αIIb expression depends on a cooperative interaction between GATA-1 (an erythroid and megakaryocytic-specific transcription factor) and Ets-1 [128]. Ets recognition sequences, which have been identified in a number of integrin gene promoters including αIIb, αv, α4, α5, αM and β2, have also been shown to act as tissue-specific gene expression regulatory sites for integrin β2 [129]. Another member of the ets family which has tissue-specific activity is the lymphoid and myeloid-specific transcription factor PU. 1, found primarily in macrophages and B cells [130]. PU.1 binding sequences have been identified in the promoters of αL, αM, αX, α2, α4, β3 and mouse β7 genes.

Integrin-dependent gene activation   A number of studies have shown that integrin-dependent adhesive interactions with the extracellular matrix can affect gene expression and regulate extracellular matrix remodelling, cell differentiation, proliferation or apoptosis. Adhesion of cells to fibronectin [131] or growth of cells in a three dimensional collagen lattice [132] causes activation of NF-κB. Integrins are considered to be critical in this process, since antibody-dependent ligation of integrins induces NF-κB activity comparable to that found following cell adhesion to fibronectin [133]. Many of the genes induced by integrins have NF-κB motifs in their upstream regulatory sequences [134], and NF-κB activity is required for the expression of a group of genes encoding vascular cell adhesion molecules, such as V-CAM-1, E-selectin or ICAM-1 (reviewed in [135]). Recent data also provide evidence that NF-κB mediates $\alpha_v\beta_3$ integrin-induced endothelial cell survival [136].

Processing and fate   During the process of synthesis, integrin αβ heterodimers are formed in the endoplasmic reticulum (ER) by the association of nascent α and β monomers, followed by the transport of the correctly folded and assembled heterodimers from the ER to the Golgi apparatus [137,138]. Transport of mutant or misfolded monomers out of the ER appears to be prevented by the heat shock protein BiP (GRP78), a chaperone protein present in the lumen of the ER, and shown to associate with truncated forms of αIIb and β3 [139]; the misfolded and retained integrin monomers are eventually degraded. The integrin α and β subunits are post-translationally processed by glycosylation. N-glycosylation of the two subunits is a prerequisite for their αβ association in the endoplasmic reticulum. With the elucidation of the precise glycosylation pattern of mature αIIb, it has become apparent that not all the potential N-glycosylation sites are actually glycosylated [140,141]. Individual α subunits also vary in their content of high mannose-type and complex-type N-linked carbohydrate chains, due to tissue-specific glycosylation. Some α subunits contain additional O-linked carbohydrate residues. The heterodimer formation itself influences further processing of N-glycosylation in the Golgi stacks. Integrin α subunits that lack the I domain also undergo post-translational cleav-

age in the Golgi, giving rise to an extracellular heavy chain covalently linked by a disulphide bond to a small membrane-spanning light chain. A common dibasic sequence followed by several hydrophobic amino acids marks the cleavage site which is encoded by a distinct exon, absent in the non-cleaved α subunits. There is however an exception for the αE subunit, which is both cleaved and contains an I domain [113]. The α4 subunit is also an exception in that it is cleaved further upstream giving rise to two fragments of similar size (80 and 70 kDa) [142]. Although the precise mechanism of this endoproteolytic cleavage is still unknown, furin present in the Golgi stacks, could be a potential candidate involved in this cleavage process.

Role in vascular biology
Physiological function   Integrins are key receptors regulating major cell interactions in the vasculature. As leukocytes and platelets circulate in the bloodstream in close contact with red cells and the endothelial cell layer of the vessel wall, their adhesive properties must be tightly regulated to prevent platelet or leukocyte adhesion to the endothelium, platelet thrombus formation or platelet-leukocyte aggregation.

Leukocyte-endothelial cell interactions   At sites of inflammation, initial contact of leukocytes with activated endothelium is mediated by a class of endothelial cell adhesion molecules called selectins, that are able to slow down flowing leukocytes, and causing them to roll on the endothelium. During this rolling process, leukocytes are exposed to activating signals, that are either released from the endothelium, such as IL-8, thrombin, C5a or TNF-α, or exposed on the activated endothelial cell surface, such as platelet activating factor or endothelial cell adhesion receptors. Each of these mediators has the capacity to increase integrin-mediated leukocyte adhesion to the endothelial cell ligands ICAM-1, ICAM-2, V-CAM-1 and fibrinogen. Activation-dependent attachment through integrin receptors converts transient rolling interactions into firm adhesion. Neutrophils flatten on the endothelium and subsequently undergo dramatic changes in shape, allowing them to diapedese through the interendothelial junctions of the vessel wall and migrate to the inflammatory sites. Lymphocytes and monocytes differ from neutrophils in that they express β1 integrin receptors including $\alpha_4\beta_1$. This integrin supports lymphocyte tethering and rolling in flow on VCAM-1 [143]. Moreover, integrin $\alpha_4\beta_7$ can support lymphocyte attachment to mucosal addressin under shear forces in the absence of selectin involvement [144].

Platelet interactions with thrombogenic surfaces   Circulating platelets are non-thrombogenic and do not interact with other vascular cells. However, following vascular injury, platelets adhere avidly to exposed sub components, where they become rapidly activated. Subsequent interactions of circulating platelets to the adherent platelets produce the hemostatic plug, that promotes thrombin generation and formation of a stable fibrin clot. Platelets adhere efficiently to the subendothelium under high shear stress present in arteries but not in. The initial contact of platelets with the subendothelium is mediated through the platelet GPIb-IX-V complex binding to immobilized von Willebrand factor (reviewed in [145]). Flowing platelets appear to attach transiently to von Willebrand factor, resulting in continuous movement of the cells along the surface [146]. In contrast, under the lower shear stress found in veins, nonactivated platelets use integrin $\alpha_{IIb}\beta_3$ to attach to immobilized fibrinogen or fibrin [147]. Under these conditions, other platelet integrins might also be involved in platelet attachment to fibronectin, laminin or thrombospondin. Once platelets adhere, they become activated by secreted products such as ADP or epinephrine, or by surface molecules such as collagen, that cross-link integrin $\alpha_2\beta_1$ and other collagen receptors. Activated platelets adhere and spread more avidly to the subendothelium, essentially through binding of activated $\alpha_{IIb}\beta_3$ to fibrinogen, which recruits additional platelets into the thrombus. Shear resistant adhesion is further enhanced by interactions of other integrins with subendothelial components. With the generation of thrombin, fibrinogen is converted into fibrin and aggregated platelets contract to strengthen the clot. Finally, signalling through integrin receptors further regulates the cytoskeletal protein redistribution allowing eventual clot retraction.

Integrins in vasculogenesis and angiogenesis Vasculogenesis, the *de novo* formation of blood vessels from differentiating endothelial cell precursors, the angioblasts, and angiogenesis, the sprouting of new capillaries from pre-existing blood vessels, are complex processes that have important connections to integrins, as they involve extracellular matrix remodelling, endothelial cell migration and proliferation, as well as functional maturation of new endothelial cells into mature blood vessels [148]. Multiple integrins are expressed in endothelial cells, and endothelial cell migration on extracellular matrix components such as laminin, collagen or fibronectin *in vitro*, and very likely also *in vivo*, relies on β_1 and β_3 integrins. Integrin function during blood vessel formation *in vivo* has been studied most extensively for $\alpha_v\beta_3$, which is not readily detectable in quiescent vessels but becomes highly expressed after stimulation by angiogenic growth factors: inhibition of endothelial cell $\alpha_v\beta_3$ with function-blocking antibodies or peptides, has been shown to block angiogenesis by inducing unscheduled apoptosis of proliferating vascular cells [149]. More recently, Senger et al. have provided evidence that VEGF-driven angiogenesis *in vitro* relies on upregulated expression of two collagen receptors, integrin $\alpha_1\beta_1$ and $\alpha_2\beta_1$ [150]. Also, the laminin receptor $\alpha_6\beta_1$ has been implicated in capillary tube formation *in vitro*. Consistent with the importance of integrin function during angiogenesis, targeted deletion of α_5 and α_v integrin subunits in mice resulted in embryonic vascular defects [151].

Pathology Adhesion of vascular cells to one another and to extracellular matrix proteins plays a crucial role during development, and is involved in the physiology and pathology of the vascular system. During vascular development in the embryo, cell adhesion receptors contribute to the morphogenesis of the heart and the blood vessels; normal physiological processes in the vasculature rely on blood cells and endothelial cells that adhere in appropriate locations or migrate to fulfil specific functions. Defective adhesion events generate diseases, such as bleeding or susceptibility to infection, while excessive adhesion can lead to thrombosis or chronic inflammation. Much of our knowledge on the functional role of adhesion receptors stems from studies of patients with genetic diseases affecting cell adhesion in the vasculature. The bleeding disorder Glanzmann's Thrombasthenia (GT) arises from genetic defects in the genes encoding the platelet fibrinogen receptor, integrin $\alpha_{IIb}\beta_3$ [152]. Similarly, the rare genetic disease Leukocyte Adhesion Deficiency 1 (LAD 1) affects the gene encoding the β_2 integrin subunit, resulting in the absence of the expression of all eukocyte-specific integrins, namely $\alpha L\beta_2$, $\alpha M\beta_2$, $\alpha X\beta_2$ and $\alpha D\beta_2$ [153].

So far, human genetic mutations affecting other vascular integrins have not been reported, most likely because they result in embryonic lethality. However, with the technical advances in genetic manipulation in mice, targeted gene knock-out or null mutations in most of the integrin genes has been achieved, providing important information on the role of integrins in embryonic development and morphogenesis of the vascular system. Among the 12 integrin subunit genes that have been inactivated in mice, 3 appear to be critical for early embryogenesis and vascular development, namely α_4, α_5 and α_v [151]. Interestingly, α_v-deficient mice develop almost normally and die only perinatally [154], while α_1, β_6 and β_7 deficient mice are viable [151].

Clinical relevance and therapeutic implications in vascular diseases Adhesion peptides such as RGD have provided a great deal of information about cell adhesion mechanisms and are serving as a basis for the development of new pharmaceuticals. When immobilized on a surface, these peptides promote cell attachment and have found a major pharmaceutical application in the development of biomaterials with greatly increased biocompatibility and tissue-regenerative potential [155]. In contrast, when presented in solution, these peptides are powerful antagonists for individual integrins, and show promise as potential drugs for the treatment of a number of diseases. By far the most active research activities concern the development of inhibitors of thrombosis. For this purpose several highly potent platelet integrin $\alpha_{IIb}\beta_3$ antagonists have been developed [156] that have successfully passed clinical trials and are already used in clinics. With the availability of orally active integrin $\alpha_{IIb}\beta_3$ antagonists, preventive anti-thrombotic therapy will become possible [157]. A second very active field of investigation is the development of antagonists to α_v integrins. The apparent dependence of angiogenic endothelium on $\alpha_v\beta_3$ has opened up promising therapeutic possibilities to suppress neoangiogenesis in the treatment of retinal neoangiogenesis in ocular neovascular diseases, as well as in cancer to suppress tumor growth [158]. Finally, progress in the development of β2 integrin inhibitors will eventually lead to a new class of anti-inflammatory drugs.

Nelly Kieffer

References

1. Hynes RO (1992) Cell 69:11-25
2. Hemler ME et al (1987) J Biol Chem 262:3300-3309
3. Hemler M (1990) Annu Rev Immunol 365-400
4. Kishimoto TK et al (1987) Cell 48:681-690
5. Larson RS et al (1990) Immunol Reviews 114:181-217
6. Fitzgerald LA et (1987) J Biol Chem 262:3936-3939
7. Tamura RN et al (1990) J Cell Biol 111:1593-1604
8. Eble JA et al (1997) In: Eble JA, Kühn K (eds) R G Landes Company, pp 199-217
9. Carrell NA et al (1985) J Biol Chem 260:1743-1749
10. Nermut MV et al (1988) EMBO J 7:4093-4099
11. Kretsinger HR (1976) Annu Rev Biochem. 45:239-266
12. Tuckwell DS et al (1992) Biochem J 285:325-331
13. Corbi AL et al (1987) ENMO J 6:4023-4028
14. Edwards JG et al (1988) J. Cell Sci 89:507-513
15. Colombatti A et al (1991) Blood 77:2305-2315
16. Michishita M et al (1993) Cell 73:857-867
17. Lee JO et al (1995) Cell 80:631-638
18. Qu A et al (1995) Proc Natl Acad Sci USA 92:10277-10281
19. Huizinga et al (1997) Structure 15:1147-1156
20. Fornaro M et al (1997) Matrix Biology 16:185-193
21. Palmer EL et al (1993) J Cell Biol 123:1289-1297
22. Rojiani MV et al (1991Biochemistry 30:9859-9866
23. Hughes PE et al (1996) J Biol Chem 271:6571-6574
24. Tozer EC et al (1996) J Biol Chem 271:21978-21984
25. Calvete JJ et al (1991) Biochem J 274:63-71
26. Reszka AA et al (1992) J Biol Chem 117:1321-1330
27. Chen WJ et al (1990) J Biol Chem 265:3116-3123
28. Bansal et al (1991) Cell 67:1195-1201
29. Ylänne J et al (1995) J Biol Chem 270:9550-9557
30. O'Toole TE et al (1995) J Biol Chem 270:8553-8558
31. Pierschbacher MD et al (1984) Nature 309:30-33
32. Ruoslahti E (1996) Annu Rev Cell Dev Biol 12:697-715
33. Pytela R et al (1994) Methods Enzymol 245:420-451
34. Leahy DJ (1997) Annu Rev Cell Devel Biol 13:363-393
35. Kloczewiak MS et al (1989) Biochemistry 28:2915-2919
36. Farrell DH et al (1992) Proc Natl Acad Sci USA 89:10729-10732
37. Main AT et al (1992) Cell 71:671-678
38. Leahy DW et al (1992) J Biol Chem 258:987-991
39. Krezel et al (1994) Science 264:1944-1947
40. Ugarova TP et al (1993) J Biol Chem 268:21080-21087
41. Tuckwell DS et al (1997) In: Eble JA, Kühn K (eds) RG Landes Company, pp 199-217
42. Altieri DC et al (1993) J Biol Chem 268:1847-1853
43. Staunton DE et (1990) Cell 61:243-254
44. Li R et al (1993) J Biol Chem 268:17513-17518
45. Li R et al (1995) J Cell Biol 129:1143-1153
46. Stanley P et al (1998) J Biol Chem 273:3358-3362
47. Underwood PA et al (1995) Biochem J 309:765-771
48. Staatz WD et (1991) J Biol Chem 266:7363-7367
49. Loftus JC et al (1994) J Biol Chem 269:25235-25238
50. D'Souza SE et al (1990) J Biol Chem 265:3440-3446
51. Smith JW et al (1990) J Biol Chem 265:2168-2172
52. Gulino DC et al (1992) J Biol Chem 267:1001-1007
53. D'Souza SE et al (1991) Nature 350:66-68
54. Schiffer SG et al (1995) J Biol Chem 270:14270-14273
55. Springer TA (1997) Proc Natl Acad Sci USA 94:65-72
56. Mould AP et al (1997) J Biol Chem 272:17283-17292
57. Landis RC et al (1993) J Cell Biol 120:1519-1527
58. Diamond MS et al (1993) J Cell Biol 120:545-556
59. KainataT et al (1994) JBiolChem269:26006-26010
60. Kern AR et al (1994) J Biol Chem 269:22811-22816
61. D'Souza SE et al (1988) Science 242:91-93
62. Bajt NIL et al (1992) J Biol Chem 267:22211-22216
63. Bajt NIL et al (1994) J Biol Chem 269:20913-20919
64. D'Souza SE et al (1994) Cell 79:659-667
65. Alemany M et al (1996) Blood 87: 592-601
66. Takada Y et al (1993) J Biol Chem 268:17597-17601
67. Loftus JC et al (1990) Science 249: 915-918
68. Puzon-McLaughlin W et al (1996) J Biol Chem 271:20438-20443
69. Goodman TG et al (1996) J Biol Chem 271:23729-23736
70. Faull RJ et al (1993) J Cell Biol 121:155-162
71. Altieri DC et al (1988) J Cell Biol 107:1893-1990
72. Bennett JS et al (1979) J Clin Invest 64:1393-1401
73. Crowe DT et al (1994) J Biol Chem 269:14411-14418
74. Frelinger AL et al (1988) J Biol Chem 263:12397-12402
75. Frelinger AL et al (1991) J Biol Chem 266:17106-17111
76. Shattil SJ et al (1985) J Biol Chem 260:11107-11114
77. O'Toole TE et al (1990) Cell Regulation 1: 883-893
78. Kovach NL et al (1992J Cell Biol 116:499-509
79. Kieffer N et al (1990) Annu Rev Cell Biol 6:329-357
80. Dustin NIL et al (1989) Nature 341:619-624
81. van Kooyk Y et al (1989) Nature 342:811-813
82. Diamond MS et al (1994) Curr Biol 4:506-517
83. Shimizu Y et al (1996) Immunol Today 17:565-573
84. Brown E et al (1996) Immunol Letters 54:189-193
85. Mould AP et al (1995) J Biol Chem 270:26270-26277
86. Bazzoni et al (1998) J Biol Chem 273:6670-6678
87. Clark EA et al (1995) Science 268:233-268
88. Meredith JE et (1996) Endocrine Reviews 17:207-220
89. Rosales CV et al (1995) Biochim Biophys Acta - Rev Cancer 1242:77-98
90. O'Toole TE et al (199 1) Science 254:845-847
91. Hughes PE et al (1995) J Biol Chem 270:12411-12417
92. de Melker AA et al (1997) Biochem J 328:529-537
93. O'Toole TE et al (1994) J Cell Biol 124:1047-1059
94. Filardo EJ et al (1995) J Cell Biol 130:441-450
95. Schaffner-Reckinger E et al (1998J Biol Chem 273:12623-12632
96. LaFlamme SE et al (1994) J Cell Biol 126:1287-1298
97. Law DA et al (1996) J Biol Chem 271:10811-10815
98. Jenkins AL et al (1998) J Biol Chem 273:13878-13885
99. Blystone SID et al (1996) J Biol Chem 271:31458-31462
100. Otey CA et al (1990) J Cell Biol 111:721-729
101. Kieffer JD et al (1995) Biochem Biophys Res Conim 217:466-474
102. Pfaff MS et al (1998) J Biol Chem 273:6104-6109
103. Horwitz AK et al (1986) Nature 320:531-533
104. Guan J L (1997) Int J Biochem Cell Biol 29:1085-1096
105. Hannigan GE et al (1996) Nature 379:91-96
106. Kolanus WW et al (1996) Cell 86:233-242
107. Schaller NID et al (1995) J Cell Biol 130:1181-1187
108. Liliental J et al (1998) J Biol Chem 273:2379-2383
109. Naik UP et al (1997) J Biol Chem 272:4651-4654
110. Chang DD et al (1997) J Cell Biol 138:1149-1157
111. Shattil SJ et al (1995) J Cell Biol 131:807-816

112. Kashiwagi HM et al (1997) J Cell Biol 137:1433-1443
113. Shaw SK et al (1994) J Biol Chem 269:6016-6025
114. Fleming JC et al (1993) J Immunol 150:480-490
115. Heidenreich et al (1990) Biochemistry 29:1232-1244
116. Uzan G et al (1991) J Biol Chem 266:8932-8939
117. Birkenmeier TM et al (1991) J Biol Chem 266:20544-20549
118. Rosen GD et al (1991) Proc Natl Acad Sci USA 88:4094-4098
119. Weitzman JB et al (1991) FEBS Letters 294:97-103
120. Lanza F et al (1993) J Biol Chem 268:20801-20807
121. Zimrin AB et al (1990) J Biol Chem 265:8590-8595
122. Pulkkinen LK et al (1997) Lab Invest 76:823-833
123. Iacovacci SL et al (1997) Mamm Genome 8:448-450
124. Jiang WM et al (1992) Int Immunol 4:1031-1040
125. Bray PF et al (1988) Proc Natl Acad Sci USA 85:8683-8687
126. Kim LT et al (1997) Proc. Soc. Exp. Biol. Med 214:123-131
127. Pugh BF et al (1990) Cell 61:1187-1197
128. Lemarchandel V et al (1993) Mol Cell Biol 13:668-676
129. Böttinger et al (1994) Mol Cell Biol 14:2604-2615
130. Klems MJ et al (1990) Cell 61:113-124
131. Qwarnstrom EE et al (1994) J Biol Chem 269:30765-30768
132. Xu J MM et al (1998) J Cell Biol 140:709-719
133. Rabinowich H et al (1995) Blood 85:1858-1864
134. Juliano RL et al (1993) J Cell Biol 120:577-585
135. Baldwin AS (1996) Annu Rev Immunol 14:646-681
136. Scatena M et al (1998) J Cell Biol 141:1083-1093
137. Duperray AR et al (1987J Cell Biol 104:1665-1673
138. Rosa JP et al (1989) J Biol Chem 264:12596-12603
139. Bennett JS et al (1993) J Biol Chem 268:3580-3585
140. Calvete JJ et al (1989) Biochem J 261:551-560
141. Miller LJ et al (1987) J hnmunol 139:842-847
142. Takada Y et al (1989) EMBO J 8:1361-1368
143. Alon R et al (1995) J Cell Biol 128:1243-1253
144. Berlin C et al (1993) Cell 74:185-195
145. Roth GJ (1991) Blood 77:5-19
146. Savage B et al (1996) Cell 84:289-297
147. Kieffer N et al (1991) 113:451-461
148. Strömblad S et al (1996) Trends Cell Biol 6:462-467
149. Brooks PC et al (1994) Science 264:569-571
150. Senger DR et al (1997) Proc Natl Acad Sci USA 94:13612-13617
151. Hynes RO et al (1997) Thromb Haemost 78:83-87
152. George JN et al (1990) Blood 75:1383-1395
153. Anderson DC et al (1987) Annu Rev Med 38:175-194
154. Fässler RE et al (1996) Curr Opin Cell Biol 8:641-646
155. Craig WS et al (1995) Biopolymers 37:157-175
156. Coller BS (1997) J Clin Invest 100:S57-S60
157. Coutre S et al (1995) Annu Rev Med 46:257-265
158. Friedlander MC et (1996) Proc Natl Acad Sci USA 93:9764-9769
159. Mould AP (1996) J Cell Sci 109:2613-2618

Vascular Permeability Factor

Definition   *same as vascular endothelial growth factor (VEGF)*

See: →Vascular endothelial growth factor family

Vasculogenesis

Definition   *Formation of blood vessels from endothelial precursor cells*

See: →Ontogeny of the vascular system; →Angiogenesis; →Vascular integrins

Vasculotropin

See: →Vascular endothelial growth factor family

Vasomotor Tone Regulation, Molecular Mechanisms of

Synonym: Regulation of blood pressure, adaptation of the vascular wall to tissue metabolic demand.

Introduction   Blood pressure is the product of cardiac output and peripheral resistance. Because cardiac output is relatively stable, peripheral resistance is the main determinant of the level of blood pressure. Peripheral resistance is dependent on the contraction of smooth muscle cells in the arteriolar wall, determining the arteriolar radius, as defined by the Poiseuille law. It is a diffuse and systemic phenomenon. The rigidity of the wall of large arteries is also a determinant of blood pressure components, mainly a determinant of the systolic part of pressure. Contraction of the smooth muscle cell is also one of the determinants of arterial wall rigidity. Therefore blood pressure is mainly determined by the level of contraction of the vascular smooth muscle cells. To understand the regulation of blood pressure one needs to analyse the intercellular communications and the intracellular signalling involved in the regulation of arterial smooth muscle cell tonic contraction and relaxation.

Arterial blood pressure is the main determinant of vascular wall tension. In accordance with the Laplace law, wall tension is an hemodynamic stress proportional to the pressure and the radius. In response to variations of tension, the arterial wall remodels itself increasing its thickness in relation to hypertrophy or hyperplasia of smooth muscle cells and to an increase in collagen content. Changes in smooth muscle phenotype are the basis of a structural increase in resistance associated with the functional determinants of smooth muscle cell contraction. Changes in smooth muscle cell phenotype is also sensitive to intercellular communications and intracellular signals similarly to the regulation of smooth muscle cell contraction and relaxation. Tension-induced chronic changes in arterial smooth muscle cell phenotype are probably also the pathological linkage between high blood pressure and arteriosclerosis development in large arteries.

Characteristics   The arterial wall is functionally and structurally compartmentalized. It associates an endoluminal compartment involving both plasma proteins, peptides and blood cells with endothelial cells; a medial compartment constituted of smooth muscle cells and of extracellular matrix, and an external adventitial compartment, interfacing metabolically active tissue with the arterial wall, and containing nerve terminals. To this structural compartmentation corresponds a functional compartmentation regulating the general and local vasomotor tone. The diffuse vasomotor tone defines the peripheral resistance and therefore blood pressure. The local vasomotor tone distributes the cardiac output proportionally to the metabolic activity of the different tissues.

The smooth muscle cells of the medial compartment are the effector cells of the vasomotor tone. The level of contraction and relaxation of smooth muscle cells defines the vasomotor tone. In physiology, the arterial smooth muscle cells are neither completely relaxed nor completely contracted. A complete relaxation or a complete contraction of smooth muscle cells are both not compatible with life. The communications generating and modulating the vasomotor tone have two directions, either centripetal communications, from the adventitia or the tissue to the media, or centrifugal, from the blood-endothelium compartment to the media. The intracellular signallings regulating the contractile function of vascular smooth muscle cells also act in two different ways, either signals generating mobilization of calcium from the intracellular stores and from extracellular environment to the cytosol, generating contraction, or signals immobilizing calcium outside the cytosol and generating smooth muscle cell relaxation. In the absence of extracellular communications and intracellular signals, smooth muscle cells are spontaneously relaxed. This implicates in an arterial system, the physiological constancy of tonic communications and signals of vasoconstriction. This contrasts with the spontaneous relaxation of smooth muscle cells in the venous territory.

The physiological communications generating vasoconstrictor signals in smooth muscle cells are mainly of a centripetal type. This is true for catecholamines, where vasoactive amines are liberated in neuromuscular junctions, situated at the interface between the adventitia, containing the terminal nervous varicosities, and the smooth muscle cells in the media [1]. This is also probably true for the angiotensinergic tone [2], where the proteins, renin and angiotensinogen, are adsorbed in the tissular interstitium, and generates vasoconstrictor angiotensins in a centripetal way. In contrast, vasodilatator communications can be both, centrifugal, generated by the endothelium [3], and/or centripetal, generated by the tissular metabolic activities, or by the non-adrenergic, non-cholinergic, probably mainly nitroxidergic, terminal nervous varicosities. The centripetal vasodilator communications predominate distally in the intratissular arterioles; whereas the communications dependent on the endothelium predominate in large conductance arteries, modulating the arterial diameter in relation to the peripheral metabolic demand. In the presence of endothelial dysfunction, vasodilatator centrifugal communications can become vasoconstrictive.

One of the characteristics of the agents of communications, such as catecholamines, peptides, gas as nitric oxide, is their lability. The extracellular diffusion spaces of these molecules of communications are very limited

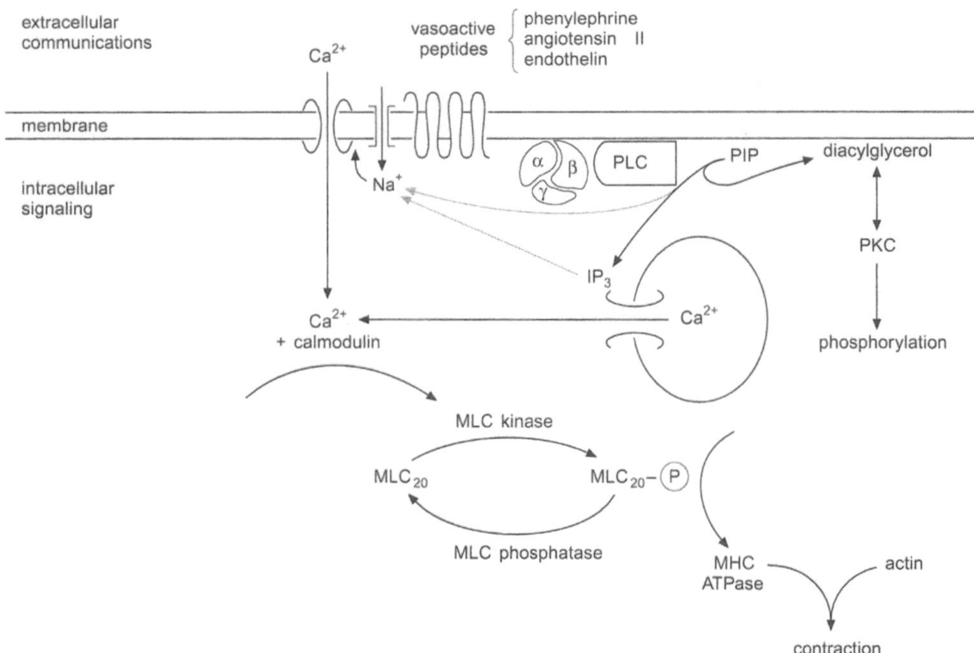

Figure 1. Extracellular communications to intracellular signalling coupling leading to contraction of smooth muscle cells. PLC, phospholipase C; PKC protein kinase C; PIP, phospho-inositides; IP3, inositol triphoshate; MLC, myosin light chain; MHC, myosin heavy chain; SMC, smooth muscle cell.

because of their quick metabolism in the vicinity of their site of production. In contrast, the syncytial organisation of smooth muscle cells allow the diffusion of intracellular signals generated by these agents, from one cell to another cell [4].

Hemodynamic stress, generated by blood pressure and flow, are able to induce signals in smooth muscle and endothelial cells [5]. Smooth muscle cells in medial position are mainly sensitive to tensional stress, the product of pressure with radius, which produce stretching of the cells. In response to this stretching, the cell contracts, and this is the myogenic tone [6]. If the heart contraction is generating blood pressure (potential energy), it is also generating kinetic energy (blood flow) which shears the arterial wall, mainly at the endothelial surface. Shear stress is proportional to the viscosity of blood, to the blood velocity, and inversely proportional to the arterial radius. As the tensional stress induces calcium signallings, phosphorylations and contraction of smooth muscle cells, shear stress induces similar signalisation in the endothelial cells.

Regulation

Genes   Formation of bridges between actin and myosin induce tonic contraction of smooth muscle cells. The smooth characteristic of the cell is defined by the absence of sarcomeric organisation of contractile proteins, which contrast with the sarcomeric organisation of cardiac myocytes (striated muscle) [7]. Preformed bridges in the sarcomeric organisation of contractile proteins, permitted the quick phasic shortening and lengthening of cardiac myocytes. In contrast, the probability of bridge frequency, between actin and myosin define the vasomotor tone in smooth muscle cells. Calcium mobilization is the first event of the contractile response of smooth muscle cells to extracellular ligand-receptor interaction. The

free intracellular calcium binds to calmoduline, and activates the myosin light chain kinase (MLCK). The phosphorylation of the myosin light chain by MLCK derepress the ATPase activity of the myosin heavy chain and therefore allows the formation of bridges between actin and myosin (Figure 1). The light chain activity is also controlled by phosphatases. Active phosphatases dephosphorylate the myosin light chain and repress contraction. Conversely, phosphorylation by protein kinase associated with small G protein Rho (Rho associated kinase) inactivates the MLC phosphatase and facilitates the action of MLC kinase. PKC also control the phosphorylation of contractile protein associated with the thin filament of actin: caldesmon and calponin [8] (Figure 2). Unphosphorylated caldesmon and calponin repress the thin filament interactions with myosin. Phosphorylations of phosphatases, caldesmon and calponin increase the duration of the contractile response to calcium. This is the phenomenon of sensitization of contractile apparatus to calcium, probably responsible for the tonic characteristic of smooth muscle cell contraction.

The interaction between extracellular vasoconstrictor agents and their cellular plasma membrane receptors leads to intracellular activation of the phospho-inositide signalling pathway, associating in biochemical cascade membranous and cytosolic events [9]. The interaction between the extracellular first messenger and its receptor modifies the allosteric conformation of the receptor and via a G-protein (GTP dependent protein) leads to activation of the phospholipase C. The G-protein is an heterotrimer of three subunits α, β, γ, associated in a quiescent state, and dissociated after activation. The αq subunit is specific of the coupling between the receptor and the phospholipase C. Phospholipase C cleaves membranous phospho-inosities in inositol phosphates, mainly IP3, and a lipid skeleton: the diacylglycerol. IP3 is

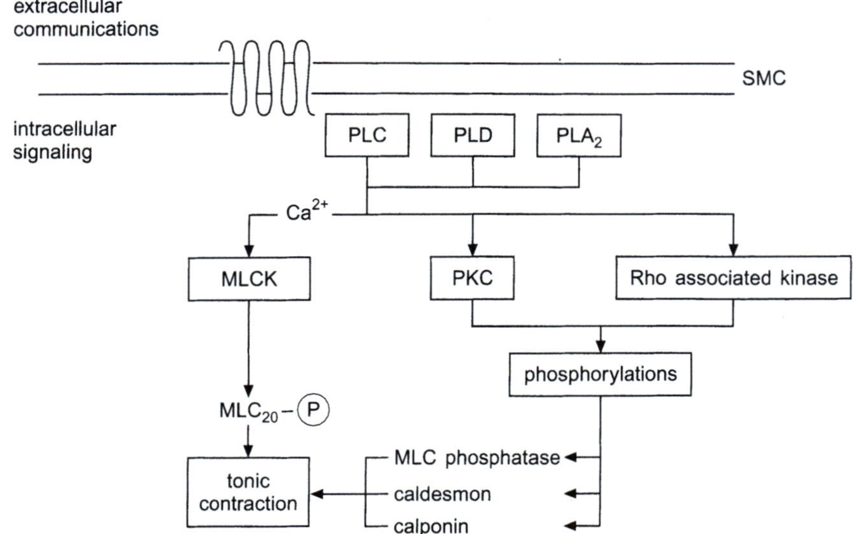

Figure 2. Intracellular signallings involving calcium mobilization and contractile protein phosphorylation leading to tonic contraction of smooth muscle cell. LC, phospholipase C; PLD, phospholipase D; PLA2, phospholipase A2; MLK, myosin light chain kinase; SMC, smooth muscle cell.

a sugar water-soluble, which solubilized in the cytosol can bind specifically a receptor-channel on the sarcoplasmic reticulum [10]. The binding of IP3 to the receptor-channel, open this later, permitting a gradient-dependent flux of calcium to be driven from the reticulum stores to the cytosol through the channel. The binding of IP3 to the receptor-channel leads to a quick increase in the free cytosolic calcium, the calcium transient. This IP3-dependent calcium transient will be followed by depolarization of the membrane and by the opening of the voltage-dependent, type L, calcium channel, driving calcium from the extracellular compartment to the intracellular cytosol.

The diacylglycerol associates with the membrane and activates protein kinase C. Protein kinase C is a cytosolic inactive enzyme in a quiescent state. In presence of free diacylglycerol at the membrane, protein kinase C translocates from the cytosol to the membrane, activates it, permitting its enzymatic phosphorylating activity towards its different targets. Activity of protein kinase C participates to the vasomotor tone by phosphorylating the caldesmon and the calponin sensitizing the contractile apparatus to calcium.

The extracellular vasoconstrictor substances which bind to seven transmembrane receptor, also interact with phospholipase A$_2$ an d D. Phospholipase A$_2$ is a cytosolic calcium-dependent enzyme which translocates to the membrane and generates arachidonic acid from membranous phospho-inositides, leading to the formation of prostaglandins by the action of cyclo-oxygenase; The phospholipase D activity generates phosphatidic acid and diacylglycerol. These different lipid pathways increase the oxydative stress within the cell leading to production of more superoxide anion [11]. The overproduction of free radicals within the smooth muscle cell has numerous consequences on extracellular communications and intracellular signalling. For example free radicals diffuse through the cell membrane and can participate in the oxidation of lipoprotein. Free radicals can decrease the efficacy of nitric oxide to vasodilate by producing peroxinitrites. Oxidative stress modulates gene expression within the smooth muscle cell, increasing the expression of genes depending on NF kappa B, and giving a pro-inflammatory phenotype to the cells. Receptor dependent oxidative stress is therefore involved in the regulation of vasomotor tone, but also in vascular wall remodelling, smooth muscle cell proliferation, and development of arteriosclerosis. Oxidative stress is probably the link by which hypertension increases the risk of vascular disease.

On the opposite side of the vasoconstrictor signalling pathway, the activity of the cyclic nucleotide pathways are to partly and permanently inhibit the vasoconstrictor tone within the smooth muscle cell. The cyclic Guanosine Monophosphate (cGMP) pathway is made up of three main components: the soluble guanylate cyclase, generating cGMP, the second messenger, and the G-kinase, an

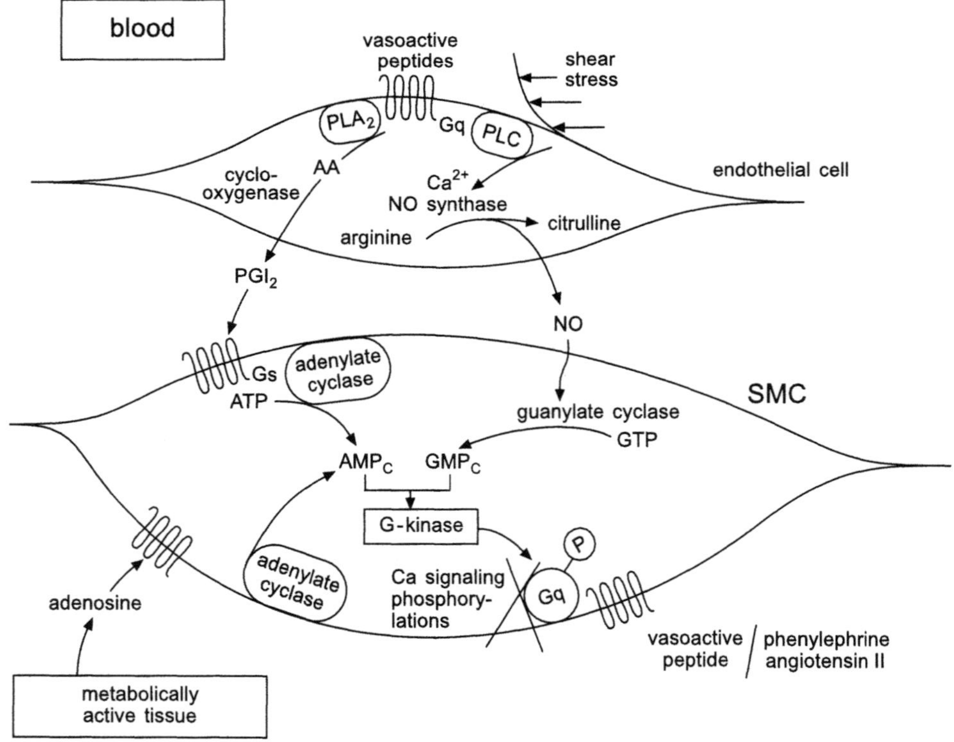

Figure 3. Centrifugal and centripetal communications and signallings leading to smooth muscle cell relaxation. AA, arachidonic acid; PLC, phospholipase C; PLA2, phospholipase A2; PGI$_2$, Prostaglandin I$_2$; Gq, G protein αq; NO, nitric oxide.

inactive cytosolic protein, activated by cGMP. The soluble form of guanylate cyclase is an heme protein, present in the cytosol of smooth muscle cells. The nitric oxide (NO) and the carbon monoxide (CO) by binding the heme radical, increase the cGMP production by the guanylate cyclase. The soluble guanylate cyclase is very different from the particulate, natriuretic peptide-sensitive, guanylate cyclases (Figure 3). The first one is a cytosolic enzyme sensitive to diffusible gas, NO and CO, whereas the second is a membranous receptor-enzyme protein. Both are product cyclic GMP from GTP. The binding of gas to the guanylate cyclase is antagonised by hemoglobin, in which NO and CO binds the heme radical, and by methylene blue which binds the heme radical of the soluble guanylate cyclase.

Cyclic GMP bind the type I isoform of the G-kinase in smooth muscle cells, whereas natriuretic peptide-generated cGMP bind the type II isoform of G-kinase in epithelial cells. Smooth muscle cells are very rich in type I G-kinase [12]. The G-kinase is a double chain protein, one chain supports the enzymatic activity, the second chain is constituted of two domains, one, the pseudo-substrate domain, inhibits the active site of the other chain, the second domain binds cyclic nucleotides. In a quiescent state, the G-kinase is inactive, the pseudo-substrate domain inhibits the active site. Binding the cGMP changes the allosteric conformation of both the chain interaction, and reveals the enzymatic activity. Therefore the activation is immediate and reversible depending on the intracellular concentration of cyclic GMP. The intracellular molecular targets of cGMP-activated G-kinase are multiple in smooth muscle cells and not all completely identified leading to smooth muscle relaxation.

Cyclic adenine monophosphate, cAMP, is the second cyclic nucleotide involved in vascular smooth muscle cell relaxation. Cyclic AMP is generated by extracellular communications able to activate receptors positively coupled to adenylate cyclase by a Gs protein. As cGMP, cAMP also bind to G-kinase in smooth muscle cells leading to relaxation.

Cyclic nucleotide metabolism is under the control of phosphodiesterase (PDE) activities. In smooth muscle cell, phosphodiesterase V is present and specific of cGMP metabolism. *In vivo* and *in vitro* inhibition of PDE V leads to the potentiation of vasodilator tone. Cyclic AMP is metabolized by rolipram inhibitable PDE IV activity and by PDE II [13].

Molecular Interactions

The physiological basis of vasomotricity and blood pressure regulation is the cross-talk between the signalling pathways of relaxation and contraction.

The G-kinase is mainly involved in relaxation by phosphorylating phospholambans associated with calcium ATPase, and by blocking the interaction between the receptors and the phospholipase C activity. Unphosphorylated, the phospholambans bind the calcium ATPase and inhibit its calcium pumping ability; phosphorylated phospholambans increase the ability of calcium pumping from the cytosol

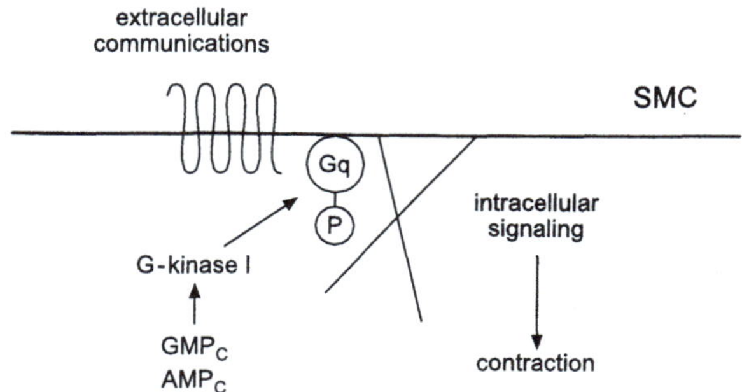

Figure 4. Cyclic nucleotides and G-kinase induced uncoupling between contractile extracellular communications and tonic intracellular contractile signalling in smooth muscle cells.

to the sarcoplasmic reticulum and decrease calcium mobilization. Mainly the G-kinase, by phosphorylating the coupling G-protein between the receptor and the phospholipases, is able to uncouple the extracellular ligand-activated receptor (1st messenger) from its intracellular signalling pathways (2nd messenger) leading to the loss of efficient coupling between vasoconstrictor agents and smooth muscle cell response [14] (Figure 4). The less the G-kinase is activated, the more the coupling between vasoconstrictive communications and vasoconstrictive response is efficient. Similarly, the more the G-kinase is activated, the less the coupling to vasomotor tone is efficient. It is probably by this cross-talk between these two signalling pathways that the vasomotor tone is permanently regulated in function of the metabolic activity.

Another example of molecular interaction is the interaction between the metabolism of the two cyclic nucleotides within the smooth muscle cell through the PDE II activity. PDE II is able to metabolize both cGMP and cAMP; cGMP has a higher affinity than cAMP for PDE II but a smaller velocity (Vmax) of consumption. In contrast cAMP has a smaller affinity but a higher Vmax. Therefore in a physiological situation, cyclic GMP competes with the PDE II activity on cAMP metabolism. Associated with cGMP deficiency PDE II becomes very efficient and predominant in the metabolism of cAMP.

Cells and Cellular Interactions   Vasomotor intracellular signallings are under the control of extracellular communications leading to activation of constrictor or relaxant signals. Vasoconstrictor and vasodilatator communications can be centrifugal from the blood to the smooth muscle cells through the endothelial cells, or centripetal from the adventitia or the surrounding tissue to smooth muscle cells.

Among the diffuse and systemic extracellular communications, catecholamines, angiotensin and sodium play predominant roles in generating vasomotor tone. Catecholamines are centripetal communications secreted by the sympathic nervous system. In the adventitial neuromuscular junctions, catecholamines bind to α_1-adrenergic receptor present on the smooth muscle membrane. From one cell the signal diffuses to other smooth muscle cells by the syncitial organisation [4]. The α_1-adrenergic receptor is coupled to phospholipase activities in smooth muscle cell. The release of catecholamine is constant. The medulla adrenal gland also releases catecholamines into circulation. Complete suppression of sympathetic activity leads to a 60 mmHg decrease in blood pressure.

The renin angiotensin system is a more complex system including two endocrine protein interaction: the active renin secreted by the juxta glomerular apparatus in the kidney, and the substrate, the angiotensinogen, mainly secreted by the liver. The secretion of renin is highly and permanently regulated by arterial perfusion pressure, ionic concentration in the macula densa of the distal tubule, β-adrenergic stimulation, and negative feedback of angiotensin II. Renin and angiotensinogen are circulating but mainly diffuse in all the vascular tissue. 20 % of the renin substrate reaction occur in the plasma, 80 % in the tissular interstitium [15]. Angiotensin I is converted into angiotensin II by the converting enzyme. Angiotensin II binds an AT1 receptor on smooth muscle cell coupled to phospholipase activities and to calcium mobilization. Angiotensin II also potentiates the adrenergic activity.

Extracellular sodium is the third communication involved in the diffuse vascular constrictor tone [16]. Excess of extracellular sodium leads to the facilitation of voltage dependent calcium channel opening. Sodium homeostasis is mainly regulated by the kidney pumps and by the endocrine system such as aldosterone. They are important cross-talks between sodium homeostasis and the renin angiotensin system: excess of sodium chloride in the macula densa decrease renin secretion and sodium depletion increase renin. This interaction is mediated by urinary ionic concentration, Na.K/2Cl. cotransport and neuronal NO synthase activity in the epithelial cells of the macula densa [17], and by specific expression of G-kinase II epithelial isoform in renin-secreting myo-epithelioïd cells of the afferent arteriole [18]. Nevertheless, the exact relationship at a cellular level, between excess of extracellular salt retention and excess of intracellular signalling leading to smooth muscle cell contraction remains to be determined.

Other peptides such as endothelin, thrombin, thromboxane are capable of generating vasoconstriction via smooth muscle cell receptors coupled to phospholipases; but these agents are more involved in the pathological local control of vasomotor tone in response to hemorrhage ischemia than in a physiological systemic regulation of blood pressure. Endothelin, thrombin, thromboxane are mainly centrifugal vasoconstrictive pathological communications locally generated by the interaction of the blood with the endothelium.

On the other hand vasodilating communications can be both centrifugal from the blood-endothelium compartment and centripetal from the metabolically active tissue to the vascular wall. Nitric oxide (NO) is the main communication from the endothelium to the smooth muscle cell. NO is produced from arginine in the endothelium by the activity of constitutive NO synthase III [19]. Systemic blockade of NO production lead to hypertension and generalized vasoconstriction [20]. Nevertheless, NO production is mainly involved in the regulation of local vasomotor tone of conductance arteries in function of the metabolic needs [21]. The activity of NO synthase III depends on the signalling pathways able to mobilize calcium and phosphorylations in endothelial cells [22]. Two great classes of communications are able to activate the endothelial NO synthase: pharmacological agents coupled to the phospholipase pathway in the endothelial cell, and hemodynamic shear stress generated by the blood flow on the endothelial cells. Seven transmembrane receptors coupled to phospholipase activities and calcium mobilization within the endothelial cells, increase the release of NO in response to circulating pharmacological active peptides such as acetylcholine on endothelial muscarinic receptor, bradykinin on the β_2 receptor, endothelin, purin, stimulation α_2-adrenergic, histamine, angiotensin II. The majority of these agents could be vasoconstrictor by acting directly on smooth muscle cells. They are ambiguous, their final direct vasoconctrictive or endothelium-dependent vasodilatator action depending on their respective bioavailability in the blood-endothelium compartment (centrifugal vasodilator communications) or in the external interstitial compartment (centripetal vasoconstrictor communications) or in the absence of endothelium. In the initial experiment of Furchgott [3], acetylcholine was vasodilatator of a precontracted vessel in the presence of endothelium and vasoconstrictor in the absence of endothelium. Growth factors for endothelium such as FGF and VEGF coupled to tyrosine-kinase receptors are also able to increase endothelial NO synthase activity by increasing the level of phosphorylation within the endothelial cells. The main factor influencing NO synthase activity is the hemodynamic shear stress of liquid moving blood on the fixed endothelial surface. The shear generates signals of calcium mobilization and phosphorylations within the endothelial cells. Blood velocity and vascular radius are the main determinants of shear stress: increased blood velocity increases shear whereas increased radius decreases shear stress. NO stimulates soluble guanylate cyclase and produce cGMP within the smooth muscle cell.

Centripetal vasodilator communications are mainly represented by tissular metabolic activity, such as adenosine, interacting with the α_2 receptor coupled to adenylate cyclase and generating cAMP in smooth muscle cells. Schematically, tissular direct metabolic-dependent centripetal vasodilator communications are cAMP dependent; whereas centrifugal endothelium dependent vasodilator communications are cGMP dependent. When the metabolic demand increases in relation with peripheral muscle activity for example, the metabolically active tissue release cAMPdependent short-living communications such as adenosine leading to strictly local vasodilation of terminal arterioles. Therefore the driving pressure and the blood velocity increase in the vasculature of this metabolically active territory, leading to more shear stress and to endothelium dependent vasodilatation of conductance arteries specific of this territory.

There are other possible vasodilator agents. Prostacyclin, produced by endothelium, is an example of a centrifugal cAMP dependent vasodilatation. Carbon Monoxide can be produced by the heme oxygenase in the smooth muscle cell itself in response to induction by heme or shear stress. CO is able to generate vasodilation by stimulating soluble guanylate cyclase as NO.

Additional Features   In summary, vasomotor tone in the arterial system can be described as a two compartment model. The first one is the extracellular compartment involving both vasoconstrictors and dilatators communications. The second one is the intracellular signallings leading to the increase or the decrease in actin-myosin bridge formation within the arterial wall. Diffuse extracellular vasoconstrictor communications and intracellular signallings control the blood pressure level via the peripheral resistances. Local vasodilatator communications, by locally modulating vasoconstrictor signalling, mainly control the repartition of blood flow in function of the metabolic demand.

Clinical Relevance   Clinical relevance of this model is shown by the characteristics of primary versus secondary hypertension. Secondary forms of hypertension are usually in relation to diffuse excess of constrictor communications. For example, renovascular hypertension is related to an excess of renin, Conn syndrome to an excess of aldosterone and sodium retention, pheochromocytome to an excess of catecholamines. Several identified genetic forms of hypertension could be also easily related to an excess of communications: Liddle syndrom, Ulick syndrome. In contrast, essential hypertension could be related to primary overfunction of intracellular vasoconstrictor signalling. Experimentally, smooth muscle cell from spontaneously hypertensive rats (SHR), in similar environmental conditions, growth faster than smooth muscle from normotensive controls. Similarly, constrictor signals are amplified in SHR smooth muscle cells independantly of the nature of the stimulus as compared to cells from normotensive controls.

Experimentally, the chronic blockade of NO production by substituted arginine antagonist lead to a model of hypertension. This model is characterized by a general decrease in blood flow which is unusual in other models of hypertension. In hypertension there is probably no true decrease in NO production. In contrast, because the overproduction of free radicals, the efficiency of NO to produce vasodilatation, is probably decreased in hypertension. Clinically, NO and adenosine generating cGMP and cAMP respectively in vascular smooth muscle cell are mainly involved in the adaptation of the vasculature to exercise rather than in the definition of hypertension. Therefore diffuse NO dysfunction is probably implicated in the peripheral syndrome of congestive heart failure [23, 24] which is characterized by a desadaptation of vasomotricity to exercise.

On the other hand, local dysfunction of vasomotor tone is also involved in pathology. For example, local excess of vasoconstrictor communications due to local overproduction of endotheline lead to cerebral vasospasm associated with subarachnoidal haemorrhage. Local defect of endothelial-dependent vasorelaxation could lead to the desadaptation of the coronary vasculature to the myocardial metabolic demand during exercise in atherosclerotic patients.

Jean-Baptiste Michel

References

1. Klemm MF et al (1993) J Compar Neurol 334:159-167
2. Danser AH et al (1995) Cardiovasc Res 29:789-795
3. Furchgott RF, Zarvadzki JV (1980) Nature 288:373-376
4. Christ GJ et al (1996) Circ Res 79:631-646
5. Owens GK (1996) Circ Res 79:1054-1055
6. Tedgui A et al (1997 Med/Sci 13:790-798
7. Horowitz A et al (1996) Physiol Rev 76:967-1003
8. Fattoum A (1997) Med/Sci 13:777-789
9. Michel JB, Arnal JF (1992) Presse Med 21:1188-1195
10. Berridge MJ (1995) Biochem J 312:1-11
11. Rajagopalan S et al (1996) J Clin Invest 97:1916-1923
12. Ecker T et al (1989) Circ Res 65:1361-1369
13. Lugnier C, Komas N (1993) Eur Heart J 14:141-148
14. Ruth P et al (1993) Proc Natl Acad Sci (USA) 90:2623-2627
15. Danser AHJ et al (1994) Hypertension 24:37-48
16. Denton D et al (1995) Nature Medecine 1:1009-1016
17. Obermuller N et al (1996) J Clin Invest 98:635-640
18. Gambaryan S et al (1996) J Clin Invest 98:662-670
19. Palmer RMJ et al (1988) Nature 333:664-666
20. Arnal JF et al (1992) J Clin Invest 90:647-652
21. Michel JB, Arnal JF (1993) Med/Sci 9:1061-1067
22. Dinerman JL et al (1993) Circ Res 73:217-222
23. Calver A (1994) Lancet 344:887-888
24. Sessa WC et al (1994) Circ Res 74:349-353

VCAM-1

Definition   *Vascular cell adhesion molecule-1*

See: →Vascular cell adhesion molecule-1

VE-cadherin

Definition   *Vascular endothelial cadherin*

See: →Cadherins

VEGF

Definition   *Vascular endothelial growth factor*

See: →Vascular endothelial growth factor family

Vitronectin/Vitronectin Receptors

Synonym: Complement S-protein, serum spreading factor, epibolin

Definition *Group of adhesive molecules that play an important role in the attachment of vascular cells and participates in the regulation of cell proliferation, differentiation and morphogenesis.*

See also: →Vascular integrins; →Extracellular matrix

Introduction Vitronectin belongs to the group of adhesive glycoproteins that play key roles in the attachment of cells to their surrounding matrix and may participate in the regulation of cell differentiation, proliferation, and morphogenesis [1, 2]. Vitronectin exerts several regulatory functions in the blood coagulation, complement and fibrinolytic systems, raising the possibility that the protein provides a unique link between cell adhesion and proteolytic enzyme cascades. Vitronectin biosynthesis may be regulated in disease states, and as a circulating and diffusible plasma protein, it is found immobilized in various tissues under (patho-) physiological conditions (Table 1). Moreover, vitronectin is a structurally labile molecule, and its biological functions are dependent and/or regulated by its conformational state(s). Predominantly multimeric forms of vitronectin are recognized by several non-related cell surface receptors including integrins, the urokinase-receptor and proteoglycans that mediate cell adhesion, migration and cellular invasion relevant for tissue remodelling particularly in the vasculature or for bacterial tropism. Although deletion of the vitronectin gene in mice is compatible with life [3], functional activities of tissue vitronectin *in vivo* may be uncovered by challenge of vitronectin knock-out mice with different pathologies.

Characteristics

Molecular Weight Human vitronectin has a molecular mass of 75-78 kD as analyzed by denaturing polyacrylamide gel-electrophoresis and circulates in single-chain and two-chain (65 kD and 10-12 kD) forms. The ratio between the 75-78 kD and 65 kD forms is variable, is genetically determined and depends on the amino acid sequence at position 381 [4]. Under non-reducing conditions, the N-terminal 65 kD and C-terminal 10-12 kD fragments are linked by a single disulfide bond. The human vitronectin cDNA encodes a molecule of 459 amino acids comprising a molecular weight of 52,400. Accordingly, the remainder of the mass of the molecule is made up by posttranslational modifications which include N-linked carbohydrate, sulfatation and phosphorylation [5]. Physicochemical parameters of vitronectin are summarized in Table 2.

Domains Vitronectin is a multifunctional protein, and different domains along the vitronectin sequence are involved in various binding interactions. The N-terminal somatomedin-B domain contains 8 of the 14 cysteine residues which are predicted to form four intrachain disulfide links. This domain is followed by a connecting region which contains the RGD sequence (residues 45 to 47), known as the versatile integrin-recognition motif, and a highly acidic region (residues 53 to 64) containing two sulfatation sites. The short acidic portion presents considerable sequence homology with regions in other proteins, relevant for haemostasis (Table 3). This part of the molecule is followed C-terminally by seven hemopexin-type repeats [6]. The hemopexin-type modules were first described in hemopexin, the heme-binding protein in plasma that has a symmetrical two-domain structure connected by a flexible hinge region

Table 1. Distribution of vitronectin

Vitronectin in body fluids:
- blood plasma (200–400 µg/ml) (reduced levels in patients with liver diseases)
- platelet a-granules (0.8% of the circulating pool)
- bronchoalvoelar lavage fluid (increased in interstitial lung disease)
- seminal plasma
- urine
- amniotic fluid
- cerebrospinal fluid

Vitronectin in tissues:
- skin (increased depositions along elastic fibres/collagen with aging and certain skin disorders)
- skeletal system, mineralized bone matrix
- associated with lamina elastica in normal vessel wall
- increased depositions in areas of sclerosis and fibrosis in various disorders (e.g. viral hepatitis, liver cirrhosis, myocardial and kidney infarction, degenerative nervous system disorders, intimal thickenings and fibrous plaques of atherosclerotic arteries, diabetic retinopathy)
- absent in most normal tissues, including basement membranes

Expression of vitronectin mRNA:
- hepatocytes as a major source of plasma form (human, rabbit, mouse, rat)
- lower levels in all other major human organs; not in blood, murine organs
- during murine development as early as day 10 in liver and central nervous system
- biosynthesis is upregulated in the liver under acute phase conditions

whereby both domains are made up of four homologous repeats. The second hemopexin-like domain contains the basic heparin-binding site which is represented by a 40 amino-acid stretch rich in Arg and Lys residues, representing two heparin-binding consensus sequences. Adjacent to the basic domain is the site for cAMP-dependent phosphorylation by protein kinase A [7], which may change the functional properties of this region.

Binding Sites and Affinity The N-terminal portion of the vitronectin molecule comprises a high-affinity binding site for plasminogen activator inhibitor-1 (PAI-1) [8] (Figure 1) and appears to be unique, since somatomedin-B-typical domains in other factors do not function as PAI-1 binding sites. Different isoforms of somatomedin-B peptides are present in human serum which were erroneously attributed to mitogenic activity [9]. Urokinase/urokinase-receptor complexes may interact with an overlapping binding site within the N-terminal region of vitronectin as well [10, 11]. Directly adjacent lies the RGD-motif that is recognized by several integrins, involved in cell adhesion and migration. This is followed by a cross-linking site (Gln93) that is required for inter-molecular linkage of vitronectin mediated by factor XIIIa or tissue transglutaminase [2], followed by a collagen binding domain (Figure 1). Within the first hemopexin repeats in vitronectin, two adjacent (conformation-dependent) regions with heparin binding capacity (residues 82 to 137 and 175 to 219) were identified by phage display. Repetitive hydrophobic sequences within the hemopexin-type repeats serve as specific binding sites for infectious group A streptococci (Figure 1).

The basic cluster towards the C-terminus has been implicated in the binding of vitronectin to heparin and other glycosaminoglcyans, including dextran sulfate, fucoidan as well as cellular heparan sulfate proteoglycans. Moreover, this site particularly exposed in vitronectin multimers is engaged in specific binding interactions with collagen (type I), osteonectin, PAI-1 and plasminogen [12].

Additional Features The identification of additional heparin binding regions distinct from the C-terminal basic cluster supports the notion that native vitronectin is heparin binding competent under certain conditions, since addition of glycosaminoglycans to unfractionated plasma resulted in the formation of covalently- and non-covalently stabilized vitronectin multimers. Likewise, in the presence of heparin, phosphorylation of the otherwise unphosphorylated two-chain vitronectin occurs indicating subtle conformational alterations of native vitronectin induced by the glycosaminoglycan. Thus, different heparin-binding domains appear to be operative both in the native and in the multimeric forms that are involved in the induction or stabilization of multimerization, respectively (Figure 2).

Structure

Sequence and Size The primary structures of human, mouse, rat, rabbit and porcine vitronectins [13-17] share more than 80 % sequence homology, whereby porcine vitronectin has a truncated compact functional form that lacks the C-terminal 10-12 kD fragment. These vitronectins contain N-linked and also O-linked carbohydrates which differ from the human counterpart. A 54 kD protein with high structural similarity to vitronectin, designated nectinepsin [18], was found in chicken and resembles the formerly described vitronectin-like factor isolated from egg-yolk.

Homologies Based on identical spacing of cysteine residues, the somatomedin-B-domain can be found in a number of other proteins [19] (Table 3). Several other extracellular matrix proteins share the RGD-motif with vitronectin, yet their specificity and selectivity of integrin recognition differs and is mostly dependent on the RGD-flanking regions in these proteins. The acidic stretch is found with appreciable sequence homology in other factors important in haemostasis (Table 2). Vitronectin may be considered as a distant member of the hemopexin family of proteins, including hemopexin itself and matrix metalloproteinases, among others. The C-terminal basic cluster contains two heparin-binding

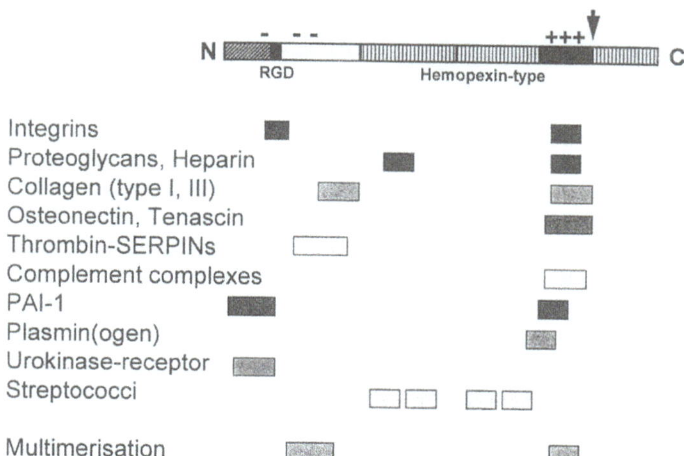

Figure 1. Vitronectin structure and ligand binding domains. A schematic representation of the domain organization of human vitronectin is shown indicating the somatomedin-B region at the N-terminus, the major cell attachment site (RGD), hemopexin-like repeats and the glycosaminoglycan-binding site (+++). The arrow points to the position of an endogenous protease cleavage site which is adjacent to the protein kinase A-dependent phosphorylation site. Binding sites in vitronectin for macromolecular ligands (listed at the left margin) are highlighted by bars according to their position along the primary structure. Note that many binding domains are concentrated around the acidic (- -) N-terminus and the basic (+++) C-terminus, both of which are believed to be partially cryptic in the circulating plasma protein and become exposed upon conformational activation (taken from [12]).

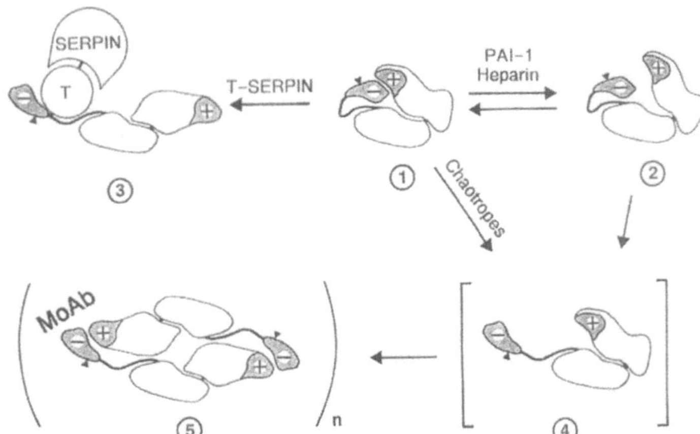

Figure 2. Hypothetical model for the interconversion of different conformational forms of vitronectin. Plasma vitronectin (1), which constitutes the predominant form of the adhesive protein in the circulation, has an RGD attachment site (arrowhead) embedded within the acidic N-terminal portion of the molecule which is at least partially cryptic. The same consideration applies to the complementary charged basic cluster at the C-terminus (heparin-binding domain), resulting in intra-molecular stabilization of the entire protein molecule. High doses of heparin that possibly interact with plasma vitronectin at accessible binding sites within the middle portion as well as reaction with PAI-1 result in (reversible) partial unfolding (2) and subsequent multimerization (5), exposing binding sites for macromolecular ligands and conformation-specific monoclonal antibodies (MoAb). A similar conformational transition of plasma vitronectin appears to be provoked by chaotropes or basic peptides, whereby an intermediate state of the unfolded molecule (4) is proposed. Upon removal of denaturants, spontaneous multimerization into high molecular weight products (5) occurs including rearrangement of the C-terminal disulfide bond. The inter-molecular stabilization may at least in part be mediated by complementary charged surface domains and hydrophobic interactions of subunits. Likewise, during blood clotting, initial reaction of thrombin-SERPIN (T-SERPIN) complex with plasma vitronectin induces profound conformational change(s) subsequently leading to a disulfide-stabilized ternary association product (3), that spontaneously forms multimers (not shown). The functional properties of these physiological complexes (3) and isolated multimers (5) are almost indistinguishable, including binary interactions with other proteins, glycosaminoglycans, cells or isolated receptors. Multimeric forms of vitronectin are also found in platelet releasate or in the extracellular matrix and contribute to multivalent interactions at sites of vascular remodeling (taken from [12]).

consensus sequences and shares this property with other glycosaminoglycan-binding proteins (Table 2). Other sites, such as the internal plasminogen-binding motif or the protein kinase A recognition site exhibit high homology with the respective general consensus sequence found in many other proteins. The fact that vitronectin- and integrin-like molecules were found in the plant kingdom, in algae, insects and lower invertebrates [20, 21], demonstrates the conservation of this important molecular system with its diverse biological activities.

Conformation   Vitronectin is a labile molecule and conformational alterations are intimately linked to changes in ligand or cell binding properties. In plasma, only 2% of vitronectin is in a conformation capable of binding to heparin-Sepharose, whereas in serum the relative amount increases to 7% [22]. Likewise, platelet vitronectin provides a structurally and functionally distinct form from that in plasma [23, 24]. In addition, complex formation of vitronectin with PAI-1, thrombin-antithrombin complex, and complement C5b-C9 induces conformational changes in the vitronectin molecule, leading to increased heparin binding as identified by functional or structural assay systems [1]. Conformational changes in the vitronectin molecule are also induced by denaturation with chaotropic

agents or detergents, heat-treatment and acidification, and these changes are accompanied by the spontaneous formation of non-covalently associated and disulfide-linked vitronectin multimers; the exact intra- or inter-molecular disulfide-linkages have not been resolved. Using a panel of conformation-sensitive monoclonal antibodies as well as synthetic peptides, evidence was provided that conformational changes in the vitronectin molecule occur in the C-terminal heparin binding domain and the N-terminal acidic portion. Within these regions, particular ligand binding sites are cryptic in native plasma vitronectin and become exposed upon denaturation/multimerization of the protein. Although both, ionic and hydrophobic interactions may contribute to subunit interfaces, multimerization is likely to include the complementary charged C- and N-terminal regions of the molecule [12, 23] (Figures 1, 2).

Additional Features   Circulating plasma vitronectin is considered to be a conformationally "leached" proform of the protein that requires induction through binding interactions (also with surfaces) to expose formerly cryptic epitopes.

Gene
Gene Structure   The human and mouse vitronectin genes (about 3.5 kb) consist of eight exons and seven introns,

resulting in 1.7 kb transcripts without indication for alternative splicing [6, 16]. The first exon of vitronectin encodes for the 19 residues of the leader peptide and the first two residues of the 459 amino acid mature protein. The second exon encodes residues 3 to 42 of the N-terminal somatomedin-B domain (residues 1-44) and is flanked on each side by a phase I intron. Well conserved tandemly repeated copies of the somatomedin-B domain are found in other proteins, indicating that this domain is an independently folding structural module which has been shuffled and duplicated during evolution [6]. Likewise, the hemopexin-type repeats in vitronectin and other factors may have been derived from a primordial hemopexin repeat.

Chromosomal Localization   The human gene (about 3.5 kb) is localized on chromosome 17 (centromeric region 17q11); the murine gene is localized on chromosome 10.

Gene Expression   Vitronectin cDNAs have been isolated from rodent and human liver cell libraries, indicating that vitronectin mRNA is present in the liver [13, 14, 16]. Using quantitative RT-PCR, significant amounts of vitronectin mRNA are also detectable at different extrahepatic sites, but vitronectin mRNA is not detectable by *in situ* hybridization in both normal and diseased vascular tissue.

Gene Regulation   Vitronectin is constitutively produced mainly in liver parenchymal cells, and expression is increased in a rodent model following stimulation with endotoxin, tumor necrosis factor-α or interleukin-6, suggesting that vitronectin is an acute phase protein [25]. *In vitro*, the expression of vitronectin in hepatocytes can be elevated by transforming growth factor-α.

Additional Features   Both, cis acting elements for the acute phase gene regulation as well as for transforming growth factor-α induction have been identified.

Processing and Fate   Vitronectin is present in normal plasma at concentrations of 200 – 400 µg/ml [26, 27] and thus constitutes 0.2 – 0.5 % of total plasma proteins. The concentration of vitronectin in plasma does not significantly differ from that in serum prepared either by recalcification of plasma or whole blood [1]. Although purified vitronectin is a substrate for transglutaminase, little if any vitronectin becomes incorporated *ex vivo* into blood clots. Reduced plasma levels of vitronectin are found in patients with severe liver failure, and changes in the plasma vitronectin levels closely parallel changes in biosynthetic markers of the liver parenchyma, suggesting that the liver is the major site of vitronectin biosynthesis. Increased plasma levels of vitronectin have been observed in patients undergoing elective orthopedic surgery and in rodents under acute phase conditions.

A second circulatory pool of vitronectin is contained within platelets in a rapid releasable form, which is endocytosed from plasma and incorporated into α-granules and which accounts for approximately 0.8 % of the circulating pool of vitronectin [28, 29]. While plasma vitronectin is mainly monomeric and lacks significant exposure of conformationally sensitive monoclonal antibody epitopes, a portion of platelet vitronectin is present in a high molecular weight, conformationally altered form [23, 24]. The ability of platelets to adhere during primary haemostasis and to increase their local concentration up to 200-fold as compared to the circulating level suggests that these multimeric forms of vitronectin will be released at the platelet-platelet, platelet-cell, or platelet-

Table 2. Physicochemical parameters of human vitronectin (modified from [2])

459 amino acids encoded by a 1.7 kb transcript		
Sedimentation coefficient, s $_{20,w}$	(monomeric)	4.2 - 4.6 S
	(oligomeric)	6.5 S
Diffusion coefficient, D $_{20,w}$		5.7×10^{-7} cm^2/s
M$_r$ (from cDNA)		52,400
M$_r$ (gel electrophoretic analysis, gel filtration)		75,000 - 78,000
Partial specific volume, v		0.711 cm^3/g
Stokes radius, r$_H$	(monomeric)	3.7 - 3.9 nm
	(oligomeric)	5.6 nm
Frictional ration, f/f$_0$		1.37
Extinction coefficient, A (1%, 1 cm, 280 nm)		13.8
Isoelectric point (8 M urea), pI		4.75 - 5.25
Carbohydrate composition (mol/mol glycoprotein)		GlcN$_{15}$, Man$_{11}$, Gal$_9$, NeuAc$_5$
Post-translational modifications		Post-translational modifications Phosphorylation (Ser^{378}) Sulfatation (Tyr^{56}, Tyr^{59})
Two to four free Cys in the native form		
Endogenous proteolytic cleavage site		Arg^{379}/Ala^{380}

matrix interface, thereby expressing a functional repertoire distinct from that in plasma.

Vitronectin also has been found in other tissue fluids including seminal plasma, urine, amniotic fluid, cerebrospinal fluid, and bronchoalveolar lavage fluid (Table 1). The amount in the lavage fluid is approximately 10-fold higher in patients with interstitial lung disease, including pulmonary fibrosis, sarcoidosis, and idiopathic pulmonary fibrosis.

Vitronectin depositions are detectable in several normal human tissues including the vessel wall and are significantly increased in areas of sclerosis and necrosis. Likewise, vitronectin depositions have been detected at sites of acute myocardial and kidney infarction. Similar staining patterns are noted in intimal thickenings and fibrous plaques of atherosclerotic arteries in association with collagen bundles, elastic fibers, and cell debris in the vicinity of elastin [30]. Specific mechanisms, which include receptor-mediated endo- and transcytosis [31, 32], appear to mediate the uptake of vitronectin into tissues. These processes are likely to regulate turnover of tissue forms of vitronectin that are part of the provisional wound healing matrix in areas of tissue injury and necrosis or during tissue remodelling. Although vitronectin depositions can be viewed as a marker of tissue injury and necrosis, they may not be causally related to the development or progression of the underlying disease, but could define a role for the adhesion protein as protective factor against tissue destruction.

Biological Activity   Vitronectin provides multifunctional properties as a potent adhesive glycoprotein and as a regulator of pericellular proteolysis at sites of cell-extracellular matrix interactions in the vascular system. In particular, vitronectin multimers, which are preferentially deposited at sites of increased vascular permeability or injury, undergo various ligand or receptor binding interactions with collagen, glycosaminoglycans or proteoglycans, β-endorphin, PAI-1, urokinase/urokinase receptor complex as well as with integrins [2]. Among these activities, binding of PAI-1 and specific interaction with thrombin-inhibitor complexes are of major importance. As a functional consequence, complex formation with vitronectin does not only increase the half-life of PAI-1 in the circulation by two- to four-fold [33], but enables the active inhibitor to become stabilized at sites of vascular injury and initial platelet plug formation. PAI-1 has also been identified as "heparin cofactor" and kinetic data indicate that both, vitronectin and heparin, drastically change the specificity of PAI-1 for thrombin, such that this pro-coagulant enzyme becomes recognized and subsequently neutralized by PAI-1. Although the overall rate of inhibition for the latter system is about 100-fold lower than for interaction of PAI-1 with plasminogen activators, thrombin inhibition by the PAI-1-vitronectin complex appears to be the major event *in vitro* when (pro-coagulant) thrombin interacts with the subendothelial cell matrix.

Upon reaction with the thrombin-antithrombin (or another inhibitor) complex during the phase of blood clotting, a ternary complex is generated that is covalently stabilized by disulfide bridge(s) between vitronectin and the thrombin component and by subsequent multimerization [34]. These ternary complexes exhibit binding to PAI-1 or heparin as well as promote cell adhesion identical to isolated multimeric vitronectin. As potent heparin scavengers in the circulation, these complexes may neutralize exogenously administered heparin.

Specific, saturable interactions of heparin-binding soluble vitronectin multimers with cultured cells involve proteoglycans, the urokinase-receptor or the gC1q-receptor, respectively [12]. In particular, reversible cell binding of vitronectin mainly involves high affinity association with the glycolipid-anchored urokinase receptor [10], whereas interaction with endothelial cells via heparin-displacable recognition can also lead to endocytosis/transcytosis of vitronectin (-complexes) [34]. The constant supply of the vessel wall-associated matrix with vitronectin by these mechanisms may explain the (age-related) accumulation of the protein along large vessels or elastic fibers in skin or during changes in vascular permeability associated with angiogenesis [35]. Similarly, processing of multimeric vitronectin by fibroblasts involves $\alpha_v\beta_5$-integrin as well as protein kinase C-dependent mechanisms [31].

A concerted action between cellular proteoglycans, urokinase receptor and α_v-integrins for vascular cell migration and invasion particularly at vitronectin-rich

Table 3. Structure-function homologies of vitronectin domains in other factors

(I) Somatomedin-B region (cysteine pattern)	megakaryocyte stimulating factor, plasma-cell expressed protein 1, placental protein 11
(II) RGD-cell adhesion motif	several adhesive glycoproteins, including: bone sialo-glycoprotein, osteopontin, von-Willebrand factor, fibronectin, denatured collagen, pro-thrombin, HIV Tat-protein
(III) Acidic region, containing sulfated tyrosines	microfibril-associated protein, hirudin, platelet glycoprotein Ib, protease-activated receptor-1 (PAR-1, thrombin receptor), selectin-ligand PSGL-1
(IV) Hemopexin-type repeats	hemopexin, several matrix-metalloproteinases (MMP, e.g. interstitial collagenase, type IV collagenase), pea seed albumin-2, nectinepsin
(V) Glycosaminoglycan-binding consensus site	several heparin-binding proteins, including: apolipoprotein E, osteopontin, lipoprotein lipase, tissue factor pathway inhibitor

matrix sites was proposed. Due to its unique ligand interactions involving components of adhesion systems and proteolytic cascades, vitronectin serves a regulatory function in urokinase-receptor-related invasiveness or differentiation of cells [36, 31]: Urokinase increases the binding affinity of multimeric vitronectin, whereas active PAI-1 but not the latent form or PAI-1-protease complexes totally abrogates urokinase-receptor-vitronectin interaction, indicating that the strength of cellular interactions can be modulated without involving matrix-lysis (plasmin-independent pathway). The ternary complex which is formed between the urokinase-receptor, vitronectin and urokinase isoforms can also be assembled on the surface of cells lacking urokinase receptor: Cell-associated vitronectin serves as specific acceptor for binding of soluble urokinase/urokinase receptor complexes, thereby initiating cellular events distant from sites of their expression. Such ternary complexes were detected in breast cancer tissue and are likely to be found in atherosclerotic vessels where all three components are abundantly present.

Vitronectin-containing substrata may direct the localization of the urokinase-receptor together with different integrins into focal adhesion areas. The co-localization of the urokinase-receptor with integrins in focal contacts which contrasts with a more scattered distribution of vitronectin-bound PAI-1 in the matrix could allow focal urokinase-mediated proteolysis to occur [38]. Moreover, increased urokinase-receptor expression and localization of the receptor to the leading edge of migrating monocytes appears to be essential for locomotion or invasiveness of (tumor) cells independent of urokinase activity [39]. Cellular migration of myeloid cells is dependent upon continued occupation of the urokinase receptor by its ligand, and vitronectin-complexed PAI-1 regulates adhesion by promoting urokinase/urokinase receptor turnover [37]. Likewise, removal of the thrombin-PAI-1 complex by lipoprotein-receptor-related proteins 1 and 2 is facilitated by vitronectin. Due to the proximity of binding sites for PAI-1 and integrins within the N-terminal portion of vitronectin, active PAI-1 but not the latent or cleaved form acts as a potent competitor for vitronectin binding to integrins $\alpha_v\beta_3$, $\alpha_v\beta_5$ and $\alpha_{IIb}\beta_3$ [39]. Likewise, active PAI-1 abrogated adhesion and migration of fibroblastic or smooth muscle cells and platelets. This indicates that excess PAI-1 favors inhibition of cellular motility at vitronectin-rich matrix-sites, whereas excess of urokinase (in the presence of plasminogen) prevents cell anchorage by efficient pericellular proteolysis. The observation that high levels of both urokinase and PAI-1 relate to a poor prognosis for patients with various types of cancer may be best explained by considering a coordination between plasmin-dependent and -independent events in cellular interactions. Support for these interactions on vascular cells stems from gene knockout experiments in PAI-1-deficient mice which are presented with characteristic vascular occlusion by cellular material following injury, very likely resulting from uncontrolled cell motility [40]. The urokinase/urokinase receptor and vitronectin-PAI-1/integrin systems are thus essential in the regulation of plasmin-dependent and -independent adhesive cellular activities particularly relevant for the dynamic remodelling of (vascular) tissue.

The recently characterized surface receptor for multimeric vitronectin, gC1q-receptor, that independently recognizes the globular heads of C1q [41], may be part of the mechanism that serves as scavenger or opsonizing complex in phagocytosis. The functional properties of vitronectin as an indirect opsonin, a participating component in apoptosis, and a major complement inhibitor are likely to be expressed in concert [42]. The distribution of vitronectin and gC1q-receptor on blood and vascular cells may very well indicate a functional linkage in processes of wound healing and immune-defense.

Role in Vascular Biology
Physiological Function   As a multifunctional protein, vitronectin not only promotes cellular adhesion and migration of a variety of cell types, but, among other functions,

Table 4. Distribution and possible functions of different vitronectin receptor types

Receptor	Distribution	(Patho-)physiological function (ligand)
$\alpha_v\beta_3$-integrin	vascular cells	cell migration, angiogenesis, restenosis (vitronectin, fibrin and other RGD-proteins) cell-cell contacts (PECAM-1, CD31) adenovirus binding and uptake (penton-base)
	tumor cells	migration, chemotaxis (vitronectin-PAI-1, osteopontin, matrix-metalloproteinase-2)
$\alpha_v\beta_5$-integrin	adhesive cells	endocytosis, phagocytosis and bacterial invasion (vitronectin, ?)
Heparansulfate-proteoglycans	endothelial cells	endocytosis, transcytosis (vitronectin-complexes)
Urokinase-receptor	vascular, blood cells, tumor cells	modulation of adhesion, cell invasion, pericellular proteolysis (vitronectin-rich extracellular matrix), receptor turnover (vitronectin-PAI-1)
Globular C1q-receptor (gC1qR)	vascular, blood cells	phagocytosis, clearance (vitronectin-complexes, foreign particles)

also provides cofactor activity for the regulation of pericellular proteolysis and serves as scavenging factor for endproducts of the haemostasis and complement systems [1, 2]. These cellular activities of vitronectin are mediated in great part by different classes of cell surface receptors expressed on vascular and other cell types (Table 4) and are of importance at vitronectin-rich sites in the body [43]. Differences in the conformation of vitronectin found in body fluids versus vitronectin immobilized within the extracellular matrix determine the recognition of these ligands by various cellular receptors that are relevant in vascular remodelling and wound healing [44]. In particular, soluble and tissue vitronectin multimers provide sufficient affinity for or induce clustering of receptors indicating that the conformationally altered protein is the cell-reactive species under (patho-) physiological conditions. Despite the widespread expression of vitronectin during embryogenesis, it should be noted that gene targeting experiments revealed that vitronectin is not required for normal development [3].

Defense mechanisms of the organism including wound healing, tissue repair, the humoral/cellular immune system, and the hemostasis system require different types of adhesive cells which all bear variable subtypes of adhesion receptors. Vitronectin-binding integrins are associated with these events and participate in signal transduction and gene regulation [43]. The integrins $\alpha_v\beta_3$, $\alpha_v\beta_5$, $\alpha_v\beta_1$ and $\alpha_{IIb}\beta_3$ are established receptors for vitronectin, but are promiscuous in recognizing the RGD motif in other adhesive proteins including fibrinogen, von Willebrand factor, fibronectin, osteopontin or denatured collagen (Table 4). The most selective vitronectin-binding $\alpha_v\beta_5$ also recognizes basic regions in HIV-Tat protein and the heparin binding domain of vitronectin, indicating that variability of integrin-ligand interaction is also found at the level of alternative recognition motifs. Together with the given lipid composition of the cell membrane, bivalent metal cations are crucial for ligand interaction, $\alpha_v\beta_3$-receptor clustering, and modulation of binding, and multivalent binding of vitronectin to $\alpha_v\beta_3$ may stabilize ligand-receptor interactions [44, 45]. With regard to mediating cell contact and motility, $\alpha_v\beta_3$ and $\alpha_v\beta_5$ serve complementary functions on vascular and tumor cells: While $\alpha_v\beta_3$ is localized to focal adhesion sites and promotes cell spreading and migration on various RGD substrata, $\alpha_v\beta_5$ supports cell attachment and is clustered in focal adhesion sites only in some cell types. However, growth factor binding to epidermal growth factor receptor appears to induce cell motility via $\alpha_v\beta_5$. Moreover, coordination with the urokinase/urokinase receptor complex appears to direct cellular migration on a vitronectin substratum [37].

While the previous information relates to cellular interactions obtained with immobilized vitronectin, the contribution of soluble vitronectin in the cell-to-cell recognition of apoptotic cells and augmentation of phagocytosis has been reported as well. Moreover, vitronectin was co-localized with $\alpha_v\beta_3$ and $\alpha_{IIb}\beta_3$ at sites of proplatelet generation in maturating bone marrow-derived megakaryocytes suggesting an important intracellular role for this integrin-ligand system during platelet formation or for translocation of vitronectin and fibrinogen between intra- and extracellular compartments. Binding of vitronectin to the luminal phase of endothelial cells is only mediated by $\alpha_v\beta_3$-integrin if highly clustered soluble forms of the adhesion protein or if vitronectin-coated microspheres were employed [44]. Interaction with vitronectin-coated particles also engages integrin-associated protein (CD47) and may result in signal transduction events.

Pathology   The fundamental role of integrin-mediated interactions in the mammalian system has been demonstrated by gene inactivation experiments in which single knock-outs of α_v- or β_1-integrins or fibronectin in mice were associated with a lethal phenotype during embryonic development at the time of vascular system formation [46, 47]. The α_v-integrins are predominantly expressed on angiogenic endothelium or migrating vascular cells and have been implicated in the formation of capillary-like structures *in vitro* as well as during retinal and tumor angiogenesis *in vivo*, where the receptor is expressed on growing but not on mature vessels [48]. The expression of receptors for vascular endothelial growth factor or basic fibroblast growth factor drives the angiogenic potential of these cells [49], and a differential molecular connection with the integrin $\alpha_v\beta_5$- and $\alpha_v\beta_3$-systems was proposed. Likewise, migrating or proliferating vascular smooth muscle cells, which significantly contribute to the process of neointimal hyperplasia at sites of atherosclerotic lesions, express predominantly α_v-integrins. Following deposition in the vessel wall, vitronectin interactions with $\alpha_v\beta_3$ not only promoted chemotaxis of smooth muscle cells, but leads to calcium-dependent intracellular signaling during haptotaxis of endothelial cells and also appears to involve calcium-dependent phosphatase(s) for neutrophil migration. Another $\alpha_v\beta_3$-ligand, osteopontin which becomes upregulated in diseased vessels, has been implicated in $\alpha_v\beta_3$-mediated smooth muscle cell migration at sites of vascular lesions [50]. Under conditions of reperfusion, adhesion of (vitronectin-coated) platelets to interleukin-1-stimulated endothelium required $\alpha_v\beta_3$-integrin, indicating its important role also in acute myocardial infarction. Cellular differentiation may be coupled to alteration in integrin expression, and cooperation between vitronectin- and fibronectin-binding integrins can influence the adhesion/migration and phagocytotic properties of cells. Both, integrin ligation and clustering on vitronectin substratum are linked to different intracellular signals that prevent endothelial cells from undergoing apoptosis [51], but soluble vitronectin was unable to reverse this event. However, in melanoma cells the expression of $\alpha_v\beta_3$ integrin prevented apoptosis in a three-dimensional collagen gel, and these cells demonstrated growth advantage compared to α_v-negative melanoma cells [52].

Clinical Relevance and Therapeutic Implications   Several molecular mechanisms related to vitronectin-cell interactions have lead to the development of potential drugs for therapy in vascular and infectious diseases. These "anti-adhesion" therapies will include antagonists that block integrin-mediated processes in (tumor-) angiogenesis, restenosis or tumor metastasis as well as novel antibiotic regimens that prevent the interaction of infectious particles with the host tissue. Together with data from animal experiments and patients' studies that indicate that α_v-integrins are upregulated or activated during vascular diseases, vascular integrins become promising targets for diagnosis and therapeutic interventions [53]. Different approaches to antagonize these receptors are under investigation using anti-integrin monoclonal antibodies as well as RGD-related low molecular weight antagonists. These were successful in blocking tumor formation and progression by inhibiting the formation of tumor vessels in skin, in preventing experimental restenosis or suppressing angiogenesis in retinopathy-related diseases [54].

Anti-microbial substances including bacterial-binding sites of vitronectin and fibronectin, complementary soluble bacterial adhesins, or antibodies thereof will provide attractive new groups of vaccines to fight bacterial infections also related to vascular diseases [55]. Approaching these issues will not only broaden our limited knowledge about the role of micro-organisms in vascular pathophysiology, but these novel "anti-adhesion"-based strategies are crucial to circumvent the emergence of antibiotic drug-resistant pathogens.

Klaus T. Preissner and Dietmar Seiffert

References

1. Tomasini BR, Mosher DF (1990) Prog Hemostas Thromb 10:269-305
2. Preissner KT (1991) Ann Rev Cell Biol 7:275-310
3. Zheng X et al (1995) Proc Natl Acad Sci USA 92:12426-12430
4. Tollefsen DM et al (1990) J Biol Chem 265:9778-9781
5. Ogawa H et al (1995) Eur J Biochem 230:994-1000
6. Jenne D, Stanley KK (1987) Biochemistry 26:6735-6742
7. Korc-Grodzicki B et al (1988) Biochem Biophys Res Commun 157:1131-1138
8. Seiffert D, Loskutoff DJ (1991) J Biol Chem 266:2824-2830
9. Heldin C-H et al (1981) Science 213:1122-1123
10. Kanse SM et al (1996) Exp Cell Res 224:344-353
11. Deng G et al (1996) J Cell Biol 134:1563-1571
12. Preissner KT and Seiffert D (1998) Thromb Res 89: 1-21
13. Jenne D, Stanley KK (1985) EMBO J 4:3153-3157
14. Suzuki S et al (1985) EMBO J 4:2519-2524
15. Sato R et al (1990) J Biol Chem 265:21232-21236
16. Seiffert D et al (1991) Proc Natl Acad Sci USA 88:9402-9406
17. Yoneda A et al (1996) J Biochem 120:954-960
18. Blancher C et al (1996) J Biol Chem 271:26220-26226
19. Merberg DM et al (1993) In: Preissner KT et al (eds) Biology of Vitronectins and Their Receptors. Elsevier, Amsterdam, pp 45-53
20. Miyazaki K et al (1992) Exp Cell Res 199:106-110
21. Nakashima N et al (1992) Biochim Biophys Acta 120:1-10
22. Izumi M et al (1989) Biochim Biophys Acta 990:101-108
23. Stockmann A et al (1993) J Biol Chem 268:22874-22882
24. Seiffert D et al (1996) Blood 88:552-560
25. Seiffert D (1997) Histol Histopathol 12:787-797
26. Preissner KT et al (1985) Biochem J 231:349-355
27. Conlan MG et al (1988) Blood 72:185-190
28. Parker CJ et al (1989) Br J Haematol, 71:245-252
29. Preissner KT et al (1989) Blood 74:1989-1996
30. Niculescu F et al (1987) Atherosclerosis 65:1-11
31. Panetti TS, McKeown-Longo PJ (1993) J Biol Chem 268:11492-11495
32. Preissner KT, Pötzsch B (1995) Histol Histopathol 10:239-251
33. Declerck PJ et al (1988) J Biol Chem 263:15454-15461
34. Preissner KT et al (1996) Semin Thromb Haemostas 22:165-172
35. Senger DR (1996) Am J Path 149:1-7
36. Preissner KT et al (1997) Thromb Haemostas 78:88-95
37. Chapman HA (1997) Curr Opin Cell Biol 9: 714-724
38. Pöllänen J et al (1991) Adv Cancer Res 57:273-328
39. Andreasen PA et al (1997) Int J Canc 72: 1-22
40. Carmeliet P, Collen D (1997) Curr Opin Lipidol 8:118-125
41. Lim BL et al (1996) J Biol Chem 271:26739-26744
42. Preissner KT, de Groot PG (1993) In: Sim E, Kerr MA (eds) The Natural Immune System: Humoral Factors. Oxford University Press, Oxford, pp 281-318
43. Felding-Habermann B, Cheresh DA (1993) Curr Opin Cell Biol 5:864-868
44. Hess S et al (1995) Thromb Haemostas 74:258-265
45. Loftus JC et al (1994) J Biol Chem 269:25235-25238
46. Fässler R et al (1996) Curr Opin Cell Biol 8:641-646
47. Hynes RO, Bader BL (1997) Thromb Haemostas 78:249-257
48. Brooks PC et al (1994) Cell 79:1157-1164
49. Risau W (1997) Nature 386:671-674
50. Liaw L et al (1994) Circ Res 4:214-224
51. Meredith JE et al (1993) Mol Biol Cell 4:953-961
52. Felding-Habermann B et al (1992) J Clin Invest 89:2018-2022
53. Mousa SA, Cheresh DA (1997) Drug Dev Ther 2:187-199
54. Hammes HP et al (1997) Nature Med 2:529-533
55. Preissner KT, Chhatwal GS (1999) In: Cossart P et al (eds) Cellular Microbiology. ASM Press (in press)

VLDL

Definition   *Very light density lipoprotein*

See: →Lipoproteins

VN

Definition   *Vitronectin*

See: →Vitronectin/vitronectin receptors

von Willebrand Disease

Definition   *Congenital bleeding disorder characterized by quantitative (type 1 and 3) and/or qualitative (type 2) abnormalities of vWF*

See: →von Willebrand Factor

von Willebrand Factor (vWF)

Synonym: von Willebrand Factor, vWF, FVIII related antigen (this last denomination should not be used anymore)

Definition *Multimeric protein required for platelet adhesion and aggregation at high shear rates.*

See also: →Thrombosis; →Bleeding disorders; →Platelet stimulus-response coupling

Introduction   von Willebrand Factor (vWF) is synthesized by endothelial cells and megakaryocytes as a large precursor, the preprovWF. It undergoes multimerization and proteolytic cleavage before being secreted in the plasma and the vascular extracellular matrix (ECM). The main function of vWF is to initiate platelet adhesion to the ECM at high shear stress, through closely regulated interactions with the platelet glycoprotein (GP) Ib-IX-V receptor. It contributes to thrombus formation, by recruiting platelets to the site of the lesion and by reinforcing platelet-to-platelet cohesion through an interaction with the platelet $\alpha_{IIb}\beta_3$ integrin (GPIIb-IIIa). In addition, it contributes to anchor the endothelial monolayer to the underlying ECM. Another important function for vWF in hemostasis is to serve as a carrier of procoagulant Factor VIII (FVIII), an essential cofactor in thrombin generation, thereby localizing FVIII on activated platelets and protecting it from proteolytic degradation. The congenital absence or dysfunction of vWF causes a moderate to severe bleeding diathesis, von Willebrand disease (vWD). This disease is now recognized as the most common inherited bleeding disorder. A large number of mutations causing several variants of vWD have been identified.

Characteristics

Molecular Weight   The relative molecular weight (Mr) of the polypeptide subunit is 250 kDa. Subunits are assembled into multimers ranging from a dimer of 500 kDa to 50 subunits of over 10,000 kDa.

Domains   The large preprovWF molecule consists of a 22-amino acid (aa) signal peptide, a 741-aa propeptide (also known as von Willebrand antigen II) and a 2050-aa vWF subunit. Numbering of the aa residues refers to respectively aa 1 to 763 of the prepropeptide and 1 to 2050 of the vWF subunit. PreprovWF contains multiple modular internal repeat domains (A to D) present in two to five copies each (Figure 1). There are three A domains, two of them containing an intrachain disulfide bond which forms a large loop of 185 aa either between Cys509 and Cys695 (A1 domain) or between Cys923 and Cys1109 (A3 domain). Some homology domains contain binding sites, so that their boundaries grossly coincide with functional domains. This is especially the case for the A1 domain, with its so-called A1 loop, which plays a key role in the interaction of vWF with platelet GPIbα, whereas the A3 loop is involved in collagen binding. The A2 domain contains two adjacent Cys (906-907) and therefore does not form a loop. The B domains are 25-30 aa long, positioned between aa 1533 and 1636 and do not contain any known binding site [1]. The C1 and C2 domains are located at the carboxy-terminal end of the subunit. There are two Arg-Gly-Asp (RGD) sequences, one in C1 domain between aa 1744 and 1746 involved in the interaction with the platelet $\alpha_{IIb}\beta_3$ and the endothelial $\alpha_v\beta_3$ integrins. The other RGD is located in the D2 domain of the propeptide, but its function remains unknown. The D1 and D2 domains constitute the propeptide and are essential for multimerization that takes place during biosynthesis. The D' domain (aa 6 to 102) containing a FVIII binding sequence and the D3 domain (103 to 478) are located in the amino-terminal end of the subunit and are also important for multimerization. The fifth D domain, called D4, is carboxy-terminal and extends to the aa 1184 to 1532.

Binding Sites and Affinity   Platelet adhesion involves an interaction of vWF with the platelet GPIbα subunit of the GPIb-IX-V complex. Platelet aggregation is dependent on vWF binding to activated $\alpha_{IIb}\beta_3$. Fluid-phase plasmatic vWF does not bind spontaneously to platelets. Two different types of vWF sequences, the regulatory sites on one hand and the direct binding sites on the other hand, are involved in binding to GPIbα and are locat-

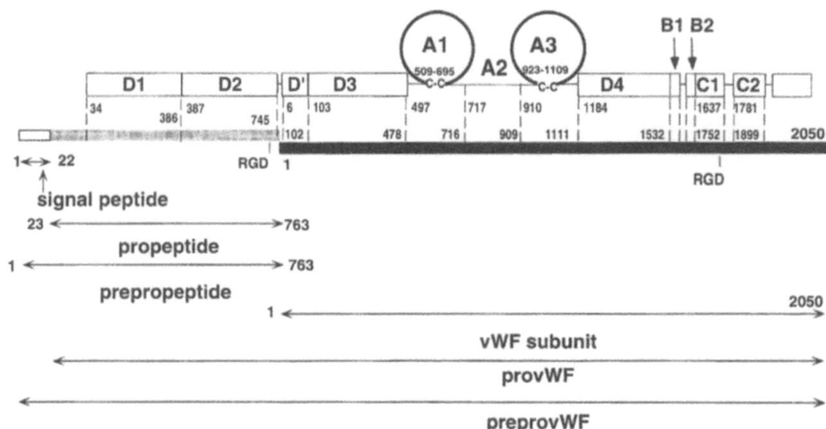

Figure 1. Homology domains of the preprovWF. Position of homology domains of preprovWF are indicated. Numbering of aa residues extends from 1 to 763 for the prepropeptide and restarts at 1 till 2050 for the vWF subunit.

ed in the A1 domain. Studies with mutated or deleted A1 recombinant fragments or full-length vWF have established that the disulfide bond Cys509-Cys695 is required for stabilization of vWF native (inactive) conformation [2-4]. The asymmetric charge distribution of the A1 domain as well as the high level of negative charges on GPIbα itself suggests that exposure of the GPIbα-binding site on vWF is dependent on electrostatic interactions. Studies with two non-physiological agents, ristocetin and botrocetin, have allowed a better understanding of vWF interaction with GPIbα. Ristocetin binds to two discontinuous, proline-rich, negatively charged regions (aa 474-488 and 695-708) flanking the disulfide bridge of the A1 domain. The inactive conformation of fluid-phase plasmatic vWF is associated with neutralisation of positive charges of the A1 loop by negative charges flanking the disulfide bond. Binding of cationic ristocetin to these negative charges could promote vWF binding to GPIbα by modification of the balance between positive and negative charges. Botrocetin, a protein isolated from the venom of the snake Bothrops jararaca, binds to predominantly positively charged sequences in the A1 loop (aa 514-542, 539-553, 569-583, 629-643) [5,6]. Binding of vWF to botrocetin has a K_d of 7.25 nM and a stoechiometry of 1.5 mol/mol vWF subunit, and botrocetin-induced binding of vWF to glycocalicin, the extracellular portion of GPIbα, has a K_d of 0.6 nM and a number of sites of 0.2 mol/mol vWF [7].

Using synthetic peptides and proteolytic fragments of vWF, the sequence 514-542 was identified as a major GPIbα-binding site [6]. More recently, Matsushita and Sadler have suggested a new complex model describing direct GPIbα-binding sites which are distinct from ristocetin- or botrocetin-dependent modulation sites. This model is depicted on Figure 2. Mutations of charged aa of the A1 domain to alanine were expressed as full-length recombinant vWF (rvWF) in order to compare their ability to bind GPIbα in the presence of ristocetin or botrocetin [8,9]. Three inhibitory sites cooperate to maintain an inactive conformation which prevents GPIbα binding

to native vWF: two acidic regions (497-511 and 687-698) and a basic region (540-578) corresponding to a cluster of mutations identified in type 2B vWD (see below § Pathology). Exposure of the GPIbα-binding site by ristocetin and botrocetin involves common binding sites. Ristocetin-binding sites on vWF (aa 520-534 and 626) are regulated by inhibitory sites that maintain the inactive conformation. Thus, ristocetin-binding sites are functionally comparable to an endogenous activating region which stimulates binding of vWF to GPIbα. Botrocetin and GPIbα both interact electrostatically with positively charged sites on vWF (aa 629, 632, 636 and 667). Upon binding to vWF, botrocetin may stimulate binding of vWF to GPIbα without a need for relieving the inhibitory sites. Ultimately, identification of vWF sequences which are directly involved in binding to GPIbα may comprise nearby sequences within aa 596-616 and particularly Lys599 [8,9]. Involvement of 596-616 and 629-632 sequences in vWF binding to GPIbα has been confirmed in another study [10]. This model is of primary importance in understanding the physiological mechanisms that allow vWF binding to GPIbα and in targeting pharmacological inhibitors that could be relevant *in vivo* (see below, § Biological Activity).

The vWF binding site to activated $\alpha_{IIb}\beta_3$ is located on the carboxy-terminal end of the vWF subunit [11] and overlaps the RGD sequence (aa 1744-1746), which is absolutely required since a mutated R1746G rvWF is unable to bind to activated platelet $\alpha_{IIb}\beta_3$ in static and in flow conditions [12,13]. The exposure of the RGD sequence is conformationally regulated and RGD appears as the main interacting site of solid-phase vWF with the endothelial $\alpha_v\beta_3$ integrin [14].

Additional Features   *In vitro*, vWF interacts with the sulfated negatively charged compounds heparin and sulfatides. The binding sites have been localized in the A1 domain [15]. Binding to heparin may reflect the ability of vWF to interact with glycosaminoglycans or other heparin-like molecules in the subendothelial ECM. The

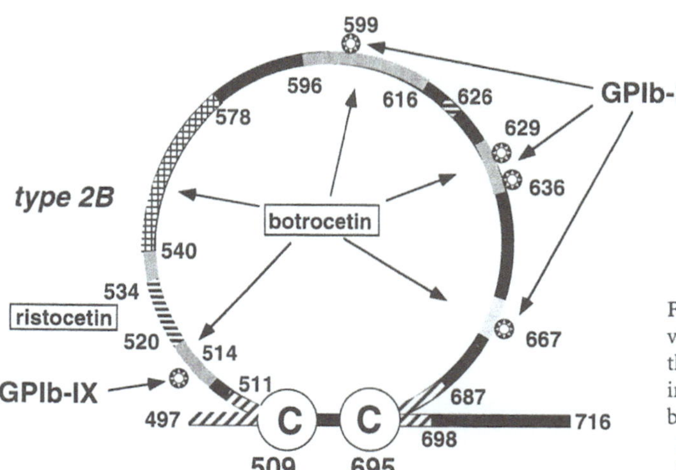

Figure 2. Schematic representation of the A1 domain of vWF Sequences involved in binding to the modulators of the vWF-GPIbα interaction are indicated by the following patterns : ristocetin-binding sites : hatched bars; botrocetin-binding sites : grey bars. Position of the Type B mutations of vWD are indicated by checked bars. ¡PIbα binding sites are indicated by an asterisk.

high affinity binding site for heparin has been localized on the positively charged sequence 565-587 of the A1 loop [16]. Binding of vWF to sulfatides involves the positively charged sequences 569-584 and 628-646 [17]. The localization of vWF binding domain to the subendothelial ECM is still debated. A tryptic fragment of vWF overlapping the A1 domain (T116, aa 449-728) has been shown to bind to the ECM and to support platelet adhesion to the same extent as vWF [18]. However, based on monoclonal antibody inhibition, additional or alternative site(s) have been described on the D4 domain [19] or on the carboxy-terminal end of the subunit [20]. A binding site for type VI collagen has been found in the A1 fragment [18]. Two binding domains to fibrillar type I and III collagen have been identified. The major site is located within the A3 loop (aa 948-998) and an accessory domain is localized within the A1 domain (aa 542-622) [21,22].

FVIII circulates as a noncovalent complex bound to vWF. The binding domain for FVIII has been localized in the amino-terminal region between aa 1 and 272 and more precisely between aa 78 and 96 [23] but is not restricted to this sequence since mutations affecting the FVIII binding properties of vWF are found outside of this region (type 2N, see below § pathology). The 1-272 aa sequence appears sufficient by itself to confer to vWF the capacity to bind FVIII and to protect it from the inactivation by activated Protein C [24]. Isotherms of vWF binding to FVIII in the fluid phase indicate a stoechiometry of 0.5 FVIII molecule per vWF monomer and a K_d of 0.32±0.09 nM [25]. However, *in vivo*, vWF circulates with many unoccupied binding sites because the molar concentrations of both proteins indicate a 50-fold molar excess of vWF in plasma.

Structure

Sequence and Size   As described above (see § Domains), the preprovWF consists of four distinct types of internal homologous domains present in two to five copies each in the following order: D1-D2-D'-D3-A1-A2-A3-D4-B1-B2-B3-C1-C2 [26,27] (Figure 1). The propeptide, composed of domains D1 and D2, directs the assembly of multimers in the Golgi apparatus during vWF biosynthesis and is proteolytically cleaved (see § Processing and Fate). Multimeric vWF is composed of identical subunits of 2050 aa associated by disulfide-bonds. The multimers can be visualized by electrophoresis on agarose gels of plasma samples followed by immunoblotting. The vWF subunit contains 19% carbohydrate with 12 N-linked and 10 O-linked glycosylation sites. The 169 Cys residues are all involved in inter- or intra-chain disulfide bonds mostly distributed in the amino- and carboxy- terminal regions [28].

Homologies   Homology searches and characteristic features suggest that the different vWF domains and various types of proteins may have evolved from common ancestors. The A domains of vWF are homologous to motifs present in proteins involved in a variety of adhesive functions such as some α and β subunits of integrin family members, component C2 and factor B of the complement activation pathway, extracellular matrix components cartilage matrix protein (CMP) and matrilin 2, types VI, VII, XII and XIV collagens and thrombospondin-related anonymous protein [29-31]. The C domains of vWF appear to be homologous to segments of thrombospondin and type I and III alpha 1-procollagen. Homologous sequences of the B, C, and D domains are also found in hemocytin, a silkworm humoral lectin implicated in the hemocyte response of the self-defense system [32].

Sequences homologous to the vWF type D domains are found in various proteins among which are vitellogenin, zonadhesin, a species-specific egg-binding protein from spermatozoa and mouse α-tectorin, a component of the ear tectorial membrane. Multiple copies of vWF type D domains are present in human mucin MUC 2 and in frog integumentary mucin FIM-B.1 in which the common structural elements including Cys-rich consensus motifs in the amino-terminal region appear to be involved in multimerization and packaging mechanisms similar to those demonstrated for preprovWF [33].

Cys-rich carboxy-terminal regions of various types of mucins, such as FIM-B.1, MUC2, MUC5AC, MUC5B exhibit striking sequence similarities to the D4, C1 and extreme carboxy-terminal domains of vWF [34]. In vWF, as described below in § Processing and Fate, this domain appears to be the only structural requirement for the carboxy-terminal to carboxy-terminal dimerization of protein monomers [35]. The similarities between the extreme carboxy-terminal domains of the mucins and vWF suggest a common role for the carboxy-terminal Cys-rich sequences in the dimerization process of these proteins. Indeed, expression studies in COS cells of the carboxy-terminal Cys-rich domain of porcine submaxillary mucin, a sequence homologous to the C-terminal region of MUC2, demonstrate a rapid conversion of the disulfide-rich monomer to a dimer [36].

Partial sequencing of preprovWF cDNA coding for either the propeptide, the FVIII binding domain or the GPIbα binding domain, were determined in various mammalian species. The deduced aa sequences of porcine, rabbit and bovine amino-terminal regions of the vWF subunits, overlapping the FVIII binding domains, can be perfectly aligned with that of human vWF. The first 175 amino-terminal aa of the bovine vWF subunit are 86% homologous to the human sequence [37]. The porcine and rabbit species show the same degree of homology for the first 116 amino-terminal aa [38]. The aa mutated in type 2N vWD, in which a defect of binding to FVIII is due to mutations inside the amino-terminal region of vWF subunit, are conserved in the four species. The bovine and porcine sequences corresponding to aa 474 to 891 overlapping the A1 and A2 domains of the vWF subunit also demonstrate a good alignment of homologous regions alternating with short stretches of non homologous domains [39,40]. The highly charged regions aa 539-553, 569-583 and 629-643 that are involved in botrocetin binding are generally well conserved (see above § Binding Sites and Affinity). Bovine and porcine vWF bind spontaneously to platelet GPIbα in the absence of ristocetin and botrocetin. Several aa in human vWF corresponding to those mutated in type 2B vWD (aa 543, 545, 550, 551), are not conserved in bovine or (and) porcine species, and it may be speculated that the overall aa changes modify the general conformation of the vWF A1 loop in these mammalian species, promoting a conformation favorable for binding to GPIbα. On the other hand, the comparison of the conserved sequences may help to determine the important role of some aa in the binding properties of vWF, such as for the conserved Asp 514, which has been demonstrated by mutagenesis to be required for the interaction of bovine and human vWF with GPIbα [8,41].

The four mammalian species share the same HRSKRS sequence overlapping the cleavage site between the propeptide and the vWF subunit, with the essential Arg at position -4 of the cleavage point and a basic (Lys) residue at position -2 [37,38]. The propeptides of porcine and rabbit species are also highly homologous for the sequences corresponding to the last 41 carboxy-terminal aa, except that the rabbit sequence lacks a stretch of four aa. The sequence of bovine propeptide is one aa shorter than that of the human counterpart and is 80 % homologous. In this animal species, the two CGLC sequences of the propeptide, promoting multimer formation are conserved, whereas the human RGD sequence (aa 698-700 of propeptide) is replaced by TGS [37].

Conformation The biological properties of vWF are related to the multimeric structure but mainly depend on finely regulated conformations at the level of the functional domains. The vWF A domains (±200 aa) appear to be major regions for macromolecular interactions between vWF and various ligands. Intrinsic fluorescence and CD spectra measurements indicate that conformational changes in the A1 domain observed upon refolding of recombinant fragments under different pH conditions result in modified affinity for binding to platelet GPIbα, showing the importance of its tertiary structure [42]. This motif is also involved in specific interactions of the members of the vWF A protein superfamily with several ligands. Analysis of the crystal structure of the recombinant A domain from the α subunit (CD11b) of the leukocyte integrin complement receptor type 3 (CR3, also known as Mac-1) revealed that the A domain is composed of five parallel and one short antiparallel β strands forming a central sheet surrounded by seven α helices [43]. The CR3 A domain has been crystallized in the presence of Mg^{2+} and the secondary structure of the metal binding site, clearly defined on the surface of the A domain at the top of the β sheet, is linked to the presence of structural motifs which are conserved in the A domain superfamily of proteins [43]. The "metal ion-dependent adhesion site" (MIDAS) motif, composed of a conserved Asp-X-Ser-X-Ser sequence, as well as non adjacent Thr and two Asp, plays an essential role for protein ligand binding to recombinant CR3 A domain. One or both of these Asp

residues are not conserved in the A domains of vWF. Protein fold predictions demonstrate that aa of the vWF A1 domain which are mutated in type 2B and in type 2M vWD are spatially clustered on two different surfaces of the A1 domain in a loop on the amino-edge of a β sheet structure [44,45]. Since other A domain proteins such as ras-p21 and flavodoxin contain their ligand binding sites on the carboxy-edge of the β sheet, it appears that the functionally important region of the A1 domain in vWF is distinct from that found in other proteins of the family [45]. The crystal structure of recombinant vWF-A3 domain (aa 916-1111), which interacts with type I and III collagen, has been recently reported [46]. The fold of the A3 domain resembles that of the A-type regions of integrins CR3 and LFA1. However, significant differences exist dealing with the nature and length of some of the α helices structures. The MIDAS motif of the crystallized A3 domain does not contain any metal ion.

Gene

Gene Structure The vWF human gene spans ~180 kilobases (kb) of DNA and contains 52 exons [47]. The length of the exons range in size from 40 to 1379 base pairs (bp), and the introns vary from 97 bp to ~20 kb [47]. The first non coding exon spans 250 bp and contains two short open reading frames. The ATG initiating translation is present at the start of the second exon. Various repetitive elements are present in the gene among which 14 Alu repeats in introns, and a TCTA repeat sequence in intron 40 which is also polymorphic in length and is useful as a marker for prenatal diagnosis. Two Alu repeats and a polymorphic poly(GT)18-26 repeat are localized in the 5'-flanking region [47,48]. The vWF gene is highly polymorphic. More than 40 polymorphisms have been described, among which nearly one half alters the encoded aa sequence (Internet WEB page: http://mmg2.im.med.umich.edu/vWF).

Chromosomal Localization The vWF human gene is located in a telomeric position on the short arm of chromosome 12 at the position 12p12-pter [49,50]. A partial unprocessed pseudogene for vWF is located on chromosome 22, spans ~30 kb DNA and overlaps exons 23 to 34 [51]. The pseudogene does not give rise to a functional transcript.

The vWF gene has been mapped on bovine chromosome 5 at q35, caprine chromosome 5 at q35, ovine chromosome 3 at q35, porcine chromosome 5 at q21 and murine chromosome 6 [52-54]. The physical assignment of vWF genes in the animal species supports the evidence of conserved synteny between the respective animal chromosomes and human chromosome 12. As shown for mouse, pig and cattle, several other genes present on the same animal chromosomes as vWF gene are also mapped to chromosome 12 in humans [52-54]. Moreover, since pig and bovine chromosomes 5 are composed of segments corresponding to human chromosomes 12 and 22, it may be hypothesized that the vWF pseudogene, which is present only in humans (chromosome 22) and higher primates [54-56] originates from a tandem duplication that has occurred before separation of chromosomes 12 and 22. Therefore, the localization of the vWF genes on chromosomes from several mammalian species illustrates karyotype evolution.

Gene Expression   vWF synthesis is restricted to endothelial cells and megakaryocytes. However, among different subpopulations of endothelial cells, the vWF gene expression is not homogenous. In adult mammals, endothelial cells show differential expression of vWF based on the location of these cells in the vascular system, and mRNA studies in porcine or canine blood vessels have shown that this heterogeneity in the expression of vWF is at the level of transcription [57,58]. These studies do not provide any evidence for a correlation between the level of vWF expression and the size of the vessel, suggesting that the synthesis of vWF is not under the control of mechanical factors. This is confirmed by recent data indicating that shear stress modulates vWF release from human cultured endothelial cells but does not act at the transcription level [59]. In mouse, vWF synthesis during early vascular development is limited to the large vessel endothelium and the heterogeneity of the gene expression in adults appears to result from the establishment of endothelial cells of different phenotypes in the embryo [60].

Gene Regulation   Expression of the vWF gene is remarkably tissue-specific and is probably controlled by the interactions between regulating transcription factors and specific DNA elements within or adjacent to the gene. Analysis of the transcriptional activity of various lengths of DNA from the 5' region of the vWF gene have recently suggested that the promoter is repressed in cell types that do not synthesize the protein, the transcription being dependent on cell-specific positive regulators able to overcome the repression.

Functional cis-acting elements implicated in the endothelial cell-specific transcription have been searched in the DNA region of the human vWF gene overlapping the sequence upstream of the transcription initiation site (nucleotide (nt) +1) and the 250 bp first exon. A promoter fragment spanning bp -89 to +19 directs expression of a linked reporter gene in transfected cells. This promoter is not cell type specific, being functional in a variety of cells [61,62]. A fragment spanning nucleotides − 90 to +155 can also function as an ubiquitous promoter in bovine aortic endothelial (BAE), bovine smooth muscle (BSM) and HeLa cells [62]. By increasing the length of the 5'-flanking sequence, it has been shown that an upstream negative regulatory region between bp -500 and -89 represses the basal (-89/+19 or -90/+155) promoter activity in all cell types analyzed [61,62]. In addition, the -487/+246 fragment exhibits a promoter activity in BAE cells and not in BSM or HeLa cells demonstrating that positive regulators counteract the repression in a cell specific manner by interacting with a DNA sequence between bp +155 and +247 [62]. The cis-acting element required for expression in BAE cells is a GATA transcription factor binding site, situated at position +221 in the first exon. Recently, a repressor of the vWF

promoter fragment -487/+155 in BSM and BAE cells has been described. This repressor acts through an interaction with a consensus NF1-binding site located between bp -440 and -470 [63]. However base substitution mutations of the NF1-binding site does not restore the promoter activity of the -487/+247 fragment in non endothelial cells, suggesting that an additional repressor is functional in these cells.

An upstream cell-specific positive regulatory element between bp -1286 and -1240 that includes an Alu sequence is able to restore a basal promoter activity to the fragment -1286/+19 in transfected calf pulmonary artery endothelial (CPAE) cells [61]. The respective importance of the -1240/-1286 and +155/+247 positive cis-acting elements for endothelial cell specific transcription *in vivo* remains to be clarified since their effects have been observed by transient transfection of human vWF promoter fragments in bovine endothelial cells. The sequence of the 5' region of bovine vWF gene is not conserved at the sites of these positive regulatory elements (see below, $ Additional Features). Therefore, the human promoter fragments could be positively regulated in transfected cells through interactions of cis-acting DNA elements with bovine endothelial cell-specific proteins according to a transcriptional mechanism being different, at least partly, from that occurring for the endogenous bovine vWF gene (see below, § Additional Features).

Transfection studies of primary cultures of human umbilical vein endothelial cells (HUVEC) showed that the -89/+19 vWF promoter fragment behaves positively as it does in CPAE cells [64]. This promoter is also negatively regulated in HUVEC by upstream sequences and, contrasting with results shown in BAE cells by Jahroudi et al, sequences in the first exon do not overcome the effect of the upstream repressor(s) in a tissue-specific manner [65].

The search for potential regulatory elements within the DNA sequence in the region flanking the vWF gene revealed a TATA box at nt -30, a CCAAT motif unusually localized at nt -18 and two DNA repeats TTTCCTTT in the -56/-36 sequence, corresponding to inverted consensus binding sites for transcription factors of the Ets family. Analysis of deletions of the -89/+19 promoter have shown that the -60/+19 region exhibits significant promoter activity in HUVEC and CPAE cells, and not in HeLa and COS cells. Transfections of mutated constructs and binding studies have demonstrated that the 5' (-56/-49) Ets binding site is essential for the activity of the -60/+19 promoter fragment in endothelial cells, through interaction with proteins of the Ets family which are not expressed in HeLa cells [66].

Besides the GATA element at +221, several potential binding sites for transcription factors are present in the first exon. Among these is a potential binding site for SP1 at nt +179. Mutagenesis studies have shown that the SP1 site does not appear to play a role in transcription [62]. A putative NF-κB binding site (at nt +80) is not functional and does not interact with NF-κB from endothelial cell extracts [67].

To examine if the -487/+246 fragment is an endothelial cell specific promoter *in vivo*, this fragment was fused to the E. coli lacZ gene and used to generate transgenic mice [68]. A limited subpopulation of yolk sac and adult brain endothelial cells expressed the LacZ construct in the transgenic animals and β-galactosidase activity was also detected in brain cells that do not synthesize endogenous vWF suggesting that the -487 to +246 human vWF gene fragment contains only a portion of the information necessary for correct spatial and temporal expression of the transgene in the endothelial cells. Recently, Aird et al have further shown that a longer promoter fragment encompassing 2182 bp of 5'-flanking sequence, the first exon and first intron could only direct LacZ expression to the endothelial lining of blood vessels in the brain, heart and skeletal muscle of transgenic mice, contrasting again with the more widespread expression of the endogenous vWF gene [69]. These authors also demonstrated that the transcriptional activation of the transgene is mediated by elements present in the local environment, providing a support for the existence of regional differences in the mechanisms of vWF gene transcription and suggesting that the cis-acting elements involved in a more widespread and specific expression of vWF are located outside of the human promoter fragments analyzed to date.

The transcription of the vWF gene therefore depends on a combination of positive and negative regulatory elements. The effect of the negative regulators would be specifically overcome in endothelial cells and in megakaryocytes by factors acting synergistically in a tissue-specific fashion. It may be hypothesized that the GATA factor bound to the first exon cooperates with Ets factors (and other endothelial proteins) and with cis-acting factors expressed locally, to promote the tissue-specific transcription by counteracting the repressors.

Additional Features   The 5' region of bovine vWF shows a perfect alignment with the 5'-flanking human sequence up to human bp -656. An Alu-type *art2* (artiodactyls 2) repeat is inserted in the bovine DNA, which also lacks the Alu element present in the human gene [70]. The bovine first non coding exon is shorter than the human counterpart and only 64% homologous. The human GATA element is replaced in bovine species by a GACA motif. The transcriptional regulation of bovine vWF gene expression has been analyzed by transient tranfections in CPAE, HeLa and COS cells with promoter fragments including 113 bp of bovine vWF gene 5'-flanking sequence and various deletions of the first exon linked to a reporter gene [71]. The bovine vWF -113/+229 fragment, spanning the entire first exon, exhibits the highest promoter activity in the three cell types at a similar level. Addition of upstream sequences from bp -113 and extending up to bp -1362 progressively represses transcription of bovine vWF in the three cell types. Therefore, the 3' end of the bovine first exon is required for the basal activity of bovine vWF gene promoter, but positive tissue-specific elements able to overcome the

negative regulation in endothelial cells could not be found in bovine vWF gene between bp -1362 and +229. These results suggest that transcription of the human and bovine vWF genes occurs via species-specific mechanisms.

The vWF gene expression may also be influenced by molecular defects of proteins involved in the regulation of the biosynthesis and processing of the precursor, provoking a partial or complete decrease of vWF *in vivo* (Type 1 or 3 vWD). Such a mechanism has been recently described in a vWF deficient mouse [56, 72]. The mouse vWD is due to the mutation of a gene, mapped to mouse chromosome 11, expressing a post-transcriptional regulator which is involved in the normal expression of vWF (the mouse vWF gene is localized on chromosome 6).

Processing and Fate   In endothelial cells, large multimers of vWF are stored in the rod-shaped Weibel-Palade bodies, while smaller multimers are secreted constitutively [73]. Platelet vWF is also located in tubular structures within the α-granules [74]. Regulated secretion of stored vWF occurs in response to several agonists of potential physiologic relevance.

The biosynthesis and secretion of vWF has been extensively studied in cultured endothelial cells and in heterologous cells transfected with recombinant wild-type and mutants vWF. Following translation, the signal peptide of preprovWF is cleaved off before entering the Golgi apparatus. ProvWF monomers are assembled into provWF dimers in the rough endoplasmic reticulum (RER), by disulfide bond formation through the carboxy-terminal region [35,75]. At the same time, a high-mannose type carbohydrate glycosylation is occurring. The initial glycosylation is required for the normal assembly of the vWF structure. In the trans-Golgi apparatus, a series of multimers is formed [76]. The provWF dimers are multimerized by formation of disulfide bonds between Cys residues located in the D' and D3 domains corresponding to the amino-terminal ends of the mature subunits. The presence of the D1 and D2 domains, constituting the propeptide, is required for multimerization and storage, since recombinant cDNA constructs encoding a vWF precursor without the propeptide region can only direct the formation of dimers but not higher multimers and mutant proteins are not stored [77]. Each one of the propeptide D domain contains the sequence Cys-Gly-Leu-Cys (aa 159 to 162 and 521 to 524) with vicinal cysteines homologous to those at the active site of disulfide isomerases that catalyze thiol protein disulfide interchange. Mutagenesis studies have demonstrated that these vicinal Cys in propeptide D domains are similarly essential for interdimer disulfide bond formation [78]. The amino-terminal D' domain, which participates in inter-dimer disulfide-bond formation and the adjacent D3 domain of vWF subunit are also required for multimer formation. Studies of FVIII binding properties of wild type vWF and of dimeric mutant deleted of the propeptide have demonstrated that the propeptide is required for the post-translational process that leads to the expression of a functional binding site on the vWF subunit [79]. Sulfation occurs and the propeptide is rapidly released, predominantly in the trans-Golgi apparatus, before storage in the nascent Weibel-Palade bodies by cleavage by specific protease(s) [76]. The "paired dibasic amino-acid cleaving enzyme" PACE (furin) of the subtilisin family which specifically recognizes aa sequences containing conserved LysArg or ArgArg sequences at position -2-1 and Arg at -4 is a candidate provWF processing enzyme ([80]; see above, § Homologies).

In cultured cells, approximately 5% of de novo synthesized vWF is stored into the Weibel-Palade bodies, whereas the remaining 95% of the protein is packaged into secretory vesicles and rapidly secreted by the constitutive pathway [73]. The stored and constitutively secreted pools of vWF are functionally and structurally different. Constitutively secreted vWF consists primarily of dimers and multimers made from both provWF and mature subunits. vWF that is stored in Weibel-Palade bodies consists of only the most biologically potent largest multimers of fully processed mature subunit which are found noncovalently associated and in stoichiometric amounts with the free propeptide [76]. Two other proteins are present in Weibel-Palade bodies, P-selectin, a cell-adhesion molecule previously referred to as GMP-140 or PADGEM and CD63, a lysosomal membrane sialoglycoprotein. Upon cell activation, P-selectin is expressed on the cell surface, where it mediates the interaction between leukocytes and platelets and the adhesion of leukocytes to endothelial cells.

The stored vWF is released by stimulation of endothelial cells. A variety of inducers such as thrombin, phorbol esters, calcium ionophores, steroids, interleukins, TNF, have been shown to stimulate vWF secretion. The secretion of vWF occurs mostly through a regulatory pathway that involves degranulation of Weibel-Palade bodies [73].

The polarity of vWF secretion by cultured human umbilical endothelial cells via the constitutive secretory pathway or by the regulated secretory pathway appears to be dependent on the culture conditions and the nature of the substratum. The constitutive secretion of cells grown on polycarbonate carbonate filters has been reported to be non polar [81] or 70% directed to the apical side [82]. Using well-defined polar growing Madin-Darby Canine Kidney-II (MDCK-II) cells stably transfected with vWFcDNA or mutants having deleted D1D2 or D'D3 domains, it has been recently shown that the preferential secretion of vWF via the constitutive pathway is apical. Moreover, since the mutant with the deleted D1D2 domains does not display polarity of secretion, it appears that determinants contained within the propeptide are involved in apical constitutive secretion [83]. Polarity of release of vWF by the regulated secretory pathway is also controversial, but the majority of reports, based on studies of treatment of endothelial cells or transfected cell models with agonists like PMA, suggest that release of vWF via the regulated secretory pathway occurs predominantly at the apical side of the

cell [83,84]. The physiological mechanism of constitutive release appears to be lacking from megakaryocytes, in which synthesized vWF is only stored within the α-granules, organelles also containing a large number of additional molecules. Electron-microscopy studies have shown that vWF multimers are tightly packed in ordered longitudinal arrays reminiscent of Weibel-Palade bodies structures placed at the periphery of the α-granules [74]. Most of the vWF stored in circulating platelets (representing approximately 20% of the total vWF present in blood), is acutely released at the site of injury upon platelet activation.

In vivo, the vasopressin analogue, 1-deamino-8-D-arginine vasopressin (DDAVP), promotes a rapid and transient release of vWF from endothelial cells, and can thereby increase the plasma levels of vWF and FVIII several fold [85]. This effect, which allows the use of DDAVP for therapy of type 1 and sometimes of type 2 vWD patients, involves an unknown mechanism. DDAVP, which does not appear to promote vWF release from cultured endothelial cells or isolated platelets, would act indirectly on vascular endothelial cells.

The potential role of the circulating propeptide is still an intriguing question. The propeptide binds to collagen and inhibits collagen-induced platelet aggregation [86]. The propeptide is also able to serve as a substrate for transglutaminase and is specifically cross-linked to laminin, suggesting that it may act as transient matrix protein upon secretion from platelets and endothelial cells at the site of vascular injury [87]. Recently, these authors have demonstrated that the propeptide promotes the attachment and spreading of melanoma cells and is a ligand for the leukocyte integrin VLA-4 ($\alpha_4\beta_1$), binding to this complex through a 15-residue sequence (aa 395-409 of the propeptide) [88]. This raises the question of the role of the propeptide as a physiological adhesion protein for VLA4-bearing leukocytes at the site of inflammation and of vascular injury. It has been recently shown that the circulating propeptide, that exhibits a half-life of less than three hours, is a valuable marker of acute secretion from endothelial cells *in vivo* [89,90].

vWF is subjected to proteolysis in the circulation at the Tyr842-Met843 bond. It is speculated that the protease would be present in plasma in an inactive form, the regulation of the proteolytic cleavage being dependent on the conformation of vWF molecules [91]. Consequently, overall conformational changes involving the site of cleavage can result in excessive susceptibility to the plasma protease. This would be the case for vWF in many types 2A vWD in which the mutations inside the A2 domain containing the Tyr842/Met843 proteolytic site, are often associated with a selective loss of large and intermediate sized multimers [92].

Biological Activity   The indirect procoagulant function of vWF is attested by the decrease in FVIII levels in plasma of patients with a severe quantitative (Type 3 vWD) or qualitative vWF defect (Type 2N vWD). vWF acts as a carrier protein for FVIII and stabilizes its activity thereby increasing the concentration of FVIII which functions as a cofactor for the generation of Factor Xa. The stabilization of FVIII by vWF appears to result from the steric inhibition of the interaction of FVIII with phospholipids. In addition, vWF protects FVIII against inactivation by activated protein C and promotes the association of secreted heavy and light chains of FVIII.

vWF initiates the hemostatic process by mediating platelet adhesion to the subendothelial ECM which is exposed when the continuous endothelial cell layer is disrupted. The GPIbα-vWF interaction is finely regulated at the molecular level (see above, § Binding Sites and Affinity), this ensures site-specific platelet adhesion to vWF associated with the subendothelial ECM of the damaged vessel wall. Subsequent platelet spreading and aggregate formation are mediated by activated $\alpha_{IIb}\beta_3$. In this section, we will address the physiological regulation of the vWF-GPIbα interaction, by discussing binding of vWF to the subendothelial ECM and the effect of high shear stress conditions, with emphasis on three main aspects: shear-induced platelet adhesion, shear-induced platelet signaling and shear-induced platelet aggregation (for a review, see [93]).

Using blood from patients with severe vWD (deficiency in vWF) or Bernard-Soulier syndrome (deficiency in GPIbα) flowing over an everted rabbit aorta placed on a concentric cylinder, it was originally reported that platelet adhesion to a vascular surface is dependent both on high shear rates and the presence of vWF. A large variety of flow chambers have been designed as *in vitro* models of arterial thrombosis, in which a coated substrate is introduced to mimick to different extents the complexity of subendothelial ECM. Single-component models comprise fibronectin, fibrillar type I and III collagen, non fibrillar typeVI collagen, proteoglycans, thrombospondin, laminin, fibrin, fibrinogen. In addition, related compounds that are not present in the vascular wall but contain some structures with an *in vitro* reactivity to vWF, such as the sulfated groups found in sulfatides and heparin, have also been studied. To determine which components may be relevant *in vivo* as substrates for vWF-dependent adhesion to the vessel wall, the ability of these compounds to support platelet adhesion in high shear conditions has to be unambiguously demonstrated. ECM-bound vWF supports platelet adhesion in high shear conditions [94]. Type VI collagen has been identified as a vWF-binding protein in subendothelial extracts [95], making type VI collagen a likely candidate as an *in vivo* ligand for vWF. However, it was found that immobilized vWF bound to isolated type VI collagen could support platelet adhesion in low shear, but not in high shear conditions [96]. Thus, the *in vitro* model of platelet adhesion to purified type VI collagen cannot fully reflect the *in vivo* situation. To date, there is ample evidence that fibrillar type I and III collagen are involved in high shear conditions. Once bound to type I or III collagen, vWF becomes firmly attached to this surface and exposes its GPIbα-binding domain. This interaction withstands the forces generated by high wall shear rates. Type IV collagen, fibronectin

and laminin do not seem to be involved in vWF-dependent platelet adhesion, but may be important for a vWF-independent interaction with platelets in low shear conditions [97]. It is clear that at low shear rates, these compounds may interact with different platelet receptors, and the $\alpha_2\beta_1$ collagen receptor as well as GPVI may be of primary importance for platelet adhesion. Fibrin may act as a substrate for vWF-mediated adhesion in both low and high shear conditions, which is interesting since fibrin is generated by coagulation [98].

High molecular weight (HMW) multimers of subendothelial matrix vWF are more effective than lower molecular weight multimers in promoting platelet adhesion in flow, probably due to their increased binding to the surface [99]. Localization of an ECM-binding domain on the vWF subunit to the A1 domain (through the T116 fragment) strongly suggests that binding to the ECM may induce a conformational change of vWF and expose the nearby GPIbα-binding site [18]. Whether both ECM-binding and shear stress act in a coordinated manner to enhance platelet adhesion by exposing the GPIbα-binding domain of vWF remains to be established. Interestingly, a recent atomic force microscopy study has demonstrated that upon immobilization on a non-physiological surface, vWF undergoes a shear stress-induced conformational transition from a globular state to an extended chain conformation [100].

The high wall shear rates found in normal circulation occur in small arterioles and range from 500 to 5,000 s^{-1} and can reach 10,000 s^{-1} in stenosed coronary arteries. The effect of high shear rates on platelet adhesion to immobilized vWF results from the fast association rate with GPIbα as well as a fast dissociation rate, so that irreversible adhesion can only be achieved when platelet $\alpha_{IIb}\beta_3$ is activated and binds to the RGD sequence of vWF. This explains why platelet adhesion in flow conditions can be inhibited by blocking both GPIbα and $\alpha_{IIb}\beta_3$. Data obtained by videomicroscopy of mepacrine-labelled platelets and analysis of superimposed video frames, indicate that platelets adhere to the surface by an initial arrest and translocation, which is dependent on the interaction with GPIbα [101]. This process is similar to the rolling of neutrophils on activated endothelial cells, that precedes adhesion and transmigration to the site of tissue damage. This interaction involves endothelial P-selectin binding to neutrophil PSGL-1. Interestingly, the molecular mechanism of vWF binding to GPIbα may resemble P-selectin binding to neutrophil PSGL-1, since an anionic sulfated tyrosine sequence has been identified in both PSGL-1 and the vWF binding site of GPIbα (aa 276-282) (reviewed in [102]).

High shear-induced binding of vWF to platelet suspensions has been measured in the fluid-phase. There is evidence that this binding is different from that observed in no (low) shear conditions, essentially as it is reversible and not saturable and only a minority of platelets binds vWF [103]. Shear-induced binding of vWF to platelet GPIbα is followed by $\alpha_{IIb}\beta_3$ activation, as indicated by an increase in binding of monoclonal antibody PAC1 which bears an epitope specific for the activated $\alpha_{IIb}\beta_3$ complex [104]. The exact mechanism for $\alpha_{IIb}\beta_3$ activation is not completely understood. Caution should be taken that in some studies, platelet activation was accompanied by secretion of platelet granules components, in particular of ADP that may contribute to $\alpha_{IIb}\beta_3$ activation. Nevertheless, several pathways for signaling have been described, some being already known for $\alpha_{IIb}\beta_3$ activation by circulating chemical agonists. The current model is that upon shear stress-induced binding of soluble multimeric vWF to GPIbα, an increase in intracellular $[Ca^{2+}]$ is associated with different signaling pathways, including protein kinase C (PKC) and other serine-threonine kinases, tyrosine kinases and phosphatidylinositol (PI)3-kinases (reviewed in [93]). The PKC pathway involves the phosphorylation of pleckstrin, but occurs independently of diacylglycerol hydrolysis. Translocation of pp60c-src to the platelet cytoskeleton suggesting an involvement of the PI3-kinase pathway has been demonstrated in the ristocetin-mediated vWF binding to GPIbα [105]. However, this process has not yet been demonstrated in shear conditions.

This shear-induced platelet activation results in aggregation, a process that is consistent with the stabilization of the initial vWF interaction with GPIbα by an interaction with activated $\alpha_{IIb}\beta_3$. This mechanism has been extensively studied in cone-and-plate or coaxial Couette viscometers [106]. Epinephrine has been shown to act synergistically with shear to increase the aggregation response. An interesting hypothesis is the generation of specific platelet signals in response to mechanical stimuli. According to a large number of studies on the effect of shear stress-related responses in endothelial cells, a role for G-proteins and G-protein-linked mechanoreceptors has been identified (reviewed in [107]). Interestingly, Chinese Hamster Ovary (CHO) cells expressing GPIb-IX-V are able to aggregate in the presence of vWF when exposed to high shaking frequencies, suggesting that aggregation a) is independent of activation of $\alpha_{IIb}\beta_3$, b) is secondary to an effect of shear on the formation of vWF-GPIbα complex and c) involves tyrosine sulfation of the GPIbα receptor [108].

Role in Vascular Biology
Physiological Function   Normal hemostasis is a coordinated sequence of cellular and biochemical interactions starting with an injury and ending with a stabilized platelet aggregate and fibrin network. In this respect, vWF occupies a central position, since it is present in the different compartments involved in this process. Namely it circulates in plasma, is stored in platelets, is present in endothelial cells and is released into the basal compartment in the subendothelial ECM. The function of platelet vWF has been demonstrated in different models including bone marrow transplantation studies in animal models of arterial thrombosis [109]. In the following section we will address the function of subendothelial ECM-associated vWF, as participating in the balance between the pro-thrombogenic properties of

the ECM and the anti-thrombotic properties of endothelial cells. Indeed, beside its function as a mediator of platelet adhesion (see above, § Biological Activity), ECM-associated vWF supports adhesion through interactions with endothelial cell receptors.

Plasma, platelet and subendothelial ECM pools of vWF are important for platelet adhesion. vWF associated to the ECM of endothelial cells has been used as a model to study the function of subendothelial vWF [94]. A shear rate-dependent increase of platelet adhesion was observed on endothelial ECM which contains vWF, in contrast to fibroblastic ECM which is devoid of it, demonstrating that ECM-associated vWF is the main effector of platelet adhesion, plasmatic vWF being involved when the vessel wall pool is no longer sufficient [99].

In contrast to the platelet, the nature of endothelial cell receptors for vWF is not entirely established. The endothelial $\alpha_v\beta_3$ integrin has been identified as a common receptor for a number of RGD-containing ligands, including vWF. Site-directed mutagenesis studies of vWF have demonstrated that the RGD sequence is an absolute requirement for endothelial cell adhesion [12]. In addition, endothelial cells may contain a GPIbα-like receptor hypothetized to function as a vWF receptor following cell stimulation by cytokines [110,111]. However, it has been established that endothelial cells adhere to vWF through a GPIbα independent mechanism. It was found that an RGD-containing fragment of vWF (SpII, aa 1366-2050) could promote endothelial cell adhesion and spreading, whereas a complementary fragment overlapping the A1 domain (SpIII, aa 1-1365) and containing the platelet GPIbα-binding region, could not [14]. Studies with rvWF defective for their interaction with platelet GPIbα or $\alpha_{IIb}\beta_3$ have indicated that endothelial cell adhesion to RGGS-rvWF is strongly impaired, whereas adhesion to rvWF deleted from the A1 domain is not different from that to wild type rvWF. Recent data providing immunochemical analysis and functional studies using a large panel of monoclonal antibodies show the lack of expression of a GPIbα-related receptor on endothelial cell surface and in cell lysates [112], but evidence for the opposite is also provided by another group so that a controversy remains [111].

Pathology   von Willebrand disease (vWD) is a frequent and heterogeneous congenital bleeding disorder which is characterized by quantitative (types 1 and 3) and/or qualitative (type 2) abnormalities of vWF. Major progress in the knowledge of vWF molecule and in the development of molecular biology techniques has led to the identification of type 2 patients with distinct anomalies of vWF gene [113,114]. A database of molecular abnormalities in vWD is available ([113]; Internet WEB page, see § Gene Structure). This review will focus on type 2 vWD which is subdivided into four subtypes: 2A, 2M, 2B and 2N [115].

Type 2N vWD refers to patients with recessive inheritance and markedly decreased affinity of vWF for FVIII. Several

2N missense mutations (Arg19Trp, Gly22Glu, Glu24Lys, Cys25Tyr, Thr28Met, Arg53Trp, His54Gln, Arg91Gln, Cys95Phe, Asp116Asn and Cys297Arg) have been reported and affect the amino-terminal part of mature vWF subunit which contains the FVIII binding site [113,116]. The expression of mutant rvWF has shown that the most frequent mutation, i.e. Arg91Gln, induces a marked decrease in vWF capacity to bind FVIII while most of the other mutations completely abolish this function. In contrast, the additive effect of Arg19Trp and His54Gln appears to be less severe [116]. The recently found mutations at position 25, 95, 116 and 297 indicate that: 1) the FVIII binding capacity of vWF does not depend only on the aa sequence of the first 272 residues corresponding to the tryptic fragment which contains the FVIII binding site [116]; and 2) that the substitution of some aa residues in the D' and D3 domains may disturb the assembly of vWF subunits since Cys25Tyr and Asp116Asn mutations are found in patients with abnormal vWF multimeric pattern [116].

Type 2B vWD refers to variants with increased affinity of vWF for platelet GPIbα. Seventeen dominant mutations localized in the 5' part of exon 28 and inducing aa changes in the A1 domain have been identified, at the heterozygous state, in numerous patients [113,116]. The Arg543Trp and Arg578Gln mutations are particulary frequent [116]. In all cases except one (Arg578Pro), the increased binding of corresponding mutant rvWF to GPIbα has been confirmed.

Type 2A vWD refers to patients with decreased platelet-dependent function associated with the absence of HMW multimers. It includes the previously described dominant IIA phenotype and the less frequent IIC to II I phenotypes among which IIC subtype is recessively inherited. Most of the candidate mutations identified so far in type 2A patients are localized in the 3' part of exon 28 and are missense mutations which induce aa substitutions clustered within a 168 aa segment (aa 742 to 909 of vWF subunit) in the A2 domain containing the Tyr842/Met843 proteolytic site [113,116]. Interestingly, mutations have also been described in the 5' part of exon 28 in patients with a 2A phenotype [116]. These mutations, Cys509Arg, Cys509Gly, Arg545His, Arg552Cys, and Cys695Tyr, correspond to aa in the A1 loop. Furthermore, one mutation, Arg545Cys, also found in patients with 2B phenotype and inducing increased affinity of corresponding rvWF for platelet GPIbα [113], was found in a patient with 2A phenotype [116]. This peculiar case appears similar to the one already reported with a 2A phenotype and a 2B (Val553Met) genotype [113]. The Cys509Gly mutation has already been expressed and, like the Cys509Arg substitution, corresponding rvWF showed spontaneous binding to GPIbα [116]. It is noteworthy that both patients with a mutation at position 509 have a 2A phenotype although the one with Cys509Gly shows a slightly decreased platelet count. The three other mutations, Arg545His, Arg552Cys and Cys695Tyr, need to be reproduced by mutagenesis of full-length cDNA in order to determine whether the corresponding rvWF shows a gain or a loss of platelet-dependent function. In the six patients with the IIC phenotype

studied, 3 missense mutations, 1 deletion and 2 insertions in exons 11 to 17 coding for the D2 domain of the propeptide have been characterized [113,116]. The patients are either homozygous for one of these defects or compound heterozygous with a IIC mutation on one allele and another defect inducing a stop codon on the other allele. Four of the IIC gene defects have been recently reproduced in rvWF and shown to induce a marked defect in multimerization [116]. In two unrelated patients with the IID phenotype, a missense mutation substituting the Cys 2010 of the vWF subunit into an Arg was shown to induce a defect in the dimerization of vWF [113,116].

Type 2M vWD refers to patients with decreased platelet-dependent function not caused by the absence of HMW multimers. The subsequent molecular defects are localized in the A1 loop. There are also gene abnormalities found in patients who remain unclassified because only a slight decrease in HMW multimers is associated to a significant decrease in platelet binding. The following mutations have been described in type 2M and confirmed by mutant rvWF studies [113]: missense mutations (Gly561Ser, Phe606Ile, Arg611Cys, Arg611His), deletions (del Arg629 – Gln639, del Lys642) and one insertion (dupl Ser 589 – Ala 623). Recently, five new mutations localized within the A1 loop, Leu513Pro, Gly561Ala, Glu596Lys, Arg611Leu and Ile662Phe, were also identified [116]. Expression studies of corresponding rvWF are under way and will help distinguish between 2A and 2M vWD among the patients with unclassified phenotypes. Except for Leu513Pro and Gly561Ala which are within or near the main cluster (aa 540-578) of type 2B mutations, the substitutions found in these 2M or unclassified (2A/2M) patients are localized in the C-terminal half of the A1 loop. It is also noteworthy that all patients within this group (except the one with the Gly561Ala mutation) have mild vWD. The existence within the A1 loop of mutations inducing either gain- or loss-of-function of vWF emphasizes that this domain has an important regulatory role in the binding to GPIbα. The study of a large number of rvWFs mimicking the various mutations found in vWD will help elucidating the aa involved in the binding of vWF to GPIbα and /or its regulation.

Acquired vWD is usually characterized by a type 2 vWD as it is associated with a loss of HMW multimers. It is characterized by an acquired bleeding diatheses without a personal or a familial past history of bleeding. Laboratory tests disclose a very low level of vWF activity while the antigen level is less affected. Acquired vWD is mostly found in patients with an underlying lymphoproliferative disease or a monoclonal gammapathy, and less frequently in malignancy, auto-immune disorder or hypothyroidism. Some drugs have been also incriminated. The pathophysiology of acquired vWD is mostly related to a circulating antibody directed against an epitope on vWF subunit. The antibody binds usually to a non functional domain of vWF subunit, although an inhibitor directed against binding sites to type I collagen has been reported [117]. The antibody-vWF complex is cleared from the circulation by the reticulo-endothelial system. The mechanism of selective loss of HMWmultimers is unknown. The second most common mechanism consists in absorption of vWF by tumor cells expressing an aberrant platelet receptor (GPIb) as shown by immunofluorescence studies on plasma cells in a patient with a monoclonal IgGκ antibody or on lymphoma cells [118]. Similar findings are confirmèd by Oleksowicz et al [119] who demonstrate that human breast carcinoma cells synthesize a protein immunorelated to platelet GPIbα, that is able to bind to vWF. This interaction could explain the ability of tumor cells to aggregate platelets via vWF and seems to be correlated with their metastatic potential.

Qualitative abnormalities of vWF have been incriminated in thrombotic microangiopathies of thrombotic thrombocytopenic purpura (TTP). HMW multimers of vWF secreted from Weibel-Palade bodies in case of endothelial damage may be involved in the pathogenesis of TTP. Supranormal multimers enhance platelet aggregation in high shear conditions that may occur in partially stenosed vessels of the microcirculation in TTP [120]. An impaired degradation of supranormal forms of multimers has been observed in acute and relapsing forms of TTP, related to the deficiency of an hypothetical depolymerase which is responsible for the cleavage of unusually large multimers of vWF to smaller polymers in normal plasma [121]. Moreover, this deficiency is transmitted in an autosomal recessive manner as only two brothers of the same family with clinical symptoms of TTP were lacking the vWF-cleaving protease activity, whereas in other symptom-free family members, a normal protease activity was found. Among patients with lupus anticoagulant, only those with arterial thrombosis associated with multiple abortions or stroke were found to have unusually large vWF multimers in plasma [122]. HMW multimers of vWF could represent an additional risk factor of arterial thrombosis in patients with lupus anticoagulant. It is not yet known if HMW multimers are released by damaged endothelial cells or secondary to an impaired degradation.

Clinical Relevance and Therapeutic Implications vWD is the most common bleeding disorder in man, although the severe form of the disease (complete absence of vWF) is rare. Recently a mouse model for severe vWD has been generated by gene targeting [123]. Those mice have no detectable vWF in plasma, platelets and endothelium of tissue sections and they have a highly prolonged bleeding time. The vWF knock-out mice will thus provide a very useful *in vivo* model to investigate the biological functions of vWF not only in thrombosis and hemostasis, fields where it has already been thoroughly studied, but also in some other pathological situations where a role for vWF has sometimes been suspected but not always completely been proven.

This includes atherosclerosis, since studies made in vWF deficient pigs have first shown that vWF deficiency makes the animals less prone to atherosclerosis. Another report failed to describe a similar phenomenom, leaving the issue unresolved. A role for vWF in human atherosclerosis has also been suggested, essentially as a marker for endothelial cell injury. Increased plasma vWF levels are found in hypercholesterolemia patients and a fall in vWF is associated with reduction of cholesterol levels [124]. However, direct evidence is missing since patients with severe type 3 vWD are not protected from the development of atherosclerotic lesions [125]. The vWF knock-out mice, after being fed a high fat diet to induce fatty streak formation and atherosclerotic lesions in the aorta, will hopefully allow the controversy to come to an end.

Cancer metastasis is another domain in which vWF is believed to play a role. Indeed, αIIbβ3-containing tumor cells may utilize soluble proteins such as vWF for binding to activated platelets. Antibodies to vWF were shown to inhibit the platelet-cancer cell interaction *in vitro* and metastasis *in vivo* [126]. In inflammatory models, vWF may play a role both directly and indirectly. The vWF propeptide has been shown to have chemotactic activity for neutrophils and therefore may be involved in the recruitment of neutrophils to the stimulated vessel wall [88]. It is also possible that in the absence of vWF, the Weibel-Palade body formation is inhibited, in which case P-selectin (also normally stored in Weibel-Palade bodies) may be stored in another compartment or directly at the cell surface. In both cases this would result in a change of adhesiveness of the vessel wall. The vWF knock-out mouse will allow those different hypothesis to be tested *in vivo*.

Acknowledgements. We wish to thank Nadine Ajzenberg and Cecile Denis for their contribution to the manuscript.

Dominique Baruch, Danièle Kerbiriou-Nabias and Dominique Meyer

References

1. Shelton-Inloes BB et al (1986) Biochemistry 25:3164-3171
2. Cruz MA et al (1993) J Biol Chem 268:21238-21245
3. Sugimoto M et al (1993) J Biol Chem 268:12185-12192
4. Siguret V et al (1996) Thromb Haemost 76:453-459
5. Sugimoto M et al (1991) Biochemistry 30:5202-5209
6. Berndt MC et al (1992) Biochemistry 31:11144-11155
7. Christophe O et al (1994) Blood 83:3553-3561
8. Matsushita T, Sadler JE (1995) J Biol Chem 270:13406-13414
9. Matsushita T, Sadler JE (1997) Thromb Haemost suppl:p364 (abstr)
10. Kroner PA, Frey AB (1996) Biochemistry 35:13460-13468
11. Girma JP et al (1986) Blood 67:1356-1366
12. Beacham DA et al (1992) J Biol Chem 267:3409-3415
13. Lankhof H et al (1995) Blood 86:1035-1042
14. Denis C et al (1993a) Blood 82:3622-3630
15. Sixma JJ et al (1991) Eur J Biochem 196:369-376
16. Sobel M et al (1992) J Biol Chem 267:8857-8862
17. Christophe O et al (1991) Blood 78:2310-2317
18. Denis C et al (1993b) Arterioscl Thromb 13:398-406
19. De Groot PG et al (1988) J Clin Invest 82:65-73
20. van der Plas RM et al (1997) Thromb Haemost suppl: p643 (abstr)
21. Cruz MA et al (1995) J Biol Chem 270:10822-10827

22. Lankhof H et al (1996) Thromb Haemost 75:950-958
23. Bahou WF et al (1989) J Clin Invest 84:56-61
24. Koppelman SJ et al (1996) Blood 87:2292-2300
25. Vlot AJ et al (1995) Blood 85:3150-3157
26. Bonthron DT et al (1986) Nature 324:270-273
27. Verweij CL et al (1986) EMBO J 5:1839-1847
28. Titani K et al (1986) Biochemistry 25:3171-3184
29. Puzon-McLaughlin W, Takada Y (1996) J Biol Chem 271:20438-20443
30. Deák F et al (1997) J Biol Chem 272:9268-9274
31. Tuckwell DS, Humphries MJ (1997) FEBS Letters 400:297-303
32. Kotani E et al (1995) Biochim Biophys Acta 1260:245-258
33. Joba W, Hoffmann W (1997) J Biol Chem 272:1805-1810
34. Keates AC et al (1997) Biochem J 324:295-303
35. Voorberg J et al (1991) J Cell Biol 113:195-205
36. Perez-Vilar J et al (1996) J Biol Chem 271:9845-9850
37. Janel N et al (1997a) Biochim Biophys Acta 1339:4-8
38. Lavergne JM et al (1993) Biochem Biophys Res Com 194:1019-1024
39. Bakhshi et al (1992) Biochim Biophys Acta 1132:325-328
40. Bahnak BR et al (1992) Biochem Biophys Res Commun 182:561-568
41. Sinha D et al (1994) Biochem Biophys Res Com 203:881-888
42. Miyata S et al (1996) J Biol Chem 271:9046-9053
43. Lee JO et al (1995) Cell 80:631-638
44. Edwards YJK, Perkins SJ (1996) J Mol Biol 260:277-285
45. Jenkins PV et al (1998) Blood 91:2032-2044
46. Huizinga EG et al (1997) Structure 5:1147-1156
47. Mancuso DJ et al (1989) J Biol Chem 264:19514-19527
48. Zhang ZP et al (1992) Hum Mol Gen 1:780
49. Ginsburg D et al (1985) Science 228:1401-1406
50. Verweij CL et al (1985) Nucleic Acids Res 13:4699-4717
51. Mancuso DJ et al (1991) Biochemistry 30:253-269
52. Barrow LL et al (1993) Mamm Genome 4:343-345
53. Janel N et al (1996) Mamm Genome 7:633-634
54. Sjöberg A et al (1996) Hereditas 124:199-202
55. Bahou WF et al (1988) Blood 72:308-313
56. Nichols WC et al (1994) Blood 83:3225-3231
57. Bahnak BR et al (1989) J Cell Physiol 138:305-310
58. Smith JM et al (1996) Am J Vet Res 57:750-755
59. Galbusera M et al (1997) Blood 90:1558-1564
60. Coffin JD et al (1991) Dev Biol 148:51-62
61. Ferreira V et al (1993) Biochem J 293:641-648
62. Jahroudi N, Lynch DC (1994) Mol Cell Biol 14:999-1008
63. Jahroudi N et al (1996a) J Biol Chem 271:21413-21421
64. Schwachtgen JL et al (1994) Biotechniques 17:882-887
65. Schwachtgen JL et al (1998) Blood 92:1247-1258
66. Schwachtgen JL et al (1997) Oncogene 15:3091-3102
67. Jahroudi N et al (1996b) Blood 88:3801-3814
68. Aird WC et al (1995) Proc Natl Acad USA 92:4567-4571
69. Aird WC et al (1997) J Cell Biol 138:1117-1124
70. Janel N et al (1995) Gene 167:291-295
71. Janel N et al (1997b) Gene 198:127-134
72. Mohlke KL (1996) Proc Natl Acad Sci USA 93:15352-15357
73. Sporn LA et al (1986) Cell 46:185-190
74. Cramer EM et al (1985) Blood 66:710-713
75. Wagner DD (1990) Annu Rev Cell Biol 6:217-246
76. Vischer UM, Wagner D (1994) Blood 83:3536-3544
77. Journet AM et al (1993) Thromb Haemost 70:1053-1057
78. Mayadas TN, Wagner DD (1992) Proc Natl Acad Sci USA 89:3531-3535
79. Leyte A et al (1991) Biochem J 274:257-261
80. Rehemtulla A, Kaufman RJ (1992) Blood 79:2349-2355
81. Sporn LA et al (1989) J Cell Biol 108:1283-1259
82. Naharara N et al (1994) Arterioscler Thromb 14:1815-1820
83. Hop C et al (1997) Exp Cell Res 230:352-361
84. Hakkert BC et al (1992) Br J Haematol 80:495-503
85. Rodeghiero F et al (1991) Blood Reviews 5:155-161
86. Takagi J et al (1989) J Biol Chem 264:10425-10430
87. Usui T et al (1993) J Biol Chem 268:12311-12316
88. Isobe T et al (1997) J Biol Chem 272:8447-8453
89. Vischer UM et al (1997) Thromb Haemost 77:387-393
90. Borchiellini A et al (1996) Blood 88:2951-2958
91. Tsai HM (1996) Blood 87:4235-4244
92. Dent et al (1990) Proc Natl Acad Sci USA 87:6306-6310
93. Kroll MH et al (1996) Blood 88:1525-1541
94. Houdijk WPM et al (1986) Blood 67:1498-1503
95. Rand JH et al (1991) J Clin Invest 88:290-296
96. Ross JM et al (1995) Blood 85:1826-1835
97. van Zanten GH et al (1996) Blood 88:3862-3871
98. Endenburg SC et al (1995) Blood 86:4158-4165
99. Baruch D et al (1991) Blood 77:519-527
100. Siedlecki CA et al (1996) Blood 88:2939-2950
101. Savage B et al (1996) Cell 84:289-297
102. Andrews RK et al (1997) Int J Biochem Cell Biol 29:91-105
103. Goto S et al (1995) J Biol Chem 270:23352-23361
104. Konstantopoulos K et al (1995) Biorheology 32:73-93
105. Jackson SP et al (1994) J Biol Chem 269:27093-27099
106. Ikeda Y et al (1991) J Clin Invest 87:1234-1240
107. Davies PF (1995) Physiol Rev 75:519-560
108. Dong Jf et al (1995) Blood 86:4175-4183
109. Nichols WC et al (1995) Proc Natl Acad Sci USA 92:2455-2459
110. Beacham DA et al (1995) Thromb Haemost 73:309-317
111. Beacham DA et al (1997) Blood 89:4071-4077
112. Perrault C et al (1997) Blood 90:2335-2344
113. Ginsburg D, Sadler JE (1993) Thromb Haemost 69:177-184
114. Mazurier C, Meyer D (1996) In: Lee CA (ed) Baillière's Clinical Haematology. Baillière Tindall, London, pp 229-241
115. Sadler JE (1994) Thromb Haemost 71:520-525
116. Meyer D et al (1997) Thromb Haemost 78:451-456
117. van Gendenren PJJ et al (1994) Blood 84:3378-3384
118. Tefferi A et al (1997) Br J Haematol 96:850-853
119. Oleksowicz L et al (1997) J Lab Clin Med 129:337-346
120. Moake et al (1986) J Clin Invest 78:1456-1461
121. Furlan M et al (1997) Blood 89:3097-3103
122. Schinco P et al (1997) Clin Exp Rheumatol 15:5-10
123. Denis C et al (1998) Proc Natl Acad Sci USA 95:9524-9529
124. Blann AD et al (1995) Thromb Haemost 74:626-630
125. Federici AB et al (1993) Thromb Haemost 70:758-761
126. Nierodzik ML et al (1995) Thromb Haemost 74:282-290

VPF

Definition *Vascular permeability factor, same as Vascular endothelial growth factor (VEGF)*

See: →Hormonal regulation of vascular cell function in angiogenesis; →Vascular endothelial growth factor family; →Tyrosine kinase receptors for factors of the VEGF family

VSM

Definition   *Vascular smooth muscle*

See: →Smooth muscle cells

VSMC

Definition   *Vascular smooth muscle cell*

See: →Smooth muscle cells

vWD

Definition   *von Willebrand disease*

See: →von Willebrand disease

vWF

Definition   *von Willebrand Factor*

See: →von Willebrand Factor

MIX
Papier aus verantwortungsvollen Quellen
Paper from responsible sources
FSC® C105338

If you have any concerns about our products,
you can contact us on
ProductSafety@springernature.com

In case Publisher is established outside the EU,
the EU authorized representative is:
**Springer Nature Customer Service Center GmbH
Europaplatz 3, 69115 Heidelberg, Germany**

Printed by Libri Plureos GmbH
in Hamburg, Germany